# Outlook for Science and Technology

THE NEXT FIVE YEARS

Eruption of Mt. St. Helens, May 18, 1980, 8:15 a.m. to 10:45 a.m. [National Oceanic and Atmospheric Administration, National Earth and Satellite Service.]

May 18, 1980 GOES-West
9:15 a.m. PDT
Eruption + 0 Hrs 36 Min

May 18, 1980 GOES-West
9:45 a.m. PDT
Eruption + 1 Hr 6 Min

MAY 18, 1980
GOES-WEST
10:15 A.M. PDT
ERUPTION + 1 HR 36 MIN

MAY 18, 1980
GOES-WEST
10:45 A.M. PDT
ERUPTION + 2 HRS 6 MIN

# Outlook for Science and Technology

## THE NEXT FIVE YEARS

A Report of the
National Research Council

Published in collaboration with the National Academy of Sciences by

W. H. FREEMAN AND COMPANY
SAN FRANCISCO

This report was prepared at the request of the National Science Foundation, under contract No. PRA-8008016.

**Library of Congress Cataloging in Publication Data**

Main entry under title:

Outlook for science and technology.

Includes bibliographies and index.
1. Research—United States. 2. Engineering research—United States. 3. Science—United States. 4. Technology—United States. I. National Academy of Sciences (U.S.)
Q180.U5094 507′.2073 81-9878
ISBN 0-7167-1345-4 AACR2
ISBN 0-7167-1346-2 (pbk.)



Printed in the United States of America

1 2 3 4 5 6 7 8 9 0 MP 0 8 9 8 7 6 5 4 3 2

**Study Chairman**

*Frederick Seitz,* The Rockefeller University

**Vice Chairman**

*John S. Coleman,* Arlington, Virginia

**Steering Committee**

*Theodore L. Cairns,* Greenville, Delaware
*John E. Cantlon,* Michigan State University
*W. Dale Compton,* Ford Motor Company
*Merril Eisenbud,* New York University Medical Center
*William K. Estes,* Harvard University
*Paul J. Flory,* Stanford University
*Herbert Friedman,* National Research Council
*Philip Handler,* National Academy of Sciences
*Alfred E. Harper,* University of Wisconsin
*William R. Hewlett,* Hewlett-Packard Company
*Mark Kac,* The Rockefeller University
*Nathan Keyfitz,* Harvard University, Ohio State University
*John P. Longwell,* Massachusetts Institute of Technology
*Gordon H. Orians,* University of Washington, Seattle
*I. M. Singer,* University of California, Berkeley
*Gary Strobel,* Montana State University
*Gilbert F. White,* University of Colorado, Boulder

**Chapter Coordinators and Contributing Authors**

**1 The Demographic State of the World**
*Nathan Keyfitz,* Harvard University, Ohio State University

*W. Parker Mauldin,* The Rockefeller Foundation
*William Petersen,* Carmel, California
*Samuel H. Preston,* University of Pennsylvania
*Ronald Ridker,* Resources for the Future

**2 On Some Major Human Diseases**
*Philip Handler,* National Academy of Sciences

*Charles L. Christian,* Hospital for Special Surgery, New York
*Oscar B. Crofford,* Vanderbilt University
*Emil Frei, III,* Sidney Farber Cancer Institute
*H. S. Kaplan,* Stanford University Medical Center

*Daniel J. McCarty,* Medical College of Wisconsin
*Russell Ross,* University of Washington School of Medicine, Seattle
*Harvey Wolinsky,* Albert Einstein College of Medicine

**3 Directions in Nutrition Research**
*Alfred E. Harper,* University of Wisconsin

*W. R. Beisel,* U.S. Army Medical Research Institute for Infectious Diseases
*Maurice R. Bennick,* Michigan State University
*Lavelle M. Henderson,* University of Minnesota
*R. E. Shank,* Washington University School of Medicine, St. Louis
*Richard J. Wurtman,* Massachusetts Institute of Technology

**4 The Science of Cognition**
*William K. Estes,* Harvard University

*Richard M. Shiffrin,* Indiana University
*Herbert A. Simon,* Carnegie-Mellon University
*Edward E. Smith,* Bolt-Beranek & Newman

**5 Ecology and Systematics**
*Gordon H. Orians,* University of Washington, Seattle

*Eric L. Charnov,* University of Utah
*W. T. Edmondson,* University of Washington, Seattle
*Gene Likens,* Cornell University
*Robert T. Paine,* University of Washington, Seattle
*Peter H. Raven,* Missouri Botanical Garden
*David Schindler,* Canadian Limnological Laboratory, Winnipeg
*Otto Solbrig,* Harvard University

**6 Plant Disease**
*Gary Strobel and Donald E. Mathre,* Montana State University

**7 Water Resources**
*Gilbert White,* University of Colorado, Boulder

*John D. Bredehoeft,* U.S. Geological Survey
*John Cairns, Jr.,* Virginia Polytechnic Institute
*Russell Christman,* University of North Carolina
*Allen V. Kneese,* Resources for the Future
*Daniel A. Okun,* University of North Carolina
*Theodore M. Schad,* Commission on Natural Resources of the National Research Council

**14 Research in Industry**
*William R. Hewlett,* Hewlett-Packard Company

*Paul Chenea,* General Motors Corporation
*Edwin Mansfield,* University of Pennsylvania
*Robert E. Naylor,* E. I. du Pont de Nemours & Company, Inc.
*Kenneth Poole,* Bell Laboratories
*John E. Steiner,* The Boeing Company
*Peter Stewart,* General Electric Company
*P. Roy Vagelos,* Merck, Sharpe & Dohme Research Laboratories

**15 Fuel Science and Technology**
*John P. Longwell,* Massachusetts Institute of Technology

**16 Issues in Transportation**
*W. Dale Compton,* Ford Motor Company

*John J. Fearnsides,* The MITRE Corporation
*Oscar Fisch,* Ohio State University
*Howard L. Gauthier,* Ohio State University
*William J. Harris,* Association of American Railroads
*Gerald Kayten,* National Aeronautics and Space Administration

**17 Prospects for New Technologies**
*Frederick Seitz,* The Rockefeller University

*Paul Barton,* U.S. Geological Survey
*Herbert Friedman,* National Research Council
*Theodore Geballe,* Stanford University
*Robert A. Huggins,* Stanford University
*Courtland Perkins,* National Academy of Engineering
*Jacob Schwartz,* New York University
*Arthur C. Upton,* New York University Medical Center
*Norton Zinder,* The Rockefeller University

## Editorial Consultants

*George A. W. Boehm,* New York, New York
*Duncan M. Brown,* National Research Council
*Elizabeth W. Fisher,* Washington, D.C.
*Gregg Forte,* American Geophysical Union
*Ruth B. Haas,* Resources for the Future
*Robert R. Hume,* National Research Council
*Robert M. Joyce,* Hockessin, Delaware
*Susan McGrath,* Washington, D.C.

*Micah H. Naftalin,* National Research Council
*Steve Olson,* Washington, D.C.
*Martin Paul,* Merrick, New York
*Kenneth M. Reese,* Washington, D.C.
*Robert C. Rooney,* National Research Council
*Dorothy M. Sawicki,* Silver Spring, Maryland
*Phillip Sawicki,* Silver Spring, Maryland
*Lydia Woods Schindler,* Germantown, Maryland
*Elizabeth S. Stephens,* Washington, D.C.
*Donna J. Turner,* Washington, D.C.
*Joseph Turner,* Washington, D.C.

## Staff

National Research Council

*Norman Metzger*
*Gary J. Midora*
*Audrey Pendergast*
*Connie K. Robinson*
*Donald C. Shapero*

# *Contents*

## 6 *Plant Disease* *233*



## 7 *Water Resources* *255*



## 8 *Radioactive Waste Management* *287*

## III RESEARCH FRONTIERS

# *Preface*

The first Five-Year Outlook for Science and Technology prepared by the National Research Council was delivered to the National Science Foundation in March 1979. This second Outlook follows the general pattern of that document in portraying selected facets of science and technology as they pertain to the disciplines themselves—from solar-terrestrial physics to cognition to mathematics—and as they affect national concerns—from water resources to the management of radioactive wastes to transportation.

The immediate purpose of this and the preceding report is to aid the National Science Foundation in responding to the Congressional request, as expressed in the National Science and Technology Policy, Organization, and Priorities Act of 1976, that Congress and the people of the United States be informed in a continuing manner of conditions which might warrant special attention within the next five years, particularly those that involve:

- Current and emerging problems of national significance that are identified through scientific research, or in which scientific and technical considerations are of major consequence.
- Opportunities for, and constraints on, the use of new and existing scientific and technological capabilities which can make important contributions to the resolution of these problems.

Frederick Seitz served as the study chairman, assisted by a resident vice chairman and a steering committee whose membership was derived largely from those responsible for the preparation of individual chapters of the report. The design of this report is premised on the desire to build on the contents of the first Outlook and to include several of the vital areas not dealt with in the first report. Therefore, to the general discussion in the first report on energy has been added a report on the management of radioactive wastes; similarly, the considerable discussion of reductionist biology in the first is joined by ecology and systematics in the second; that of the structure of matter by solar-terrestrial physics; that of U.S. demography by world demography; that of a description of the health care system by a report on biomedical research on a limited list of major afflictions. In like manner, some of the vital areas not reviewed in the first Outlook are now in the second, while others have been deferred. This report, thus, includes chapters on chemistry and mathematics, but, for lack of space, does not deal with, for example, economic science and agricultural research. These and other topics will be considered in future reports on the Five-Year Outlook for Science and Technology.

Like the first report, this second Outlook has undergone several formal reviews. Its early contents were evaluated and given mid-course corrections by the steering committee. Each chapter was subsequently reviewed in detail and revised in response to comments from those expert in its subject. In addition, the Governing Board of the National Research Council reviewed the entire manuscript prior to its final revision before delivery to the National Science Foundation on March 31, 1981.

The names in this volume of those who wrote it and those who reviewed it testify that its preparation engaged a wide spectrum of the very best of American scientists and engineers. That this volume exists at all is due directly to the knowledge, thought, and care they brought to its preparation.

# *Observations*

"Inflation, unemployment, regulation, energy, productivity, disarmament, balance of payments, innovation. . . ."
"Migration, world population growth, food shortages, changing age distribution, mineral supplies. . . ."
"Disease prevention, environmental protection, consumer activism, radiation, weather prediction. . . ."
"Recombinant DNA, nuclear wastes, cancer, cholesterol. . . ."

Each of the above cue words has been made familiar by repetition in the news media. Each relates in greater or lesser degree to some aspect of science or its applications to technology. Each is a matter of major consequence to our society to be addressed by appropriate, informed public policymakers. It is our purpose in these Five-Year Outlook reports to provide current summaries of various aspects of science and technology which may be useful to those involved in formulating such policies. The observations below are offered as a perspective from which to view these several summaries.

## THE ENDLESS FRONTIER

Scientific research and the development of new technology already had been important aspects of American life for at least a century, but it was the success of science and technology in World War II that finally stimulated the development of explicit policy by the federal government. In the 1945 report, *Science, The Endless Frontier,* our government was encouraged to share the faith of scientists that the advancement of science is unquestionably in the public interest, that scientific knowledge should be cultivated in its own right and, legitimately, be fertilized by public funds. That report supported the belief that applications of scientific knowledge would make the nation militarily more secure, increase the food supply, improve the public health, expand the economy, and in diverse ways enhance the quality of daily life. Recognizing that both the findings and the fruits of science are unpredictable, it averred that the support of science by the very best scientists in all disciplines would redound in time to the national interest. Acceptance of these beliefs, as embodied in the charter of the National Science Foundation, was a historic landmark, the formal marriage of science and the government, a bond that has dramatically and permanently altered the history of both.

What a remarkable period it opened! Science flourished as never before. Armed with powerful accelerators, physicists studied the nucleus of the atom to reveal structures never anticipated. Improved optical telescopes and spaceborne sensors using the entire electromagnetic spectrum revealed a wondrous universe filled with previously unimaginable celestial bodies, while geophysicists, reading the clues in cores drilled in the sea floor, offered us "plate tectonics"—a grand synthesis that provides an improved understanding of the position and contours of continents, of earthquakes and volcanism, and of ore body formation. Chemists developed a family of powerful synthetic techniques, giving us pesticides, herbicides, new drugs, new textiles, polymers and plastics, and a safer food supply. Experiments in quantum physics gave birth to the laser. The invention of the transistor transformed electronics, made possible the modern computer, and revolutionized the communications industry. Military requirements and a buoyant aircraft industry paved the way for commercial jet aircraft. All the while, biological understanding was moving to the molecular level, acquiring extraordinarily powerful insights into the nature of life, the functioning of living cells, tissues, and organs, the nature of the genetic apparatus, the process of biological evolution, and those aberrations of human biology known as disease.

As the relationship of government to the research and development process expanded, the idea was widely accepted that henceforth it would be in the public interest for government to support fundamental research, defined in the broadest possible context. However, no single agency would be uniquely responsible for this mission. Certain agencies of government would support those areas of science and technology for which government itself was intended as the principal consumer, for example, the Department of Defense, the Atomic Energy Commission, and the National Advisory Committee for Aeronautics, the predecessor agency of the National Aeronautics and Space Administration. The Departments of Health, Education, and Welfare and of Agriculture developed powerful facilities and programs to support research that is clearly in the public interest and for which there is no logical, strong, nongovernmental mechanism of support.

A special agency for the broad support of fundamental research was created—the National Science Foundation (NSF). Universities increasingly became the principal loci for the conduct of fundamental research, whereas most of the applied research and development financed by the mission agencies used the facilities and resources of America's technological industries, the products of which, for a period, became dominant in world markets. And we prospered as never before.

## CHANGES

For almost two decades, the unique position of the United States in world science, in the world economy, and as the world's political and military leader was essentially unchallenged. Thereafter, a series of circumstances acted collectively to erode that unique global position: the war in Vietnam; the beginning of uncontrolled inflation; the sudden rise in oil prices; the inability of American petroleum production to keep pace with domestic consumption; growing American dependence upon a variety of other nations for diverse nonfuel minerals critical to the economy; the combination of extended life spans and reduced birth rates which markedly altered the age distribution of the American people; and the consumer and environmental movements which gave rise to legislation formulated to reduce hazards in the food supply, in air and water, workplace and home, while maintaining the amenities of the physical environment. Simultaneously, there arose powerful competition with both American science and technology as Western Europe and Japan emerged from the ashes of World War II while the Soviet Union, committed to a strong national scientific endeavor, embarked on a major buildup of strategic and tactical weapons capability. It is in the combination of these and other circumstances, rather than in any one alone, that are to be found the roots of current anxiety for the future of the American economy.

In very rough terms, American fundamental research now constitutes somewhat more than one third of the world total, with the remainder divided approximately equally between Western Europe and Japan on the one hand and the nations of the Soviet bloc on the other. As these Five-Year Outlook reports make plain, American science remains at the forefront in virtually every discipline. That is to say, Americans continue to publish a considerable percentage of all significant scientific papers and give a fair share of the major papers at international meetings. Numbers of Americans are internationally recognized as leaders in all fields and American laboratories operate many of the more powerful instruments whose capabilities pace the progress of science. In general, that is a valid estimate, but it does not afford one a sense of the direction of current trends in the relative strengths of American and overseas science. Neither does an examination of simple tables of annual expenditures for research afford much insight.

To get a reading on such trends, about 200 particularly knowledgeable American scientists were asked to compare their perceptions of the quality of American research, in their own disciplines, with that done elsewhere. The respondents work in such areas as condensed-matter physics, chemical synthesis, photosynthesis, neurobiology, child development, fusion, cognitive science, and macroeconomic modeling, as well as in two applied areas, new drug development and large integrated circuits. For most of these areas of science, informed impressions—for that is what they were—proved to be much the same. Earlier American dominance has been succeeded by increasingly impressive international competition which, in the main, American scientists welcome for the broader resources and collaboration it makes possible.

Numerous significant foci of first-rate research have developed in Western Europe and Japan. In a few instances, European laboratories moved to the forefront by acquiring unique, powerful new instruments as yet unavailable in this country. However, of no discipline was it said that leadership had clearly left the United States behind; in several fields, American research remains the world standard of excellence. If these assessments describe a trend, it is not clear how long American science will continue to be at the forefront of virtually all disciplines, particularly if support for research as a fraction of Gross National Product continues to decline in the United States while it rises in some of the countries which are our most successful competitors.

Science itself is an open-ended endeavor. Information and understanding gathered anywhere can be used by scientists everywhere. Rarely can scientific information that is not classified for military purposes long be held proprietary. Nevertheless, over time, economic rewards come to the country that is first to discover, since it is also likely to be first to invent and to apply.

But the history of the last decade bears witness to the fact that open-endedness does not hold equally true for technology. Markets can be saturated and the huge U.S. domestic and export economy appears inviting to the producers of technology everywhere, many of whom, in fact, design expressly for the export market. Although the U.S. balance of trade for manufactured goods has been affected very seriously thereby, few, if any, major innovations generated elsewhere underlie the successful invasion of the American market by foreign technologies. Some have represented the marriage of purchased U.S. expertise with cheaper foreign labor. Some, more serious, have been made possible by cost reductions based on improved manufacturing technologies, and others have come about through improvements in the quality, reliability, and price of technologies which in many cases were initially of U.S. design.

This circumstance has led to assertions of a decline in the innovative capabilities of American industry. When innovation is defined as the entire process from the conception of a new or modified technology to its ultimate implementation or marketing, there would appear to be some justification in that charge. Whether this partial innovative failure has its origin in inflation, in lack of investment capital, in deflection of company resources or of the attention of management from the innovation process to the requirements for compliance with regulation, or in yet other institutional barriers to innovation is not a judgment which will be attempted in this volume. What can more appropriately be asserted here is that, from the standpoint of current industrial research activity, there is every reason to believe that American industry is potentially as fully innovative as ever, that the research end of the technology pipeline continues to be full of exciting new ventures and possibilities which—given opportunity—could assure America's place in the forefront of innovative technology.

If the American economy is to grow in a socially desirable way, the scientific-technological sector of the economy must be given its full share of opportunity in the future. The technological competition now offered by the entire industrialized world will remain a fixture of the global economy. Indeed, that competition surely will become even more intense. Moreover, just behind current sources of competition is a set of newly emerging nations deliberately preparing to secure places in the same system, for example, Brazil, Mexico, India, as well as such recognized achievers as Korea, Taiwan, Hong Kong, and Singapore. For the most part, their manufactured products will enter potentially saturable markets whose expansion depends on the growth of the world economy.

Only a wholly new technology, such as television or the digital computer, can create a new open sector of the world economy. America's place in that competitive arena can be assured not by quota-type protectionism but by maximizing its competitive edge, by returning to the forefront of the technological innovative process, and by utilizing fully the unique fundamental research capacity and the huge industrial research capability already in place.

In our mixed system of private/public enterprise, the development of technology to be marketed and used in the private sector is a function of the private sector itself. The government may assist, as it has in the development of standards, in logical spinoffs such as those in the aircraft industry or, occasionally, when it is apparent that the costs of development may exceed the resources that can be brought to bear by entities in the private sector, as is true for nuclear energy. Otherwise, the role of the government with respect to industrial research and development is that stated by Congressman George Brown:

> The federal government has generally taken the position that, in an essentially capitalist economy, it is not its role to directly support efficiency and innovation in the process of production. Rather, these should result from the financial incentives provided to industry. However, when this process weakens and the nation's economic position is affected, it becomes government's role to find ways to either increase incentives or decrease disincentives toward a healthy industrial economic climate.

In other nations engaged in technological innovation, government policy is highly supportive of industry, particularly that segment seeking export markets. Yet, to quote Congressman Brown again:

> In the United States, an adversarial relationship has developed between industry and government. If we are to succeed in enhancing innovation and productivity, we must follow the lead of other nations and develop more cooperative attitudes and behavior between industry and government. Our economic planning, both on the national and individual business level, must be long term, and must not be directed at just what is politically hot today or just what shareholders want today. We must plan and act for the future.

There have been many attempts to calculate the economic return on the national investment in fundamental research. All such calculations necessarily must remain somewhat suspect. But all do agree that that economic return has been enormous. Indeed, it is at least arguable that, over time, other investment modes would lose their growth potential were it not for the expansion of opportunity ultimately made possible by fundamental research. In an ever shrinking, overpopulated world of finite mineral resources, where arable lands may be reaching their carrying capacity for human populations, further improvements in the human condition will require that we live by our productive imagination and wits, using scientific understanding for the development of ever more effective and efficient technologies. The one resource which is still unlimited is man's ability to seek out, to comprehend, and to find the paths and institutions through which his new comprehensions can be translated productively into social purposes.

The question for the federal government is not whether to support fundamental research; that decision lies behind us. Rather, we require a means of determining the appropriate magnitude of that support. We need a comprehensive reevaluation of the mechanisms by which support is administered and provided, by which choices are made, and by which resources are allocated. In this regard, it at least should be remarked that the federal support of fundamental research, expressed in constant dollars, grew from World War II until fiscal year 1967, after which it dropped significantly. Since fiscal year 1975, federal research support has climbed to slightly more than the 1967 level. But allocations for fields of research have changed dramatically within that time with little debate concerning the associated rationale. Mechanisms for decisionmaking and for research support remain much the same. This is a circumstance that warrants careful scrutiny and consideration, now that it is understood that fundamental research is not a luxury for an economy that can afford it—although that may have been the historic fact—but a vital necessity if we are to achieve the social and economic goals to which we aspire.

## THE SECOND FIVE-YEAR OUTLOOK

As indicated in the Preface, the present volume and the first Five-Year Outlook report should be regarded as successive stages, with this volume complementing its predecessor. Thus, the presentations of Planet Earth, The Living State, and The Structure of Matter in the first volume are followed now by presentations on Sun and Earth, The Science of Cognition, Chemical Synthesis of New Materials, and Some Recent Developments in Mathematics. The Demography of the United States in the first volume is succeeded here by The Demographic State of the World. A general treatment of the Health of the American People is followed up by a more detailed consideration of Some Major Human Diseases and of Directions in Nutrition Research; to appreciate these fully, it may be necessary to refer to some of the material in the earlier presentation of The Living State. To the earlier chapters on Computers and Communications and on Materials are added chapters on Prospects for New Technologies, Research in Industry, and Issues in Transportation. A general treatment of problems of Energy is supplemented by treatments of Fuel Science and Technology and of Radioactive Waste Management. Toxic Substances in the Environment is paralleled by treatments of Water Resources, Ecology and Systematics, and Plant Disease. Finally, the summary of Academic Science and Graduate Education is followed here by a comparison of the modes of support and conduct of research in Europe with those in the United States. It must be recognized, however, that large elements both of the scientific endeavor and of American technological enterprise remain untreated in these two volumes. It is our hope that they will be considered in future Outlook reports.

## ON THE STATE OF AMERICAN SCIENCE

The two Outlook reports make it evident that American science, which was relatively underdeveloped at the turn of the century, is approaching maturity. Several streams have fed this advance: the inherent interest and the enlightenment provided by science; the creation of graduate and medical schools by American universities; the needs of increasingly sophisticated industry for well-trained scientists and engineers; the intense national interest in improving the public health; the challenges and accomplishments of applied science in wartime; the use of science as an instrument for establishing national prestige in the international competition for the minds of men; and the possibilities generated by the great rise in national wealth which came with the development of what once seemed our limitless natural resources.

As stated earlier, U.S. scientists are now effectively engaged in research and development in essentially all of the major areas of science and technology. The tremendous success of these endeavors is evident in the remarkable panorama of scientific accomplishment portrayed in these two volumes.

The chapter on Sun and Earth describes the remarkable ingenuity, careful planning, and clever instrumentation that has been used to gain further insight into the physics of the sun and the manner in which the sun's radiations govern the properties of the gaseous layers surrounding the earth, determining their chemical composition and the nature of both the magnetosphere and the ionosphere—with important implications for radio communication. Solar energy, a fraction of which is directly used for plant photosynthesis, is largely absorbed at the earth's surface, from where it ultimately stokes the engine which creates weather in the lower atmosphere. Nevertheless, the extent to which meteorological phenomena are determined by variations in the emanations from the sun remains uncertain.

The chapters Chemical Synthesis of New Materials and The Science of Macromolecules suggest that chemists have the gifts of wizards, capable of creating in the laboratory any stable chemical structure that can be imagined. The chemical properties of such newly synthesized materials are fairly predictable, but predictions of their physical properties are still somewhat chancy. Surprise is still the order of the day. Nevertheless, a sufficient basis of understanding has been established to suggest that synthetic chemistry will have an ever more productive future in tailoring molecules to order for the diverse purposes of man.

Like so many other aspects of our civilization, however, the future of the chemical industry will be affected markedly by the supply and price of energy and feedstocks, particularly of hydrocarbons. Natural gas and petroleum fractions are the normal feedstocks for much of the chemical industry, including agricultural chemicals. Until recently, the price of the raw material has been only a small fraction of the total cost of the final product, but that circumstance is changing rapidly. The future of the entire chemical industry may turn significantly on the availability and prices of methane and higher hydrocarbons. We can turn back to cotton, wool, and leather for our apparel, but not for the myriad industrial uses of synthetic polymers or for the thousands of other synthetic compounds now deployed in our diverse economy.

A recurring theme in the chapters on basic science and on technology is the way in which the findings and the instrumentation developed in one area of science find application elsewhere.

The first Five-Year Outlook portrayed the manner in which the development of solid state physics and the computer and communications industries have gone hand in hand, each stimulating and making the other possible. A not dissimilar circumstance has existed in the field related to synthetic polymers and their various applications in fibers and plastics.

As indicated in the chapter Directions in Nutrition Research, much of our current understanding of the functioning of vitamin D is traceable to decades of work on the stereochemistry of organic molecules, on the kinds of molecular rearrangements occasioned by the absorption of light, and on the diverse physical instrumentation now employed to help decode the structure of molecules. Or note the statement in the chapter On Some Major Human Diseases describing the kaleidoscopic interchange of ideas from one research area to another:

> Who could have predicted . . . that studies of the genetics of skin transplantation in mice would provide a principal clue to understanding rheumatoid arthritis in man; that a variant in the structure of the sulfonamides developed as antibacterial agents would make possible management of glaucoma of the eye, or that the combination of a viral infection and inappropriate formation of an antibody to some structure on the surface of one's own cells could give rise to a family of diverse diseases?

Such interweavings occur no less in the physical sciences and in the development of new technology. For example, the theory of plate tectonics provides a meaningful framework for resource exploration of commercially useful minerals and hydrocarbons. Yet this great new synthesis of geophysical understanding owes its creation to the curiosity of paleontologists about the shell structures of almost microscopic creatures found in cores drilled in the ocean floor, to painstaking surveys over the ocean floor to detect magnetic polarities in seabed rocks, to inquisitiveness about the geography of the Hawaiian Islands, to matching the flora and fauna on different continental borders, to a maze of work on the properties of chemical isotopes, and to the imaginative application of that work to the dating of ancient rocks and sediments.

It is an inspiring fact that lasers, invented out of the insights afforded by quantum physics, have spawned new arts and technologies. They enable much more precise alignment of untold different physical arrangements, including tunnels built under riverbeds or drilled through mountains. Lasers are the basis for one approach to controlled fusion, are used to repair damaged retinas, and are at the heart of the instrumentation which makes possible the detection of fleeting intermediates in chemical syntheses. They also are used to drive photochemical synthetic processes in the laboratory and may soon find similar commercial application.

The chapter On Some Recent Developments in Mathematics reveals the delight of mathematicians that the most abstruse mathematics occasionally finds application in extremely practical circumstances. Elsewhere in this volume (Ecology and Systematics), ecologists express their fervent hope that mathematicians will continue with their recent successes in the treatment of nonlinear equations in order to help them study living communities that are not at equilibrium but move from state to state when recovering from large ecologic disturbances such as earthquakes, fires, or droughts.

In sum, the scientific enterprise will be seen as an extraordinarily dynamic system. The practitioners of each field successively attack in increasingly sophisticated fashion the layer of questions revealed by previous research. In the process, they find new surprises and unsuspected arrangements which generate more questions as well as more opportunities for applications to human welfare.

In the best American tradition, the American scientific enterprise has been highly pluralistic. Fundamental research is performed in universities, in nonprofit research institutes and hospitals, in industrial laboratories, and in the government's own laboratories. The work has been supported by a multiplicity of federal agencies, by state governments and a few municipalities, by private foundations and voluntary public giving, as well as by industry. Furthermore, the entire system has provided an example of the classical American compromise between egalitarianism and elitism. That compromise was entirely acceptable when the sum of such support enabled the continuing healthy growth of the scientific endeavor. However, the system is now under considerably greater strain than it was when the current federal expenditures for basic research were developed several decades ago.

In the decade and a half since the support of research ceased to grow, the fully trained scientific community has more than doubled. The intrinsic costs of doing science also have doubled more or less, largely as a result of the increased costs of essential instrumentation and the need for the larger teams required to address more complicated experimental situations. Under these circumstances, the project grant system not only has been unable to provide reasonable numbers of the major, relatively expensive (for example, $100,000-$1,000,000) research instruments that are needed at the leading edge of most disciplines, but it has markedly reduced the mean period of project support and frequently offered support in amounts insufficient to enable the successful conduct of a proposed project. It also has denied support to many investigators with successful track records and has significantly reduced support for the training of the next generation of scientists. At the same time, the drain on university resources created by the expanded academic scientific endeavor, along with difficulties in the funding of higher education, has required the expenditure of almost one third of the total funds otherwise available for the support of research to defray the associated indirect costs of research—including those necessitated by increasing government regulation.

The most precious component of the entire system for research is the relatively small group of generally acknowledged leaders—those whose insights and creativity fashion the intellectual structure of the scientific disciplines. These, the architects of science, are easily identified by their peers and by the managers of the grant and contract programs of the federal agencies. Supporting them and their most promising associates at an adequate level is the *sine qua non* of the entire enterprise. The thousands of other scientists who constitute the bulk of the enterprise are, collectively, no less essential. Indeed, only if the entire system is operating effectively can future leaders receive training, initiate their careers, and be identified.

Science-supporting agencies with closely defined missions, such as space and defense, may utilize external peers to assist in decisions concerning research priorities and research performers, but it is clear that ultimately their decisions must be based primarily on quality. However, those agencies with rather broader senses of mission, such as the National Science Foundation and the National Institutes of Health, whose programs in the main are fashioned by selecting among proposals voluntarily submitted by the scientific community—largely its academic component—have had to function in the atmosphere of continuing tension between egalitarianism and elitism that has characterized the Republic since its founding.

While support for the scientific enterprise was growing, those agencies could finance not only the best and the next best but also relatively lesser science that was widely distributed geographically. Generally, this was justified on the ground that it would improve the quality of science education and thus inspire some students in what are, from the research standpoint, lesser institutions and thereby satisfy our egalitarian views concerning the use of tax-derived funds. But, as indicated above, financial support of scientific endeavors effectively stopped growing in fiscal year 1967, while the intrinsic costs of research have continued to rise. This circumstance may necessitate that the science-supporting agencies sacrifice their egalitarianism in sufficient measure to assure that the very best, whose contributions will determine the quality of the national future, are fully supported.

These complex circumstances occasioned the comparative examination, reported herein in the chapter on Research in Europe and the United States, of the patterns and mechanisms of research support now operating in several European nations. This chapter makes no recommendations concerning which of the arrangements found in the United Kingdom, France, and Germany might be usefully transplanted to the American scene, but it does provide a mirror with which to examine ourselves in order to consider how we might contemplate remodeling aspects of the American scientific support system in the light of our changing financial circumstances and demography.

Meanwhile, it should be feasible to protect the scientific community from the unnecessary drain of an ever increasing load of paperwork, to use funds in such fashion as to increase significantly the mean period of project support, to provide adequate stable support to distinguished productive scientists, and to protect the course of basic research from well-meaning but short-sighted pressures for early applicability or "relevance." If American science, which has been so extraordinarily productive for three or four decades, is to maintain its pace as world leader, our society also must find ways to provide the scientific community with the advanced instrumentation which is at the very heart of scientific progress.

## EDUCATION IN SCIENCE

This Outlook has not treated explicitly the status of education in science. Nevertheless, the reader should be aware of certain large and highly relevant problems.

A series of recent reports has rendered evident the deterioration of school science. Public attention has been dramatically called to the functional illiteracy—the limited ability to read and write—of a significant fraction of secondary school graduates. But a far greater fraction is illiterate with respect to science and the trend is in the wrong direction. It may be that the more gifted students will somehow find their way to become successful scientists, but an increasingly technical world creates demand for very large numbers of at least minimally scientifically literate persons—as citizens and as members of the labor force. The latter point is particularly compelled by comparison of our circumstances with the intensive secondary school experience in Japan, now our most effective industrial competitor, and in the Soviet Union, our potential competitor or adversary in other senses. Surely the circumstances warrant the full attention of responsible persons in each dimension of the education system, local, state, and federal.

The previous Outlook noted that our demographic trends are leading to declining numbers of college-age students, and will until the mid-1990's, which have resulted in a lack of suitable opportunities for academic employment for young scientists because of declining student enrollments. This foreshadows a shortage of research scientists two decades hence. Some acceptable interim solution is urgently required.

A rather more acute problem of opposite character is developing with respect to highly trained engineering manpower. Shortages are now evident in industry. Moreover, there are numerous vacancies on engineering faculties and much of the engineering graduate student population consists of foreign students intending to return home. If not reversed, this shortage may prove to be a limiting factor that will pace the revitalization of the U.S. industrial economy.

Of yet another character is the changing mood of young physicians concerning the attraction of a career in academic medicine as clinician-teacher-investigator. Whereas, two and three decades ago, more than one third of medical students aspired to such careers, that fraction has now diminished very markedly. This is evident in vacancies on medical faculties and in the steady decline in the number of applications for research support sent to the National Institutes of Health by persons with M.D. degrees. The causes for this trend are poorly understood. They appear to include an increased desire for immediate social relevance as compared with the more intellectual and long-term rewards of research; the large discrepancy in personal income as compared with private practice; the large personal financial requirements of medical graduates who are often heavily in debt when they leave their medical training; the instability of research funding; and the difficulties of clinical research engendered by some irritating aspects of regulation, for example, the need for "informed consent" before the investigator may examine tissue routinely taken at surgery, much less a sample of blood or excreta. Much of the great progress in medicine for three decades is owed to the work of dedicated academic physician-investigators. And there is yet much for them to do. Reversing the current trend by whatever means are required is an imperative for our society.

## ON SCIENCE-BASED TECHNOLOGY

Concern for adequate energy supplies is a central fact of American life, as indeed it must be for the whole world. The first Five-Year Outlook scanned this problem area, summarizing the constraints and some of the opportunities lying before us. That essay on energy emphasized the great potential and desirability of conservation measures, i.e., more efficient end use of available energy in order to minimize demand. However, that assessment concluded that such measures alone would not suffice and that serious attention must be given to expanding the energy supply.

What makes the short-range problem so severe, beyond the violent increase in the price of imported petroleum and the potential hazard of a sudden, politically motivated cessation of petroleum supplies, is the simultaneous awakening to a restatement of Lord Action's aphorism, "All energy pollutes." Growth of a nuclear power supply has been inhibited by public concern for associated hazards, both real and imagined. In view of the immediate and probably continuous constraints on the supply of petroleum, it would seem offensive that oil continues to be burned under boilers in large utility or industrial plants; yet the practice persists. The use of coal is limited by concern for emissions that may include carcinogens, radioactivity, and sulfur and nitrogen oxides that contribute to air pollution over cities. Coal combustion has been regarded as the principal contributor to the phenomenon of acid rain, although it appears that there are also other major contributors to this process. Only less immediate is a concern for the buildup of atmospheric carbon dioxide and the possibility of a "greenhouse effect" which, in the next century, might alter regional climates significantly. The timing of such a phenomenon will depend upon the rate of increase in the worldwide combustion of wood, coal, oil, and natural gas. If estimates adjusting that rate downward prove valid, the "greenhouse effect" could be deferred beyond the middle of the next century, but it remains a serious prospect.

One solution to a part of these perplexing problems would be the development of a domestic synthetic fuel industry on a scale sufficient to reduce, if not eliminate, the need for petroleum imports. This, however, would present its own set of environmental hazards. Even apart from political and economic considerations, our country must appreciate that it cannot assume that the petroleum we may wish to import will necessarily be available. Indeed, a recent report from the International Institute for Applied Systems Analysis, the most comprehensive analysis published to date of future world energy requirements and supplies, suggests that by the year 2000 North America must become self-sufficient in fossil fuels. Yet most forecasts indicate that total energy use in the United States, currently about 78 quads (quadrillion British thermal units) per year, will increase by 15 to 20 quads in the next two decades. Staying within such bounds and achieving something like self-sufficiency will require vigorous efforts to conserve energy and improve the efficiency of its use, increasing use of natural gas, enhanced oil recovery techniques, and the development of a commensurate capability for the extraction and refining of shale oils and for the gasification and liquefaction of coal. The technologies for these purposes, along with their prospects and constraints, are reviewed in the chapter Fuel Science and Technology. It asserts that, whereas present methods for the conversion of coal and oil shale are now relatively unsophisticated, these methods will evolve rapidly with continuing national need—to become cleaner, more efficient, and perhaps even less costly.

The principal source of demand for liquid fuels will continue to be the nation's transportation system. The chapter Issues in Transportation portrays, among other things, the response of the U.S. automobile industry to the challenges of foreign-made automobiles. Considering the dominant position of the automobile in our economic life, much turns on the outcome.

One aspect of automobile manufacture, herein mentioned only in passing, is the inevitable growing use of robots in this and many other manufacturing industries. Introduced both for economy and for precision of manufacture that contributes to the quality of the product, such processes are being adopted widely here and abroad. Highly automated plants also can raise productivity and thus effectively offset cheaper labor costs in other countries. Such developments if applied domestically will help to expand the U.S. economy in desirable ways. But they also generate the specter of what has been termed "jobless economic growth." The resultant dislocations—only temporary, it is hoped—will demand serious attention both by industry and by the public sector.

The previous Five-Year Outlook explained why one must take a conservative position when estimating the time at which fusion or active solar energy transforming systems may make large-scale contributions to the American energy economy. Nuclear energy based on fission has been demonstrated to be practical, but it is now, in effect, being withheld from further development in our country. This current national posture, which is not based on formally stated policy but is nevertheless operating, has arisen, ostensibly, from two sets of concerns: the possibility of a catastrophic reactor accident, and the hazards alleged to be associated with the ultimate disposition of radioactive waste.

The report of the NRC's Committee on Nuclear and Alternative Energy Systems indicated that, taken overall, a nuclear energy system based on light-water reactors has the smallest environmental impact and the lowest risk of any of the alternative major energy-providing systems to which we may now turn. Even if a catastrophic accident were to occur—the chance of which is exceedingly small, although not zero—the total number of lives lost might well be less than that associated with the direct use of coal for similar purpose aggregated over a number of years. Choice between them, therefore, is a political, not a scientific, decision.

The second concern, management of the radioactive wastes, appears to be publicly less well understood, and is a source of public worry primarily for that reason. To help provide enlightenment, the present volume offers the chapter Radioactive Waste Management. The chapter concludes that:

> The available evidence supports a substantial degree of confidence that the technical aspects of geologic isolation can be managed in a manner that will protect the public.

Moreover, it indicates that:

> As with defense wastes, there is no technological urgency about immediate selection of a permanent isolation option for high-level waste from the nuclear power industry. . . . Nevertheless, because of widespread public concern about the hazards of radioactive waste, it remains in the national interest to proceed expeditiously with plans for permanent isolation [of the wastes].

Thus, it would seem that we have a sufficient body of understanding in hand to permit an early political decision concerning the future of nuclear energy.

The question of reprocessing spent nuclear fuel is not treated in the chapter. Such reprocessing would involve the isolation, preparation, and transportation of essentially weapons-grade plutonium. As a result of current concern for the dangers of proliferation and for possible theft by illegal groups, current U.S. policy forbids reprocessing, a policy which also would act to prevent commercialization of fast breeders, if they were developed to the point where they might be used. However, it should be recognized that the entire problem of the management of permanent isolation of radioactive wastes would be simplified if the spent fuel from light-water reactors were reprocessed—as it probably will be in other countries.

## Risk/Benefit Considerations

Concerns which permeate both Five-Year Outlooks and are prominent in public discussions of technology are those for safety and for avoiding adverse effects on the public health. In place is the Delaney Amendment to the Food and Drug Act which renders it illegal to add to any foodstuff any material known to be carcinogenic in any species of animal and at any dosage. This law, in effect, places an infinitely negative value on the fact of carcinogenesis regardless of the magnitude or location of its incidence. The original terms of the water quality legislation, which called for zero levels of pollutants in the effluents from factories and cities discharged into streams, were enacted in a similar spirit.

Discussions of these matters hinge on the quality of data that relate some form of toxicity to the presence of a chemical species in the environment and the reliability of the data that relate the dose to the incidence of such adverse effects. Unfortunately, there are few instances in which the data are of sufficient quality to permit reliable extrapolation to entire populations under reasonably predictable conditions. Yet, once the safety of a given material has been called into question, that fact itself demands resolution and decision by government action. Each such decision has proved troublesome. The associated risks and benefits are generally incommensurable; they may affect entirely different populations at different times or places. All too frequently, the risks are known only as a result of the exposure of a modest number of animals to relatively high doses; rarely is there reliable information concerning the effects of exposure to the low doses that might actually be experienced in practice by human beings. There is wide disagreement concerning the validity of extrapolations from high to low doses and across species. Finally, translation of adverse effects on the public health into economic terms comparable with those of the dollar costs of risk abatement remains fraught with difficulty.

No simple technique is available to enable decisions to be made under most such circumstances. The best one can ask at the present time is that both risks and benefits be assayed by panels of competent, qualified scientists and that the results of their findings, in turn, be delivered to an appropriately authorized body, appointed to make regulatory decisions with the full understanding that what is to be made is in part a political—not a scientific—decision. This is not to suggest that scientists or groups of scientists need avoid advocacy; it is to argue that there must be a clear demarcation of the processes of analysis and of advocacy.

The acceptability of a given risk and its social weight, as compared with the associated benefits, is necessarily a political decision into which may be factored moral, ethical, social, and economic values and even considerations of the national security. It should be clearly understood as such. There can be no better illustration of this circumstance than the status of the risks associated with radiation, which are known with greater precision than are those of any of the chemical pollutants that are sources of concern. This knowledge derives not only from studies using thousands of experimental animals subjected to varying dosage levels, but also from the data available from the medical histories of the irradiated survivors of the bombs at Hiroshima and Nagasaki. Although those data do not permit direct evaluation of the consequences of exposure to very low doses of radiation—such as those that might be encountered outside a nuclear power plant—they do come close enough to permit reasonable extrapolations. Since the theory of how radiation induces cancer is itself somewhat controversial, there is disagreement concerning the manner of that final extrapolation. However, barring truly catastrophic accidents, both current schools of approach to that extrapolation agree that the increased cancer incidence due to possible irradiation of the general public, arising from the operation of nuclear power plants, is so low as to be essentially without statistical meaning. The effect, then, may be said to represent a slight and practically indeterminate increment to the current mortality due to cancer.

Perhaps the best news yet to arrive from Hiroshima and Nagasaki is the report that, in a biochemical genetic study of 18,946 children of irradiated survivors, involving electrophoretic examination of 28 different proteins present in normal blood, only one somewhat questionable indication of a mutation that would not have been observed in any other comparable population of Japanese children was detected. These facts are here recorded not so much in support of an expanded nuclear energy program, but as an indication of the need for an institutionalized procedure in our society for separating the analytical process for the assessment of risks and benefits from the unavoidably political process of decisionmaking.

## Technological Innovation

Two chapters of this Outlook are devoted specifically to the vitality of our technological resources and our abilities to use them—Prospects for New Technologies and Research in Industry. Both chapters are sources of encouragement in that they offer vistas of a technological future involving exciting, apparently feasible new technologies that are working their way through the system. Both will repay the reader with heightened appreciation for the breadth of the applied research and development enterprise and its continuing strength.

However, neither edition of the Outlook treats the history of some specific, innovative technology already in the marketplace. It would be useful to understand how the germ of the original idea arose; what decisions were required within the innovating company in order to initiate the research trail; when, how, and by whom a decision was made that the matter had genuine promise and warranted deployment of the relevant resources of the company; when the decision to enter the development stage was reached and what factors influenced that decision; when that information was made available to the marketing division of the company, and what difference that fact made to the course of development itself. Such a tale should be told in some future Outlook, if indeed such a history can be reconstituted.

The chapter Research in Industry describes both accomplishments and problems. Basic research, as defined in industry, is found to be a healthy, vigorous enterprise, although there are major industries which, it can be argued, do not support a sufficiently vigorous research enterprise, e.g., automobiles, mining, metallurgy. Each of the areas discussed involves recognition of the serious effects of rises in the price of energy or of feedstocks. In several of the contributions, concern is expressed for what is depicted as the rather heavy hand of regulation. This problem is strikingly evident in the discussion of research in the pharmaceutical industry, where regulation has increased markedly the cost and the time required to bring a new drug to the market and, hence, has raised the market price of new drugs. It is not clear how much additional protection the current level of increased regulation has afforded the public, but it has slowed down the flow of new drugs, reduced the number of companies with financial resources adequate to engage in the development of new drugs, and rendered it effectively impractical for a pharmaceutical company to consider the development of drugs for any but the major diseases. Would somewhat less stringent controls be more compatible with the overall public interest?

The difficulties of translating research into marketable products and processes vary for each industry. Still, there are some rough parallels—higher energy costs, the effects of inflation, the need to improve productivity while simultaneously achieving higher quality in the product line itself, and responding within governmental regulation to competition from companies in other countries which can operate under different rules. For example, it now seems appropriate to reexamine aspects of antitrust legislation and its implementation with respect to the research and development sector.

Taken as a whole, the picture that emerges is one in which it is not the pace of research that limits the apparent innovative capability of these industries. Accordingly, an appropriate concern at the governmental level for the next several years would appear to be to seek federal policies that might encourage research and development by industry and facilitate the processes whereby capital-intensive industry is able to exploit its innovative ideas in the marketplace.

## HUMAN RESOURCES

The Demographic State of the World is an engrossing chapter which carefully states its premises with regard to future demographic projections. It is evident that population growth will continue to be relatively slow in the industrialized nations for some years to come, whereas relatively rapid population growth is inevitable across the developing world. This situation gives rise to a complex set of future realities which will exert pressures on the land, water, energy, and mineral supplies, as well as on what remains of the world's forests, particularly the tropical forests.

One matter which emerges with particular force from a reading of this chapter is the danger of overaggregating or averaging data. As the chapter explains, the gross water resources of the planet probably are ample to manage a global population of nine billion, but that does not mean that we will avoid severe water shortages in specific geographic locales still subject to rapid population growth, for example, in Egypt. Similarly, the aggregate total of grain production in the world could suffice to feed us all, but that does not mean that there will not be acute food shortages in many quarters of the globe, some of which may well become chronic and serious.

It has become evident that the industrialized nations of the world cannot conceivably engage in capital transfer to the developing nations on a scale sufficient to manage these huge problems, even with great sacrifice. If, however, they wish to minimize the possibility that burgeoning populations elsewhere will generate irresistible pressures that will destabilize the world political situation, the industrialized nations must engage in extensive efforts to assist the developing nations to learn how to manage their problems and opportunities. This means assistance in the development of adequate infrastructures, including educational systems, as well as indigenous agriculture and industry. Particular emphasis should be given to the support of light industry and research and development programs which will, one day, enable the developing nations to manage their own affairs more adequately. Some such technical assistance programs are in existence today, funded and managed by the United Nations and by a series of bilateral arrangements. But their sum is far from adequate for the task. This circumstance constitutes a huge challenge to the scientific and technical communities of the industrialized world. It also requires the sympathetic understanding, the wisdom, and the courage of the peoples and governments of the nations of the already industrialized world.

A principal lesson derived from these Outlooks is that science and its application repeatedly solves problems—only to generate new ones. That, of course, is the principal conclusion to be drawn from the chapter on demography. In the developing nations, a modicum of sanitation, agricultural improvement, and nutritional understanding has had a profound effect on death rates, particularly of infants and children, but without a commensurate feedback effect on birth rates. Ironically, therefore, the humanitarian application of that modicum of knowledge of sanitation, nutrition, and agriculture that has reduced infant death rates has acted to increase population size and thereby denied to those same populations much of the benefit of the growth of their own industrial economies. Indeed, whereas agricultural production in some of these nations, taken in the aggregate, appears to have kept pace with population growth, the numbers are frequently misleading since a small fraction of the population in each country is eating much better than in the past, while the circumstances of much larger numbers of persons may actually be deteriorating.

Yet another example of the creation of problems by solving other problems is evident in the chapter On Some Major Human Diseases. This chapter makes it obvious that the extraordinary progress achieved in the United States and other industrialized nations in bringing under control the infectious processes—diseases which, until only recently, were the principal causes of death—has reduced infant mortality rates and extended life expectancy at all ages. Tuberculosis, pneumonia, typhoid fever, syphilis, and infantile diarrhea, for example, have been succeeded by heart disease and cancer as the principal causes of death. As a result, a significant fraction of the population spends much of the last decade or two of life disabled in greater or lesser degree by chronic ailments. It is these diseases which now place so great a burden upon the health care delivery system; and precisely because there are no quick cures and no truly definitive therapy, these diseases have increased health care costs markedly. Current care can provide, in the main, only supportive management and symptomatic therapy. As the previous Outlook noted, this increased life expectancy coupled to a reduced birth rate has also altered dramatically the age distribution of the American population, with profound social consequences. Consider, for example, the fact that, when the present Social Security system was enacted, there were 11 workers in the labor force for each person over 65; today, there are less than 4.

As the chapter stresses, in almost no instance is there as yet definitive understanding of the causes or course of development of the major disorders of later life (their etiology and pathogenesis). Indeed, although there have been extremely heartening developments in the management of these illnesses, and some progress has been made toward their prevention, any hope for definitive preventive or curative measures must await the acquisition of sufficient understanding of fundamental human biology to make rational, effective approaches possible. Fortunately, progress in such understanding, summarized in this chapter, has been proceeding apace. That progress gives us reason to hope.

Somewhat surprisingly, perhaps, the chapter Plant Disease affords a striking parallel to that on human diseases. Agricultural progress has been made possible by the development of extraordinarily productive strains of those crops that are planted for food and lumber. Such selection, however, leads to monocultural practices which, in turn, can render entire crops vulnerable to the onslaught of disease. The previous Five-Year Outlook described the manner in which recent understanding of photosynthesis, cell biology, and genetics has been utilized for the development and selection of superior producing plant strains. The current Outlook chapter summarizes groups of diseases to which plants are vulnerable, and discusses the mechanisms being developed to contend with them. The principal stratagem available is to breed new plant strains which are genetically resistant to a given invading microorganism. This has been accomplished without first gaining a complete understanding of the basis for this resistance, that is, what the gene does that makes the resistance possible. This practice is now being supplemented by a variety of applications of current understanding of biochemistry and cell biology. Although this approach is in its infancy, it appears to offer great promise.

We end where we began: science and technology in the United States are mature and highly productive, judged by any standard. With understanding and support, these endeavors not only can expand further and generate new opportunities as yet unimagined, but they also can be of inestimable value to our country in dealing with complex problems now and in the future. The scientific and technological communities, and the institutions in which they work, recognize that challenge and their responsibilities in meeting the problems ahead. To do so, they must communicate with other sectors of our society, especially those responsible for formulating and implementing the national policies which will govern our ability to utilize the results of scientific and technological progress for the benefit of all society. This volume, we hope, is a step in that direction.

*Frederick Seitz*
Study Chairman,
Five-Year Outlook Report

*Philip Handler*
President,
National Academy of Sciences

April 1981

# I

# *HUMAN RESOURCES*

# 1

# *The Demographic State of the World*

## INTRODUCTION*

Most of the more developed countries[1] have birth rates too low to replace their populations. Their low rates of birth are associated with high proportions of women in the labor force, the diminishing stability of marriage, and a high-consumption society that emphasizes the production of goods rather than the reproduction of people. Americans see no great need to arrest the approach of a stationary population; the governments of France, Austria, and especially the countries of Eastern Europe are doing what they can to reverse the downturn of fertility. The population problems in the more developed countries are not too few or too many people, but include such factors as a prospectively disadvantageous age distribution, some effects of internal and external migration, and a spatial distribution not everywhere congruent with economic opportunities.

These problems are slight compared with those that face the two thirds of the planet counted as the less developed countries.[1] Many of these show rapid technological and economic progress but, at the same time, the number of their poor is increasing; this certainly applies to India and possibly to Brazil. Population growth handicaps the ability to accumulate capital and presses on resources. Recent energy and materials shortages

---

* Demographic terms are defined at the end of the chapter.

---

◀ "The 15 million or so alien residents of northwestern Europe in the early 1970's comprised what was often called the tenth member of the Common Market, larger in total population than several of the member countries." [International Labour Office.]

have brought this aspect to general consciousness. But population also hurts in another way that did not arise in the nineteenth century, when masses of workers were important for the early stages of industrialization. In the more advanced and automated production of the late twentieth century, human muscles lose their economic value. Unemployment becomes a major problem of the poor countries in particular.

Only a small part of the population problems reveal themselves in the planetary figures. On the optimistic supposition that bare replacement will be reached early in the twenty-first century and that mortality will continue its present decline, the ultimate world population will be 9 billion, attained late in the twenty-first century. It will thus double from the present total. Current progress in agriculture and industry suggests that world food and other production also could double in the next century. On a global view, one could be moderately sanguine about the population-resource relation and even the population-capital ratio for an ultimate 9 billion people.

Such optimism is not justified when individual countries are considered. At present growth rates, Bangladesh has little prospect of attaining a satisfactory balance between its population on the one side and land and capital on the other. Not many of its 90 million people are likely to be able to sustain themselves by selling television sets, automobiles, and ships to the rest of the world. Egypt is creating a modern industrial machine and sending out skilled labor to the oil-producing countries but the growth of Cairo presents intractable difficulties, as does the narrowly circumscribed tillable area of the Nile valley. Mexico, despite even more advanced industry as well as oil wealth, has some of the same crushing difficulties as Egypt; both require increasing food imports for their rapidly expanding urban populations. Brazil is more advanced industrially than either Egypt or Mexico, but the energy price increases of the 1970's have deferred the hope that its phenomenal economic advance would soon reach the poor part of its population. Brazil's plans to use alcohol derived from sugarcane for fuel raise in sharp relief the conflict between expanding the industrial sector and feeding the poor. Nations of rising income from Poland to Tanzania suffer chronic balance of payments deficits.

The increase in life expectancy, at the rate of half a year per year in less developed countries, while births have remained high, has produced crowding in areas that up to World War II had abundant space. Thus, Burma, with 15 million people before World War II, could export 3 or 4 million tons of grain; today, with over 30 million people, it has no surplus for the world grain market; add another 15 million people and it probably will have to import food. In a further stage, one can imagine Burma's forests overcut and its oil reserves exhausted, so that it will have no funds to pay for food imports. This is what can happen with dynamic population numbers and a static economy.

Therefore, it is inevitable that a realistic discussion of population will refer to individual countries. The compactness and simplicity of a global overview have to be relinquished if the implications of today's population trends are to be appraised. A fresh water shortage for the world as a whole could not occur for centuries, but it is here now for parts of the United States, not to mention Egypt. If the forests of the U.S.S.R. were accessible to the population of Bangladesh, there would be no firewood problem in either country. With free movement of grain around the world, no one would be hungry and the reserves needed to cover year-to-year fluctuations could be small.

If the problem disappears when one considers issues globally rather than nationally and regionally, it is in some sense created artificially when one disregards possible technical changes. Just as the world's population will at least double before it becomes stationary, so the world's resources will be multiplied by advancing technology in the century ahead. The technical progress of recent centuries is going to be maintained in some degree, although difficulties in spreading the technology to poor countries can be expected. The trouble is that it is easier to visualize population growth than it is to picture technical and economic change. In this chapter, an effort is made to take both into account.

No one studying population should forget that technological advances could overcome all of the concerns about a population growing from 4 to 9 and perhaps to 12 billion. If nuclear or solar energy on a sufficiently large scale were to become available, or if tomorrow someone were to find a means of extracting energy inexpensively by controlled fusion, then cheap fertilizer and hence cheap food would be provided. Splitting of water would bring forth a hydrogen economy, with indefinite amounts of clean-burning fuel for transport, heating, lighting, and industry. Materials could not possibly be scarce once fuel was cheap, since much of the earth's crust could be used. Alternatively, many billions of people could live prosperously with present-day technology if capital were accumulated with single-minded determination.

The disappointments of the last decade have demonstrated the inadvisability of counting on technical leaps. Such leaps are hard to plan and, if they come, they can have unexpected unfavorable consequences. It is possible to learn to extract coal cheaply and use it within acceptable limits of pollution, but it is not possible to learn to burn coal without producing carbon dioxide. Given that the distribution of population over the earth's surface has taken account of local food productivity, however roughly, any major change in global temperatures which might result from high concentrations of carbon dioxide could be disastrous even if the world's total food supply were increased—which there is no reason to assume would result from a rise in temperature.

To have faith in science and technology, and to pursue promising leads that would relieve shortages of energy and materials and enable larger populations to live better is one thing; to count on near miracles that would solve the population problem at one stroke is quite another. The serious student of population wants to see intense activity in both basic science and its technological applications, as well as in the operation of the economy, that would make the globe fit for more people to live on; to pursue single dramatic solutions is not as promising a strategy as making smaller changes along many lines, with emphasis on those lines that are labor intensive.

Applications of science and technology are too large a subject for the present chapter; they will be treated in the succeeding chapters of this volume. Here it is enough to say that it is never sufficient to ask whether the earth can sustain 4 or 9 or 12 billion people. The question is always whether the advances of technology can keep extending resources at rates that will accommodate population increases and permit a rising standard of living, country by country, in an imperfectly coordinated world.

An encouraging feature of the post-World-War-II period is the rapid increase of human capital in virtually all countries. The spread of education at several levels and the accumulation of industrial skills are probably more important than any shortages of physical capital or resources. In this chapter, emphasis is placed on the components of population change. Thus, the chapter begins with mortality—the present levels and differentials among countries, between the sexes, and among social classes. It continues with fertility, where the range of uncertainty is greater and the implications for future population are more difficult to forecast than for mortality. Migration then is treated in some detail. A few years ago, it was thought that, with the filling up of the Americas, the last great international movement—across the Atlantic from east to west—was over. That has proved to be wrong. Refugees from East to West Germany and from Vietnam to the United States, guestworkers from southern Europe and western Asia to northern Europe, legal and illegal movement—across the U.S.-Mexican border, from Bangladesh to northern India, from ex-colonies to ex-metropolises—these are a few of the migration currents since World War II. It would be idle to suppose that the total movement will diminish in the future, although the directions of the streams may be expected to change. Finally, the relation of population to resources is considered briefly.

## MORTALITY

Deaths are one component of population change; the others are births and migrations. Changes in the death rate have been responsible for most of the change in human population growth during the twentieth century. At the turn of the century, the death rate of the human population was on the order of 30-35 deaths per 1,000 population per year. By 1950-55, it had declined to about 19 per 1,000 and, by 1975-80, to about 11 per 1,000.[2] Most of the recent change has occurred in less developed regions, where crude death rates declined from approximately 23 per 1,000 in 1950-55 to an estimated 12 per 1,000 in 1975-80. During the same period, more developed countries as a group experienced a decline from only 10 to 9 per 1,000, their decline being impeded by the aging of their populations. Only a small fraction of the decline in death rates in less developed regions has been offset by declines in birth rates during this century; the result is that the world population grew at unprecedented rates in the period 1950-80.

Other measures of mortality are better indicators of a population's health and longevity. Life expectancy at birth has increased from approximately 47 years in 1950-55 to about 57 years by 1975-80 for the human population as a whole. For less developed regions, the corresponding figures are 43 and 55 years; in more developed regions, 64 and 71 years. The magnitude of these achievements is suggested by the fact that in 1900 no country in the world had a life expectancy as high as the world average of 57 years in 1975-80.

These mortality improvements have been distributed unevenly. As indicated by the above figures, improvements have been slower in the more developed countries as they approach the limits to longevity under current medical knowledge and practices. Male longevity has improved more slowly than female in the post-World-War-II period in these countries. For example, in the United States, male life expectancy grew by 3.7 years between 1950 and 1977 (from 65.6 to 69.3), while female life expectancy grew by 6.0 years (from 72.2 to 77.7 years).[3] Among less developed countries, Latin American populations appear to have experienced a slower than average mortality decline in the last decades but, once again, the explanation may lie at least for some countries (Cuba, Costa Rica, Uruguay, and the commonwealth of Puerto Rico) in their location at the upper end of the range of observed life expectancies, where progress is slow.

For countries in sub-Saharan Africa, neither trends nor current levels are known with precision. The evidence on mortality levels in the region supports only the crudest assessment of current conditions. The United Nations estimate of life expectancy of 47.1 years in sub-Saharan Africa in 1975-80, which is probably the most reliable one available for the region, must be viewed in this light. All one can say is that mortality, especially in West Africa, is very high.

There is even less information about mortality levels in China, for which the United Nations assigns a life expectancy of 64 years in 1975-80. A more recent, unpublished analysis of data covering a part of the Chinese population in 1975 indicates for that part a life expectancy of 60-65 years and a crude death rate (total deaths divided by total population) of

8-10 per 1,000. The largest less developed country with reasonably reliable data on mortality is India. Its nationally representative Sample Registration System (SRS) probably includes about 90 percent of deaths that fall within the sample frame. It yielded a crude death rate of 16.5 per 1,000 in 1970-72 and 15.0 in 1976. SRS data have been combined with those from other sources to produce an estimate of life expectancy in India for the 1961-71 period of 45.5 years for females and 47.9 for males.[4] Life expectancy probably has increased by several years since that period.

## Factors Influencing Mortality Trends in Less Developed Regions

The factors responsible for declines in mortality vary from time to time and from place to place. Much of the research that attempts to identify these factors for less developed countries in the postwar period has focused on the island of Sri Lanka. The reasons for this concentration are a dramatic decline in mortality and the availability of good regional data on both mortality and presumptive causal factors, particularly the prevalence of malaria before and after a successful antimalarial campaign. The current consensus is that the antimalarial campaign of 1946-47 reduced the island's crude death rate by 4-5 per 1,000, representing about 40-50 percent of the decline that occurred in crude death rates between 1936-45 and 1946-60.[5] Other important contributors were improved nutrition and better organization of health facilities. This order of magnitude for the effectiveness of antimalarial campaigns is probably not seriously in error for Mauritius, Venezuela, Guatemala, and India, other countries with moderate malaria endemicity and successful antimalarial campaigns.

Longevity is being increased rapidly in many poor countries and regions. Kerala state in India, Sri Lanka, and Cuba all had life expectancies in the upper sixties or above in 1970-75 despite quite low levels of income (less than $200 per capita in U.S. dollars in Kerala and Sri Lanka). Each of these populations has a health system oriented toward delivering basic services to the entire population, with emphasis on rural areas; each has a large-scale, government-sponsored nutritional supplementation program; and each has achieved unusually high levels of literacy.[6] But each is also a relatively small population with a well-organized administrative structure commanding popular support. It is not clear how readily their success can be transferred to other areas.

With an average life expectancy of about 55 years, less developed countries have completed about two thirds of the transition from preindustrial mortality levels to those characteristic of a more developed country. But they still show enormously high death rates among their poorest population sectors, as revealed by recent information on socioeconomic differences in mortality. Their white collar or well-educated urbanites enjoy life expectancies similar to those of more developed countries, while rural illiterate groups have levels that are 15-30 years lower.[7] Multivariate analyses of several bodies of data suggest, but cannot prove, that the literacy, or educational attainment, of mothers is one of the most important factors in a household's mortality level.[8]

***Mortality at an Early Age*** The death rate from diarrheal diseases in infancy and the second year of life continues to be high. These diseases are more prominent than they were in more developed countries at equivalent mortality levels, and they may help to explain why mortality below the age of five is often high relative to mortality above the age of five in less developed countries.[9] Problems become particularly acute around the time of weaning, when a child often first comes into contact with contaminated food and loses immunities transmitted through breast milk or acquired *in utero*. These problems sometimes result in a local peak in the age-curve of the death rate around the time of weaning, as in Guatemala and Senegal.

While Latin American and Asian countries have reduced malaria death rates to a small fraction of their earlier levels, the same is not true of tropical Africa. Malaria probably shortens life expectancy in tropical Africa by some four to eight years; infective mosquitoes continue to be many times more common in parts of Africa than in other regions. Low incomes, a widely dispersed population, and weak health services contribute to the malarial death rate. What can be done is vividly illustrated by an insecticide spraying campaign in a village in Kenya, where the crude death rate was reduced from 24 per 1,000 to 13.5 per 1,000 in two years. No change in mortality occurred in the control village.[10] But international initiatives against malaria have lost much of their momentum since the 1950's.

Continued progress in reducing mortality in less developed countries does not require scientific or technical breakthroughs but only the spread of relatively few basic preventive and curative procedures throughout entire populations. Oral rehydration for diarrheal disease, boiling of drinking water where piped clean water cannot be made available, construction of pit privies, immunization, and insecticide spraying are probably the keys to success over the next five years and beyond.

## Mortality in More Developed Countries

Recent declines in fertility in more developed countries have greatly increased the leverage that mortality conditions exert on the size and structure of the population. These countries have reached the stage where approximately 93 percent of those born survive to their 50th birthday.

Therefore, subsequent declines in mortality will not increase greatly the number of births occurring in these countries, a number that is projected to be steady or to decline in many of them. Without the youthful bias typically imparted to age distributions by high fertility, what happens to life expectancy at the age of 50 obviously has major implications for future populations.

This observation gains consequence from the fact that important changes seem to be under way in mortality rates at older ages, particularly in the United States. For example, life expectancy at age 50 has increased from 24.5 years in 1949-51 to 27.6 in 1978. Much of the recent improvement is attributable to declining death rates from diseases of the heart as noted in the next chapter. The age-adjusted death rate from heart diseases declined slowly from 3.08 per 1,000 in 1950 to 2.86 in 1965. Thereafter, the decline accelerated rapidly and the rate reached 2.16 in 1976.[11]

Several factors are involved in this rapid recent advance against heart disease mortality. There is solid evidence that Americans are increasingly heeding warnings about the adverse effects of practices such as cigarette smoking and high consumption of animal fats. For example, the proportion of the adult male population that smokes cigarettes has declined from 52.4 percent in 1965 to 41.9 percent in 1976. Department of Agriculture figures show large declines between 1963 and 1975 in per capita consumption of animal fats.

Perhaps more important than these changes in personal habits is the diffusion of improved methods of treating hypertension. The percentage of people with hypertension who are receiving treatment has increased greatly over the last decade. This increase is spread throughout all race-sex groups and may account for the fact that improvements in life expectancy among blacks from coronary heart disease mortality have exceeded those among whites in the last decade.[12]

The United States is still behind several other developed countries. Table 1 displays estimates of life expectancy at birth for the latest available years in five developed countries. United States males fall three years short of male life expectancy in Japan, Sweden, and Norway. The U.S. female deficit is about half as large as the male. In general, the national differences shown in this table (as well as the sex differences) are dominated by differences in heart disease mortality. Some of these differences may have genetic components; it is instructive, for example, that Japanese-Americans have even higher life expectancies than the population of Japan itself.[13]

The relatively rapid decline in U.S. mortality from heart disease in the last 15 years is more than an erratic fluctuation, but there is still scope for additional progress. Much of this progress probably will occur through "risk factor intervention," in which large groups are persuaded to modify

**Table 1** Ranking of selected more developed countries according to life expectancy at birth (1970's)

| Country | Year | Life expectancy at birth | |
|---|---|---|---|
| | | Males | Females |
| Japan | 1977 | 72.69 | 77.95 |
| Sweden | 1976 | 72.12 | 77.90 |
| Norway | 1975-76 | 71.85 | 78.12 |
| United States | 1976 | 69.1 | 76.7 |
| U.S.S.R. | 1971-72 | 64 | 74 |

SOURCE: United Nations Population Division, drawn from official publications and files of the United Nations Statistical Office.

their behavior in longevity-enhancing ways. (The next chapter discusses these risk factors and the changing rates of cardiovascular disease in considerable detail.) Some community programs of this type have been successful.[14] Whether advances in understanding and slowing down the aging process will occur in the next decade remains to be seen.

Infant and youthful mortality still demand attention. Although infant mortality has fallen in recent years in the United States, a number of countries still do much better than we do. Accidents, homicides, and suicides are not causes of death on which medical advances will have much effect but one can hope for various kinds of nonmedical alleviation.

One can expect some major social adaptations to current and prospective mortality trends:

- As the life cycle lengthens, the need for individual adaptability increases. The lengthening of life is associated with more changes in the course of the life cycle: changes in careers, spouses, and residences, as well as reeducation. Life expectancy at birth for females in 1900 was 48 years, the same as life expectancy at age 32 today. Instead of being terminated by the death of a spouse, many marriages end in divorce. Education of adults is useful in proportion to the length of life remaining to them.
- Mortality declines at older ages will change the ratio of the retired to the active labor force. According to official Census Bureau projections that employ conservative mortality assumptions, the number of persons aged 65 and above is slated to increase by nearly 50 percent between 1975 and 1995. The ratio of the population 65 and over to that aged 15-64 will increase by more than 50 percent before the end of the first quarter of the twenty-first century. Faster-than-expected mortality declines will put upward

pressure on the age at retirement, a pressure that will be accommodated partly by the increased vitality of the elderly.

- More older people will have to concern themselves with support for and care of a parent. Under recent mortality and fertility conditions in the United States, 34 percent of persons aged 60 would have a living parent; if cancer were eliminated as a cause of death, this figure would increase to 42 percent.
- The huge sex difference in mortality is creating very different life cycles for males and females. Under recent U.S. conditions, a newborn American girl can expect to spend nine years of her life as a widow and has a 60 percent chance of dying in that state; for males, the corresponding figures are two years and 22 percent. Recent mortality changes show no sign of narrowing these sex differences.

Mortality conditions at older ages will stand close watching in the years ahead. In 1977, the Bureau of the Census modified its population projections from those produced in 1975 to incorporate several more years of data on mortality declines. The result was that the projected number of persons over the age of 65 in 2000 grew from 28.8 million to 31.87 million, a 10 percent increase in the projection over the course of two years.[15] But the 1977 projection already appears extremely conservative in its assumption that male life expectancy will increase from 69.1 years in 1976 to 71.8 years by 2050; three countries have surpassed the latter figure already and, between 1960 and 1977 alone, U.S. males gained the 2.7 years that are projected for the next 75 years. Mortality conditions among the elderly will become particularly significant when the baby boom cohort reaches retirement age.

## FERTILITY

The world's crude birth rate (total births divided by total population) was about 36 births per 1,000 per year in 1950; it has been reduced by about 20 percent during the past 30 years and is below 30 today. That is perceptible progress toward the goal of a stationary population, where a crude death rate of 15 would correspond to a crude birth rate of 15.

The more developed countries, with one quarter of the world's population, had an average crude birth rate of about 23 in 1950. Since then, there has been a decline of 32 percent, and the level of crude birth rates today is about 16 (see Figure 1).

A few of the more developed countries, including both Germanies, have more deaths than births, and many have rates of reproduction that, if continued, would result in a decline from present population levels. Population growth among the more developed countries poses no serious

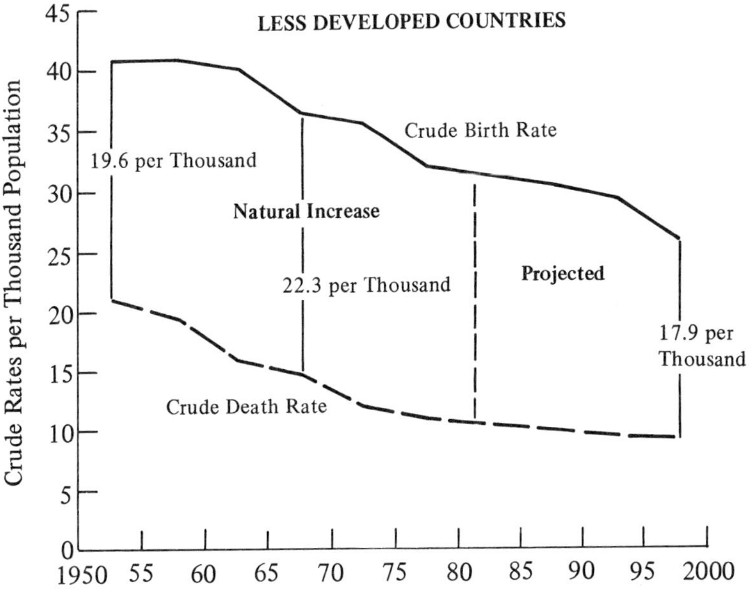


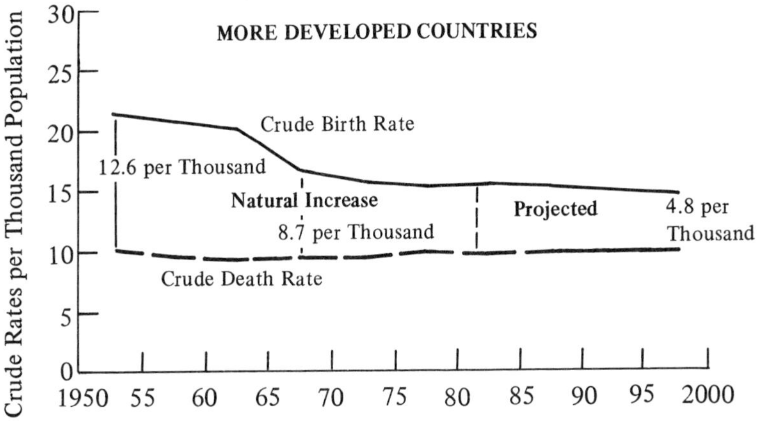


**Figure 1** Crude birth and death rates and percentages of natural increase (1950-2000). United Nations medium assumptions. [SOURCE: United Nations Population Division.]

problem of total numbers, but there are many problems of urbanization and distribution.

The less developed countries present a different picture. In 1950, their populations numbered 1.7 billion, and death rates were high—more than 23 per 1,000. The revolution in mortality has reduced the crude death rate almost by half to 11 or 12. Fertility was high in 1950, about 42 per 1,000. Thus, the rate of growth was slightly less than 2 percent per year in 1950. The crude birth rate has declined from 42 to about 33—a decrease of 9 points, or about 20 percent. Changes in fertility have been dramatic in Asia and the Pacific, substantial in Central and South America, and hardly noticeable in Africa. In Asia, where the bulk of the population lives, the crude birth rate declined by 25 percent from more than 41 to 31. In the Americas south of the Rio Grande, the decline was about 16 percent—from more than 41 to less than 35. The data for many African countries are inadequate, but what information exists suggests a crude birth rate of about 48 in 1950, perhaps dropping a couple of points to 46 by now. Only two countries in Africa, Tunisia and Mauritius, have shown appreciable changes in fertility over the 30-year period. Egypt's crude birth rate declined in the late 1960's and very early 1970's but, since that time, it has increased, to 41 in 1979, about the same level as in 1965.

## More Developed Countries

With the sole exception of Ireland, all of the more developed countries experienced declines in fertility from 1950 to 1978. Declines ranged from about 15 percent in Spain to more than 25 percent for 24 countries (see Table 2) and were common to large and small countries, to capitalist and socialist nations, to European and non-European populations, to Catholics and Protestants. Most of western and northern Europe, Canada, the United States, and Japan have below replacement levels of fertility.

A number of more developed countries, chiefly in eastern Europe (where legalized abortion came to be associated with, or indeed produced, the low birth rate), are making determined pronatality efforts, using both negative and positive means. Negative measures include tightening restrictions on induced abortion and reducing the availability of contraceptives. Positive measures are a liberal increase in wages for couples having three or more children, interest-free loans for the purchase of an apartment or house and household furnishings with, for example, 20 percent of the loan being canceled at the birth of the first child, 30 percent for the second child, and the remaining 50 percent for the third child if born within eight years. Czechoslovakia, East Germany, Bulgaria, Hungary, and Romania have shown that such policies are not without effect.

In addition, a number of other developed countries have pronatalist ambitions or mild policies: Israel, for obvious internal and regional reasons of a politico-ethnic character; France, with a target since 1975 of attaining replacement fertility or slightly higher; Finland, with quantitative targets to prevent a decline in population in any of its counties; Greece, with a target of ensuring a population growth rate not much lower than one percent per year; and Argentina, with various pronatalist

**Table 2** Population (1980) and changes in crude birth rates (1950-78) in selected more developed countries

| Country | Population (thousands) 1980 | Crude birth rate 1950 | Crude birth rate 1978 | Percent change 1950-78 |
|---|---|---|---|---|
| The Americas | | | | |
| Canada | 24,073 | 27 | 15 | -44 |
| United States | 222,159 | 24 | 16 | -33 |
| SUBTOTAL | 246,232 | 24 | 16 | -34 |
| Asia and the Pacific | | | | |
| Australia | 14,487 | 23 | 16 | -33 |
| Israel | 3,950 | 35 | 25 | -27 |
| Japan | 116,364 | 28 | 14 | -50 |
| New Zealand | 3,268 | 26 | 17 | -35 |
| SUBTOTAL | 138,069 | 28 | 15 | -47 |
| Europe and the U.S.S.R. | | | | |
| Austria | 7,481 | 16 | 12 | -26 |
| Belgium | 9,920 | 17 | 12 | -27 |
| Bulgaria | 9,007 | 25 | 16 | -38 |
| Czechoslovakia | 15,336 | 23 | 18 | -21 |
| Denmark | 5,105 | 19 | 12 | -35 |
| Finland | 4,818 | 25 | 14 | -45 |
| France | 53,450 | 21 | 14 | -32 |
| Germany, Dem. Rep. | 16,864 | 17 | 14 | -17 |
| Germany, Fed. Rep. | 60,903 | 17 | 10 | -42 |
| Greece | 9,329 | 20 | 16 | -22 |
| Hungary | 10,761 | 21 | 15 | -28 |
| Ireland | 3,307 | 21 | 22 | +2 |
| Italy | 56,959 | 20 | 13 | -36 |
| Netherlands | 14,082 | 23 | 13 | -45 |
| Norway | 4,080 | 19 | 13 | -34 |
| Poland | 35,805 | 31 | 19 | -38 |
| Portugal | 9,856 | 24 | 17 | -31 |
| Romania[a] | 22,268 | 26 | 19 | -27 |
| Spain[a] | 37,378 | 20 | 17 | -15 |
| Sweden | 8,262 | 17 | 12 | -30 |
| Switzerland | 6,310 | 18 | 11 | -38 |
| United Kingdom | 55,888 | 17 | 12 | -25 |
| U.S.S.R. | 266,666 | 27 | 18 | -32 |
| Yugoslavia | 22,328 | 30 | 17 | -43 |
| SUBTOTAL | 746,163 | 22 | 16 | -27 |
| TOTAL | 1,130,463 | 23 | 16 | -30 |

[a] The latest figures available for Romania and Spain are for 1977.

SOURCE: W. Parker Mauldin, The Rockefeller Foundation.

measures such as cash subsidies, housing and medical benefits, and others. The Netherlands reports the goal of achieving a stationary population, and Japan, that a stationary population is "estimated and expected."

Very soon in the more developed countries—and ultimately in the less developed countries—the inevitable approach of the stationary population will bring important changes, not all of them desirable. There will be far more old people in relation to the labor force; the projections of the Bureau of the Census go from 11 persons 65 and over in 1940 for each 100 persons 18-64 years old to 17 by 1960, then rise slowly to 19 by the end of the century. Even the slow rise is causing trouble in the United States social security system. What, then, can be expected when the baby boom cohort comes of age and the number of persons 65 and over per 100 persons 18-64 years old goes to 24 in 2020 and 29 in 2030? A lowering of benefits, or people working to older ages, are the most obvious means of avoiding tax increases.

In a sense, the costs associated with people living to older ages will be offset by the community spending less on raising and educating children; however, in practice, the savings on the young are not available to the old.

In industry and business, the labor force will be older than in a time of rapid population increase, and this could have an effect on innovation. We do not know to what degree the creation and implementation of new ideas calls for new people.

## Less Developed Countries

There are 92 less developed countries with populations of a million or more, but 2.6 billion, or 80 percent of the population of all less developed countries, live in 16 countries with over 35 million people each. Population data, crude birth rates, and percentage changes for 1965 and 1975 for these countries (except Vietnam) are given in Table 3.

Less developed countries vary enormously in culture, level of development, population policy, and changes in fertility. Brazil and Nigeria do not seek to reduce their rates of population growth but, in recent years, have adopted a policy of supporting family planning for reasons of health and as a human right. Iran has tried to reduce its rate of population growth, but it is unlikely that the present government will continue that policy. Burma is frankly pronatalist. It recognizes family planning on maternal and child health grounds, but contraceptives are not easily available and family planning clinics are banned. All of the other countries have adopted policies intended to reduce the population growth rate.

There has been no significant reduction in fertility in Bangladesh,

**Table 3** Population (1980) and declines in crude birth rates (1965-75 or later) in less developed countries with populations of 35 million or more

| | Population (millions) | Crude birth rate | | Percent change |
|---|---|---|---|---|
| Country[a] | 1980 | 1965 | 1975[b] | 1965-75[b] |
| China | 975 | 34 | 20 (1978) | -41 |
| India | 694 | 43 | 36 | -16 |
| Indonesia | 152 | 46 | 36 | -22 |
| Brazil | 122 | 42 | 33 (1977) | -21 |
| Bangladesh | 89 | 48 | 48 | [c] |
| Pakistan | 82 | 47 | 47 | [c] |
| Nigeria | 77 | 49 | 49 | [c] |
| Mexico | 70 | 44 | 37 (1978) | -16 |
| Philippines | 51 | 44 | 34 (1977) | -23 |
| Thailand | 48 | 44 | 33 | -26 |
| Turkey | 45 | 41 | 34 | -16 |
| Egypt | 42 | 41 | 41 (1979) | [c] |
| Iran | 38 | 45 | 45 | [c] |
| South Korea | 38 | 33 | 23 | -31 |
| Burma | 35 | 40 | 40 | [c] |
| TOTAL | 2,523 | 40 | 31 | -23 |

[a] Excludes Vietnam with an estimated population of 52 million; information on vital rates over time is thought to be unreliable.
[b] Figures are for 1975 unless a later year is specified.
[c] No significant change.

SOURCE: W. Parker Mauldin, The Rockefeller Foundation.

Burma, Iran, Pakistan, Nigeria, or Egypt, but there is evidence of change in each of the other 9 countries of 35 million or more population. The decline has been particularly strong (more than 25 percent) in China, South Korea, and Thailand, and also has been impressive (more than 20 percent) in Brazil, Indonesia, and the Philippines.

The rate of population growth in less developed countries as a whole is down only slightly from its recent peak (see Table 4). It is now about 2.2 percent per year with a country range from about 1 percent to well over 3 percent.

After 1965, fertility declined more rapidly than mortality in Asia and Latin America, and the rate of natural increase was smaller in Latin America and much smaller in Asia during the last five years than in 1960-65. Fertility and mortality declines were about equal in North Af-

**Table 4** Percentages of natural increase in population (1950-1980)

| | 1950-55 | 1960-65 | 1975-80 | Percent change 1960-65/1975-80 |
|---|---|---|---|---|
| World | 1.77 | 1.99 | 1.81 | - 9.0 |
| More developed countries | 1.28 | 1.19 | 0.67 | -43.7 |
| Less developed countries | 2.00 | 2.35 | 2.21 | - 6.0 |
| Africa | 2.16 | 2.49 | 2.91 | +16.9 |
| Latin America | 2.72 | 2.77 | 2.66 | - 4.0 |
| Asia | 1.88 | 2.06 | 1.37 | -33.5 |

SOURCE: W. Parker Mauldin, The Rockefeller Foundation.

rica and the Middle East, but in sub-Saharan Africa fertility did not decline, whereas mortality probably did; so the rate of natural increase there continued to climb to an estimated 2.9 percent per year.

The net effect of these differing trends in the less developed countries was a modest decrease of about 6 percent in the rate of natural increase, from 2.35 percent in 1960-65 to 2.21 percent in 1975-80 (see Table 4).

A summary of the eight largest developing countries follows. These countries have a population of almost 2.25 billion—about half of the world's population. Together, they contain about two thirds of the population of all of the less developed countries.

***China (975 million)*** China has managed to reconcile Marxist tradition with population control. The official goal is to lower growth from 1.2 percent in 1978 to 0.5 percent in 1985 and to zero by the year 2000. Population growth is viewed as detrimental to capital accumulation; it hinders the elevation of scientific and cultural levels and improvement of the level of living.[16] Third and higher order births are to be reduced and then eliminated, and termination of childbearing after one rather than two children is to be promoted. The means are mobilization of party committees on all levels, strengthened propaganda and education, a system of rewards and penalties, and improved family planning services.

Most estimates of the population of China fall within the range of 950-1,000 million, but a reasonably firm figure must await the 1981 census. The government has announced a crude birth rate of 18.34 and a death rate of 6.29, and thus a growth rate of 1.2 percent in 1978. These figures are said to be estimates derived from incomplete data and there may be some undercounting. In the absence of accurate knowledge, the crude birth rate is shown as 20 in Table 3. Some scholars speculate that the crude birth rate might be as high as 22 and the death rate as high as

8 per 1,000 population. Whatever the precise figures, it is evident that China has achieved remarkable improvements in health conditions and in the reduction of fertility.

***India (694 million)*** India has a long history of population censuses and its total population is known with moderate accuracy, but the registration of births and deaths is incomplete. Prior to the mid-1960's, most estimates of vital events were necessarily based on the censuses. Now, and for the last 15 years, the Sample Registration Scheme (SRS) has produced estimates of birth and death rates for most of the states of India and for the country as a whole.

Analysts estimate that the average annual crude birth rate in India was about 45 per 1,000 population for the decade 1951-61.[17] The latest estimate from the SRS is 34 for 1977, but the SRS probably undercounts events, and a figure of 36-37 is more commonly accepted. These figures suggest a decline in the crude birth rate of about 10 percent from 1951-61 to 1961-71, and another decline of 10 percent from 1961-71 to about 1977. In 1976-78, the crude death rate was 14.5 per 1,000.[18]

There is great diversity in fertility in India. Rural rates average 5-7 points higher than urban rates, and rates in states range from the mid-20's to a high of 40. Six states report rates below 30 and four have rates of 35 or more. Uttar Pradesh, with a population of more than 100 million, has the highest crude birth rate, 40, and Maharashtra, with a population of 60 million, has a crude birth rate of 26.

A distinctive feature of the Indian family planning program has been its reliance on vasectomies, particularly in some states. Starting late in 1970, sterilization "camps" were organized in Kerala. These camps brought temporary facilities for sterilization into rural areas; they were accompanied by considerable publicity and usually by incentives for acceptors, partly in cash and partly in simple gifts. The cash value of incentives was of the order of $15-20. These camps became quite popular and were introduced into 16 states by 1972-73. But a number of problems arose, primarily because of a lack of adequate postoperative follow-up. There were some cases of tetanus and death following vasectomies in one of the camps and the program was discontinued.

During the emergency period, family planning was pushed to the point of coercion. The number of sterilizations increased to 8 million during 1976-77. With the change of government in 1977, the number of sterilizations decreased dramatically to 950,000 in 1977-78, and then increased to 1.6 million in 1978-79, still below the early 1970's.

Although India's crude birth rate has fallen significantly in recent years and, according to United Nations projections, may fall to about 26 for the period 1995-2000, a decrease of almost 29 percent from the estimated figure of 36.9 for 1975-80, the population will continue to grow rapidly. The population base is about 694 million; an increase of just under one-half by the end of the century would take the population past the billion mark—1,037,000,000.

***Indonesia (152 million)*** Indonesian fertility changed little from 1950 until the late 1960's. A large-scale family planning program, initiated in 1968 and well organized since 1970, has brought substantial change during the 1970's. This is reflected by a drop in the crude birth rate from 46 per 1,000 in 1965 to 36 in 1975.

The success of family planning has led to optimism about future fertility. The United Nations projects a decline from 37.9 in 1975-80 to 24.5 for the period 1995-2000. Although mortality is expected to decrease substantially (the crude death rate was 15 per 1,000 in 1978[19]), the Indonesian authorities are considering a target of replacement level fertility by the end of the century. The United Nations figures project a 35 percent increase in the population from just over 150 to just over 200 million.

***Brazil (122 million)*** Until recent years, Brazil has had high fertility and a rapidly growing population. During the past decade and a half, however, fertility rates have decreased markedly. The crude birth rate declined from 43 to 37 between 1960-70, and more rapidly since 1970; the crude birth rate is estimated to be in the range of 31-33 for 1976-77.[20] Other estimates are even more specific, with one statistician stating that Brazilian fertility fell 26 percent from 1970 to 1976.[21]

In spite of recent fertility declines, the momentum of population growth in Brazil remains strong. Contributing to it is a low crude death rate of 8-9 per 1,000 (1978).[22] The country's population is projected to be about 200 million in the year 2000,[23] with replacement fertility to be reached in the year 2015, according to the World Bank. If that materializes, Brazil's population could stabilize at 341 million. Thus, Brazil is likely to become the most populous country in the Americas sometime during the twenty-first century.

***Bangladesh (89 million)*** Although there is no evidence that fertility rates have begun to fall in Bangladesh, the United Nations projections (medium assumptions) assume a decrease in the crude birth rate from 47 to 35 by the end of the century. Its crude death rate is about 16.

Projecting these figures, the population would exceed 150 million by the year 2000. If Bangladesh does not achieve replacement fertility before the year 2035, as the World Bank projection assumes, the population would ultimately stabilize at 334 million. These hypothetical figures illustrate the problem ahead.

***Pakistan (82 million)*** Like Bangladesh, Pakistan has a low per capita income, a low adult literacy rate, and a relatively low life expectancy at birth. Although there are differences of opinion on the level of the crude birth rate, it is clearly above 40. In 1979, the crude death rate was 16.[24] The World Bank projects a population of 139 million in the year 2000, and an ultimate population of 335 million.

***Nigeria (77 million)*** Nigeria's crude birth rate is around 50 and its death rate 17-20.[25] The World Bank assumes that Nigerian replacement fertility will not be reached until 2040 and that the population would be 157 million in the year 2000—with a minimum of more than 200 million and a possible maximum of 435 million ultimately. One can be skeptical whether such a population is possible; the fear is that, if fertility does not decline, mortality will increase.

***Mexico (70 million)*** Mexico, with a population of 70 million, had high fertility rates and high rates of population growth, at least until the early 1970's. Since then, the crude birth rate has fallen by 13-15 percent to 37 in 1978 (the age structure acts as a brake on the crude birth rate decline in a high fertility society). In 1977–79, the crude death rate was 8.[26] The major decreases in fertility did not occur until a large-scale national family planning program was launched.

## Causes of Fertility Decline

Low fertility is always found in societies that have become industrialized and modernized, but a high degree of modernization does not seem to be a necessary condition for achieving low fertility, as the case of China shows. Some degree of consensus is emerging to the effect that:

- Marital patterns (age at marriage, proportion of reproductive years spent within marriage) account for a substantial part of the recent decline, perhaps one third in developing countries, but marital fertility itself is the major component.
- Social setting, especially health and educational status, has a close relationship to fertility decline.
- Family planning programs have a significant independent effect, at least in less developed countries with favorable social settings, including the three largest: China, India, and Indonesia.

## Prospects

Population growth rates in less developed countries are high. Although they will moderate somewhat during the next two decades, their momentum will remain high. This momentum will lead to larger and larger absolute increases in the populations of the less developed countries during the remainder of this century, even though fertility rates will continue to decline. A typical country with past high fertility and a current crude birth rate of about 40 would increase its population by 60 percent even if replacement fertility were achieved immediately.

Population projections for the year 2000 center on 6 billion. Thus, it is projected that the population of the world will increase by about 1.5 billion during the next 20 years.

Africa is expected to experience the most rapid growth, more than 75 percent, and Latin America the next most rapid, with a growth rate of about 65 percent. South Asia would grow by 55 percent and East Asia by 24 percent, according to these projections. North America and the U.S.S.R. would grow by 17-18 percent. Europe would grow very slowly, by only 7 percent.

No one knows when fertility will begin to decline, and at what rates, in sub-Saharan Africa, Bangladesh, and Pakistan. Nor is there consensus as to what the size of the world's population will be when (and if) it ceases to grow. If replacement fertility were reached in every country of the world by the year 2000, the world's population would not exceed 8.5 billion. But such a rapid decline in fertility seems unlikely. If replacement fertility were achieved by the period 2020-25, the population of the world would grow to 10.7 billion.

## MIGRATION

Each year, the United Nations *Demographic Yearbook* supplements a general review of population statistics with a more detailed survey of one particular topic. In 1977, for the first time, it featured international migration. The data are poor, however, and not easily interpreted. For example, the number of recorded immigrants from Economic Commission for Europe (ECE) countries[27] into other ECE countries was 57 percent greater than that of the corresponding emigrants, partly because of time lags, but mainly because "emigrant" and "immigrant" are not defined consistently. With that much inaccuracy in the documentation of "normal" *legal* migration within the region with generally the best population statistics in the world, one must be wary of conclusions about refugees and illegal, or quasilegal, migrants.

Statistics on internal migration are usually even poorer, particularly in a country like the United States with no identity cards. Data are collected in sample surveys and, inferentially, from questions in the census on where respondents were living one or more years previously. The "mobile population," defined as those who resided in different houses within the United States on two successive dates, is divided by the Bureau of the Census into "movers" within a single county and "migrants" from one county to another. The latter category is divided further according to whether the migration was within the same state, to a contiguous state, or to a noncontiguous state, with international migrants classified separately. This differentiation is intended to distinguish shifts in residence that are and are not accompanied by changes in job, school, type of neighborhood, and so on; but it is not possible to mark these socioeconomic boundaries clearly by a geographical index.

Perhaps the best estimate of the number of refugees in the world is given in the annual report of the U.S. Committee for Refugees, a private organization. By its count, the wars since 1945 have resulted in the flight of more than 60 million persons, of whom about 15 million still were not resettled in the mid-1970's. Although the figures obviously are not accurate enough to make precise comparisons, there has been a discernible trend upward. The total can be compared with the entire emigration from all of Europe between 1800 and 1950—also 60 million. Refugees are apparently the largest category of migrants in the world today.

Clandestine movements are widespread. According to a working estimate used in the late 1970's, one tenth of the alien workers in European Common Market countries,[28] or a total of some 600,000 persons, were at their current locations illegally. Following the decision of an international commission, anyone who knowingly organized illegal movements or employed illegal immigrants was to be subject to a punishment to be specified by each member country. Such controls are hampered by the citizenry's typical ambivalence toward illegal aliens, and often by an ambiguity of the illegality itself. Britain, for example, reacted to the influx of unskilled former colonials by imposing new controls without entirely abandoning the prior goal of free migration within the Commonwealth.

A somewhat similar situation developed in the region of the United States bordering on Mexico, where "illegal" immigrants are spoken of as merely "undocumented." Although the western hemisphere was formally excluded from the Immigration Acts adopted in the 1920's, consuls in Mexico were instructed to enforce rigorously the existing laws limiting visas to literate persons who would not engage in contract labor or become public charges. But, over the longer run, the influx of Mexicans continued to respond to the demand for labor in the United States.

By definition, the number of undocumented migrants is unknown. The number of undocumented Mexicans apprehended rose very slowly from a few thousand in the mid-1920's to well over a million in 1954, then fell to several tens of thousands in the following decade, rising again from under 100,000 in 1966 to over 500,000 in 1973. No useful estimates of the total flow or numbers present in the United States are available.

As far back as we have records, some migrants have been in temporary status. Chinese, as a prime example, generally left their home villages with the intention of returning eventually; the large numbers who settled throughout southeastern Asia and elsewhere did so more or less reluctantly. In Africa, as another example, a villager would spend some years working in the mines in order to acquire the means of establishing himself in village life. Even at the height of immigration to the United States, from around 1880 to 1914, a sizable proportion of immigrants—and of some nationalities, even a majority—came as sojourners and, after amassing modest sums, returned home and were able to move up a notch in their village or small town.

Temporary migrants in Europe, usually called by their German designation of *Gastarbeiter* (guestworker), have received labor permits of varying duration, with which they have moved usually from one of the Mediterranean countries to one of those in northwestern Europe. By 1974-75, West Germany had a resident alien population of 4.1 million, over 60 percent male, scattered through most of the country but concentrated in urban-industrial regions, and of diverse nationalities. Because of their favorable age structure, the aliens' birth rate was almost three times that of the natives, and their death rate was less than a fifth of the native one. Moreover, although juridically the aliens were temporary, they were in fact becoming, or so it seemed, a part of the society. The 15 million or so alien residents of northwestern Europe in the early 1970's comprised what was often called the tenth member of the Common Market, larger in total population than several of the member countries.

The *Gastarbeiter* institution, however, proved to be more flexible than some critics believed it to be, as well as a more serious problem than its proponents had anticipated. In Switzerland, with a higher proportion of temporary aliens than any other western European country, the world-famous amity among language and religious groups was seriously damaged. *Überfremdung* (in French, *hyperxénie*; in a possible English translation, "hyperforeignization") became a large issue in several elections, which resulted in some major revisions in immigration law. In Switzerland and in Germany, however, it was possible both to import labor when it was needed and to reduce the flow substantially when the need dried up. That there was more difficulty in France and Britain was due mainly to the fact that those countries imported labor from former colonies, persons who claimed—in some cases with a degree of success—full equality with the natives. The question is whether the economic benefit of the *Gastarbeiter* program outweighs its social cost.

Among permanent legal migrants, as stressed in the United Nations *Demographic Yearbook*, there also has been a marked change from the past. The major new fact is that emigration is from the world's less developed areas. In mid-1974, an estimated 9.5 million emigrants from those countries were residents of northwestern Europe, northern America, and Oceania. Latin America, which once had been second only to the United States as a target of immigration, has become a region of net emigration.

Between 1960 and 1974, the international migration from Asia more than tripled.

The United States is still a significant country of immigration, with nearly 400,000 newcomers per year. After the national quotas enacted in the 1920's were abandoned by the Immigration Act of 1965 (which went into effect in mid-1968), the sources of immigration shifted enough to affect the composition of the foreign stock almost immediately. Even when the nationality is the same by bureaucratic count, there may be significant differences between earlier and more recent immigrants. Virtually all of the Chinese living in the United States before the change in law, for instance, spoke Cantonese, reflecting the province from which their forebears had come. The new immigrants from (or often through) Hong Kong speak Mandarin, and they often feel as alien to the Chinese-American community as to the general American culture.

The advisory committee to President Nixon on population, known as the Rockefeller Commission, recommended more stringent measures to block illegal immigrants and that every two years Congress consider whether "the impact of immigration on the nation's demographic situation" indicated that the number admitted should be cut. Implementation of this will raise complex issues.

As a country develops, its urban sector grows by its own natural increase, by the annexation of new territory, and by the massive migration of countrymen out of agriculture into city occupations. Those who make this move successfully are rewarded with better jobs and incomes, and more varied and comfortable living. Although the rapid urban growth of less developed countries also derives in part from the better-paying jobs available in cities, the prevalent rural stagnation provides a strong impulse. Nonindustrial countries have enjoyed a remarkable decline in rural mortality; medicines, technicians, and death-control measures, disseminated from advanced nations through international agencies, have shattered the prior balance between the traditional economy and the rural population. The effects of the growing disparity between the countryside's resources and the numbers dependent on them, moreover, have been aggravated by a rise in expectations so rapid as to outstrip any conceivable improvement in the actual situation.

A very large proportion of this new type of migrant pushed out of the countryside lacks access to the most elementary urban facilities. Almost all large cities in less developed countries are ringed by squatter settlements, shantytowns of self-constructed huts which, in Latin America, for instance, may constitute a third or even half of the city's population. That as many as 50 percent of the Latin Americans in cities live in these conditions does not imply the shift in the economy, the degree of literacy, and the overall urbanity that these data once would have suggested.

One of the most significant elements of cities' innovating history—the

Lima, Peru. [International Labour Office.]

small-family pattern—arose there and then diffused among other sectors of society. In the historic West, urban fertility was lower everywhere than that in the countryside; for, as a part of their effort to move up the social ladder, most of the villagers who came to cities reduced their family burden. The lower urban fertility that has been routine in the West, however, is only one of several patterns now to be found in less developed areas. In various countries, according to the data available, the differences between urban and rural fertility are very small.

According to several sample surveys by the Bureau of the Census, migration within the United States retains a characteristic behavior pattern. Most migrants are, as before, young adults, and the proportion of young children has become almost as high as that of persons in their twenties. In other words, there is now greater family migration in the United States. Although the population in retirement areas is growing very fast, the persons who move to them are only a relatively small proportion of the country's older people. From 1965 to 1978, there was a net movement of 1.4 million persons out of the northeast and the north central regions to the South and West; that is, the historical trend of westward movement has continued, and the migration out of the South has been reversed. This combination is typically explained as a drift to "the sun belt"; along with the attraction of climate in the southern and southwestern states are their relatively favorable attitudes toward business, their frequently lower taxes, and their proximity to sources of energy.

By now, it has become a commonplace that urban areas have been growing in a new way. From about 1920 on, there has been a backflow from the central cities of metropolitan areas both to residential suburbs and to smaller industrial or commercial cities, the so-called satellites in the standard metropolitan statistical areas[29] (SMSA's). By 1970, two thirds of the national population lived in SMSA's but, over the period 1970-77, the annual growth of the metropolitan population was 0.7 percent, while that of the nonmetropolitan population was 1.2 percent. Within the SMSA's, central cities lost population and suburbs grew at a substantially higher rate than the nation as a whole. That some central cities have become black is well known; but, in spite of the stereotype of Mexicans as field workers, Hispanics are more metropolitan than either whites or blacks, with an estimated 84 percent of the Hispanic population living in SMSA's in 1978.

The in-migration to American cities has changed in a way analogous to that of the urban centers of less developed countries. In many cases, the rural-urban migration of blacks and Hispanics has been based less on qualifications for city-based occupations than on such factors as displacement from cotton farming by the mechanical picker, a lag behind the rest of the country in the South's recognition of civil rights, and growing population pressures in Puerto Rico, other Caribbean and Central American countries, and especially Mexico.

In sum, both international and internal migration have been undergoing fundamental change. The models derived from past experience, usually at least implicit in the interpretation or even the perception of what is happening today, are at best deficient, at worst obsolete. The most common migrant in the world today is the refugee, and generalizations derived from economically motivated movements hardly apply to this politically generated flow.

The illegal flow of immigrants, partly overlapping with that of refugees, is not a temporary phenomenon that can be disposed of with either an absolute ban or total relaxation. Although in trouble, the American economy still provides a life which is so much better than in most parts of the world that, given unlimited access, the world's destitute would submerge this country. What limit to place on access will continue to preoccupy the American government and public.

## RESOURCES

From a rudimentary viewpoint, population and resources present two sides of the same problem. With double the resources, including capital, double the population can be sustained at a given level of comfort. If resources and the techniques for converting them are fixed and population grows, then a declining level of consumption is certain. If technology expands indefinitely, then so can both population and the standard of consumption. It is the prospects for technological and economic growth that distinguish those who see a 12 billion population living comfortably at equilibrium on the earth's surface from those who see the present population at most doubled and largely continuing in poverty, with the nations engaged in a bitter struggle for the world product.

Since resources are unevenly distributed over the surface of the planet and by no means in proportion to population, the division of the world into partially closed national states adds a political factor to economic and technological ones. Trade is the means by which availabilities and needs can be made to coincide. Political and social obstacles to trade and the rational use of resources ought in some sense not to exist; and yet they are no more to be wished away than physical limitations.

With only a moderate degree of optimism, we may grant that the planet *ultimately* will be able to hold comfortably 9 to 12 billion people, but the immediate question is whether the pace of the growth of population can be matched by the pace of the advance in materials, energy, and capital. Any realistic discussion must be largely concerned with the bridge between the present state of affairs and the ultimate stationary one. Is agriculture expanding fast enough so that the 90 million Bangladeshis now in place (not to mention the 150 million expected in the year 2000) are moving toward a satisfactory diet? Will new sources of energy be available before oil supplies are exhausted? Energy is the key to the release of materials; will energy supplies and the efficiency with which they are extracted and used increase fast enough to provide the aluminum and steel that 6 billion people will need by the end of the century, and that 9 to 12 billion will require in 2100? The following discussion concerns prospective rates of change rather than absolute levels of resources.

Global per capita production of cereals has continued erratically upward. But it has grown more rapidly in a few more developed countries than it has in the majority of the less developed countries. The latter have compensated by increases in imports which make them more vulnerable to shifts in trade, balance of payments problems, and shortfalls abroad. Whether this situation can continue in countries faced with mounting energy import bills and limited export potential depends on the availability of foreign credit and grants. The uncertainties in the situation are perplexing and worrisome. While they are exacerbated by population growth, they would be there even if population growth were to cease. Given the fluctuations of output in the United States and abroad and the demand for our cereals in countries with strong economies, the dependence of some poor countries on American agriculture could give rise to some difficult ethical and political problems in the next few years.

Agricultural production can be efficient in a variety of ways. The Japanese have achieved high productivity levels by using relatively little land but a great deal of labor and chemicals. In contrast, the United States uses little labor but much land and mechanization to achieve similar productivity levels.[30]

Both approaches, however, are based on a highly developed scientific infrastructure and the substantial use of energy. Given these two ingredients, there appears to be sufficient flexibility in agricultural production techniques to achieve reasonable levels of productivity in most countries.

Roger Revelle[31] has asked how many people could be fed and fed well if yields comparable to those obtained by Iowa corn farmers—which are not the highest in the world—were achieved globally on all arable land. His answer is a large multiple of the present population. Such production levels may never be feasible because of organizational and institutional limitations, but his analysis is useful in making it clear that the sheer availability of land is not the central problem in feeding the world's growing numbers. Rather, it is the persistent disparity in the distribution of people on the one side and production capacity on the other. Production capacity depends on all of the factors contributing to the production of food—capital, scientific expertise, trained manpower, and adequate institutions, as well as land, energy, water, and other physical elements.

The National Research Council has made contributions to knowledge through the several volumes of its report on food resources.[32] One service has been given by calling attention to the possibilities of Leucaena, a leguminous tree that can fix up to 500 pounds of nitrogen per acre per year, with some strains growing to 65 feet in height. It puts down deep roots that enable it to tolerate drought. Cattle feed on its leaves; its wood is used for timber and fuel; it holds down soil on slopes that have been subject to erosion; its pods and seeds are used as human food in Central America and Indonesia. When cut, the stumps produce shoots that in some varieties reach 18 feet in 12 months. There are few more effective converters of sunlight. The chief problem is to get this valuable tree disseminated.[33]

Worldwide per capita production of petroleum appears to have peaked in 1974. This is the result of the four-fold increase in prices in 1973-74 which dampened demand, rather than a sign of impending global depletion. Since World War II, depletion of older working fields in the United States and elsewhere, along with discoveries in the Middle East, led to concentration of production in a small group of countries which, for political as well as economic reasons, banded together to raise the price. There are serious dangers in the current world energy situation, stemming from the fact that importing countries have become increasingly dependent on supplies from a politically unstable area of the world, rather than from the likelihood of imminent exhaustion.

According to H. E. Goeller and Alvin M. Weinberg, a reasonable standard of living for billions of people would require 60 terawatts of energy.[34] This figure compares to about 7.5 terawatts used today and is about 1/2,000 of the energy absorbed from the sun. Can a 60-terawatt world be reached and sustained?[35]

It can, insofar as the sheer availability of energy is concerned. Coal alone could supply the energy needs of 10 billion people for several hundred years. Nuclear and solar sources could do the same for even longer eventually. But each of these sources has potentially serious problems. Coal dirties the environment. Continued and expanded combustion of fossil fuels along with deforestation could lead to increased atmospheric carbon dioxide levels which could increase world temperature levels. Carbon dioxide in the concentration that would follow the spread of industrial civilization based on coal could warm the planet, change the windstreams of the atmosphere and the currents of the oceans, raise the temperature of some cold countries, and make some wet countries dry. While the effects on agriculture would be negative in some regions and positive in others, there is little doubt that severe social disruptions could ensue. If the worst fears of some are realized, we will not be able to use all of the coal and fossil fuels we have available to us.

Nuclear fission avoids increasing carbon dioxide, but it raises a further set of difficult and dangerous issues, from the risks of catastrophic accidents to the diversion of fissionable materials into atomic weapons. If the secrets of nuclear fusion can be unlocked and applied at reasonable cost, most of these concerns will be resolved (although some radiation and waste disposal problems will remain). But, for practical purposes, fusion is several decades away.

Solar energy creates only minor environmental problems; the large land areas it requires would be available in deserts useful for little else. It is the one source that gives promise of production by the household for its own use. But to use it in large quantities will require a substantial lowering of cost, especially in the direct generation of electricity, as well as storage devices not now available. To produce significant quantities of biomass for energy use with today's technology would bring energy production into direct competition with agriculture for arable land. Still, solar energy in its various forms, plus fission, are the best hopes we have for a sustainable energy future.

Nonfuel minerals are not likely to run short. While current plus prospective reserves of some important minerals would be exhausted in a few decades at present relative prices if there is no technological change and no change in recycling rates, these conditions probably will not hold. In the worst case, global prospective reserves of bauxite (from which aluminum is derived), for example, might be exhausted around 2040, but there are many other sources of aluminum. If they had to be used today, production costs would rise by at least 20 and perhaps as much as 80 percent; with prospective technological improvements, the crossover price is likely to decrease significantly over time, certainly before bauxite is exhausted. For some other nonfuel minerals, substitutes can be found easily at costs only marginally higher than current levels. In the case of most minerals, the odds are against concerted cartel action.

Can more abundant minerals be substituted for less abundant ones without substantial increases in cost? With three exceptions, Goeller and Weinberg argue that they can—that societies can turn to nearly inexhaustible minerals with little loss of welfare.[36] Detailed studies of cadmium, zinc, lead, copper, tin, and mercury indicate that a number of materials such as iron, aluminum, silicon, magnesium, titanium, and others can be substituted in many uses. Other studies find that these abundant materials can be acquired from lower grade ores which are in nearly inexhaustible supply, provided that sufficient energy is available at reasonable costs, both economic and environmental. In addition, glass and cement from virtually unlimited sources (sand, soda, clay, limestone), plastics, and some materials derived from renewable wood and plant sources can be substituted for many of the minerals that might be exhausted eventually. (A section of the last chapter in this volume, "Prospects for New Technologies," offers a detailed description of new methods of resource exploration.)

The exceptions—important resources that are not in unlimited supply—are fossil fuels, phosphorus, and a few elements essential in trace amounts for agricultural production, such as copper, zinc, and cobalt. While reserves and prospective reserves of coal, phosphorus, and these trace elements are adequate for several hundred years, they are not inexhaustible. Ultimately, phosphorus and some portion of the trace elements may have to be reserved for agricultural use and returned to the soil by recycling agricultural and animal wastes, and perhaps even bones. But fossil fuels cannot be recycled; eventually, a replacement must be found, even for coal.

Recycling can stretch supplies of nonfuel minerals. It has the advantage that the energy required to put materials back into productive use is generally less than that needed to reduce and refine ores. Goeller and Weinberg point out that, for magnesium, the remelt energy is only 1.5 percent of the energy required to win the metal from virgin ore; for aluminum, it is 3 to 4 percent; and for titanium, it is 30 percent.[37]

There are signs that forests and fisheries are being overexploited. One sign is the downward trend in per capita wood production since 1965; another is the decline in the surface area of the globe covered by forest. Most of the deforestation is occurring in less developed countries where energy price rises, population growth, and the need for additional agricultural land are intensifying pressures on fuel wood and forested land, especially in the delicate ecosystems of tropical forests. But production techniques are changing rapidly in forestry as well as in agriculture. Settled farming techniques are beginning to be applied. The change is most marked in the forestry sectors of more developed countries but, even in less developed countries, isolated examples of successful reforestation and village woodlot programs can be found. The fishing industry is further behind in applying farming techniques, in large part because the technological problems are more difficult; but research and commercial activities in this area are increasing. Some changes are clearly necessary; there has been no increase in the world fish catch since 1970, despite great increases in fishing efforts.

Water, air, and land pollution pose serious problems for those countries that have not mounted the effort necessary to avoid or offset environmental deterioration. Both population and economic growth tend to have adverse effects on the environment. But environmental protection programs are halting many forms of deterioration at affordable costs.

Possibilities for the more sparing use of materials, and for conservation by both producers and households, are extensive. The answer to the question of whether growth of demand is outstripping supply is inconclusive. There are some signs of deterioration on a global level but also some signs of improvement.

The key to the long-run sustainability of life on earth is energy. If there is sufficient low-cost energy and it can be used without adverse environmental effects, land can be reclaimed, seawater can be desalted, fertilizers produced, low-grade ores mined, metals recycled, and substitutions made between minerals.

We are told that: "If present trends continue, the world in 2000 will be more crowded, more polluted, less stable ecologically, and more vulnerable to disruption than the world we live in now."[38] Other statements go much farther:

> . . . our planet is grossly overpopulated . . . the limits of human capability to produce food by conventional means have very nearly been reached . . . more likely than extinction [of the human race] is the possibility that man

> will survive to endure an existence barely recognizable as human—malnourished, . . . surrounded by the devastation wrought by an industrial civilization that could not cope with the results of its own biological and social folly.[39]

Yet, on the other side, ". . . the world has been exhausting its exhaustible resources since the first cave man chipped a flint, and . . . the process will go on for a long, long time."[40] It is argued that energy and materials are only a small part of the cost of production and their prices could double or treble without much effect on the prices of finished goods; that resources under the ground are much more extensive than any exploration so far has revealed; that impending exhaustion of any particular material would be signaled by a rise in its price and there would be adequate time and opportunity to adapt; that technical advances permit existing materials to be used more effectively and the operation of the market constantly substitutes more abundant for scarcer materials; and that technical and administrative advances are more than equal to the task of safeguarding the environment.

The citizen may well be puzzled by such contrary opinions provided by authoritative writers. And he will hear more in the coming years, for the debate will not be brought to an end soon because one side or the other runs out of facts to support its position. Yet facts that would decide the overall issue of whether resources are the constraining factor on our industrial civilization are conspicuously lacking.

Our knowledge is an island in a great sea of ignorance. We do not know how much crude oil and gas lie under the surface of the land and sea because exploration is expensive and companies carry on only enough of it to assure the production of the next few years. The prospects of workable fusion by the twenty-first century cannot be estimated. It had been hoped that nodules of magnesium and other minerals would be brought up from the mid-ocean floor, but this may be much more expensive than at first calculated. Food production in poor countries has been increased by the complex of modern practices known as the Green Revolution, but the key question is whether the supplies will keep ahead of population increases. And no one knows how those pushed out of agriculture in the process can be put to effective work that would enable them to earn their share of the food.

Because of these and many other unanswered questions, this report does not attempt to assess the degree to which the further progress of the United States and the world will be held back by resource scarcity, as against limits set by shortages of capital and labor.

In the face of uncertainty on so vital a matter as resources, one would be ill advised to choose a middle point between the optimistic and pessimistic views. Prudence suggests asymmetry, taking more account of possible bad outcomes than of equally likely good ones.

## CONCLUSION

Population problems take very different forms in the more developed and the less developed countries. The less developed countries still face much infectious disease, to which malnutrition makes many of their citizens more susceptible; in the more developed countries, the question is rather how to bring heart disease and cancer under control. For less developed countries, the prime question on fertility is when childbearing will come down to levels at which the population would just replace itself; more developed countries are mostly below long-term replacement already.

The stability and wholesomeness of the environment are prominent in the minds of those in comfortable economic circumstances; for the poor of the world, such considerations are subordinate to sheer survival—an issue that reveals itself in different attitudes to the use of DDT against insect pests and in many other ways. For the more developed countries, the ocean's minerals represent an immediate opportunity to supplement expensive terrestrial resources; less developed countries would like to see those minerals left where they are until they themselves are in a position to join in their exploitation. The list of real differences is a long one and finds expression in lively ideological and political exchange.

The less developed countries have shown rapid progress in extending life expectancy over the post-war period, many increasing it by six months per year. Among more developed countries, progress has been slower; the United States, for instance, gained 26 years since 1900 to reach 73.5 years by 1977. Differentials are still found everywhere; for the United States, the most favored sex-race group is white females, with 77.7 years' life expectation at birth and a median age at death of 81.1 years. This means that half of the white girl children born would live to 81.1 years or more, according to the mortality rates of 1977.

The resistance to further advances in life expectancy can be illustrated by the effort to eradicate cancer in the United States. The utmost that complete eradication of cancer could do is to increase the expectation of life by about two years. Increase in some forms of cancer in recent years has come about in part because people who are saved from heart disease are of the age to fall victim to cancer. Aside from this, the basic biology of cell reproduction is not well enough understood that one can proceed confidently to specifics for the prevention or cure of cancer. (The unknowns are described in detail in the second chapter of this report.) How to allocate research resources between increasing knowledge of the basic biology on the one side and devising specific therapies on the other will continue to be a problem of scientific administration.

Although no exact figures are available, a recent Presidential Commission on World Hunger reports that the number of hungry people in the world is increasing. Mortality increase through famine is a possibility, but more to be feared is the susceptibility to disease brought about by chronic malnutrition. Malnutrition is only one of the conditions that raise the possibility of a reversal of the downward trend in deaths for the world as a whole. The complex ecology of snails and of the anopheles mosquito, resulting in schistosomiasis and malaria, may be another; the interaction of these diseases with malnutrition is especially threatening.

Unprecedentedly low birth rates have started to appear in the industrial countries. The Federal Republic of Germany shows 9 per 1,000 population in 1978 (against a death rate of 12); Austria and Sweden show 11; the United Kingdom 12; and the United States 15. Whether such low rates are a temporary dip or are a durable feature of advanced industrial societies, no one can yet say. It is known that, with increasing wealth in families, children come into conflict with other expenditures, and parents see them as too expensive. But the apparent cost of children is only one element in the decline of the birth rate.

The falling birth rate is associated with important changes in the institution of the family, in which the United States may be the leader. Between 1960 and 1975, two-worker, husband-wife families without children increased from 23 percent to 30 percent of all American households; one-worker, husband-wife households dropped from 43 percent to 25 percent; one-worker households of other types—those with female heads, and men and women living alone—increased from 14 percent to 20 percent of households. It can be expected that the one-worker household containing husband and wife will continue to fall throughout the remainder of the century.

Women increasingly work outside the home and reduce childbearing. The divorce rate is high and the prospect of divorce is itself a deterrent to childbearing. A child is a handicap, both for work and for remarriage, to whichever member of the couple has to look after it. Divorce is only one reason for the numerous single-person households; another is children leaving the parental home at young ages, before they contemplate marriage. Recently, more people have had the means to live alone, and evidently the wish to do so is widespread. Households consisting only of a woman and her one or more children are increasing, partly because of divorce, but also because of a great rise in illegitimate births.

Should the United States have a national population policy? Some industrial countries do—particularly those of Eastern Europe; others, for instance West Germany, do not. If present tendencies continue, the native United States population will taper off at only 20 to 40 million more than we have now, and then will decline slowly. A variety of incentives—income tax exemptions, housing loans, subsidies for child-care centers—are now in effect, but they do not constitute a population policy; they are far too mild to motivate childbearing at the expense of jobs and careers. Very large sums would be required to counteract the cost of rearing and educating children and the wages foregone in looking after them. In the Russian part of the U.S.S.R., despite enormous state expenditures on day-care for children, free medicine, pregnancy leave, and so forth, birth rates are about as low as those of the United States.

When, during the late 1950's, the United States was showing birth rates that would have led to nearly four children per couple, U.S. promotion of family planning in less developed countries was unconvincing. Considerable controversy developed in the press around the world, with frequent use of such words as genocide and cannibalism. Now that the United States is below replacement, our example is noted, and this kind of criticism of our efforts is seldom heard. The fear of excessive urbanization in the less developed countries has helped make birth control more acceptable to citizens and to governments; overpopulated cities are more conspicuous and more politically exigent than an overpopulated countryside.

The Committee on Demography and Population of the National Research Council and its staff are seeking more precise information on levels and changes of fertility in the world. With World Fertility Survey data and refined techniques for making indirect inferences on birth rates where direct information is unavailable or incomplete, we should attain more knowledge of current changes on the several continents.

Demographers watch those numbers with great concern. When an increase in income comes along, two opposite responses are possible. One, which might be called the Malthus effect, is to turn the increase in income into more children. The other, which might be called the demographic transition effect, turns the increment in income into a higher standard of living, and then goes on to reduce births in order to preserve and increase the higher standard. Current fertility rates tell how far the demographic transition effect is taking over from the Malthus effect.

Europe's fertility decline was accomplished not only without governmental encouragement but mostly without contraceptive apparatus. A sufficiently strong motivation can overcome material shortcomings. Conversely, if people want to reproduce—for instance, in intertribal or international competition, or for old age security—no amount of contraceptive equipment will make much difference. Yet the incipient motivation to reduce fertility in the less developed countries can be strengthened by technical advance. A safe, once-a-month pill for women, and something corresponding for men, would accelerate the decline of the birth rate in many countries. Improving the technology of contraception is one of the ways in which the United States can make a contribution.

As the number of births in the more developed countries falls toward the number of deaths, migration will constitute an increasing fraction of their population increase. In the United States, the present excess of births over deaths of about 1.5 million will last only a generation or so, unless current tendencies are reversed. As migration comes to provide the largest part of the increment to the United States population, it will receive increasing attention.

Nowhere do the dilemmas of policy show themselves more clearly than in migration. Migration has changed from a free movement in search of a better life, one in which the ancestors of most Americans were welcomed to these shores, to a movement dominated by refugees and illegal entrants. Illegal migration is a response to genuine economic need—not only the need of the migrants but of the receiving economy. To keep illegal migrants out would require either a severity of border control for which Americans are not prepared, or else making employers responsible for identifying their workers, which again seems politically unacceptable. Even if measures severe enough to be effective were congenial to American public opinion, they would be found offensive by the countries from which the migrants come.

On the other hand, the forces in place do not point to any easy legalization of the migrant flow. Some of the migrants' economic utility to employers would be lost if they were legalized and subject to taxation, unionization, and minimum wage standards. Thus, the Comptroller General heads a current report, "Prospects Dim for Effectively Enforcing Immigration Laws."[41] The many constituencies involved impede any clear-cut solution to the problem of illegal migrants.

The size and persistence of the flow of illegal migrants may well depend less on our government than on how rapidly Mexico and other countries expand their economies and control their births. And yet American imports of scarce fuels and materials go up with increases in population, legal or not.

The future population of the United States depends on births, deaths, and migrations. Extrapolating on the basis of considerations such as the preceding, the Bureau of the Census gives for the year 2000 a middle estimate of 260 million, and low and high figures of 246 and 283 million respectively. The low, medium, and high estimates assume that we are moving toward 1.7, 2.1, and 2.7 lifetime births per woman from the present level of 1.8.

For the more developed countries, a rise is likely from the present 1.1 billion to an ultimate 1.4 billion or so, predicted to be reached within the next two or three generations. The major variation among estimates is with respect to the less developed countries; for these, everything depends on how soon they go through demographic transition. If they have dropped to bare replacement by 2000-05, they will stand at 7.4 billion in 2075 (from their 1975 2.8 billion); if replacement is 15 years later, they will reach 9.8 billion by 2075. That puts world population in the range of 8.8 to 11.2 billion.

The above estimates take the continuing decline of mortality for granted and vary only on fertility. But it is legitimate to suppose that mortality will rise if the higher birth rates persist. There are already signs of a slowdown in the fall of mortality in some countries. Some parts of the world are making real progress in birth control; those that are not could be facing a major increase of mortality.

The population problem is not one and the same for the world and for separate countries; each country has its own. Bangladesh has the classical problem of too many people on too little arable land. Burma has a good deal of land and rich agricultural production, but a population that is rapidly filling the land; the combination of a dynamic population and a static economy will bring a difficult situation within a very few years. Some have problems of population distribution: two thirds of Indonesia's population is concentrated on Java and Bali. The Soviet Union has a fast increasing Asian population; in the last census, the Russian part had dropped to less than half of the total, at least in children under ten, and this is of concern to the regime. France sees itself threatened by its low birth rate.

It is easy to rationalize away the population problem. If the poorest countries would show as high rates of economic advance as Japan and Korea have done, they would quickly approach their demographic transition. If there were no national restrictions on trade, then even at the present time food is adequate for all. If the Javanese would only eat corn rather than rice, they could get enough carbohydrate per acre; if they could attain the yields of Japan or Iowa, they could eat plenty of whatever they want.

It is fair to say that the population problem consists of the fact that these things are not done easily. Nationalism, cultural preferences in food, inefficient agriculture, and urban-centered development are very persistent. So is the large-family culture in many parts of Latin America, Asia, and especially Africa. To wish these things away is as unrealistic as to ignore shortages of physical resources. The control of oil supplies by a dozen countries, to promote what they conceive to be their national interests, is as much a part of the energy problem as the ultimate exhaustion of the oil fields.

Many of the series bearing on the physical capacity of the earth to sustain population have tended to level off during the 1970's, including the world grain yield per hectare and the world fish catch. The increases of national economies that were taken for granted in the 1960's are no longer in prospect. Famines or near famines have occurred in Ethiopia, Bangladesh, Haiti, India, and the Sahel (involving six countries just south of the Sahara). After a long period of decline, relative prices of raw materials have on the whole increased in the 1970's. No one knows to what extent these changes will prove to be temporary, or whether the growth of the 1960's will resume. This may be the beginning of a long epoch of hard times; on the other hand, the world's numerous resource, environment, and population problems may be solved quickly, and growth rates of 10 percent return.

It is not even certain that the climate will allow present grain yields to continue. Climatic records suggest that, in the perspective of historical and geological time, conditions in the present century have been exceptionally favorable to cereal crops. And overall climate could be favorable but soil and water in short supply in particular areas. Increasing carbon dioxide could raise the global temperature and make things worse for many countries by changing the main currents of the atmosphere and oceans. Unstable tropical soils are being pressed into service for needed subsistence crops; in this situation, more people work the land harder and produce more desert, which results in further pressure on the ecology of the remaining good land. There are large areas where crops are limited more by water than by land; in parts of the United States, the use of ground and fossil waters is lowering the water table at an alarming rate, as Chapter 7, "Water Resources," explains, while in Pakistan and elsewhere irrigation and poor drainage are waterlogging and salinizing the soil.

One should not make these statements—least of all in a report on science and technology—as though the techniques of agriculture and the knowledge of ecology were going to be frozen in their present position. Holding technique constant and allowing population to rise is a way of creating spurious problems. New sources of energy would free oil and gas to become feedstock for synthetics and fertilizers. Cereal plants that fix their own nitrogen, as do present legumes, are being developed. Oil-based energy to distill the ocean waters for agriculture is now unlikely; solar energy will have to continue to produce rainfall for that, as it has from time immemorial. However, the timing and place of rainfall to increase yields might be influenced by cloud seeding, now under investigation.

The question is not whether such things can be done, but whether they can be done at a rate that will keep up with population and provide the minimum increase in welfare that is expected. On the population side, there will be stability sooner or later; on the resources side, technology will produce plenty, sooner or later. The question is whether technical improvements can occur at a pace that will bridge the time to this balanced condition.

This is where the United States can bring enormous power to bear—that of its scientific and technological establishment. It is no longer the world's richest country in per capita income, but its research capacity is unequaled. It could develop new technologies that would multiply resources and make the present surplus labor of the poor countries fruitful.

# Summary and Outlook

The mismatch of population, resources, and capital will continue to hold the world's attention for the next five years and much longer. A supposed population excess can only be relative to the means of sustaining population. It is this linkage that makes the problem so controversial. Those who think that the earth contains rich resources waiting to be discovered do not foresee population pressures. Those who think that capital can be multiplied easily with a different social organization see no reason for population restraint.

Yet, recent experience has dampened the optimism that would let population grow without limit because new resources and capital can be found or created easily. We have lived for eight years now in a general energy crisis. It is difficult to extract coal and burn it without polluting the atmosphere. Some 23 percent of the earth's land surface is desert, and an additional 25,000 square miles become desert each year. The local impact of this in sub-Saharan Africa and Asia, south of the Himalayas, is catastrophic. Arable land per capita is declining everywhere and the energy needed to raise yields is more and more costly. Demand for lumber and firewood causes overcutting and a potentially disastrous shrinkage of the world's forest area. The world's fish catch seems to have peaked and could well have entered a long-term decline, despite more ships engaged.

Even if resources and capital were ultimately unlimited, there is a clear ceiling to the pace at which they can be expanded. That is what gives importance to statistics of the increases in population, resources, and capital, country by country. One way of making the problem disappear is by taking a global view and showing, for instance, that over two decades cereal production has been going up by 2.7 percent per year, and population increases by only 1.7 percent per year. Detailed data not only for the world as a whole, but also for individual countries, are needed. By recognizing that there are independent national states and an unequal distribution among them of population and resources, the center of the problem can be approached.

Thus, this chapter takes up in some detail the components of population increase—births, deaths, and migrations—for major countries. A principal finding is that the populations of the more developed countries have mostly dropped below replacement levels, but with considerable variation among them. American birth rates were higher than those of western Europe during the 1950's, but they have fallen more sharply since then. American mortality was discouragingly level during the 1950's and 1960's but, during the 1970's, it took a turn downward. Soviet mortality has been rising.

At first, birth rates in the less developed countries did not decline as their death rates dropped. There seemed to be an impasse—no fall in births before development, but development held back by high birth rates and large populations. Experience of the late 1970's suggests that there may be an escape. In China, Indonesia, and several other countries, effective government action along with social changes are resulting in a dramatic decline in birth rates. The fall is recent and the means of measuring it are inadequate, but evidence shows that birth rates can be made to decline before full economic development occurs. One of the factors that brought about the declines in recent years has been American-aided programs.

However, even if things continue to go well on the population control front, the world's population will be 6 billion by the end of this century, and 9 billion by the latter half of the twenty-first century. Population projections are notoriously uncertain but, assuming continued progress in the spread of contraception, simple arithmetic shows that the present annual increase, now 70 million or so, will exceed 90 million by the year 2000. The transition to a happier balance between the human population and the earth that is its habitat will not be easy. If the Chinese can reach their stated objective of a one-child family, already virtually attained in West Germany, and the example is followed elsewhere, a stationary world population living in security and prosperity may indeed be possible by the late twenty-first century—but one must not underestimate the effort and restraint needed.

The United States has had some part in the progress made to date and should have a larger part. Yet, the public and its leaders are disillusioned about the benefits of outright gifts of money to the less developed countries. In what is the largest (if unintended) social experiment of all time, enormous transfers have gone to a dozen countries through the operations of OPEC (Organization of Petroleum Exporting Countries), but they have not produced a proportional gain in the welfare of the people of those countries. An investment in their own human capital is far more decisive for development than any physical or financial capital can be. American assistance generates the highest returns in the welfare of the poor countries when educational, scientific, and technical leadership, rather than money transfers, are provided. This country's capacity to devise better methods for activities—ranging from village industry to contraception to satellite systems that reveal new resources—is unparalleled. Improved ways of using biogas and solar energy, new plant sources of protein, textiles, rubber, firewood, and other necessities will help to raise living standards in poor countries, especially if they can be applied with local capital.

Through the United Nations and elsewhere, the less developed countries have communicated a strong nationalism. They have reminded rich countries of the principle of noninterference, and have stressed that only domestic effort can solve population and other developmental problems. The principle is a sound one, and it will be severely tested by growing populations, increasing urbanization, fuel shortages, environmental deterioration, and rising foreign debts. Again, the help that the United States can provide which will be most effective as well as most consistent with the independence of the poor countries is in the form of new discoveries and inventions contributed to the worldwide pool of science and technology. That contribution is the subject of the remaining chapters of this volume.

## DEMOGRAPHIC TERMS AND CONVENTIONS

### Mortality

Demography has developed ways to avoid the misunderstanding that can easily arise because populations have different or changing *age distributions*. Thus, the *crude death rate* (total deaths divided by total population) of the United States is higher than that of Venezuela: 9 per thousand against 6 per thousand. To make a proper comparison between the two countries, *age-specific death rates* (for example, the deaths to people aged 20-24 at last birthday divided by the population aged 20-24) are compared; it will be found that, at each age taken separately, the United States rate is lower. A convenient summary of the death rates is the *expectation of life* for a particular year. That for the United States in 1977 at age zero was 73.2; for Venezuela, less than 70. Hypothetical individuals subject to U.S. 1977 mortality would live 73.2 years. On the average, real individuals will live longer than this because they will be subject to mortality declining from 1977 levels. The hypothetical expectation is a good way of comparing countries as well as different times in the same country.

To see the effect of particular diseases, we calculate *cause-deleted life tables*; one of these shows that, if there were no cancer deaths and all other mortality were the same as the current rates in the United States, the expectation of life would be increased, but only by about two years. That the amount is so small is due to *competing causes*: people whose lives are saved by eliminating cancer are old enough that they soon would die of heart disease. Such competition among causes applies less to accidents because these occur in large part to younger people; it does not apply to infectious diseases, whose virtual elimination in the United States, combined with other public health measures, has raised the expectation of life in this country by 20 or more years since the beginning of the century.

## Fertility

Some of the same considerations apply with respect to *fertility*. Thus, the *crude birth rate* (total births divided by total population) has been rising in the United States over the last few years largely because an increasing proportion of the population happens to be in the marrying and childbearing ages. (These are the baby boom cohorts of the 1950's.) Demographers examine *age-specific birth rates*, for example, births to women aged 20-24 years at last birthday divided by the total number of women aged 20-24. They combine mortality and fertility to calculate the *net reproduction rate*, which is the expected number of girl children that would be born to a girl child with the life table and the age-specific birth rates of a given year. This is the *replacement ratio* of the population; it tells how many girl children a girl child is replaced by in the next generation. In the United States now, it is about 0.9. The total fertility rate is a little over double this, or about 1.8 children of both sexes per woman—not enough to maintain the population in the long run. That means that the U.S. population is failing to replace itself, even though it is still increasing by virtue of *population momentum* that reflects the age distribution resulting from previous high fertility. If present age-specific rates of birth and death continue, the native population would ultimately drop off 10 percent per generation.

Sooner or later, all populations must become *stationary*, meaning that on the average they would neither increase nor decrease; over the long run, their net reproduction rate would be unity. A long-run stationary condition would take the form of fluctuations in births around a net reproduction rate of unity. All of this applies to the *closed population* consisting of the present inhabitants of the United States and their descendants, and disregarding migration; migration at present rates would offset any shortfall of natural increase.

Very different is the condition of the less developed countries, many with a net reproduction rate of 2 or more. This means that they double in a generation where the *length of generation* (approximately the mean age at childbearing) is usually 25 to 30 years. The rate of increase of Mexico a few years ago, and of some African countries now, is as high as 3.5 percent, which implies a *doubling time* of $70 \div 3.5 = 20$ years. A country that doubles in 20 years multiplies by 32 in a century. In the *demographic transition*, the decline of deaths is followed at a longer or shorter interval by the decline of births; each decade of delay in the fall of births can increase the population by fully one-third. For the poor countries, the rapid population increase is made even more difficult by *internal migration*, specifically, the flow of population to the cities, or urbanization. The cities of many poor countries are increasing at 4 percent or more per year, which is to say that they double in less than 18 years.

*Population forecasts* used in this report and elsewhere are essentially *projections*, the working out of consequences of certain sets of assumptions. If past forecasting experience is a guide, we cannot know the world population by the end of the century even within half a billion, nor that by the middle of the twenty-first century within 2 or 3 billion. The projections provided here are hypothetical or illustrative only.

## REFERENCES

1. The regional classification system used by the U.S. Bureau of the Census groups countries and regions into either "more developed" or "less developed." The system is based on the assumption that, within regions, levels of such factors as industrial development, literacy rate, gross reproduction rate, per capita income, etc., are somewhat constant. Countries in North America and Europe plus Japan, Australia, New Zealand, and the U.S.S.R. are "more developed"; all others are "less developed."

2. United Nations Population Division. *Demographic Estimates and Projections for the World, Regions, and Countries as Assessed in 1978.* January, 1979, Tables 2A and 2B.

3. U.S. Bureau of the Census. *Statistical Abstract of the United States.* Washington, D.C.: U.S. Government Printing Office, 1979, p. 70.

4. Drawn from Timothy Dyson, "A Working Paper on Fertility and Mortality Estimates for the States of India"; and Registrar General of India, "Selected Indicators of Mortality and Fertility in India, 1976." Both papers presented at Workshop, Panel on India, National Academy of Sciences' Committee on Population and Demography. New Delhi, November 1979.

5. The bibliography of studies of mortality decline in Sri Lanka is quite lengthy. The most authoritative analysis is Peter Neurman, "Malaria and Mortality," *Journal of the American Statistical Association*, Vol. 72(June 12, 1977), pp. 257-263.

6. On Kerala, see John Ratcliffe, "Social Justice and the Demographic Transition: Lessons from India's Kerala State," *International Journal of Health Services*, 1978, Vol. 8(1), pp. 123-144; and *Poverty, Unemployment, and Development Policy: A Case Study of Selected Issues with Reference to Kerala*, New York: United Nations, 1975. On Sri Lanka, see Paul Isenman, "The Relationship of Basic Needs to Growth, Income Distribution, and Employment: The Case of Sri Lanka," Manuscript, Policy Planning and Program Review Department, The World Bank, May 18, 1978. On Cuba, see Sergio Diaz-Briquets, "Income Redistribution and Mortality Change: The Cuban Case," paper presented to Population Association of America annual meeting April 13-15, 1978, Atlanta, Georgia.

7. These relations are best documented in Latin America. See the series United Nations, Centro Latinoamericano de Demografía. *La Mortalidad en los Primeros Años de Vida en Países de la America Latina.* Santiago, 1976-79.

8. J. C. Caldwell. "Education as a Factor in Mortality Decline: An Examination of Nigerian Data," *Population Studies*, Vol. 33(November 1979), pp. 395-414.

9. Samuel Preston. *Mortality Patterns in National Populations.* New York: Academic Press, Inc., 1976.

10. D. Payne, B. Grab, R. E. Fontaine, and J. H. G. Hempel. "Impact of Control Measures on Malaria Transmission and General Mortality," *Bulletin*, World Health Organization, Vol. 54(1976), pp. 369-377.

11. U.S. Bureau of the Census. *Statistical Abstract of the United States, 1979.* Washington, D.C.: U.S. Government Printing Office. U.S. Department of Health, Education, and Welfare, National Center for Health Statistics. *Health: United States: 1978.*

12. U.S. Department of Health, Education, and Welfare. *Proceedings of the Conference on the Decline in Coronary Heart Disease Mortality.* Bethesda, Md.: National Institutes of Health Publication No. 79-1610.

13. Evelyn M. Kitagawa and Philip M. Hauser. *Differential Mortality in the United States.* Cambridge, Mass.: Harvard University Press, 1973.

14. Lester Breslow. "Risk Factor Intervention for Health Maintenance," *Science*, Vol. 200(May 26, 1978), pp. 908-912.

15. U.S. Bureau of the Census. *Projections of the Population of the United States: 1977 to 2050.* Current Population Reports Series P-25, No. 704, July 1977. Washington, D.C.: U.S. Government Printing Office, 1977.

16. C. Muhua. "Birth Planning in China," *Family Planning Perspectives*, Vol. 11, No. 6(November/December 1979).

17. A. K. Jain. "An Appraisal of Family Planning in India," *Population in India's Development 1947-2000.* Edited by A. Bose et al. Indian Association for the Study of Population, 1974.

18. *Sample Registration Scheme Bulletin*, Vol. XIII, No. 2(December 1979). Office of the Registrar-General, India, Ministry of Home Affairs, New Delhi, India.

19. U.S. Bureau of the Census. *International Population Dynamics, 1950-79.* ISP-WP-79(A). 1980.

20. U.S. Bureau of the Census. *Country Demographic Profiles: Brazil.* (Forth-

coming Publication). U.S. Bureau of the Census. *International Population Dynamics, 1950-79.* ISP-WP-79(A). 1980.

21. Elsa Berquo. *Veja.* São Paulo, October 24, 1979, p. 139.

22. U.S. Bureau of the Census. *International Population Dynamics, 1950-79.* ISP-WP-79(A). 1980.

23. *World Development Report.* Washington, D.C.: World Bank, 1979.

24. U.S. Bureau of the Census. *Country Demographic Profiles: Pakistan.* ISP-DP-24. March 1980.

25. U.S. Bureau of the Census. *International Population Dynamics, 1950-79.* ISP-WP-79(A). 1980.

26. U.S. Bureau of the Census. *Country Demographic Profiles: Mexico.* ISP-DP-14. September 1979. U.S. Bureau of the Census. *International Population Dynamics, 1950-79.* ISP-WP-79(A). 1980.

27. The 34 member countries of the Economic Commission for Europe are Albania, Austria, Belgium, Bulgaria, Byelorussian S.S.R., Canada, Cyprus, Czechoslovakia, Denmark, Finland, France, German Democratic Republic, Federal Republic of Germany, Greece, Hungary, Iceland, Ireland, Italy, Luxembourg, Malta, Netherlands, Norway, Poland, Portugal, Romania, Spain, Sweden, Switzerland, Turkey, Ukrainian S.S.R., U.S.S.R., United Kingdom, United States of America, and Yugoslavia.

28. The nine Common Market countries are Belgium, Denmark, France, Federal Republic of Germany, Ireland, Italy, Luxembourg, Netherlands, and the United Kingdom.

29. The basic criteria for standard metropolitan statistical areas (SMSA's), defined by the Office of Management and Budget, are that each SMSA must include at least:

1. One city with 50,000 or more inhabitants, or
2. A city with at least 25,000 inhabitants which, together with contiguous places (incorporated or unincorporated) having population densities of at least 1,000 persons per square mile, has a combined population of 50,000 and constitutes for general economic and social purposes a single community, provided that the county or counties in which the city and contiguous places are located has a total population of at least 75,000.

As of December 31, 1978, there were 283 SMSA's (including 4 in Puerto Rico).

30. Vernon Ruttan. "The Prospect for Agricultural Growth," paper presented at the Conference on Economic and Demographic Change: Issues for the 1980's, Helsinki, 1978.

31. Roger Revelle. "Will the Earth's Land and Water Resources be Sufficient for Future Populations?", *The Population Debate: Dimensions and Perspectives, Papers of the World Population Conference*, Bucharest, 1974, Volume Two. New York: United Nations, 1975.

32. Roger Revelle. "Flying Beans, Botanical Whales, Jack's Beanstalk, and Other Marvels," *The National Research Council in 1978: Current Issues and Studies.* Washington, D.C.: National Academy of Sciences, 1978.

33. *Leucaena: A Promising Forage and Tree Crop for the Tropics.* Washington, D.C.: National Academy of Sciences, 1977.

34. H. E. Goeller and Alvin M. Weinberg. "The Age of Substitutability," *Science*, Vol. 191, No. 4228 (February 20, 1976), p. 688.

35. For a discussion of energy, see *Energy in a Finite World: Volume One: Paths to a Sustainable Future. Energy in a Finite World: Volume Two: Global Systems Analysis.* Both volumes edited by Wolf Hafele. Cambridge, Mass.: Ballinger Publishing Co., 1981.

36. Goeller and Weinberg, op. cit., p. 687.

37. Goeller and Weinberg, op. cit., p. 686

38. Council on Environmental Quality and the Department of State. *The Global 2000 Report to the President*. Volume One. Washington, D.C.: U.S. Government Printing Office, 1980.

39. Paul R. Ehrlich, Anne H. Ehrlich, and John P. Holdren. *Human Ecology: Problems and Solutions*. San Francisco: W. H. Freeman and Company, 1973.

40. Robert H. Solow. "The Economics of Resources or the Resources of Economics," *Economics of the Environment, Selected Readings*. Second Edition. Edited by Nancy Dorfman and Robert Dorfman. New York: W. W. Norton & Co., Inc., 1977.

41. U.S. General Accounting Office. *Prospects Dim For Effectively Enforcing Immigration Laws*. Publication No. GGD-81-4. Washington, D.C.: U.S. Government Printing Office, 1980.

## BIBLIOGRAPHY

Harrison Brown. *The Challenge of Man's Future*. New York: Viking Press, Inc., 1954. An early statement of problems of population in relation to the world habitat.

Ansley J. Coale and Edgar M. Hoover. *Population Growth and Economic Development in Low-Income Countries. A Case Study of India's Prospects*. Princeton, N.J.: Princeton University Press, 1958. Classic treatment of the way population handicaps the accumulation of capital.

*Economics of the Environment. Selected Readings*. Edited by Robert Dorfman and Nancy S. Dorfman. Second edition. New York: W. W. Norton & Co., Inc., 1977. A collection of important papers with an illuminating introduction.

Paul R. Ehrlich, Anne H. Ehrlich, and John P. Holdren. *Ecoscience: Population, Resources, Environment*. San Francisco: W. H. Freeman and Company, 1977. An extensive presentation of the resource base of population.

Nathan Keyfitz. *Applied Mathematical Demography*. New York: John Wiley & Sons, 1977. A relatively advanced treatment of demographic theory and its applications.

George Masnick and Mary Jo Bane. *The Nation's Families: 1960-1990*. Cambridge, Mass.: Joint Center for Urban Studies of Massachusetts Institute of Technology and Harvard University, 1980. An account of the current and prospective condition of the American family that has attracted attention far beyond sociology.

Judah Matras. *Introduction to Population:A Sociological Approach*. Englewood Cliffs, N.J.: Prentice-Hall, Inc., 1977. Among the numerous short textbooks on population meant for introductory classes, this one seems to be as good as any.

William Petersen. *Population*. Third edition. New York: Macmillan Publishing Co., Inc., 1975. An extended and scholarly treatment of population from a social scientist's viewpoint.

Henry S. Shyrock and Jacob S. Siegel. *The Methods and Materials of Demography*. Two volumes. Third printing (rev.). Washington, D.C.: U.S. Government Printing Office, 1975. (Also published in a hard-cover, one-volume edition.) A very detailed and careful, but elementary, introduction to the handling of demographic data.

U.S. Bureau of the Census. *Statistical Abstract of the United States*. 100th edition. Washington, D.C.: U.S. Government Printing Office, 1979. An indispensable collection of data on the United States and the world.

# 2

# *On Some Major Human Diseases*

## INTRODUCTION

Continuing improvement in the public health has been one of the triumphs of human progress. The average human lifetime has extended concomitantly with improved living standards, the institution of relatively simple sanitation measures, improved nutrition, and the growth of general scientific understanding. In the twentieth century, these several processes, in concert, have resulted in remarkable increases in life expectancy at all ages, largely because of dramatic declines in the incidence of and mortality from nutritional deficiencies and acute bacterial and viral infections (see Figure 1).

As many as 175,000 cases of pellagra in the American South were annually *reported* in the years between the first and second world wars; this disease is no longer to be seen in the United States. Such microbial infections as tuberculosis, syphilis, typhoid, diphtheria, streptococcal septicemia, and pneumonia, which collectively accounted for about 40 percent of all deaths in the United States in 1900, represented no more than 4 percent of all deaths in 1970. Perhaps most dramatic was the decline in the death rate during the first years of life. In 1900, only 80 percent of all babies born reached the age of five; in 1977, 98.3 percent so survived. Accordingly, gross life expectancy at birth increased from 47 to 73 years

◀ "With changes in longevity and the marked reduction in mortality due to acute infectious disease has come a steady rise in that fraction of humanity which ultimately succumbs to 'degenerative' diseases." [Administration on Aging, U.S. Department of Health and Human Services.]

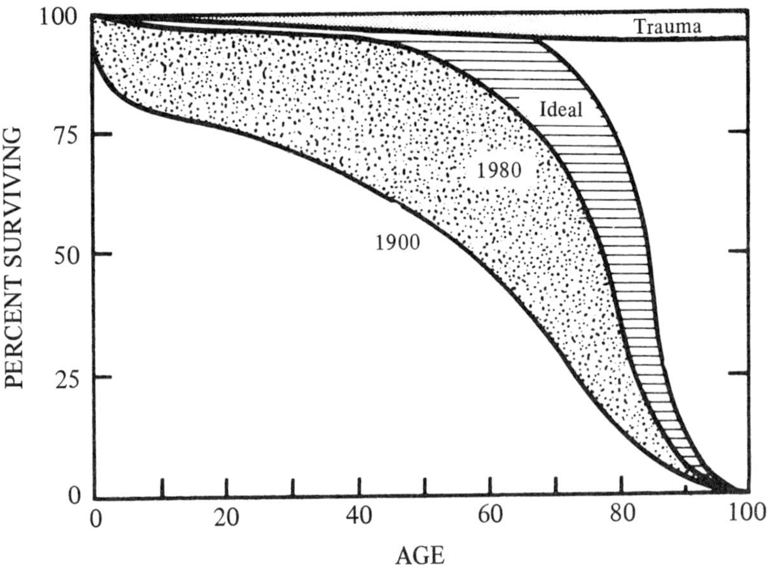


**Figure 1 Survival of the American population.** The curve for 1900 depicts the percentage of all people born before 1900 who in that year would have been 10, 20, 30, 40 years old, etc., and who, in fact, were alive in 1900. The curve for 1980 is constructed similarly. The "ideal" curve is a hypothetical extrapolation, with trauma (externally generated injury) the dominant cause of death in early life. Since 1900, life expectancy has increased significantly at all ages. Because of the remarkable increase in survival during early and middle years, the survival curve of Americans today need be improved relatively little to approximate the "ideal." [SOURCE: James F. Fries. "Aging, Natural Death, and the Compression of Morbidity," *New England Journal of Medicine,* Vol. 303(1980), p. 131. Reprinted by permission.]

during the same period. But this increase has not been due entirely to the reduction of childhood mortality. Since 1940, the average life expectancy for 20-year-old white males has increased by four years while that for white females has increased by almost eight years. Expressed somewhat differently, more than 70 percent of all white males now live to the age of 65 and beyond, as do 84 percent of white females. The figures for 1940 were 58 and 68 percent, respectively, and for 1900, 39 and 35 percent, respectively.

Acute infection still comprises a large fraction of all illnesses; it is the outcome that has been so dramatically altered. Moreover, as will be discussed later in this chapter, some infections may leave, as a residuum, permanent changes that initiate the development of chronic diseases.

With these changes in longevity and the marked reduction in mortality due to acute infectious disease has come a steady rise in that fraction of humanity which ultimately succumbs to "degenerative" diseases—diseases that are several years in developing and that characteristically appear in the later years of life. In 1978, diseases of the heart, most prominently coronary artery disease, accounted for almost 38 percent of all deaths. Related cerebrovascular diseases, primarily strokes, accounted for 9 percent of all deaths, and cancers accounted for somewhat more than 20 percent of all deaths. At the same time, the aging American population exhibits an increased incidence of a variety of debilitating or crippling disorders, which, while responsible for a relatively small death toll, nevertheless cause a large fraction of human pain, disability, economic loss, and tragedy.

A principal goal of biomedical research is to reduce this vast burden of illness and prevent premature mortality. But the task is scarcely simple. In considerable measure, the etiology (the causative factors) of these diseases of the later years remains obscure, their pathogenesis (the development of disease) is not well enough understood to permit regularly successful intervention, and much of the clinical care of their victims consists merely of supportive management. Successful future prevention and definitive therapy will probably rest on an ever more detailed understanding of the fundamentals of human biology. Until such understanding is in hand, much of clinical medicine must be content with "halfway medical technologies," costly supportive measures that cannot be definitive even though they may extend life.

This chapter will offer a brief survey of the current frontiers of knowledge for the two major classes of fatal disease, cancer and atherosclerosis, the most common metabolic disease, diabetes mellitus, and one of the most common of all debilitating disorders, arthritis. Patently, these research efforts are of immense importance in their own right, but in the present context they also serve as prototypes of how current biomedical science addresses the entire panoply of human disorders.

## ATHEROSCLEROTIC DISEASE

Among the American population at any given time are more than four million individuals who have overt evidence of coronary heart disease, manifest by a history of angina pectoris or an acute heart attack. Managing this population is a major challenge to the American medical system, and preventing such occurrences is a major goal of those concerned with the American public health.

In the history of science, it has always been considered appropriate to attempt to explain—and deal with—a given phenomenon in terms of such relevant knowledge as may be available. Seen in retrospect, such behavior may appear to be that of accepting a series of popular fads, each prevalent when belief in its fashionable concept of causality is at its height. At some future date, the history of attempts to prevent and treat atherosclerotic disease may strike a reader similarly.

Atherosclerosis is the name given to a process wherein the inner wall of an artery becomes thicker and thicker due to the formation of a plaque, which as it grows impedes the flow of blood. The plaque may finally occlude the vessel, thereby depriving the tissue (which it should nourish) of oxygen and nutrients, leading to rapid death of cells in the tissue so affected (an infarction).

Early in the history of studies on this disorder, it was noted that such plaques may contain unusual quantities of cholesterol. Since a disease syndrome somewhat resembling human atherosclerosis can be produced in rabbits by feeding them diets extremely high in cholesterol, much clinical and experimental attention has been given to how the concentration of cholesterol and other fatty materials (lipids) in the blood plasma relates to the genesis of coronary heart disease. The possibility that cholesterol plays a central role in atherosclerosis has been strengthened by the discovery of a protein in the membranes of normal cells that serves as a receptor for cholesterol and is required for cholesterol to pass into the interiors of cells. The very rare individuals who lack both of the genes that specify this receptor protein exhibit very high concentrations of serum cholesterol and generally die of coronary occlusion before the age of 20. The cells of individuals who have one of the two genes contain a subnormal amount of the receptor. Such individuals have *familial hypercholesterolemia*; they have elevated levels of serum cholesterol and develop severe coronary artery disease relatively early in life.

In the years since World War II, epidemiological approaches, focusing particularly on diet, have dominated much of the research on this disorder. Extensive surveys were conducted comparing the incidence of and mortality from coronary heart disease among populations eating diets high or low in fat, diets varying in their ratio of unsaturated to saturated fat,* and diets with different cholesterol contents.

In a general way, several of these major studies appeared to support the thesis that diets generally high in fat content, particularly those much richer in saturated than in unsaturated fat and those containing greater quantities of cholesterol, correlate with an increased incidence of and

* Long carbon chains of saturated fatty acids are joined by consecutive single bonds:

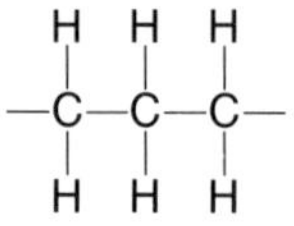

Unsaturated fatty acids contain one or more double bonds:

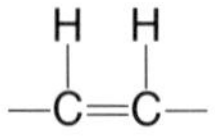

Polyunsaturated fats contain many double bonds.

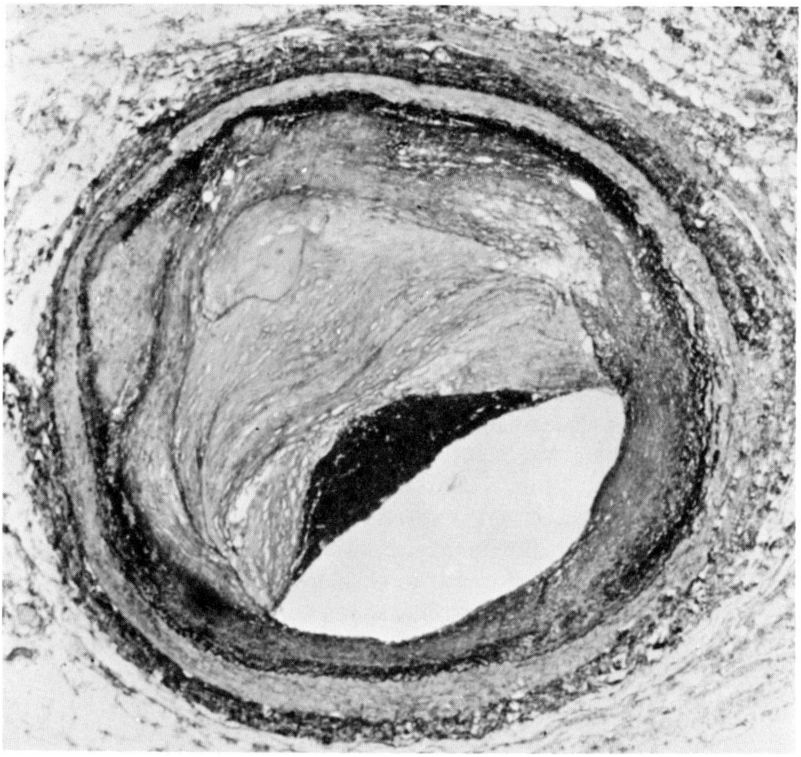

Cross section of the coronary artery of a 65-year-old man who suffered an acute myocardial infarction six days before his death. A thrombus, or blood clot, has formed on top of an old atherosclerotic plaque. [National Heart, Lung, and Blood Institute, National Institutes of Health.]

mortality from coronary artery disease. Such studies have done much to give direction to current trends in preventing and managing the disorder. However, from the beginning of such studies, it also became apparent that there were numerous exceptions to the seeming rules—individuals who failed to develop coronary artery disease while living on diets considered to be atherogenic, or individuals who succumbed to coronary artery disease despite a lifetime on a low-fat diet.

## Risk Factors

Without an understanding of the prime etiological factors that initiate the development of atherosclerotic plaques, attention turned to other phenomena that correlate with the appearance of atherosclerosis. These correlates, now called *risk factors*, include male sex, age, hypertension, elevated serum cholesterol concentration, diabetes, cigarette smoking, sedentary life style, being overweight, emotional stress, and a "type A" personality (marked by an overdeveloped sense of urgency, by drive, by competitiveness). A family history of numerous individuals who have succumbed to heart attacks relatively early in life is also a risk factor, although only a tiny fraction of all coronary heart disease arises from specific monogenic disorders. The presence of these risk factors, presumably operating in concert, appears to account for perhaps one half of all deaths due to coronary artery disease; hence, other significant factors must be at play.

For a decade and a half, physicians in practice, public health agencies, and such organizations as the American Heart Association have emphasized these risk factors. Americans have been urged to quit smoking cigarettes, to exercise at least in moderation, to have their blood pressures checked and, where hypertension exists, to follow the advice of their physicians, to reduce their intake of salt and table sugar, and to bring their weights down to accepted norms. They have also been advised, on the basis of less compelling evidence, to reduce the total fat content of their diets to about 30 percent of total calories (the American norm has been about 40 percent), to substitute sources of unsaturated fats (vegetable oils) for saturated fats (those of animal origin), and to reduce their intake of cholesterol (which requires reduced consumption of animal fats and particularly of egg yolks).

Beginning about 1966, as shown in Figure 2, the death toll due to coronary heart disease began to recede. From a peak age-adjusted death rate of 243 per 100,000 in 1963, there occurred a steady decline, the rate falling to 177 in 1977.

The death rate due to stroke declined from just over 100 in 1940 to just less than 50 in 1977, while the death rate due to hypertensive disease declined from 69 in 1940 to less than 9 in 1977. These two trends must be attributed in considerable part to increased awareness of hypertension, earlier diagnosis of elevated blood pressure in the population at large, and the availability of effective measures. Among these measures are reducing blood pressure through both reduction of body weight by dietary means and marked restriction of sodium intake, and a battery of drugs that have proved effective in managing a large fraction of all hypertensive disease. These medications involve combinations of drugs that are vasodilators (causing relaxation of the muscles of blood vessels and, therefore, an increase in the caliber of small arteries and arterioles), drugs which block the receptors for the nerve transmitters epinephrine and norepinephrine, and drugs that prevent blood clotting. It is estimated that about 35 million Americans are hypertensive in significant degree (that is, they have casual systolic pressures greater than 160 millimeters of

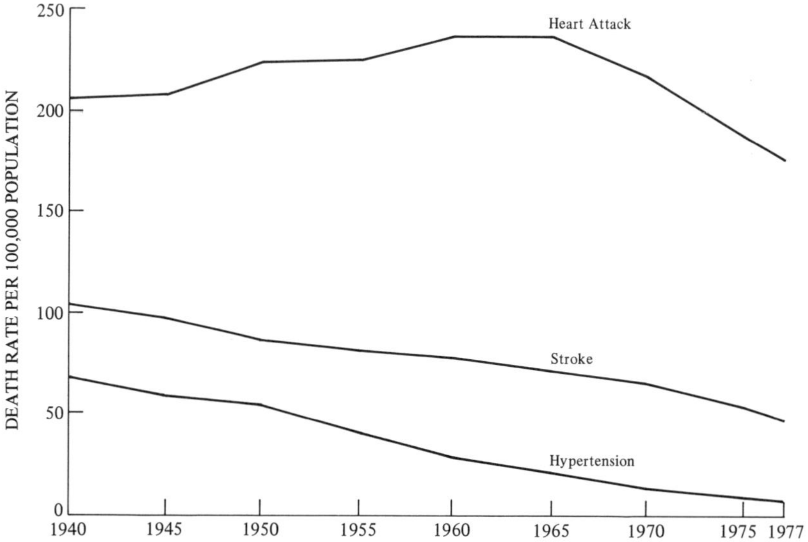


**Figure 2 Mortality due to heart attack, stroke, and hypertension.** *Note:* All rates are age-adjusted to the 1940 U.S. population. [SOURCE: *Adjusted Death Rates, United States, 1940-1977.* Dallas, Texas: American Heart Association, National Center, Division of Planning and Evaluation, 1978 (updated yearly).]

mercury or diastolic pressures greater than 95 millimeters of mercury when untreated).

While this generalized campaign was in progress, hospitals around the country developed intensive care and coronary care units staffed by knowledgeable teams and equipped to respond instantly to the emergency aspects of coronary care. Increasing numbers of lay citizens were taught the rudiments of such care, and emergency medical services around the country were upgraded.

There are experts who ascribe the welcome downward trend in mortality from coronary artery disease to the success of the campaign urging people at risk—and, indeed, all people—to take the recommended preventive measures. Certainly, large numbers of adult Americans have stopped smoking; the consumption of dairy products has declined while the consumption of vegetable oils for such uses as salad dressing, margarine, and general cooking has increased. The mean level of cholesterol in the blood serum of the American population is said to have declined by more than 5 percent. Many Americans, including individuals who understand that their risk of coronary disease is above average, have taken to programs of physical exercise. And, of course, emergency medical services have rescued the lives of thousands who otherwise would have died.

All of the evidence indicates that multiple factors contribute to an unfavorable outcome in atherosclerotic disease. The question is whether relatively small improvements in each of a number of risk factors can collectively result in a significant net decrease in untoward outcomes of the atherosclerotic process. As observed in several large population groups in the United States, the mean concentration of cholesterol in the blood of middle-aged men has fallen from 233 milligrams per deciliter to 217 milligrams per deciliter, a 7 percent decline. By accepting the epidemiologically positive association of atherosclerotic disease with elevated blood cholesterol, it has been calculated that an average decline of 5 milligrams per deciliter could result in a 4.3 percent decline in coronary deaths over a six-year period, all other factors remaining equal. Similar reckoning indicates that this 4.3 percent decline in cardiovascular mortality would be trebled by reducing the average diastolic blood pressure among the same population group by a mere 2 millimeters of mercury. If they were also to cease smoking cigarettes, the three factors together might lead to a net decline in cardiovascular mortality of about 18 percent. This result may be compared with the 21 percent decrease actually observed in the entire population of the United States from 1968 to 1976.

Clearly, such constructs cannot establish cause and effect relations among the variables; they are offered here to counter the facile argument that small changes in these life style factors cannot make much difference in a health problem of this magnitude. Should the line of reasoning behind these constructs indeed prove valid, it would constitute a startling demonstration of how narrow is the margin between health and disease with respect to these variables.

Even the more ardent advocates of the importance of risk factors, particularly of the composition of one's diet, have attributed only a part of the entire decline in the death rate to the favorable response to the campaign urging that all adult Americans take "preventive" measures. There are several reasons for this reservation. The decline in the death rate started well before the campaign advocating preventive measures had taken hold in the United States; the number of people who continue to smoke heavily (two packs a day or more) appears not to have declined; the blood pressures of a large fraction of all hypertensive individuals remain uncontrolled; the exercise fashion did not commence until the declining trend in the coronary artery death rate was very well along; and the diabetes, a contributing factor to cardiovascular disease, of most persons so afflicted is little better controlled today than in the middle 1960's. All of these observations suggest that the beneficial behavioral changes that have occurred cannot fully account for the magnitude of the favorable trend in the death rate due to coronary artery disease. Patently, if some other major influence is at play, knowledge of it would be most welcome.

"Exercise in moderation brings a sense of well-being." [Library of Congress.]

There are additional powerful motivations, in addition to preventing coronary artery disease, for following most of the recommendations stemming from a consideration of the principal risk factors. Reducing the blood pressure of hypertensive individuals is highly desirable of itself; cessation of cigarette smoking markedly reduces the risk of lung cancer and emphysema as well as of coronary artery disease; exercise in moderation brings a sense of well-being, which is its own reward; careful management of diabetes reduces the intrinsic risks from this disease and may delay the onset of the serious secondary afflictions that occur in the later years of the chronic diabetic; and maintaining a reasonable weight is both its own reward and a means of reducing the possibilities of hypertension and diabetes as well as of atherosclerosis.

## The Diet-Heart Hypothesis

One cornerstone of the set of measures recommended for the prevention of atherosclerosis is the diet-heart hypothesis, the concept that a reduction in dietary fat and cholesterol will reduce the concentration of cholesterol in the blood serum, which will in turn reduce the likelihood of severe coronary artery disease. However, apart from the weak case that can be drawn from the general downward trend of mortality due to coronary artery disease in the last decade and a half, the bulk of other recent evidence neither supports nor denies the diet-heart hypothesis. Some of that evidence is worth noting.

A comparison of the "butter eaters" of northern Europe with the "olive oil eaters" of southern Europe was among the original bases for the diet-heart hypothesis. However, in that study, the correlation between dietary cholesterol, serum cholesterol, and coronary artery disease rested in rather considerable measure on the population of the Karelian peninsula of Finland. If the data concerning that specific population are removed from consideration, the correlation among diet cholesterol, serum cholesterol, and coronary artery disease found in these classic studies becomes decidedly less statistically significant. In the four years after the beginning of a campaign to reform diet in Sweden, one similar to the campaign in the United States, the death rate due to coronary artery disease increased by 30 percent.

Israelis consume a diet that is higher in polyunsaturated fats than that of any other western society. Indeed, the Israeli diet is quite similar to that recommended by the American Heart Association, the Senate Select Committee on Nutrition, and the United States Department of Agriculture, yet mortality from coronary heart disease in Israel is about three quarters of that in the United States and is still rising. Neighboring Bedouins in the desert are alleged to have relatively little coronary artery disease. Certainly, they eat frugally and are rarely obese. But when they settle in Israeli towns and accept the general Israeli diet, their depot fat begins to resemble that of their Israeli neighbors and their susceptibility to coronary heart disease increases markedly.

A reduction in dietary fat, particularly saturated fat and, to a lesser degree, cholesterol, induces significant reductions in the levels of cholesterol in the blood serum of persons rigorously maintained in metabolic wards, but it appears to be only half as effective in free-living individuals. In intervention trials conducted with several thousand middle-aged men, such dietary modification has resulted in significant decreases in serum cholesterol levels and in the associated levels of low-density lipoproteins (see "Atherogenesis" below), but it has led to only marginal decreases in the incidence of coronary artery disease and to no decrease in mortality therefrom. Similarly, in a series of large-scale trials, using drugs that interfere with either the absorption of cholesterol in the intestine or its synthesis in the liver, serum cholesterol levels were reduced, but the incidence or severity of coronary artery disease was not.

Thus, it would appear that the admonition that to avoid coronary artery disease people who do not show overt signs of atherosclerosis, who do not exhibit an elevated blood level of cholesterol, and who exhibit none of the other risk factors, except perhaps age and male sex, should eat diets significantly restricted in cholesterol and rich in polyunsaturated fatty acids rests on somewhat shaky foundations. At this juncture, the following recent statement by the president of the American Heart Association appears to be apt:

> Our Committees conclude that the evidence supports our existing statements recommending modest reductions in saturated fats and cholesterol. As a practical matter, given the current American diet, it would be difficult for an individual to maintain desirable weight without cutting down on total fats. . . . Overall, our diet advice urges balance, prudence, and selections from the four basic food groups to ensure good nutrition.

Such advice cannot be faulted. However, some advocates of diets very high in polyunsaturated fat and low in cholesterol go on to suggest that, in any case, such a diet can do no harm. But that is not certain. There are opposing suggestions from animal experimentation: monkeys fed diets rich in peanut or coconut oil were found to develop considerable proliferation of the intimal cells of their arteries, with subsequent scarring, while butterfat in similar amounts produced only harmless "fatty streaks," rather like the lesions seen in infants that appear not to be precursors of atherosclerotic plaques in adults. Further, it has been suggested that hydrogenated vegetable oils (such as margarine) contain some of the trans forms of the unsaturated fatty acids that normally occur as cis isomers. (These terms are explained in "Chemical Terms and Conventions" at the end of Chapter 11.) Esters, particularly cholesterol esters of unsaturated fatty acids in the trans form, are intensely sclerogenic (scarring) when implanted into tissues. Finally, several recent studies have reported that the segments of a given population, in New Zealand and in Hawaii, for instance, with the lowest levels of serum cholesterol have the highest overall mortality rates, particularly from cancer. These facts are cited here merely to indicate that it may be too facile to suggest that diets of the sort being recommended as preventive measures against coronary heart disease are entirely innocuous.

In summary, the controversy over the diet-heart hypothesis remains unsettled. What is uncontroversial is the great boon to health that results from the combination of measures required to lose excess weight and to maintain normal weight thereafter.

## Atherogenesis

Extensive research into the nature of the arterial wall and the changes occurring during atherosclerotic disease has made it increasingly obvious why so many risk factors exist. New findings are beginning to shed light on how some of these may operate. Much of the progress has hinged on several developments in modern molecular and cell biology: the advent of electron microscopy; methods for the culture of pure cell lines *in vitro*; enlarged understanding of intermediary metabolism, protein chemistry, and the biology of such specialized cells as blood platelets, macrophages, endothelium, smooth muscle, and the protein molecules bound to cell membranes that function as highly specific receptors for a great variety of materials from the cells' surroundings; detailed knowledge of the blood clotting mechanisms; recognition of the tissue hormones called prostaglandins; appreciation of the manner in which a variety of hormones affect normal and aberrant tissues; and understanding of the molecular modes of action of a variety of drugs.

The locus of the atherosclerotic process is the interior wall of the artery. As indicated in Figure 3, the innermost structure is a sheet of endothelium, a layer only one cell deep, below which lies the working part of the structure—densely packed, smooth (involuntary) muscle cells. These muscle cells share some of the metabolic capabilities of the cells known as fibroblasts, which are normally responsible for constructing connective tissue. Thus, these muscle cells can produce some of the principal macromolecules of connective tissue, such as collagen, the primary fibrous protein component of connective and scar tissue, or glycosaminoglycans, complex, long-chain carbohydrates that are usually linked to special proteins.

Perhaps the first great surprise was the discovery that, whereas clinical atherosclerotic disease is largely evident in the middle or later years of life, signs of its origin are to be found much earlier and are almost universally present in the vessels of young males dying of trauma (e.g., casualties of the Korean War). The beginning is sometimes visible as a "fatty streak," an indurated area in which can be seen a milky-looking deposit of mixed fatty materials, the mixture having the same general composition as that of fatty materials in the blood. More frequently, however, no lipid deposits may be discerned. Instead, there is a thickening of the smooth muscle cell layer, which bulges slightly into the lumen of the vessel where the integrity of the endothelium has been disrupted in some manner. This suggests that the primary event is the endothelium's injury, which seemingly occurs in response to some mechanical stress, perhaps just the normal water hammer of the hemodynamic cycle, or turbulent blood flow in the region of a break in the endothelium.

With the passage of time, the smooth muscle cells continue to proliferate and secrete about themselves an increasing amount of the components of scar tissue, particularly if cholesterol levels are elevated. Slowly, the muscle cells and the scar tissue begin to fill up with fatty materials, particularly cholesterol. The structure bulges farther and farther into the lumen of the vessel, becoming increasingly obstructive. Toward the end, it may begin to calcify. Should the material effectively occlude the vessel, clinical symptoms become apparent—angina pectoris when a coronary artery is involved, or intermittent claudication (leg pain upon exertion) when the lesion affects one of the distributary arteries of the lower ex-

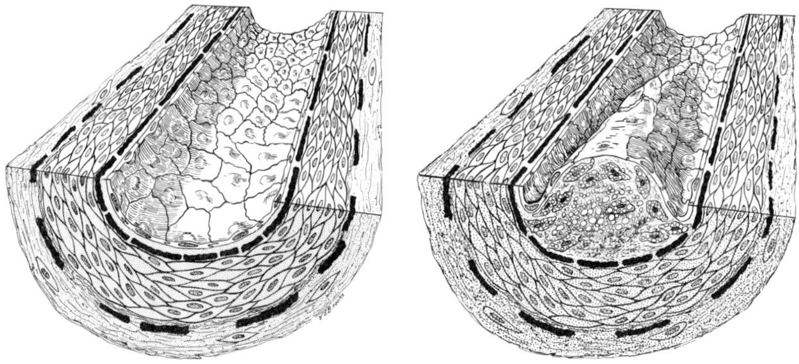

**Figure 3 The atherosclerotic plaque.** A normal artery (left) compared with an artery containing a lesion of atherosclerosis (right). The lesion is located entirely inside the innermost layer of the artery, the intima, and contains newly proliferated smooth muscle cells, which in the diagram are shown to contain lipid droplets in the form of small spherical deposits. Extracellular lipid deposits are also seen. Normally, the endothelial cells of the artery form a continuous monolayer, as seen on the left. In the diagram on the right, some endothelial cells are missing at the apex of the lesion, suggesting that these cells are altered in atherogenesis. Three principal changes occur in the formation of these lesions: smooth muscle cells multiply within the intima, these cells form new connective tissue proteins, and lipids are deposited both in the cells and in the connective tissue proteins. [SOURCE: Russell Ross and J. A. Glomset. "The Pathogenesis of Atherosclerosis," *The New England Journal of Medicine,* Vol. 295(1976), p. 371. Reprinted by permission.]

tremities. Should complete obstruction occur, so that blood cannot flow, then an infarction may occur beyond the lesion. More frequently, a blood clot may form on the partially occluded vessel, and the clot then completes the occlusion.

The mechanism of blood clotting seems to play a key role in atherosclerosis. Circulating in the blood at all times are the blood platelets, formed in the bone marrow from large precursor cells called megakaryocytes. The normal role of platelets is to prevent leakage of blood from the vascular tree by lodging at a ruptured vessel and discharging their contents, which initiates the cascade of events that is the blood clotting mechanism. It is imperative, therefore, that such platelets move freely in the circulation and not adhere to the vessel walls, since, should they begin to clump, inadvertent clotting might occur. On the other hand, they must be able to adhere to the site of a ruptured vessel to begin the clotting process.

The stickiness of platelets to the surfaces of the blood vessels appears to be based upon a balance between two related compounds, thromboxane $A_2$ and the prostaglandin known as prostacyclin. (See the next page.) Platelets themselves produce thromboxane $A_2$, which makes the platelets more able to aggregate to each other and to blood vessel walls. (Platelets from poorly controlled diabetics seem to produce an excess of this sub-

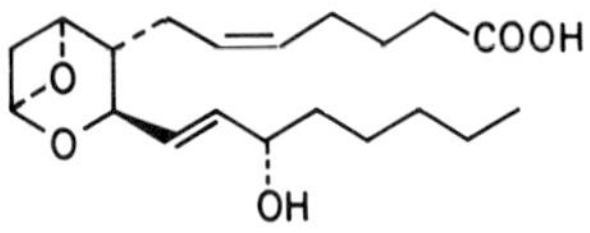


THROMBOXANE $A_2$

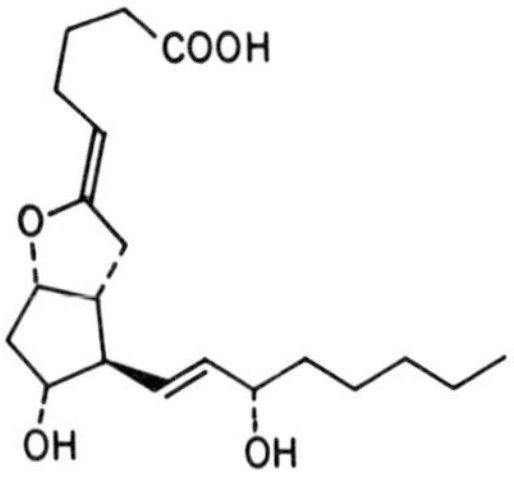


PROSTACYCLIN

stance, and for reasons not yet clear this results in platelets aggregating in the vessels of the lower extremities.) On the other hand, the cells of the endothelial lining predominantly produce prostacyclin, which renders the platelets less sticky to each other and to the vessel wall. The molecular mechanisms of these effects are unknown. Aspirin appears to inhibit the synthesis of thromboxane $A_2$, an action that presumably accounts for the finding that daily low doses of aspirin may prevent temporary attacks of reduced blood flow (transient ischemic attacks) and the occurrence of stroke.

Where the endothelium has been damaged and is deficient in prostacyclin, platelets may be expected to aggregate. As the lodged platelets decompose, they appear to release a specific protein, the normal role of which is unknown, that serves as a powerful stimulus to the smooth muscle cells; the cells increase both in number and size and also produce increasing amounts of scar tissue.

Early in this process, the endothelium may repair itself. Endothelial cells in tissue culture continue to divide until they make a single smooth sheet, whereupon division ceases. Presumably, broken endothelial lining can heal itself in this way. But if enough of the plaque forms, its growth may become effectively irreversible.

There is evidence suggesting that cholesterol in the blood plasma, if present in abnormally high amounts, can harm the endothelial lining and perhaps even initiate the entire process. But it seems unlikely that this is the most frequent primary causal event. Once the entire process is initiated, however, cholesterol, when bound to specific carrier proteins to form what are called "low-density lipoproteins," can attach to the receptor protein on the surface of the smooth muscle cells and enter into the cells to accumulate. Further, cholesterol bound to low-density lipoproteins is surrendered to the glycosaminoglycan component of the plaque's scar tissue, where it also accumulates. Plaque, therefore, is not so much an accumulation of lipid materials as it is a cellular extravasation into the blood vessel that acts as a depository for lipids, particularly cholesterol, in the blood.

Finally, one should note that blood plasma also contains high-density lipoproteins, which appear to function in the opposite sense—they accept and remove cholesterol from tissues in which it has concentrated and transfer that cholesterol to the liver for further metabolism. Therefore, even while the cellular aspects of plaque formation proceed, the net deposition of cholesterol represents a dynamic balance between the two classes of lipoproteins. Accordingly, people with high concentrations of low-density lipoproteins tend to have more cholesterol deposited into plaque, once the plaques begin forming, while people with high concentrations of high-density lipoproteins would be expected to be protected from that process. Estrogens appear to promote the formation of high-density lipoproteins and suppress the formation of scar tissue by the hyperactive smooth muscle cells, in keeping with the differences in the occurrence and severity of atherosclerosis between the sexes.

## Cholesterol Metabolism

In view of the centrality of cholesterol to atherosclerosis, some understanding of its metabolism in the body is in order. Cholesterol is a sterol required for the structural integrity of the plasma membrane of every cell in the body. It is also the specific starting material for the synthesis of male and female sex hormones and the important steroids manufactured and secreted by the adrenal cortex; clearly, it is of vital significance to normal life.

Cholesterol itself is not a dietary requirement; the body can synthesize all of the substance it requires. Cholesterol is made in the liver by a complex series of enzymically catalyzed reactions beginning with intermediates that arise in the normal metabolism of both glucose and the common fatty acids of the diet. Its rate of synthesis is regulated at the step in which the intermediate 3-hydroxy-3-methylglutaryl coenzyme A is transformed into mevalonic acid by an enzyme sensitive to the concentration of cholesterol in its environment. In a general way, higher levels of cholesterol will inhibit the synthesis, while lower concentrations will enhance it. Once made, the cholesterol binds to a low-density lipoprotein, which is also made in the liver, and in this form it departs from the liver cell to all of the other tissues in the body.

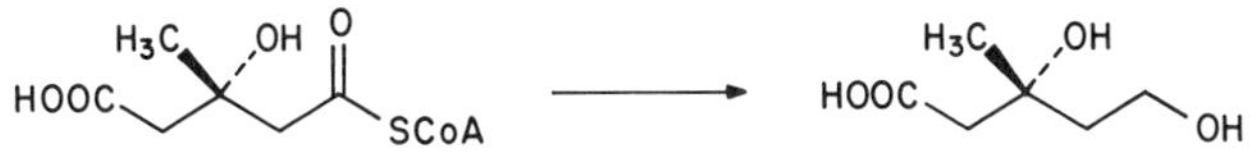


3-HYDROXY-3-METHYL-GLUTARYL COENZYME A

3,5-DIHYDROXYL-3-METHYL-VALERIC ACID (MEVALONIC ACID)

The process, however, is dynamic. Some of the cholesterol returning to the liver attached to a high-density lipoprotein is secreted by the liver into the bile and excreted into the intestine. Some of the cholesterol is oxidized to form the bile acids required in digestion. The bile acids and cholesterol both "recycle" to some degree; they are reabsorbed in part lower in the intestine and return to the liver for reuse.

Ingested cholesterol can be absorbed from the intestine by the mechanism that reabsorbs the recycled cholesterol, so a significant increase in dietary cholesterol will result in an increased absorption from the intestine. Humans normally absorb one fifth to two thirds of the cholesterol that passes through the intestine. (The fraction absorbed varies among animal species, as does the efficiency of excretion through the bile. It is for this reason that the rabbit can stand as a model of human atherosclerosis, since it absorbs ingested cholesterol with greater efficiency than do humans and is decidedly less efficient at excretion. In contrast, the rat absorbs ingested cholesterol relatively poorly and is extremely efficient at excreting cholesterol when its level in the blood rises.)

In the light of this partially self-regulated system, it is apparent that modest changes in the amount of dietary cholesterol are unlikely to have profound effects on the level of circulating cholesterol in essentially normal human adults.

## Summary and Outlook

The degree of understanding of atherosclerosis portrayed above is heartening. Few of these details were available ten years ago, and the process is much more complex than was anticipated. That complexity indicates why there exist a variety of risk factors which contribute in varying degree to the development of atherosclerosis in a given individual. By the same token, it is in that very complexity that hope for the future may reside. There is reason to believe that in the near future the initiating events that damage the arterial endothelium in vulnerable areas of the body may be understood. Much more will also be discovered about the role of platelets and the behavior of smooth muscle in atherogenesis.

While, understandably, much attention is focused on preventing coronary artery disease by reducing risk factors, there have been encouraging developments in prolonging the lives of individuals with established disease. For example, mild exercise programs can lead to the establishment of small new arterial networks; highly effective drugs can prolong clotting and dilate arteries, thereby reducing blood pressure; and experience with other drugs intended to lower the levels of cholesterol in the blood serum is growing, though it is not yet sufficient for evaluations to be made. Angiography, which requires the injection of large quantities of a contrast agent opaque to X-rays, has permitted the coronary artery tree to be visualized directly and hence major occlusions to be localized. This procedure may yet be replaced by observations using reflections of ultrasound, a method that would be relatively innocuous and noninvasive. Techniques such as these permit precise diagnosis and make possible the increasingly frequent performance of coronary artery by-pass surgery, which shunts blood flow around one or more major occlusions. This procedure relieves the oppression of angina pectoris in a large fraction of individuals so afflicted, restoring the quality of their lives, but whether it brings an attendant increase in life span remains uncertain.

Even now, an era of explosive development in understanding the prostaglandins and their role in these and other processes is under way. Undoubtedly, much more light will be shed on the processes regulating the synthesis, metabolism, transport, and excretion of cholesterol; the structures of the low-density lipoproteins; the factors regulating the formation of these lipoproteins and the manner by which they accept and deliver up their burden of cholesterol to and from other cells; and the manner in which high-density lipoproteins remove cholesterol from tissues, particularly from developing plaques, and deliver it to the liver for excretion.

With that increased understanding should come increasingly attractive opportunities for intervening in the atherogenic process: repair of the endothelium; inhibition of the proliferation of smooth muscle; inhibition of the ability of platelets to aggregate locally; inhibition of the process by which lipids, particularly cholesterol, are delivered to the growing plaques; acceleration of the removal of such materials; and inhibition of the formation of scar tissue by the smooth muscle cells. If, as seems likely, the blood level of cholesterol proves to be critical to atherogenesis, inhibitors of cholesterol formation could be assessed, particularly those affecting the regulating enzymic step. Discrimination of the genetically controlled differences among individuals with respect to various aspects of these processes should help to identify those individuals at high risk and to develop specific corrective measures.

Improved regulation of hypertension can be of major importance, given its significance in the pathogenesis of atherosclerosis in hypertensive individuals. Research concerning the role of salt in the regulation and elevation of blood pressure and the mechanisms by which hypertension becomes established is entering a new phase of sophistication. More physiological approaches to treatment will capitalize on reversing well-defined physiological aberrations, in distinction to today's rather broadly based pharmacological approaches. An example is the clinical use of drugs to inhibit the enzyme that converts angiotensin I to angiotensin II, the peptide that is the pressor material responsible for hypertension of renal origin. As such efforts become successful, one can expect fewer undesired side-effects from drug therapy and, it is hoped, better adherence to treatment programs by hypertensive individuals.

In all likelihood, however, individuals will still have to regulate their own behavior in order to minimize the chance of atherosclerosis. Gross risk factors will remain: heredity, being overweight, hypertension, diabetes, cigarette smoking, lack of exercise, life under pressure, diet. Only with sufficient motivation can an adequate response to each of these risk factors be expected, and far better motivational techniques than those presently available are required, particularly to meet the challenge of cigarette smoking.

Blood cholesterol and its deposition into the growing plaque will probably remain of central importance. In the next few years, a definitive judgment will be needed concerning the usefulness of dietary approaches in regulating the deposition of cholesterol, since a rigorous modification of diet, which may prove necessary to affect that process, will require high motivation in view of its effect on the pleasures of the table. Moreover, if a drastic change should be found desirable for all Americans, it will have a profound effect on American agriculture. It is imperative, because of the possible effects on the quality of life, the nature of the American economy, and the moral burden, to be reasonably certain of one's ground before going forward with an immense experiment with the nutrition of the entire American people. The requisite understanding must therefore be obtained as soon as possible.

Meanwhile, other mechanisms may be brought into play which will continue the welcome downward trend in mortality from coronary artery disease. Various forms of drug treatment can be used to lower the number of out-of-hospital deaths in individuals with recognized coronary heart disease. Community-based resources can be better organized through mass education and minicourses. The American people can be directly involved in reducing the enormous toll that is now exacted outside medical facilities. No one should ever be more than a few seconds away from informed, effective emergency support, primarily in the person of appropriately educated fellow citizens.

A convinced, educated populace would be willing to accept certain prudent approaches to its life style and to take these matters into its own hands. The gains already made can be expanded when the currently obscure risk factors that contribute to atherosclerosis are revealed. In time, it should also become clearer to what degree each of the pieces of the last 15 years' assault is responsible for the decline in mortality. The government can accelerate this process by supporting and conducting research and by developing effective health care delivery systems. It may be too much to expect that atherosclerosis will go the way of poliomyelitis or pellagra. But it is no longer reasonable merely to assume that atherosclerosis is an inescapable concomitant of aging.

## CANCER

### Introduction

Cancer is not a single disorder but a generic category of diseases which share certain fundamental biologic attributes that distinguish them from other types of disease.

Collectively, cancers are the most feared of the major life-threatening forms of disease. Approximately one person in four now living in the United States will develop cancer, and one in six will die from cancer if present incidence and mortality rates remain unchanged. Recent mortality rates from various forms of cancer are shown in Table 1. In 1979, some 765,000 new cases of cancer were diagnosed and 395,000 Americans died from the disease. The direct cost that year of caring for people with cancer was on the order of $9 billion. Obviously, cancer is a vast human problem, fully warranting every effort to understand, prevent, and cure it. About a third of all serious cancers are now being treated successfully (five-year remissions), and there is some reason to believe that this rate will continue to improve.

It would be of considerable value to have reliable data on long-term trends in the incidence of and mortality from each major site-specific form of cancer. Regrettably, few such data exist. A number of local or regional tumor registries have kept such data for as long as four decades, but their completeness is open to question. A Ten Cities Survey was conducted in 1937, and National Cancer Surveys—each representing about 10 percent of the U.S. population—were performed in 1937, 1947, and 1969-71. Since 1973, such data, collected in a systematized way, have been made available annually through the Surveillance, Epidemiology and End Results (SEER) program of the National Cancer Institute.

The data do not suffice to reveal small or subtle changes, but the larger changes are amply evident. Incidence data are suspect but highly desirable. Mortality data are affected both by the slowly but steadily improving cure rate for certain cancers and by the inaccuracy of hospital death reports. Nevertheless, certain large trends are clearly evident, for example, a dramatic decline in stomach cancer, some increase in cancer of the colon and rectum among males, an increase in breast cancer, a large increase in cancer of the lung, and small declines in several but not all other cancers. In summary, the aggregate death rate due to cancer has been rising primarily because (1) an ever larger fraction of people survive to that age when cancer takes its toll, and (2) there is a considerable increase in the age-corrected incidence of those forms of cancer associated with cigarette smoking.

## Nature of Cancer

Cancers involve a defect in the regulation of cellular growth such that one or more cells, and their progeny, grow excessively, that is, at rates which are not suited to the needs of the entire organism. The resultant tissue of this disordered, unchecked growth is known as a neoplasm. The tissues of adults are normally in a state of equilibrium; the numbers of newly formed cells equal the numbers that become senescent and die (except in the adult nervous system, where, after maturity, cells that die appear not to be replaced).

Only a very small fraction of cells in any given tissue, designated as "stem cells," normally have the capacity to undergo an indefinitely sustained series of mitotic divisions. As the progeny of these cells differentiate to perform their specialized functions, to serve as liver, muscle, bone, kidney, etc., their proliferative activity diminishes and ultimately ceases

**Table 1** Death rates from cancer, by sex, age, and type of cancer (1940–1978)

| Age at death and selected type of cancer | Male | | | | | | Female | | | | | |
|---|---|---|---|---|---|---|---|---|---|---|---|---|
| | 1940 | 1950 | 1960 | 1970 | 1975 | 1978 | 1940 | 1950 | 1960 | 1970 | 1975 | 1978 |
| Total U.S. rate | 114.1 | 142.9 | 162.5 | 182.1 | 192.3 | 203.5 | 126.4 | 136.8 | 136.4 | 144.4 | 152.1 | 161.4 |
| 25–44 years | 24.6 | 31.0 | 34.2 | 33.9 | 28.7 | 26.9 | 50.3 | 49.1 | 45.8 | 40.3 | 33.4 | 31.0 |
| 45–54 years | 135.3 | 156.2 | 170.8 | 183.5 | 187.4 | 192.2 | 204.1 | 194.0 | 183.0 | 181.5 | 176.6 | 177.1 |
| 55–64 years | 352.2 | 413.1 | 459.9 | 511.8 | 512.3 | 522.0 | 384.1 | 368.2 | 337.7 | 343.2 | 357.7 | 369.7 |
| 65 years and over | 896.2 | 968.9 | 1,066.2 | 1,221.2 | 1,301.1 | 1,357.0 | 792.3 | 755.7 | 709.1 | 708.3 | 725.2 | 759.0 |
| Persons 45–54 years old | | | | | | | | | | | | |
| Respiratory system | 23.3 | 39.3 | 54.7 | 72.1 | 78.2 | 81.3 | 6.2 | 6.7 | 10.1 | 22.2 | 28.1 | 33.3 |
| Digestive organs, peritoneum | 70.7 | 59.9 | 53.2 | 45.9 | 44.8 | 45.0 | 58.6 | 47.0 | 38.9 | 32.5 | 29.7 | 29.0 |
| Breast | 0.3 | 0.4 | 0.2 | 0.4 | 0.2 | 0.3 | 47.5 | 46.9 | 51.4 | 52.6 | 50.4 | 51.0 |
| Genital organs | 5.6 | 4.8 | 4.0 | 3.4 | 3.3 | 3.3 | 69.2 | 59.0 | 44.8 | 34.4 | 29.9 | 26.4 |
| Lymphatic and hematopoietic tissues, excluding leukemia | N/A | 9.7 | 11.6 | 12.8 | 11.0 | 10.3 | N/A | 5.7 | 7.1 | 8.3 | 7.4 | 7.0 |
| Urinary organs | 9.1 | 8.7 | 8.3 | 8.0 | 8.1 | 8.3 | 4.5 | 3.8 | 3.5 | 3.5 | 3.3 | 2.9 |
| Mouth, throat, and pharynx | 6.0 | 6.3 | 7.9 | 7.9 | 8.2 | 8.4 | 1.6 | 1.8 | 2.8 | 2.8 | 3.0 | 2.7 |
| Leukemia | 5.6 | 6.5 | 7.2 | 6.6 | 5.9 | 6.2 | 4.6 | 5.1 | 5.3 | 4.9 | 4.4 | 4.4 |
| Persons 55–64 years old | | | | | | | | | | | | |
| Respiratory system | 45.7 | 94.2 | 150.2 | 202.3 | 214.1 | 223.5 | 12.6 | 15.4 | 17.0 | 38.9 | 58.3 | 69.4 |
| Digestive organs, peritoneum | 195.3 | 174.8 | 153.7 | 139.0 | 129.5 | 128.9 | 149.1 | 125.1 | 102.2 | 86.0 | 83.5 | 81.0 |
| Breast | 0.8 | 0.8 | 1.0 | 0.6 | 0.8 | 0.6 | 74.9 | 69.9 | 70.8 | 77.6 | 79.2 | 83.0 |
| Genital organs | 30.7 | 26.2 | 23.5 | 22.8 | 23.3 | 23.3 | 100.9 | 87.4 | 73.0 | 58.2 | 53.0 | 51.8 |
| Lymphatic and hematopoietic tissues, excluding leukemia | N/A | 18.2 | 23.1 | 27.1 | 26.7 | 25.8 | N/A | 11.5 | 15.7 | 17.7 | 17.5 | 17.9 |
| Urinary organs | 23.7 | 26.9 | 25.9 | 26.4 | 24.8 | 24.1 | 11.8 | 11.2 | 9.3 | 9.4 | 9.0 | 9.2 |
| Mouth, throat, and pharynx | 16.1 | 16.2 | 16.3 | 20.1 | 18.9 | 18.2 | 3.0 | 3.3 | 3.8 | 6.2 | 6.8 | 6.1 |
| Leukemia | 10.6 | 14.1 | 16.1 | 15.4 | 14.9 | 13.8 | 7.5 | 10.1 | 10.6 | 9.0 | 8.6 | 8.4 |

*Notes:* Deaths given per 100,000 people in the United States. N/A = not available.

SOURCE: *Statistical Abstract of the United States, 1980.* Washington, D.C.: U.S. Government Printing Office, 1981.

entirely. Collectively, the progeny of a single stem cell constitute a clone. The total numbers of cells in a clone may vary considerably, depending upon the numbers of cell divisions genetically programmed to occur during the successive steps in the differentiation pathway.

It is against the background of this system of carefully regulated compartments of proliferating stem cells and their differentiating clonal progeny that the essential features of neoplastic growth may be understood. *Any cell capable of mitotic division can undergo neoplastic transformation.* In most instances, the initial change appears to involve a single cell, the clonal expansion of which generates the population of tumor cells. In a neoplastic clone, many of the progeny of the initially transformed cell are also clonogenic, that is, they also can divide indefinitely, with the result that a neoplastic clone, unlike a normal clone, is "immortal."

As a neoplastic clone continues to grow, it gives rise to subclones that diverge from the parental clone. Whereas these subclones are also "immortal," they may differ in various ways. It is now possible to distinguish crudely their degrees of differentiation, the kinds and quantities of antigens they produce, and their varying capacities to spread, or metastasize, to other parts of the body. For example, a specific subline of an original tumor cell may develop in such a way that it has a high capacity to metastasize to one specific organ, e.g., the lung. Such tumor invasiveness relates not only to the tumor's intrinsic capacity for growth but also to the production of mediators which, in some manner, modify the cellular microenvironment to the tumor's advantage. For example, some tumors produce collagenase, an enzyme that adversely affects the supporting structure of a normal tissue, thereby helping the cancer invade that tissue. The growth behavior of malignant neoplasms differs from that of benign tumors in the capacity of neoplasms to penetrate adjacent tissues and invade normal tissue barriers. Presumably, such information may someday afford a useful lead for managing the spread of invasive tumors.

The nature of this fundamental neoplastic process has major clinical implications. For example, the emergence of many independently evolving subclones greatly increases the likelihood that one or more of these will, by chance, be relatively resistant to any given drug or combination of drugs. Since the survival of even a single clonogenic tumor cell can lead to renewed growth and recurrence of the tumor, successful therapy must currently aim to either extirpate or destroy every last cancer cell.

## Experimental Induction of Cancer

Clearly, the neoplastic cell differs significantly from the stem cell from which it originated. It can multiply indefinitely, it is easily separated from its neighbors, it can invade other tissues, and it fails to differentiate into the mature form of its tissue of origin. It is unclear, in this entire process,

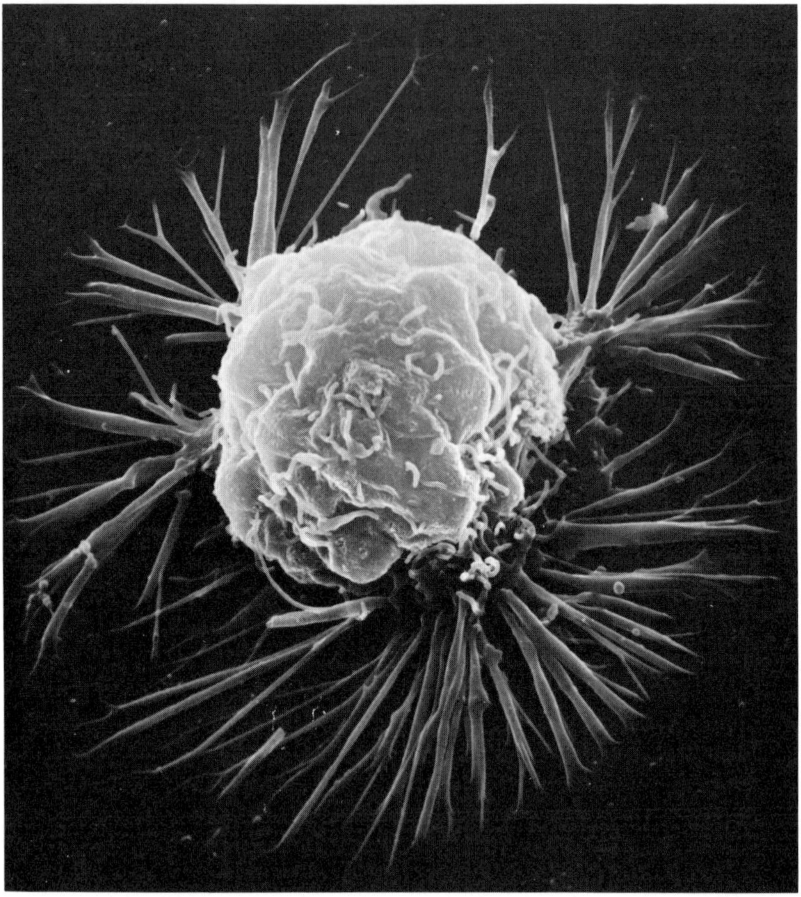

Scanning electron micrograph of a cell from human breast cancer tissue. The neoplastic cell "can multiply indefinitely, it is easily separated from its neighbors, it can invade other tissues, and it fails to differentiate into the mature form of its tissue of origin." [National Cancer Institute, National Institutes of Health.]

exactly what change takes place in the genetic structure of the cell, how many distinct biochemical changes are involved in this profound transformation, how many new enzymes, if any, are formed, how many proteins that normally should be formed fail to form, or how many biochemical entities appear or disappear from the cell's surface and interior.

Since all generations descended from the initially transformed cell (the entire clone) behave in essentially similar fashion, the entire process, in a formal sense, appears to spring from an alteration in the genetic apparatus of the cell. By definition, such an alteration is a mutation. Mutations include changes in single base molecules of DNA (deoxyribonucleic acid), deletions of single base pairs or of larger segments of DNA, or diverse rearrangements of the DNA on a somewhat larger scale.

Identification of agents that cause neoplastic transformation and studies of their mechanisms of action have been pursued via two quite different but complementary approaches: experimental investigations using laboratory animals under carefully controlled conditions, and epidemiological studies of human populations. Most laboratory investigations have used mice and rats, although limited data also are available from studies involving chickens and other avian species as well as mammalian species ranging from guinea pigs to monkeys. Many different types of tumors have been induced in animals, ranging from lymphomas and leukemias that resemble the corresponding neoplasms in humans to species-specific tumors that appear to have no human counterparts. More recently, much of this research has been conducted with pure lines of animal or human cells maintained or grown in culture; this hastens the neoplastic process and facilitates the ability to observe the transformation of noncancerous cells as it occurs.

Most of the agents studied consistently induce one or a few types of tumor, but a given tumor type can be induced by a considerable variety of seemingly unrelated agents. Identified carcinogens include viruses, physical agents (ultraviolet and ionizing radiations and impermeable foils and films implanted into tissue), and approximately 500 chemical agents belonging to several widely differing classes of chemical structures. Reliable dose-response data are available only for ionizing radiation and a few chemicals and are limited to the induction of two types of tumors: skin cancers, and lymphomas or leukemias.

There seems to be firm evidence that carcinogenesis is a multistage process. A sufficient dose of most carcinogens causes the entire process of neoplastic transformation to proceed, leading, for example, to carcinoma of the skin. Lower doses, however, may induce only the first stage, designated initiation—the stem cell is neoplastically transformed but does not begin unrestricted growth. Under these conditions, succeeding stages in the process can proceed in the presence of another class of agents, known as promoters. Promoters cannot themselves initiate the process, but in some manner they permit the rest of it to proceed once initiation has occurred. This usually leads to benign papillomas of the skin rather than to carcinoma. Just which substances in nature, if any, actually serve as promoters is not known. Phorbol esters, widely distributed in nature, serve effectively in the laboratory, but it seems unlikely that they so serve *in vivo*.

Initiation and promotion have been observed best in studies of the transformation of animal cells in tissue culture. Such studies also indicate that dose-response relationships for carcinogens cannot easily be predicted. Seemingly similar strains of mammalian cells may differ by a factor of 100 in their sensitivity to the same carcinogen. Moreover, the genetically determined susceptibility of the species and strain of an animal, as well as its age, sex, nutrition, hormonal balance, and other constitutional factors, contributes significantly to the ultimate outcome of the initial carcinogenic exposure. This complexity makes it difficult to establish a "safe" dose of a known carcinogen or even whether a given agent is, or is not, a carcinogen.

Experimental chemical carcinogens are not a single class of chemicals. Indeed, their enormous diversity seems almost bizarre. Some carcinogens are not themselves active in the form in which they are administered but are converted into the active carcinogen, which appears to react with DNA, by their metabolism in the liver. Other carcinogens are active originally but to reach the sensitive tissue must escape deactivation by the liver.

In a general way, studies of chemical carcinogens have supported the concept that neoplastic transformations are mutations. However, systematic analyses comparing the carcinogenicity of chemical agents in animals and their mutagenicity in bacteria have given much less than a one-for-one correlation; some classes of carcinogens are powerful mutagens, while others appear to be almost inactive in the bacterial test for mutagens now conventionally used. Conversely, many substances active as bacterial mutagens have been observed to have no carcinogenic activity. Thus, having a simple bacterial procedure for screening for mutagens is extremely useful, but it will not suffice for identifying carcinogens.

## Viral Carcinogenesis

There is as yet no conclusive evidence that a specific virus, acting alone, causes any type of human cancer, but there is evidence that viruses contribute to certain cancers. For instance, there is a strong association between a human herpes virus, the Epstein-Barr virus, and two forms of human cancer, endemic Burkitt's lymphoma and nasopharyngeal carcinoma. In a recent field survey of 20,000 Ugandan children, 14 cases of Burkitt's lymphoma were diagnosed. In 12 of these 14 cases, the levels of antibodies against antigens on the protein coats (the capsids) of Epstein-Barr viruses were significantly higher, months prior to the appearance of the tumor, in the blood serums of tumor-bearing patients than in the blood serums of control children. Epidemiological studies in Chinese populations in the United States, Hong Kong, Singapore, and the People's Republic of China have all revealed strikingly high levels of antibodies to Epstein-Barr virus antigens in patients with nasopharyngeal carcinoma; these titers diminish after successful treatment and increase again if the tumor recurs. The paradox is that this virus is ubiquitously distributed in human populations throughout the world; indeed, it is the causative agent of the common disease infectious mononucleosis. Yet the association of Epstein-Barr virus with Burkitt's lymphoma is rare except in certain endemic regions of equatorial Africa and New Guinea. Some coexisting circumstance must hold the explanation.

The genome for the Epstein-Barr virus is absent from the tumor cells of most lymphomas arising outside the above regions. However, in a few instances, permanent cell cultures of such lymphomas and related leukemias have released viruses similar to those known to induce corresponding types of neoplasms in lower mammalian species. The biological significance of these viral isolates remains to be established. Recent immunological studies have suggested that human breast cancer tissue often contains antigens that react with antibodies induced by the mouse mammary tumor virus. This seems a potentially important lead which warrants further investigation. It is of particular interest because the mouse virus is a very small DNA virus which is integrated into the mouse's own genome, including the genome of the cells of the germ line. Hence, it can be transmitted to the next generation without need for reinfection. Of itself, this virus is not necessarily tumorigenic, its action being highly dependent on the concurrent effects of hormones and, perhaps, other factors.

Epidemiological studies have also revealed an association between carcinoma of the cervix and herpes simplex virus type 2. Antibodies to this virus are detected in the blood plasma of a significantly greater proportion of women with invasive carcinoma of the cervix than in the blood of matched controls. Moreover, herpes simplex virus type 2 has been shown to have the capacity, under certain experimental conditions, to transform hamster and mouse cells in culture. However, it is not yet possible to assess whether the herpes simplex virus is actually an oncogenic agent (an initiator) in cervical cancer or is nonspecifically linked to the disease through other common exposure factors.

Finally, epidemiological studies on four continents suggest an association between primary cancer of the liver and chronic infection by hepatitis B virus. (Only a very small fraction of those infected develop liver cancer.) This viral DNA has also been found to be integrated into the DNA of tumorous liver cells in the host. Again, however, such integration appears to be necessary but insufficient to transform normal liver cells into tumor cells; some unidentified cocarcinogen must also be involved. The development of laboratory methods for propagating this virus and for studying its molecular biology should shed further light on the nature of this association.

Only a few years ago, it seemed possible that there might be a general relationship between viruses and cancer, that there might even be a "human cancer virus." This now seems quite unlikely, but as we have seen there remain some associations that appear to be meaningful.

Whereas only a minute fraction of all human cancer has been found to be associated with viral activity, studies of viral carcinogenesis in animal tissues have been extraordinarily revealing. Such activity has been known since 1911, when Peyton Rous demonstrated that a virus was responsible for a sarcoma (tumor of the connective tissue) in the subcutaneous tissue of rabbits. Since then, a great many viruses of various classes have been found to be able to induce tumors in some organ of an inoculated animal—frequently a species that is not normally a host to the virus—as well as in certain tissues in culture.

Ordinarily, when a virus invades a host cell it multiplies many times and emerges to invade other cells, destroying the original cell and giving rise to its specific disease (for example, the respiratory diseases caused by adenoviruses). The formation of a tumor, however, involves a somewhat different process. The DNA of the virus is totally integrated, as a unit, into the DNA of the host cell and duplicates only as the host cell divides. The same thing happens with tumor viruses that consist of RNA; the viral RNA codes for an enzyme that catalyzes the synthesis of a DNA copy of the viral RNA, and this copy is then integrated into the host genome. The formation of a tumor, therefore, must be an "expression" of this viral genetic material within the host DNA.

The "expression" of genes is the synthesis of the proteins for which they code. Such proteins may, for example, be enzymes that catalyze some chemical process. In some cases, they are regulatory proteins that bind to some specific site on the cell's DNA, thereby "turning on" or "turning off" the expression of an adjoining segment of DNA that may code for one or several enzymes. The fact that the most thoroughly studied tumor viruses are relatively short strands of nucleic acid (very short as compared with the full genome of the host) provides an opportunity to examine the actual role of viral DNA. For example, the virus SV40 and the mouse polyoma virus each consist of only about 5,200 base pairs, which comprise a total of five genes. In these and several other instances, the activities of only two of the genes are necessary to transform the host cell into a tumorous cell. In at least two instances, one of these genes is expressed as a "kinase," an enzyme capable of catalyzing the transfer of phosphate groups from ATP (adenosine triphosphate) to a protein in the cell (a process known as phosphorylation):

$$\text{ATP} + \text{Protein-OH} \longrightarrow \text{ADP} + \text{Protein-O-P}$$

In another case, the expressed protein is an ATPase, an enzyme that catalyzes the hydrolysis of ATP.

These simple processes appear to be all that is necessary to transform the host cell! The protein upon which the kinase acts—if, indeed, there is only one such protein—seems to be a vital component of the cell's inner skeletal structure, that which affixes the protein strands of the inner skeleton to the cell membrane. Its phosphorylation presumably accounts for the observed structural deformation of the cell but does not account for the failure of growth control. It remains to be learned whether tumorigenesis by viruses generally or by other agents can result from a variety of enzymes or whether only a very restricted number are involved. If only a limited number of enzymes are involved, then inhibitors of such enzymes might serve as highly specific cancer therapeutic agents.

Yet more remarkable is that, in the few instances tested to date, the specific cancer gene of the viral DNA (that expressed as the protein kinase) has been found to be identical with a gene that is normally present in the host genome. Introduction of the gene isolated from either the virus or the host results in transformation of host cells in culture. Viral carcinogenesis stands revealed, therefore, as probably an aberration of the process of cellular differentiation. Normally, the malefactor gene functions—is expressed—only for some brief period as the cells cloned from a stem cell go through their programmed series of alterations en route to becoming fully differentiated, mature tissue cells—liver cells, bone cells, lymphocytes, etc. Thereafter, the activity of the gene is largely or completely repressed. But no such control is imposed on the virally introduced gene; it functions more or less freely, and with disastrous consequences.

Thus, viral carcinogenesis as seen in these experimental systems appears to be a surprisingly simple process. It remains to be determined whether this process is a useful model for other forms of carcinogenesis.

## Radiation Carcinogenesis

Both ultraviolet and ionizing radiations have been convincingly identified as human carcinogens. Sunlight is the major cause of cancer of the skin, which is usually not metastatic. It has also, however, been implicated as a causative factor in malignant melanoma. Individuals exposed to heavy doses of ionizing radiation for therapeutic reasons, as well as the irradiated survivors of the atomic bombs at Hiroshima and Nagasaki, exhibit a significantly increased incidence of acute and chronic myeloid leukemia. More recently, the atomic bomb survivors have been shown to suffer lesser, but nonetheless significant, increases in the incidence of lymphomas and several types of epithelial cancers. Of related interest is the fact that no increase in the number of genetic mutations has yet been observed among the offspring of the irradiated survivors of the two atomic bombs. Whether they will show an altered incidence of any form of cancer remains to be determined.

Some controversy persists concerning the dose-response relationship for the induction of leukemias by ionizing radiation. This is in large part an inevitable consequence of the insufficiency of the data available at the low-dose end of the scale, data available directly only through the use of hundreds of thousands of animals. It had been held that data at higher dose levels were consistent with the "linear hypothesis," which postulates that each increment of dose yields a proportionate increment in the incidence of leukemia and, hence, that any dose of radiation above zero should yield some leukemias in a sufficiently large population. Recently, this view has been sharply challenged; current analyses favor interpreting the dose-response data as indicating that at low doses the incidence of leukemias should be some fraction of that expected from the linear hypothesis. This alternative view has major implications with respect to public policy, since the number of cases would be expected to decrease to negligible levels at very low doses; if so, there may be doses of ionizing radiation appreciably greater than zero that, for practical purposes, are unimportant.

## Chemical Carcinogenesis

Only a few dozen of the roughly 500 chemical compounds known to be carcinogenic in animals have been implicated as carcinogens in man. They range from inorganic materials such as arsenic, chromium, nickel, and asbestos to diverse classes of organic compounds, including benzene, $\beta$-naphthylamine, vinyl chloride, and chloromethyl methyl ether. Each of these is significant primarily with respect to people working in special environments and poses little threat to the population at large. In sum, the established human chemical carcinogens, like ionizing radiations, are known to account for only a very small fraction of all human cancers.

Epidemiological studies reveal that the incidence rates for different forms of cancer vary extraordinarily from country to country (see Table 2) and, in some instances, among regions within a single country, suggesting that major environmental factors are at work. Further, when large populations have migrated from a country of origin where the incidence of a particular cancer was unusually high (e.g., gastric carcinoma in Japan), the incidence of that cancer has typically decreased sharply among the migrants in their new geographic environment and has further decreased in their first-generation progeny. (Presumably relevant is the fact that gastric carcinoma was among the most frequent forms of cancer in the United States 50 years ago and, unaccountably, has now all but disappeared.)

The totality of such observations has led to the suggestion that as much as 85 percent of all human cancer may be "environmentally" determined. Although there are a number of naturally occurring carcinogens, for instance, the aflatoxins that can occasionally be found on fungus-infested peanuts, corn, and other plants when stored in a hot humid

**Table 2** Variation in worldwide incidence rates for some common cancers

| Site of cancer | High-incidence area | Incidence in high-incidence area (percent of all cancer) | Low-incidence area | Highest rate divided by lowest rate |
|---|---|---|---|---|
| Esophagus | Iran | 20 | Nigeria | 300 |
| Penis | Uganda | 1 | Israel | 300 |
| Skin | Australia | 20 | India | 200 |
| Liver | Mozambique | 8 | England | 100 |
| Nasopharynx | Singapore | 2 | England | 40 |
| Bronchus | England | 11 | Nigeria | 35 |
| Corpus uteri | United States | 3 | Japan | 30 |
| Stomach | Japan | 11 | Uganda | 25 |
| Mouth | India | 2 | Denmark | 25 |
| Pharynx | India | 2 | Denmark | 20 |
| Cervix uteri | Colombia | 10 | Israel | 15 |
| Colon | United States | 3 | Nigeria | 10 |

SOURCE: Richard Doll and Richard Peto. "Quantitative Estimates of Avoidable Risks of Cancer in America Today," *Journal of the National Cancer Institute,* May 1981.

climate, their amounts in nature and their dose-response curves make it unlikely that they could account for more than a trivial fraction of all cancer. If there are particularly important naturally occurring carcinogens, ones capable of accounting for a large fraction of all cancer, they have not been identified.

What, then, is the mechanism of nonoccupational, "environmentally induced" cancer? There appears to be no one explanation; indeed, much of the burden must be laid at the door of "life style." There is evidence that diet may affect the incidence of cancer, not through the prevalence of carcinogens but through nonspecific factors such as the caloric, fat, and fiber content of the diet. Obesity certainly increases the incidence of cancers of the endometrium and of the gall bladder in women. Failure to ingest pentose-containing fiber, a substance that promotes the formation of soft bulky stools composed largely of colon bacteria seems related to a high incidence of colonic cancer. How these and diverse other dietary factors operate is unclear.

From such observations, one might be tempted to label a considerable fraction of all cancer as potentially avoidable. But such avoidance would require major alterations in eating habits, and avoidance of one form of risk seems almost necessarily to engender another. There is also little solace to be found in the fact that, whereas the pattern of site-specific cancers varies so dramatically around the world, the gross age-corrected rate of incidence of cancer is much the same the world over.

Social habits such as smoking cigarettes and chewing betel nuts have been linked strongly to the development of cancer of the lung and the buccal mucosa, respectively. Careful studies of the numbers of cigarettes smoked daily by lifelong smokers suggest that the dose-response function for the induction of lung cancer is quadratic rather than linear (see Figure 4). Carcinogens may also interact synergistically to increase the incidence of cancer; examples include the combination of alcohol and cigarette smoking in carcinoma of the esophagus, with neither alone being a serious carcinogen for this tissue, and the combination of asbestos exposure and cigarette smoking in cancer of the lung. What is abundantly clear is that cigarette smoking is by far the most serious of all "environmental" factors operating to cause cancer and that it is most readily eliminated from the human environment.

Behavioral patterns of individuals and populations may also influence the incidence of cancer . Women who have their first child at an early age are partially protected against breast cancer. Conversely, early sexual activity increases the risk of cancer of the cervix. Certain occupational exposures are appreciably more dangerous in males, for example, exposure to specific chemical carcinogens such as asbestos, vinyl chloride, and $\beta$-naphthylamine. In these situations, effective preventive measures can be instituted, whereas, with the still largely unspecified processes associated with diet, social habits, and life styles, the difficulties of effective prevention appear rather formidable.

## Causes of Cancer

From all of these considerations, the question that emerges is this: granted that some forms of cancer in man have a clear pattern, granted that natural promoters do exist and operate, granted that the absolute numbers of people developing and dying of cancer have risen considerably, is the great bulk of cancer caused by exposure to unidentified exogenous initiators, or can much of cancer incidence be the consequence of otherwise normal metabolic events? At this writing, that question has no definitive answer. The age-corrected gross incidence of cancer in the United States actually declined somewhat for a quarter century after World War II. That decline would have been more dramatic had it not been for the striking rise in cancers associated with cigarette smoking, first in males, more recently in females. For decades, the median age of all deaths due to cancer has remained constant, at close to 60 years worldwide.

As people have become more aware of the immense increase in the amount and number of industrial chemicals in our civilization, including those substances usually regarded as "pollution" to which all of us are variably exposed, concern has grown that such exposure may contribute

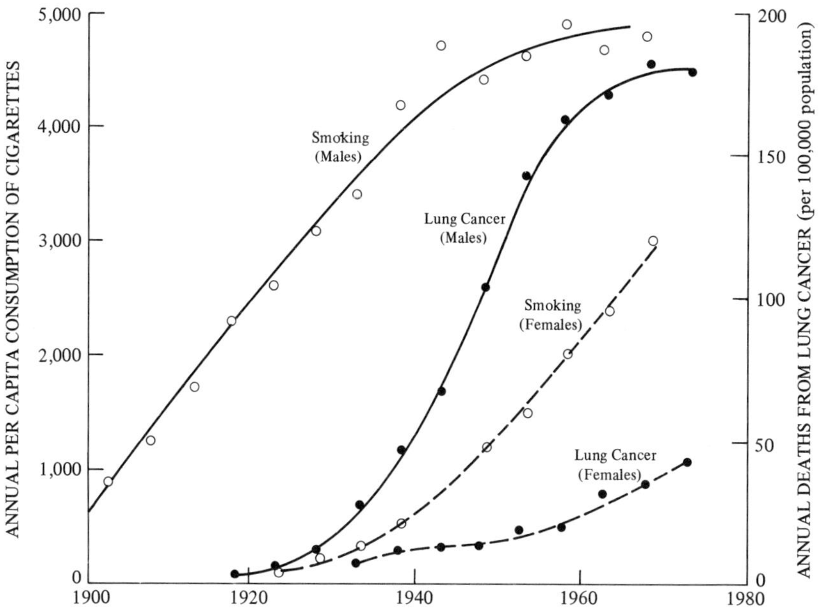


**Figure 4 Comparison of cigarette smoking with incidence of lung cancer deaths.** Cigarette smoking and lung cancer are unmistakably related, but the nature of the relation remained obscure because of the long latent period between the increase in cigarette smoking and the increase in the incidence of lung cancer. The data are for England and Wales. In men, smoking began to increase at the beginning of the twentieth century, but the corresponding trend in deaths from lung cancer did not begin until after 1920. In women, smoking began later, and lung cancers are only now appearing in large numbers. [SOURCE: John Cairns. "The Cancer Problem," *Scientific American,* Vol. 233(November 1975), p. 72. Copyright © 1975 by Scientific American, Inc. All rights reserved.]

significantly to the burden of "environmentally induced" cancer. However, although such exposure may, in theory, constitute a chemical cancer time bomb already set to go off, it is clear that if it exists it has not been detonated yet, even though it would have started ticking decades ago. As noted above, were it not for cigarette smoking the gross age-corrected rate of cancer incidence would have fallen markedly since World War II. Moreover, neither gross incidence rates nor site-specific rates show any correlation with the nation-by-nation use of energy per capita, although energy use should roughly correlate with the extent of chemical industrialization. In any case, only a few percent of all current cancers can be accounted for in such terms.

Given these circumstances, attention seems warranted to the possibility that one or more "initiators" arise in the course of normal metabolism, the amounts of which, in different tissues, are differentially altered by "life style factors" and even by some "chemical carcinogens." The validity of this alternative hypothesis remains to be appraised.

In any case, current knowledge of the details of viral carcinogenesis offers a compelling guide to future studies of carcinogenesis by chemical and physical agents. Central to this understanding will be determining the nature of the alteration in the genetic apparatus that these agents produce. Meanwhile, the principle "causes" of cancer—if such there be—remain somewhat obscure, as does the molecular nature of the initiating events.

## Carcinogen Testing and Regulation

It has long been known that certain carcinogens, such as ionizing radiation and alkylating agents (chemicals capable of transferring methyl or ethyl groups to the purine and pyrimidine bases of DNA), can also induce mutations in DNA. However, the mutagenicity of other chemical carcinogens remained masked until it was appreciated that many carcinogens must be activated during metabolism and that these activated derivatives are also mutagenic. Thus, the correlation between carcinogenicity and mutagenicity is not sufficiently rigorous to provide a firm basis for regulatory actions. Nonetheless, mutagenicity testing has played a useful role as a screening procedure in selecting chemical compounds of particular interest for carcinogenicity testing.

There are major experimental difficulties in identifying chemical agents as carcinogens in animal tests. Such tests are time consuming and expensive, and their sensitivity is both limited and variable from species to species. To maximize sensitivity, investigators have typically used mice and other species of relatively high general susceptibility and have administered test chemicals at maximally tolerated doses. Many chemical compounds for which no incriminating evidence exists in humans have been identified as carcinogens under these admittedly artificial circumstances.

Although all but one of the agents known to be carcinogenic in humans are also carcinogenic in animals, no firm scientific foundation exists for the converse, namely, that a chemical which elicits certain types of tumors in animals when given at maximally tolerated doses is likely to induce similar tumors in humans at the dose levels prevailing in the workplace or environment. Finally, animal tests of the carcinogenicity of chemicals in the low-dose range, the range of importance from the standpoint of public health, require extremely large scale, costly experiments that are fraught with logistical difficulties so severe as to render them essentially impractical.

In recent years, carcinogenicity tests that use animal cells in culture have been devised which offer hope for the future, but they will require much more extensive study and validation before they can be recommended for adoption. Meanwhile, there is a growing need to develop a more sophisticated set of regulatory policies. It has been estimated that some 40,000 chemicals are used by industry, agriculture, or medicine and that 1,000 new chemicals are added each year. Yet the capacity to assay carcinogens in the United States is currently limited to a few hundred compounds per year, testing only at very high dose levels, and there is an extensive backlog of compounds waiting to be tested. The sense of crisis has been compounded further by enactment of the Toxic Substance Control Act, which mandates obtaining such information for all substances that may be suspect.

A separate problem is raised by the Delaney Amendment, which makes it unlawful to add to any foodstuff any substance known to be carcinogenic in any amount in any species. Such legislation appears, at first reading, to be rational and defensible. Who would wish deliberately to add a carcinogen to the food supply? But the problem arises with materials, such as saccharin, that have only been shown to be carcinogenic at very high doses and whose carcinogenicity at approximately normal doses is unknown but must be exceedingly small. Such examples suggest abandoning a zero-risk policy and adopting an approach wherein the anticipated risks are weighed against the anticipated benefits, a decision much more political than scientific. The problem applies to other areas besides food additives, bearing, for example, on the minute traces of pesticides or of diethylstilbestrol that might find their way into the food supply even if their use is appropriately regulated and monitored.

## Early Detection and Pretreatment Diagnostic Evaluation of Cancer

Most cancers are unlikely to cause symptoms until they have reached an appreciable size, by which time there is a significant probability that they will have invaded neighboring structures or distant parts of the body, thus greatly diminishing the likelihood of successful therapy. It is therefore highly desirable to be able to detect cancers while they are still small and without symptoms. This may readily be accomplished with cancers of the skin or lip which are visible to the naked eye, but it is a far more formidable task with cancers arising deep within the body.

By far the most successful screening technique is the Papanicolaou exfoliative cytology test (the "Pap smear") for detecting cancer of the uterine cervix. The widespread introduction of this test is held by some scholars to have been an important factor in the more than 50 percent decrease in the death rate from this type of cancer observed among American women. Another screening test, mammography, for the early detection of cancer of the breast, has been much more controversial. There is little question that the procedure can detect minute breast cancers. But epidemiological studies suggested that repeatedly exposing women at intervals of approximately six months would have resulted in cumulative radiation exposures that would have been likely to induce more new cancers of the breast than they would have detected. Hence, the procedure fell into disfavor. However, it is currently being advocated once again following the introduction of technological advances that have sharply reduced the radiation exposure required. Certainly, women over 50 have much to gain and little to lose from this procedure.

New types of diagnostic equipment have been introduced in recent years. The fiber-optic endoscope, an instrument that permits the inside of a hollow organ to be directly observed, is much more flexible than previously available instruments and has facilitated the search for early cancers in the lung and colon. Two new techniques, computerized axial tomographic (CAT) scanning and ultrasound scanning, can detect small tumors in otherwise relatively inaccessible sites, such as the brain and upper abdomen. They represent the most striking of recent advances in cancer detection. Radioisotopic imaging procedures for detecting lesions in bone, liver, and kidney have been greatly improved with radionuclides having optimal characteristics. (Chapter 17, "Prospects for New Technologies," offers further details on these and other imaging procedures.)

These procedures, however, are far too time consuming and costly to be considered for mass screening. Instead, they are likely to play an increasingly important role in the pretreatment diagnostic assessment of how far cancers have spread that have already been detected and proven by biopsy. Although these procedures cannot guarantee against the presence of microscopic metastases, they can materially assist in the selection of optimal treatment by distinguishing those cancers that still appear to be regionally localized from others that have already spread to distant sites.

The fact that most, perhaps all, malignant tumors bear distinctive antigens on their cell membranes, and that some of these antigens may be released into the circulation, provides new opportunities for the early detection of cancer and for the detection of microscopic metastases. Highly sensitive radioimmunoassays have been introduced as screening procedures for certain types of cancer. Limited but encouraging progress has also been made in the imaging of cancers deep within the body using antitumor antibodies tagged with radioactive labels.

Until now, these approaches have had to use a complex mixture of antibodies generated by immunizing animals with human tumor tissues, a technique that has been less than satisfactory. The "hybridoma" technique, whereby a specific line of antibody-producing B lymphocytes is rendered immortal by fusing them to myeloma cell lines, yields "cell factories" that are permanent sources of molecularly homogeneous antibodies. (A given B lymphocyte makes only one kind of antibody.) The hybridoma technique, first developed for mouse cells a few years ago, has recently been successfully adapted to human cells, thus opening for the first time the possibility of using human antibodies in diagnosing and treating cancer. If human hybridomas can indeed be generated against specific antigens that distinctively mark a given type of cancer, the resultant antibodies should be well tolerated and quite safe even in repeated administrations. The extraordinary specificity of these antibodies and the ease with which they can be attached to radioactive labels that can easily be imaged outside the body may make diagnostic detection of asymptomatic cancers in such sites as the lung, breast, and gastrointestinal tract a reality during the next few years. This approach also holds promise for detecting and perhaps even eradicating microscopic metastases. These capabilities should be reflected in improved cure rates for the relevant types of cancer.

## Treatment of Cancer

Surgery and radiation therapy, which have a very high curative potential for solid cancers that are still regionally localized, are the treatments of choice. However, when cancers invade adjacent vital structures or spread to distant sites within the body via blood or lymph channels, curing them with surgery becomes impossible and the chance of curing them with radiation therapy, although not entirely eliminated, is profoundly reduced. Limited degrees of local invasion by a tumor may be dealt with successfully by the combined use of radiation therapy and surgery. Localized zones of hypoxia (low concentrations of molecular oxygen) within tumors appear to be an important factor in rendering some cancer cells radioresistant, permitting them to survive doses of radiation which destroy the rest of the tumor. Two promising approaches to circumventing this problem are entering clinical testing. The first involves the use of new, less toxic chemical radiosensitizers which can penetrate hypoxic tissues and substitute for oxygen. The second, hyperthermia, involves increasing the temperature of tumors to about 43°C (Celsius), a temperature at which hypoxic cells are selectively killed. Either or both of these approaches may augment significantly the efficacy of radiation therapy for solid tumors.

The success of such procedures depends on the stage of the disease. Thus, for relatively small breast tumors that have not invaded the overlying skin, the underlying muscle, or the adjacent lymph nodes, surgery and/or radiation will effect a cure in 75 percent of the cases. If the primary tumor is larger and has spread to the lymph nodes, only 25 percent of the patients will be cured, due to the metastatic spread of the cancer to many other sites.

A principal problem in formulating therapeutic regimens to manage cancer has been to devise ways of killing the neoplastic cells while doing minimal damage to normal tissue. This problem besets radiotherapy and is at least equally difficult in chemotherapy. The strategy with radiation and most chemotherapeutic agents is to take advantage of those particular aspects of cellular life that make the growth and division of cells possible. Any agent so restricted should, therefore, have comparatively little effect on normal nondividing mature cells. Bone marrow and intestinal mucosa are two tissues where cellular proliferation is the norm; hence, they are particularly likely to be damaged by such therapy.

Figure 5 shows some of the chemotherapeutic strategies that use agents which interfere with one or another of the metabolic steps required to form nucleic acids, an absolute prerequisite to cell division. A few drugs interfere with the mitotic machinery that causes two distinct cells to form and separate. Certain tumors deriving from particularly hormone-sensitive tissues have responded well to such hormones. The great advantage of chemotherapy is that the administered agent, distributed all over the body by the circulation, can find its way to the most disseminated microcancers and, if they are sensitive, destroy them.

Hundreds of different chemical compounds have been screened for their potential as chemotherapeutic agents. At this time, over 30 different such agents are in use, each because it has been found to be somewhat effective in treating various human cancers, that is, it significantly shrinks solid tumors or reduces the number of abnormal white cells in the blood in leukemia.

A major problem in this process is the development of cellular resistance to the chemical agent. A variety of mechanisms have been observed to account for this resistance—decreased cellular uptake, an increase in the amount of a target enzyme, decreased activation of the drug into the form in which it actually operates, and increased activity of DNA repair enzymes. Largely because of this effect, but also to increase the chances of success, it has been common practice to use not one but a number of chemotherapeutic agents simultaneously, usually of two different classes.

Chemotherapy has achieved profound success in the treatment of acute lymphocytic leukemia in children, Hodgkin's disease, and cancer of the testes, as well as lesser success with a variety of other forms of cancer. It is this approach which offers most hope in treating disseminated, multiple metastases from some primary lesion. Moreover, chemotherapy applied immediately after surgical or radiotherapeutic control of a local cancer will improve survival by controlling microcancers at distributed sites.

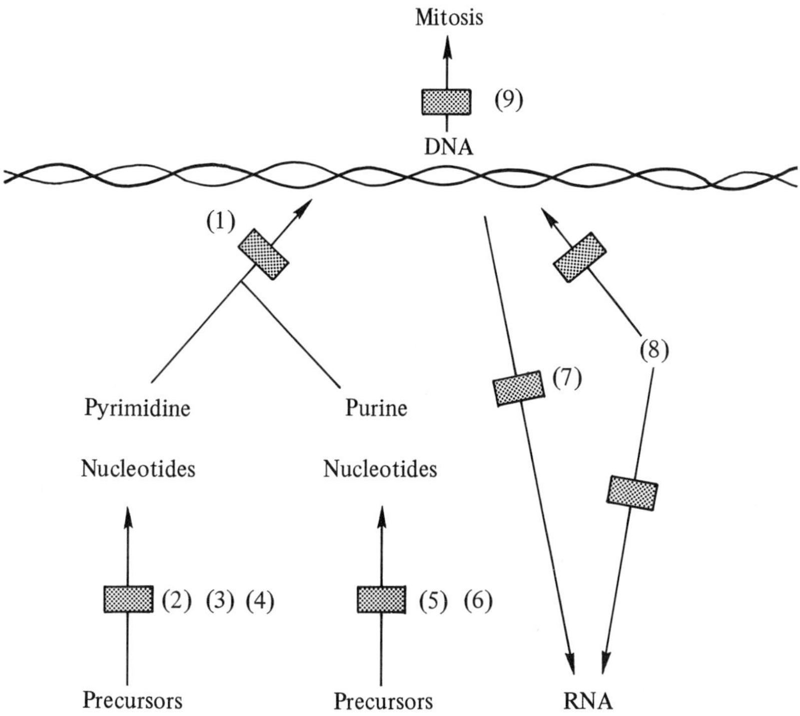


**Figure 5 Strategies employed in cancer chemotherapy.** Chemotherapeutic agents have been selected for their ability to inhibit certain metabolic steps necessary to cell division. Various agents prevent the synthesis of DNA from its precursor molecules (1-6), interfere with the mechanism responsible for cell division (9), or prevent DNA from producing proteins (7-8). (1) Arabinosyl cytosine inhibits DNA polymerase, which is responsible for the synthesis of DNA. (2) Arabinosyl cytosine also blocks the reduction of cytidylic acid to deoxycytidylic acid. (3) Hydroxyurea acts like arabinosyl cytosine. (4) Methotrexate blocks the methylation of deoxyuridylic acid to thymidylic acid. (5) Methotrexate prevents formyl folate from taking part in purine synthesis. (6) Mercaptopurine blocks the interconversion of purines. (7) Various antibiotics (e.g., adriamycin) bind with DNA and prevent the synthesis of RNA. (8) Diverse alkylating agents (e.g., chlorambucil) alkylate the bases of DNA and produce crosslinking, so that neither RNA nor DNA synthesis is possible. (9) A series of plant alkaloids (e.g., vincristine and vinblastine) react with the fibers of the mitotic apparatus and prevent mitosis, freezing the cell in metaphase.

A few varieties of tumor, cancer of the breast and prostate, for example, are hormone-dependent and can be manipulated with hormones or antihormones. Thus, estrogen will markedly shrink 60 to 70 percent of breast cancers if estrogen receptor proteins are present in the cytoplasm of the tumor cells, but only 10 percent of such cancers if the receptor proteins are absent. For premenopausal women, estrogen deprivation, as by the removal of one or both ovaries, is often done if estrogen receptors are present. For postmenopausal women, common treatments include an antiestrogen, progesterone-related hormones, male hormone, or diethylstilbestrol. With these regimens, the responses relate to the presence or absence of estrogen receptors as indicated.

It has been the combined use of several classes of drugs, for example, an adrenal steroid, a mitosis blocker, and an alkylating agent for leukemias and lymphomas, that has slowly increased the success of chemotherapy. But an agent that is effective against the carcinomas, the common tumors of epithelial tissues, is still badly needed.

The presence of specific antigens on the surfaces of tumor cells would appear to afford yet one more avenue for intervention. Lacking knowledge of the chemical identity of these antigens and their function within the tumor cell, one is left only with the possibility of immunotherapy. As noted above, the availability, in abundance, of homogeneous antibodies specific to a given form of tumor may become a powerful therapeutic weapon. It is not clear whether such an antigen-antibody reaction will, of itself, result in the destruction of the tumor, although this is highly possible if it renders the cell susceptible either to killer lymphocytes or to the complement system in the blood, which disables the cells. In addition, affixing a highly radioactive material to the antibody, iodine-131, for example, could enable a radioactive agent to seek out a neoplastic cell and destroy it from close range.

Although there have been striking successes, there have also been sad disappointments. The chemotherapeutic agents that have done so well in some instances have afforded relatively little benefit in treating the more common epithelial tumors arising in the lung, the breast, the gastrointestinal tract, the bladder, and the prostate. The problem of distant metastases continues to dominate these major types of cancers.

***Interferon*** Much public attention has focused on the prospect that interferon may prove to be a useful addition to the cancer-chemotherapy armamentarium. At this writing, one can merely state that the available body of information supports belief that this possibility warrants the most serious attention. But it is still too soon to develop a sense of where interferon will fit in the total scheme of cancer therapy.

Interferon is a medium-sized glycoprotein that an animal cell releases when invaded by a virus. When absorbed to the surface of another animal cell, interferon markedly increases the cell's resistance to invasion by a virus. Interferons are produced in small quantities by most virus-infected animal cells; the actual proteins differ somewhat both between animal species and among various tissues of the same animal.

The interferon that can be collected from human white blood cells has received particular interest. The mechanisms by which interferons increase resistance to viral infection are being intensively studied; it is evident that several different aspects of cell biology are affected, among them the properties of attachment receptors on the cell surface, a "second messenger" effect, immune response, and the release of a small nucleotide that inhibits one or more intracellular enzymes. Indeed, interferon could prove to be a valuable tool for probing details of cell biology.

Normal cells treated with interferon become larger, divide only very slowly, and change structurally. Thus, it seemed possible that interferon might beneficially affect cancer cells, whether or not the cancer were induced by viruses. Preliminary observations of both transformed cells in culture and small numbers of patients with osteogenic sarcoma, lymphocytic lymphoma, and myelomatosis have all appeared to be sufficiently promising as to warrant more full-scale trials. But such trials are presently limited by the very small available supply of human interferon from white blood cells or fibroblasts. That stringent limitation underlies the current commercial competition to prepare interferon using recombinant DNA technology. At this point, no final judgments are appropriate; one can only say that the stage is set for an interesting, perhaps exciting phase of the endeavor to improve and extend the lives of those afflicted with cancer.

## Summary and Outlook

Striking progress in biochemistry, molecular biology, cell biology, and immunology, combined with the ever greater sophistication of techniques, is certain to shed light—definitive light, it is hoped—on the nature of neoplastic transformation and on the differences between normal and cancerous cells. Just what specific event in the genome is critical to that great change and what its immediate consequences are should become clear. All of the tools required appear to be at hand. In turn, such knowledge should permit a much better understanding of the *modus operandi* of chemical carcinogens and of the role of life style factors, hence illuminating such possibilities as there may be for prevention. This understanding should also reveal the rational possibilities for therapeutic intervention, of which many must surely exist in addition to those already noted. What is abundantly evident is that no shortcut was possible. Cancer is a disturbance of cell biology. Its management, by means other than surgical extirpation and brute irradiation, had to await the wondrous and still accelerating accretion of fundamental understanding of all aspects of the normal living cell.

Whether there will be striking improvements in the prevention, detection, or treatment of cancer remains unpredictable. But, if the last decade is an indicator, one may look forward to steady improvement. Meanwhile, it is no longer true that the diagnosis of cancer is necessarily a death sentence. Almost every form of cancer is amenable to some form of intervention if it is diagnosed early enough, although some forms remain more amenable to treatment than others. Despite the great gaps in understanding, there has been steady clinical progress. The understandings to which one may confidently look forward indicate just how much more clinical progress we may justifiably expect.

## DIABETES MELLITUS

Diabetes mellitus is a disease characterized primarily by an elevated concentration of glucose (sugar) in the blood and a relative intolerance to ingested glucose—ingested glucose rapidly causes a marked further rise in the blood glucose level, which subsides only after several hours. All other aspects of the disease appear to reflect these circumstances.

For 60 years, it has been clear that the disease is due to an insufficiency of insulin, a small protein hormone that is manufactured and secreted by the beta-islet tissue of the pancreas. Commercially available ever since that discovery, insulin has been the principal means for clinically managing the disorder. This effort has been highly successful. The life spans of diabetic individuals have been markedly extended. Death in the acidotic coma that formerly claimed the lives of juvenile diabetics with the fulminating form of this disorder is now a great rarity. Ironically, however, much as the great triumph over many infectious diseases has freed us to succumb to atherosclerosis and cancer, effective clinical management of acute diabetes has freed diabetics to succumb to its unfortunate later sequelae.

In the past 40 years, the reported prevalence of this disease in the United States has increased more than sixfold. The current diabetic population exceeds 5.5 million individuals, and perhaps another 4 or 5 million Americans have potential or actual abnormalities that resemble diabetes. In some part, this increase in prevalence probably reflects better screening tests, changes in definitions, and increased awareness among physicians. However, the upward trend in prevalence has affected all ages and socioeconomic groups, occurring in twice as many women as men and with somewhat greater frequency among black and low-income populations. Now the fifth leading cause of death, diabetes directly claims over 35,000 lives annually and is a contributing factor to many more deaths each year.

Although treatment with diet, exercise, and various drugs as well as insulin has extended the average life expectancy of diabetics significantly, there is still no *cure* for this disease. Its secondary complications adversely affect the blood vessels, the heart, the kidneys, the eyes, and the nervous system. Diabetics are more than twice as prone to coronary heart disease and stroke as nondiabetics; they account for nearly 20 percent of all the patients with end-stage kidney disease who enter hemodialysis programs; diabetic retinopathy is the leading cause of new blindness among adults in the United States today; and the peripheral vascular disease that accompanies chronic diabetes results in an amputation rate almost 40 times higher than that among the nondiabetic population. Moreover, the management of diabetes during pregnancy remains somewhat difficult, although the numbers of stillbirths and defective infants have been sharply reduced while perinatal deaths have declined from 12-13 percent to 6-7 percent.

The totality of this morbidity, as large-scale disability, the inability to work, hospitalization, and the other costs of care, is brutal. An estimated $7 billion a year goes toward the health care of diabetics. But there is no way to estimate the human toll exacted from the quality of life of those individuals so afflicted.

## On the Nature of Diabetes Mellitus

Diabetes mellitus has served as a window through which to observe virtually all bodily functions. Sixty years of research, motivated in part by the search for a better understanding of this disease, has not only markedly enhanced such understanding and improved the clinical management of the diabetic. It has also revealed a great deal about such diverse phenomena as the mechanisms of protein synthesis; the process by which a cell makes and releases a hormone; the mechanisms by which hormones operate in their target cells; the manifold details of carbohydrate metabolism; the interrelationships among carbohydrate, lipid, and protein metabolism; the structure of cell membranes; cellular transport processes; the mechanism of nerve conduction; the structure of basement membranes in diverse organs; the factors involved in maintaining the integrity of the vascular epithelium; and the structure and metabolism of the lens of the eye. While this effort has been in progress, a remarkable panoply of diverse hormones has been revealed; the interrelationships among these hormones are, even now, under intense scrutiny.

Further increasing this diversified understanding of the human body offers hope that significantly improved clinical management, prevention, or even cure of the disorder can be achieved. Obviously, review of these many facets of diabetes is beyond the scope of this chapter. Instead, it will essay a brief summary only of what appear to be the most relevant highlights of such understanding, presenting the rationale for further pursuit of such information.

***What Insulin Does*** Glucose is the primary fuel of most body cells. When ingested, it is temporarily stored in the liver in a polymeric form, glycogen (animal starch), and is subsequently released to the blood at such a rate as to keep the blood sugar level constant while all of the other body tissues accept glucose from the blood according to their own energy requirements. When the liver's store of glycogen runs out, some hours after a meal, the liver starts to synthesize glucose from other materials available, primarily amino acids, and continues to release glucose to the circulation.

The other important form of fuel for the body is fat. Fat is synthesized from blood glucose by all body cells, but the major loci of this activity are the fat cells widely distributed within the connective tissue of the body and the liver cells. When the concentration of glucose in the blood declines, the fat cells release fatty acids to the circulation, and the fatty acids, in turn, are used as fuel, together with glucose, by diverse other tissues, including the liver itself.

The membranes of living cells are not freely permeable to glucose molecules. At normal blood glucose concentrations and in the absence of insulin, glucose does not enter muscle, fat, or other cells at an adequate rate. The principal function of insulin is to bind to the surface of such cells and cause a change in permeability so the entry of glucose, as well as of most amino acids, is facilitated. In the absence of insulin, entry is possible only at decidedly elevated glucose concentrations in the surrounding fluid, and then at what may be an inadequate rate.

The full details of this mechanism remain to be established. Central to it is the presence in the cell membrane of specific receptor proteins that on their outer surfaces offer a binding site for insulin. When an insulin molecule binds to such a receptor protein, the structure of the membrane is altered in such a way that glucose molecules more readily enter the cell. The receptor protein has been isolated in pure form, and the manner of its binding to insulin is being studied with X-ray crystallography. Detailed knowledge of such binding is eagerly awaited since this process is now considered insulin's primary mechanism. An understanding of its details conceivably may lead to the synthesis of a relatively simple chemical compound with the important structural features of insulin, one that could appropriately attach to the receptor protein and thus serve as a "synthetic insulin."

The entry of glucose seems to depend upon special proteins that serve as "transport structures." Before the arrival of insulin at the surface of a fat cell, for example, most of these protein structures are scattered in the interior of the cell. As the insulin molecules bind to the receptors, the transport structures rapidly migrate into the cell membrane, where they somehow serve as conduits for glucose molecules.

The binding of insulin to the receptors is not permanent. In due course (minutes to hours), a segment of the cell surface invaginates, forming a tiny sealed sphere inside the cell. Its entire complement of receptors, bound to insulin, is drawn into the cell interior. There, released insulin may occasion yet other metabolic effects. Ultimately, the insulin molecules are drawn into the subcellular bodies called lysosomes, where the insulin is hydrolyzed into its component amino acids which are then released back to the cell cytoplasm. The receptor proteins also may be thus degraded, but in large measure they are recycled back into the membrane. It is this process, therefore, that creates the requirement for a continuing supply of fresh insulin to the cell's surface.

Diverse other phenomena, in addition to easier entry of glucose, also occur when insulin binds to surface receptors. For example, the intracellular activities of a number of enzymes required for the further metabolism of glucose increase rapidly. Whether this is due, somehow, to a direct effect of internalized insulin or to the release from the cell surface into the interior of some second messenger—quite possibly the calcium ion ($Ca^{++}$)—is not yet established.

Insulin is but one of a family of hormones that serve as regulators of these metabolic processes. Indeed, several hormones operate in the opposing direction; that is, in various ways they result in an increase in blood glucose concentration. Among these, for example, are glucagon (another protein hormone of the pancreas); the growth hormone of the anterior pituitary (somatotropin); the adrenocorticotrophic hormone (ACTH) of the same gland; the steroid hormones of the adrenal cortex; and the catecholamine hormones (e.g., epinephrine) of the adrenal medulla. Indeed, it has been known for almost 50 years that the experimental diabetes that may be induced by pancreatectomy can be alleviated by removal of the anterior pituitary. Thus, in normal individuals, a sensitive system of multiple checks and balances keeps the blood sugar concentration almost constant, regardless of other circumstances, thereby ensuring a supply of glucose to body cells commensurate with their requirements.

The major tissue which had been thought to be independent of insulin is the brain. Drastic diminution of the blood sugar concentration, by administration of an excess of insulin, results in profound neurologic disturbances (insulin shock), presumably indicating that the brain, which requires a continuing supply of glucose, is not among the body tissues that accept glucose from the circulation through the action of insulin. However, research has recently revealed that at least some cells in the brain manufacture their own insulin. Indeed, it appears that all body cells make very small amounts of insulin. This was neatly demonstrated by showing that, whereas the guinea pig pancreas secretes an insulin quite different from that of other mammals, the insulin in the guinea pig brain is much like that of mammals generally. It is not clear whether this intracellular insulin serves a function, but it may be released to the local environment in the brain, in the manner of various other brain hormones, thereby obviating the need for insulin from the pancreas, which cannot traverse the blood-brain barrier.

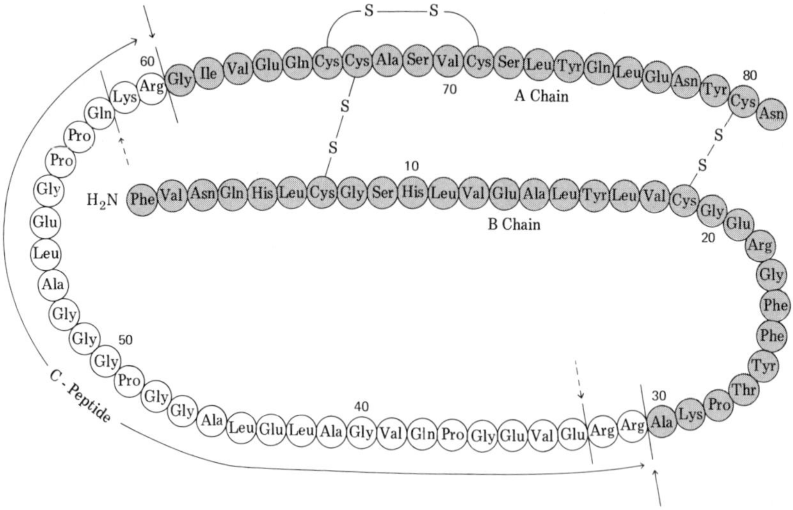


**Figure 6 The structure of bovine proinsulin.** Proinsulin is the intact chain of 81 amino acid residues with all three disulfide structures (—S—S—) in place. Hydrolysis, catalyzed by appropriate enzymes, cleaves the bond between amino acids 30 and 31 and between amino acids 60 and 61, releasing a 30-membered C peptide (positions 31-60). This forms the two-chain structure of insulin, which is bridged by two disulfides. Insulin is shown as the darkened portion of the proinsulin structure. [SOURCE: Abraham L. White et al. *Principles of Biochemistry.* Sixth Edition. New York: McGraw-Hill Book Co., 1978, p. 1266.]

***Insulin Synthesis*** As shown in Figure 6, insulin is a protein constructed of two polypeptide chains linked by two disulfide bridges. The principal region involved in the binding of insulin to the receptor proteins in cell membranes appears to be the hydrophobic amino acid residues (which repel water and attract hydrocarbons) of the B chain.

This structure is in the final stage in a synthetic process that commences very much like that of all other proteins in animal cells. The messenger RNA for insulin is translated on the surfaces of ribosomes in the usual way to form a polypeptide of 109 consecutive amino acid residues, termed preproinsulin. This molecule is released from the ribosomes and migrates through the complex Golgi apparatus of the cell, where 23 amino acid residues are removed, leaving proinsulin. The structure of bovine proinsulin is shown in Figure 6; the C peptide segment of the chain shown is five amino acid residues shorter than that of human proinsulin. Proinsulin spontaneously assumes a specific three-dimensional conformation such that three sulfhydryl (-SH) pairs are appropriately aligned and then linked to form the final disulfide bridges. After this stage, the internal C peptide segment is enzymically cut away, leaving the final two-chain, linked structure.

***Insulin Secretion*** The signal that produces the release of insulin from the cells in which it has been synthesized is the concentration of glucose in the fluid just outside those cells—in effect, the concentration of glucose in the blood plasma. The beta-islet cells are extremely sensitive to the glucose concentration and normally release insulin as required.

Although this general mechanism was recognized many years ago, the details of the process remain obscure. At the minimum, it appears to involve specific glucose receptors on the beta-cell surface, an increased concentration of $Ca^{++}$ just inside the cell membrane, and an increased rate of glucose metabolism within the beta cell itself. Moreover, other factors modulate this response. In the pancreas are at least three (perhaps more) other types of islets, each responsible for the synthesis of a separate, distinct hormone. These hormones, when released, can variously affect the blood sugar concentration. They also appear to affect the beta islets themselves in varying degrees, as do several hormones that come to the pancreas from the gastrointestinal tract. One of the latter, the "gastric inhibitory polypeptide," which enhances the beta cells' response to glucose, is released only when carbohydrates are present in the intestine; hence, the response of insulin to ingested glucose is larger in certain individuals than it is to intravenously administered glucose. Despite this complexity, the primary signal to which the beta islets are sensitive is the blood glucose concentration.

## The Causes of Diabetes

The abbreviated description of the blood sugar system given above makes it apparent that a variety of conceivable aberrations could give rise to superficially similar forms of diabetes, for example, an absence or insufficient number of beta-islet cells; failure of the islet tissue to respond to the blood glucose concentration; the absence or diminished numbers of insulin receptors in peripheral tissues; genetically faulty receptors; mutated forms of insulin unable to bind to receptors; the failure to process properly proinsulin to insulin; or the excessive production of one or more of the opposing hormones, such as adrenal steroids or glucagon. Indeed, each of these aberrations, and others, may occur to some extent within the human diabetic population. For example, at least one individual has been detected whose own insulin is defective, due to a mutation that resulted in an amino acid substitution in the B chain, and two families of diabetics, whose blood contains significant quantities of proinsulin, were found to have amino acid substitutions in the C peptide segment.

In the main, however, two major classes of diabetics may be recognized. The first class includes virtually all juvenile diabetics, plus cases that may appear decidedly later in life. It consists of individuals whose islet tissues appear to be damaged or scarred or so impaired as to be unable to provide normal amounts of insulin, probably as the residuum of an infectious process. This has been documented in a few human diabetics whose pancreas contained a virus that induces diabetes when injected into mice; such damage has been produced experimentally in rodents using the mumps, Coxsackie B, and reo viruses. The actual damage to the islet cells appears to result from the operation of the immune system (antibodies plus attacking lymphocytes) on the islet cells while they are infected with viruses. There is also evidence to suggest that, in some instances, islet damage reflects an autoimmune reaction in which some compound on the surface of the islet cell wall serves as the autoantigen.

Clearly, it is of primary importance to learn whether the immune system plays a role in this form of diabetes. If it does, the hereditary determinant would not directly specify inadequate insulin secretion; it would specify, rather, which individuals are predisposed to having islet tissue damaged by viral infection. In any case, diabetes in such individuals is the consequence of their being unable to respond to an increased load of glucose by releasing a compensatory amount of insulin; the result is a prolonged large increase in the circulating blood glucose concentration, which never returns to the normal range.

The second class of diabetics almost invariably acquire their diabetes somewhat later in life, and it is, almost equally invariably, associated with chronic obesity. In such individuals, the islet tissue appears to be normal, but, for reasons which are unknown, it does not respond adequately to the glucose signal; at the same time, the peripheral tissues (muscle and fat cells) are somewhat resistant to the insulin released by the pancreas. As a sedentary life style and obesity have increased in the American population, so, too, has this form of diabetes. But the nature of the mechanisms linking obesity with diabetes remains obscure.

Whereas the diabetics of the first class can be managed effectively only through the use of insulin, people in the second class—whose diabetes is indeed sensitive to administered insulin—are most effectively managed through dietary restriction, weight reduction, and mild exercise. Hence, this form of the disease is referred to as "noninsulin-dependent diabetes." Indeed, a significant loss of weight frequently eliminates the need for insulin. It is noteworthy that such individuals need not reduce back to "ideal weight." In fact, they appear to do best if they remain moderately overweight.

## The Consequences of Diabetes

Some of the consequences of longstanding diabetes were noted earlier. Impaired conduction in peripheral nerves appears to reflect faulty carbohydrate metabolism in these tissues. The kidney defect begins with a thickening of the basement membrane of the kidney glomeruli, followed by a series of other changes, frequently including the early appearance of hypertension. When present, careful early treatment of the hypertension with one of the drug regimes now available ameliorates the rise in blood pressure and can protect the kidneys against further deterioration.

Cataract of the lens is currently considered to follow upon an aberrant form of carbohydrate metabolism in the lens generated by a high internal glucose concentration. Peripheral vascular disease consists of damaged arterioles with consequent impairment of blood flow, ultimately enough to result in gangrene. The reason for the damage to the arterioles is not clear, but substantially similar vascular changes underlie the retinopathy which can lead to blindness. Photocoagulation of small areas of the retina, using lasers, appears to afford considerable protection against the further progress of this otherwise irreversible, destructive process.

It is not clear that chronic diabetes, of itself, hastens the atherosclerotic process. But diabetics tend to have a significantly increased concentration of low-density lipoproteins in their blood, which exacerbate the atherosclerotic process once it commences.

It is now widely considered that these various consequences of diabetes, which occur despite seemingly successful regulation of the disease, may all reflect the fact that current regimens for managing diabetes exercise relatively "loose," rather than "tight," control of the blood sugar concentration. These regimens suffice to maintain the blood sugar, on average, at levels such that, for example, minimal amounts of glucose escape into the urine. But they do not really smooth out the surges in blood glucose concentration that result from the ingestion of food. Even in diabetics under "successful" treatment, transient episodes of mild hyperglycemia occur frequently.

In experimental trials with diabetic animals, mechanical systems for achieving "tighter" control have prevented the appearance of secondary consequences and, indeed, have reversed early changes in the kidneys and nervous system. It is for this reason that much attention is now focused on seeking means to achieve tighter control of the blood sugar concentration in diabetic individuals and, thus, to test this hypothesis.

## Efforts to Achieve Tighter Control of Diabetes

Several alternative approaches are now under consideration. It has been demonstrated that islet tissue from rats, cultured for a week at 24°C, can be implanted into the livers of diabetic mice, take up residence, and achieve control of the diabetes. Islet tissue from one rat fetus will suffice to this end, indicating that the amount of material which must be implanted is but a small fraction of the normal complement of islet tissue in an adult animal. The intervening period in culture at lowered temperature somehow obviates the problem of the immune system rejecting the transplanted islet cells. Moreover, it is such models that have produced the reversal of nephropathy and impaired nervous conduction. Such demonstrations offer considerable hope not merely for "tight control" but for an actual "cure" for the disease.

Meanwhile, an alternative approach is a mechanized system that can infuse into the animal a steady trickle of insulin at a rate somewhat comparable with that of a normal animal—in contrast to the one or more intramuscular insulin injections per day that human diabetics now require. Superior still would be a system equipped with a sensor capable of monitoring the blood sugar level, thereby governing the rate of insulin administration in a feedback system analogous to that of the normal islet tissue.

Experimental systems of the former type have been constructed and are reasonably effective in experimental animals. One such system can effect relatively tight control in human diabetics for several weeks, but such devices are still bulky and clumsy and are not suitable for general use. If, however, these experiments reveal that tight control will, in fact, avert the catastrophic secondary consequences of chronic diabetes as currently managed, an intensive effort should be mounted to design systems that would permit diabetics to live unhindered, normal lives. Critical to this effort would be the development of an effective sensor for blood glucose. Of interest is the fact that, as tight control continues, the requirements of tissue for insulin appear to decline.

Commercial insulin has been prepared from the pancreases readily obtainable from stockyards. Occasionally, individuals develop antibodies to such foreign insulin, which may differ at one or two amino acid positions from human insulin. It is then feasible to switch to insulin from another animal species, but resistance due to antibodies may appear again. Obviously, the ideal circumstance would be to use human insulin itself. It now appears that, in due course, human insulin will become available through large-scale production by bacteria into which the genetic information required for synthesizing human insulin has been transferred by recombinant DNA technology. (Chapter 17, "Prospects for New Technologies," discusses the details of this technique.) The feasibility of such a process has already been demonstrated, but it remains to be established that highly purified human insulin can be prepared on a sufficient scale and with reasonable economics.

***Oral Insulins*** Considerable effort has gone into the search for a drug which, unlike insulin, could be taken by mouth and which would tend to stabilize blood glucose concentration within the desirable range. The sulfonylureas (oral insulins), which apparently increase the amount of insulin released by a given sugar stimulus, do so operate, but only in persons whose beta cells are viable and functional in some degree. Thus, they have found a place in the management of late-onset diabetes, particularly for individuals who resist the thought of insulin injections. It is not known whether the degree of regulation so achieved affords protection against the consequences of chronic diabetes, and there is concern that use of such drugs may serve, in some, as a crutch for avoiding the greater rigors of dietary care and exercise, thereby degrading somewhat, rather than improving, the quality of their care. In any case, the long-term role of this class of compounds remains to be established.

## Summary and Outlook

The vast expansion of general biological understanding and the increasing sophistication of experimental biology ensure continuing progress in understanding the fundamental aspects of the biology of diabetes: insulin synthesis and secretion, the control of these processes, the molecular mode of action of insulin and its metabolic consequences, the detailed interplay of insulin and a dozen other hormones, and the role of insulin in the central nervous system. Progress should be made in distinguishing the genetic factors that appear to contribute to the genesis of insulin-dependent diabetes in various subgroups of patients.

A clinical test of the relationship between "tight" control of the blood sugar and the risk of developing diabetic complications (especially retinopathy and nephropathy) probably will be initiated during 1981. This trial seems feasible now because of the development of instruments for the home monitoring of blood glucose and the imminence of programmable devices for continuously delivering more natural insulin doses. The results of such a trial will have enormous implications for the intensity with which all health care providers should strive to correct the metabolic abnormalities of diabetics. The potential that "tighter" control may reduce substantially the incidence of blindness, end-stage kidney disease, heart attacks, stroke, amputations, and other morbidities associated with diabetes warrants a full and intensive effort.

The next five years should witness the beginning of large-scale production, through recombinant DNA technology, of human insulin and its testing in human diabetes. An alternative to the insulin from hogs and cattle now used will offer the obvious advantage of avoiding a shortage of the world's supply of insulin (a potential problem of little importance, perhaps, in the United States but of considerable import in low-income developing countries where livestock production per capita is much below the U.S. level). It may also, because of a lesser potential for immunologic reactions, reduce the incidence of diabetic complications.

Progress should surely be made in identifying specific viruses and perhaps other environmental factors that can produce insulin-dependent diabetes in genetically predisposed individuals. If such agents are found, programs could be established to immunize susceptible individuals against the malefactor viruses and to reduce the exposure of susceptible individuals to diabetogenic environmental toxins.

The likelihood that diabetes can be treated, perhaps cured, with beta-cell transplantation depends largely on progress in managing the problem of immunologic rejection. Should this problem be solved, as one set of experiments has suggested it can be, large quantities of insulin-producing beta cells can probably be grown in tissue culture and used for such purposes. Fortunately, since endocrine tissue does not require an "organ structure" to function adequately, treating diabetics with transplants will not involve the "one donor-one recipient" problem inherent in organ transplant procedures.

The most discouraging aspect of the diabetes problem is the lack of progress in halting the rampant growth of obesity and a sedentary life style in our society. Obesity is either a primary or contributing cause of 75 percent of all diabetes. Reversal of these trends will require fundamental changes in the eating habits of a majority of the population. More than one generation may go by before such changes are well established. Until such changes are made, the incidence of noninsulin-dependent diabetes will continue to increase and the complications of this form of diabetes will continue to be among the nation's most significant health problems.

## ARTHRITIS

Strictly defined, the term arthritis refers to inflammation or inflammatory changes in the joints of the body. However, the term also is used to refer to dysfunction or inflammation of structures and tissues surrounding the joints, including muscle, bone, cartilage, tendon, and ligament. Taken together, these conditions constitute the rheumatic diseases (from rheumatism).

The terms arthritis and rheumatism are often used interchangeably, but it is important to recognize the distinction, as many of the syndromes classified as "arthritis" are expressions of generalized diseases that affect the joints or surrounding tissue. Some patients with hepatitis B, for example, may develop a particular form of joint pain and rash. Conversely, some forms of primary "arthritis" may affect other organ systems as well. For example, patients with juvenile arthritis may develop a unique form of eye disease.

Estimates of the prevalence of individual rheumatic diseases in the United States, derived from surveys conducted by the National Center for Health Statistics, indicate that 27.5 million people were affected by "arthritis, rheumatism, or gout" in 1976; 3.8 million people reported that their activity was limited because of rheumatic disease. Functional impairment occurs most often in individuals over the age of 65, but it is estimated that juvenile rheumatoid arthritis affects more than 250,000 children in the United States.

Direct expenses for the care of people with rheumatic diseases, including the cost of hospitals, nursing homes, physicians' and other professionals' fees, and drugs, totaled $2.5 billion in 1976. Perhaps $4 billion was lost from unemployment due to incapacitation in the same period. If all forms of "arthritis" were considered, the total number of afflicted persons would increase by 50 percent, as would the economic costs.

## The Biology of Rheumatoid Arthritis

Skeletal joints are lined with a thin membrane of connective tissue (the synovium), and the joint space contains a very small amount of fluid (synovial fluid). The membranes must present a smooth surface so that articulation is as frictionless as possible, while the fluid serves as a lubricant. The membranous material and the fluid are both products of the special connective tissue of the membrane and underlying structure is collagen, a triple-stranded protein molecule organized into sheets by cross stranding. (Collagen is the principal component of all connective tissue and is the most abundant single protein in the body.)

The major lubricating component is called a proteoglycan. The exact structure of these lubricants varies from site to site, but its general plan is that of the huge organized molecular complex shown in Figure 7. The total complex has an effective molecular weight that may be as much as 200 million. The hyaluronic acid—a linear polymer of certain sugars—is as long as 4,000 nanometers. The chains of protein to which keratan sulfate or chondroitin sulfate (smaller polymers of other sugars) are affixed occur at intervals of about 25 nanometers along the hyaluronic acid chain.

At this writing, it is not possible to offer a consistent, coherent account of the etiology or pathogenesis of the rheumatoid diseases. However, the accumulated data are all compatible with the working hypothesis that infectious organisms or other agents initiate the disease, in genetically susceptible individuals, and that the process centrally involves an inappropriate immune response that leads to chronic disease.

Rheumatoid arthritis is expressed as painful stiffening of the joints, followed by swelling and distortion. A significant fraction of juvenile arthritis is rheumatoid. Characteristically, the proteoglycans of the synovial fluid are found to be depolymerized and the collagen of the supporting tissue is partially degraded, due to two enzymes, a hyaluronidase and a collagenase, released into the joint fluid.

However, rheumatoid arthritis—manifested prominently as a joint disease—is a systemic disorder which may involve lesions of the eye, heart, lung, subcutaneous tissue, and reticuloendothelial system as well. Systemic lupus erythematosus, a potentially fatal disease primarily affecting women of childbearing age, is a related disorder, frequently evi-

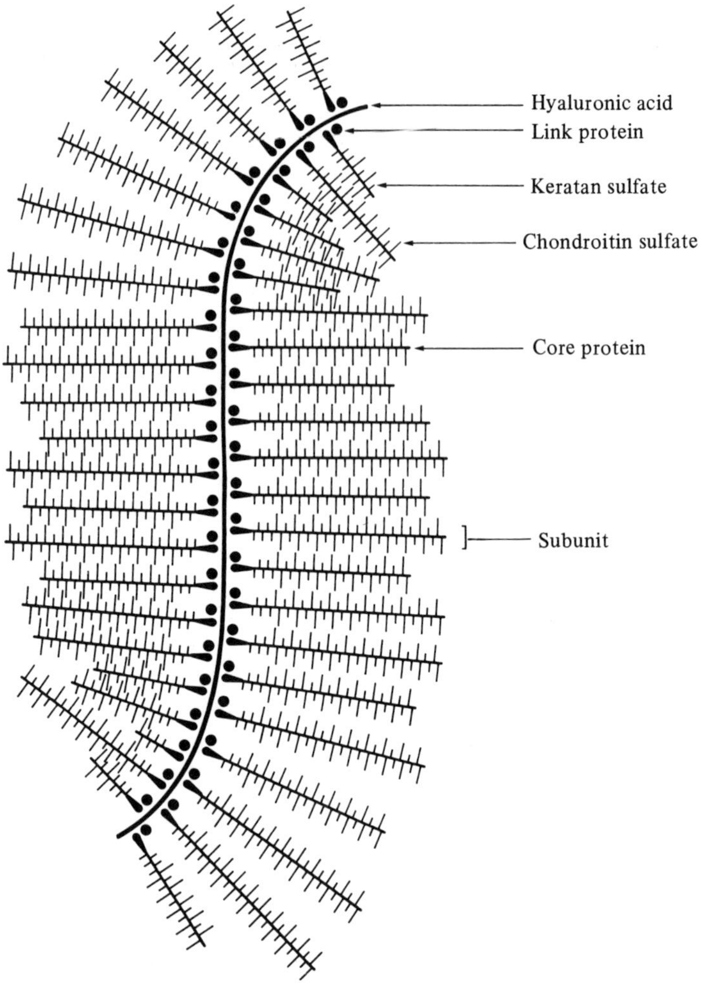


**Figure 7 The bottlebrush structure of proteoglycan aggregates from cartilage.** Proteoglycan subunits are bound by many weak bonds to a long filamentous hyaluronic acid molecule with the aid of link proteins. The length of the hyaluronic acid may vary considerably, but the proteoglycan subunits are evenly spaced along its length. These subunits consist of a polypeptide backbone (core protein) to which are covalently attached keratan sulfate and chondroitin sulfate chains. The lengths of the chondroitin sulfate chains have been reduced to avoid overlapping neighboring subunits, but the remainder of the diagram is drawn to scale based on electron micrographs. [SOURCE: L. Rosenberg. "Structure of Cartilage Proteoglycans," *Dynamics of Connective Tissue Macromolecules*. Edited by P. M. C. Burleigh and A. R. Poole. New York: American Elsevier Pub. Co., 1975, p. 107.]

dent first as redness of the skin, particularly a butterfly rash spanning the nose and cheeks. It later involves fever, pleurisy, pneumonia, and heart disease, as well as arthritis itself, and frequently terminates in fatal kidney disease. In 1955, the survival rate was less than 50 percent four years after initial diagnosis; now, supportive therapy achieves 80 percent survival after ten years. About 50,000 new cases are diagnosed annually in the United States.

***Immunological Aspects*** The immune system may be conceived of as having two main effector arms: serum antibodies, and certain formed elements (e.g., lymphocytes, macrophages) in the blood. The cell populations that contribute to this system consist of B lymphocytes (derived from bone marrow) and T lymphocytes (activated in the thymus). The latter are subdivided further into helper T cells and suppressor T cells, according to both structural and functional characteristics.*

In normal individuals, the helper and suppressor T cells work in concert to produce a suitable level of immunological response. The rheumatic diseases, as a group, are characterized by abnormalities in this delicate balance. Patients with systemic lupus erythematosus have altered levels of suppressor lymphocytes, and this may correlate with the activity of the clinical disease.

The serum of people with rheumatoid disease appears to contain two unusual types of antibodies. The first type consists of antibodies directed against some of the lymphocyte population, as well as against cells of other body tissues (e.g., platelets, red blood cells). This means that the serum contains antibodies directed against normal components of the surfaces of these cells. This condition, known as autoimmunity, violates the normal rule that one's immune system "learns," very early in life, not to make antibodies against any component of one's "self." Regrettably, that rule is not absolutely inviolate, and breaches result in disease. For example, myasthenia gravis seems to result from the formation of antibodies directed against the receptor protein for acetylcholine in the nervous system. The inflammatory damage seen in rheumatoid arthritis also may involve abnormalities in lymphocyte number, function, or both, as well as the production of autoantibodies.

The second unusual antibody in the blood serum of people with rheumatoid arthritis is a protein called "rheumatoid factor," an antibody that, surprisingly, reacts with two major classes of antibodies in blood serum (denoted as IgG and IgM). However, the binding constant is low and the rheumatoid factor may be a much more specific antibody against an as

---

* For a further explanation, see pp. 85-95 in the National Academy of Sciences' *Science and Technology: A Five-Year Outlook.* San Francisco: W. H. Freeman and Company, 1979.

yet unrecognized antigen. The rheumatoid factor appears to be made by mature plasma cells (antibody synthesizers) located in the synovium and in other affected tissues. This antibody is thought to react with another antibody already attached to whatever cell surface material has served as the autoantigen. It is this large surface-bonded "immune complex" that attracts T cells and the large polymorphonucleocytes which engulf and digest them. At the same time, this process releases the enzymes that degrade collagen and the proteoglycans, touching off the inflammatory process. This sketchy picture may be profoundly modified as more information is gathered. Knowledge of these phenomena, though substantial, remains in its infancy, and much more intensive work is necessary to clarify the complex cellular processes that influence the recognition and control of immune responses.

While much information is available about qualitative and quantitative abnormalities in the regulation of the immune response, little is known of the causative factors that lead to the clinical features of the rheumatic diseases. The search for antigens possibly involved in immunological injury has focused both on the endogenous immunological regulatory systems and on several types of infectious organisms suspected of playing an etiological role. The evidence is conclusive that, in persons with systemic lupus erythematosus, their own nucleic acids, as well as other nuclear components, serve as antigens. In addition, the serums of patients with rheumatoid arthritis react with a cell surface antigen that is associated with the Epstein-Barr virus, an observation that has restimulated interest in the possible role of this viral infection in the pathogenesis of rheumatic disease. The simplistic explanation of such a circumstance suggests that a cell surface must offer in close proximity both the autoimmune antigen and some material made in the same cell because of the virus infection that also acts as an antigen, so that both attract their specific antibodies before killer T cells attack.

***Genetics*** A huge body of information supports the idea that a system of closely linked genes, located on human chromosome 6, is responsible for the genetic regulation of the immune response. At least five distinct groups of genes are involved, each of which codes for the synthesis of unique surface markers that were originally detected on the surface of lymphocytes but are present on most other nucleated cells.

This major histocompatibility genetic complex is designated as HLA (for human leukocyte antigen); the corresponding marker series are designated by A, B, C, D, and DR (for "D-related"), respectively, each of which represents a group of closely related genes, physically arranged on the chromosome as shown in Figure 8. The chemical structures on cell surfaces that correspond to these markers have yet to be identified; at this time, the markers are recognized by their antigenic activities, that is, their

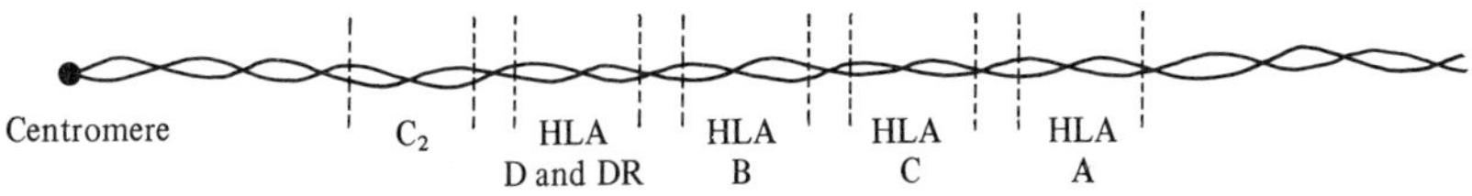


**Figure 8 Relative positions of gene sites in the major histocompatibility gene complex on human chromosome 6.** The centromere is the clear, constricted portion of the chromosome that attaches to the spindle during cell division. HLA stands for human leukocyte antigen; A, B, $C_2$, D, and DR are groups of closely related genes arranged on the chromosome in the order shown.

reactions with the specific antibodies that can be raised against them in other animals of the same species. These markers were initially used to assess the compatibility of donor and recipient for organ transplantation. (An individual whose leukocytes contain surface marker A will not generate antibodies against tissues transplanted from another individual with the same A-series marker, but will raise antibodies against transplanted tissues bearing markers B, C, D, etc., and thus initiate rejection of the transplant.)

In the last few years, much research has tried to correlate the appearance of each of these genetic markers with the occurrence of diverse diseases. The antigen HLA-B27 occurs in over 90 percent of patients with ankylosing spondylitis, a form of arthritis involving spinal and sacroiliac fusion which is particularly prevalent in Great Britain. This same antigen also occurs in the majority of patients with certain other rheumatic conditions, suggesting a common genetic base for these afflictions. It is of great interest that this cell surface antigen also appears in the majority of patients who develop arthritis following infection with certain bacteria, notably shigella, salmonella, and yersinia. Significant associations also have been found for rheumatoid arthritis (with the antigen HLA-DR4), systemic lupus erythematosus (with DR2 and DR3), Sjögren's syndrome (with DR3), and psoriatic arthritis (with B27).

The recognition that particular markers of this HLA system are associated with certain rheumatic diseases and syndromes of altered immunoreactivity has significantly advanced the genetic and epidemiological understanding of these conditions. The ability to recognize individuals "at risk" for these conditions will permit studies that may lead to the identification of environmental factors (such as infectious organisms and toxic chemicals) involved in their etiology.

***Infection*** The search for microbial agents as pathogens for illnesses such as rheumatoid arthritis and systemic lupus erythematosus has not met with much success. Reports of the isolation of mycoplasmas, viruses, bacteria, or chlamydial species from subjects with rheumatic disease have generally lacked confirmation. Other observations, however, have supported the hypothesis that infectious factors may be involved in the etiology or pathogenesis of rheumatic syndromes. Infection with hepatitis B virus is associated with two rheumatic diseases: an acute remittent polyarthritis, and polyarteritis. Lyme arthritis, a newly defined rheumatic syndrome, seems due to a virus or other infectious agent carried by ticks. The occurrence of the antigen HLA-DR2 is unusually frequent in the lymphocytes of patients with this illness.

***Models of Disease*** Striking similarities between the manifestations of systemic lupus erythematosus and the characteristics of an inbred mouse strain (NZB) have furthered enormously the exploration into the immunological and genetic aspects of autoimmune disease. Diseases in mice and other species that resemble systemic lupus erythematosus and rheumatoid arthritis in their immunological and pathological features are providing correlative studies equivalent to those provided by NZB mice. Insights gained from the study of experimental infectious arthritis have guided equivalent studies seeking an infectious basis for human rheumatoid arthritis. A recent experiment with chronic arthritis initiated by mycoplasmal infection in swine, which showed that immune complexes are deposited in various tissues despite the absence of any identifiable microbial antigen, may have relevance to the study of human rheumatic disease. Withal, to date, most of the available knowledge derives from studies of affected humans.

***Inflammation*** Although knowledge of the etiological factors responsible for rheumatic syndromes is less than satisfactory, understanding of the biochemistry of inflammation, the final stage in the pathogenesis of most forms of arthritis, has expanded markedly. The destruction of bone, cartilage, and other skeletal tissues results from the production of enzymes such as collagenase by specialized cells in joint tissue and by the phagocytic white blood cells. The extent to which these processes occur is, in turn, influenced by a variety of chemical mediators, including prostaglandins and unidentified factors produced by lymphocytes and monocytes.

Activation, by the immune complexes, of the complement system, which generates products that are potent inducers of inflammation, both locally and systemically, is a prominent feature in rheumatic diseases such as rheumatoid arthritis and systemic lupus erythematosus. Unraveling the interrelationships among the cellular and humoral factors that initiate and modulate the inflammatory response is currently among the most rewarding areas of biomedical research. The various prostaglandins and complex mixture of vasoactive and chemotactic factors involved in the complement system are clearly important, although most of the details remain to be established. However, there are promising prospects for the development of pharmacological agents that may selectively alter those specific processes that are deleterious.

Supportive therapy for arthritis has largely focused on the inflammatory process. Aspirin is not only an analgesic but an inhibitor of one enzymic step in the synthesis of the form of prostaglandin that activates the inflammatory sequence. It remains the most common and useful form of therapy; other available drugs appear to act in the same manner. Corticosteroids are general inhibitors of the inflammatory process, dissolving some of the operative white blood cells and partially suppressing the mechanisms that lead to the degradation of collagen and proteoglycans. However, their effects are yet more general and can be excessive; some agent more specific than those now available could be very helpful.

## Osteoarthritis

Research during the past several years has altered the older concept that osteoarthritis (degenerative joint disease) is the simple, inevitable effect of aging and trauma on articular cartilage. It is now evident that articular cartilage in patients with osteoarthritis has several consistent features. These include increased water content; partial disappearance and depolymerization of proteoglycans; disruption of collagen; increased levels of degradative enzymes; and an increase in the rate of synthesis of nucleic acids, protein, and glycosaminoglycans. The causes and mechanisms of these changes are not entirely clear. However, it is apparent that more than one biochemical abnormality may initiate osteoarthritis, since it may develop following a number of quite unrelated diseases. Accordingly, it is hoped that defects will be characterized in some patients that will be as specific and as amenable to therapy as is hyperuricemia in persons with gout.

## Gout

Gout is an ancient, much studied form of arthritis, now known to be the specific consequence of the accumulation of crystals of uric acid or its salts in the joint space. These crystals attract polymorphonucleocytes, which ingest the crystals and release one or more factors that induce the painful, destructive local inflammatory process. Uric acid itself is the normal end product of the metabolism of the purine bases (adenine and guanine) that are normal constituents of nucleic acids. Its concentration in blood plasma is characteristically elevated in gouty persons. This could result from circumstances that lead to either overproduction or diminished excretion of uric acid. The latter seems the dominant factor in most gouty individuals. However, in a small fraction of afflicted individuals, overproduction occurs because of one of several heritable disorders of metabolism.

In any case, inhibiting the enzyme (xanthine oxidase) responsible for the final step in the synthesis of uric acid has proved to be highly effective. Available drugs for this purpose, such as allopurinol, reduce the concentration of uric acid in the blood, lower the frequency of attacks, and provide such persons with relative comfort. Although there seem to be few undesirable complications of such therapy, the underlying cause of the disease remains to be established, and research directed at this gap in knowledge should be rewarding. One should also note that colchicine, a substance that inhibits cell mitosis, has long been known to relieve the pain of an attack of gout rather rapidly; it could not affect the body's burden of uric acid in so short a time. Although, unlike the new drugs, it does not modify the course of the disease, the mechanism of its rather specific effect warrants exploration.

### Orthopedics and Biomaterials

The development that has revolutionized the management of patients with severe chronic arthritis is an example of applied research: the development and application of surgical techniques for replacing joints. The great success of the low-friction (Charnley) hip prosthesis has led to analogous surgical reconstructions of other joints severely damaged by arthritis. Refinement of these techniques is a vigorous, continuing process involving research into such areas as detailed biomechanics, the development and testing of biologically inert materials, and studies of the structure and metabolism of skeletal tissue. Current endeavors are directed at developing metal and plastic materials with surfaces yet more suitable for implantation, improving the fixation of implanted materials to bone, producing innovations that would reduce the incidence of the occasional serious infections, and exploring techniques for the transplantation of living cartilage and bone. (See Chapter 17, "Prospects for New Technologies," for more information on prosthetic and orthotic devices.)

## Summary and Outlook

From the vantage point of 30 years ago, today's methods for preventing disability from acute infectious arthritis, rheumatic fever, and gout would be viewed as revolutionary. They were made possible by the fact that, for each of these three forms of rheumatic disease, a key step in pathogenesis has proved to be amenable to successful intervention. Similarly, major improvements in the management of rheumatoid arthritis, systemic lupus erythematosus, and osteoarthritis will almost certainly require discovering some missing parts in the puzzle of how these illnesses originate and progress. Truly effective control of disease is usually based on knowledge of its etiology, but acquiring new information short of this goal can also bring significant success. For example, the uses of allopurinol for gout and insulin for the mature onset of diabetes mellitus are successful over long periods, although the ultimate defects in these diseases remain somewhat obscure.

Some research areas that hold high promise in the search for either etiologies or elements in pathogenesis that may be amenable to control include:

1. *Immunology* Each of the following, all of which are the subjects of intensive research, should be richly rewarding: (a) further elucidation of the cellular and humoral mechanisms that influence the initiation and control of the immune response; (b) new information regarding HLA and other genetic factors that affect susceptibility to disease; (c) characterization of the immune reactants in immune complexes, with attention to the immunospecificity of antibodies and the nature and source of antigens contained in complexes; (d) determination of the deposition in tissue of immune complexes and their role in initiating and perpetuating the inflammatory response, and exploration of the potential for immunological therapy or prevention.
2. *Microbial studies* (a) Experimental clinical models of rheumatic disease in which infectious stimuli are explicit would enable the insights gained to be applied in the study of the more common rheumatic syndromes. (b) Epidemiological studies that seek to identify microbial agents or other environmental factors involved in the etiology of disease may yet reveal critical missing links in these puzzles.
3. *Biochemistry of inflammation* The complex processes underlying inflammation are only partially understood. Attention to steps that can be specifically and selectively controlled by pharmacological intervention should prove clinically rewarding.
4. *Biology of connective tissue* Further knowledge of the biochemical changes in cartilage and other connective tissues before or during degenerative diseases should be enlightening and may offer the opportunity for intervention.
5. *Orthopedics and biomaterials* New materials and new methods for fixation designed to reduce problems that currently mar success in joint replacement are almost certain to be developed, along with improved techniques and procedures in the transplantation of bone and cartilage.

## POSTSCRIPT

It is no longer known who first used the term "biomedical science"—perhaps an early clinical investigator desiring to cloak his relatively crude arts with the mantle of precise science, or maybe a fundamental biologist seeking to attract funds more readily available for distinctly medical research. Be that as it may, this chapter is a testimonial to the vitality and enormous utility of "biomedical science"—a spectrum of research extending from the most esoteric explorations of the diverse manifestations of life to astute observations made at the clinical bedside.

It includes studies of nerve conductivity in the giant axon of the squid, mutations of the pigments of the fruit fly, the light-driven proton pump of bacteria that live in brine, virus infections in bacteria, the synthetic activities of the bread mold, the aging of the rotifer, the ultrastructure of the mouse myelin sheath, the fusion of cells of different tissues from different species, the crystal structure of proteins, the kinetics of enzyme reactions at liquid-nitrogen temperatures, and so forth. It includes the study of a crippled child who engages in self-mutilation, of fetal development in a hypertensive diabetic pregnant mother, of a child who develops diarrhea each time it drinks milk, of an aging woman whose distorted hands must be pried apart in the morning, of a middle-aged man who loses his breath and whose chest is agonizingly constricted after climbing a flight of stairs, and of human cells in tissue culture. And it includes epidemiological studies that seek correlations between specific aspects of the environment and the incidence of diverse diseases. All of these and many, many more constitute the continuum of biomedical science.

The very fact that this is an interactive continuum supports the conviction that clues to understanding the etiology and pathogenesis of the diseases considered in this chapter and of the many other disorders that afflict mankind, clues necessary to improve the management, or perhaps to cure or even prevent, these diseases, will come from observations gathered by those working at widely separated parts of the continuum. Periodically, voices arise to assert publicly that numerous items of scientific understanding, already gathered in the laboratory, could be of material benefit in the clinic and yet have not been transferred. Little evidence exists to substantiate that concern. Significant new understandings are transferred remarkably rapidly from laboratories elsewhere in the continuum to scientists addressing each specific disease. If hazard there be, it may arise from premature attempts to apply the as yet not quite applicable.

As noted, a remarkable array of diseases have been brought under effective control. A considerable list of disorders for which, only yesterday, pronouncements of diagnosis were, in effect, death sentences either have disappeared due to preventive measures or are treated effectively in the clinic. What is particularly noteworthy is that these successes were made possible, in whole or in part, by the fact that, for each, the etiology had been clearly established before control was achieved: identification of the specific vitamins that were lacking in deficiency diseases; knowledge of the specific consequences of an excess or deficiency of a given hormone; clear identification of the organisms that caused tuberculosis, pneumonia, syphilis, typhoid fever, whooping cough, or subacute bacterial endocarditis and of the viruses that caused smallpox, measles, mumps, or poliomyelitis.

No such circumstance yet obtains for the chronic diseases considered in this chapter or for such others as schizophrenia, multiple sclerosis, and muscular dystrophy. There can be no guarantee, furthermore, that, when all the details of etiology and pathogenesis are in hand, truly effective management, prevention, or cure will become self-evident. We can only hope blindly that persistence combined with serendipity will bring us closer to the desired goals.

Not all biomedical understanding is relevant to all disease states. But which clues will illuminate which pathological states remains unpredictable. Who could have predicted that a deficiency in the enzyme hypoxanthine-guanine phosphoribosyltransferase would be the primary cause of a form of gout in which children mutilate themselves? Who could have predicted that a deficiency in the enzyme ornithine aminotransferase would be the cause of the gyrate atrophy of the choroid and retina that leads to blindness; that studies of the genetics of skin transplantation in mice would provide a principal clue to understanding rheumatoid arthritis in man; that a variant in the structure of the sulfonamides developed as antibacterial agents would make possible management of glaucoma of the eye; or that the combination of a viral infection and inappropriate formation of an antibody to some structure on the surface of one's own cells could give rise to diverse diseases? The list goes on and on.

What stands out in such histories is that each new major technique or procedure enables a leap to unanticipated new understandings and insights, that each new broad biological understanding illuminates a host of pathological circumstances never even considered by the original investigators. As the understandings sought become more sophisticated, more recondite, and more detailed, the efforts demanded become ever greater, and the tools and instruments ever more elaborate, more powerful, and more expensive. There is little choice but to pursue this path.

Fortunately, clinical progress need not await the great breakthroughs. Small progressive steps, applying whatever can be applied, using halfway measures, have made possible very substantial differences in the life expectancy, and particularly in the quality of life, of people afflicted with each of the diseases treated in this chapter. Regrettably, such measures contribute significantly to the ever rising national bill for health care. There is no reason to believe such progress will abate, just as there is no reason to hold that these diseases are intrinsically unavoidable concomitants of the human condition. Because the pace of biological progress ensures that genuine etiological understandings cannot remain elusive much longer, investigators grappling with these diseases have never been so universally optimistic over the prospects for future success. Meanwhile, we may rejoice in the expanding, ever more detailed panorama of the nature of life, of the nature of human beings, being laid out before us.

In the words of Pierre Charron in *Traité de la Sagesse* (Book I, 1601), "La vraie science et la vraie étude de l'homme, c'est l'homme."

## BIBLIOGRAPHY

Carlo M. Croce and Hilary Koprowski. "The Genetics of Human Cancer," *Scientific American*, Vol. 238(February 1978), pp. 117-125.

Department of Health, Education, and Welfare. *Report of the National Commission on Diabetes to the Congress of the United States*. Washington, D.C.: U.S. Government Printing Office, 1976.

Raymond Devoret. "Bacterial Tests for Potential Carcinogens," *Scientific American*, Vol. 241(August 1979), pp. 40-49.

Isaiah J. Fidler. "Mechanisms of Cancer Invasion and Metastasis," *Cancer: A Comprehensive Treatise*. Volume Four. Edited by Frederick F. Becker. New York: Plenum Publishing Corp., 1975.

W. Henle and G. Henle. "The Immunological Approach to Study of Possibly Virus-Induced Human Malignancies Using Epstein-Barr Virus as an Example," *Progress in Experimental Tumor Research,* Vol. 21(1978).

David Koffler. "The Immunology of Rheumatoid Diseases," *CIBA Clinical Symposia*, Vol. 31(1979).

J. F. Mustard et al. "Platelets, Thrombosis and Atherosclerosis," *Progress in Biochemical Pharmacology,* Vol. 13(1977), pp. 312-325.

National Health and Lung Institute Task Force on Arteriosclerosis. *Arteriosclerosis*. Washington, D.C.: U.S. Department of Health, Education, and Welfare, 1971.

Abner Louis Notkins. "The Causes of Diabetes," *Scientific American*, Vol. 241(November 1979), pp. 62-73.

*Proceedings of the Conference on the Decline in Coronary Heart Disease Mortality*. Edited by Richard I. Havlik and Manning Feinlieb. Washington, D.C.: U.S. Department of Health, Education, and Welfare, 1979.

[Joe Baker, Medical World News.]

# 3

# *Directions in Nutrition Research*

## INTRODUCTION

Nutrition is both a science and a practice. The science of nutrition, as a branch of biology, is concerned with acquiring knowledge about the sources of nutrients and the need for them. It seeks to determine the factors that influence the efficiency with which nutrients are used as well as the effects of deficiencies, excesses, and disproportionate amounts of nutrients on growth and development, on behavior, and on physiological and pathological processes. Finally, it tries to find out how nutrient needs are altered by environmental factors such as disease, stress, and changing physical conditions. The practice of nutrition uses the knowledge gained through these efforts to maintain health and to treat disease.

Nutrition has developed as an independent science only since the end of the nineteenth century. Until the early part of this century, foods were thought to provide just three major nutrients—carbohydrates, fats, and proteins—and a few minerals. Between 1906 and 1912, foods were found to contain both fat- and water-soluble substances (vitamins) needed for survival, and certain amino acids were shown to be essential dietary constituents.

The past 75 years have seen remarkable progress. Some 45 substances, ranging from polyunsaturated fatty acids to the mineral element selenium, have now been identified as essential nutrients for mammals. Quantitative studies have established, with varying degrees of accuracy, the amounts of 35 nutrients that are vital for human growth and development, maintenance of health, and reproduction and lactation.

The physiological functions of many nutrients have been successfully identified. We know, for example, that thiamin plays a central role in the chemical reactions through which the body obtains energy from the oxidation of carbohydrates, and that essential fatty acids are precursors of a family of hormonelike substances, prostaglandins, which are involved in the functioning of the nervous system, the gastrointestinal tract, and the reproductive organs.

Advances in the science of nutrition have made it possible for seriously ill patients to subsist for months on parenterally (intravenously) administered fluids that contain only known purified or crystalline substances while they undergo and recover from surgery to restore major body functions. Even persons who have lost all gastrointestinal function can live this way with portable infusion units, leading reasonably active lives for many years. The development of therapeutic diets has made it possible to treat patients for a variety of metabolic problems. Phenylketonuria is one of the best-known heritable metabolic diseases among many that can be treated by diet modification. Others are lactase deficiency, galactosemia, and several vitamin dependency diseases.

## Diet

Nutrition and health surveys indicate that 85 percent of the U.S. population surveyed, even in low-income counties, shows no evidence of nutritional deficits. Of the rest, most show marginal but nondebilitating anemia, presumably the result of insufficient iron in the diet. (Iron is the nutrient that is most likely not to occur in sufficient amounts in American diets.) A proportion of the population, usually 1 to 2 percent, but varying from one age group to another and from one locality to another, was classified as having marginal intakes of some vitamins, but this cannot be attributed to inadequacies of the food supply.[1] Most Americans are obviously well nourished. There are a variety of reasons, nevertheless, for nutritional deficits in populations that have access to a nutritionally adequate food supply. Many elderly people do not consume enough food because they do not expend much energy and thus have small appetites; many people restrict their food intake to lose weight; others who are ill consume little food, and surveys in hospitals have shown that many patients arrive for treatment malnourished. There are always individuals who have bizarre food habits either from choice or ignorance; social instability in families may lead to neglect of children; alcoholism is commonly associated with inadequate diets.

A healthful diet is one in which the combination of foods provides an appropriate balance of the various nutrients. Using the food group approach (i.e., eating appropriate proportions of dairy products, meats and legumes, fruits and vegetables, and cereal grain products), nutrient needs can be met with diets that provide from 1,200 to 1,800 kilocalories of energy per day.[2] Individual foods differ greatly in the amounts of nutrients they provide. Despite these differences, however, no single food is complete, nor does any food possess unique health-promoting properties. It is not scientifically sound to assume that diets should be composed only of foods that have high nutritional value. A food that is rich in energy but low in nutrients can be valuable in the diet of an active youngster, whereas a large quantity of such food may be inappropriate for a sedentary, overweight adult. In devising regulations and food policies, and developing nutrition education programs, the relationship of individual foods to the total diet should be a major consideration.

Although most of the population of the United States is generally well nourished, there is still widespread malnutrition and undernutrition throughout the world. The causes lie in complex political, socioeconomic, and general public health problems. In dealing with these, it is important to recognize the limits of nutrition intervention and to assess realistically the role of nutrition in comprehensive programs for health, and agricultural and economic development.

## New Directions

The remarkable advances in the science of nutrition and its application since World War II, combined with an improved standard of living for most of the population, have all but eliminated nutritional deficiency diseases as a serious public health problem in the United States. In addition, the development of vaccines and the discovery of antibiotics have made it possible to control many of the major infectious diseases. During this time, overweight and obesity have become more prevalent, and chronic and degenerative diseases have become the major causes of death and disability in this country.

The possibility that modification of the fat components of the diet might be an effective treatment or preventive measure for managing heart disease—now a major medical problem—served as an impetus to change the direction of much nutritional research toward investigation of problems that are not exclusively or even primarily nutritional in nature. These include the chronic and degenerative diseases—cancer, cardiovascular disease, osteoporosis, hypertension, and the general process of aging—as well as trauma induced by burns, accidents, or surgery. We only have indicators of the role that nutrients play in these conditions and of the way that the body responds to them. The potential rewards of this change in direction are exceeded only by the possibilities for controversy. If we are to continue to make real progress, we must proceed carefully, using the most rigorous scientific methodology. Our prescriptions for treatment must be based upon fact, not hope. Only in this way can nutritional research continue in the twenty-first century the dramatic improvements in individual health that it began in the twentieth.

## NUTRITION AND HEALTH

### Introduction

In the United States, the major nutritional deficiency diseases had been eliminated by the early 1940's as public health problems and there was little concern about the nutritional needs of the population. The food supply was adequate, and cereal grain products were in large measure fortified with iron and the vitamins—thiamine, niacin, and riboflavin—that were considered most critical for the prevention of nutritional deficiencies. *Over*nutrition was beginning to receive attention as a major public health problem. Then, in the 1960's, reports of undernutrition and malnutrition, particularly in some of the southern states, raised questions about the nutritional status of the population and the adequacy of the food supply in the United States. The problems uncovered by nutrition surveys (for example, the Ten-State Nutrition Survey[3]), with the possible exception of iron deficiency anemia, could not be attributed to shortcomings in the food supply. The results of the Health and Nutrition Examination Survey, undertaken in 1971-72, led to similar conclusions.[4] Both surveys also revealed that substantial numbers of people in the United States were overweight.

These observations shifted the focus of attention toward the nature of major medical problems and the possible relationships between changes in diet composition and changes in the major causes of death and disability. A series of reports released by the Senate Select Committee on Nutrition and Human Needs drew attention to the fact that the composition of the food supply had changed.[5] Analysts in the Department of Agriculture had established that since the turn of the century there had been a 10 percent increase in total daily calories derived from fat and a 10 percent decrease in those derived from starch (see Figure 1).[6] The Senate Committee hearings emphasized what other reports had revealed: that the major causes of death were chronic and degenerative diseases, with heart disease and cancer accounting for close to 70 percent of all deaths; and that a large proportion of the population was substantially overweight.

If obesity is defined as being 20 percent or more above desirable weight, then about 40 percent of American women aged 40 to 49 years and about 32 percent of American men in this age category are obese.[7] The National Center for Health Statistics has estimated that, on the basis

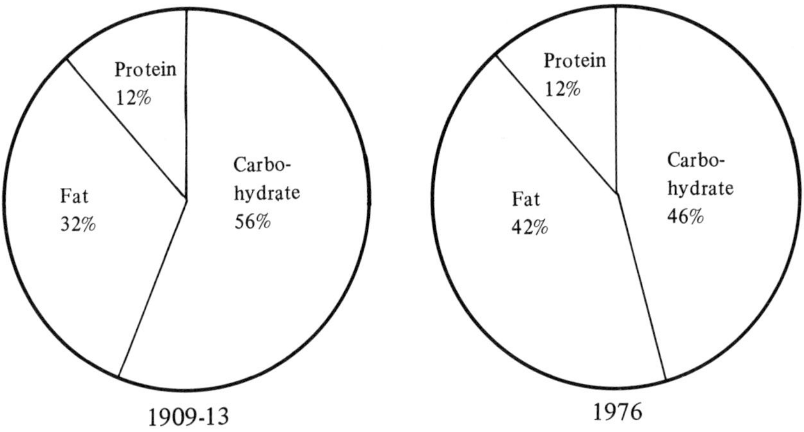


**Figure 1 Calories from energy-yielding nutrients (per capita civilian consumption).** [SOURCE: Louise Page and Berta Friend. "The Changing United States Diet," *Bio-Science,* Vol. 28:1(1978), p. 632. Copyright 1978, American Institute of Biological Sciences. Reprinted by permission.]

of the standards for appropriate body weight developed by the Metropolitan Life Insurance Company, slightly over seven million Americans are extremely obese. There has been a substantial increase in average body weight over the period since 1900, despite a decline in caloric intake.

These changes in diet composition, body weight, and the major medical and health problems have been accompanied by other changes in health indicators in the United States. Infant mortality has fallen to one tenth of what it was in 1900; childhood and maternal mortality have declined even more sharply, and life expectancy has increased from 47 to 72 years. With 70 percent of the male and 84 percent of the female infants born reaching age 65, the proportion of elderly people in the population has increased from 4 percent to over 11 percent.[8] The incidence of chronic and degenerative diseases increases with age so, since most people now live beyond 65, it is to be expected that disability and death from these diseases also will increase. However, there is debate about how much this incidence is affected by factors other than aging.

## Overweight and Obesity

Excess body fat is accumulated when the major sources of energy in the diet—carbohydrates and fats—are consumed in quantities that provide more calories than are expended. The body has only a limited number of ways of disposing of excess nutrients. Extra sodium, vitamin C, and many water-soluble nutrients are excreted in the urine; extra calcium, iron, and some other minerals are not absorbed efficiently from the intestine, but excess carbohydrates, alcohol, and fats are absorbed efficiently and are not excreted except in certain pathological conditions. The excess is removed from the circulation and stored as fat. This ability to store surpluses of energy sources has great value for the survival of organisms that may find food plentiful during certain periods of time and scarce during others, conditions that undoubtedly prevailed for the human population throughout most of its existence.

An excessive intake of carbohydrate initiates a series of coordinated responses that lead to conservation of the unneeded calories as fat. Most food sources of carbohydrate are digested to yield glucose or glucose precursors which, on absorption, stimulate release from the pancreas of the hormone insulin. Insulin facilitates the uptake of glucose by muscle and adipose tissue. It causes a rise in the activity of the liver enzyme glucokinase, which catalyzes the conversion of glucose to an intermediate product. This in turn undergoes a series of chemical transformations, resulting in a portion of the glucose being converted to the glucose polymer glycogen, a storage form of carbohydrate which is retained in the liver. Another portion of the glucose is oxidized completely to carbon dioxide and water, with the release of energy for the work of the liver. The remainder is converted to small units, acetyl coenzyme A, a derivative of acetic acid, which serves as a precursor for the synthesis of fatty acids, the major constituent of body fat.

While these events are taking place, the amounts of enzymes involved in the synthesis of fatty acids increase and the amounts of those involved in the breakdown of fat decrease. The body thus becomes adapted to a high carbohydrate intake by developing a high capacity for the synthesis of fat from carbohydrate. In humans, fat synthesized in the liver is transported through the blood in lipoprotein molecules to other tissues and organs, most notably adipose tissue, where it is stored for future use and where additional fat may be synthesized.

Although this adaptive system is highly advantageous for people who may have to face alternating periods of food surplus and deprivation, it may be disadvantageous for persons in a modern, industrialized society. The high incidence of obesity in the United States suggests that substantial numbers of Americans have great difficulty regulating their body weight (see Figure 2). At the same time, the proliferation of popular articles on how to lose weight testifies to the fact that in America fat is not beautiful. It also is unhealthy, as statistical studies show; obesity is associated with a high incidence of diabetes, hypertension, and gall bladder disease, and is thought to be a risk factor in heart disease.

***Appropriate Body Weight*** One question that arises from all of this is, "What is appropriate body weight?" For years, the guides for desirable body weights have been based on tables prepared by life insurance actu-

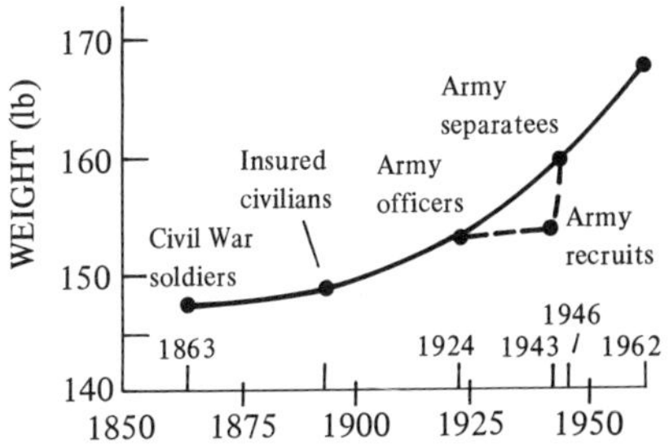


**Figure 2 Average body weights for American men 5′8″ tall, aged 30-34 (since 1850).** [SOURCE: Theodore B. Van Itallie. "Obesity: The American Disease," *Food Technology,* Vol. 33, No. 12 (December 1979), p. 43. Copyright 1979 by the Institute of Food Technologists.]

aries, largely on the assumption that body weight should remain essentially the same throughout later life as it is at age 25. The average weight of American men aged 40 to 59 is 15 percent above their weight at age 25.

Interest in the relationship between obesity and health has stimulated further examination of appropriate body weight and the degree of health hazard associated with excess weight. An analysis of the accumulated information on the subject has been reported recently in relation to life expectancy and the probability of death from heart disease among men aged 40 to 59 years.[9]

The examination of these relationships, corrected for elevated blood pressure, which is a known risk factor for heart disease, used relative weight or a body mass index (weight in kilograms divided by height in meters), instead of simple body weight. This gives a more accurate indication of adiposity. For men at middle age, the probability of death was lowest for those who were somewhat above the average in relative weight, which would correspond with a weight of between 20 and 30 percent above the recommended desirable weight found in life insurance tables. The probability of death increased considerably both for those who were greatly over and for those who were greatly under the average. Also, in the absence of hypertension, overweight appeared not to be a risk factor for heart disease.[10] These observations agree with those of other researchers and, most recently, with results obtained in a long-term study, sponsored by the National Institute on Aging, on the effects of aging in men.[11]

Despite the comforting information that some overweight is not particularly detrimental, excessive overweight is associated with increased health risk and thus is undesirable. In addition, obesity affects people's lives in countless ways in addition to curtailing normal physical activity, such as walking and climbing stairs. In spite of all these undesirable effects, it is still difficult for fat people to achieve and maintain weight loss. The low rate of long-term success in weight control programs points up the need for improved understanding of the reasons body weight is not regulated appropriately. Despite the progress of recent years, our knowledge of the complex of systems involved in regulating body weight is still inadequate.

## Internal Regulators of Body Weight

Body weight is essentially a reflection of energy balance, which can be represented by the equation

$$\text{Energy balance} = \text{energy intake} - \text{energy expenditure}$$

When the body is growing, the balance is positive and the energy content of the body increases. After mature body weight is achieved, if the amount of energy consumed as food exceeds the amount expended, the extra energy is stored as fat. If this continues, the end result is obesity. To maintain constant body weight, the amount of food consumed must be controlled so that the energy it provides just balances the amount needed for physical activity and maintaining body functions. In order to understand body weight regulation and its failure, it is necessary to understand the way in which food intake is controlled and the factors that influence energy expenditure and the efficiency with which it is expended. Weight loss can be achieved by reducing intake below expenditure, increasing expenditure above intake through increased physical activity, and by a combination of the two. The last of these is considered the most effective as there is some evidence that the control of food intake is more accurate when energy expenditure is high than when it is low.[12]

Control of food intake in man has been studied intensively. As might be expected of a biological or metabolic process that is critical for survival, it has evolved as a multifaceted system so that the failure of one component rarely results in the failure of the entire system. The features that emerge, and they are only incompletely understood, are those of a system of nerve tracts in the brain that serve as receptors for signals arising in other parts of the body which provide information about the environment, the expenditure of energy, and energy reserves. The information received is integrated by this system in such a way that another signal is elicited which either initiates or terminates food consumption.

Many of the critical nerve tracts in this system have been identified by cutting or damaging specific nerve fibers in animals and then observing feeding behavior. Several of the substances—called neurotransmitters—that are necessary for transmission of the signal also have been identified.[13] They are for the most part derivatives of some of the amino acids that are constituents of dietary proteins—serotonin, for example, which is derived from tryptophan, and norepinephrine, which is derived from tyrosine. Their importance has been established by injecting animals with drugs to block or specifically inhibit the action of the individual neurotransmitters. Many of the properties of this feedback system have become known through this type of research, but the complexity of the process is such that anything approaching a complete understanding of it will require continued intensive and imaginative investigation.

The signals that provide information about the energy needs of the body to the control system in the brain arise from a variety of sources. There is evidence that sites in the upper part of the spinal column sense external temperature changes and elicit signals to increase food intake when environmental temperature falls. Nerve responses can be detected in the feeding control system in the brain when nutrients are being absorbed. Feeding responses are observed when animals are injected with inhibitors that block the ability of the tissues to oxidize glucose. Also, feeding tends to occur when little glucose is being used, for instance, after a prolonged period without food when fatty acids released from adipose tissue are the main source of energy. Satiety and cessation of feeding tend to occur when glucose is in plentiful supply. All of this suggests that the extent to which glucose is being used for energy is sensed by the brain.

After body fat has been depleted, animals and human subjects will consume extra food until the fat stores have been replaced. Also, after being overfed, animals will reduce their food intake until body weight has fallen to what it was before overfeeding. Distension of the stomach will depress food intake temporarily but, if this occurs because the diet has been diluted with inert material, such as cellulose, the animal will adapt and consume gradually increasing amounts of food until it is obtaining as many calories as it needs to prevent weight loss. Feeding is often initiated also by the sight or odor of food and may be excessive when the taste is highly appealing. Social environments in which overeating is encouraged at an early age may be important in setting life-long patterns of eating behavior that lead to obesity. In sum, a variety of information is processed and integrated to determine food intake.

***Metabolic Differences*** The other side of the energy balance equation is that of energy expenditure. The inability to understand the tendency of some people toward obesity, the readiness with which they regain excess weight after having lost it, and the fact that some persons can maintain their body weight with low caloric intake, have aroused interest recently in the possibility of differences in the efficiency of energy-requiring processes in the body. The processing of food in the body after a meal increases heat production, so that a certain amount of energy is lost. Experimental studies with mice and some with people suggest that the obese may lose less energy in this way than their lean counterparts.

Basal metabolic rates differ among individuals; some people and animals require more energy to maintain basic body functions than do others. Although it is by no means a general phenomenon, some obese people are found to have low basal metabolic rates. Much energy is used in the body for the transport of nutrients across membranes between cells, particularly for the transport of sodium, which is continuously pumped out of cells by an energy-requiring process. Studies on obese mice indicate that they use less energy than lean mice of the same strain to maintain a low intracellular sodium concentration. In other obese mice, overeating was found to be associated with high insulin production, suggesting that hormonal differences may account for some differences in the tendency toward obesity.

These are a few among an accumulating number of observations which suggest that obesity and the overeating that leads to obesity may have a variety of causes, including metabolic or endocrine differences that alter the efficiency with which individuals use energy or that influence their food intake. Some of the differences observed between lean and obese strains of mice are inherited, suggesting that a genetic component may be responsible for the tendency of some people to eat beyond their basic needs.

The primary abnormality in obesity is an excess of adipose tissue. This can occur either through an increase in the size of fat cells, an increase in the number of fat cells, or both. Underfeeding or overfeeding of rats during the preweaning period can influence the number of fat cells they have at maturity. Observations that rats underfed during early life had fewer fat cells and were leaner than others that were allowed to eat freely during early life have led to the concept that a propensity toward obesity may be determined by the number of fat cells in the body. In humans, fat cells proliferate mainly during the first two years of life and during puberty. Individuals who become obese in childhood usually have higher numbers of fat cells than those who become obese later in life. Nevertheless, there are observations indicating that even in adulthood, once obesity reaches a certain level and fat cells have enlarged to a certain point, fat cell proliferation may begin again. Individuals with the greatest number of fat cells tend to be the most obese, presumably because there is a limit to the size to which fat cells will expand. There is active debate over the importance of large numbers of fat cells induced early in life as determinants of later obesity.

Our increased knowledge of the processes involved in the control and use of food intake unfortunately still does not enable us to provide much better advice to the overweight than to increase their physical activity and reduce their food intake. New clues that provide a better understanding of the regulation of body weight are coming to light, but the promise of new approaches to preventing obesity or stimulating weight loss is not likely to be fulfilled without further knowledge of the basic processes that determine energy balance. This information can be obtained only through painstaking research.

## NUTRITION AND DIET IN THE MANAGEMENT OF CLINICAL PROBLEMS

Steady progress in understanding the metabolism of carbohydrates, fats, and proteins and the functions of the essential nutrients—vitamins, minerals, and amino acids—has made it possible to use diet modification as a therapeutic measure in diseases that are caused by genetic defects of metabolism and in others that result in organ failure. Treatment of such diseases has provided an interface between basic science and clinical medicine and has contributed to an understanding of metabolism and the function of nutrients. Observations of the effects of malnutrition, whether originating from the diet or from disease, have created an awareness of the importance of nutritional support for hospitalized patients and led to effective ways of providing needed nutrition for patients who are unable to eat enough food.

Geographic differences in the incidence of many diseases and differences in susceptibility to certain diseases among groups within populations have aroused renewed interest in the possibility that the type of diet consumed may be an important factor predisposing many people toward developing specific diseases. This subject is dealt with in the previous chapter "On Some Major Human Diseases." Therefore, it is not treated here.

## Genetic Defects of Metabolism

Some 100 genetic disorders are known in which modification of the diet or the intake of specific nutrients is a useful part of the treatment. This treatment reduces the severity of the signs and symptoms, but does not provide a cure for the basic defects causing these diseases. Most of them are characterized by the loss of a single highly specific function, such as the ability to degrade a particular amino acid, to convert one dietary carbohydrate into another, to synthesize a specific protein for the transport of a nutrient, or to use a vitamin efficiently to form a functional coenzyme or enzyme. These diseases are rare but are often devastating. Ironically, successful management of them may increase the gene pool of these diseases. Thus, progress in one direction may have the potential for increasing problems in another.

Congenital lactase deficiency is caused by a lack of the enyzme lactase in the small intestine. This enzyme is needed to hydrolyze lactose (milk sugar) into its component simple sugars. When the enzyme is missing, lactose accumulates in the intestine where it undergoes fermentation and causes bloating, abdominal distress, and diarrhea. In the severe form of lactase deficiency, the signs are evident immediately after birth and, unless the disease is treated quickly, the continuous diarrhea that results will lead to a degree of dehydration that can be catastrophic for the infant. Since the enzymes for digestion of carbohydrates other than lactose are present in normal quantities, elimination of milk from the diet and substitution of a formula diet containing other carbohydrates is an effective treatment.

A milder form of lactose intolerance is unique among genetic metabolic disorders because it is widespread among many populations, particularly Oriental, African, and Middle Eastern ethnic groups. In these populations, the ability to synthesize lactase falls sharply within the first few years of life and remains low thereafter. Traditionally, most of these populations do not include milk in their diets, or use it only to a limited extent after early childhood.

Lactose is broken down during digestion to glucose and galactose. An excess of galactose in the blood, or galactosemia, results from a deficiency of the enzyme galactose-1-phosphate uridyltransferase in the liver, kidney, and small intestine. This enzyme is needed to convert galactose to glucose, the normal blood sugar. The accumulation of galactose in the blood and a conversion product, galactose-1-phosphate, produces hypoglucosemia (low blood sugar). Unless it is treated, cataracts, liver and renal disease, and mental retardation eventually occur. If the hypoglucosemia is severe, coma and death may follow. As with congenital lactase deficiency, the removal of lactose—the source of galactose—from the diet is a highly effective treatment.[14]

During the past two decades, a group of genetic diseases classified as vitamin-dependency diseases have been recognized. One of these is characterized by convulsions that occur within hours to a few days after birth, and by an abnormal electroencephalographic pattern in affected infants. These signs can be prevented by administering pyridoxine—vitamin $B_6$—in amounts that exceed the recommended dietary allowance by 30- to 100-fold.

Infants with another hereditary disease are unable to degrade the branched-chain amino acids, leucine, isoleucine, and valine, and thus cannot convert them to glucose or fat that can be oxidized for energy. This causes the amino acids and derivatives of them to accumulate in the body and to be excreted in large amounts in the urine. The disease varies considerably in severity, but usually results in mental retardation. Diets that contain only the minimum required amounts of branched-chain amino acids have been devised for treatment of this disease. These diets are expensive and monotonous because protein intake from foods must be curtailed severely, but they provide a means of reducing the probability of severe mental retardation. In some of these infants, a defect in the dehydrogenase enzyme required for the metabolism of these amino acids reduces the enzyme's ability to interact effectively with its thiamin-containing coenzyme. This can be partly overcome by administering large doses of the vitamin thiamin.

Another type of hereditary disease that can be treated by modification of the diet is $\beta$-lipoproteinemia, which is characterized by a defect in the ability to synthesize a protein needed to transport fat from the intestine into the lymphatic vessels that carry it to other organs. This disease results in poor absorption of fat and fat-soluble vitamins (A, K, D, and E) and is accompanied by diarrhea and severe loss of fat in the feces. Weight loss occurs and progressive neuromuscular disease develops. This condition can be improved by replacing most of the usual dietary fat with fats that contain medium-chain fatty acids. These are less dependent on $\beta$-lipoprotein for absorption.

## Organ Failure

Organ failure that results from the ingestion of toxic substances or infections, or autoimmune reactions, can lead to impaired metabolism of nutrients. In some of these instances, dietary modification or manipulation of specific nutrients can be an important part of the clinical management of the patient. Hepatic (liver) failure, in which alcoholism is often a causative factor, and renal (kidney) failure, often the result of infection, are two examples in which the potential for dietary treatments has been explored extensively. Illustrations of the research that underlies therapeutic advances in nutrition follow.

***Hepatic Failure*** Liver damage can result in abnormalities in the metabolism of protein, carbohydrate, vitamin, and mineral intake. It often is accompanied by low food consumption, which leads to malnutrition, creating further complications of the original disease. Diet therapy is an important part of the management of patients with liver diseases.

The liver plays a critical role in the metabolism of amino acids. Degradation of most amino acids and conversion of the nitrogen from amino acids into urea, a relatively nontoxic substance, are major functions of the liver. When these functions are sufficiently impaired in liver disease, there are toxic effects from the accumulation of the end products of the metabolism of amino acids, and cerebral function may be affected.

Despite the impaired ability of the liver to metabolize amino acids, patients with liver disease still need protein, which is essential for repair of the damaged liver. If they consume excessive amounts, their systems become overloaded, the rate of synthesis of urea fails to increase appropriately in response to the increased load of amino acids, and ammonia accumulates. This ammonia, together with the accumulated amino acids or other products of amino acid metabolism, can cause hepatic encephalopathy, in which cerebral function is impaired. Confusion, apathy, and personality changes ensue; muscle spasms may occur and the patient may go into coma.

Changes were observed in the concentrations of amino acids in the blood of patients with liver disease many years ago, but there were few clues to their possible significance. Blood concentrations of several amino acids, notably tyrosine and tryptophan, precursors of brain neurotransmitters, were elevated; concentrations of the branched-chain amino acids were depressed. As knowledge of the transport of amino acid into tissues expanded, it was realized that all of those amino acids depended on a common carrier for entry into the brain. The various amino acids would then compete with each other for entry. Those in high concentration in the blood would suppress the brain's uptake of those that were in low concentration.

Dogs in which the blood supply is diverted to bypass the liver serve as models for studying liver disease. Such dogs show many of the features of patients with severe liver disease, including an altered blood amino acid pattern, and will go into a coma resembling that associated with hepatic encephalopathy. It has proved possible to prevent dogs with hepatic portal bypasses from dying in coma by infusing them with a mixture of amino acids that was designed to correct their abnormal blood pattern. Subsequently, the mental state of some patients with liver disease was improved when they were treated in this way.[15]

The results of these and similar studies suggest that, besides the adverse effects from the accumulation of ammonia, there is increased uptake by the brain of the precursors of neurotransmitters, owing to low blood concentrations of the branched-chain amino acids, which otherwise would compete with them. This may contribute to the development of hepatic encephalopathy. The possibility that neurotransmitters such as serotonin may be produced in abnormally large amounts, or that other false neurotransmitters arising from phenylalanine may block the action of the true neurotransmitters, opens up new possibilities for understanding some of the severe adverse effects of cirrhosis of the liver and of hepatic failure generally.

***Renal Failure*** Renal failure causes a syndrome known as uremia, which develops gradually and insidiously as the kidneys cease to function and metabolic waste products, especially of amino acid or nitrogen metabolism, accumulate in blood and tissues. The syndrome is complex and reflects the loss of excretory, endocrine, and metabolic functions of this organ. The symptoms are many, including loss of appetite, nausea, vomiting, muscle cramps, weakness, emotional irritability, and decreased mental comprehension. Many of the symptoms can be controlled with diet therapy or dialysis but, if untreated, the disease can lead to coma and death.

There are many abnormalities in renal failure that require adjustments in the diet but, since dietary protein is a major source of waste products that cannot be eliminated efficiently, efforts to limit the accumulation of these have received particular attention. A general principle in devising diets for renal patients is to use a low amount of high-quality protein, such as meat, milk products, eggs, or fish, which contains a high proportion of essential amino acids. This reduces the load of waste products, primarily urea, that must be excreted by the malfunctioning kidney, and also reduces the quantities of a wide variety of metabolites that accumulate in the body. There is little information about the toxicity of the specific compounds found in increased quantities in the blood in uremia, but several of the nitrogenous end products and amino acid metabolites are thought to contribute to the undesirable consequences of the disease. Careful dietary management, even if it is not adequate to control all of the adverse consequences of the disease, can reduce the frequency of the need for dialysis.

Some of the $\alpha$-ketoacids that are formed during the degradation of amino acids are readily reconverted into the corresponding amino acids in the body. Nitrogen that ordinarily would be converted to urea and excreted must be used for this conversion and thus is recycled in the body. If, in the diets of renal patients, $\alpha$-ketoacids are substituted for the amino acids for which they serve as precursors, the amount of nitrogen needed is reduced and the work of the kidneys is correspondingly reduced. Whether this offers a practical approach for the routine treatment of renal patients is debatable, but it holds promise as an adjunct to therapy, not only for patients with renal failure but also for the treatment of infants who are unable to make urea because of genetic defects of urea cycle enzymes.

Even when patients with renal failure are maintained by diet or dialysis or some combination of the two and live for many years, some of the metabolic disorders from which they suffer persist or progress. One of these, a disorder of bone metabolism known as renal osteodystrophy, leads to softening of the bones and loss of bone mineral. Complex skeletal changes occur, accompanied by diffuse bone pain. A part of the disorder is due to the inability of the diseased kidney to convert vitamin D to 1,25-dihydroxycholecalciferol, its active form. Since the final step in this conversion occurs solely in the kidney, vitamin D deficiency develops despite the presence of an adequate supply of vitamin D itself in other parts of the body. Recent research has established the pathway for the synthesis of the active form of vitamin D in the body.[16] This information, combined with the availability of synthetic 1,25-dihydroxyvitamin $D_3$, promises to provide a treatment for this disorder.

## Nutrition and Resistance to Infection

Malnutrition is not limited solely to people in underdeveloped nations who face famine and economic deprivation; it occurs, although infrequently, in this country as well. Also, it is an all-too-common secondary problem in hospitalized patients who have cancer and other chronic degenerative diseases that suppress the desire for food. Severe malnutrition increases the susceptibility of a patient to infectious agents by depressing generalized host resistance and reducing the effectiveness of specific defensive measures by the immune system. It contributes frequently to the development of pneumonia, bacterial sepsis, and other forms of infection that occur as a final, irreversible complication in these illnesses.

Malnutrition not only predisposes to infection, but the reverse is also true; infection may initiate or worsen malnutrition.[17] The most noticeable effect of an infectious illness is the breakdown or wasting of body tissues and the depletion of stores of nutrients. This occurs particularly as the result of intestinal infections that cause diarrhea and impair the absorption of nutrients. At the same time, however, all host defensive mechanisms, including immunological defenses, require that body cells manufacture new proteins in order to eliminate or control the infectious process. Nutrients must be obtained from whatever body stores remain, from the diet, or through special feedings in order to maintain the biochemical functions of the cells that supply energy and the free amino acids needed to build the new proteins required for host defenses. When an infection occurs in a patient who is already malnourished, susceptibility to other infections is increased and a vicious cycle may be generated.

Although nutritional and immunological interactions have been studied most frequently in man during generalized protein-energy malnutrition, immune systems also can fail to function as a result of deficiencies, imbalances, or excesses of single nutrients. Generalized protein-energy deficiency is usually accompanied by deficiencies of one or more individual nutrients. A deficiency of vitamin A, iron, zinc, folic acid, or vitamin $B_{12}$ can occur in an individual patient. Each of these single nutrient deficiencies has produced immune dysfunction in man, and each of them has proved to be reversible. Other hints have emerged to suggest that deficiencies of pyridoxine, riboflavin, ascorbic acid, or excesses of iron, vitamin E, cholesterol, or polyunsaturated fatty acids also may be detrimental to some aspects of immune function in man. Animal experiments suggest that, if a single nutrient normally is essential for preserving or contributing to the functional activity of a metabolic process within body cells, it also will have an effect on the competence of immune system cells. Unfortunately, no individual deficiency has been studied comprehensively enough to determine its potential effects on all measurable immunological functions. Few details are known about how individual nutrient deficiencies exert their effects on specific immune functions or the extent to which these are the result of the generalized malnutrition that accompanies any specific nutrient deficiency.

Iron deficiency is the most common form of single-element malnutrition in this country. Because of its importance as a component of many cellular enzyme systems, a lack of iron will lead to the dysfunction of lymphocytes and phagocytic cells even before anemia can be detected. On the other hand, an excess of iron can be deleterious to immune system functions. In addition, an excess of iron in the body allows this key nutrient to become available to microbial organisms, which need it for luxurious growth and replication and, in the case of some bacteria, for the production of toxins as well. A therapeutic attempt to correct a deficiency in body iron before a coexisting deficiency of body protein is corrected can, therefore, be potentially dangerous.

Since much of the zinc in the body cannot be mobilized rapidly, this element can become depleted readily. Zinc deficiency affects many aspects of immune function. Acrodermatitis enteropathica, a severe congenital disease of infants and young children which includes skin, intestinal, and immunological abnormalities, recently has been recognized to result from a markedly depressed ability of intestinal cells to absorb zinc from ingested foods. Feeding large doses of zinc to patients with this disease can reverse the immunological defects and skin lesions and restore the ability of the patients to resist previously fatal infections.

Great advances have been made during the past decade in the identification and characterization of different components of the immune system and in understanding how these components must function together to be effective. It is becoming increasingly possible to identify immune system components that are not functioning properly in a malnourished patient and to improve disease resistance by appropriate nutritional therapy.

## Parenteral Nutrition

Total parenteral nutrition for the debilitated patient is a recent development that has stimulated renewed interest in nutrition in clinical medicine. Its primary purposes are to maintain an adequate nutritional state or to correct undernutrition or malnutrition in patients by supplying adequate amounts of all nutrients through intravenous feeding. Fluids, glucose, and protein hydrolysates were given intravenously for many years prior to the mid-1960's, but efforts to provide total nutritional support by such procedures were not consistently successful. However, between 1965 and 1970, puppies were shown to maintain a rapid growth rate when they received their entire supply of nutrients through a catheter fixed in a major blood vessel, such as the superior vena cava.[18] A large blood vessel with a rapid rate of flow is essential when concentrated glucose solutions are used because they exert high osmotic pressure and cause smaller veins to deteriorate around the insertion point of the catheter. Where fat emulsions can be used to provide a part of the energy for the patient, the lower osmotic pressure of the solution permits the use of a smaller peripheral vein.

This technique for meeting total nutritional needs intravenously was of particular interest initially because it enabled the physician to prevent weight loss and depletion of energy reserves and nutrients in patients, particularly infants and young children, who had to undergo extensive gastrointestinal surgery and therefore could not eat. Early experience with this technique showed that it was possible, in the case of patients with severe lesions in the gastrointestinal tract, to nourish them adequately for months while they underwent corrective surgery. Similar demonstrations showing that malnourished infants could be rehabilitated rapidly to withstand surgery to correct intestinal malformations firmly established the usefulness of the procedure.

The potential of this technique for increasing the probability of survival where life is at risk from nutritional depletion and from the direct consequences of disease or trauma has been demonstrated. Its value in patients who must undergo gastrointestinal surgery is obvious. In patients with extensive burns, major trauma, or severe infections, where energy needs may increase from 20 to 80 percent above normal, and who may be unable or unwilling to eat, it provides a way of preventing or reducing the severe nutritional depletion that otherwise would occur. In cases of severe malabsorption or diarrhea, it can be used to prevent dehydration and at the same time maintain patients in a satisfactory nutritional state. More recently, it has been tested as a means of preventing nutritional depletion in cancer patients who lose their appetite while undergoing chemotherapy. There are indications that in some persons tolerance for chemotherapeutic agents is increased and immune response is improved, but the final answer as to how generally effective it may be remains to be established.

When it is considered that most cases of severe trauma and burns are the result of accidents, that accidents are the major cause of death among people under 45 years of age, and that accidents are responsible for the greatest loss of time from work, the ability to maintain and rehabilitate accident victims through nutritional support has been an important medical advance in reducing mortality and speeding the recovery of traumatized patients. Improvements in methods of providing this support and a better understanding of the nutritional needs of persons who suffer trauma and burns should increase further the rate of success in treating them and reduce the duration of their hospitalization.

The ability to maintain for years patients who have no gastrointestinal function, mainly with portable infusion systems that enable them to lead reasonably active lives, has opened up an opportunity to investigate human nutrient requirements. It also has provided a chance to study differences in nutritional needs and in the metabolism of nutrients when they enter the body without first being absorbed through the gastrointestinal tract, and when water-soluble nutrients do not pass through the liver before being delivered to other organs and tissues.

In one such study, a patient who had been maintained for more than five years by this procedure developed a condition resembling diabetes, with an accumulation of glucose in the blood.[19] She also developed impaired nerve conduction in the limbs. Insulin did not restore the glucose tolerance test to normal. Other tests showed that chromium concentration in the blood was less than 10 percent of normal. Chromium deficiency in animals was known to cause impairment of glucose removal from the blood. Administration of this element to the patient restored the glucose tolerance test to normal and improved the peripheral neuropathy. With chromium added to the infusion fluid, the patient remained well. This investigation provided a convincing demonstration of the need for chromium in humans and an opportunity to study this requirement quantitatively. It also provided evidence that a lack of chromium causes deterioration of normal nerve function, either directly or indirectly by impairing glucose metabolism.

Studies of this type open up the possibility of establishing whether all of the nutrients that are essential for the human species have been identified. In view of the prolonged periods during which patients have been able to stay healthy on parenteral infusions, it seems likely that any nutrients that remain unidentified must be contaminants of known substances and required in minute amounts. In connection with this, animals kept in an environment in which there was strict control of diet and environmental contamination have developed signs attributable to deficiencies of nickel, vanadium, and silicon. It is possible that these elements may be required in trace amounts in human diets.

## Nutrition and Aging

Reliable knowledge about the effects of nutrition on aging is very limited. Nevertheless, there has been much speculation throughout human existence about the possibility of discovering a food or diet that would prove to be the elixir of life. It is not difficult to understand the appeal of myths that promise to prolong youth and delay death by some simple expedient such as a nutritional nostrum. Despite the fact that there is no scientific evidence to support the possibility of this, there is no lack of customers for those who promise that the ravages of age can be prevented by nutritional supplements.

Within a species, an individual cannot expect to live longer than the maximum life span that can be authenticated. For humans, this is 114 years, although there have been reports of people surviving beyond this. Much evidence indicates that the life span of mammals is basically a species characteristic that is genetically determined. The life expectancy of an individual within a species is influenced by many factors throughout life, so that for most persons the genetic potential at birth is not likely to be achieved. Among these factors is the increasing susceptibility to death from disease, particularly from chronic and degenerative diseases, with increasing age. For a large number of diseases, the mortality rate grows exponentially with increasing age beyond 20 years. It is not clear whether or not increasing susceptibility can be attributed directly to the process of aging. Nevertheless, the relationship holds, whatever its underlying basis. It might be noted that life expectancy at age 65 differs very little between countries such as the United States and Japan, where diets differ substantially and the major causes of death are distinctly different.

The efficiency of many physiological processes changes with age (see Figure 3).[20] To indicate just a few: heart rate declines; distensibility of the arteries decreases; maximum breathing capacity falls; renal blood flow and glomerular filtration rate decrease; there is a reduction in the ability of parietal cells of the gastric mucosa to secrete hydrochloric acid (in other words, some digestive procedures slow down); handgrip strength and muscular function generally decline; the response of the nervous system slows. There are reductions in metabolic responses as well. Tolerance for cold is reduced; response in glucose tolerance tests is prolonged. In general, there is a reduced ability to adapt to changes in the external environment and to achieve homeostasis after a disturbance in the internal environment. It should be emphasized that the rates of decline of these processes differ greatly among individuals.

There are thus two broad questions we might ask about relationships between nutrition and aging. The first is whether, by modifying diet or feeding practices, it is possible to lengthen the life span. The second is whether, in view of the physical and metabolic deterioration that occurs with advancing age, dietary or nutritional modifications can improve the quality of life for the elderly. This becomes particularly important when it is considered that 11 percent of the U.S. population is over 65 years of age and that more than half of the infants born today can expect to survive beyond 70.

Although there is considerable evidence that sound nutritional practices can prolong life expectancy, there is very little to suggest that nutritional manipulation can retard the aging process and increase the maximum life span. The only evidence that the life span of mammals can be prolonged by altering feeding practices comes from studies with rodents—mice, rats, and hamsters. The major finding of such studies is that dietary restriction—usually quite severe dietary restriction of the order of 20 to 50 percent of normal food intake—increases survival over that of animals allowed to eat freely. In most of these studies, restriction was begun early in life, and hence it retarded growth and development. Food restriction also tended to delay the onset of diseases from which the animals died, but for a few types of tumors the incidence increased and a higher proportion of the tumors that developed were malignant.

Efforts have been made to determine whether restriction of food in-

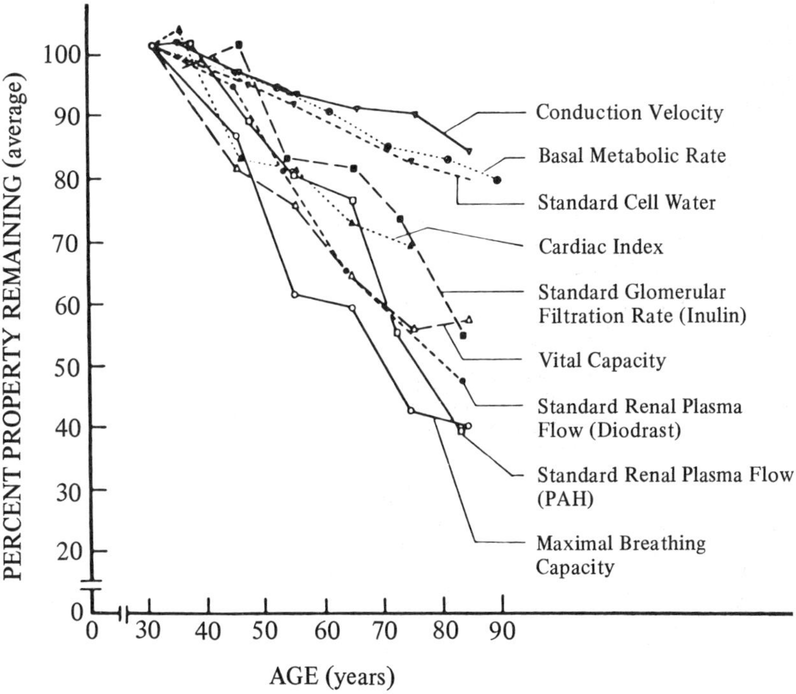


**Figure 3 Age changes in physiologic functions (percent of mean value at age 30).** [SOURCE: N. W. Shock. "The Science of Gerontology," *Proceedings of Seminars 1959-61 Durham, N.C.: Council on Gerontology.* Edited by E. C. Jeffers. Duke University Press, 1962, pp. 123-140. Data are derived from cross-sectional studies.]

take has these effects only if it is begun early enough in life to retard growth and development. The answer from the studies is not clear. It appears to depend within narrow limits on the age at which restriction is initiated and the duration of restriction. The same is true of studies of the effects of diet modification.[21] A high protein intake early in life was observed to exert a beneficial effect, whereas a high protein intake late in life was detrimental.

Consumption of a high-fat diet, which produced obesity in the animals, shortened survival. Analysis of the available information indicates that interaction between variables characterizing growth and those characterizing diet is complex and that no one factor alone exerts an inordinate effect on longevity.

Measurements of certain functions provide evidence that chronic underfeeding, which lengthens the life span of rodents, also tends to delay changes associated with aging. The ultimate question is whether these observations apply to humans. In assessing their significance, several

"There is a dearth of nutritional information generally about the needs of the elderly for the various essential nutrients." [Administration on Aging, U.S. Department of Health and Human Services.]

things should be kept in mind. The animals were kept in isolation and did not interact socially. The environment was clean and as free of disease agents as possible. Most of the treatments impaired growth and development. The diets fed were invariable through periods of many months. Physical activity was limited because the animals were housed in small cages. Finally, the assessment of what is the desirable body weight in man should be recalled.

The most striking change that is known so far to occur in nutritional needs in the aging process is a gradual decline in caloric requirements. This is associated with a declining basal metabolic rate and declining activity. There is a dearth of information generally about the needs of the elderly for the various essential nutrients. Studies of protein and a few vitamin requirements indicate that the need for these substances changes little, if at all, with advancing age. Perhaps a more important need than information on nutritional requirements of aging healthy people is the need for more information about the effects of the various chronic diseases from which the elderly suffer and the drugs they take to control such diseases.

It is perhaps appropriate to conclude a discussion of nutrition and the elderly with a word about the possible existence of Shangri-La's, where people live to an extremely old age while retaining the vigor of youth. One place among three or four cited from time to time in newspapers is in Ecuador, where highly active people well over 100 years of age are reported to be common. During a study of the bone structure of this population, a discrepancy in the ages reported by the same individuals over a period of years was noted. On comparing birth records with reported ages, it was found that after age 65 to 70 the elderly population systematically exaggerated their ages. Nearly all of those reporting their ages as being 100 or more were actually in their 80's to 90's. This was a consequence of the increasing status accorded the elderly in the social structure of the community. As with claims for nutritional miracles, reports of small populations with uniquely long life span seem to be just another myth.

## NUTRIENT REQUIREMENTS

As knowledge of the essential nutrients expanded, as each in turn was identified, and as methods for measuring them were developed, it became possible to make quantitative estimates of human requirements for them. This was done initially by measuring the amounts present in the diets of populations that developed deficiency diseases and comparing them with the amounts consumed by people who remained healthy. Later, when the pure compounds could be isolated or synthesized, experiments were done on human volunteers who consumed diets that were deficient in the nutrient until they developed signs of the deficiency disease. By returning incremental amounts of the item to the diet, it was possible to determine how much was needed to relieve the signs or to prevent them from developing. Such studies established the human requirements for the vitamins for which deficiency diseases were known, for protein, and for the more important mineral elements.

Experiments of this type provided the basic information needed to establish standards for evaluating the nutritional adequacy of diets and for planning food supplies that would meet basic nutritional needs. The standards used in the United States are the Recommended Dietary Allowances (RDA) of the Food and Nutrition Board of the National Research Council.[22]

These standards are based on the available scientific knowledge of the nutrient requirements of healthy people. They are considerably higher than the average requirements because individual variability in nutrient needs is taken into account. In experiments in which the requirements—for energy, protein, thiamin, and a few other nutrients—of a substantial number of individuals have been determined, the variability ranges from about 50 percent below to 50 percent above the average. The RDA are therefore set high enough to cover the needs of most of those with the highest requirements. In addition, consideration is given to the nature of the food supply, with the RDA being increased further for nutrients that may not be released efficiently during digestion or may not be completely absorbed in the form in which they are available. The RDA thus provide a standard for the amounts of nutrients needed to ensure that the requirements of most, if not all, healthy persons will be met.

The RDA serve as the basis for interpreting the results of food consumption surveys designed to assess the nutritional adequacy of diets. They provide guidelines for planning and procuring food supplies for hospitals and other institutions. Food guides for nutrition education programs are designed to ensure that the combination of foods recommended will meet the RDA. They serve as standards for modifying diets for the treatment of disease. They are used as guides by regulatory agencies that have the responsibility for nutrition labeling of foods and for developing regulations to ensure the nutritional quality of foods.

Despite the value of reliable dietary standards for these many purposes, they do have shortcomings. An individual's intake is not necessarily inadequate when it is below the RDA. When intake is severely inadequate, clinical signs of deficiency disease are evident. Detecting marginal deficiency or depletion of the nutrient before any evidence of disease occurs requires biochemical assessment; for this, a knowledge of the functions of nutrients is needed. Also, the degree of reliability with which nutrient needs are established varies from one nutrient to another, often because some have been studied very little, owing to the cost and difficulty of experimenting on human subjects. It should be pointed out that, although it is necessary to have estimates of human needs for essential nutrients, such as the RDA, and that these must be sufficiently accurate to serve as guidelines for many practical purposes, it is unrealistic to assume that continuous reinvestigation of requirements by classical methods will provide a set of single values that are of unique utility.

Estimates of requirements for some nutrients are controversial because of the limitations of the standard methods for determining them or differences of opinion in interpreting results. The requirements for calcium, vitamin C, and protein fall into this category. Requirements for some of the trace minerals have not been established, but research on the significance of trace minerals in human nutrition has developed rapidly in recent years. One stimulus for this has been the observation that zinc deficiency can occur in some human diets.[23] A type of dwarfism that was partially reversed by zinc supplements was encountered in the Near East, and mild zinc deficiency has been reported among some children in Colorado. This deficiency appeared to result from a diet containing a high proportion of cereal products from which zinc was not readily released during digestion.

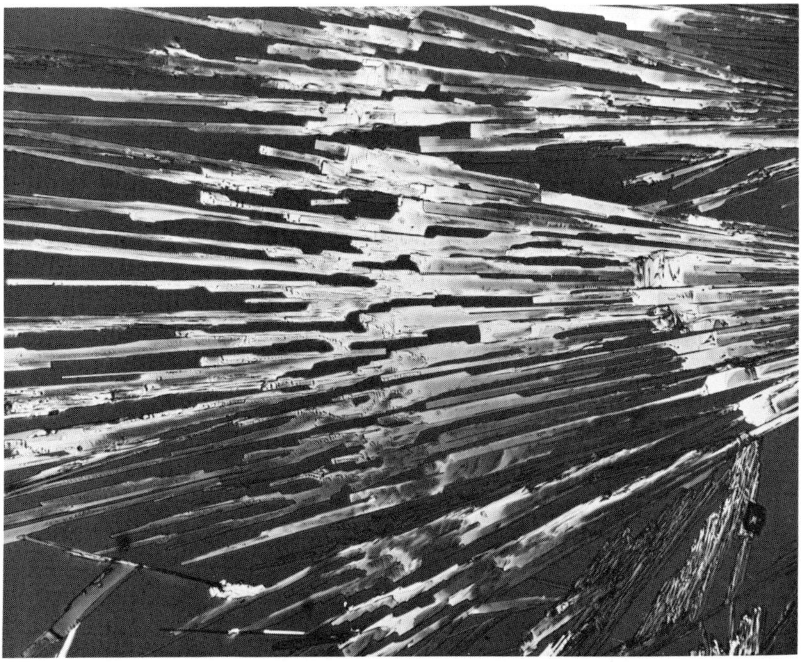

Crystals of vitamin B6. [Merck Sharp & Dohme Research Laboratories.]

Such observations highlight two areas which need much more attention with regard to meeting nutrient requirements. The first of these is the need for new methods and approaches for establishing requirements. With improved techniques for measuring nonradioactive isotopes, it is now becoming possible to determine quantitatively the rate of loss of nutrients from the body without exposing human subjects to the hazards of radioactive isotopes. This approach is being used to estimate the rate of oxidation of individual amino acids in an effort to determine the amounts lost daily through degradative processes. The availability and development of such methods and modifications of them provide an opportunity to reinvestigate human nutrient requirements for which the available methods have been less than satisfactory. They also should simplify the determination of those nutrient requirements that are needed in only minute amounts.

The second need is for much more knowledge about interactions between nutrients and among dietary constituents that may alter the availability or assimilation of nutrients from diets. Research in this area holds promise for explaining discrepancies between the anticipated response to a diet based on direct chemical analysis and the actual response observed when the diet is fed. Dietary interactions that influence the availability of iron for absorption are an example. It has been known for a long time that iron is poorly absorbed from foods; yet, although the amount of iron in the U.S. food supply is considered marginal in relation to need, 85 percent of the population shows no evidence of iron inadequacy and few of the rest are severely affected. Investigations of iron absorption have revealed that iron from meat is much more readily available for absorption than that from most plant products. More important, if other foods are eaten with meat, absorption of the iron they contain is greater. Ascorbic acid, vitamin C, also has been shown to improve iron absorption from a meal.[24]

Interactions between calcium retention in the body and protein consumption provide another example of nutrient interactions that can influence nutritional needs. Calcium loss in the urine increases if protein intake is high, indicating that the requirement for calcium depends upon the amount of protein consumed. On the other hand, the loss appears to be less if phosphate intake is increased concomitantly. The nature of these interactions is not clearly established, but the observations suggest that, in considering calcium needs and the effectiveness of diets in meeting them, it may be necessary to consider not only the amount of calcium in the diet, but also the amounts of phosphate and protein, as well as, perhaps, those of other dietary constituents.

There are suggestions of many such interactions, particularly among the trace elements, whereby the amount of one element in the diet influences the need for another. For instance, investigations of zinc and copper requirements indicate that the amount of one in the diet influences the efficiency with which the other is used. Understanding the general nutritional significance of interactions of this type will require painstaking research, but eventually it will make it possible to assess more accurately the contributions of various foods toward meeting nutritional needs and to devise diets that should be uniquely effective for clinical therapy.

## Summary and Outlook

Advances in knowledge of the functions of nutrients and their roles in metabolic processes have led nutrition research in new directions. The evidence that the debilitating effects of many metabolic defects can be relieved by dietary modification has stimulated interest in the potential of this type of treatment for preventing or delaying the onset of other diseases not caused by nutritional deficiencies. Great effort has been, and continues to be, expended in studying associations between diet and the development of chronic and degenerative diseases.

Recent investigations in animals of the effects of deficiencies of some of the mineral elements that are required in minute amounts have revealed that these may have a number of unique and physiologically important functions not previously recognized. Information about the control of food intake and its relationship to obesity indicates that the causes of obesity are more complex than has heretofore been assumed, and that some of them may be metabolic or genetic. In addition, questions have been raised as to whether appropriate body weight for persons over 25 may not be higher than currently accepted norms.

Observations that diet and specific nutrients can influence nerve function have provided new approaches to some of the diseases caused by malfunction of the nervous system. Recent knowledge of the way in which immune systems respond to malnutrition has increased appreciation of the importance of nutrition in the medical treatment of surgical cases and diseases, such as cancer, in which chemotherapy is required. Improved knowledge of relationships between diet and chronic degenerative diseases is essential for establishing clearly the extent to which dietary modifications may influence the development and the treatment of such diseases.

Food-drug interactions are an important area for future research because knowledge of such interactions is in a primitive state.[25] It is known that some drugs can affect taste and the acceptability of food. This is particularly true of some of the drugs used in cancer chemotherapy. Others, such as antibiotics, affect the absorption and use of nutrients. Some antagonists of folic acid are used in cancer chemotherapy because they inhibit tumor growth by suppressing the conversion of folic acid to its active coenzyme form. Some anticonvulsant drugs appear to impair the metabolism of vitamin D and folic acid. Penicillamine, which is used for treating certain inflammatory conditions, can induce vitamin $B_6$ deficiency. The myriad potential interactions of this type raise a serious question about the adequacy of our knowledge of the nutritional consequences of drug treatment generally and particularly in the case of the elderly.

Little is known about whether the gradual deterioration of organ function that occurs during aging can be delayed by diet modification. Some animal experiments suggest that restriction of food intake reduces the rate of deterioration of organ function, but observations from human studies suggest that body fat stores that are somewhat above average may be advantageous for the elderly. These are among the frontier areas in nutrition research that promise new information for the betterment of human health.

Public concern with using diet modification to control the diseases that are major causes of death has led to unrealistic expectations from nutrition research and to impatience with scientific caution in applying new findings. It is understandable that fears about health abound today just as they have in times past, that the desire for a long and active life is just as compelling, and that magical cures and promises of an active old age are still alluring—even more so, perhaps, if they come wrapped in modern-day scientific garb. It would be unfortunate, however, if public perceptions based on beliefs that have not been evaluated critically were to become the predominant basis for public policy decisions in a modern scientific and technological society. Information that is to be used as the foundation for public policy deserves even more critical scientific evaluation than that prepared by scientists for their colleagues.

## REFERENCES

1. *Ten-State Nutrition Survey, 1968-70.* Pub. No. (HSM) 72-8230-8134. Washington, D.C.: U.S. Department of Health, Education, and Welfare, 1972.

2. American Medical Association. "Concepts of Nutrition and Health," *Journal of the American Medical Association*, Vol. 242(1979), pp. 2335-2338.

3. *Ten-State Nutrition Survey, 1968-70.* Pub. No. (HSM) 72-8230-8134. Washington, D.C.: U.S. Department of Health, Education, and Welfare, 1972.

4. *Preliminary Findings of the First Health and Nutrition Examination Survey, U.S. 1971-72.* Pub. No. (HRA) 74-1219-1. Washington, D.C.: U.S. Department of Health, Education, and Welfare, 1974.

5. Senate Select Committee on Nutrition and Human Needs. *Dietary Goals for the United States.* Second Edition. Washington, D.C.: U.S. Government Printing Office, 1977.

6. L. Page and B. Friend. "The Changing United States Diet," *BioScience*, Vol. 28(1978), pp. 192-197.

7. T. B. Van Itallie. "Obesity: The American Disease," *Food Technology*, December 1979, pp. 43-47.

8. A. E. Harper. "What Are Appropriate Dietary Guidelines?" *Food Technology*, September 1978, pp. 48-53.

9. A. Keys. "Overweight, Obesity, Coronary Heart Disease, and Mortality," *Nutrition Review*, Vol. 38(1980), pp. 297-307.

10. Ibid.

11. R. Andres. "Effect of Obesity on Total Mortality," *International Journal on Obesity*, Vol. 4(1980), pp. 381-386. See also G. A. Bray, "To Treat or Not To Treat—That Is the Question," *Recent Advances in Obesity Research.* Volume Two. Edited by G. A. Bray. London: Newman Books, Ltd., 1978, pp. 248-265.

12. National Research Council, Food and Nutrition Board. *Recommended Dietary Allowances.* Washington, D.C.: National Academy of Sciences, 1980.

13. J. F. Marshall. "The Role of Central Catecholamine-Containing Neurons in Food Intake," *Recent Advances in Obesity Research.* Volume Two. Edited by G. A. Bray. London: Newman Books, Ltd., 1978, pp. 6-16.

14. R. H. Herman. "The Interaction Between the Gastrointestinal Tract and Nutrient Intake," *Human Nutrition.* Volume Four. *Nutrition, Metabolic and Clinical Applications.* Edited by R. E. Hodges. New York: Plenum Publishing Corp., 1979, pp. 105-140.

15. J. E. Fischer. "Amino Acid Derangements in Hepatic Failure," *Clinical Nutrition Update—Amino Acids.* Edited by H. L. Greene, M. A. Holiday, and H. N. Munro. Chicago: American Medical Association, 1977, pp. 174-182. See also R. J. Wurtman and J. D. Fernstrom, "Control of Brain Monoamine Synthesis by Diet and Plasma Amino Acids," *American Journal of Clinical Nutrition*, Vol. 28(1975), pp. 638-647.

16. H. F. DeLuca. "Some New Concepts Emanating from a Study of the Metabolism and Function of Vitamin D," *Nutrition Review*, Vol. 38(1980), pp. 169-182.

17. W. R. Beisel. "Metabolic and Nutritional Consequences of Infection," *Advances in Nutrition Research.* Volume One. Edited by H.H. Draper. New York: Plenum Publishing Corp., 1977, pp. 125-144.

18. M. E. Shils. "Parenteral Nutrition," *Modern Nutrition in Health and Disease.* Sixth Edition. Edited by R. S. Goodhard and M. E. Shils. Philadelphia: Lea and Febiger, 1980, pp. 1125-1152.

19. K. N. Jeejeebhoy, et al. "Chromium Deficiency, Glucose Intolerance, and Neu-

ropathy Reversed by Chromium Supplementation in a Patient Receiving Long-Term Total Parenteral Nutrition," *American Journal of Clinical Nutrition,* Vol. 30(1977), pp. 531-538.

20. E. J. Masoro et al. "Nutritional Probe of the Aging Process," *Federation Proceedings,* Vol. 39(1980), pp. 3178-3182.

21. M. H. Ross, E. Lustbader, and G. Bras. "Dietary Practices and Growth Responses as Predictors of Longevity," *Nature,* Vol. 262(1976), pp. 548-553.

22. National Research Council, Food and Nutrition Board, op. cit.

23. T-K. Li and B. L. Vallee. "The Biochemical and Nutritional Roles of Other Trace Elements," *Modern Nutrition in Health and Disease.* Edited by R. S. Goodhart and M. E. Shils. Philadelphia: Lea and Febiger, 1980, pp. 424-428.

24. National Research Council, Food and Nutrition Board, op. cit.

25. *Nutrition and Drug Interrelations.* Edited by J. N. Hathcock and J. Coon. New York: Academic Press, Inc., 1978.

[Roger Landrum.]

# 4

# *The Science of Cognition*

## INTRODUCTION

Cognitive science has two faces, one turned toward the nature of human mental abilities and the processes that bring them to bear on problems, and the other turned toward the general nature of intelligence, whether embodied in man or machine. One direction is identified with cognitive psychology, the other with the branch of computer science termed artificial intelligence. This review will deal mainly with the first aspect, which represents quantitatively the greater research effort, but will devote some attention to the second aspect and the interrelationships between the two.

### The Problems and Goals of Cognitive Science

Intellectual and social forces similar to those that have driven us to understand human biology drive us to understand human cognition. In the newer discipline of cognitive science, we seek explanations of the acquisition of language, knowledge, and skills; the development and nature of reasoning abilities; and the ways that objects and events are perceived and represented in memory. But, as with other sciences, some areas of potential application are especially relevant. Among the most notable of these for cognitive science are education, the psychological aspects of medicine and health, and the human element in advanced technology.

Attempts to solve two general problems of education have drawn on cognitive science in the past and, as the discipline matures, may be expected to do so more strongly. One is the assessment of intellectual capacities, the other the learning processes basic to instruction. A preliminary definition of human intellectual capacity, or intelligence, was the contribution of the French psychologists Binet and Simon, who were commissioned near the turn of this century to identify children of school age who could profit from instruction. In the United States, the problem of measuring mental abilities became one of national scope almost overnight when, at the onset of World War I, it suddenly became necessary to assess swiftly the abilities of large numbers of recruits to absorb the special training needed to prepare them for innumerable tasks peculiar to a rapidly expanding military effort. The challenge was met, and the residue is a technology of ability testing still widely applied in schools and industry, as well as in the military. In recent years, however, changing social values have led to strong demands for a shift of emphasis from the assessment of abilities for purposes of assigning people to tasks to the diagnosis of disabilities for the purpose of providing remedies and helping the tested individuals to meet societal demands.

With regard to instruction, the pioneering American psychologist Edward L. Thorndike applied the learning theory of the early 1900's to developing drill methods for teaching basic arithmetic and spelling skills. His methods long constituted a major part of the basic elementary school curriculum. Much more recently, the development of computer-aided instruction has been strongly influenced by cognitive psychologists who saw the possibilities of implementing more modern learning theories through the use of computer-aided management of individual instructional routines. This form of instruction is still experimental with regard to mass application in the schools, but it is already widely used in specialized industrial training.

Currently, perhaps the most difficult and challenging applied problems for cognitive science in education concern the need for contributing to a multidisciplinary attack on the social problems associated with widespread mental disabilities. Some of these problems are very large and call for correspondingly massive efforts at amelioration. It is estimated that there are at least six million severely handicapped school-age children in this country. The three largest categories of disability are mental retardation, speech defects and related problems of communication, and learning dysfunctions connected with reading.

With regard to these problems, perhaps the most extensive efforts over the last two decades have been directed at applying learning theory and its associated methodology to the training of the mentally retarded. These efforts have proven valuable, for example, in developing techniques to help even severely retarded individuals form habits that reduce their dependence on others. There is now reason to hope that, within the foreseeable future, a fuller understanding of learning processes may help to uncover the causes of disabilities and retardation, permitting earlier treatment and even prevention of disabilities.

Among the newer branches of cognitive psychology, some of the most striking recent advances have occurred in the investigation of speech perception and production, opening up the possibility of greatly improved methods of dealing with problems of speech and communication. At the same time, several strains of research and theory are converging on the perceptual and cognitive processes underlying reading.

Special problems of education calling for substantial research are not limited to disabilities. For example, during the late 1970's, Title I funds under the Elementary and Secondary Education Act were expended at a rate of some $1.5 billion annually, with large sums devoted to such special programs as Head Start, aimed at improving early education among disadvantaged groups. One of the important lessons learned from these efforts is that across-the-board, empirical evaluations of such programs are likely to yield almost meaningless results. The effects of such interventions depend on the social backgrounds and learning capacities of the individuals being educated and on characteristics of the learning situations. Thus, meaningful interpretation of the effectiveness of educational intervention programs requires contributions from a number of disciplines, among them the aspects of cognitive science that deal with learning processes and abilities.

In health science, cognitive research is perhaps most importantly involved as a contributor to the multidisciplinary attacks on mental illness. Major research programs are directed at such matters as increased understanding of cognitive processes in schizophrenia, essential to obtaining maximum benefit from breakthroughs in biochemical or pharmacological treatment of such disorders. Also, a new and rapidly expanding application of cognitive science exploits the vast memories and computational powers of digital computers in the service of medical diagnosis. Progress depends on understanding the mental processes of human diagnosticians and thus the ways in which human intuition can best be complemented by computers.

Some of the most novel and intriguing problems for cognitive science arise in advanced technology. In the first major development of engineering psychology, during and immediately following World War II, consideration of cognitive processes was only a secondary element. A typical problem was the design of instrument arrays and control devices such as joy sticks to fit human perceptual-motor capacities; a related problem was the design of training devices and programs for operators of aircraft and other complex equipment.

However, cognitive processes of a different order are becoming recognized as important aspects of the man-machine interface in the current development of systems for the teleoperation of mechanical devices under remote control. These systems, important in dangerous or inaccessible environments, had their origin in the handling of radioactive materials and now are being extended to the manipulation of objects under the sea, on satellites, or in more remote explorations of space. Manual skill in the operation of equipment is relatively unimportant in these systems, since the operator is far removed from the device and can communicate with it only symbolically over telecommunication channels. Here, basic research is required to advance understanding of human capacities for symbolically controlling intricate manipulations under heavy information load.

## Cognitive Science and Neural Science

When we observe a human being solving a problem, all we observe is information being taken in by eye or ear and, after some time, an action. What happens in the meantime is one of nature's principal mysteries. The information is passed in some form from the sensory organs into the uncharted regions of the brain, where mechanisms, possibly akin to those in a computer but perhaps very different, carry out the operations that produce the action. It is a prime objective of psychology and neural science to fill the gaps between the observed sensory input and the observed action.

The problems facing efforts to trace the causal chain of events through the brain and nervous system have proven wholly refractory to approaches that, on the psychological side, draw only on a general knowledge of the way people acquire, retain, and use information. It has been the task of experimental psychology to bring the results of everyday observation into the laboratory, refine them with suitable controls, and proceed to the formulation of descriptive and predictive laws, sometimes quantitative in form, sometimes embodied in computer simulations of behavior. Although this discipline is still young, in some instances psychological laws and principles and models so derived enter into the direction of research in much the same manner as has become familiar in older sciences. Cognitive science thus should complement neural science in the arduous but increasingly fruitful effort to find the specific bases of psychological functions in the brain.

## Methods of Research on Cognition

For the reader unfamiliar with this area, it may not be obvious just what an experiment on cognition or a model of a cognitive process would be like. A few examples may provide a useful concrete background for a more general discussion of problems and results.

***Rote Learning*** The first genuine experiment on human learning and memory was contrived by the German philosopher Herman Ebbinghaus about 100 years ago. He set himself to obtain a quantitative account of the way in which a person memorizes a list as, for example, items to be ordered from a pharmacy. Ebbinghaus recognized that, to provide the basis for scientific understanding, the experiment would have to be conducted under carefully simplified conditions, with the observer (subject) protected from distractions and with physical conditions such as illumination and time intervals controlled so that the experiment could be accurately described and, if need be, replicated. Also, he realized that, if the items to be memorized were familiar words, a subject might come to the experiment with existing associations between various items, strong for some word pairs and weak for others. Thus, in order to be able to follow the whole course of learning, he devised what came later to be known as "nonsense syllables," that is, unfamiliar trigrams that can be pronounced readily but do not resemble words—for example, TAV, XOQ.

He constructed lists of different numbers of these syllables and presented the items of a list one at a time to the subject, usually himself, at a constant rate. On the first cycle through the list, the subject simply viewed the items; on the second cycle, he tried to anticipate the first item at the start signal, the second item on seeing the first, and so on, until the list was mastered. Ebbinghaus' work, and much that followed with variations of the paradigm, produced a fairly complete account of the way in which the speed of simple associative learning and its rate of forgetting are influenced by such factors as the time intervals between items on lists and the familiarity of the items. Some important principles emerged, for example, that the forgetting of a list is influenced strongly not only by the interfering effect of other lists an individual might learn later (retroactive inhibition), but also, rather counterintuitively, by the persisting effects of lists learned earlier (proactive inhibition).

***Tachistoscopic Perception*** A tachistoscope is a device for presenting visual displays for very brief controlled intervals. In experiments conducted with this device, it has been shown that a person can perceive a surprisingly large number of letters, or even words, at exposures as short as a few thousandths of a second. Optimal conditions are needed, however, for the amount perceived can be reduced drastically by the "forward masking" or "backward masking" effects of other displays presented a fraction of a second before or after the one being reported. A problem not yet entirely resolved is the degree to which masking reflects the limited temporal resolution of the peripheral visual system rather than more central, probably cortical, processes related to attention.

***Judgments of Probability*** In one current line of research, an experimental situation is presented to subjects as a simulation of public opinion polls. On each of a series of trials, the subjects are shown, by means of the display screen of a computer, the results of polls simulated by the computer for a number of hypothetical election candidates. The purpose is to see how people form mental representations of event frequencies, which then influence their expectations. After having had an opportunity to observe tallies of respondents favoring each candidate, the subjects are asked to predict the results of elections involving the same candidates. One outcome of such studies is the demonstration that in normal adults the buildup of memory for relative frequencies of events (wins and losses in this example) is an automatic process that can operate with great precision, and that on the average probability judgments based on these memories tend to match the true probabilities. Whether this probability-matching tendency applies outside the laboratory depends on the conditions of observation. The same learning theory that accounts for matching in the simulated environment predicts, correctly, that the impressions individuals build up concerning probabilities of such events as crimes or airplane accidents are often extremely wide of the mark. The reason is that one observes incidents or news accounts only on occasions when these events occur; one has no way of identifying the specific occasions when the events might have occurred but did not, and such "negative instances" are therefore underrepresented in memory.

***Models*** Perhaps the most important purpose of the experiments is to generate models that summarize compactly the facts obtained from previous experiments and to help anticipate the results of new ones. In many cases, continuing testing and refinement have led to formalization of mathematical or computer simulation models with properties similar to those that hold for these kinds of models in other sciences.

***The Constant Ratio Rule*** A simple mathematical model whose predictive validity has been well established for a considerable range of circumstances reflects a property that has been termed "independence of irrelevant alternatives." The gist of the model is that, in a situation in which an individual makes a choice from among a number of alternatives, the relative probabilities of selection for any two alternatives are independent of the number of other alternatives that may be available at the point of choice. For example, in the experiment on simulated opinion polls, suppose subjects were asked to predict the winner of an election in which A, B, and C were the candidates. If, separately, the subjects' probabilities of expecting A to win over B had been determined for a two-way contest in which those candidates were pitted against each other, then, from the constant ratio rule, one could predict exactly the relative probabilities of subjects' expecting A to rank higher than B in the three-way election. An implication of the model is that, if one is interested in pair-wise choices among a number of alternatives (which can be large), as often occurs in situations ranging from preferences for foods to preferences among potential election candidates, it is necessary to collect data on only $n - 1$ of the much larger number of pairs $[n(n - 1)/2]$ that can be chosen from n alternatives. Then one can predict, by way of the model, choice probabilities for all possible pairs. The constant ratio rule has been incorporated into choice theories that have wide application in psychology and may prove a link between psychology and economics.

***General Problem Solver*** This model, which has reached a considerable degree of sophistication and generality, takes the form of a computer program that accepts a specification of any of a wide variety of problems,* selects and carries out a sequence of operations that produces a solution, and prints out a record of the sequence and the result. By successive cycles of programming, comparing the printouts with human performance, and reprogramming, what was originally simply a computer program to simulate a limited form of performance becomes a model that can predict how human beings deal with many kinds of problems, predict the relative difficulties of problems, and identify points of special relative difficulty within problems.

## Basic Questions

As cognitive research proceeds with relative autonomy at the psychological (as distinguished from the neurophysiological) level of analysis, it seeks answers to a sequence of questions roughly paralleling those one might raise in the course of trying to understand the functioning of a digital computer:

- How does the mechanism work? What are the mental operations that enable accomplishment of intellectual tasks?
- What are the form and mode of storage of the information that feeds cognitive operations? How is the external world represented in memory?
- How does this information accrue and how do skills develop as a function of a person's past experience?
- How are cognitive operations and the products of learning brought to bear on problems of adjustment to one's environment?

---

*An example from elementary physics would be a specification of parameters, "A man of mass M wishes to lower himself to the ground from height H by holding on to a rope passing over a frictionless pulley and attached to an iron weight of lesser mass M'," and the question "With what speed would the man hit the ground?"

In the remainder of the chapter, these questions will be addressed in sections on information processing, mental representation, learning, and cognitive skills.

## INFORMATION PROCESSING

### The Sources of New Vigor in Cognitive Psychology

Although the study of human mental processes has been a part of psychological research since the beginnings of experimental psychology about a century ago, the increased vigor and activity in this area during the past decade or two have been truly dramatic. The basis is to be found in influences from several fields—notably computer science and linguistics—converging on the mainstream of experimental and mathematical psychology.

The advent of large, high-speed digital computers capable of manipulating enormous amounts of information in solving mathematical, statistical, or logical problems has had a revolutionary impact on cognitive psychology. Perhaps most importantly, the analogy between human and computer problem solving suggested a useful way of partitioning the unmanageably complex task of comprehending simultaneously the mind and brain in all their aspects.

The partition stems from the distinction between computer hardware and software. One can seek to understand how a computer works by studying its wiring diagram, the way patterns of electrical input from a teletype can be recorded in the polarities of magnetizable elements, and so on—the hardware. But such knowledge is not essential for the programmer, who needs to know rather the informational capacity in logical units, how the units can be addressed, the elementary arithmetical and logical operations that can be performed on them, the logical and temporal constraints on the sequencing of operations—the software.

This distinction suggested a tractable approach to human mental function. One might hope to construct a useful (though of course not complete) theoretical picture of the structures and processes implicated in mental activity at the level of software—the logical form and organization of encoded information, the cognitive operations performed on it, and the functional rules governing the sequencing of operations. Implementation of this strategy is a main theme of contemporary cognitive psychology.

Similarly important has been the growing interaction between linguistics and psychology. In an earlier period, linguistics was regarded, properly at the time, simply as the study of properties of recorded language.

However, beginning especially with the seminal contributions of Noam Chomsky in the mid-1950's, it has become increasingly well understood that an adequate theory of the properties of spoken and written language provides a picture of what the human mind accomplishes in dealing with language. Linguistic theory thus defines with increasing precision the problems for experimental and developmental research in tracing out the joint influences of innate and experiential factors that make possible the production and use of grammatical language by the adult human being. Clues to innate factors come from the discovery of attributes that hold universally for all languages, independently of the particular experiences of people in different cultures or environments.

Contributions from linguistics and computer analogies indicated the kinds of research needed to advance understanding of comprehension and semantic memory; technical advances in research methods made it possible for the research to be done. A key advance has been the development of increasingly precise methods for measuring the time required for mental processes. Contrary to the common impression that the speed of thought is akin to the speed of light, the conduction of messages in nerve fibers is relatively slow. Speeds in human peripheral nerves are about 60-70 meters per second, and there is no reason to believe that transmission times within the brain can be appreciably faster. Since storing and retrieving items of information in memory must depend in part on the transmission of messages over nerve fibers, mental operations that depend on the use of information in memory must take measurable times.

This supposition has proved well founded, and methods for carrying out these measurements within the framework of evolving theories of the memory system are providing a major empirical basis for testing and refining models of cognitive processes.

The effects of these interactions with linguistics and computer science can be seen in current research on cognitive operations, processes of attention and memory, and mental representation. Concepts emanating from the research have contributed to the interpretation of speech, reading, and problem solving.

## Cognitive Operations

How the mind works can be investigated at several different levels of analysis, each level pertaining to different questions. Two levels are particularly important: the level of elementary information processes and the level of cognitive strategies.

Elementary information processes include such steps as recognizing a visual stimulus, scanning lists in long-term memory for the presence or absence of information, and comparing a newly presented item with a representation of one retrieved from short-term memory. Characteristically, their duration ranges from a few milliseconds to a few hundred milliseconds. Today, the generally accepted picture of how the processes are organized hypothesizes a short-term memory of very limited capacity that accounts for the narrow limits of attention; a long-term memory in which are stored the recognition capabilities, the schemas incorporating organized knowledge, and the motor skills and strategies; and the sensory and motor systems.

The entire system can be viewed as a symbol manipulator that reads symbols through its sensory organs, writes them through its motor organs, and copies, stores, and compares them in its memories. When the mind compares two symbols, it can behave differently, depending on whether the two are judged to be the same or different; this is known as conditional branching. In these information-processing theories, symbol means simply any pattern, capable of assuming a rich variety of forms, that can be stored and operated on in a physical system, whether a brain, a computer, or a blackboard.

The basic hypothesis that guides research in cognition is that it is precisely these kinds of symbol-manipulating abilities, strategically organized, that enable a physical system, computer or human, to address tasks intelligently. Of course, opinions differ greatly on how wide a span of the activities we call intelligent can be accounted for by the hypothesis. However, the hypothesis is empirically testable, and determining the limits to which it can be pushed provides a powerful and challenging research program for cognitive science.

Research at the second level, on cognitive strategies, has made the most progress in elucidating problem-solving processes, strategies for concept attainment, and strategies for memorizing. Also, a start has been made toward describing the strategies that people use for understanding natural language text, although many complexities of language remain to be unraveled.

## Selective Attention

A characteristic of many complex tasks is the need to divide attention among a number of active sources of information. A pilot about to land an aircraft must listen to instructions coming over a radio channel while monitoring several visual indicators. A member of a surgical team must attend without fail to messages directed to him or her by other members, sometimes simultaneously, while ignoring still others. These tasks are difficult because we have only limited capacity to cope with problems requiring many simultaneous decisions. According to current theories of selective attention, our mental apparatus includes the equivalent of the executive program of a computer time-sharing system, its function being

"There have been a number of startling demonstrations of abilities to carry out two complex independent tasks simultaneously." [The Bettmann Archive, Inc.]

to allocate attention to different sources of information in accord with task demands.

Optimal allocation of attention in taxing situations appears to require a process termed automatism—a consequence of specific forms of learning. Given a consistent distinction between relevant and irrelevant aspects of a task during a period of training, some uniformly relevant items of information come to evoke appropriate action sequences automatically, without the need for conscious decisions. There have been a number of startling demonstrations of abilities to carry out two complex independent tasks simultaneously, when sufficient amounts of practice are provided. For example, some individuals have been trained to read text while simultaneously writing down different spoken sentences. After much training, reading rates and comprehension approach normal levels, while the taking of simultaneous dictation is quite accurate and rapid. It is even possible to draw implications from the dictated sentences while simultaneously reading at normal performance levels. A reasonable interpretation is that the performers have automatized their dictation skill to a large degree, reducing its demands on the limited central processing capacity and allowing simultaneous tasks to be done. Virtually all complex cognitive skills can be carried out only because many subcomponents of the task performance are delegated to automatic mechanisms that do not require control and attention. The study of automatism and the more general problem of cognitive resource allocation is just beginning, but may grow into a major research area in the next few years, with potential applications to the demanding forms of skilled performance required of, for example, aircraft controllers and pilots.

## Control Processes

Detailed experimental exploration of the strategies, decision rules, and mental techniques that are termed control processes, in distinction to more fixed structural processes or capacities, represents a major new direction taken by cognitive psychology in the last 15 years. The experimental tools necessary for the study of control processes have been developed only recently; the result has been a variety of new lines of research.

Control processes are based on information that is stored in long-term memory, but is brought into active, short-term memory when the processes are activated, occupying a certain amount of the available capacity. Much of the intentional, conscious portion of human cognition is carried out by control processes.

One type of control process that has received intensive study in psychology is that involved in detecting and recognizing unclear, ambiguous, or noisy stimuli, as in detecting a distant aircraft on a radar screen or recognizing the symptoms of a lesion on an X-ray plate. The Theory of Signal Detection (TSD) has provided a quantitative model of performance in such situations that is often remarkably accurate and has become a standard tool. The model was originally developed in the engineering of auditory communication systems to give a basis for the detection of signal stimuli embedded in noise that is "normally" distributed, but it has been applied in numerous other settings.

The basic idea can be understood by reference to Figure 1. It is assumed that the sensory states generated in an observer, either by a signal or by noise that might be mistaken for a signal, can be ordered along a single dimension representing the likelihood that the state was generated by a signal. The x-axis of the figure represents this dimension and the curves represent frequencies (technically, probability densities) with which states generated by signals or by noise alone fall at various points along the dimension. Because the curves overlap, one can see that on some occasions stimuli will give rise to sensory states that provide ambiguous information to the observer, since they might have arisen either from a signal or from the noisy background. In those instances, should the observer decide that a signal was present or not? TSD assumes that the observer selects, not necessarily consciously, a criterion (that is, a point on the x-axis of Figure 1) and labels any observation to the right of the criterion as a signal and any to the left as a nonsignal.

The theory provides a basis for choosing a criterion that takes account of prior odds on signal occurrence and the costs of various types of errors. Most importantly, however, it provides a means of obtaining two quantities: the observer's criterion and the distance ($d'$) between the means of the signal and noise distributions. The first of these is a control process, and highly variable; the second is an invariant, representing the discrimi-

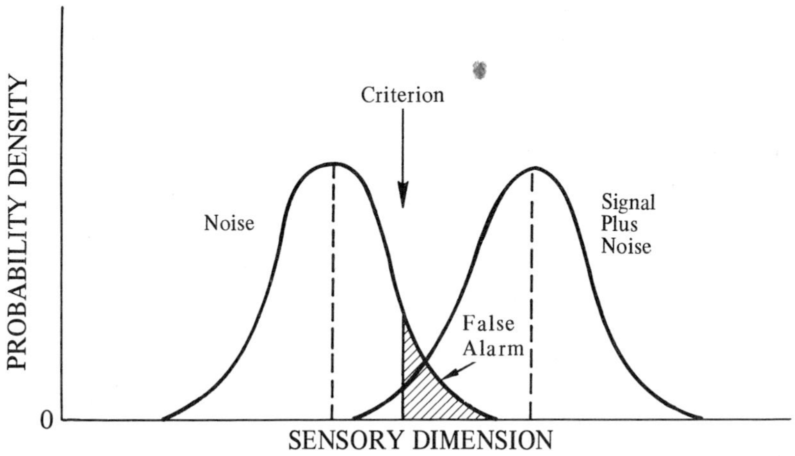


**Figure 1 Theory of signal detection.** Results of observations made in a task requiring signal detection give rise to states of the observer that can be ordered along a single dimension. The curves represent probabilities with which values on this dimension arise when the observer is actually responding only to noise and when he is responding to a signal in a noise background. When an observation yields a value to the right of the criterion, the observer reports a signal. Some of these responses will be correct detections (the area under the signal-plus-noise curve to the right of the criterion), but some will be "false alarms" (the crosshatched portion of the noise distribution).

nability of signal from noise (signal-to-noise ratio) specific to the stimulus situation and task. The ability of TSD to separate the decision component from the invariant component of detection is the central reason for the importance of this theory in modern psychology.

## MENTAL REPRESENTATIONS

### The Importance of Representations

Human beings, viewed as processors of information, are clearly knowledge-based systems. Even everyday activities require a vast amount of knowledge—an ordinary conversation, for example, calls on the meanings of thousands of words. Thus, a central problem for cognitive psychology is to discover how knowledge is stored in memory and how it is retrieved.

Since any storage device, be it biological or of human design, has particular ways of representing knowledge, understanding a memory system is partly a matter of describing its permissible representations. Computer representations, for example, can be described at "molecular" or "molar" levels—either as configurations on magnetic tape or as semantic networks. So, too, human representations can, in principle, be described at a molecular, or neurophysiological, level, or at a more molar one, the level of mental representations.

## Representing Lexical Knowledge

Knowledge about the meanings of individual words, which psychologists refer to as lexical knowledge, is particularly important because it can provide the building blocks of more complex units.

The study of lexical knowledge in contemporary psychology has been influenced greatly by other disciplines. For example, the traditional study of meaning by linguists and philosophers has made psychologists aware of certain kinds of lexical knowledge that people using a language must have: knowledge that enables us to determine, for example, that two words are synonymous (as, illness = disease), or that certain words license certain inferences (as, anything that is a bird is therefore an animal). Another major influence on psychology is artificial intelligence, particularly attempts to design programs that understand natural language. This line of research has contributed concrete ideas about how lexical knowledge is represented and how such representations are accessed and used.

The link between artificial intelligence and psychological studies of lexical knowledge is particularly direct. An early development, proposed about 1960, was a method for representing knowledge about object terms, that is, words designating concrete objects, in computer programs. The next step was a joint effort by computer scientists and psychologists to test this proposal as a psychological theory.

In recent versions of the theory, the basic idea is that each object term is represented by a knowledge schema, which characterizes it in terms of its supersets, its subsets, and its distinctive properties. The schema for "bird," for instance, lists "animal" as its superset; "robin, sparrow, chicken," etc., as subsets; and "has feathers," "can fly," etc., as distinctive properties. Another critical assumption is that the different schemas are connected into a hierarchical network through their supersets; for example, the schema for robin is connected to that for bird, which in turn is connected to that for animal. The total meaning of an object term, then, is given by its unique knowledge schema together with all other schemas to which the latter is connected. It is assumed that, when a word is presented, some or all of the components in its schema are activated. This activation then spreads by superordinate connections to other schemas, but weakens with distance so that ultimately the entire process is damped. A sentence expressing, for example, a relation between an object term and a superordinate category ("Measles is a disease") is recognized as true if activation spreads from the object to the category representation. It would be recognized as false if there were no acceptable connecting pathway, as might be the case for "Aging is a disease."

Experimental tests of the psychological validity of this theory have been based on a verification task, which allows use of response time measures to draw inferences about underlying representations. On each trial of the experiment, subjects are presented a sentence that expresses either a true or false superset relation (for example, "A penguin is a bird" versus "A bat is a bird"), or a true or false property relation (for example, "A porpoise has skin" versus "A porpoise has scales"). The subject's task is to decide as rapidly as possible whether the statement is true or false, with the decision time being the main datum of interest. For sentences expressing true superset relations, the theory predicts that decision times should increase with the number of connections intervening between the two schemas for the object terms, because the more intervening connections, the longer it should take the activation emanating from one representation to reach the other. For the same reason, the theory predicts that decision times for sentences expressing true property relations should increase with the number of connections between the representation of the object term and that for the property. Both predictions have been confirmed.

The concept of spreading activation in associative networks provides a basis for explaining the speed and efficiency with which people locate words in memory while speaking, listening to speech, or reading. The vocabulary of the average adult includes upwards of 50,000 words and, to produce a sentence, one must locate and activate the representations of precisely the words needed. Since the transmission of nerve impulses is quite slow, it seems scarcely possible that words could be located with the necessary speed by the kind of search of memory locations employed by a computer program for the same purpose.

Reaction time studies indicate that people are able to bring the task within their capacities by taking advantage of the fact that the words occurring throughout a brief conversation or a passage of text are generally related in meaning. Analyses of a phenomenon termed "priming" show that the time required to access a word in memory is reduced if it has been preceded recently by a semantically related word, the facilitation being inversely related to the inferred distance between the words in the network. Therefore, when a word occurs in any linguistic task, the activation of its memory node must spread to those of words related in meaning. It appears, then, that at any point in the course of speaking, listening to speech, or reading, a search of the temporarily active segment of the associative network (which can be accomplished in much less time than a search of the whole network) is likely to locate the words that are needed for continuing production or comprehension of meaningful speech.

There are, however, a number of problems with the network-activation theory. Perhaps the main one is that it does not account for variations in decision times for sentences expressing false superset relations. One might think that such decision times should be longer the farther apart the two schemas are in the network (for example, the decision time should be longer for "A penguin is a shark" than for "A penguin is a duck"), but in fact the opposite occurs. This anomaly accounts for much current research aimed at revealing mechanisms other than network activation that may be involved in accessing lexical knowledge.

## Imaginal Representations

The representations considered thus far are all symbolic descriptions. An alternative is an imaginal representation, which depicts rather than describes its referent. It seems that a part of our knowledge about objects is represented by visual images. Terms with concrete referents are remembered better than terms with abstract referents, presumably because the concrete terms can be represented both imaginally and symbolically, while abstract terms are limited to symbolic representations.

More unequivocal evidence for imaginal representation again hinges on results obtained in the verification task. Usually, the time to verify a true property statement decreases as the distinctiveness of the property increases. Thus, "The station has a clock" would typically be verified faster than "The station has walls" (presuming both statements to be true). However, this inequality would be reversed if an individual were instructed to imagine the station and inspect the image, for the time to verify a true property statement decreases as the size of the object requiring attention increases. The fact that size matters strongly suggests that the representation includes spatial extent, which is a hallmark of a visual image. Still more definitive evidence is obtained from experiments in which, in terms of the example, an individual who has viewed a station or other building earlier is asked unexpectedly to report from memory the number of windows on the front wall. The typical result is that reaction time is linearly related to the number of windows, as though the individual had to scan a mental image of the structure and count the windows.

Finally, the alternative hypothesis that memories of this sort are actually verbal in form is virtually ruled out by a phenomenon termed mental rotation. A person is shown a stimulus, for example, a printed letter or an irregular polygon. After an interval during which that stimulus is out of view, the person is shown a test stimulus, either the same or different in form from the first stimulus and displayed either in the same orientation or rotated to a new orientation (for example, a letter A rotated to appear as ⊳ ), and is asked to judge as quickly as possible whether the original and test stimuli are the same or different. The critical observation is that the reaction time varies linearly with the number of angular degrees between the orientation of the original stimulus and the test stimulus. This observation confirms the assumption that the individual mentally rotates the image of the first stimulus to the orientation of the second and then determines whether they coincide.

## Representing New Knowledge

Since much of our new knowledge is transmitted in spoken or written sentences, we need to know how sentences are represented. While it seems obvious that a sentence is at least partly represented by its meaning, not just by its exact wording, the extent to which we rely on meaning is quite dramatic. Experiments indicate that, about 30 seconds after a sentence is presented, memory of its exact wording is nil, as subjects perform at chance when trying to discriminate between the presented sentence and a close paraphrase. As a 30-second interval may mark the end of a short-term memory representation, these results suggest that long-term memory representations of sentences carry information only about meaning.

Such findings, together with theoretical notions derived from the work of modern linguists and the tradition of the philosophy of language, have led to the idea that sentences express propositions, symbolic descriptions of single thoughts, such as "Botany is a science."

More rigorously, a proposition may be partly defined by three features. First, it is abstract, in that it is independent of the specific means by which it is presented (words or pictures, a particular sentence or its paraphrase). Second, it is either true or false. Third, it obeys conditions, such as including a relation; for example, "botany library science" does not express a proposition.

Since these properties capture part of what is meant by "meaning," psychologists have tried to show that propositions are the units of long-term memory. Supporting evidence comes from such observations as that the time needed to read and understand a sentence increases with the number of propositions it expresses, independently of the number of words in the sentence, and that the time one needs to learn a multiproposition sentence is less when one has already learned one of the component propositions in another context. All told, then, there is substantial evidence that sentences are represented symbolically by the propositions they express.

## Interactions Between Old and New Knowledge

Characteristically, learners try to integrate new information with prior knowledge. Indeed, in many situations when one is reading or listening to supposedly new information, so much is left out by the author or speaker that one could never understand the message without prior knowledge. This point is illustrated in the following vignette: "At the security gate, the airline passenger presented his briefcase. It contained metallic objects. His departure was delayed." Although the incident may appear commonplace, readers or hearers need much prior knowledge about air terminals to understand it. They must know, for example, that the briefcase would have to pass an inspection and that certain metallic objects may not be carried into the cabin of an airplane.

Presumably, this kind of prior knowledge is represented by a schema. The schema for air terminals would specify: the roles played by various people in a terminal, the objects typically encountered in a terminal, and the actions that typically ensue in a terminal. Similarly, there must be schemas for other recurrent situations, and in each case the situation schema would be organized in terms of the roles, objects, and actions that typically occur in that situation.

Schemas are thought to play a major role in comprehension. Essentially, people understand a new proposition by accessing the most relevant schema and then fitting the new proposition into the framework so provided. Thus, without a schema for voting in an election, for instance, it would be difficult to understand most stories about this topic.

The picture that emerges, then, is that comprehension is based on knowledge. People understand the new in terms of the old, and memory organization is thus a cornerstone of comprehension. This principle's major implication for education is that new information should be presented in a way that ensures maximum contact with prior knowledge. Educators must determine what knowledge schemas individual students already have, and structure new information so that it best matches these old schemas.

## LEARNING

Learning has been studied in such a variety of organisms and situations that it is hard to see common characteristics. Learning is usually defined as any modification of behavior resulting from an individual's previous experience. However, the effects of a learning experience, especially in human beings, may be complex and not immediately apparent in observable activity. Thus, for theoretical purposes, learning is the acquisition of information in any form that ultimately can influence behavior. Major research problems revolve around the need for finding ways of measuring acquired information and tracing its storage, transformation, and decay in the course of learning and forgetting.

Because the observable characteristics and manifestations of learning vary so enormously over phylogenetic levels, from one-celled organisms to human beings, a central question has been that of the kind and extent of generality to expect with regard to concepts and principles. The great variation in the nervous systems of different animal forms suggests little commonality in processes, like learning, that directly reflect neural function. On the other hand, common evolutionary pressures may yield similar functional principles at different phylogenetic levels, in spite of differences in the responsible neural mechanisms, as has been observed in respiration, reproduction, and the sensory systems.

After the first century of experimental research on learning, the overall picture appears to be that some elementary forms of learning, notably conditioning, exhibit important common properties over a very wide range of phylogenetic levels. Conditioning constitutes a major part of the adaptive repertoire of lower organisms, but in higher organisms it is overlaid and dominated by more complex processes, largely verbal in human beings. Even in humans, conditioning is important in the regulation of bodily processes and the acquisition and modification of motives and emotions. Consequently, an understanding of conditioning may be implicated in the treatment of some forms of illness (for example, high blood pressure, or recovery of movement following a stroke) and behavioral disorders such as phobias and addictions. Currently, however, much more research is being directed at more complex and characteristically human forms of learning.

The state of research on learning can be illustrated best by considering first some elementary forms of learning presumed to be common to both human beings and lower animals, and then some varieties of exclusively human learning.

## Basic Learning

Classical conditioning, made a household word by the work of Russian physiologist Ivan Pavlov and his followers, is the prototype of a basic learning process: a mechanism whereby organisms learn to anticipate biologically significant events such as the appearance of food or water. In a form of conditioning widely studied in American laboratories, a stimulus that has preceded pain subsequently evokes behaviors symptomatic of fear and distress, even after long intervals. This conditioned emotional response in the laboratory rat is used widely in assessing the effects of tranquilizing drugs and other means of alleviating fear or anxiety.

In the ordinary life of human beings, one observes classical conditioning only in the case of some defensive reflexes and, more importantly, emotions and attitudes. Conditioning of visceral functions, such as changes in heart rate or blood pressure, intestinal function, and glandular discharges, can be demonstrated under carefully controlled conditions, often in connection with medical procedures. But why does such conditioning not occur constantly in everyday life, leading to a chaotic state in which bodily functions are conditioned to innumerable external stimuli? The answer was a mystery until the discovery was made that, once an organism has learned that some cue or signal reliably signifies the occurrence of an event, other stimuli that may occur later in the situation will not undergo conditioning—their conditioning is said to be blocked. This result is a special case of the more general principle that the initial experiences of a person or an animal in any new situation may determine which cues or aspects of the situation will enter into learning and which will come to be ignored or ineffective.

The implications that experiences early in life are especially important for later learning are borne out by such phenomena as imprinting, the almost instantaneous learning by fledgling birds to follow their mothers. Some investigators of phobias in human beings find evidence that these, too, result from early conditioning experiences.

In recent decades, theories of conditioning have progressed from the qualitative formulas of Pavlov and his followers to mathematical models capable of describing in detail the course of conditioning under varying circumstances and even predicting such novel phenomena as blocking. The mathematical form of a major class of these models has been shown recently to be identical to a family of equations describing properties of saturation and stability in electrical networks. This discovery and other related findings support a growing belief that properties of basic forms of learning reflect rather directly the properties of neural networks, which in turn are governed by principles common to still broader classes of communication networks and adaptive systems.

A substantial body of research and theory indicates that some forms of elementary human learning, such as the development of motor skills, can be interpreted in terms of processes closely resembling those of conditioning. However, efforts to open up the analysis of more complex, characteristically human forms of learning evoke more interest and excitement.

## Language Acquisition

A preeminent task for a theory of learning is to account for the acquisition of language by children. Perhaps more than any other faculty or skill, language distinguishes human beings from other animals; it makes possible the communication basic to civilization, shapes our thoughts, and preserves our intellectual heritage. To a linguist, language is a most complex and intricate assemblage of sounds (phonological rules and representations), meanings (semantic rules and representations), and mappings between the two (syntactic rules and representations). Yet language is generally learned by children almost effortlessly, with little dependence on formal instruction and with speeds far surpassing those of the acquisition of less complex skills and knowledge in the schools.

How can one study early language acquisition scientifically? The problem is difficult to address directly because it is not feasible or desirable to experiment on young children; to do so might disturb the natural learning of language. A strategy considered appropriate for many decades, during which the work of Pavlov and Thorndike evolved into a body of learning theory, was to uncover laws and principles by experimenting on animals and then apply the results to human language learning. The best-known effort following this strategy was the work of B. F. Skinner, who developed a body of "operant conditioning" principles for shaping the behavior of animals in both experimental and practical situations. In 1957, Skinner extended his principles to the learning and use of language, in his volume *Verbal Behavior*. In essence, he proposed that a child learns to make appropriate utterances in various situations by hitting on them in the course of more or less random spontaneous vocalizations, and then selecting those utterances for further use on the basis of their rewarded, or reinforced, consequences.

The linguist Noam Chomsky aroused a major controversy by attacking vigorously all of the premises of Skinner's theory. Chomsky held that knowing a language implies the ability to produce an infinite variety of novel sentences, a skill far too complex to be learned by reinforcement or reward on the basis of the small amount of adult speech that a child hears. Chomsky's claim, in its essentials, was that language acquisition is not a matter of learning at all, but rather an unfolding of an innate program to develop a mental representation of the grammar of a language. He found support in evidence for what were termed linguistic universals—abstract properties common to the grammars of all languages but not deducible from samples of sentences such as those that children hear.

It was difficult to settle the controversy decisively without critical experimentation on young children, but controlled observations by psycholinguists such as Roger Brown helped to undermine the reinforcement position. It was observed, for example, that many kinds of grammatical errors that ought to arise frequently in young children's speech if they were producing utterances more or less randomly simply do not occur. Furthermore, competence in grammar could not be learned as a consequence of adults selectively rewarding acceptable grammatical utterances, since it was discovered that adults reward children for speech that is correct in meaning rather than for its grammatical properties.

On the other hand, the idea of language learning being a matter of releasing elaborately organized innate capacities to understand and produce grammatical speech has by no means been fully supported by studies of children's early speech. Much observation intended to bear on the issue has yielded little compelling evidence that children's early efforts at combining words reflect knowledge of the grammar of the language. Perhaps at first a nonreinforcement learning process is at work. There is accumulating observational support for the idea that children begin by learning simple rules for producing acceptable combinations of words to communicate specific items of information, such as short sentences with fixed word orders denoting relations between agents and actions or between actions and objects in familiar situations. Only after considerable experience of this sort do they begin to notice more abstract grammatical regularities in the speech of adults or older children and to incorporate them into their own linguistic abilities.

Such ideas do not, to be sure, constitute a theory. Consequently, a major current enterprise is the effort to embody the ideas in mathematical or computer simulation models which, if the hypotheses entering into the models are sound, will actually infer the grammatical rules of a human language after processing a sample of sentences of the sort that children hear. This approach cannot be evaluated yet, but it offers important possibilities for constructing useful theories of language acquisition that would be wholly out of reach otherwise. At the same time, the attention of some theorists is turning to language itself and the properties of grammars that make them learnable in principle. Part of this work is purely mathematical and logical in character, already yielding some interesting formal results about what sorts of innate knowledge are necessary and sufficient for language acquisition.

Optimism about the practical results to be expected from theories of language acquisition relying on formal analysis and computer simulation must be tempered by the observation that, with regard to the problem of acquisition, the theories have been limited to restricted or artificial languages. Further, most of them have made little contact with the body of established empirical facts about the stages of children's linguistic development. This situation should be remediable, but the remedy will require a synthesis of the more empirical approaches and the more abstract ones.

## Acquisition of Knowledge and Cognitive Skills

Developments in cognitive science during the last decade have made it possible to open up the study of complex forms of adult human learning—for example, the acquisition of factual knowledge or problem-solving strategies.

A principal research strategy for analyzing such learning has been to examine in detail the tasks performed by skilled practitioners in various fields, to detect the processes that experts and novices use to perform those tasks, and to model the performers' knowledge, strategies, and other skills. These models, which often take the form of computer programs that simulate the skilled behavior, have reached respectable levels of competence in skills as diverse as playing chess, synthesizing chemical compounds, and solving college-level physics problems. Current research is aimed at deeper questions: by what process does a novice become an expert? What has to be learned, and how is it done?

One kind of learning that has been modeled successfully is acquisition of the expert's ability to make fine perceptual discriminations among the wide variety of situations that are encountered in practice, and to recognize most of them instantly from a small number of cues. A computer program, EPAM (Elementary Perceiver and Memorizer), which employs simple mechanisms to accomplish such perceptual learning, has been used to simulate the chessmaster's intuitive grasp of the essential features of chess positions. The program also has been used to estimate the variety of familiar chunks of information an expert must be able to recognize—some 50,000 recurring patterns of pieces in the case of the chessmaster. In addition, the program has been used successfully to interpret the findings of a wide range of classical experiments on rote verbal learning.

The EPAM program and similar efforts are concerned mainly with acquiring the index to the body of expert knowledge—the recognition cues that are used to access that knowledge. Other research in learning is concerned with the content of the knowledge and its organization in memory. It is now widely agreed that long-term memory can be modeled as a network of nodes, with associative structures representing the packets of information, termed schemas, stored at the nodes and connected by numerous pathways. An important research question is how information presented in natural-language text is transformed so that it can be incorporated in such structures, augmenting and enriching the information contained in them.

Several research efforts have achieved some success in modeling this acquisition of knowledge from natural-language text. One computer simulation, for example, interprets the language of physics problems in college textbooks and creates a representation in memory of each problem. Using this representation, it sets up the appropriate equations and solves them. This program does only a moderate amount of new learning, for it is already provided with schemas to represent the common kinds of objects, such as levers and masses, which it will encounter in the problems; it must simply assemble them correctly.

In the next few years, other learning processes are likely to be identified and modeled. The modeling efforts themselves are still at an early stage of development, but the time should not be far off when educational psychologists can begin to develop and adapt the basic findings for application to problems of teaching and learning in the schoolroom. Already, ideas emerging from the research are being used to improve instruction; for example, organization of the material in a manner compatible with the memory structure that the student brings to the task is being emphasized. These and similar ideas will need careful empirical testing before their usefulness in education can be evaluated. However, we can expect a continuing stream of such ideas from basic research now under way.

## COGNITIVE SKILLS AND INTELLECTUAL FUNCTION

A major step toward applying the results of cognitive research to human problems is the analysis of the cognitive skills and capacities basic to communication, reasoning, and decisionmaking. Progress can be cited both for processes underlying speech and reading and for the relatively complex forms of reasoning customarily associated with intelligence and intelligent behavior.

### Speech Perception

The stimulus with which one must deal in understanding speech is well understood to be borne by air upon which a speaker's vocal apparatus has impressed wavelike disturbances of varying frequencies. How the hearer extracts information from this flow of auditory energy could not be understood until the introduction of the sound spectrograph, which permits the energy spectrum carrying a speech message to be passed through frequency analyzers and then plotted on a screen or in a photograph as a temporal record of the varying frequencies of constituent sound waves.

Inspection of a sound spectrogram shows immediately that the information received by the hearer is not "packaged" in successive discrete units corresponding to the letter sounds or syllables making up the spoken message. Rather, the auditory information on which the identification of successive vowel and consonant sounds depends overlaps in a complex manner. Vowel sounds are relatively simple to deal with, being associated with formants, concentrations of acoustic energy in frequency bands that are modulated over time in characteristic ways. Consonant sounds are added mainly by very rapid modulations or discontinuities in the formants (for example, delay of onset, change in the characteristic frequency, brief noise bursts, or gaps).

The first step in "cracking the speech code" was to discover these correspondences, a task made especially difficult by the fact that the information responsible for a particular consonant sound cannot be localized at any point or in any narrow band in the speech stream. If, for example, the consonant sound "s" is produced by adding a brief burst of noise at the beginning of the formant for the vowel, then a noise burst and the formant together, in proper temporal relationship, are perceived as the syllable "sa", but the noise burst alone would be heard as only a chirp. If, in a computer-controlled speech synthesizer, one quantitatively varies the duration of a silent interval between the noise burst and the beginning of the formant over the range of values, a listener hears the syllable "sa" at the shortest durations and the syllable "sta" at a duration of about 50 milliseconds.

Although the physical parameter basic to identification of the consonant, duration of the silent interval in this case, is varied over a continuous series of values, the listener does not hear a continuous series of changes in speech sound but rather an abrupt transition at some point from "sa" to "sta". This phenomenon, termed categorical perception, is the key to the capability of a listener to reduce the infinitely varying patterns of auditory information produced by different speakers under different circumstances to a small number of phonemes, or basic speech sounds, that are sufficient to allow identification of all of the syllables that can be distinguished in a given language.

Categorical perception may be unique neither to human beings nor to speech, but it is certainly most conspicuously developed in the case of speech. The human auditory system is especially sensitive to just those variations in the auditory stream that are produced when a speaker varies the manner and place of articulation, the voicing, the nasality, etc., of spoken sounds in order to produce recognizable syllables. The listener responds sensitively to these significant dimensions of variation and is capable of ignoring—literally not hearing—the enormous variation in other irrelevant aspects of the auditory stream. The difficulty adults characteristically have in learning to speak a new language is doubtless related to the fact that, in the adult's categorical perception, speech sounds that must be discriminated in the new language may in the native language fall in the same category and thus be undifferentiated.

Clearly, continuing research on the conditions of acquisition and modifiability of the speech code is essential to improving the treatment of speech disorders and the teaching of second languages. However, the advances in analyzing speech perception at the phonological level constitute only a step toward understanding how we comprehend spoken language. One of the most active current lines of cognitive research (as indexed, for example, by the program of the 1980 International Congress of Psychology) is directed to the next step of analyzing the perception of grammatical sequences of words.

## Cognitive Processes in Reading

Because of the obvious practical importance of understanding reading, this topic has had a long history of research dating back to the early 1900's. Throughout much of this period, controversy has centered on the basic approaches to the teaching of reading; one approach is to teach directly the correspondences of printed and spoken letters (the phonics approach), and the alternative is to teach pattern recognition of words as units (the look-say approach). The debate remains unresolved, due to our limited understanding of the processes underlying reading. However, there are signs that the situation may improve, in part as a consequence of the increasing demand for some remedy for the widespread incidence of reading disabilities, and in part as a consequence of the fact that reading provides a natural medium in which to study in combination new models for processes of perception and memory ranging from pattern recognition to comprehension of meaning.

The early laboratory studies of reading uncovered several important facts that could not have been obvious to the layman. For example, one's impression that reading progresses by a continuous intake of information as the eyes move smoothly across lines of print is an illusion. Actually, the eyes move at high velocity from one fixation point to another, and information is taken in only during the intervals, typically one tenth to one fifth of a second, in which the eyes are at rest. Further, it was demonstrated in early psychological laboratories that during such a fixation pause the visibility of letters surrounding the fixation point falls off rapidly toward the periphery, and since normally only a few fixations occur per line, many letters are never seen clearly. At the same time, other investigations showed that readers acquire familiarity with the statistical structure of the language (that is, probabilities with which different letter combinations occur) and use this knowledge to fill in gaps in what is perceived.

The researchers who uncovered these facts essentially exhausted the methods available in the early 1900's, and little further progress was made until the last decade or so when concurrent advances in technical methods (such as eye movement monitors and computer-controlled visual displays) and in cognitive theory permitted a new wave of progress.

The present theoretical picture of the reading process can be outlined as follows. First, there are clearly two main routes from print to the brain, one direct and one branching off to involve phonological encoding and articulation. In the direct route, information from the printed message is processed through a series of analyzers that filter out extraneous visual information. The relevant information is transmitted in progressively more highly encoded forms that are resistant to masking (disturbance by subsequent events), and that provide appropriate inputs to the pattern-matching systems that yield recognition successively of letters, letter groups (spelling units), and words.

The system is normally highly reliable in operation. However, it has the major limitation that, once representations of letters or words in the memory system have been excited by information from a printed message, their level of activity in these units declines quickly as the analyzers are captured by new inputs. Since associative mental processes are relatively slow compared with those responsible for sensory and perceptual analyses, some means is needed to keep the output of the perceptual system in an active state long enough for information from a meaningful unit of text, such as a sentence, to be simultaneously available for recognition and other forms of processing.

This need is met by the phonological route. This path branches off from the one just described. Following the initial identification of letters in the printed message, the input from combinations of letters activates memory representations of syllables, which in turn are articulated, subvocally if not aloud. The importance of this step, even when the task does not call for speech, is that the engagement of the articulatory system leads to the entry of the syllables into a short-term memory system termed an articulatory loop, in which up to six or eight syllables can be maintained in a high state of activity while further inputs are processed. This process permits the assembly of enough information to activate a still higher order structure, such as the memory representation of a concept or proposition.

Because skilled readers obviously do not mumble words to themselves as they read, it was once thought that the role of speech should be eliminated, insofar as possible, from reading. However, numerous lines of research (including electromyographic recordings from speech musculature, various techniques to suppress vocalization, and comparisons of normal readers with individuals suffering from forms of brain damage that lead to dyslexias or specific reading disabilities) make it clear that skilled reading requires some participation of the articulatory system. Skilled readers simply become so efficient in its use that overt signs of articulation during reading are not apparent to casual observation. However, if articulation is artificially suppressed (by a concurrent task that requires activity of the vocal muscles), they have difficulty reading anything but simple material and, especially, remembering what they have read. Clearly, the neural and psychological systems involved in reading and in speech are closely related, and skilled reading involves coordination of the two, not suppression of one at the expense of the other.

In view of these dual aspects of reading, its learning must begin with training in the recognition of the visual patterns corresponding to printed letters, and then must proceed to a mapping of these onto phonological units. However, taking together the results of research on reading and on speech production, it is clear that this step cannot be taught by sounding out the successive letters encountered in reading a word, as is often misguidedly considered to be a basic part of the teaching of reading. Since syllables are the smallest phonological units that can be articulated independently, learning to read must involve associating visual representations of letter groups with spoken syllables. Research on methods of implementing these principles in the practical teaching of reading has just begun, but already training programs growing out of the research have begun to show advantages over traditional methods.

## Human Problem Solving

The nature of intelligence in cognitive science, or in common speech for that matter, is not easily captured by formal definitions, but is connected in general with efficiency in acquiring, generating, and using information in solving problems. Vigorous efforts to understand intelligent problem solving are being pursued in both psychology and computer science. Research on the psychological aspects of intelligence is aimed at measuring and understanding the conditions of use and the development of problem-solving abilities. Research on computer programs and computability relate to more abstract properties of intelligence. A significant contemporary trend is interaction between these approaches to intelligence, following decades of quite autonomous development.

One of the first applications of the new information-processing concepts that arose after World War II was research on problem solving. The initial emphasis was on puzzle-solving tasks in the psychological laboratory, but research is moving steadily to problems whose solutions require professional skill and extensive knowledge of subject matter.

Even the simple puzzles (for example, getting the three missionaries and three cannibals safely across the river) are hard enough for intelligent adults, often requiring half an hour for their solution. Research on how subjects solved them in the laboratory, while their verbal protocols were recorded, revealed a few basic strategies that appear to be central to almost all human problem solving. Problem solving characteristically involves highly selective search through large problem spaces; the selectivity is essential to a successful search in a reasonable amount of time. Frequent use of means-ends analysis is necessary, that is, solving problems by comparing present situations with goals, recognizing one or more differences between them, and applying operators stored in memory that may be relevant to reducing the differences.

Beyond identifying these processes of selective search and means-ends analysis, an important task has been specifying exactly what is involved in carrying out these processes and demonstrating empirically that such processes are sufficient for dealing with problems intelligently and successfully. The precision is achieved by incorporating the processes into computer programs, the basic operations in the program corresponding to the elementary information processes revealed by research. The proof of sufficiency is achieved by using the computer programs to simulate the actual human behavior, and showing that the programs can solve the

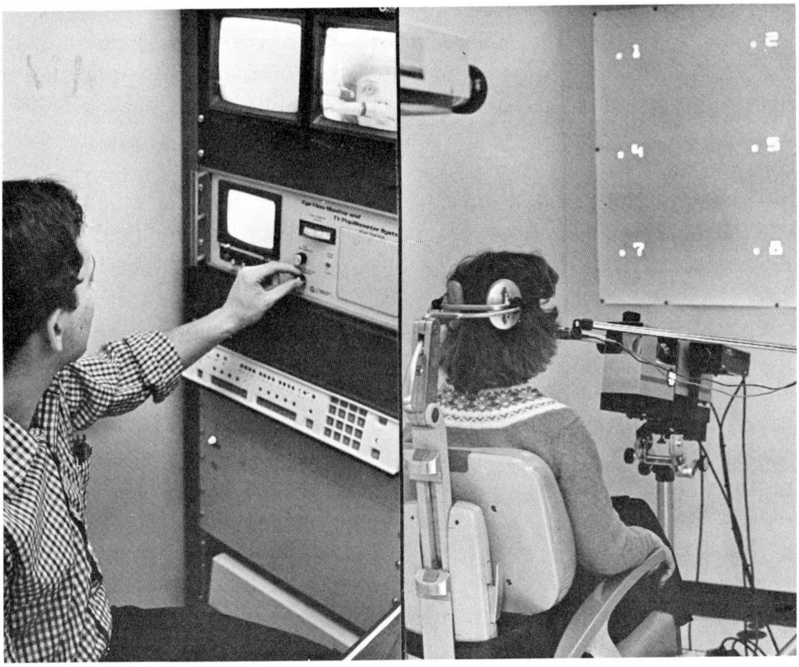

Studying eye movements during problem solving. "A vital question in understanding problem solving is how search efforts are organized and guided by strategies." [Copyright 1980, Learning Research and Development Center, University of Pittsburgh.]

same problems as their human counterparts, organizing their efforts in much the same way that verbal protocols and other evidence have shown human problem solvers to organize theirs.

A vital question in understanding problem solving is how search efforts are organized and guided by strategies. We have learned a great deal about both stimulus-driven and goal-driven search, and know that most problem solving is guided by a combination of these. The key idea is that problem-solving behavior consists of a sequence of actions, each initiated, or fired, by recognition of some clues that indicate its relevance to the current situation. The clues may consist of perceived features of the environment (sometimes simply the problem solver's worksheet) or goals and subgoals held in short-term memory. When environmental clues predominate, we say the behavior is stimulus-driven; when short-term memory clues predominate, we say that it is goal-driven. The strategic elements, each consisting of a set of stimulus or goal conditions linked to an action, are known technically as productions, and a strategy organized in this way as a production system.

Most contemporary programs for simulating human problem-solving behavior, simulations of learning processes, and artificial intelligence programs for performing complex tasks are constructed as production systems. All productions have the same basic form, and new productions can simply be added to the system. This simplicity has been a major factor in recent successes in devising programs that can learn and that can simulate human learning.

The same processes that have been identified as fundamental to problem solving also have figured prominently in research on other cognitive skills, in particular, schema formation, concept attainment, and the discovery of patterns in data.

The concept of a schema, important in models for memory representations, reappears here as a plan for acquiring, storing, and using knowledge. It appears that an important early stage in solving a problem is bringing to mind a schema comprising certain general knowledge that must be true of a given type of situation, together with an outline of the information structure that must be filled in before a problem can be solved. Thus, diagnosing a malfunctioning machine may begin with the mechanic's bringing to mind a schema comprising certain concepts and relations that must necessarily be involved in the type of machine, together with a flow chart whose more detailed entries must be filled in as the diagnosis develops. An even more common example is comprehension of a spoken sentence by a listener. Here, the initial stage of comprehension appears to be guided by a schema comprising the grammatical case structure that is to be anticipated in the given context, again, with information concerning the particular grammatical units and their interrelations to be filled in by information forthcoming as the sentence is heard.

In concept attainment, the individual must learn to discriminate among a set of items that belong to different categories, for example, picking out all of the adverbs from a passage of text. In pattern discovery, the individual must discern a regularity and formulate a rule to describe and apply it, as in discovering the regularity in the sequence 1, 2, 4, 8 . . . and predicting the next few numbers. The strategies used to perform such tasks, which form an important component of intelligence tests, are now fairly well understood. Among the abilities required, beyond general problem-solving strategies, are skills in recognizing symbols and regular successions of symbols in lists like the alphabet or the natural numbers, and in categorizing these patterns in meaningful ways. An important outcome of this research is to demonstrate that a small set of rather simple skills, differently organized for different tasks, suffices to account for behaviors that are commonly used as indicators of human intelligence.

A growing number of college-level courses intended to develop general problem-solving abilities have been designed on the basis of this research. Such courses seem to be well received, although they have not been clearly demonstrated yet to be cost-effective uses of students' time. They raise the age-old pedagogical issue of transfer: must people be trained in the specific skills they need, or can they be trained instead in more general skills that can be applied to specific situations? The correct answer must be "some of each," and current advances in cognitive science provide us with perhaps the first opportunity to attempt to specify the optimal combination. This question is central to the design of curricula and will almost certainly receive increased attention, with considerable illumination from the new research tools.

## Artificial Intelligence

Originally, computers were designed to solve some specific kinds of problems (especially mathematical ones) that are difficult or tedious for human beings. It was natural, therefore, for computer scientists to attempt to program these machines to solve also the kinds of problems that challenge human intelligence (intellectual games such as chess and go, and the types of reasoning problems that appear on intelligence tests) and to branch off into efforts to program computers to comprehend and produce language. These extended efforts are subsumed under the term "artificial intelligence."

Artificial intelligence is obviously germane to efforts to let computers take over difficult tasks, and the value of these efforts is well known. Less easy to answer are questions about the relevance of artificial intelligence to human cognitive science.

One source of relevance is the common objective of computer scientists and cognitive scientists in developing general theories of information-processing systems that will help in understanding the limitations and potentialities of all systems, whether machines or human beings, capable of problem solving. Despite their conspicuous differences, human beings and computers share common characteristics as information-processing systems. In each case, information must come into the system through channels of finite capacity—electrical cables in computers, nerve trunks in humans. These have limited capacities for information transmission, must cope with similar problems of distinguishing signals from noise, and must use codes to transform information in ways that increase the efficiency of transmission and achieve error-correcting properties. Substantial progress has been made over several decades in developing general theories of information transmission. One function of artificial intelligence is to relate this theory to the kinds of information processing essential to intelligent human behavior.

An especially active and promising line of investigation in artificial intelligence is the development of programs for analyzing visual space, an aim being to enable computers to describe and understand visual scenes much as human beings do. One motivation is the indirect value such work may have in advancing the theory of human vision. More directly practical are the potential applications to such problems as programming computers that can be carried in space vehicles to increase both the amount and the reliability of transmitted information about the visual properties of landscapes on other celestial bodies.

## Augmentation of Human Problem-Solving Abilities

The past quarter of a century of research has shown that many powerful ideas for artificial intelligence programs can be borrowed from human cognition, as that begins to be understood. Conversely, solutions to problems of designing artificial intelligence programs have offered many hints about the human processes. There continues to be a close and fruitful partnership between these two lines of research.

But there is another connection between human cognition and artificial intelligence. Computers and human problem solvers in concert may be able to accomplish tasks that, individually, neither could do as well, or at all. Although we have already become used to thinking of the computer as an indispensable ally in tasks in science, engineering, and business, we are just beginning to see the potential of human-computer cooperation over a much wider range of nonnumerical, symbol-manipulating tasks.

One such cooperative effort that is already highly visible is word processing. In using word-processing systems for editing manuscripts, for example, the computer provides a large and flexible, but for the most part passive, memory. It allows us to examine the text we have written, to revise it and to reproduce it, all without repeated retyping. Yet the computer, in this application, does not really provide any of the intelligence; that comes from the human user.

The next step is for the computer to participate in the problem-solving process. Automatic programs for analyzing spectrograms of materials (for example, DENDRAL) and for medical diagnosis (for example, INTERNIST and MYCIN) can help to uncover molecular structures and diagnose diseases. The program MYCIN almost literally talks with the physician faced with diagnosing and recommending treatment for bacterial infections when symptoms are confusing and conflicting, and estimates the reliability or certainty of the diagnoses and recommendations. This program is even capable of assimilating new knowledge as it operates and of responding to objections or suggestions from the physician so as to gain from his or her experience and intuition. For a long time to come—perhaps forever—we will want to leave the final diagnosis to the physician, but the diagnostic program can be a powerful additional source of information and insight.

Similarly, a number of papers published in the chemical literature in recent years report the automatic identification of organic molecules by the DENDRAL computer program, which draws its inferences from mass spectrograph data. MOLGEN, a program not quite so far along, provides advice for the design of research experiments in molecular genetics. Programs for the automatic synthesis of chemical compounds have been in the literature for nearly a decade and are now used in the chemical industry.

Interest in industrial robotry is undergoing a considerable revival in the United States, Europe, and Japan. Much of this activity is concerned with extending familiar techniques of factory automation. A part, however, is concerned with building robots that can protect human beings from various kinds of dangers by replacing them in difficult environments, as in space or in areas where there is danger from radioactivity or the like.

The boundary between what human beings do and what machines do will shift continually as our knowledge about what would be truly helpful in the solution of human problems increases, and as we achieve new technical understanding of how to provide that help. Computer intelligence will come to supplement human intelligence in many ways (for example, appraising both the social and the environmental effects of alternative routes of energy development), just as mechanical power came, in the first Industrial Revolution, to supplement human and animal muscle. (Artificial intelligence is also discussed in Chapter 17.)

## RESEARCH SETTINGS

Following the pattern laid down by William James at Harvard University nearly a century ago, cognitive research in this country has developed largely in universities. However, the rather slow pace of development during the first half of the century, and the great acceleration in the second, may be due to some extent to the need for major complementary contributions by a number of disciplines, something not often easily accomplished in traditional academic environments. Thus, the emergence of modern psycholinguistics and the human information-processing movement in the 1950's and 1960's was sparked to an important extent by work in a few institutions where the situations happened to be favorable for the interaction of linguists, psychologists, and computer scientists. Also, disproportionate contributions came from a few independent research organizations where such interactions are the rule, as at the Haskins Laboratories, now in New Haven, Connecticut, and the Learning Research and Development Center at the University of Pittsburgh. The former has been the source of major developments in speech perception, the latter of major segments of basic cognitive research aimed specifically at education.

In western European countries and Japan, the pattern has been much the same, with even more conspicuous contributions from independent research organizations. Important among these are the Applied Psychology Unit of the British Medical Research Council in Cambridge; the Institute for Perception Research in Eindhoven, The Netherlands; and the Max Planck Institute at Nijmegen, The Netherlands. The first of these has been the source of some of the most important theoretical ideas about cognitive factors in communication and human engineering; the second has been the focus for fruitful international conferences on perceptual factors in reading and communication; a major program in the third is at the forefront of contemporary research on cognitive factors in the comprehension of natural languages.

In the Soviet Union, cognitive psychology was until very recently almost solely the province of the Institute of Pedagogy and its ramifications in schools of education. The most important results in that tradition appear to be connected with the training and education of the mentally retarded. Currently, research in cognitive psychology is largely the province of the Institute of Psychology in Moscow and smaller institutes in Tbilisi and other cities. Soviet psychology at the level of human behavior only recently has become a recognized discipline, and research that would be considered current by western standards is largely concentrated in psychophysiological studies of basic sensory capacities on the one hand, and large-scale, computer-simulation models of complex decision processes on the other. Examples of the latter are simulations of the decisionmaking in a segment of an industrial organization, or in moving ships through the locks of a canal. The inaccessibility of most Soviet cognitive research literature to investigators in this country makes it difficult to evaluate these developments or the degree to which they depend on concentrated efforts in the institutes as distinguished from the programs slowly gaining ground in some Soviet universities.

# Summary and Outlook

A conspicuously active, and theoretically productive, area in the behavioral and social sciences is the study of human cognitive processes and capacities. Part of the reason lies in the magnitudes of the social problems in which these processes and capacities are implicated. It has long been apparent that efforts to cope with the prevalence of illiteracy and learning disabilities, the psychological aspects of health and environmental problems, the information explosion, and the educational demands of a technologically advanced society have all made only limited progress by means of research aimed directly at practical applications. Major advances in dealing with these problems may be initiated only by corresponding advances in our understanding of the human mind and its functions—that is, in scientific theories of human learning and cognition.

The recent burst of scientific activity that gives promise of beginning to meet these needs resulted from a convergence of influences from theoretical developments in linguistics, experimental psychology, and the computer revolution. In particular, demonstrations that computers can accomplish tasks hitherto the province solely of human intelligence, such as playing chess or answering factual questions, suggested the possibility that computers can "think" in ways similar or superior to human reasoning. The analogy has proven more than superficial, providing theoretically suggestive concepts and fruitful methods for the study of human perception, memory, and problem solving. In return, new insights into human cognition are leading to more effective ways of programming computers to accomplish tasks, such as the recognition of speech, which seem to be on the verge of important practical applications.

- The formulation and application of computer simulation models that have had a revolutionary impact on our conception of how human problem solving is accomplished.
- The development of a family of mathematical models of elementary decision processes, originally derived from signal detectability models of communication engineering. These models permit the appraisal of the separate contributions of a human observer's biases and discriminative capacity (signal-to-noise ratio) to performance on such tasks as the perception of simple visual and auditory stimuli near threshold, the detection of blips representing aircraft on radar screens, or the discrimination of shadows representing disease conditions on radiographic plates.
- Analyses of speech perception, revealing how phonologically significant items of information are extracted from the spectrum of sound waves generated by a speaker, and analyses of cognitive processes in reading, elucidating the distinct but complementary visual and auditory routes for the transmission of information from printed text to the brain.
- Hierarchical network models for the organization of semantic memory, accounting for the speed with which recall of one word or concept leads to recall of another as a function of similarity and categorical relationships.

Current work that appears likely to come to fruition during the next few years includes:

- Mathematical models relating basic learning processes to the properties of neural networks.
- Computer-implemented models of language acquisition.

- Efforts to apply the results of current research on communicative processes to the amelioration or prevention of disorders of speech and reading.
- Development of methods for teaching general problem-solving skills and for computer augmentation of human capacities for dealing with extremely complex problems.

In general, the near future seems likely to bring increasingly effective exploitation of the complementary strengths of cognitive psychology and related fields in forging a discipline that may play a significant role as a basic science to many facets of education and research on human factors in advanced technology.

## BIBLIOGRAPHY

J. R. Anderson. *Cognitive Psychology.* San Francisco: W. H. Freeman and Company, 1980.

G. H. Bower. "Cognitive Psychology," *Handbook of Learning and Cognitive Processes.* Volume One. Edited by W. K. Estes. Hillsdale, N.J.: Lawrence Erlbaum Associates, 1975.

G. H. Bower and E. R. Hilgard. *Theories of Learning.* Fifth Edition. Englewood Cliffs, N.J.: Prentice-Hall, Inc., 1981.

W. K. Estes. "On the Organization and Core Concepts of Learning Theory and Cognitive Pscychology," *Handbook of Learning and Cognitive Processes,* Volume Six. Edited by W. K. Estes. Hillsdale, N.J.: Lawrence Erlbaum Associates, 1978.

R. Lachman, J. L. Lachman, and E. Butterfield. *Cognitive Psychology and Information Processing.* Hillsdale, N.J.: Lawrence Erlbaum Associates, 1979.

A. Newell and H. A. Simon. *Human Problem Solving.* Englewood Cliffs, N.J.: Prentice-Hall, Inc., 1972.

M. I. Posner and G. L. Shulman. "Cognitive Science," *The First Century of Experimental Psychology.* Edited by E. Hearst. Hillsdale, N.J.: Lawrence Erlbaum Associates, 1979.

*Cognition and Categorization.* Edited by E. Rosch and B. B. Lloyd. Hillsdale, N.J.: Lawrence Erlbaum Associates, 1978.

*Cognitive Processes in Choice and Decision Behavior.* Edited by T. S. Wallsten. Hillsdale, N.J.: Lawrence Erlbaum Associates, 1980.

# II

# *NATURAL RESOURCES AND ENVIRONMENT*

# 5

# *Ecology and Systematics*

## INTRODUCTION

In a general sense, ecologists study the relationships between organisms and their surroundings. But any field so broad must consist of a number of more limited, somewhat artificial disciplines. Thus, individual ecologists may investigate the evolutionary significance of the shape and behavior of an organism, the interrelations among a group of similar organisms (a population), the functioning of all of the organisms in a specific area (a community), or the interactions of these organisms with their nonbiological environment (an ecosystem). This division of ecology into separate levels of organization is one way to partition the field; many other organizational schemes have also been proposed and are in use.

Crucial to almost any study of ecology is the work of systematists, who classify organisms by analyzing the differences and similarities between them and attempt to determine their evolutionary relationships. The roots of systematics, as well as of ecology, lie in natural history, and modern ecologists and systematists draw heavily upon this field in seeking out key problems to investigate, in determining the best species and systems for examining specific problems, and in assessing the accuracy of observational data.

Because ecologists and systematists often try to understand and predict the characteristics of large systems, they confront a problem faced by all scientists who deal with complex systems having many interacting

◀ Western tent caterpillars (*Malacosoma californica*) on the leaf of a red alder. All are descended from a single egg mass and, hence, are full siblings. Recent theory predicts that in a population of this kind females should outnumber males. This species and others in the genus are notable for major fluctuations in population. [D. F. Rhoades.]

parts. These separate parts must be grouped, and to form such groups certain relationships must be emphasized while others are deemphasized. The problem is that it is difficult to know if one has made the best groupings. Moreover, the very process of setting up aggregate units channels thinking about problems in subtle but powerful ways. Thus, forming a group such as herbivores emphasizes the source of food common to those organisms, plants, while deemphasizing such attributes as size, color, behavior, or tolerance to temperature changes. Clearly, the category of herbivores will be useful in discussing flows of energy in ecological communities, but it will not be appropriate for many other problems.

Ecology and systematics passed an important milestone in the 1930's when they welcomed into their conceptual structures the "neo-Darwinian" synthesis of Mendelian genetics with the theory of evolution. This union, forged by R. A. Fisher and J. B. S. Haldane in England and Sewall Wright in the United States, made the study of evolutionary adaptation a central concern of ecology.[1] In G. Evelyn Hutchinson's apt phrase, it was a matter of the "ecological theater and the evolutionary play."

More recently, the fields of ecology and systematics have been profoundly affected by three events, two of them technical, the other political. The first technical event was the development of modern computational devices that made it possible to integrate large quantities of field data and to obtain solutions to complex modeling problems that previously were intractable. The second event was the development of new equipment, such as the mass spectrophotometer, the atomic absorption spectrometer, and the gas chromatograph, that can detect chemicals that previously were impossible to measure.

The political event was the sudden growth of environmental awareness in the United States. Because ecological knowledge is a key to solving important environmental problems, ecologists found themselves thrust into the political arena. Initially, the expectations of the environmentally concerned public and the hopes of many professionals tended to be unrealistically high. Now a pattern is emerging in which the modest but useful contributions of ecologically related sciences are being appropriately applied to solve environmental problems.

From the wide range of topics under active investigation by ecologists and systematists, this chapter will treat a small number that illustrate the scope of the field, that reflect the intensity of current activity, and that have applications in environments dominated by humans. Many equally important topics have, of necessity, been left out.

Because no organism is an island, ecology has always been concerned with the overall interactions among the living and nonliving parts of specified systems. After a brief look at systematics, this chapter will describe research on two different aspects of total ecosystems—their energy flows and nutrient cycles. Successive sections will focus on progressively more specific levels of organization, from interactions of species to characteristics of populations to individual life histories. At each stage, the question will be posed, "What processes and patterns appear to be most interesting at this level of organization?" Understanding how these processes work and how they might be manipulated is essential to make proper decisions about environmental problems.

## SYSTEMATICS

The classification and naming of the animals, plants, and microorganisms of the world provide an orderly basis for accumulating and communicating information about them. Tens, if not hundreds, of millions of dollars are spent directly or indirectly in the United States in an effort to understand the environmental impacts of public and private works on communities of plants and animals. Studies of this sort seek to determine not only the direct effects on living creatures of construction projects, such as dams, highways, and subdivisions, but also the indirect effects of pollution, such as atmospheric sulfur, acid rain, and sewage runoff. These efforts are of little value if the organisms are not properly identified.

### Classification

In the United States alone, there are an estimated 20,000 kinds of native and naturalized plants that maintain themselves without cultivation. These are the plants of the forests, deserts, and grasslands of the country, the habitats that are modified or replaced during agricultural, suburban, industrial, or urban development. Because virtually every organism on earth depends upon the photosynthetic productivity of green plants and microorganisms, knowing the numbers and kinds of these organisms and the ways in which they are distributed is of general importance.

Despite popular belief to the contrary, the kinds of plants, animals, and microorganisms that inhabit the world, even those in the United States, are not well known or understood, and information about many of them is not readily available in museums or books or other publications. Unlike many other developed countries, the United States has no national inventory of its plants or of any group of animals, except birds and mammals, that is relatively complete and national in scope. There is no reference work or map that one may consult to find the nationwide distribution of a given species of plant, and no standard authority for the names of plants or animals except birds and mammals. Of course, some state and local surveys are useful in finding particular information, but they are very uneven in format and coverage.

Various pieces of federal legislation, such as the National Environmental Policy Act and the Endangered Species Act, mandate that such information be gathered. A comprehensive national biological survey intended to catalog existing information and make this catalog available in a form that is readily understandable would be far less expensive than the piecemeal efforts we are now undertaking.

Our lack of information about plants, animals, and microorganisms outside the United States, particularly in the tropics, is far worse. There are probably at least 4.5 million kinds of organisms on the earth, about two thirds of which occur in the tropics. Of the roughly 1.5 million kinds of organisms that live in temperate regions, more than two thirds have been cataloged and given a name. But of the approximately three million kinds of organisms found in tropical regions, perhaps only one in six is known and cataloged.

The world's tropical forests are being cut down at a rate that will lead to the conversion of essentially all of them, principally to agricultural purposes, by early in the twenty-first century.[2] This massive conversion of tropical forests is expected to cause an extinction of plants and animals at a rate unparalleled in the history of the world. Up to one million of the three million kinds of tropical organisms may become extinct by the end of this century, and an additional million kinds of tropical organisms will meet the same fate during the next century. By the year 2000, if current population growth rates continue, approximately three fifths of the world's human population will be living in the tropics. These people will depend for their livelihood either upon the communities made up of these largely unknown plants and animals or upon the agricultural and industrial systems that replace them.

Given the great utility of many organisms for human welfare, and the many ways in which human economies are based directly or indirectly upon them, ignorance of the magnitude outlined above is serious indeed. It is important not only to identify and classify organisms of proved or potential economic importance, but also to investigate the potential usefulness of these organisms. For example, many tropical plants synthesize chemicals that may be useful as drugs or pesticides.

## New Techniques

The past two decades have witnessed a remarkable revitalization of systematics. Developments in many other fields, ranging from molecular genetics to statistics, have provided a wide range of data concerning relationships among organisms, more powerful ways to compare large data sets, and improved capacities to store and retrieve data.

These developments have had an especially important effect on the field of population biology. For instance, population biologists are intensely investigating the factors that control genetic variation in populations. Using electrophoresis to distinguish different proteins, which are coded for in genes, they have demonstrated that natural populations are more genetically variable than had been suspected. Among the proteins, enzymes have been selected for study, in part because they are simple and in part because they are known to be highly responsive to selective pressures from the environment. Proteins whose functions can be directly related to specific environmental factors are especially promising targets for analysis.

One of the most striking developments of the past decade has been the wide application of chemical data to systematic and evolutionary problems. New techniques have made it possible to compare organisms chemically in many different ways, and new journals have been formed (*Photochemistry, Journal of Molecular Evolution, Biochemical Systematics, Chemical Ecology*) to accommodate the increased flow of information in this and related areas.

Chemosystematics comprises two main areas: the study of macromolecules, such as proteins, and of simpler compounds, such as flavonoids and terpenoids. In a major recent development, macromolecular systematics has begun investigating the evolutionary relationships of organisms by comparing the proteins and genetic sequences of contemporary organisms. DNA and DNA-RNA hybridization techniques, which reveal how similarly the genes on chromosomes of different organisms are ordered, are receiving particular attention, as are immunological and electrophoretic approaches to determining protein structure.

The extensive use of computers during the past 25 years to compare many traits of organisms has, at the same time, led to the development of well-established methods for classifying, ordering, and representing data. Indeed, many of the current leading questions in numerical taxonomy are mathematical and methodological ones, such as measuring the distance between the characteristics of organisms when these characteristics are defined in multidimensional spaces, developing techniques to measure the significance of clusters of similar organisms, and discovering ways to compare classifications. The application of these and other techniques developed by systematists extends far beyond the field; they are being extensively applied to subjects ranging from anthropology to psychology, from economics to grants management.

## STUDIES OF TERRESTRIAL ECOSYSTEMS

Because human beings exploit the world around them so intensively, ecologists are especially interested in the productive capacity of ecosystems. Consequently, studies of energy flows and nutrient cycles have dominated research in the past few decades. Complete measurements of flows of energy and nutrients in ecosystems, however, are notoriously difficult to obtain, and the large differences between years mean that data from one or two years may not reflect long-term averages.

## Energy Flows

All current models of ecosystem energetics are empirical. The most common conceptual approach is to group organisms according to their sources of energy. For example, organisms may be categorized according to their trophic levels as photosynthesizers, herbivores, carnivores, and detritivores (eaters of detritus). Studies of ecosystem energetics then try to determine the amount of energy contained within, and flowing through, each of these groups and the factors most strongly influencing those quantities.

Such information can be used to develop management plans that will increase yields in certain components of ecosystems. For instance, total photosynthesis, which determines the overall rate of supply of essential nutrients, seems to be most strongly related to the potential evaporation of water from the plants at a site, although this factor is modified by such variables as soil fertility, the seasonal distribution of precipitation, and the distribution of stressful conditions such as droughts, unusually cold weather, and high winds. As another example, the fraction of energy entering one trophic level that is consumed by the subsequent level is higher in communities where the most abundant plants produce no wood. In systems dominated by woody plants, much more of the primary production accumulates as wood, which is difficult for organisms to digest.

Another way to know how much energy is flowing between organisms is to know how much carbon moves between them, a fact that has led to the development of what are called carbon budget models. These models also establish compartments, but they use different units than do trophic models. Carbon budget models mostly employ groups such as wood, leaves, roots, and fruits because these different plant tissues have very different lifetimes and therefore different patterns of carbon movement. These models are useful in studying interactions between leaf- and root-eaters or between fruit- and seed-eaters.

Growth and succession models are similar in structure to carbon budget models, but they focus on changes in the ecosystem over time. The more complete growth and succession models employ compartments for the dominant plant species in the system. Several recent models of forest succession in the eastern United States are detailed enough to predict changes in the abundances of species.

Ecosystem energetics may also be viewed from an evolutionary perspective. As species in a particular system evolve, they allocate their energy reserves in different ways, and these allocation patterns can influence the properties of the entire system. For example, the ratio of energy bound up in persistent tissues to the energy bound up in short-lived tissues varies in different parts of the world.[3] Ratios are higher in most temperate forests than they are in tropical ones, perhaps because the stresses and strains of the physical environment in the temperate zone favor a relatively greater investment in roots and wood. Also, temperate-zone trees invest much less energy in fleshy fruits than do tropical ones, partly because the pollen and seeds of nearly all tropical trees are dispersed by animals attracted to the energy-rich parts of the fruit provided for them, while many temperate trees use the wind as a pollinator and seed-dispersing agent.

A very interesting but as yet little explored area, because of the recency of theoretical developments, is the application of insights about the evolutionary defenses plants have developed against herbivores to energy models of consumption of plant tissues by herbivores. As yet, it is impossible to predict overall patterns of consumption from existing theories about the chemical defenses of plants, partly because many of the theories are untested and partly because many other factors besides plant defenses influence the actual abundances and dietary preferences of herbivores in nature. It is known, however, that monocultures of genetically similar plants are especially susceptible to herbivores; consequently, some aspects of energy flow in communities may soon yield to predictive analyses, and it may be possible to apply some of these results to agricultural systems.

Energy flow through carnivores seems to be highly similar in all ecosystems so far studied (that is, they get about the same fraction of total energy in all ecosystems). This is probably because the quality of food in animal tissues is less varied than is the food in plant tissues. With carnivores, it is the quantity of food that is important, rather than the quality, as with herbivores. (Herbivores can die of starvation with full stomachs.) Thus, carnivores cannot survive unless their prey are more abundant than some lower population threshold. Such thresholds prevent predators from completely exterminating their prey.

## Nutrient Cycles

Unlike energy, nutrients can be used many times within a system. One way to measure nutrient cycling through an ecosystem is to build a weir in a stream that emerges from a small drainage basin and measure the materials that pass out of the system. The best drainage basins for such experiments are those underlaid by impervious rocks, so that all materials that leave the system pass through the weir and are not lost through seepage into soil and rocks below. Materials contributed to the system by wind, precipitation, and animal migrations can also be monitored.

With these movements of nutrients under careful scrutiny, the system's response to natural variations and to experimental manipulations, such as fertilization or clear-cutting, can be measured accurately. Analysis can also reveal the amounts of various minerals stored within components of the systems and how those quantities vary over seasons and in response to perturbations.

The few such studies carried out to date indicate that chemicals in precipitation contribute significantly to the nutrient budgets of natural ecosystems. In undisturbed forested ecosystems, precipitation generally contributes more inorganic nitrogen and phosphorus than is lost in stream water. Calcium, magnesium, sodium, and potassium, however, register net losses that must be made up by weathering from the underlying rocks. The availability of these elements depends upon the nature of the geological base, or substrate, and the number of millennia over which it has been weathering. In general, young soils—such as those characteristic of much of the temperate latitudes—are better sources of new elements than are the deeply weathered soils of the tropics.[4]

Disturbances to ecosystems, such as clear-cutting and burning, unleash massive losses of some nutrients from watersheds, most notably calcium and nitrogen. The time it takes to recoup these losses from precipitation and weathering varies widely, depending on the chemical composition of the underlying rock, the amount of weathering that has already occurred, temperature, and moisture. When key nutrients are removed faster than they can be restored (because of harvesting, for example), growth rates of vegetation will decline, unless artificial processes can replenish these nutrients.

These nutrient cycling studies have also revealed that temperate and tropical ecosystems differ substantially in the relative amounts of nutrients held in different compartments within the system. For example, in most temperate forests well over half the total nitrogen in the system is found in partly decomposed plant and animal remains on the forest floor; in tropical forests, almost all the nitrogen is located in the living plants themselves. Cycling of nitrogen and other elements in tropical forests is very rapid because the roots of plants actively penetrate freshly fallen leaves, branches, and fruits on the forest floor and extract nutrients before those nutrients enter the soil and are leached into ground water by heavy rains. Temperate soils are more likely to contain reserves of nutrients that can support agriculture for many years.

Sustaining high agricultural production is difficult on tropical soils because cutting and burning, the usual prelude to cultivation, releases nutrients and plowing the soil destroys the intricate structure of fine roots that normally recapture them. Accordingly, large losses of nutrients occur quickly, and usually crops can be grown for only a few years before yields become very low.

Many important relationships among the components of ecosystems remain to be elucidated. For example, the role and regulation of nitrogen fixation and denitrification in natural ecosystems are still poorly understood. At present, data are scattered, but a thorough understanding of these processes is vital to better management of disturbed ecosystems. Given that more and more ecosystems are being disturbed—by logging, agriculture, mining, and other processes—knowing how to disturb them in ways that will minimize losses of valuable materials, as well as knowing how to guide the recovery of damaged systems, is vital to human welfare.

## AQUATIC ECOLOGY

For a number of reasons, aquatic ecologists have pioneered the study of whole systems and how the system's interacting parts determine its properties. This is in part because aquatic ecosystems are often simpler and more clearly delimited than terrestrial ones, and in part because it is not possible with aquatic ecosystems to observe directly many of the behavioral interactions among organisms that tend to absorb the attention of terrestrial ecologists.

Much of the most interesting work going on in aquatic ecology today concerns the structure and functioning of lake and riverine communities and examines the evolutionary aspects of interactions among the species that inhabit them. These two types of communities are very different because in flowing waters most resources pass by relatively stationary organisms and are available to them only briefly.

In the past, considerable progress was made comparing lakes that differ strikingly in certain respects, for example, fresh water versus salt water, shallow versus deep, large versus small, fast versus slow flowing, and polar versus tropical. These comparative studies revealed a number of important patterns. For instance, the abundance of algae, and therefore of total chlorophyll, can be predicted by knowing the total amount of phosphorus. In many cases, this knowledge can indicate the total amount of photosynthesis, which is useful because phosphorus is much easier to measure than photosynthesis.

Merely measuring conditions in natural bodies of water, however, no matter how extensively and carefully these conditions are selected, does not suffice to understand the factors that control productivity and community structure. One must manipulate or alter the system in some desired manner and then measure the system's responses. Ecosystems are complex, and the measurements required to untangle complex causal relationships are usually correspondingly complex. Hence, such studies require many persons with diverse skills and interests, such as aquatic chemists, experts on different groups of plants and animals, and modelers, working together toward a common goal. Furthermore, the response times of complex systems are often long, requiring that experiments be conducted and monitored for periods sometimes as long as decades. The recovery of Lake Washington in Seattle, following the diversion of municipal sewage away from the lake, has been going on for 20 years but is still not finished.

Sometimes relatively simple manipulations of ecosystems can be telling. For example, the important controlling roles of higher level predators have been demonstrated by removing starfish from the rocky intertidal zone of the Pacific coast of North America. Starfish prey preferentially on the dominant competitor for space, the California mussel, making it possible for many other species to invade and survive where they would otherwise be eliminated from the system through competition with mussels. When this starfish is removed, mussels thrive and these competing species disappear or become very rare. Similarly, when alewives were introduced into New England lakes, the dominant species of zooplankton were replaced by other species of different sizes, shapes, and behavior because alewives selectively ate the larger zooplankton.

Other useful insights have come from experiments designed to help restore lakes degraded by pollution. The manipulations that have been used in these cases include reducing the input of sewage; adding phosphorus, which causes immediate rises in photosynthetic productivity; and reducing the biomass of rooted plants by mechanical harvesting, by adding herbivorous fish, or by using herbicides.

One exciting aspect of these studies is the very interesting linkages they reveal between biochemistry and ecology. Lake experiments have uncovered chemical interactions that standard laboratory experiments would never have revealed. For example, aquatic ecologists studying the nutrient requirements of planktonic freshwater algae are finding that different algal species need very different amounts of various nutrients and trace metals. Within a single drop of water, one species will be limited in its growth because there is not enough silicate, another because there is not enough phosphate. In addition, many species produce chemicals that enhance or inhibit the growth of other species. By taking into account the effects of nutrients, the differential effects of light and temperature, and spatial and seasonal patchiness, important insights concerning the shapes and structure of algae are being attained.

Another interesting chemical interaction has to do with the spines and helmets that animals such as water fleas and some rotifers develop in response to certain physical conditions. For many years, the function of the spines was not understood, but studies have now shown that rotifers develop these spines in response to a substance liberated by their predators.

It is insights such as these that are vital to understanding why natural communities have the number of species they do, why the species possess certain properties, and why interactions among them produce the patterns of flow of energy and materials characteristic of these systems.

Sometimes it takes ecosystems so long to respond to perturbations that even long-term experiments will yield only preliminary results. To extend time horizons, paleoecologists are studying lake sediments, which contain a detailed record of past aquatic communities and an indication of the surrounding terrestrial community, since pollen and seeds are blown into the water. Recently, important advances have been made in recognizing remains of organisms and in chemically analyzing sediments. Specific investigations of some organic molecules, such as carotenoids, many of which are highly specific to certain groups of algae species, are developing rapidly, enabling better reconstruction of the details of past communities. Such insights are valuable because the present is but one moment in time, tied to the past by a web of connections. Knowledge of past communities can reveal how they responded to disturbances and, therefore, how present communities are likely to respond to similar disturbances in the future.

## COMPETITION, PREDATION, AND DISTURBANCE

From 1935 to 1970, population biologists dedicated much of their efforts to understanding competition. They demonstrated that species living together use environmental resources in different ways, and that these species change their use of resources if competitors are absent.

During the past decade, more attention has been given to predation. The first studies were carried out to test ideas that predators may influence such important features of ecological communities as the number of species living together, the distribution of species, and how commonly certain communities occur.

Disturbance of communities, whether generated internally by biological forces (particularly predation) or externally by some physically imposed stress, is now perceived to be as important a phenomenon as competition itself. With disturbance, however, the scales of time and space are often expanded.

Enormous strides have recently been made in integrating predation-disturbance viewpoints with those based on competition. These three areas of current research have broad applications (as in pest control and forestry practices), have been studied using innovative mathematical techniques, and are susceptible to experimental testing.

## Environmental Patchiness

The organic world is heterogeneous, yet most older ecological theory dealt with closed systems that were internally homogeneous and assumed to be near equilibrium. Many newer approaches have discarded these restrictions. Disturbances such as earthquake-generated landslides, volcanic eruptions, forest fires, the activities of badgers on short-grass prairie, or wave action along intertidal rocky shores all produce patches of environmental change. These patches are generated at different rates, seasons, and scales, but in all cases they open up the affected communities for invasion by new species. Thus, these communities are not at an equilibrium but move between different states as they are disturbed and recover from disturbances.

Recognizing this fact and its biological implications, the U.S. Forest Service has changed its policy of actively suppressing all fires. Indeed, planned disturbance is being used to maintain particular habitats for endangered species. For example, Kirtland's warblers breed only in jack pine woodland in Michigan where the trees have reached a height of 5 to 20 feet following a burn. Without a careful program of burning, the species would probably become extinct.

Conceptualizing the world as a mosaic of changing patterns emphasizes the importance of learning more about how species disperse and about their capacity to invade and successfully compete for resources in disturbed sites. Such information is basic to testing models of dynamic biogeography and will help to establish a framework with which to understand the extraordinary variability of species that are early invaders of disturbed sites, be they tropical forest trees, insects, or marine algae. There is now real promise that an array of adaptations for coping with a capricious world, including an organism's dispersal, selection of a habitat, use of altered environments, and reproductive strategies, can be compared across a wide variety of habitats.

Further study of dynamic biogeography promises to provide new insights into several pressing and important issues, including the question of why some natural communities are fragile whereas others are robust. As noted earlier, damage to a crop by herbivores may be greater if the crop is grown as a single-species stand rather than interspersed with other species, apparently because mixtures make it harder for species-specific herbivores to find the appropriate plants. For example, insect damage to collards is less when collard plants are grown among plants of other species than when they are grown in pure stands.

Dynamic biogeography also has important implications for planning the optimum size, shape, and dispersion of parks and nature reserves.[5] The number of species in an area depends upon a balance between the rate of immigration of new species and the rate at which species become locally extinct. Theory is not yet well enough developed, however, to enable us to say whether a single large reserve, which should have the lowest extinction rate, is better than a series of smaller ones, where extinction rates are higher but opportunities for exchanges of species between units are better. Nonetheless, these ideas may help in the design of management techniques to maintain the largest possible number of species in the reserves and parks we have set aside.

## Temporal Changes in Ecosystems

The term harlequin environment has been coined to describe a landscape of patches in which two or more species are competing for resources. For example, the spaces created by logs bumping against rocky shores are invaded by algae and a variety of different sessile invertebrates. These species differ in their abilities to disperse into an area and in the seasons when they disperse, and once they reach an area they differ in their rates of growth.

Early theoretical studies suggested that species could coexist in such systems even if they used the same resources, such as space and food. Later research added both complexity and verisimilitude to this view. It appears that the times at which competitors arrive at a gap or patch may influence the ultimate composition of the community, because once an individual species has prospered in an area it is harder to displace it. Experimental work in forests and marine communities has shown that invasion, even by a competitively superior species, can be severely altered or restricted by species already there. For instance, on pilings along the Atlantic coast of the United States, the presence of the bryozoan moss animal *Schizoporella* inhibits the establishment of all or most other species, yet it in turn essentially cannot invade space already monopolized by another species.

When prior residence is important in determining what other species may invade, the natural pattern of a community's change over time can become stalled, and alternative states of a community may be stable for long periods. Thus, succession of organisms over time after a disturbance may not lead to a single final stage that is independent of the nature of the disturbance and the species that first colonized the site. Rather, a variety of communities may persist simultaneously until the next disturbance occurs, and we cannot understand the pattern of these communities without knowing their histories.

## Dynamic Stability

A developing alliance of theoretical and applied ecology is making possible fundamental advances in understanding complex interactions involving many species, such as pollination systems, host-parasite interactions, and symbioses. In each of these systems, ecologists are becoming better able to understand how natural selection acts on traits of the participants and how adaptations of interacting species—flowers and pollinators, for instance—constrain their mutual evolution. Knowledge of the specificity of a host to a parasite has permitted pests to be controlled with little impact on other species in a system. For example, St.-John's-wort, a serious pasture weed in California, was reduced to very low levels by introducing a beetle that eats that plant and no other. Such techniques will become more important as the use of broad-spectrum pesticides is reduced.

The sum of the interactions among species living together imparts a dynamic order to natural communities. Recent experimental studies mirroring human intervention—in which one species, for example, is removed or added—show that such a change often disrupts the distribution and abundance of some associated species. For example, when sea otters were reintroduced into places along the Pacific coast where they had been exterminated by hunting, they caused sharp reductions in populations of one of their preferred prey—sea urchins. Sea urchins are, in turn, the most important grazers on large kelp; without predation by sea otters, sea urchins become common enough to keep kelp at very low population densities. In the presence of sea otters, kelp increase in abundance and overpower many other types of algae and sessile invertebrates that are common when kelp are rare. One would not have suspected that a single predator could have such an important effect on the populations of so many species that it does not actually eat!

Studies of the whale-krill-phytoplankton community of the southern oceans highlight the complexities involved in attempting to manage a single species so as to produce a maximum sustainable yield.[6] The great whales feed predominantly on krill, and overexploitation of the former has generated a 150-million-ton increase of the latter in the southern oceans. This in turn appears to have contributed to increased populations of seals, penguins, and squid, all of which feed on krill. Krill harvesting, which is being seriously contemplated, may decrease stocks of both whales and seals. Thus, when species share a limiting resource, exploiting one that is commercially or aesthetically valuable affects the entire system. Understanding and successfully managing multispecies stocks is at the frontier of a combined theoretical and applied ecology.

All studies of how species interact impinge directly on an issue that pervades most ecological subjects: to what extent are natural assemblages stable enough to resist, or recover from, change? The multiplicity of species, the nature and strength of their interactions, how they eat and are eaten by one another, and how they invade and succeed in different kinds of sites are all believed to relate to this fundamental problem. The ques-

Rocky intertidal zone on Tatoosh Island, Washington. Predation by the starfish is responsible for the sharp lower boundary to the bed of mussels and goose-necked barnacles. [R. T. Paine.]

tion is central because the ability to effectively predict the behavior of disturbed communities is important for many environmental problems. Questions of stability will dominate, either directly or indirectly, most studies of ecological communities in the decade to come.

## PLANT POPULATION BIOLOGY

In the last ten years, plant population biology has emerged as an important field that attempts to integrate concepts and insights of ecology, genetics, and physiology. All sources of food, as well as fossil fuels, ultimately derive from plants. However, our understanding of plant behavior is still incomplete, especially in the areas of how plants affect the environment, how pollutants affect plants, and how the richness and forms of plant species affect other organisms and the dynamic stability of ecosystems. Given the importance of plants, it is essential that we thoroughly understand the way they behave.

The goal of plant population biology is predictive, testable theory to explain the distribution and the functioning of plant populations.[7] For this, ecological, physiological, and genetic data and theory must be integrated and reconciled. The stage is set for a rapid development of concepts and ideas, and the next ten years should see very important progress toward understanding why different species occur in different environments and why some species are more abundant than others.

## Plants and the Environment

Botanists have tended to believe that plant species and populations vary from place to place because of physical factors, especially moisture. Indeed, there are well-known correlations between the physical environment and the form, abundance, and diversity of plants. For example, chaparral-type vegetation occurs in all areas of the world with a Mediterranean climate. However, direct information on how environmental factors, such as wind, cold, and drought, affect plant mortality has been lacking.

Detailed demographic studies and reliable, affordable technology to measure microenvironments now make it possible to field-test the precise responses of plants to their physical environments and to quantify the causes of mortality. When populations are studied in such detail, it becomes apparent that biological interactions, especially competition and predation, play an important role in the structure and functioning of plant communities. The success of many programs of biological control of weeds using insects or microorganisms illustrates the point.

## Plant Physiological Ecology

Conceptual and technical developments are making it possible, for the first time, to formulate realistic models of plant functions, especially photosynthesis and water movement. Using solid-state technology, and microelectronics in particular, plant biologists can now directly measure gas exchange rates in photosynthesis, respiration, and transpiration of intact leaves in the laboratory and the field under varying conditions of temperature, light, and humidity. These studies are demonstrating, as one might expect, a close relationship between environmental parameters and plant performance. Plants that are indigenous to high-light and high-temperature environments, such as many summer-active desert shrubs and herbs, have their highest rates of photosynthesis at high levels of temperature and light; those that grow at low temperatures have their photosynthetic optima at low temperatures.

Most species investigated are able to adjust their behavior, within certain limits, to fit changing circumstances. For example, plants can change the temperature at which they photosynthesize best if they are grown for a while at temperatures either higher or lower than the ones to which they were previously adjusted. Little is known regarding the genetic control of such physiological variations. However, now that the inheritance patterns of single enzymes and their biochemical and physiological roles in photosynthesis can be analyzed simultaneously in natural populations, investigators confront a great and exciting opportunity to analyze the genetic and biochemical machinery of adaptive mechanisms in plants.

## Adaptive Value of Plant Structures

Although in evolutionary terms form is just as important as function, we do not understand it as well. Nevertheless, theoretical models have been developed that examine the profits and costs (in terms of calories they produce by photosynthesis or calories they expend in construction and maintenance) of plant structures and reproductive patterns. For example, models have been proposed that predict reasonably well how the shape of leaves and the structure of canopies vary in different environments. It is now understood why it is better for a plant to have its leaves concentrated in a dense upper layer if it grows in the shade, but to have them spread over a number of more open layers if it grows in the full sunlight. Other models measure the significance of rewards that plants offer, in the forms of nectar and pollen, to pollinating organisms.

Although this approach is still in its infancy, it has already shown considerable predictive capabilities, and it should become an important element of population biology theory. Particularly promising is the possibility of predicting the best properties for new plant varieties in agriculture. It was once thought that the larger the individual plant, the greater its yield. But as the use of dwarf varieties for high-yielding grains and fruits demonstrates, growth entails a series of trade-offs; vegetative growth occurs at the expense of reproduction, and larger plants have less favorable ratios of photosynthesis to support tissues. Advances along similar lines can be expected in studies of how plants allocate energy reserves to the production of fruits and seeds.

## Genetic Analysis of Plant Populations

The theory of evolution by natural selection is based on the implicit assumption that there is a relationship between an organism's outward characteristics (its phenotype) and its genetic material (its genotype). For 100 years after Darwin enunciated the idea, our knowledge of the mechanics of heredity was so incomplete that the assumption could not be tested. Once DNA was shown to be the genetic material, it became possible to address the question. But only in the last five years, with the development of relatively rapid and reliable methods for sequencing genes, has it become possible to envision testable hypotheses of the way genes are affected by natural selection.

Some of the recent discoveries in this field have been startling and unexpected, and they pose questions with serious theoretical and practical implications. One of the most important findings is that many genes are found in hundreds or even thousands of repeated copies. In these cases, the general model that selection proceeds by selecting between alternative forms of genes cannot be correct, because it requires all those copies to mutate simultaneously. How selection operates in these cases has profound implications for both natural and artificial selection.

The technique of isozyme analysis, which analyzes those sequences of DNA that are actually translated into proteins, has revealed a great deal of variation within and between plant populations in the products of genes. Indeed, the variation is so great that it cannot be explained by current models of population genetics. This has given rise to alternative explanations, the most popular of which is that a great deal of genetic variation is neutral; that is, it is not affected by natural selection and accumulates by random mutation. This still needs to be proven experimentally.

If neutral genes really exist, evolution may be more complex than heretofore assumed, and adaptations to environments may be less precise than they otherwise would be. It is too soon to tell, however, how much of genetic variation is in fact neutral.

## SOCIAL ORGANIZATION

Animal social systems have been vigorously studied since 1920, when the British amateur ornithologist Elliott Howard drew attention to the existence of territoriality in many species of birds. For many years, most observers simply documented the basic social patterns in a large number of species and attempted to interpret the adaptive significance of those patterns. The last decade, however, has seen a fundamental shift in perspective.

Social systems currently tend to be viewed as associations of individuals with unique but partly overlapping interests. Due to differences in age, sex, experience, and size, each individual in a social group is affected differently by the actions of other members of the group and gains benefits in different ways. Among the benefits that a social existence confers are protection from predation (sometimes), communal defense and exploitation of resources, and better success in competition for mates (females, for instance, may ignore a solitary male in favor of competing males from which they can choose). Because the value of these options is highly specific to situations and species, the social systems found among living organisms exhibit great diversity. It has also become apparent that there are inevitable disadvantages to associating in groups: the necessity of sharing food, conditions that favor the spread of disease, greater susceptibility to predation, and competition for mates. That most animals are solitary much of their lives suggests that the costs of being social often exceed the benefits.

Because each individual is unique, ecologists need extensive data on groups in which all individuals can be identified, and preferably in which genetic interrelationships are known, to understand the real dynamics of social systems.[8] As a result, most of the important insights and results have sprung from intensive long-term studies of specific social groups, such as those that have been made of Florida scrub jays, Arabian babblers, and several species of monkeys. Without a thorough background knowledge of a group, the interactions of its members are very difficult to interpret.

## Choice of Associates

All individuals who live in groups choose associates, and the most fundamental choice is that of a mating partner. The relationship between parents is unique, because usually parents are the only individuals in a family group who are unrelated.

The choice of a mate commits the chooser's genetic investments and parental efforts. A very important insight is that members of the sex that invests the greater biological energy in its sex cells, usually the female, have the most to gain from good choices of mating partners and the most to lose from poor choices. Not surprisingly, females in most species exercise greater discrimination in choosing mating partners than do males.

Choosing a mate is a complex process. The parental contributions made by a spouse differ greatly among species, and decisions must often be made quickly using incomplete information about the spouse's qualities. At the one extreme, males may provide only a set of genes to their offspring; all subsequent parental care is provided by the female. At the other extreme, as with the small sandpipers known as phalaropes, the females desert their mates soon after laying their eggs, and the offspring are cared for by their fathers.

Which partner deserts first, and when, is one of the central issues of social organization. Basically, an individual is expected to desert when its overall reproductive success is higher if it deserts and leaves its offspring in the care of its spouse than if it remains with its current spouse and assists in raising those offspring. This is difficult to measure in the field. Many factors affect this decision, including the ability and willingness of the deserted spouse to continue caring for offspring, the probability that additional mating partners can be found during the current breeding season, and the likelihood of success with the new spouse. The prevalence of monogamy in many vertebrates (especially birds, in which about 90 percent of all species are monogamous over a single breeding season) is apparently linked to the great difficulty that deserted partners have in raising offspring unassisted, combined with the limited opportunities that the deserter has to find new mates.

## Altruism in Animal Social Systems

To what extent do individuals behave in ways that lessen their own chances to survive and reproduce but that boost the survival and reproductive success of their associates? It is generally thought that such behavior more readily evolves in groups of closely related individuals, where the recipients of aid share many genes in common with the "altruist." In this way, individuals may enhance the frequency of their genes not only by producing their own offspring but by helping their relatives to survive and reproduce. This realization, developed mathematically in the 1960's, has been the most important advance in our understanding of natural selection since the theory was given its rigorous formulations in the 1930's.

An area of considerable current interest that relates to the concept of altruism between relatives, or kin selection, is the role of individuals other than parents in caring for offspring. One of the most striking findings in recent years has been the discovery that more than two individuals care for offspring in a much larger number of species than previously suspected. In nearly every case for which detailed information is available—in scrub jays, Mexican jays, and Arabian babblers, for example—these helpers remain with their social groups and help rear later siblings or half-siblings. A key question is whether they are primarily helping or whether they are gaining experience which may enhance their own future reproductive success. In some intensely studied species, parents with helpers evidently do rear more offspring, but in other cases they do not. Evidently the benefits and risks from helping and accepting help are not the same in all species, though it is not understood why this is so.

In colonies of social insects, individuals have complex and variable genetic relationships that have helped to mold the evolution of the species. For instance, ants, bees, and wasps have a sex-determination system in which males are haploid (having half the full complement of chromosomes) and females are diploid (having a full set of chromosomes). As a result, if the queen mates only once, females (all of which are sterile workers except for the queen) are more closely related to their female siblings than they would be to their own offspring if they could bear them (sharing three quarters of their genes in the former case and half in the latter). Females are even less closely related to drones (sharing just a quarter of their genes). However, if the queen is fertilized by more than one male, as occurs in many species, sterile workers are not necessarily more closely related to their female siblings than they would be to their own offspring. These genetic relationships play an important part in determining the relative proportions of males and females in insect colonies and the interactions between members of different castes.

Despite the success of kin selection in accounting for features of some social systems, it has proven difficult to determine the full extent to which this force has influenced the evolution of social systems, particularly among vertebrates. Evidence is accumulating from a few bird species that patterns of social organization do not differ significantly between groups in which the individuals are closely related and groups in which relationships are much more distant. The relative importance of the genetic component in shaping the interactions among members of vertebrate social groups is generally not known.

## What Do Group Members Say to One Another?

Social systems are maintained by frequent behavioral interactions among their members. Communication can occur via several senses (vision, sound, smell, touch) and can carry very different messages (information about predators, food, the intentions of the signaler). The predominant theory (though the theory is not without its dissenters) holds that for social signals to evolve they must influence the behavior of the receivers of those signals in ways that enhance the fitness of the senders. At the same time, receivers of signals are expected to respond to them in ways that enhance their own fitness. Thus, if a bird hears an associate sound a hawk alarm, it should try to seek cover in the fastest and safest way possible. These requirements on interactions between signaler and receiver impose substantial constraints on the kinds of signals that can evolve.

They also provide a way of looking at the evolution of honesty and deceit in communication. Potentially, an individual could benefit by sending false signals concerning its desirability as a mating partner, its ability to defend space, and so on. However, receivers should evolve to be good detectors of deceitful signaling.

A satisfactory theory of deceit is still to be developed, but some powerful ideas and data are now accumulating. For example, wintering Harris sparrows have a highly variable amount of black on the feathers of their faces and throats. The quantity of black is strongly correlated with the dominance and actual fighting abilities of the individuals in winter flocks. When the feathers of dominant individuals are bleached, making these birds appear subordinate, they are challenged by individuals that would otherwise not do so. Nevertheless, the bleached birds successfully subdue the attackers, and in fact may become despots as a result of these frequent challenges.

## EVOLUTION OF LIFE HISTORIES

All organisms pass through a series of physiological and behavioral stages over the courses of their lifetimes. Sometimes, these stages of life are reflected in profound changes in the form of an organism: for instance, a butterfly begins life as an egg, hatches into a larva, and goes through a pupal stage before emerging in its adult form. Other creatures undergo less drastic physical changes.[9]

The study of life histories takes as its subject these general patterns in the lives of individual organisms. It is a powerful tool with which to understand the compromises that organisms have evolved to deal with the many physical and biological factors that affect them during their lives.

Most of the important concepts in the study of life histories can be framed in terms of trade-offs. For example, Pacific salmon of the genus *Oncorhynchus* spend two to three years of rapid growth at sea and then travel up a river or estuary to spawn, after which they die. Trout, of the genus *Salmo*, are closely related to salmon and also come in from the sea to spawn, in many of the same streams. But trout do not die after a single spawning; they return to the sea and breed again the next year. Assuming that these two patterns were produced by natural selection operating upon the reproductive advantages to an individual organism, why should any fish die after one spawning? Why not return, as trout do, to spawn many times?

In this case, the trade-off relates to the fact that a fish must reduce the number of eggs it produces during a single spawning if it is to have enough energy to return to the sea and survive until the next spawning season. Trout produce fewer eggs each year than they would if they invested all their energy in reproduction and died soon after spawning. With salmon, this trade-off is made differently—they put all of their energy into reproduction and in this way increase the chances that more of their offspring will survive.

The concept of fitness—an organism's genetic contribution to future generations—is central to analyses of such problems. Almost all current models use as a criterion for fitness the reproductive success of an individual, that is, the number of its offspring that survive to reproduce. But reproductive success is in turn determined by several different factors.

The general problem was first clearly posed by Darwin in 1871:

> Thus the fertility of each species will tend to increase, from the more fertile pairs producing a larger number of offspring, and these from their

> mere number will have the best chance of surviving, and will transmit their tendency to greater fertility. The only check to a continued augmentation of fertility in each organism seems to be either the expenditure of more power and the greater risks run by the parents that produce a more numerous progeny, or the contingency of very numerous eggs and young being produced of smaller size, or less vigorous, or subsequently not so well nurtured. To strike a balance in any case between the disadvantages which follow from the production of a numerous progeny, and the advantages (such as the escape of at least some individuals from various dangers) is quite beyond our power of judgment.[10]

The balance spoken of by Darwin is the basis of current life history theory. For example, there is a trade-off between a parent's reproduction and its survival—helping offspring to survive reduces the chance that the parent will survive to reproduce again. A simplified representation of this trade-off is shown in Figure 1. The exact form of the trade-off curve depends on the species and its particular environment. To determine this form, one must know the extent to which producing more eggs or providing more care for offspring reduces the parent's chances of reproducing in the future.

An interesting feature of this trade-off is that parents and offspring have different genetic interests in how much effort parents invest in current reproduction. This is because the offspring is 100 percent related to itself but shares only 50 percent of its genes with any future offspring (assuming that the parents stay together). Each of the parents, on the other hand, is only 50 percent related to all of the offspring it produces. Parents therefore try to reduce their current investment in reproduction in favor of future reproduction, while offspring try to make them do the opposite (see Figure 1). The common parent-offspring conflict revolves around attempts by the offspring to gain more resources from its parents than the parents are selected to give.

## Life Histories and Changing Ecological Conditions

The classic examples of broad correlations between life histories and ecological conditions are the patterns of larval development in bottom-dwelling marine invertebrates. At high latitudes, most of these species develop directly from large eggs into larvae that eat the same food as do the adults and are miniature adults in their form; at low latitudes, most species have very small eggs, and the larvae spend some time among plankton, feeding in ways very different from those of adults.

The identification of such patterns is the first stage in analyzing the causal mechanisms that relate life histories to environmental factors. Much more detailed knowledge of these patterns is needed, and alterna-

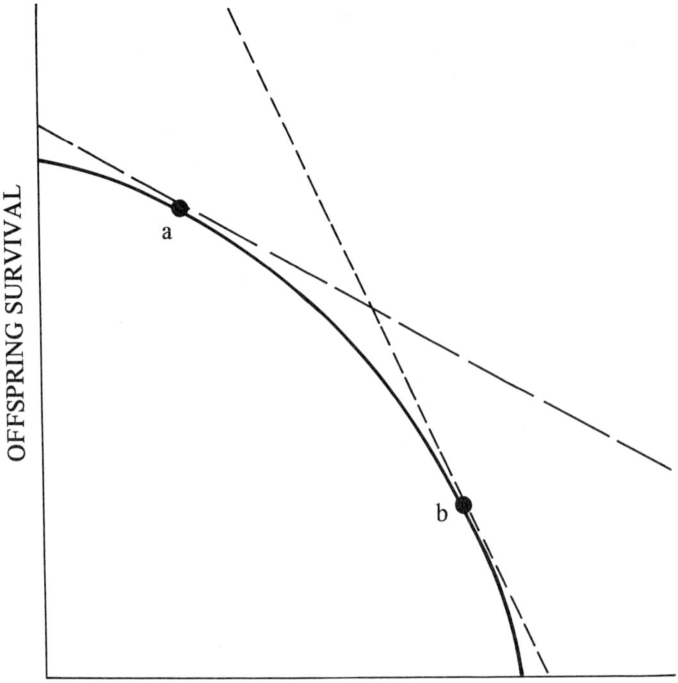


**Figure 1 Life history trade-offs.** In many species, there is a trade-off between a parent's own chances of survival and its efforts to help its offspring survive. In salmon, the parents invest all of their energy in reproducing and die shortly after the female gives birth. In trout, the parents survive to return to the sea and spawn again. Using standard population genetics, it is straightforward to show that parents and offspring have different genetic interest in how this trade-off is made.[11] From the viewpoint of the parent, the optimum trade-off is the point on the curve with slope $-2$ (point b). From the viewpoint of the offspring, the optimum point is that of slope $-\frac{1}{2}$ (point a). Between a and b, parent and offspring conflict. Genes in the offspring will influence the trade-off toward point a; those in the parent will move it toward point b. [SOURCE: B. Charlesworth. *Evolution in Age-Structured Populations.* New York: Cambridge University Press, 1980.]

tive hypotheses of their causes must be posed and tested. The fastest progress is likely to be made by choosing situations in nature where a single environmental variable affects a major feature of an individual's life history (for example, age of first reproduction). One interesting study is of a species of coral reef fish in which individuals are functional females first and then change into functional males. The time of sex reversal turns out to be related to the size of the local reef and the number of males present on the reef.

## Current Problems

A number of important questions in the theory of life histories have to do with compromises that govern certain stages of an organism's life. Among these are the following.

- How do reproduction and death rates change with age of the individual? The age at which an individual first reproduces, and how often it does so thereafter, greatly influences the representation of that individual's genes in future generations. For example, long prereproductive periods reduce the total number of offspring produced much more than do decreases in the size of litters.
- Why does development take the form it does? For instance, why do some marine organisms, like mussels and barnacles, have tiny larvae which they broadcast into the open water where they feed for a time before settling, while others, like some snails and starfish, lay large eggs that develop directly into bottom-dwelling juveniles and adults?
- In which species do parents care for their offspring and which parent provides it? In most species of birds, both parents feed the offspring; in most mammals, parental care is provided entirely by the female. In some species of fish, the males guard the eggs, while in others only the females do so. Some of these patterns are easy to understand. The role of male mammals in feeding young is limited by their inability to produce milk; it is largely among carnivores that males make substantial contributions to the feeding of young. In other cases, reasons for the differences are much less obvious. In certain birds, for example, only females feed the young, while males devote their time to defending the territory or attracting additional mates; yet those males have no physiological traits that prevent them from feeding the young.
- Why do higher organisms reproduce sexually rather than asexually? Until recently, it was assumed that the value of sexual reproduction was that it generated variability among offspring, which enhanced the population's ability to survive future environmental change. Whereas sexual reproduction undoubtedly does generate such variability, the problem is to discover an immediate adaptive value of variable offspring to their parents. Current research deals with how variability may enhance survival in patchy environments where rapid changes are occurring in parasite and disease organisms that may preferentially kill offspring that are very similar to their parents.
- How is reproductive effort allocated between male and female

offspring? This relates to the evolution of sex ratios and, for hermaphroditic plants, the resources allocated to pollen versus seeds.

- What is the appropriate measure of fitness? In order to fully understand patterns in life histories, investigators must be able to combine life history traits into a measure of fitness that tells how natural selection has affected a particular organism. As noted above, almost all life history models now use the individual's reproductive success as a criterion of fitness, as opposed to some benefit to the population, such as its size or future chances of survival. However, there are exceptions. For example, current sex ratio theory about the role of local competition for mates involves a form of selection on populations of individuals. Other cases can be expected to arise as more precise models are tested in the laboratory and in the field.

One way to analyze life history traits is to experimentally manipulate the situations faced by individuals and measure their responses. One example, drawn from the study of sex ratios, used the wasp *Nasonia,* which has haploid males and diploid females and parasitizes blowflies. When a single wasp invades a blowfly, all of its offspring mate among themselves (that is, brothers compete with one another to mate with their sisters). In such a population, because all mating is between brothers and sisters, a female gains by producing few sons, since one male can mate with many females. However, as the number of females colonizing an area increases and their male offspring compete for access to many groups of females, theory predicts that the proportion of males should also increase.[12]

Through a clever use of genetic markers, the offspring from the first female to lay eggs on a blowfly host were distinguished from the offspring produced by the second female. As predicted by theory, the first female wasp produced many more daughters than sons, while the second female produced relatively more sons.[13] In addition, the second female adjusted the proportion of her sons according to the number of competing males already present, as estimated by the relative number of eggs laid by the two females. In experiments involving a large number of female wasps, all producing offspring that compete for mates, the relative numbers of males and females approached equality, a result also predicted by theory.

Another way to analyze life histories is to look for correlations between life history traits and environmental variables.[14,15] This process is somewhat simplified by the fact that life history traits tend to come in groups; for example, long lives tend to be associated with few offspring per litter while short lives are associated with many offspring per litter.

Cases in life histories where a trade-off is straightforward are important for analysis. The trade-off is extremely direct for sex ratios—more sons means fewer daughters. Similarly, producing larger seeds means producing, for the same investment of energy, fewer seeds. In other cases, however, such as how reproduction reduces growth and raises mortality, or how early reproduction affects total length of life, very little quantitative information about the trade-offs is known.

Since breeding for life history traits is the basis for much agricultural improvement, agricultural literature is a likely source of valuable information about how changes in one trait influence other traits. For the most part, little use has been made of this literature. Life history theory may also be able to contribute to animal and plant breeding and to understanding sex- or age-specific changes in populations subject to extensive human predation.

## Summary and Outlook

The domain of ecology is the relationships between organisms and their environments. Within this wide field are subjects ranging from the properties of complex assemblages of species with their inanimate environments to the behavior of individual organisms. Several concepts are central to pursuing a better understanding of these interactions. The first is the theory of evolution by natural selection, which provides the basis for models that predict organisms' physiological, morphological, and behavioral responses to their environments. Because all organisms are influenced by many different factors, simple adaptations to any one of them seldom occur. Rather, an organism's characteristics represent a compromise among conflicting selective pressures. In addition, selective pressures seldom remain constant for very long. Therefore, a key problem is to understand the trade-offs involved in adaptations to combinations of varying influences.

Determining the evolutionary relationships among organisms and developing a rational scheme for classifying them are basic pursuits upon which nearly all other aspects of biology depend. For a few groups, the process of naming and describing species is nearly complete; but for others, including some of the most important ones, such as insects and nematodes, most living species have not yet even been described.

Studies of whole ecosystems have proceeded farthest for aquatic systems, because of their relative simplicity and clear boundaries. These studies are throwing new light on chemical interactions between species and on factors that regulate total productivity. These insights suggest methods by which lakes can be restored and managed.

For terrestrial systems, some of the best results have come from studies that use a weir installed at the mouth of a small watershed to monitor the total output of water and mineral elements from an area. These figures can then be compared with the quantities entering the system from elsewhere and arising through local weathering. Such studies have highlighted the importance of nutrient cycling mechanisms in intact ecosystems and the ease with which they can be disrupted by various types of disturbance. To use the services and products of ecosystems on a sustained basis will require better understanding of these cycling mechanisms.

Population dynamics, a field initially supported by the need of insurance companies for more accurate mortality tables, can now provide valuable contributions to the management of exploited animal and plant populations. These ideas are also central to current efforts to understand better the interactions among competitors and between predators and their prey. Future studies will attempt to extend ecologists' predictive powers to complex systems with multiple predators and multiple prey. Some of today's most difficult management problems involve these complex systems.

During the coming decade, ecological research will undoubtedly continue to investigate adaptation. Recent advances in the theories of sex ratios, allocation of energy, reproductive rates, and patterns of parental investment have developed to the point where we can understand how exploited animal and plant populations change as a result of human harvesting practices. However, adaptations of organisms must be better understood in the context of their total life histories, because selective pressures vary over the stages of an organism's life and organisms seldom achieve ideal solutions to the problems of each stage.

In addition to their intrinsic interest, studies of adaptation are central to predicting the properties of organisms in different types of environments. As European biologists traveled around the world during the last century, they discovered ecological communities that appeared very similar to the ones they knew at home. For example, in all five regions of the world that have a Mediterranean climate (the Mediterranean basin, California, central Chile, South Africa, and southwest Australia), with mild wet winters and hot dry summers, they found vegetation dominated by evergreen shrubs with thick leathery leaves. Some of the birds they found in those areas reminded them of familiar European birds, even though they turned out not to be closely related. Ecologists now understand why thick-leaved shrubs are best at using the resources of those areas and why similar vegetation favors similar birds. However, it is not known how much and in what ways living organisms converge in their attributes if they evolve in similar physical environments.

If adaptations can be predicted from a knowledge of the physical environment, other properties of ecological communities can be predicted as well, because many of these properties are the result of interactions among the component organisms. Nonetheless, a large gap exists between adaptations of individuals and properties of ecological communities, such as energy flow, nutrient cycling, and species richness and stability; a major goal of future research will be to bridge this gap. Although the task is a difficult one, there is reason to be optimistic. Energy flow is the result of organisms capturing sunlight through photosynthesis or eating one another. Substantial progress is being made in understanding how organisms forage and what prey they select. What remains to be done is to explore the connections between large numbers of individuals foraging in different ways and the total pattern of energy flow and storage in ecosystems.

Similarly, the number of species living together is in part determined by the range of resources available in a community and in part by how similar two species can be and still coexist in the same environment. Therefore, once we know how species change their use of resources when supplies are reduced by competitors, it will be possible to predict not only the number of species in an area but also something about their use of resources. Obviously, such a task will be easier for organisms such as birds, with relatively few, easily studied species, than it will be for insects, where thousands of species may be present at a single site.

Even highly technical societies depend heavily on resources and services supplied by ecosystems. These systems provide food and fiber; they process wastes; they enrich human lives aesthetically. We can continue to enjoy these benefits only if it is understood how ecosystems work, how they respond to disturbances, and how their capacities to carry out certain functions are limited. Abuse of these systems can only result in a decline in the quality of human life.

## REFERENCES

1. R. A. Fisher. *The Genetical Theory of Natural Selection.* New York: Oxford University Press, 1930.
2. Norman Myers. *Conversion of Tropical Moist Forests.* Washington, D.C.: National Academy of Sciences, 1980.
3. C. F. Jordan. "A World Pattern in Plant Energetics," *American Scientist*, Vol. 59 (1971), pp. 425-433.
4. F. H. Bormann and G. E. Likens. *Pattern and Process in a Forested Ecosystem.* New York: Springer-Verlag, 1979.
5. B. A. Wilcox. "Insular Ecology and Conservation," *Conservation Biology. An Evolutionary Perspective.* Edited by M. E. Soule and B. A. Wilcox. Sunderland, Mass.: Sinauer, 1980.
6. R. M. May et al. "Management of Multispecies Fisheries," *Science,* Vol. 205(1979), pp. 267-277.
7. J. L. Harper. *Population Biology of Plants.* New York: Academic Press, 1977.
8. J. L. Brown. "Avian Communal Breeding Systems," *Annual Review of Ecology and Systematics*, Vol. 9(1978), pp. 123-156.
9. S. Rohwer. "Status Signalling in Harris Sparrows. Some Experiments in Deception," *Behavior*, Vol. 61(1977), pp. 107-129.
10. Charles Darwin. *The Descent of Man.* New York: D. Appleton and Co., 1971.
11. B. Charlesworth. *Evolution in Age-Structured Populations.* New York: Cambridge University Press, 1980.
12. W. D. Hamilton. "Extraordinary Sex Ratios," *Science,* Vol. 156(1967), pp. 477-488.
13. J. H. Werren. "Sex Ratio Adaptations to Local Mate Competition in a Parasitic Wasp," *Science,* Vol. 208(1980a), pp. 1157-1159.
14. T. Clutton-Brock and P. H. Harvey. "Mammals, Resources, and Reproductive Strategies," *Nature,* Vol. 273(1978), pp. 191-195.
15. G. E. Hutchinson. *An Introduction to Population Ecology.* New Haven: Yale University Press, 1978.

# 6

# *Plant Disease*

## INTRODUCTION

Research on plant diseases involves the study of the interaction of two systems—the host and the pathogen. Pathogens are divided into infectious and noninfectious types. Infectious pathogens are living agents, such as fungi, bacteria, nematodes, and viruses; noninfectious pathogens are such things as acid rain, nutrient deficiency, and air pollution. People and insects commonly act as agents in transmitting pathogens to the host plants.

Research in plant pathology, plant genetics, and related plant sciences has made tremendous gains in protecting crops and other plants by revealing the methods that protect plants against pathogens, and by developing plant varieties resistant to disease. Recent research on the physiology of host-pathogen relationships has demonstrated that there are also dynamic mechanisms in plants that provide defenses against attack by pathogenic bacteria and fungi. Factors such as weather can greatly influence the host-pathogen interaction and therefore the severity with which a disease affects a plant population. This makes it difficult to estimate accurately the annual losses in agriculture and forestry that result from plant disease. Nevertheless, it is estimated that plant diseases cause about a 30 percent loss in potential yield of major crops each year. When a primary food crop is seriously damaged, political, economic, and social disruptions may occur. The epidemics of potato blight in the mid-1840's in Ireland, a country that had come to depend very heavily on this food

◄ Leaf rust on wheat. [Office of Information, U.S. Department of Agriculture.]

source, resulted in massive starvation and the emigration of hundreds of thousands of Irish to the United States. Millions starved in the great Bengal famine of 1943, caused, in large part, by an epidemic of brown spot of rice.

Other notable plant disease epidemics include the chestnut blight, which destroyed every important stand of this valuable forest tree within 50 years, and Dutch elm disease, which is currently decimating one of the most popular shade tree species in our country.

Plant diseases have taken and continue to take their toll of American agricultural production. In 1916, red rust of wheat destroyed two million bushels of wheat in the United States and another million bushels in Canada. Stem rust destroyed about 60 percent of the wheat crop in Minnesota in 1935. Northern leaf blight of corn can cause up to 50 percent yield loss in some parts of the Corn Belt. The Department of Agriculture has estimated that in 1976 $4 billion in U.S. crops was lost because of plant disease. Unexpected disease problems can explode at any time, resulting in a serious loss to agricultural productivity.

Devastating epidemics do not normally occur in wild plant communities. Rather, they are caused primarily by the accidental importation from other parts of the world of new pathogens to which the native plants have little, if any, resistance, or they result from the breeding of genetically uniform crops. While uniformity may improve yields and harvestability, it may also make the crops vulnerable to disease epidemics.

The story of southern corn leaf blight illustrates the genetic vulnerability of crops bred for uniformly high yields and convenience of harvest. Corn varieties bred for maternally inherited pollen sterility, which eliminates detasseling by hand, proved extremely susceptible to a new race of fungus. By the time this was discovered, the genetic trait for pollen sterility and susceptibility to the disease had been bred into more than 85 percent of the American corn crop. In 1970, leaf blight broke out in the United States and spread from Florida to Wisconsin in one season. Many cornfields in the southern states were killed entirely, and U.S. losses nationally amounted to 600 million bushels of corn, worth more than $1 billion.

When a crop variety is genetically uniform, disease control is all the more important because any breakdown of resistance can lead to loss of the entire crop. This chapter is confined primarily to the impact, development, and prospects of plant disease research in addressing the problems of control.

## DISEASE RESISTANCE AS A KEY TO PLANT HEALTH

Although quarantines are sometimes effective in preventing the introduction of foreign pests, for diseases already present in this country, growing disease-resistant varieties is easier, less expensive, and often more effective than other methods of control. Even where pesticides are essential for adequate disease management, resistance is important because it makes other control methods more effective. The value of resistance is measured in terms of its durability and its effect on rates of disease increase. The critical questions are: will the resistance reduce the spread of disease to a manageable level, and will the resistance remain effective for the normal useful life span of a crop variety?

The 1972 National Academy of Sciences study on genetic vulnerability emphasizes that our major crops are composed of very few plant varieties. For instance, 77 percent of the total U.S. peanut acreage and 96 percent of the pea acreage consist of only two varieties of the respective plants; 72 percent of the potato acreage consists of just a few varieties. However, recent efforts have produced new varieties of vegetables, and durable resistance to several diseases has been incorporated into single varieties of cucumbers, cabbages, and tomatoes.

Resistance to disease in plants is more common than is susceptibility, but what factors actually control resistance? Plants, unlike animals, do not produce antibodies in response to bacterial, fungal, or viral infections. The underlying biochemical and physiological mechanisms of resistance are controlled by genes, and manipulation of these genes by a variety of techniques has produced some dramatic successes.

Disease-resistant genes are obtained from induced mutations, from wild species, from relatives of domestic crop plants, and from old, discarded varieties. The lack of disease-resistant genes in plant populations that are otherwise suitable for breeding with a specific variety may in some cases be overcome with recombinant DNA techniques. These and other genetic technologies may allow barriers to gene transfer between distantly related plants to be broken, and may even allow the synthesis and introduction of genes for resistance. However, enormous further advances are required before this direct approach becomes practical.

## Durability of Resistance

Plant pathogens are more capable of genetic change and adaptation than are their host plants. There is therefore no guarantee that resistance to a given pathogen will provide permanent protection against it; there is always the possibility that the pathogen will adapt to the resistance and overcome it.

Disease resistance may be either of two basic types: that in which high levels of resistance are simply inherited (governed by one or two genes), and that in which lower levels of resistance are inherited as complex traits based on the combined effects of many genes in the plant. Understandably, early plant breeding efforts concentrated on simply inherited resistance controlled by one or two genes. Such resistance was easier to breed and, initially, more effective than resistance derived from complex inheritance.

The problem with simply inherited resistance is that it is often associated with simply inherited virulence in the pathogenic organisms. That is, resistance controlled by one new gene in the host plant may be overcome by one or more new genes for virulence in the pathogen. In many diseases, there is a gene-for-gene relationship between resistance in the plant and virulence in its pathogens. This type of resistance is referred to as specific resistance because it is effective only against races of the pathogen that lack the corresponding genes for virulence.

The exclusive use of specific resistance favors the selection of pathogen races that can overcome it. Specific resistance lacks durability, but resistance does not have to be very durable to be useful. Since the early 1900's, the American and Canadian wheat crops have been protected from rust diseases by specific resistance, but there have been some notable failures. In 1952 and 1954, new races of the wheat stem rust fungus caused the loss of about 70 percent of the durum wheat crop and 25 to 35 percent of the bread wheat crop. In most years, however, the resistance held up.

The key to the success of specific resistance in wheat has been a continual search for new genes for resistance and the development of an extensive monitoring program to detect new fungus races before they become prevalent. This system has become more effective with better understanding of the genetics of rust fungi, but it is expensive to maintain. It requires a central laboratory with several scientists devoted almost exclusively to studying the changes in the population of a single fungus. It also requires periodic expeditions to explore and sample the dwindling populations of wild wheats in the Middle East in a search for new genes. Furthermore, it detracts from programs to develop higher yielding varieties of wheat because breeders must devote much of their time to the manipulation of genes for rust resistance.

Two changes in the use of specific resistance have been advocated to increase its durability, and thus reduce its cost. One is to regulate the geographic distribution of specific plant genes. The wheat stem rust fungus cannot survive winters north of Texas or Oklahoma, but each spring and summer it spreads north to the prairie provinces of Canada. The useful life span of a gene for specific resistance in Canadian varieties could be prolonged by keeping it out of the wheat in Texas and Oklahoma. This would prevent selective changes in the fungus population before it reaches Canada. Division of the North American continent into wheat rust zones and the distribution of distinct sets of genes for specific resistance in those zones could be highly effective in frustrating adaptation by the fungus.

The second approach to increasing the durability of specific resistance involves a simulated "return to nature," and is based on studies that show that wild plants and their pathogens have evolved relatively stable associations in which diseases are not ordinarily debilitating. For instance, in the wild oat populations in Israel and surrounding areas, many types of specific resistance are found and there is a corresponding diversity of pathogen races within the population of the fungus that causes oat crown rust. The plant and fungus populations seem to have reached a stable equilibrium in which most of the host plants are highly resistant to most races of the fungus.

Multiline varieties of crops are designed to reproduce this kind of balanced protection. Each multiline variety, such as those developed for wheat and oats, is composed of a collection of lines or varieties that are similar in appearance but different in their specific resistance to races of the rust fungi.

The gradual increase of general resistance by crop selection over several generations is an old method for developing durable resistance. Recently, this approach has aroused new interest because it provides long-lasting resistance. General resistance is often difficult to work with because its effects may be subtle and its inheritance is usually complex, but it has the advantage of being more durable than simply inherited specific resistance. Crops such as corn, which are readily cross-pollinated, have traditionally depended on complex, general resistance to diseases. For these crops, there has been no need to set up extensive programs for surveying the races of pathogens. It is now apparent that general and specific resistance are compatible. General resistance increases the durability of specific resistance—hence the renewed interest in improving general resistance in such self-pollinated crops as wheat, which are more difficult to breed than corn.

The recent emphasis on general resistance has led to improved techniques for measuring subtle, quantitative differences in plant responses to pathogens. The emphasis now is on how resistance affects the pathogen's reproductive ability. Most epidemics depend on multiple cycles of pathogen reproduction. What may seem a relatively slight reduction in reproduction in each cycle, when multiplied over several cycles during the crop season, is often sufficient to prevent the otherwise explosive buildup of the pathogen; complete inhibition is not essential. The most efficient use of disease resistance will be based on results of basic research into the population dynamics of the organisms that cause disease.

## New Methods for Developing Resistant Crops

The classical approach in plant breeding is to combine disease resistance with desirable horticultural or agronomic characteristics. While this approach has been successful in many crops, in others it has been impossible to achieve this combination. Protoplast culture is a promising new method for acquiring disease resistance traits in plants.

Although tissue culture techniques at the protoplast level have led to the regeneration of a number of different plants, until recently none has involved a major food crop. In the case of the white potato, plants regenerated from protoplasts commonly have traits not observed in the mother clone from which they are derived. Furthermore, the newly acquired traits seem to be stable and are maintained in subsequent generations of plants that are propagated by tubers. Thus, protoplast regeneration has the potential to improve disease resistance and other horticultural properties of vegetatively propagated food plants. The results of research seeking to ascertain the underlying genetic mechanisms that control these traits could be of enormous value to crop improvement programs.

Another mechanism for acquiring disease resistance in plants is through the use of mutagens. Chemical agents or X-rays have been effective in some cases in inducing resistance not previously known in the plant population. A notable example is the development of mint varieties with resistance to *Verticillum* wilt. It is possible that induced mutations can be combined with the protoplast regeneration system described above.

## CHEMICAL CONTROL OF PLANT DISEASE

Chemicals have long been used successfully as disease control agents. For example, in the late 1800's common stinking smut regularly destroyed 10 percent or more of the Kansas wheat crop; today, instances of this disease are rare because of the successful use, first, of copper-containing fungicides and, more recently, of organic mercury-containing fungicides and hexachlorobenzene as seed treatments. Because of its toxicity to humans, however, the use of mercury for seed treatment was banned in 1978. As a result, several diseases of small grain cereals, particularly bacterial leaf streak and leaf stripe of barley, are beginning to reappear. Whereas non-mercury fungicides are acceptable substitutes for control of the smuts, a major effort is now under way to find controls for other diseases of barley.

Other chemical compounds, particularly those widely used for control of leaf diseases of fruits and vegetables, are currently being examined by the Environmental Protection Agency. If they are rejected, and growers are denied these chemical controls, new compounds, new resistant varieties, or other control measures will need to be developed.

In 1979, an epidemic of blue mold devastated the Cuban tobacco crop, causing losses of more than 90 percent. The same disease struck tobacco fields in the United States and Canada, resulting in losses estimated at more than $250 million. Why this disease, which had not been seen for years, suddenly caused such tremendous losses is not known. In 1980, growers had limited access to a new fungicide called metalaxyl, which dramatically reduced losses from blue mold in fields where it was used.

Metalaxyl also promises to be valuable in controlling other serious plant diseases, including late blight of potato, downy mildew on vegetable and grain crops, and several root rots on a variety of plants.

Systemic fungicides, such as metalaxyl, which are absorbed by plant tissues and which move through the plant to control disease, have been developed largely within the past 10 to 15 years. The first such compound, carboxin, was described in 1964 and is now marketed worldwide as a seed treatment for control of different smuts of wheat, barley, and oats. Another systemic fungicide, benomyl, controls a wide variety of diseases of fruits and vegetables. There is currently a major research effort to develop other compounds that act systemically.

One difficulty that has arisen in using certain systemic fungicides is that the fungi being controlled develop resistance to the fungicide. This has been the case with benomyl and metalaxyl. However, strategies for slowing down or preventing the development of resistant strains of the pathogen are being used. These involve the use of fungicides that have different modes of action and that are applied as mixtures with benomyl or alternately with benomyl. Thus, if strains resistant to benomyl develop, they may be eliminated by the other fungicide. However, this approach has failed in situations where the pathogen is resistant to several chemicals.

## BIOLOGICAL CONTROL METHODS

Possible environmental and health hazards associated with the use of chemicals for plant disease and insect pest control have aroused considerable public concern. Therefore, as an alternative, biological control has been examined, with some encouraging results. An ingenious new biological treatment of young fruit trees for crown gall disease caused by *Agrobacterium tumefaciens* is now commercially available. In this instance, root stocks are treated with a suspension of the bacterium *Agrobacterium radiobacter*, a nonpathogenic strain of *Agrobacterium*. This bacterium produces an antibiotic that prevents growth of the crown gall bacterium.

Many bacteria associated with plants produce specific antibiotics called bacteriocins. Thus, diseases caused by bacterial pathogens may be controllable either with the appropriate antibiotic or with avirulent strains of the pathogenic organism. Some likely candidates for control by these methods are fire blight of apples and pears, bacterial wilt caused by *Pseudomonas solanacearum*, and numerous bacterial diseases that are borne on seeds.

In Australia, fungal root rot of avocados has been controlled by intensive cover cropping and applications of chicken manure and dolomitic limestone. Just why these measures keep the destructive fungus under control is not completely understood. In any case, microbial activity in soil treated in this manner is enhanced, whereas germination and development of pathogen spores are inhibited. The challenge for biological control, particularly of soil-borne pathogens, is to understand better the interactions that occur among soil organisms and the ways that these interactions can be manipulated to control plant pathogens.

Many species of *Pseudomonas* produce antifungal agents, only some of which have been chemically characterized. Some consideration has been given to altering the microflora of perennial plants (trees, vines, and shrubs) by introducing bacteria that are antagonistic to the most troublesome fungal pathogens. For instance, injecting cultures of certain strains of *Pseudomonas syringae* into young elm trees can decrease the chances that symptoms of the dreaded Dutch elm disease will appear. The effectiveness of this approach depends on the extent to which the bacteria: (1) produce antifungal agents that are effective against a wide range of isolates of the fungal pathogen, or trigger defense responses in the host, (2) can establish and distribute themselves in the plant, (3) do not cause adverse effects in the plant, and (4) are effective in controlling the disease for a prolonged time.

Plants may have latent (unexpressed) mechanisms of disease resistance. For instance, some races of *Colletotrichum lindemuthianum* (the cause of bean anthracnose) inoculated into bean hypocotyls activate a defense mechanism that protects the plant from disease when challenged by pathogenic races of the fungus. Such protection seems to be systemic, since inoculation of the first true leaf of the plant with the nonpathogenic strain confers resistance on the entire plant. Protection of tobacco from the bacterial wilt organism with avirulent heat-killed cells of *Pseudomonas solanacearum* has also been reported. Additional research is needed to discover the extent of this phenomenon and to determine how it can be manipulated to provide economical disease control.

Some fungi offer promise in controlling other destructive fungi. For example, *Fomes annosus* is a fungus which is responsible for heavy losses of timber in American and European pine plantations. This fungus grows on the stumps of harvested trees; from these stumps, spores and mycelial threads spread rapidly to nearby healthy pines. However, if a harmless fungus, *Peniophora gigantea,* has already colonized the stumps, *F. annosus* cannot become established. This technique for protecting pines has been adopted in Great Britain, where tablets containing the reproductive structures of *P. gigantea* are placed on freshly cut stumps, thus allowing for the initial colonization of this organism to the exclusion of *F. annosus.*

One of the more interesting observations of biological control concerns chestnut blight, caused by the fungus *Endothia parasitica.* A hypovirulent (nonpathogenic) form of the fungus will render harmless normally virulent (pathogenic) forms of the fungus in culture. Thus, when the hypovirulent form is injected into actively growing cankers on limbs, the cankers become quiescent. Apparently, RNA from the hypovirulent fungus makes the virulent fungus impotent. So far, this method of controlling chestnut blight appears to have been more successful in Europe than in the United States. Research to discover other biological relationships of this type and to delineate the mechanisms by which they operate may yield important new techniques for the control of plant disease.

## EPIDEMIOLOGY

In plant pathology, relatively little attention is given to therapy of diseased individuals; instead, the primary concern is with the conditions that govern the spread of diseases through populations. Early study of plant epidemiology focused on the influence of such climatic conditions as temperature, relative humidity, and precipitation. For instance, fungus spores usually require a film of water in which to germinate before they can penetrate and infect the host leaf. The time required for spore germination depends in turn on the temperature. When temperatures are too low or too high, or when the leaves are dry, infection will not occur. Careful studies of these conditions have led to some relatively simple but effective methods for determining when infection is likely to occur in an orchard or field and when a fungicide should be applied.

Comparative epidemiology involves mathematical analyses of the different patterns by which epidemics develop in different types of host plants beset by pathogenic organisms. The so-called disease progress curve is the cornerstone for analyses and comparisons. Epidemiologists plot relative amounts of disease (expressed in units, such as percentage of leaf area damaged, number of disease lesions, proportion of roots infected, and so on) against time in order to analyze the effects of weather, pesticides, and other factors, on the progression of a disease.

Diseases that build up through a series of many cycles of pathogen reproduction are commonly referred to as compound-interest diseases. The progress of such epidemics—for example, potato late blight, wheat stem rust, or rice blast—increases exponentially over time. Each increment of disease generates a larger increase until the supply of healthy plants is exhausted.

Other diseases, such as those that affect the roots of plants, build up at a slower rate and are commonly called simple-interest diseases. The rate of increase of these diseases within any crop year is closer to being steady than exponential. Simple-interest diseases are less explosive but are often more persistent than compound-interest diseases. They are caused by organisms that compensate for their slower reproduction by an increased ability to survive from one crop to the next. Simple-interest diseases are especially destructive to perennial crops and forest trees.

Comparisons of disease progress curves for different types of epidemics provide knowledge of the importance of, for example, initial infection levels, length of the growing season, and suitability of weather. It is then possible to base disease control strategies on the components of the epidemic that are most vulnerable to disruption.

### Monitoring Disease Epidemics

Plant epidemiology has greatly benefited by the development of the high-speed digital computer. Detailed data on microclimates can be collected, processed, and merged with complex programs to simulate crop growth and parasite reproduction. It has become possible to elucidate the complexities of plant disease interactions at levels undreamed of a few years ago. Details of thousands of interactions among host plant, pathogenic organism, and weather can now be stored and integrated. Plant pathologists with broad training in such diverse areas as advanced mathematics, computer science, plant ecology, and meteorology are using their skills to produce mathematical models and simulations of plant disease epidemics. Apple scab and potato late blight have been modeled so successfully that the computer simulations now serve as rapid, accurate training guides to help students and extension agents learn optimal disease management systems that are useful in integrated pest management.

## INTEGRATED PEST MANAGEMENT

Integrated pest management involves fitting together information about the biology of pests, the environment, and the status of the host to obtain maximum results from the application of biological and chemical control methods. This management technique not only reduces the environmental load of pesticides but is also cost effective. Customarily, a given situation is monitored and specific control measures are instituted only after pest populations exceed a certain tolerance level. Control involves such measures as applying effective amounts of pesticides at the appropriate times, crop rotation, use of microbial agents, or whatever other manipulations minimize loss. Integrated pest management relies heavily on knowledge of the biology and life cycles of pests, and on trained personnel who can apply the information.

For some crops, resistant varieties or other methods of disease control are unavailable, so growers must resort to pesticides. The disadvantages of using fungicides on crops may be mitigated in the future by more accurate forecasting of disease. In Pennsylvania, for example, a forecasting system know as BLITECAST has been developed for late blight of potatoes. This program involves placing weather instruments on a grower's farm. The weather data are then phoned in periodically to a computer center, and a forecast of the probability of a late blight outbreak is

No-till corn in a double-cropping system quickly grows above the wheat stubble, which serves as a moisture-holding mulch throughout the growing season. [Soil Conservation Service, U.S. Department of Agriculture.]

quickly relayed to the grower, who then uses this forecast to determine whether he should apply a fungicide or delay doing so. It has been shown that smaller amounts of fungicides are needed to control late blight if forecasting is combined with the use of potato varieties that have some degree of genetic resistance. This approach needs to be tried with other crops and diseases to see if it has general application.

## ENERGY SUPPLY AND PLANT DISEASE

As energy costs soar, agricultural producers are looking for ways to reduce energy consumption and yet maintain productivity. One approach is to decrease tillage or even to adopt a no-tillage system, whereby weeds are killed with herbicides rather than by being plowed under or disced. This tillage method (known as conservation tillage) has the added advantage of significantly reducing erosion—by 50 to 90 percent. There is some concern, however, about the effect that changes in tillage practices will have on the incidence of plant disease since many bacterial and fungal pathogens overwinter in and on crop residues. Experience thus far has not been sufficiently extensive to indicate whether this concern is justified. However, some examples are encouraging.

In Nebraska, a cropping system of this kind seems not to have increased the incidence of disease; indeed, stalk rot of corn and sorghum has actually decreased. This particular system involves a rotation of winter wheat, grain, sorghum, and fallow; fields are kept weed-free by herbicides and the crop residue is left undisturbed on the surface. Yields of wheat have increased 8 to 10 percent and sorghum yields have increased 40 to 50 percent over those obtained with conventional tillage. On the other hand, in some areas barley and wheat leaf diseases have increased in severity in fields where conservation tillage has been practiced.

There is little doubt that agriculture in America, and perhaps elsewhere in the world, will change markedly in the coming decades unless alternative, cheap energy sources are found. If they are not, research on the impact of changes in tillage practices on plant diseases will assume new importance.

## BIOCHEMISTRY OF HOST-PATHOGEN INTERACTIONS

The biochemistry of host-pathogen interactions is an important aspect of plant pathology. Research in this field is based on the assumption that all diseases, whether in plants or animals, are the result of molecules in the host interacting with those of the pathogen. With the advent of such techniques as recombinant DNA, gene sequencing, affinity chromatography, high-pressure liquid chromatography, and molecular spectroscopy, the tools for making advances in biochemical plant pathology are at hand.

The major goals of this field are to elucidate the mechanisms by which pathogens function and are specific to their hosts, and to explain the biochemical nature of disease resistance and defense responses. Many plant pathogens, especially fungi and bacteria, release phytotoxins that are, in part, responsible for the effects of plant disease. Some of the most biologically interesting toxins, known as host-specific toxins, are those that affect the specific plants or varieties that are attacked by the pathogen that produces the toxin. The host's specificity seems to be related to receptor sites that are proteins found in or on the cells of the plant host. When the fungal toxin contacts the receptor protein, a series of events is triggered that leads to cellular death, and eventually the entire leaf or plant may die.

A common response of plants (especially legumes) to an invasion by a pathogen is the synthesis of phytoalexins, which are low molecular weight antibiotics capable of inhibiting the growth of microorganisms, including plant pathogens. About 100 different phytoalexins have been isolated and characterized. A single plant species is capable of producing several structurally related phytoalexins, but the mechanisms by which they inhibit the growth of pathogens have not been determined. Healthy plants do not normally contain detectable amounts of phytoalexins. These compounds are synthesized at the sites where the tissues are invaded, in response to certain molecules, called elicitors, in the invading pathogen. One such elicitor is a complex carbohydrate present in the cell walls of a plant pathogenic fungus. Not only will this substance induce phytoalexin formation in the plant, but small fragments of the complex carbohydrate also act as elicitors. Applying an elicitor molecule prior to invasion by a pathogen may confer resistance to disease through the formation of phytoalexins. The use of these compounds, or their related synthetic analogs, as a means of conferring protection on plants is currently in its infancy, but holds promise for the future.

Viruses are noncellular parasites that have their own genetic system but lack the machinery necessary to synthesize their constituent proteins. Outside the cell, they are inert aggregates of nucleic acids, proteins, and other molecules. Once a virus invades a suitable cell, interactions between it and its host are at a very intimate level—the level of the gene. Hence, a practical antiviral agent must act very subtly, interfering with the replication or pathogenesis of a virus while at the same time doing minimum damage to the host cells. Because of this stringent requirement, progress in the chemotherapy of plant virus disease has been slow. Rather, the emphasis has been on preventing the virus from reaching the crop by controlling such virus vectors as insects and by developing virus-resistant crop varieties.

Enhancing present strategies against plant viruses and developing new ones will require more complete knowledge of how they replicate than is currently available. In recent years, significant progress has been made in understanding the biochemistry and molecular genetics of plant viruses. The entire complement of genes has been identified in only a few plant viruses. Most of these results have been derived from nucleotide sequence data or by analyzing complex cell-free systems in which viral nucleic acids have been caused to function *in vitro*. Often, the more obvious approach of studying infected tissue has been impractical. Furthermore, in only a few instances have the functions of the viral genes been determined, even though their functions must be known in order to trace step-by-step how a virus is able to induce the host cell to produce new virus particles.

## PLANT PARASITES AS BENEFICIAL AGENTS

*Agrobacterium tumefaciens* causes tumors known as galls to form in plants, apparently through a plasmid harbored in the bacterium. (A plasmid is a piece of circular DNA that exists and replicates independently of the bacterial DNA.) After the bacterium attaches itself to the plant cell wall in an injured area, the plasmid DNA enters the cell by a mechanism not yet fully understood. The initial steps in this interaction are heat sensitive. About 10 percent of the total plasmid DNA becomes incorporated into the DNA of the plant cell, most likely the nuclear DNA of the host plant. After the uptake of the plasmid DNA, the host plant begins neoplastic growth.

The process by which the DNA of one organism affects the biology of another in this manner is called transformation. It occurs because the plasmid DNA has unique sequences of bases which allow it to become inserted into the plant DNA. Thus, the new genes confer new properties or new regulatory functions on the plant. Unraveling the mysteries behind these events ultimately may allow us to engineer plants in desired ways.

The plasmids of *Agrobacterium* are among the more promising tools for applying recombinant DNA techniques to plants. This approach involves acquiring genes from the DNA of a given organism by digesting the DNA with specific enzymes. These genes are then joined with a plasmid or other vector to produce a recombinant. Finally, the recombined plasmids are placed into a host and the host is tested for the expression of new characters, such as drug resistance or synthesis of a metabolic product. This technology has potential not only for understanding host-parasite interactions but for genetically manipulating plants to be disease resistant, to make new products, or to fix nitrogen.

Sometimes the mere classification of an organism as a pathogen dramatically influences the way that its importance is gauged. For instance, a number of plant-associated bacteria have biological properties that are potentially useful. One such example is the hairy root syndrome caused by *Agrobacterium rhizogenes* (see Figure 1). The primary roots of plants infected by this organism develop numerous functional secondary roots. If the etiological agent, a plasmid of relatively large size, is lost from the bacterium, it becomes nonpathogenic, but transfer of the plasmid to non-infectious strains of *Agrobacterium* renders them infectious. Presumably, the change in the growth of the root is similar to what happens in crown gall disease. A matter of interest is the possibility that *A. rhizogenes*, its plasmid, or its genetically modified DNA might be useful in causing certain plants to produce more roots, which might increase their ability to tolerate drought. Genetic manipulation might also succeed in transferring this rooting characteristic to species of *Rhizobium*, which are the organisms responsible for symbiotic nitrogen fixation.

Mycorrhizae exemplify a unique relationship between fungi and plants. This symbiotic association of fungi with roots is beneficial both to the plant and to the fungus itself. The host benefits primarily through greatly increased uptake of nutrients, particularly of phosphorus. Two

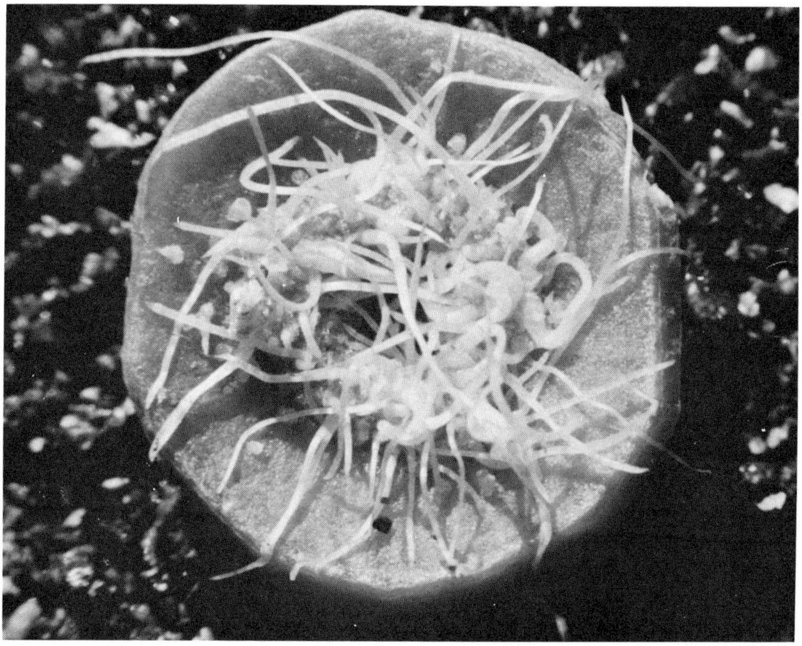

**Figure 1 Hairy root syndrome in carrot.** A carrot disk, having been inoculated with a culture of *Agrobacterium rhizogenes*, begins to show the hairy root syndrome.

basic types of mycorrhizal fungi exist; the ectomycorrhizal forms that live on the outer tissues of roots, and the endomycorrhizae that invade and live primarily inside root tissues. Ectomycorrhizae have been shown to be particularly important to forest trees, especially for the establishment and growth of new seedlings. These fungi can be grown in culture and are currently being used to inoculate seedlings before they are transplanted to the field. The endomycorrhizae are found primarily in the roots of annual and perennial grasses and shrubs. They cannot as yet be grown in pure culture; however, their value to plant growth has been clearly shown. The challenge now is to produce or grow them so that nurseries, seedbeds, seeds, and land areas with little or no native population of endomycorrhizal fungi can be inoculated. The reclamation of strip-mined areas, for example, may be improved by the addition of these fungi. Indeed, this has been shown to be true in the eastern United States.

## PATHOGENS AS BIOLOGICAL CONTROL AGENTS

Plant pathogens that attack weeds offer the possibility of effective and environmentally safe herbicides. Weeds are attacked by a number of pathogens, each of which interacts with the host through the production of one or more phytotoxic compounds. These phytotoxins, especially those from bacteria and fungi, may provide a new starting point for investigators seeking to synthesize biologically effective herbicides.

A case in point is the Canadian thistle, which invades croplands in the northern part of the United States. It is susceptible to several fungal pathogens, including two species of *Alternaria*, a *Septoria*, and a *Fusarium*. Fermentation broths of *Alternaria* contain substances that kill leaves upon application. Some plant specificity has been demonstrated when more purified preparations of these compounds are used. Conceivably, compounds from biological sources can be produced which are biodegradable, plant specific, and nontoxic to animals. Such compounds might be made by fermentation or by chemical synthesis. A substantial worldwide effort is needed to acquire these weed pathogens and preserve them in culture collections. The isolation and characterization of phytotoxins from such organisms has just begun.

In a few cases, plant pathogens have been used directly in the biological control of weeds. Two approaches are being used. The first is the classical biocontrol approach, in which a pathogen is imported. For example, a rust fungus from Italy was released in Australia in 1971 to control skeleton weed, a major pest in wheat fields and grasslands. A similar program is under way in the western United States. Great care must be taken when this method is used to test thoroughly any imported plant pathogen to ensure that it will not harm native plants.

A second approach to biological control of weeds makes use of pathogens already present in the country. A good example is the program against northern joint vetch in rice fields in Arkansas. Here, spores of a pathogenic fungus, a species of *Colletotrichum*, are mass-produced in fermentation culture and then sprayed over the rice fields. The fungus is harmless to rice, but a single application of spores is sufficient to kill the weeds. While plant pathogens may never fully replace chemical herbicides, they can be extremely useful in certain situations. Fungi such as skeleton weed rust may be effective in rangelands where chemical control is too expensive. Weed pathogens will be valuable in situations where a high degree of specificity is needed, or where chemical residues may be a special problem, as they are in and around waterways.

## NEW ASPECTS AND CAUSES OF DISEASE

Several new types of plant pathogens have been discovered in recent years, including viroids, mycoplasmas, spiroplasmas, and rickettsialike organisms. Furthermore, certain plant viruses, though themselves pathogens, have been shown to be colonized by other viruses, known as satellite viruses. A satellite virus can replicate only in plant cells that also are

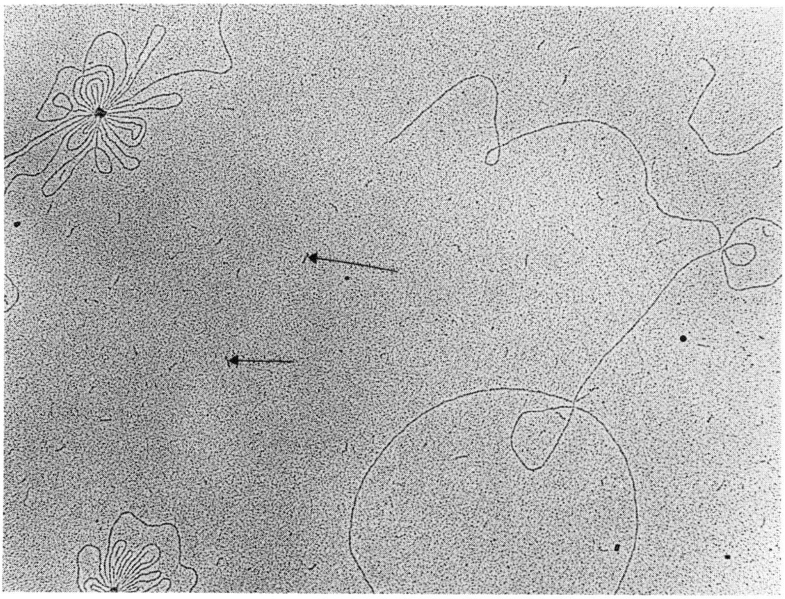

**Figure 2 Potato spindle tuber viroid.** The extremely small particles of potato spindle tuber viroid (arrows) in reference to the DNA from a virus (large winding strands) magnified 55,000 times. [SOURCE: T. O. Diener, T. Koller, and J. M. Sogo, Swiss Federal Institute of Technology, Zürich.]

infected with the particular plant virus on which it depends, referred to as the helper or supporting virus.

The satellite virus inhibits replication of its companion virus, often drastically, and may mitigate the severity of pathogenic symptoms. It is therefore conceivable that certain satellite viruses can be developed as control agents of plant viruses, especially for fruit, grapes, and other perennial crops. The RNA content of some satellite viruses is so small that only one or at most two very small genes could be encoded, and it is possible that some satellite viruses do not encode at all in the usual sense. The simplicity of the RNA in satellite viruses makes them especially amenable to manipulation.

Another newly discovered pathogen of plants is the viroid. The simplest known biological agent of disease, it is an RNA molecule that is roughly the size of the RNA in the smallest satellite virus (about 360 nucleotide residues). However, unlike RNA from a satellite virus, it is not encapsulated (surrounded by protein) and it replicates independently. The viroid causing potato spindle tuber disease is a circular RNA molecule for which the entire sequence of nucleotide residues is known (see Figure 2). Although viroids are only known to occur in higher plants,

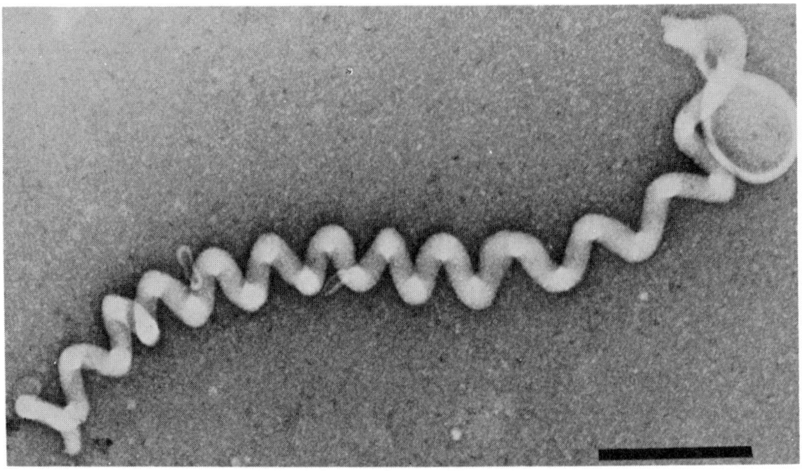

**Figure 3 Corn stunt disease spiroplasma.** The bar represents 1 micrometer. [SOURCE: T. A. Chen.]

certain slowly developing diseases of animals and humans may be induced by similar agents.

In the late 1960's, mycoplasmalike organisms were first shown to be agents of plant disease. These organisms resemble bacteria but are generally smaller and lack a cell wall. They have been shown to cause over 60 diseases of various plant species, the most common of which is aster yellows disease of ornamental plants. The identification of the causal agent of these diseases was all that was needed for scientists to realize that certain antibiotics, such as tetracycline, could be used to control plant diseases because mycoplasmas are extremely sensitive to them. Other agents of plant disease were recognized in the 1970's, including rickettsialike organisms that cause Pierce's disease of grapes and the spiroplasma that causes citrus stubborn disease and corn stunt (Figure 3).

Some of the newest threats to plants have come from acid rain, atmospheric deposits, and air pollution. These conditions are caused mainly by industrial and transportation activities and by the burning of coal for electricity and heat. Breeding plants resistant to various pollutants would help alleviate the losses sustained as a result of these human activities.

## PLANT PATHOGENS AND ENVIRONMENTAL DAMAGE

Each year, somewhere in the United States, major damage to agricultural crops occurs from untimely frost in the late spring or early fall. Recently, plant researchers discovered that certain bacteria, particularly some important plant pathogenic forms, can cause the nucleation of ice (frost); ice crystals form around these bacteria at temperatures at which water would not normally freeze. This work also showed that corn plants, which are very sensitive to frost damage, could withstand much lower temperatures if they were grown in the absence of these ice-nucleating bacteria.

The practical value of this discovery to agriculture was recently shown in California, where some citrus groves were protected from late spring frosts by applying antibiotics to eliminate the ice-nucleating bacteria normally present. It now becomes important to determine how many other frost-sensitive plants could be similarly protected.

## Summary and Outlook

Plant diseases will continue to damage the fields, forests, and gardens on which all humans depend. However, some new developments in genetic manipulation show promise as weapons in the fight against plant disease.

The concept of planting seeds of diverse genetic backgrounds (multilines) in order to suppress pathogen populations will gain wider acceptance. Incorporating several genes, rather than depending on only one or a few traits for disease resistance, has already received general acceptance as a goal for research on many of the world's food crops. Increased work in genetics will be needed to incorporate resistance against disease and pests into major food plants. The technique of cloning plants through the regeneration of whole plants from protoplasts, initially explored in work on white potatoes, could be applied to other important food species. Finally, recombinant DNA techniques suggest a means for mobilizing and moving resistance traits (and other traits as well) from one plant to another. Demonstrations of the feasibility of this approach will undoubtedly be attempted in the near future.

Much emphasis is now being placed on the biological control of plant pathogens. The use of antibiotic-producing organisms to control pathogens will undoubtedly receive more attention. Cropping systems will change as energy supplies dwindle or take other forms, influencing the type and amount of disease that develops. Techniques to encourage microorganisms that discourage the establishment of pathogens in these new cropping systems will be needed.

Post-harvest diseases and hazards such as mycotoxins, which develop as plant products decay in storage, are of major economic importance. In many cases, their control could provide the quickest gain in net productivity.

Integrated pest management will become more widely accepted. Its continued development will require more knowledge about the life cycles, growth requirements, and biochemistry of pathogens, about vectors that transmit pathogens (insects, other animals), and about pathogen hosts. It will also require new, more effective, environmentally safe pesticides and a cadre of professionals trained to integrate and apply this knowledge.

An understanding of resistance or susceptibility to disease at the biochemical level has the potential of adding new approaches to disease control. It already appears that elicitors may be used to protect plants and that toxins may act as screening agents and models for herbicides. There is much still to be understood about the myriad biochemical events that surround the transformation processes occurring in plant disease.

The study of plant pathogens and the diseases they cause can be expected to produce some fundamental information on the etiology of disease in general, including diseases that affect animals and human beings.

## BIBLIOGRAPHY

K. F. Baker and R. J. Cook. *Biological Control of Plant Pathogens.* San Francisco: W. H. Freeman and Company, 1974.

J. A. Browning and K. J. Frey. "Multiline Cultivars as a Means of Disease Control," *Annual Review of Phytopathology*, Vol. 7(1969), pp. 355-382.

T. O. Diener. "Viroids: Structure and Function," *Science*, Vol. 205(1979), pp. 859-866.

B. Doupnik, Jr. and M. G. Boosalis. "Ecofallow—a Reduced Tillage System—and Plant Disease," *Plant Disease*, Vol. 4(1980), p. 31.

G. J. Green and A. B. Campbell. "Wheat Cultivars Resistant to *Puccinia graminis tritici* in Western Canada: Their Development, Performance, and Economic Value," *Canadian Journal of Plant Pathology*, Vol. 1(1979), pp. 3-11.

A. Kerr. "Biological Control of Crown Gall Through Production of Agrocin 84," *Plant Disease*, Vol. 4(1980), pp. 24-30.

D. R. Knott. "Using Race-specific Resistance to Manage the Evolution of Plant Pathogens," *Journal of Environmental Quality*, Vol. 1(1972), pp. 227-231.

J. M. Krupinsky and E. L. Sharp. "Reselection for Improved Resistance of Wheat to Stripe Rust," *Phytopathology*, Vol. 69(1979), pp. 400-404.

G. B. Lucas. "The War Against Blue Mold," *Science*, Vol. 210(1980), pp. 147-157.

D. H. Marx. "Ectomycorrhizae as Biological Deterrents to Pathogenic Root Infections," *Annual Review of Phytopathology*, Vol. 10(1972), pp. 429-454.

L. Moore et al. "Involvement of Plasmid in the Hairy Root Disease of Plants Caused by *Agrobacterium rhizogenes*," *Plasmid*, Vol. 2(1979), pp. 617-626.

K. O. Muller. "Studies on Phytoalexins. I. The Formation and Immunological Significance of Phytoalexin Produced by *Phaseolus vulgaris* in Response to Infections With *Sclerotinia fructicola* and *Phytophthora infestans*," *Australian Journal of Biological Sciences*, Vol. 2(1958), pp. 275-300.

National Research Council, Committee on Genetic Vulnerability of Major Crops. *Genetic Vulnerability of Major Crops.* Washington, D.C.: National Academy of Sciences, 1972.

R. E. Phillips et al. "No-tillage Agriculture," *Science*, Vol. 208(1980), pp. 1108-1113.

J. Shepard et al. "Potato Protoplasts in Crop Improvement," *Science*, Vol. 208(1980), pp. 17-24.

R. D. Shrum. "Forecasting of Epidemics," *Plant Disease—An Advanced Treatise*. Volume Two. Edited by J. G. Horsfall and E. B. Cowling. New York: Academic Press, 1978, pp. 223-238.

J. E. Van der Plank. *Disease Resistance in Plants.* New York: Academic Press, 1968.

H. E. Waterworth et al. "CARNA 5, the Small Cucumber Mosaic Virus-Dependent Replicating RNA, Regulates Disease Expression," *Science*, Vol. 204(1979), pp. 845-847.

R. F. Whitcomb and J. G. Tully. "Plant and Insect Mycoplasma," *The Mycoplasmas.* Volume Three. New York: Academic Press, 1979.

# 7
# *Water Resources*

## INTRODUCTION

Although the amount of water available to the nation as a whole appears adequate for its needs, water shortages at regional and local levels are becoming more common, and water quality problems are increasing in complexity and scope. As the population continues to shift to arid and semiarid regions of the country, the competition for water intensifies, leading to conflicts over allocation and complicated negotiations for scarce supplies. With a finite ground- and surface-water resource, reuse and recycling will become more and more necessary, exacerbating the effects of man's disturbances of the natural water cycle.

At the same time, the increased volume and variety of wastes generated by a highly industrialized society are reducing the usability of the water supply in some areas and precluding recycling in others. In still others, where the water is not obviously unfit for use, refinements in analytical techniques have made it possible to detect smaller and smaller amounts of contaminants, raising questions of health effects that cannot be answered quickly or easily. The institutional and technological problems that result from all of this are enormous, but some are less difficult to resolve than others.

The problems of obtaining near-term water supplies are important but can be solved in most instances by institutional and social changes. The changes may be difficult to achieve, but they are within the realm of

---

◄ Los Angeles aqueduct. "The political, legal, and social problems involved in reallocation of water, while posing difficult obstacles, have been solved in the past." [EPA Documerica—Charles O'Rear.]

possibility. For example, we know how to use water more efficiently in irrigation and industrial cooling. The difficulties lie in overcoming the political and institutional barriers that prevent application of the scientific knowledge we have.

The problems of water quality are of a different order. In almost all areas, there is a dearth of data, sometimes even of monitoring or testing techniques, that are critical for making intelligent decisions about water management in the near as well as the long term. It is these problems that will be the most difficult to resolve. Accordingly, it is water quality rather than water quantity that is discussed in this chapter.

The nation will face some hard decisions about water in the years ahead. We will have to decide on "safe" levels of contaminants when we have inadequate data for assessing health risks or the benefits of controls. We will have to decide which water quality problems need immediate attention and which are less urgent. These decisions often will have to be made in an atmosphere of regional pressures for action and uncertainty about the full long-term effects of changes. As the last section of this chapter indicates, new institutional arrangements and new research organizations must be developed to deal with water quality issues. We need new strategies for management in the face of uncertainty and better methods for monitoring their results. In order to develop these strategies and improve current approaches, some important scientific problems must be attacked. Several of these problems are taken up in this chapter. Of necessity, the choices are selective. Not all of the water quality problems can be discussed here. Because of limited space, we have included only those that are high on the list of problems requiring attention in the near future.

## WATER SUPPLY

In its second national assessment of the nation's water resources, over the period 1975-2000, the U.S. Water Resources Council drew attention to ten important problems created by intensive use and competition for water: inadequate surface-water supply, overdraft of ground water, pollution of surface water, pollution of ground water, quality of drinking water, flooding, erosion and sedimentation, dredging and disposal of dredged material, wet-soils drainage and wetlands, and degradation of bay, estuary, and coastal waters.[1]

Although the council's list has led some policy officials, writers, and others to suggest that a "major water crisis," comparable to the "energy crisis," is inevitable by the year 2000, there is substantial evidence that this is not the case. In the first place, water is a renewable resource; it is not used up as are deposits of coal, natural gas, petroleum, or uranium, except where ground-water supplies are exhausted. It can be used and reused over and over again if its quality is protected. Also, the total amount of water available in an average year is very large in comparison with present and foreseeable demands. The hydrologic chart in Figure 1 shows that the system is essentially a self-renewing one.

The quantity of water withdrawn from surface- and ground-water sources for public water supply systems in the entire United States in 1975 (the latest year for which figures are available) was about 27 billion gallons per day, or about 6 percent of the average annual flow of the Mississippi River at New Orleans.[2] Less than one fourth of the water withdrawn for public water supply systems is actually consumed (evaporated, transpired, or incorporated into a product); therefore, the remainder is available for reuse after appropriate treatment.

Irrigation consumes a much greater percentage of the water it withdraws. In 1975, consumption by irrigation was estimated to equal about 86 billion gallons daily in the contiguous 48 states.[3] Even this is only about 50 percent of the average annual flow of the Columbia River at its mouth. It should be obvious, therefore, that the United States as a whole has within its boundaries sufficient water for all foreseeable needs, although there have been and will be local shortages.

Therefore, the view that the United States is about to run out of water must be based on the erroneous assumptions that the water resource is nonrenewable, that it cannot be reused, or that no mechanisms exist, or are likely to be developed or improved, for *reallocating* water supplies from less valuable to more valuable uses or for *relocating* water from areas of surplus to areas of deficiency. Many states already are reallocating water to ameliorate shortages resulting from inadequate surface-water supply. Hundreds of interbasin transfers already move excess water to areas that are deficient.

The Upper Colorado and Upper Missouri/Yellowstone river basins are the regions in which the most dramatic adjustments in water use probably will be required in the near future because of the demand for water for the development of synthetic fuels.[4] Recent reports show that even in these basins, however, there is unappropriated local water that could be made available, although the adverse environmental consequences resulting from changes in instream flow are not fully understood.[5] In only a few areas, according to the U.S. Geological Survey, do local water supplies appear to be insufficient for the development of synthetic fuels.[6] Interbasin transfers to remedy these deficiencies are well within the scope of available technology. The political, legal, and social problems involved in reallocation and relocation of water, while posing difficult obstacles, have been solved in the past and should not be considered to be insoluble in the future.

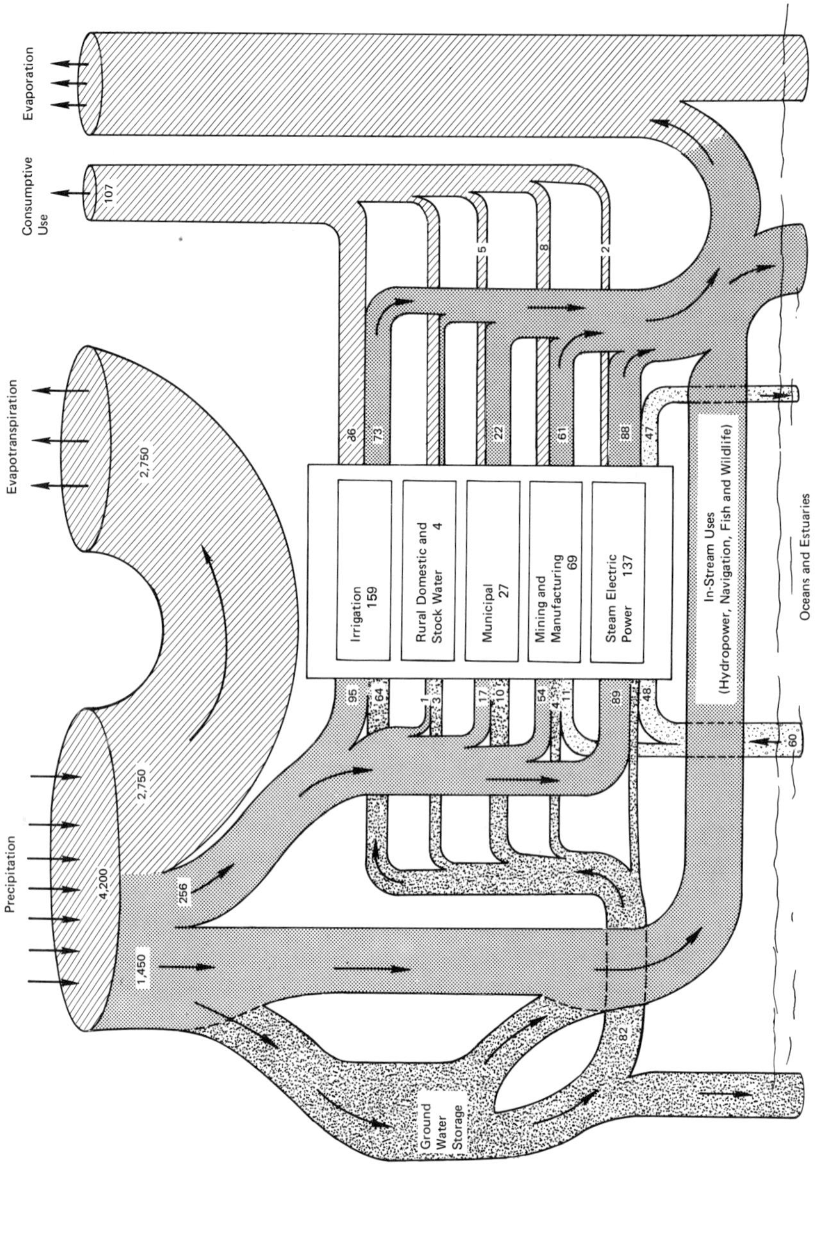


**Figure 1 Water supply and use in the United States in billions of gallons per day (1975 estimates).** [SOURCE: Data taken from the U.S. Water Resources Council, Second National Assessment, December 1978.]

## WATER QUALITY

### The Search for Solutions

Experience gained over the past 50 years has shown that relying on a single technology or institutional device can be counterproductive. Massive attacks on what appeared to be obvious problems have been launched and vast sums of money have been expended, only to learn that the wrong approach was being taken. For example, although the nation has spent some $15 billion on flood control structures such as dams, levees, and channels over the past five decades, average annual flood damage expressed in constant dollars is continuing to increase. While these investments have yielded benefits in making some floodplains usable, it is doubtful that the objective of reducing net dislocations from floods has been reached or can be reached without also using other measures, such as land use planning and building design. Even larger expenditures since 1972 for the control of point sources of water pollution through the construction of waste treatment plants have not achieved the improvements expected because the effects of nonpoint sources of pollution often are dominant and do not respond easily to technological solutions.

Similarly, the nation's fragmented approach to environmental problems has legislated independent attacks on air and water pollution without proper consideration of how such approaches might affect the land or, more important from the viewpoint of water resources, how they will affect the quality of ground-water reserves. For instance, prohibition of the disposal of wastes in water or by burning has led to disposal on land, with concomitant buildup of pollutants in the soil, or to underground injection. Because of the slow movement of underground water, the full effects of some of the control policies that have been set may not be felt for decades or even centuries.

The need for an interdisciplinary, integrated approach to pollution problems is apparent. In the near future, four important water quality issues will require attention and would benefit from such an approach. They are: (1) the quality of drinking water, which is being questioned; (2) the increased contamination of ground water; (3) the control of nonpoint sources of pollution; and (4) the development of measures that could be used to test the deterioration or enhancement of aquatic ecosystems.

## DRINKING WATER QUALITY

The United States has long prided itself on having a relatively safe supply of drinking water for its city dwellers and most farm families. The threats of typhoid fever and cholera are almost forgotten and, while the quality

There is "a growing awareness that hundreds of different organic compounds may be present in public water supplies and that some of the compounds may be harmful." [U.S. Environmental Protection Agency.]

of waterworks design and management sometimes is deficient, particularly in small towns, domestic water supplies generally are abundant and potable. During the 1960's, however, complacency was disturbed by two developments: the increased ability to identify and measure contaminants in very small amounts such as parts per billion, and the possibility, whether or not supported by unambiguous evidence, that some of these contaminants, particularly the manmade organic compounds that have been produced in ever growing volume since 1946, may have serious toxic effects.

The Safe Drinking Water Act (Public Law 93-523) of 1974 recognized these concerns and authorized significantly increased research activity on organic contaminants found in surface and underground waters used for public water supplies. The act provided the first federal regulation of all public water supplies and established an iterative process for the evolution of standards. This involves national analytical surveys, a review of earlier Public Health Standards, review by the National Research Council, and determination of safe levels for contaminants that affect human health by the administrator of the Environmental Protection Agency (EPA).

## Organic Compounds

The principal result of the research effort to date has been a growing awareness that hundreds of different organic compounds may be present in public water supplies and that some of these compounds may be harmful. It is not known how many compounds actually may be present, nor is it known when and where they may occur, although chloroform is the most pervasive. We also do not know yet the nature of all the contaminants that may be present. Some of the compounds (e.g., trihalomethanes) may be formed by the reaction of complex natural materials with chlorine used for purification, although many of the others are obviously of industrial origin. Confidence in the data we do have is limited because, in the search for volatile compounds that can be identified through gas chromatography and mass spectroscopy techniques, other compounds may be overlooked or not detected by these methods. In addition, national sampling programs are not entirely adequate and current data may not include all of the reaction products that exist. Therefore, it is not possible to say with certainty to what extent specific organic reaction products may present hazards for human health.

Only a small percentage of the compounds that have been identified in water systems have been tested adequately for carcinogenicity. An evaluation of the significance of the hazards of any environmental contaminant is complicated by ambiguous data on human health effects and by conflicting views held by scientists on the validity of the experiments. This problem is accentuated in the case of drinking water because of unique features of the Safe Drinking Water Act which require the EPA administrator to regulate "*any* contaminant which may have *any adverse* health effect on people." [Emphasis added.]

Chloroform provides an excellent example that shows the magnitude of the scientific problems that remain even when sampling data reveal the presence of a virtually ubiquitous chemical that has been demonstrated to be carcinogenic in animals. The National Cancer Institute (NCI) has concluded that chloroform produces a dose-related response in the kidneys of male rats and that highly significant dose-related increases in liver cell (hepatocellular) carcinomas occur in mice of both sexes. In addition, liver disease has been reported in male rats exposed to dosages too low to produce hepatocellular carcinoma. It seems fairly certain that mice get cancer from chloroform. The question—and here is the rub—is: can we deduce from this that man will? This is not a new problem in the testing of toxic or carcinogenic substances, but the answer is still not clear. Extrapolating experimental results from one species to another involves relative differences in metabolic factors, in biotransformation and systemic distribution, and in rates of cellular repair. There are other uncertainties, not the least of which is the possibility that a carcinogenic response in test animals is a result of the high dosages that have to be used to obtain measurable results. For example, in the NCI chloroform study, the test animals were administered oral dosages of 90-400 milligrams of drug per kilogram of body weight, although consumers of public water supplies typically ingest less than 0.001 milligrams per kilogram per day from that source.

The limitations are not restricted to extrapolations of laboratory data. Epidemiologic studies of human exposure are faced with enormous statistical constraints because the studies seek to analyze relatively small effects. To specify the exposure variable properly, it would be desirable to segregate test populations into one group consuming contaminated water and another consuming clean water. The difficulty is the high correlation between contaminated surface sources and dense urban populations in which other factors may cause cancer. These sorts of problems are among the reasons animal studies are employed. The dilemma is that most humans consuming the dosages of chloroform normally found in drinking water live in urban areas where observable small effects could be due to other causes but, in order to obtain statistically valid effects in a reasonable number of test animals, much higher dosages would have to be employed. Our understanding of the organic contaminants in drinking water is far from complete and our ability to establish the risks they pose for human health is based at present on unverifiable extrapolations from animal studies. We expect to learn considerably more about organic compounds found in water, but it still will be difficult to demonstrate their effects on human health.

## Other Problems

The search for alternative disinfectants that has been stimulated by the discovery of trihalomethanes in water supplies has reawakened interest in the resistance of certain pathogens to disinfection, as well as in the treatment techniques themselves. The roles of cysts, protozoa, and viruses in water systems as contributors to background levels of contagious disease should receive further attention within the next few years. In general, the problem of chlorine and trihalomethanes has focused attention on the potentially undesirable effects of a wide range of additives used in treating water. Aluminum, which is frequently added as a treatment chemical, may be involved in human health effects—namely, senile dementia. Preliminary research in this area has generated many questions but few results that can support confident regulatory action.

The effects of the distribution system itself on water quality have not been fully evaluated; to the extent that they remain speculative, public choice as to appropriate action must be based on a crude assessment of risk. Reactions initiated in a water treatment plant may not stop at the treatment site, and entirely new reactions may take place in the distribution system. It is known, for instance, that the trihalomethane reaction continues after purification agents are added to water, producing nearly twice as large a quantity of trihalomethanes at the end of the distribution system as that leaving the plant. The effect of corrosive waters on pipe coatings, particularly those of asbestos cement pipe, and the adhesives used in bonding plastic pipes may be important influences on the quality of the water reaching the consumer. The chemicals used in plastic pipes (tetrahydrofuran and methylethyl ketone) are known to be toxic to the nervous system, and some epoxy resins used to coat storage tanks and pipes may contain epichlorohydrin and other curing agents, such as phenylenediamine, which have unknown consequences for human health. The risks involved need to be quantified to determine whether corrective measures are needed and, if they are, how extensive they should be.

Drinking water contains a number of naturally occurring compounds that may affect human health. Among these are dissolved metals; mercury and arsenic are of particular concern. Most of the available data are based on inhalation studies, and the effects of ingested forms cannot be predicted from them. The role of water hardness in cardiovascular disease and the effects of trace elements on this debility have not been elucidated yet and will continue to be explored.

Research developments in drinking water quality in the next few years will be dictated in large degree by how regulatory agencies decide to allocate their research resources. Pressures from water-short regions of the country for increased treatment to make indirect reuse possible will heighten the importance and urgency of research on methodology and contaminants. Although we may expect refinements in analytical techniques for identifying organic contaminants, there seems to be little reason to expect rapid improvements in our ability to assess the significance of these compounds for human health. Thus, there will be increased uncertainty as to how much treatment water should receive and what types of treatment are acceptable to society.

## GROUND-WATER CONTAMINATION

For millennia, man has used subsurface water as a domestic water supply, particularly in locations remote from large surface sources. However, during the past three decades there has been an increased awareness that ground water can be a major source of water for municipal, industrial, and agricultural development. To some extent, the increased use of ground water has been brought about by the development of more efficient and powerful pumps and by cheap energy. At the same time, the volume and variety of wastes generated and the prohibitions on using streams for waste disposal have led to greater reliance on underground waste disposal.

During the 1970's, there was an increased awareness of the fact that shallow ground water is being polluted by man's activities. This is particularly disturbing because of the slow rates of flow through the ground-water system—rates typically measured in centimeters or meters per year. Contaminants entering the system are flushed out slowly, if at all. Because of the nature of subsurface flow, remedial measures usually are so costly they are not economically feasible, although they may be politically desirable. Once contaminated, ground water may remain contaminated for tens, hundreds, or even thousands of years, and its use as potable water may be precluded for all practical purposes.

Problems of ground-water quality fall into four broad categories:

1. General water quality (natural background).
2. Point sources of contamination (including most waste management problems).
3. Nonpoint sources of pollution.
4. Use of the subsurface for storage.

Natural ground-water quality has been the subject of longstanding investigations by hydrologists and geochemists and has a well-developed data base. It is the latter three types of problems, brought on by man's activities, which we believe will concern us in the 1980's.

Although we do not know how far contamination extends in many areas, there is an increasing public awareness and concern about the hazards of toxic chemicals found in ground-water sources that supply wells, springs, streams, or lakes. Contamination reports are increasing, partly because of more frequent sampling and more comprehensive chemical testing of ground water. At the same time, improved analytical techniques are making it possible to detect pollutants (particularly organic solutes) at significantly lower concentrations than was possible earlier.

## Status of Scientific Understanding

Ground-water quality is determined by a large number of chemical, physical, and biological processes and factors. Many of these are fairly well understood individually and theoretically. However, in actuality, several processes and factors interact in a complex manner to affect water quality. The whole subject of chemical reactions in contaminated ground water is an area of research that needs to be advanced in the next few years if suitable measures are to be designed to handle increases in waste

Farm children at a spring in Roanoke County, Virginia, in 1926. "For millennia, man has used subsurface water as a domestic water supply." [National Archives.]

production. Even though some processes and parameters are known in a theoretical sense, the factors controlling the values involved are not well defined. For example, in many pollution problems, changes in water quality are dominated by the effects of convective transport and hydrodynamic dispersion, which in turn are related to flow velocity.

Mathematical models can be developed for the flow of ground water through a porous medium and for the transport of chemical constituents, although the empirical data base which supplies the necessary information for modeling transport is much smaller than the data base for flow. When chemical concentration affects the flow and transport equations, however, the specific processes often are not well understood.

In most field situations, the degree of mixing produced by hydrodynamic dispersion is surprisingly large. The data suggest that dispersion in the field is two to four orders of magnitude (100-10,000 times) larger than that observed in the laboratory. This implies that the heterogeneity of geologic deposits and soils introduces a mixing phenomenon which has the effect of greatly increasing the dispersion. The size of the dispersion increases as the flow path lengthens, posing problems in extrapolating the effects of small-scale tests to realistic field scales.

Because of this, it is desirable to increase our understanding of the relationships between primary geologic factors (such as tectonic stresses and original depositional environments) and the hydraulic and geochemical properties of the aquifer system (such as porosity and permeability, mineral content and variability, and aquifer boundary conditions). An increase in our knowledge along these lines also will aid in interpreting geophysical and geochemical surveys when estimating aquifer parameters and when defining patterns of ground-water quality.

Using numerical methods to solve partial differential equations usually offers an economical and reliable technique with simulation models of ground-water movement. However, coupling the equations that describe flow and transport with those describing chemical reactions produces equations that are extremely difficult to solve numerically. Further advances in numerical analysis are needed to develop models that represent complex contamination problems efficiently and accurately.

The "restorability" of a contaminated aquifer is highly dependent on its hydrogeologic and geochemical properties and on the chemical and physical properties of the contaminant. In some cases, such restoration is neither technically nor economically feasible. However, in response to public or governmental demands for positive action in clearly documented cases where ground-water contamination threatens public health, aquifer cleanup programs are being required more and more frequently. One possible approach to this problem is to induce *in situ* chemical or biological reactions that would neutralize or immobilize the contaminant. Additional research is required to evaluate the viability of this approach.

Problems of ground-water pollution may be localized or may spread over a large area, depending on the nature and source of the pollutant and on the nature of the ground-water system. A problem of growing concern is the cumulative impact of pollution on a regional aquifer from nonpoint sources as well as from many small point sources. The former is illustrated by the potential impact of atmospheric deposition on ground-water quality. The latter is exemplified by the situation on Long Island, New York, where approximately three million residents of Nassau and Suffolk counties rely on wells as their sole source of water supply. Domestic wastewater seeping from thousands of septic systems and leachates from landfills or industrial waste disposal sites have contaminated the shallow ground water in many developed parts of Long Island. We do not know enough about how the contamination from such diffuse sources is attenuated by the buffering capacities of ground-water systems to evaluate the long-term integrity of threatened ground-water resources.

In order to design water systems based on ground-water supplies, it is necessary to predict the future response of the aquifer. Inherent in any such prediction is a degree of uncertainty which increases as the predictions are extended into the future. In many engineering problems, the period of observation is commonly about equal to the period of prediction; for example, for a ten-year forecast, one usually has at least ten years of observations of the system's behavior on which to base the forecast. For predictions of waste movement in underground aquifers, the time horizon is so distant that the period of system observation will not be meaningful in predicting long-term effects. Scientists and engineers planning the use of ground water are put into a difficult position when they are required to forecast what will happen over the period of time involved in typical ground-water movements.

## Waste Disposal

The earth has certain obvious advantages for waste disposal. Most experts would agree that, properly used, it can provide a safe repository for toxic wastes. The challenge is to design disposal systems that have negligible potential for contaminating the biosphere and that permit monitoring and correcting faulty designs. This involves predicting geologic behavior and the movement of contaminants for long periods, then observing what happens. The problem is perhaps best exemplified by the current search for a geologic repository for high-level radioactive wastes, an issue discussed in Chapter 8. As with the disposal of many other wastes, the opportunities and limitations of marine disposal of radioactive waste are controversial and are not discussed in this chapter.

While toxic and hazardous waste dumps have resulted largely from failures in regulations or in their enforcement, many waste disposal sites are a result of decisions by local and federal authorities. For example, Congress has specified land disposal as an innovative alternative for disposing of wastewaters from sewage and has authorized a 10 percent increase in construction grants for such projects. Certain contaminants in wastewaters, such as synthetic organic chemicals, are not readily removed in the treatment given prior to land disposal and thus may be dispersed on the land and find their way into underground and surface waters. Added to this are the contaminants from illegal dumping of toxic wastes, a practice that in thousands of situations has allowed liquid wastes to percolate into ground water and thence into surface waters. Such situations become far more serious when storm-water runoff saturates the area and increases the rate at which the wastes spread.

Sludge, which is produced in wastewater treatment and contains many of the contaminants present in the wastewaters, may be appropriately applied to land. In proper circumstances, it can enrich the soil with nutrients and humus and other organic materials important to the soil. In other circumstances, it can contaminate the land and the subsurface aquifer.

Most communities in the United States dispose of refuse in dumps or sanitary landfills. Where these are properly sited and operated, they need not impair the quality of nearby waters. Where improperly sited, built, or operated, they have posed major pollution problems. At times, toxic materials produced by man find their way to such disposal sites, and the problems that can result when these are leached out are numerous and may be intractable.

The most frequently used land disposal methods are septic tanks and the tile fields that dispose of effluents from the tanks. Where these septic tanks are properly sited, such as in truly rural areas with appropriate soil and water table conditions, they may pose no hazard whatsoever. Frequently, however, they are improperly installed by developers who provide such facilities, together with individual wells for water supply, for housing developments. In such situations, many fail within five years. Even where these tanks operate successfully, they may pollute ground water. Recent studies suggest that as many as one out of every three septic tank installations is not operating properly, and that the consequent pollution both above and below ground is substantial. The number of household water supplies that have been deleteriously affected by failed septic tank systems is legion. The solution, beyond better engineered onsite facilities or improved maintenance, may lie in better land use control and in imposition and implementation of effective regulations for septic tank installation.

Waste disposal activities undoubtedly are introducing viruses into a variety of ground-water sources, but the persistence and movement of these viruses only rarely have been the subjects of scientific inquiry. We still can only speculate on the extent to which such viruses pose a public health hazard.

Waste disposal is undertaken in several rock types which have not been sources of water and which hydrologists generally have not investigated. These are the more nearly impermeable rocks—the crystalline rocks (granites) and the argillaceous rocks (shales). These rock types pose particularly difficult investigative problems.

Ground-water flow in crystalline rocks is mostly through fractures, and research on fracture flow and transport is active. Debate continues about how to treat theoretically flow and transport in such media, about which parameters are most significant, and about how to collect meaningful data in the field.

Perhaps one reason that hydrologists have difficulty predicting flow in fractured rocks is that not enough is known about the processes that create fractures, affect their orientation and spacing, control aperture distance and permeability, and allow their occurrence to be predicted quantitatively.

In the arid and semiarid regions of the United States, the vadose or unsaturated zone above the ground-water level may be an appropriate place to store toxic wastes. However, before this can be undertaken with confidence, a number of scientific questions concerning transport and reactions in this zone must be answered. The transport of water and vaporized contaminants in natural porous media is poorly understood and rarely predictable but, in some subsurface environments (particularly the unsaturated zone), transport may be a significant factor and an important link between ground-water pollution and human health.

Some types of argillaceous rocks have the spaces between grains closed up by clayey particles and their permeability is so low that it is difficult to measure it with standard techniques. At the same time, if these rocks are located deep within the earth, pressure from overlying strata may fracture them. Research continues on how to treat ground-water flow in this class of rocks.

### Prospects

The nation faces a period when the need for waste disposal will increase rapidly while the capacity to estimate the effects of wastes will grow slowly, given present rates of investigation. Once polluted, ground-water supplies can be cleansed only over long periods, and public management agencies will have to decide in place after place how much risk to take in permitting any discharge into aquifers.[6a]

The level of scientific understanding is adequate to evaluate and interpret a few problems of ground-water quality, but the required site-specific data are frequently too inadequate for an evaluation to be made with reasonable certainty. In some places, geologic and hydrologic knowledge will permit confident decisions about the consequences of underground disposal and safe ways to achieve it. Elsewhere, concern about threats to water quality and about the difficulties in controlling them will heighten. The realization that these threats are long-term should stimulate increased efforts in the 1980's to reduce the amount of hazardous wastes that must be disposed of into or on the earth by avoiding their production, recycling them, neutralizing them before disposal, destroying them, or improving the efficiency of waste-generating processes, as well as by conserving resources.

## NONPOINT SOURCES OF POLLUTION

Nonpoint sources of pollution include, among others, runoff from farms; urban runoff, much of which reaches water courses from storm sewerage systems or from combined-sewer overflows; wastes disposed of on land which are then leached out into soil and water; and airborne wastes. Whereas money and technology can be applied readily to conventional pollution problems, the control of nonpoint sources requires politically difficult strategies, including proper land management.

It is now apparent that, even if all wastewaters disposed of in sewers were provided with treatment directed to the goals of the Federal Water Pollution Control Act Amendments of 1972 and the 1977 Clean Water Act, pollutants from nonpoint sources would continue to affect water quality, often masking any improvement that wastewater treatment of point sources could accomplish. As already noted, such nonpoint sources are a continuing threat to ground-water resources, making some of them unsuitable as a public water supply.

Research on control of nonpoint sources of water pollution needs to be high on the agendas of regulatory agencies and of the industries and local authorities which contribute to the problem and inevitably must be part of the solution. Because of the diverse nature of the pollution sources and their highly variable impact on underground and surface waters, and because of the exceedingly high costs involved in controlling these sources, any appropriate management strategy will have to establish priorities.

## Agricultural Runoff

Erosion of topsoil by water runoff from agricultural land, whether such land is used for crops, pasture, or other purposes, has been going on for ages. The major fertile deltas of the world have been built up from topsoil carried into rivers by rainwater and deposited at the mouths of these rivers. Inland, this process, particularly when accelerated by human action, may have a deteriorating effect, sometimes filling reservoirs created for water supply, power, or recreation while denuding the surface of the land. The EPA has estimated that sediments from nonpoint sources are 360 times greater than those discharged from municipal and industrial treatment plants, with agriculture contributing more than 50 percent of all sediment loadings.[7] Work in numerous areas, including the Tennessee Valley, has shown that such erosion can be reduced by sound agricultural land management. Soil losses in other areas continue to be significant.

The pesticides and fertilizers that are widely used in modern agricultural practice are less amenable to control than is erosion. These chemicals were relatively inexpensive until recently and have been used heavily, with the result that waters from agricultural lands have been affected significantly. Thus, pesticides from farms kill fish in the same waters that wastewater treatment plants have been designed to clean. Fertilizers supply nutrients to rivers and lakes, leading to accelerated eutrophication and countering the efforts to remove such pollutants at expensive advanced wastewater treatment plants.

Agriculture in certain areas increases the nutrient burden in streams some ten times more than forested lands do. Where concentrations of nutrients are sufficiently high to render waters eutrophic and trigger algal blooms, a partial reduction in concentration, such as might be effected by reducing phosphorus or nitrogen from point sources, may have no discernible effect on water quality.

The degree to which pesticides affect water is a function primarily of how much is used. Pesticides have become a part of the food chain to the extent that some of them are ubiquitous both in the environment and in man and animals—so much so that all of the synthetic organics covered by the EPA drinking water regulations are pesticides.[8] Other types of pesticides are now either banned or controlled but, as long as stable synthetic organic chemicals are used for pesticides, they will constitute a major part of the problem in the management of pests.

Where irrigation is widely practiced, as in most of the western states, agricultural use and return flows are responsible for sharp increases in concentrations of dissolved solids, mainly salts, which have seriously affected downstream uses of water. As a result of this effect in the Colorado River basin, the U.S. government has begun construction of a major and costly desalination plant to render the waters of the lower Colorado River suitable for agricultural use in Mexico, and is involved in controversy over the efficacy of other corrective measures upstream.

## Construction Runoff

On a per-acre basis, construction and its immediate aftermath contribute far more pollution, primarily sediment, than do any of man's other activities. Construction sites for highways, airfields, shopping centers, housing developments, and industry all yield vast loads of sediment following rainstorms. Most states and many local authorities have enacted sediment control legislation for construction sites, but staffing of the regulatory agencies is so limited and the number of sites so large that the regulations have been virtually unenforceable.

## Urban Runoff

Runoff from urban and industrial areas includes fertilizers and pesticides from residential lawns, park areas, and other public greenswards; the runoff from construction sites; and the myriad other contaminants that emanate from urban centers, including those from vehicular traffic and the stack discharges of industry, power plants, and large buildings.

The contribution of urban development to contaminants in stormwater runoff can be appreciated from the units of measurement that are often used to identify this pollution: pounds of pollutant per curb-mile of street or per square mile of development. Sediments and associated toxic materials, biodegradable organic material, nutrients, oils and greases, asbestos, and microorganisms, all have been identified in runoff water.

Contaminants washed out of the air are particularly important in urban areas because they fall on impervious surfaces and are readily washed into storm sewers. Dustfalls on major cities amount to hundreds of tons per square mile per year, which translates into concentrations of solids on the order of hundreds of milligrams per liter. The increases in the concentrations of heavy metals in streams draining from urban areas are on the order of two to ten times above normal background levels.

Urban runoff reaches water courses in two ways:

1. In older cities, where storm-water collection involves the use of combined sewers that carry both storm-water runoff and household, commercial, and industrial wastewaters, the intercepting sewers never can be adequate to carry much more than the peak dry-weather flow to the treatment plants. Accordingly, overflows are provided so that, when the sewers are surcharged with storm-water runoff, they spill directly into water courses carrying not only the storm-water runoff but also significant volumes of untreated domestic and industrial wastewaters. The latter amounts to 5 to 10 percent of the total of these wastewaters.
2. In more modern cities, where sanitary and storm-water sewers are separated, storm water is collected and discharged to the nearest water courses while the domestic and industrial wastewaters are conducted to treatment plants prior to discharge.

Although extensive studies have been made of methods to control combined-sewer overflows and separate storm-water runoff, no economically feasible techniques have emerged yet. In some instances, programs of sewer separation have been instituted at vast cost. Even if an effective approach were found that ends the problem of combined-sewer overflows, it would not begin to address the problem of pollution resulting from the storm-water runoff itself.

The 1978 Needs Survey of cost estimates for the construction of publicly owned wastewater treatment facilities conducted by the EPA indicates that, in January 1978 dollars, while an estimated $35 billion will be required for additional wastewater treatment plants to meet municipal needs between 1978 and 2000, an additional $85 billion will be required to prevent or control the discharge of untreated wastes from combined-sewer overflows and to treat storm-water runoff itself. The latter alone is estimated to cost almost $62 billion, more than the cost of all of the treatment plants and control facilities for combined-sewer overflows.[9]

The cost estimates for control of urban nonpoint sources are far less certain than those for treatment facilities, and there is no guarantee that investing such funds will, in fact, accomplish what is intended. Structural technical solutions such as treatment plants are likely to be of little avail. Solutions to the urban runoff problem will require major modifications in land use policies (including the imposition of a wide range of regulations upon individual property owners), the construction of permeable rather than impermeable parking areas, the construction of retention basins for storm-water runoff which can be integrated with park and recreational areas, and other so-called nonstructural approaches to reducing flood loss. At present, the EPA provides funds for constructing interceptor sewers and treatment facilities but not for improved land management. Less costly or local measures may be ignored because construction grants for them are not available.

## Atmospheric Deposition

Pollutants carried by winds eventually reach water bodies when rain washes them out of the atmosphere or when they are directly deposited. They may fall on water courses or they may fall first on the land and then travel along its surface or underground, eventually reaching rivers, lakes, and the ocean.

A particular example of atmospheric deposition, acid rain or, more accurately, acid precipitation, results from the emissions of gaseous sulfur and nitrogen oxides which react with atmospheric water and, when oxidized, form dilute solutions of sulfuric and nitric acids. The sources for these oxides are diffuse: fossil-fueled power plants, industrial manufacturing plants, cars, and so forth. Of these, the stationary sources are more amenable to control, particularly for sulfur dioxide, through the construction of scrubbers in large plants. Emission controls for cars and smaller furnaces will be slower in coming; the control of nitrogen oxides will be particularly difficult because the requisite technology is insufficiently developed (see the section on "Environmental Effects" in Chapter 16, "Issues in Transportation," for a discussion of highway vehicle, aircraft, rail, and marine emissions).

Early evidence of the impact of acid rain in the United States was found in the Adirondacks, where the excess acidity killed off life in many lakes; the problem is more acute during spring runoff when snow-melt floods the lakes with highly concentrated acids just as vulnerable minnows and fingerlings are hatching. In addition to increasing the acidity of lakes, acid rain, contrasted with normal rain, tends to leach from the soil substances that may be of health significance in drinking water supplies.

The phenomenon of acid rain has spread rapidly across the continental United States. In 1955-1956, acid precipitation was recorded in 12 northeastern states. By the early 1970's, it had intensified in the northeast and extended westward; it now affects all of eastern North America, except for the southern tip of Florida and the northern areas of Canada. In the western United States, acid rain affects major urban centers on the coast and the mountain areas of the continental divide.

Research to explore the full manifestation and significance of the increasing acidity of surface waters has not yet begun; thus, no meaningful attempts have been made to control it.

### Control Strategies

Some scientific observers have concluded that the equity approach implicit in the Clean Water Act results in tackling all problems with equal fervor, with the result that some of the more serious problems continue to be troublesome while major investments are made in addressing problems of only marginal consequence. They argue that the nation cannot afford this expensive and slow process with nonpoint sources of pollution. Under this view, the most serious situations must be identified and efforts must be made to alleviate them. Then, the lessons learned can be applied to problems of lower priority. It is inappropriate to invest billions of dollars in combined-sewer overflow controls that may make little discernible improvement in the discharge waters, while neglecting the Love Canals. Rendering water suitable for swimming in one place while it continues to threaten life in another appears to be an abdication of common sense in the name of administrative and bureaucratic conformity. These problems will require corrective action in the years immediately ahead. As noted later in this chapter, the design of new institutions, economic incentives, and regulatory surveillance may be as important as new technology.

## AQUATIC ECOSYSTEMS

The problems of water quality discussed in this paper affect not only human health but also the health of the water systems that support a wide variety of plant and animal life—lakes, rivers, swamps, and estuaries. It is in these bodies of water that changes in quality are most evident. The sterile lakes and streams of northern New York State are perhaps the most dramatic demonstration of how man's activities can have unexpected and far-reaching consequences. Deposition of material from the atmosphere is only one way in which aquatic ecosystems can be affected. Changing nutrient levels, introducing new species or toxic chemicals, dredging, or filling can all upset the delicate communal web that keeps an ecosystem alive and productive.

Unfortunately, it is not yet possible to measure or predict with any degree of certainty when and how this balance will be upset by man's activities. Aquatic ecologists still have to address some fundamental scientific questions about these systems, including those concerning what standards and methods should be used to measure the effects of human intervention or, indeed, what things should be measured. Aquatic biologists need to develop a system analogous to that of the medical profession, in which blood pressure, temperature, etc., indicate whether the patient is healthy or ill. Such a system allows diagnoses to be verified and cases to be discussed among colleagues in all parts of the world. While many water quality problems are local, aquatic ecosystems often cross jurisdictional boundaries, and a uniform language for discussing problems would go a long way toward solving them.

While there has been some progress in selecting standard methods for measuring characteristics, it is still not clear what characteristics should be measured. Only when we have reached some agreement on what characteristics indicate changes in the health of an ecosystem can we begin to predict changes validly and deal with degraded ecosystems.

Our inability to predict changes means, among other things, that the Toxic Substances Control Act cannot be implemented effectively because there is no way to determine accurately the hazards to the environment (as required by the act) posed by chemicals. It means that the environmental impact statements required by the National Environmental Policy Act contain no scientifically based estimates of effects on aquatic ecosystems.

The lack of agreement among aquatic ecologists on how impacts should be assessed is reflected in criteria documents prepared by the EPA. Among other things, the data on which the documents are based have not been collected systematically; a variety of methods, species, and time limits are used; statistical analyses are almost totally lacking; and biological information is fragmented and is not coupled to chemical and physical information.

It is one thing to test toxicity in the laboratory—as can be seen from the chloroform studies described earlier—and another to extrapolate the results to the real world. Aquatic ecosystems, with their complex collections of plant and animal species that respond to many fluctuating variables—water temperature, salinity, acidity, and so on—are even more complicated than human beings.

One way to bridge the gap between laboratory and field is to set up a fragment of an ecosystem in the laboratory—a microcosm of the larger system—and examine a particular ecological relationship or function. In order to do this successfully, it is necessary first to determine the basic processes of the ecosystem and then to discover which are useful in evaluating particular effects of pollution. Once this has been done, different types of microcosms can be developed to study specific relationships or functions. The hazard to an aquatic ecosystem from a particular course of action (for example, use of a new chemical) can be estimated crudely in laboratory tests by using only one species. However, the accuracy of the estimates should be confirmed in laboratory microcosms rather than in the "real world," so that outside influences can be eliminated.

While a promising beginning has been made in using this technique, present methods are inadequate for the needs and are not being developed rapidly enough to keep up with man's ingenuity in creating new hazards for ecosystems—such as new chemicals. Development of this predictive technique will have to be accelerated if we are to avoid wrong decisions and ill-conceived corrective programs.

## Recovery of Damaged Ecosystems

In spite of the lack of scientific agreement on how to evaluate aquatic ecosystems, there have been some notable successes in the recovery or rehabilitation of damaged ecosystems. For example, Lake Washington, a large body of fresh water lying to the east of Seattle, was polluted by effluents from a number of small sewage treatment plants as well as by runoff from city streets. This pollution has been virtually eliminated by a concentrated attack on the problem, leading to the construction of interceptor sewers to convey wastes to a large regional sewage treatment plant.[10] Likewise, the remarkable recovery of the Thames and Clyde rivers in the United Kingdom provides compelling evidence that present methodologies and waste treatment technologies, however imperfect conceptually, can produce remarkable results in a relatively few years in favorable economic circumstances. Furthermore, the steps that led to the recovery of these ecosystems were not financially ruinous to the industries using the water and have, at least in the case of the Thames, turned an ecological eyesore into a multiuse recreational facility. The success of the overall program on the Thames is seen in the rising demands for recreational uses that may conflict with each other (for example, sailing and fishing).

In an era when energy costs are skyrocketing, it is quite likely that recreation close to one's residence will become increasingly desirable. In light of this, it seems prudent to consider the rehabilitation of aquatic ecosystems near large population centers.

Rehabilitation may be defined as a pragmatic mix of nondegradation, enhancement, and restoration (returning an ecosystem to its initial state). If restoration or rehabilitation is to be effective, there are several areas that require attention:

1. Greater precision is needed in determining how much ecological improvement results from specific reductions in pollution loads at point and nonpoint sources.
2. The ability to estimate recovery time once pollution sources are removed needs to be substantially improved.
3. Improvements are needed in monitoring quality during the rehabilitation or recovery periods and better ways must be developed to maintain the desired quality once it has been achieved.

Any plans for rehabilitation on a national scale should:

1. Systematically identify ecosystems where rehabilitation is desirable.
2. Establish priorities for attention.
3. Determine the degree of rehabilitation that would provide amenities attractive to the public.
4. Estimate the degree of ecological improvement that would result from reductions in waste discharges and nonpoint source discharges into the system.
5. Develop either a professionally endorsed set of parameters that will permit the improvements to be tracked as they occur or a monitoring system that would ensure quality control measures to protect the rehabilitated system.

### Growing Importance of Instream Uses

In recent years, it has become increasingly important to maintain minimum flows in surface streams in order to protect aquatic ecosystems and improve aesthetic values. Several of the western states in which the use of water is allocated under the appropriation doctrine have passed laws that provide for the establishment of minimum flows to protect fish and wildlife.[12] The statutes are silent as to how such minimum flows are to be established, and many law suits have placed the decision before the courts. Since we do not have a predictive skill based on an understanding of the assimilative capacity of aquatic systems, it is very difficult to establish a rational basis for determining the instream flows required.

## CRITICAL INSTITUTIONAL AND ECONOMIC ISSUES

### The Need to Reexamine Approaches to Water Quality Management

In the 1930's and 1940's, a bold attempt was made to deal with the continental waters of the United States through drainage basin development plans, such as those for the Tennessee and the Columbia basins. These plans went a long way toward managing large rivers for irrigation, navigation, and hydropower. Some other aspects, such as improved water quality, received less attention.

Notwithstanding these efforts, and those of the 1960's and 1970's, responsibilities for the overall management of water resources in the United States have not been defined clearly since the federal government assumed responsibility for improving water courses for navigation in the

1820's. The missions of federal agencies are diverse and in partial conflict. The roles of federal and state governments with respect to water quality remain in controversy. Even where the physical and biological problems and the possible avenues to their solution are well defined, it is difficult to marshal the public capacity to deal with them.

Most of the water quality problems referred to in this chapter have been apparent for many years but received little attention until the 1960's. This section examines the institutional attempts to improve water quality that have been made in the past two decades and suggests that a reevaluation of our approaches is needed.

## Regulatory Strategies

When the Congress started to make a serious effort to improve the quality of the nation's water courses in the mid-1960's, it had a choice of several strategies in pursuing its aim. Its basic choice was reflected in the Federal Water Pollution Control Act of 1965. Simply put, the strategy was to design an applied research program that would permit an understanding of how major water courses would respond to control actions (so-called "river surveys") and then to develop a plan for achieving prescribed standards in the water course itself (*ambient* standards). At the same time, a major program of subsidies for the construction of municipal waste treatment plants was initiated. Several of the surveys were completed and, in the process, some notable methodological improvements in analyzing the quality of water courses were produced. The interdisciplinary modeling work done on the Delaware estuary was especially outstanding.[13]

Based upon these analyses of water courses, the states were then to produce water quality plans to meet the prescribed ambient standards. The states, however, were slow to act and, when the plans were produced, they usually had no discernible link to the ambient standards that were to be achieved.

A few fledgling regional water management agencies (for example, the Ohio River Valley Sanitation Commission and the Delaware River Basin Commission) were in existence at the time and many of their supporters hoped they would become effective and efficient regional water managers that would use the tools being developed in the surveys. By then, research also had shown that regional management approaches using a wide range of technologies could achieve a given set of ambient standards at much lower cost than the conventional regulatory approaches. However, as noted, the approaches taken under the 1965 act failed to yield quick results and the quality of the nation's water courses continued to worsen while, at the same time, environmental concerns were mounting rapidly. Congress attacked the problem by means it regarded as more direct and decisive.

This resulted in the 1972 amendments to the Federal Water Pollution Control Act of 1965, the basic law in force today. The new amendments relied on uniform, technologically based *effluent* standards, with specific timetables for the application of technology, the threat of enforcement actions if the regulations were violated, and substantial federal grants to encourage cooperation. Thus, attention largely turned away from understanding conditions in the water course itself and focused on reducing and eventually eliminating all effluent, whatever its impact upon a river basin. Regional and state agencies were reduced to being implementers of federal law. While many of the objectives of the act have not been met and may never be met, it has resulted, with the aid of federal subsidies, in large capital investments in control facilities by both municipalities and industries.

Experience in attempting to implement the 1972 amendments taught a variety of lessons. While the legislation established the symbolic goal of committing the nation to water quality, actual accomplishment has come slowly. Although progress has been made in cleaning up some waters, the cost has been enormous and the effort has not been especially efficient. Clearly, the cooperation of state and local governments is crucial to implementation, and limitations upon state resources and the lack of state commitment to federal programs that show only modest benefits have made implementation difficult.

Currently, the problems mentioned in earlier sections, along with the continued relevance of some of the considerations stressed by advocates of regional approaches, suggest the need for careful reconsideration of the approach adopted in the 1972 amendments. This is the case at least in the more heavily developed industrial and agricultural river basins in the United States. Such reconsideration requires both analytical studies and the encourgement, rather than the displacement, of regional institutions for water quality management. The nation faces a reassessment of its water pollution abatement policies during which new calls will be made on science and technology to develop cost-effective strategies to meet realistic goals.

For example, there is much interest in developing advanced-level treatment of wastewater to remove plant nutrients and various trace contaminants from effluent streams discharged from point sources. Such treatment would be very expensive and it is not known how much it would benefit water quality. Nor is it clear in many areas how much emphasis upon point sources is warranted in comparison with other types of measures. It seems wise, therefore, to make renewed efforts to try to understand river basins quantitatively as aquatic ecosystems—that is, through models. The problem is difficult and a solution would require a collective research effort, but the alternative may be to undertake huge further investments which, in the final analysis, are likely to yield little or no return.

One reason for this lack of success might be that nonpoint sources, including, as noted earlier, inputs from the atmosphere, may be contributing as many polluting substances as do point sources. If this is so, the problem takes on an even more distinct regional quality, including not only the river itself but the watersheds as well, and possibly even larger regions. If the problem is manageable at all, the control of contributions from nonpoint sources would have to depend on measures taken on a regional scale, such as changes in agricultural and other land use practices, possible modifications of water courses, and special measures to control air pollution.

Another frontier problem, especially in highly developed areas, is keeping hazardous wastes out of surface and underground water supplies. As indicated in the section on drinking water quality and groundwater contaminants, this problem is extremely troublesome in some areas. It seems likely, and limited research suggests, that this problem is more efficiently and effectively handled on a regional scale than in a dispersed manner. Some institutional innovation is occurring in this area; for example, the Gulf Coast Waste Disposal Authority is in the process of establishing a regional-scale management system for hazardous industrial wastes.

It is therefore an appropriate time to reexamine the experience that has accumulated with regional approaches to water quality management, especially in the light of a new generation of problems. There is a range of experience to be assessed, although most of it deals with the more conventional quality problems. Long-standing examples are the Genossenschaften in the Ruhr area of West Germany and the Ohio River Valley Sanitation Commission. More recent are the federal-interstate compacts in the Delaware and Susquehanna river basins and the Gulf Coast Waste Disposal Authority. The British water authorities, which are responsible for water supply as well as wastewater disposal, represent the ultimate in regional ownership and management at present.[11] Other regional wastewater management authorities have been created in metropolitan areas (for example, in the Twin Cities area of Minnesota). These various enterprises have been in operation long enough to have accumulated experience. Questions that should be examined include why some of these groups have been so frustrated in their efforts to mount regional management programs; what distinguishes the more successful ones; how they have been affected by national policy; what are the appropriate interrelationships among local, state, regional, and national authorities; and how the authority of such agencies should be arranged to take account of nonpoint sources of pollution and hazardous wastes.

Human activities interact with a complex natural system that usually does not conform to jurisdictional lines. The accumulated experience under the Federal Water Pollution Control Act Amendments of 1972 and the water problems outlined in this chapter may have increased the incentives for states to cooperate in forging regional institutions that will be politically more viable.

## Economic Incentives in Water Resources Management

Water supplied by federal agencies has, for historical and political reasons, almost always been subsidized (sometimes heavily, usually in the form of subsidized capital investments), as in the case of agricultural water provided by federal agencies, particularly the Water and Power Resources Services. In recent years, there has been heavily subsidized investment in water pollution control facilities for municipalities and industries. Another "subsidy" has been in the form of free use of environmental common property resources (i.e., air, land, and water) for waste disposal. These various forms of subsidies have led to the unnecessarily large use of water, excessive generation of waste materials, and degradation of the environment.

Some economists have been calling attention to this problem for a long time, but interest in management approaches using economic incentives has now spread to public agencies at all levels of government. For example, for the first time there is a substantial group of people in the Environmental Protection Agency looking into economic incentives as a supplement to the present command and control regulations and some municipalities have experimented with incentive pricing of municipal water supplies. The latter calls for pricing, at least for the larger users, not on the basis of subsidized or even average cost, but at the cost of development of the next increment of water supply. The idea is that, unless this is done, users will consume additional water for which they would not be willing to pay if they were charged its true cost. This idea has gained attention because many municipalities are finding the development of additional water supplies difficult and costly.

Large cutbacks in water use achieved by cities like Tucson through public information programs suggest that other incentives in addition to price also may be effective. Water shortages in some areas and periodic droughts, along with the environmental movement, have made the idea of water conservation more attractive. As stresses on the nation's water resources continue to increase in the 1980's, it will be necessary to reconsider carefully the management policies of public agencies in the light of research concerning the practicability and implications of alternative pricing policies, along with other incentives for reducing water use and encouraging its reuse.

Effluent fees have been proposed frequently as a means of dealing with the damage caused by the free use of common property resources for waste disposal. Effluent fee systems are in effect, or shortly will go into effect, in several countries. In this plan, a fee is levied on discharges to public water courses that takes account of the fact that other uses are precluded or damaged by the discharge.

It often has been argued that a system of effluent fees could make noncompliance expensive for the laggards and therefore could be a powerful stimulant to action. Three additional arguments are made for fees:

1. Fees could be designed to provide incentives for good operation and maintenance of facilities which otherwise would suffer from poor management.
2. Fees could help to stimulate innovation and technological improvement in processes and products that would reduce the generation of polluting wastes and improve the treatment of wastewater once it is generated. They also would help to finance the administrative aspects of water quality management, i.e., planning, maintaining, sampling, and applying sanctions.
3. Fees might provide an efficient source of finance for research to understand the natural systems involved and for planning and administering the regional approaches to water quality problems discussed previously.

On the other side, it is argued that effluent charges would be impractical. It would be difficult to set the fees. Complex systems of monitoring might be required. Public understanding is lacking, and effluent charges are sometimes looked on as licenses to pollute.

A major problem for applied research in the coming five years is to find if and how practical economic incentive schemes can be designed to supplement existing or modified regulatory schemes and how the incentives provided at all levels of government can be made more consistent. Without such research and testing, involving a wide range of interdisciplinary study, the nation will have advanced little in solving water policy issues—which grow more acute as demands on water resources increase.

## Research Organization

Many of the problems scientists need to address if we are to have effective and efficient management of our water resources do not fall neatly into the single-discipline boxes that have characterized research in this area to date. The more difficult and more interesting problems of ground-water contamination and aquatic ecosystems involve complex interactions between man and the environment. They draw on chemistry, physics, and biology as well as on the social and behavioral sciences for solutions. For example, advanced-level waste treatment at the Blue Plains wastewater treatment plant in Washington, D.C. is being considered to remove nutrient nitrogen from the effluent discharge. The initial investment for this facility is large, on the order of $100 million. The purpose of removing this nutrient is to improve the quality of the Potomac River estuary, but it is necessary to assess the impact that this removal will have on the ecosystem in the estuary.

In order to approach this problem scientifically, we need information on:

1. Flow in the estuary and its variability.
2. Chemical fluxes, especially those of the major nutrients which, in the benthic zone (i.e., the bottom layer of water), play an important part.
3. Sediment transport in the system.
4. The interaction of the aquatic ecosystem with the flows of nutrients and sediment.

A fuller study would examine the costs, benefits, and social returns of achieving alternative levels of water quality in the estuary, thus greatly increasing the range of pertinent data and analysis.

Research on problems such as this one requires an interdisciplinary team. However, it is not easy to assemble such teams and, once assembled, they are difficult to orchestrate in such a way that the end products of their work address the total problem. Frequently, each team member ends up working independently and the research product becomes nothing more than a collection of smaller projects. In addition, a team effort requires a funding commitment that is considerably larger and usually longer in duration than that for an individual investigator. An investigation of an estuary requires large-scale logistic support simply to collect the data. Many funding institutions, including many research institutes, are reluctant to take on commitments of such scale and duration; they tie up too much of the available resources for too long.

In the studies that have been attempted, the failures outnumber the successes. Emphasis on traditional disciplines and the competitive nature of the university world make it difficult to organize an interdisciplinary team of more than two or three principal researchers. A few universities have organized institutes or centers to deal with the interdisciplinary nature of water problems, but these also have not been very successful since they are outside the traditional disciplines and therefore not in accord with the academic reward structure.

The interdisciplinary teams that have been successful appear to have the following features in common:

1. Stable and adequate funding for at least five years. Each investigation has a certain minimum funding level below which the effort is unlikely to succeed.
2. A clearly identified leader who has some leverage over all of the members of the team and who can inspire team enthusiasm.

3. Continuous and close communication across disciplinary lines.
4. An environment in which the individual researcher can publish in his own discipline while participating in the team effort. This requires that the individual investigator have a degree of freedom and autonomy.
5. A finite life for the effort, or at least finite phases of the effort, so that there are definable target dates for synthesizing results.

These features should be taken into account by government agencies funding research in water management. Either an in-house team that meets these conditions should be assembled or research grants to institutions should be monitored closely to ensure that the results are integrated and relevant to the problems under study.

## Summary and Outlook

While water shortages and increased demands for water will continue to affect some parts of the nation in the next five years, the more urgent problems will be those of water quality. A lack of scientific knowledge is leading to decisions affecting water quality that may not be soundly based. Vast sums of money are being spent on programs that have effects that cannot be measured scientifically. Increases in our knowledge are needed and can be obtained over the next five years. The most immediate scientific problems that must be solved are found in the areas of drinking water quality, ground-water contamination, nonpoint sources of pollution, and deterioration of aquatic ecosystems. Research to find a basis for better methods of managing water quality is needed as problems are tackled increasingly on a regional scale. These deserve the higher priorities. The search for solutions to these and other problems will be facilitated by better research organizations that are capable of dealing with them on an interdisciplinary basis.

The United States has public water supplies of better quality than most other countries, and the spread of waterborne communicable diseases has been almost entirely eliminated. There are, however, questions concerning the possible carcinogenicity of certain organic contaminants that have become widespread in recent years and that show up under improved analytical techniques. The long-term effects of these contaminants are not fully understood, but the expenditure of large sums of money to remove them is mandated by regulatory agencies.

One of the ways in which more contaminants are reaching the population is through ground water, which serves as a source of drinking water for about 40 percent of the nation's population. This contamination occurs partly because the statutory prohibitions against the disposal of pollutants in the nation's surface waters or by burning are encouraging land disposal, without consideration of its long-term effects. More recently, statutory encouragement for land disposal of sewage effluents may have enlarged the potential hazards. Knowledge of how contaminants react in underground aquifers is imperfect and efforts to avoid or deal with further pollution will at best be crude, except where detailed studies are undertaken.

The nation's approach to pollution abatement has concentrated on point sources, but the gravity of pollution from nonpoint sources, including agricultural runoff, erosion from construction sites, storm-water runoff from urban areas, land disposal of wastes, and atmospheric deposition, suggests that it must be controlled in the years ahead if the quality of the nation's waters is to be improved significantly. New institutions that can operate on a regional, interdisciplinary basis, economic incentives, and better regulatory surveillance are needed if water quality and aquatic ecosystems are to be maintained. The effects of pollutants on aquatic ecosystems are not at all well known, partly because of the lack of agreement on parameters to measure the health of the ecosystems and on the analytical methods to be used. Such methods need to be developed in order to predict the impacts of human activities on ecosystems and to formulate plans that will facilitate the recovery or rehabilitation of damaged ecosystems. The partial recoveries of Lake Washington in Seattle and of the Thames and Clyde river basins show that much can be and is being done.

## REFERENCES

1. U.S. Water Resources Council. *The Nation's Water Resources 1975–2000: Second National Water Assessment.* Volume One. Washington, D.C.: U.S. Government Printing Office, December 1978.

2. Ibid.

3. Ibid.

4. Constance M. Boris and John V. Krutilla. *Water Rights and Energy Development in the Yellowstone River Basin.* Baltimore, Md.: Johns Hopkins University Press for Resources for the Future, 1980.

5. U.S. Geological Survey. *Synthetic Fuels Development, Earth Science Considerations.* Washington, D.C.: U.S. Government Printing Office, 1979.

6. John B. Weeks et al. *Simulated Effects of Oil Shale Development on the Hydrology of Piceance Basin, Colorado.* Geological Survey Professional Paper 908. Washington, D.C.: U.S. Government Printing Office, 1974.

6a. *Report of the Consulting Panel on Health Aspects of Wastewater Reclamation for Groundwater Recharge.* State of California State Water Resources Control Board, Department of Water Resources, Department of Health, June 1976.

7. U.S. Council on Environmental Quality. *Environmental Quality.* Ninth Annual Report. Washington, D.C.: U.S. Government Printing Office, 1978, p. 119.

8. U.S. Environmental Protection Agency, Office of Water Supply. *National Interim Drinking Water Regulations.* Washington D.C.: U.S. Government Printing Office, 1976.

9. U.S. Environmental Protection Agency. *1978 Needs Survey, Cost Estimates for Construction of Publicly Owned Wastewater Treatment Facilities.* Washington, D.C.: U.S. Government Printing Office, 1979.

10. W. T. Edmondson. "Lake Washington and the Predictability of Limnological Events," *Archives of Hydrobiology*, Vol. 13(1979), pp. 234-241.

11. D. A. Okun. *Regionalization of Water Management.* London: Applied Science Publishers, 1977.

12. For example, see Rev. Code Washington, par. 90.22.010 and 90.22.020 (1969) and Oklahoma Stat. 82 Sec. 1451-1459 (1971).

13. Walter O. Spofford, Clifford S. Russell, and Robert A. Kelly. *Environmental Quality Management: An Application to the Lower Delaware Valley.* Research Paper R-7. Washington, D.C.: Resources for the Future, 1976.

# 8

# *Radioactive Waste Management**

## INTRODUCTION

This chapter is concerned primarily with the current state of technical knowledge in the management of radioactive wastes and with identifying uncertainties in this area that can be expected to yield to scientific and technical advances over the next five to ten years. But the scientific and technical aspects exist within a larger social context of even greater complexity. For that reason, even though treatment of broad social questions is not and cannot be our primary focus here, we begin with some brief remarks regarding the social context created by public perceptions and institutional constraints which directly affect and complicate the management of radioactive waste. As technical matters are reviewed, the social context and its difficulties should be kept in mind, and we will touch upon them explicitly again from time to time.

---

*Although primary responsibility for this chapter rests with the chapter coordinator, it is based on contributions from members of the Committee on Radioactive Waste Management (CRWM) of the Commission on Natural Resources of the National Research Council, and the discussion owes much to the suggestions and criticisms of the CRWM membership. The document has been approved by the CRWM for inclusion in this report on science and technology.

---

◀ Inside Gable Mountain, Washington. Two canisters of spent fuel and one of vitrified waste placed in this cavern will be part of a test to qualify basalt as a medium for the storage of radioactive wastes. [Richland Operations Office, U.S. Department of Energy.]

## NONTECHNICAL ASPECTS OF RADIOACTIVE WASTE MANAGEMENT

### Public Perceptions

Within the last decade, the public has shifted partly away from its once strong support of the use of nuclear power to generate electricity. The shift has been associated with a lack of confidence in the ability of appropriate agencies to develop a safe method of transporting and isolating radioactive wastes. Opinions concerning nuclear waste management must now be seen in the context of more general attitudes toward nuclear power, for the two matters have become linked in public discussion. After the accident at Three Mile Island, support for nuclear power declined to only a narrow majority, while concern increased that something is also amiss in the management of nuclear wastes. By mid-1979, this was considered to be a major problem by more than 80 percent of the public.[1] Yet, by a wide margin (more than two to one), those polled were willing to accept risks of waste disposal if the best safeguards available are used. A condition voiced by significant minorities was that the disposal options would be more acceptable if they involved temporary storage of wastes at the nuclear plants and permanent disposal "not in [one's own] state."[2]

In any event, a number of questions present themselves. Where do people get their information, and how much do they actually know? Is the risk from low-level radiation as perceived by some segments of the public comparable with the real risk? How does the public respond to arguments involving probability? What is the public's perception of the dangers of radioactive wastes relative to other risks they take, both voluntarily and involuntarily? How much does the public know about the natural radioactive background and its variations? What does the public regard as an acceptable risk? To what degree does the public's appreciation of these matters affect their acceptance of radioactive waste management practices? Has the public been adequately informed about waste management problems? All these questions are relevant to waste management planning.

### Institutional Constraints

While three governmental agencies—the Department of Energy (DOE), the Environmental Protection Agency (EPA), and the Nuclear Regulatory Commission (NRC)—share the major responsibility for nuclear waste management, numerous other agencies—the Council on Environmental Quality, the Office of Science and Technology Policy, the Department of Transportation, and the Department of the Interior—play important roles as well. Disagreements among these agencies and in Congress over the adequacy and direction of the current program were reflected in the difficult search for consensus experienced by an interdepartmental review group established by President Carter.[3] These disagreements have hampered efforts of the federal government to define a coherent waste management policy.

Regulatory gaps remain in the waste management program. The Nuclear Regulatory Commission's mandate to regulate radioactive waste, as defined in the Atomic Energy Act of 1954 and the Energy Reorganization Act of 1974, specifically exempts many activities of the Department of Energy and its contractors from the Nuclear Regulatory Commission's licensing mandate. The result is continuing disagreement as to whether some of the required radioactive waste storage facilities should be licensed. It is clear that there is an urgent need for a thorough review of the regulatory process.

A third source of managerial and regulatory uncertainty lies in the difficulty of meshing complex developmental and regulatory activities in an environment of rapidly escalating programs and changing policies, missions, and assumptions. Criticism has been directed, for example, at the Nuclear Regulatory Commission for its failure, as of 1980, to promulgate site suitability criteria despite the fact that the national search for repository sites has been under way since 1976. One must take into account that, because of technical complexities, selection and characterization of the initial sites will take several years. Moreover, the eventual NRC site selection criteria are likely to be based on the experience gained in the first round of site investigation. Although President Carter chose in favor of a more deliberate schedule, the announced timetable still appears optimistic and may not take adequate account of the procedures required by the National Environmental Policy Act and of the realities of public participation.[4]

A fourth problem lies in the challenge by local governments to federal preemption of control over radioactive waste management, as indicated by the temporary closing of low-level waste repositories in 1979 and the numerous instances of state and local legislation dealing with radioactive wastes over the past several years. Some states have prevented the Department of Energy from undertaking geologic studies aimed at evaluating possible sites.

Behind these intragovernmental conflicts over radioactive waste management lie the deeper issues of federal-state relationships and public acceptability of nuclear power. It is not clear how current efforts will resolve these issues. But without their resolution, it will be exceedingly difficult to implement any waste management plan, however technically feasible.

There are two important implications of current public perceptions and institutional uncertainties. First, even if it were possible to prove that current technology is adequate for the problems of handling radioactive waste, such proof in itself would not entirely remove public apprehension or solve the difficulties of institutional cooperation. Evidence of this kind would play a part, but public acceptance of a solution to the radioactive waste management problem will be influenced by other factors as well. Second, institutional constraints depend upon the social and political environment and require institutional action rather than more research on the physical nature of wastes in varying repository environments. But the more complex problem of how the public perceives the interplay of costs and benefits is, in contrast, poorly understood. The problem pervades the fashioning of technical solutions, and it should receive adequate attention in the nation's allocation of its research.

Our discussion turns now to the state of scientific and technical knowledge of radioactive waste management, and we will look first at sources and effects of exposure to radiation. We will review the current state of our knowledge about the effects of low-level radiation and the present status of criteria for protection of the public health. Next, the characteristics of radioactive wastes will be described along with options available for waste management. Finally, the most advanced of those options, isolation in mined cavities, will be discussed in detail.

## SOURCES OF EXPOSURE TO ENVIRONMENTAL RADIATION*

An understanding of the natural and manmade sources of radiation to which humans are normally exposed[5] provides a useful perspective on any discussion of radioactive wastes (see Figure 1). Whatever the source of the radiation, our concern is mainly with its effects on humans, and we will turn to those effects in the next section.

By far the largest sources of human exposure are from the practice of medicine and natural radiations from either the earth's crust or outer space. Each of us is irradiated both from within (as in the case of carbon-14 produced by cosmic rays and ingested with foods) and from without (as in the case of radiation emitted by naturally radioactive soils, rocks, and construction materials).[6] Potassium, an element essential to all life, is sufficiently radioactive to impart readily measurable radioactivity to the body. The radioactive element radium, which is present in soils, is dissolved into drinking water and finds its way into foods we eat. Radon, a radioactive gas, is produced by radium and becomes part of the air we breathe.

---

*The word radiation is used here in a narrow sense. It includes only radiations produced by radioactivity or cosmic rays. It does not apply to other kinds of radiation, such as visible light or ultraviolet radiation, that play no part in the waste problem.

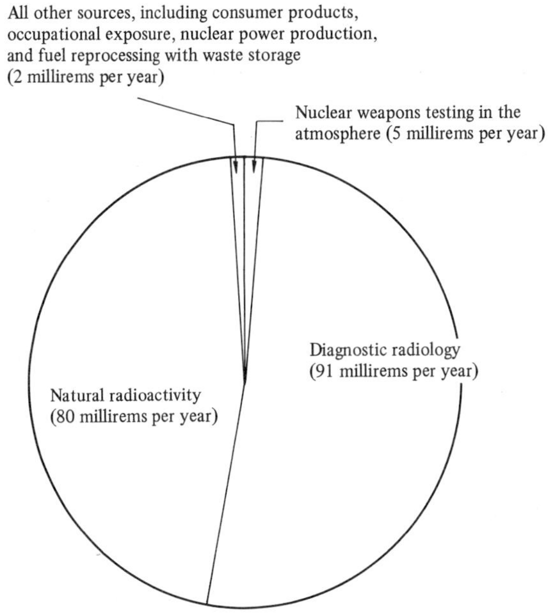


**Figure 1 Estimates of U.S. per capita radiation dose to bone marrow from all sources of exposure.** [SOURCES: "All other sources" from National Research Council, Committee on Biological Effects of Ionizing Radiations, *The Effects on Populations of Exposure to Low Levels of Ionizing Radiation.* Third Edition. Washington, D.C.: National Academy of Sciences, 1980. *Estimates of Ionizing Radiation Doses in the United States, 1960-2000.* ORP/CSD 72-1. Rockville, Md.: U.S. Environmental Protection Agency, 1972.]

The average person in the United States and most parts of the world receives a dose to the whole body of about 80 millirems per year from all sources of natural activity, but this may vary depending on geographical location and life style. (The millirem, one thousandth of a rem, is a measure of radiation dose.) Some individual organs, such as bone and the lungs, receive a higher dose than does the whole body from natural sources. An important variable is the dose received from cosmic radiation, which averages about 28 millirems per year at sea level but increases with altitude because of reduced atmospheric shielding. In Denver, one mile above sea level, the annual dose from cosmic radiation is more than 50 millirems per year.[7]

At some places in the world, exposures are much higher than normal. In the city of Guarapari, Brazil, more than 10,000 inhabitants receive a whole body dose of about 600 millirems per year, and in the Indian state of Kerala, about 100,000 residents of coastal fishing villages receive a somewhat higher dose. In both places, the sources of radioactivity are local mineral deposits containing thorium.[8] The exposure to natural radioactivity is also elevated, though to lesser degrees, in parts of the United States, such as western Colorado and central Florida.

Specific parts of the body sometimes receive higher doses from localized exposures to radiation. The lungs, for example, are exposed directly to radon and to a number of radioactive decay products that tend to attach themselves to the inert dust of the atmosphere. The radon concentration in the air within buildings is often higher than it is outdoors and varies with the materials of construction. Weather stripping and other measures that reduce ventilation in buildings also reduce the dilution of radon that diffuses from construction materials. Recent measurements in a number of countries indicate that the lung dose from indoor exposure to radon may range as high as 1,000 to 2,000 millirems per year.[9-11] (In this case, "lung dose" refers to the dose to the basal cells of the bronchial epithelium, which is the tissue from which most lung cancers originate. Radiation is often highly specific in its anatomical site of action.)

Both oil and coal contain small amounts of radioactive uranium, thorium, and their decay products. Some of these substances are discharged to the atmosphere in measurable amounts, and although the dose received from oil and coal emissions is minuscule, it is nevertheless greater than the doses from many nuclear power plants of comparable capacity under normal operating conditions.[12] Additional exposure is received from cosmic rays when people fly at high altitudes: the dose received during a transcontinental trip across the United States is about 2 millirems.

The use of ionizing radiation in medical diagnoses and therapy is the largest deliberate source of human exposure, estimated to be about 91 millirems per person per year in the United States.[13]

Another source of human exposure has been radioactive fallout from nuclear weapons tests, varying with the time and location of the tests. Residual strontium-90 from tests conducted before the weapons-testing moratorium of 1963 is currently resulting in an annual average dose of about 4 millirems.[14-15]

## EFFECTS OF EXPOSURE TO LOW LEVELS OF RADIATION

If radioactive wastes are not effectively isolated, the public may be exposed to more radioactivity than normal, particularly from contaminated food, water, and, to a lesser extent, air. Such contamination could provide an internal source of potentially harmful radiation.[16-17]

The main concern is the likelihood of an increase in the probability of developing cancer. Chromosomal changes may also be produced that can genetically cause malformations or disease in the offspring of the persons exposed. During the past two decades, however, research has shown that the genetic effects of low levels of radiation are slightly less than originally estimated, in contrast to the risk of cancer induction, which has proved to be greater than first thought.[18]

An underlying reason for public difficulty in gaining confident understanding about radioactivity and radioactive wastes may be due in part to the enormous range of quantities that must be considered. Discussion of radioactive waste can involve units as low as picocuries (a curie is a measure of radioactivity equal to 37 billion disintegrations per second in any nuclide; a picocurie is $10^{-12}$ curie) and as high as many hundreds of megacuries (a megacurie is one million curies). This is a range of 20 orders of magnitude, comparable to the ratio of the diameter of an atom to the distance from the earth to far beyond the sun. Such a range is not easy to comprehend.

Information about the biological effects of radiation comes both from experiments with laboratory animals and from human experience, such as that of the survivors of the World War II bombings of Hiroshima and Nagasaki, of patients irradiated in the course of diagnostic or therapeutic procedures, and of industrial workers exposed to X-rays or radium early in this century. The animal experiments and the human experiences involved exposures that were often more than 1,000 times greater in dose and in dose rate than those received from nature.

Guidelines for maximum permissible exposure of workers at their jobs and members of the general public are developed by organizations such as the International Commission on Radiological Protection (ICRP) and its U.S. counterpart, the National Council on Radiation Protection and Measurements. These two organizations were founded in the late 1920's to protect users of X-rays and radium, which were then the principal artificially produced sources of ionizing radiation. Standards for radiation protection are issued in the United States by the Nuclear Regulatory Commission and the Environmental Protection Agency, which use the basic information provided by the United Nations Scientific Committee on the Effects of Atomic Radiation, the International Commission on Radiological Protection, and the National Council on Radiation Protection and Measurements.

Theoretical and experimental evidence suggests that there may be no threshold for radiation effects; in other words, there is no "safe" dose of radiation, and any exposure, however slight, may ultimately produce harmful effects. For this reason, the National Council on Radiation Protection and Measurements and the International Commission on Radiological Protection have cautioned that exposures should be reduced to the lowest practicable level.

If in fact long-term effects are produced at levels of exposure on the order of the natural background level, they occur at so low a frequency as to be difficult or impossible to detect. In round numbers, if everyone in the United States were to receive throughout life an additional whole body dose of 100 millirems per year, the increase in cancer mortality (now about 500,000 per year) would at most be about one percent, or about 5,000 additional cases per year. There are currently no epidemiological techniques by which an increase of this magnitude could be detected. The actual effect might be less than this estimate, and the possibility that there would be no effect at all cannot be excluded.[19] Thus, the dilemma concerning the effects of low levels of radiation is not likely to be resolved in the foreseeable future.

There is evidence that the biological damage per unit of certain types of radiation is somewhat reduced at low doses delivered at low dose rates.[20] Furthermore, there is evidence that, when the dose is reduced, not only is the incidence of tumors reduced but the time between exposure and occurrence of tumors is increased. One cannot rule out the possibility that, at sufficiently low doses, the latent period for some may exceed the life span, thus providing what has been termed a practical threshold.[21-22] Research to elucidate how low total doses and how administering low doses at different rates affect the form of the dose-response curve deserves high priority during the next five years. Until the low-dose dilemma is resolved, it will be necessary to continue to act on the conservative assumption of proportionality.

Although there is still much to be learned about the effects of low-level radiation, it should be noted that our ignorance of the effects of many chemicals to which the public is exposed is even greater. It must also be emphasized that the current assumptions concerning dose-response relationships place upper bounds on estimates of the effects produced, except possibly for alpha particles.

## CRITERIA FOR PROTECTING THE PUBLIC HEALTH AND SAFETY

The basic requirement in protecting the health and safety of the public is that radioactive wastes be managed in a way that prevents any substantial direct external and internal radiation damage and any significant contamination of the biosphere, i.e., the earth's life-supporting systems.

The radioactive waste products from nuclear power production can be in gaseous, liquid, or solid form. The radioactivity of the gases in spent fuel diminishes to near zero within a few years after the fuel is removed from a reactor. Most of the gaseous fission products are short-lived and decay within a few days.* Two exceptions are krypton-85, which has an

*Fission products are the radioactive elements produced within the fuel by the splitting of uranium nuclei during the production of nuclear power. Activation products are radioactive forms of cobalt, iron, and other elements produced by the bombardment by neutrons of nonradioactive elements within the reactor system. Transuranic elements, which include plutonium, neptunium, californium, curium, and americium, are produced by neutrons interacting with uranium.

11-year half-life (i.e., the time it takes for the radioactivity of a nuclide to decay to 50 percent of its original value) and tritium, which has a 12-year half-life; but the chemical and physical characteristics of these nuclides are such that they are two of the least hazardous fission products.[23]

Contamination of food or water is possible if isolation of the wastes is not successful. Several nuclides have long half-lives—ranging from about 30 years to more than 25,000 years—and they could be carried by water from their storage site to places where they could contaminate the biosphere.

The hazard presented by a nuclide is determined by a number of its properties, including its half-life, the type of radiation it emits (for example, when absorbed into the body, alpha emitters are generally more hazardous than are nuclides that emit beta or gamma radiation), the energy of the radiation, and the chemical properties of the nuclide. The chemical properties determine the ease with which the nuclide will be assimilated by living organisms and, once absorbed, the anatomical site of deposition.

The first and most important criterion for protecting the public is that all nuclides must be isolated for a sufficiently long time to permit their decay to acceptable levels.

How long is long enough? The question is complex. The answer depends in part on the half-lives of the radionuclides involved and on their toxicities, but in practice it depends on many other factors as well. Figure 2 shows the volumes of water required to dilute the reprocessed radioactive waste from one year's operation of a 1,000-megawatt reactor to the maximum permissible concentration in drinking water as recommended by the International Commission on Radiological Protection. To understand this concept, consider that the total quantity of radionuclides has been dissolved in enough water to dilute the radioactive material to the concentration considered permissible for drinking water. This volume of water is the "required water dilution volume"; it is plotted in Figure 2 as a function of time from 1 to 100 million years. The rapid decay of most of the fission products is seen during the first 1,000 years, after which the residual radioactivity is due to the long-lived fission product iodine-129, the transuranic elements, and two nuclides of radium.*

Thus, for the first two centuries the risk is mainly from fission products, in particular strontium-90 and cesium-137, which have half-lives of about 30 years. After about 600 years, when the strontium-90 and cesium-137

---

* Data[23a] published after this chapter was completed suggest that the nuclide neptunium-237 may have a greater influence on the long-range radiotoxicity of the wastes than was heretofore believed. One effect of a larger contribution from neptunium would be to increase the required water dilution volume for the total mixture. This would invalidate statements that have often been made that in about 600 years the water dilution volume for reprocessed wastes would fall below that of the ore from which the waste arose.

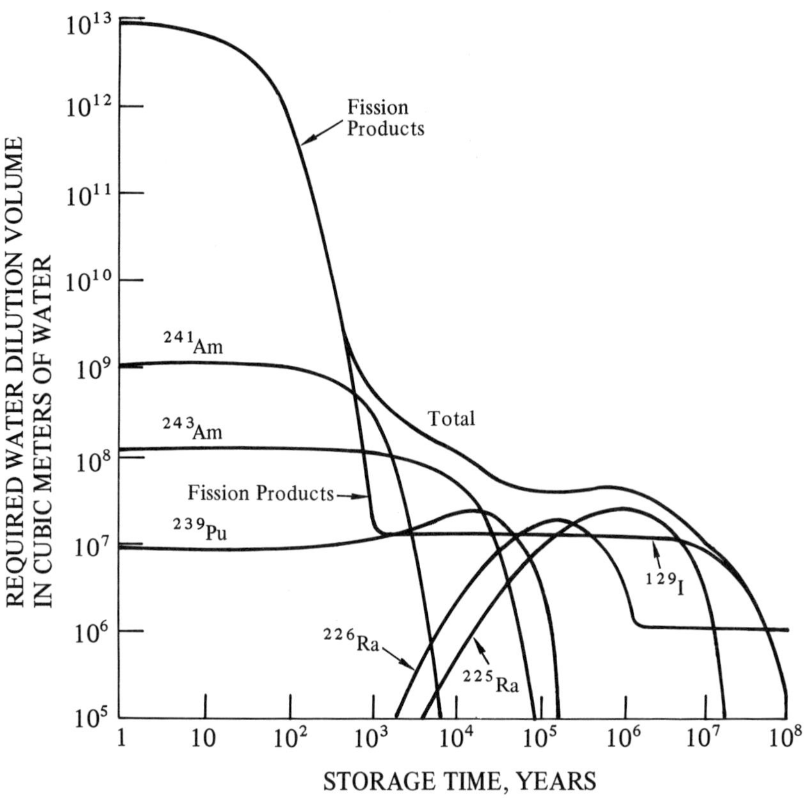


**Figure 2 Water dilution volumes for radionuclides in wastes.** Required water dilution volumes for radionuclides in high-level wastes produced by reprocessing the fuel from a 1,000-megawatt light-water reactor operating for one year. [SOURCE: T. H. Pigford and J. Choi. "Report to the American Physical Society by the Study Group on Nuclear Fuel Cycles and Waste Management," *Reviews of Modern Physics,* Vol. 50, No. 1, Part II(1978), p. S110.]

have almost completely disappeared, the residual hazard is due mainly to isotopes of plutonium and other members of the actinide family of elements.

However, two isotopes of radium—radium-226 and radium-225—form in the wastes after several thousand years and eventually contribute significantly to the required water dilution volume. The radium was originally present in the ore, was removed in the uranium refining process, and re-forms partially in the waste from the slow decay of residual actinide radionuclides. The complexities of the long-lived decay chains and the biological effects of the nuclides produced should be more fully examined in the next five years so that the question "How long is long enough?" can be answered more definitively.

What approach can be taken if it is necessary to consider long periods of time, such as hundreds of thousands of years? How can we predict the geochemical behavior of radioactive wastes over periods of time that are without precedent in human experience? Here we can in part look to nature for guidance. A large deposit of thorium on a hilltop near Pocos de Caldas, Brazil, is believed to have been in place for 80 million years. The thorium, which is chemically similar to plutonium in this kind of environment, shows little change of location despite an average annual rainfall of 170 centimeters.[24]

In a uranium mine called Oklo, in the West African country of Gabon, there is a remarkable geologic anomaly—the remains of a natural fission reactor that was apparently active periodically for several hundred thousand years, beginning about two billion years ago. The radioactive fission products have long since decayed, but most of the resulting stable nuclides can be located, and most, though not all, of the radioactive elements have migrated only insignificant distances over that period of time.[25]

Many mineral deposits are known that have remained in place for tens of millions of years, including a number of deposits containing radioactive elements. Research that increases our knowledge of the conditions required to maintain such stability will provide an additional basis of assurance when selecting sites and waste forms for geologic repositories.

If isolation can be achieved for a sufficiently long period of time, protection of the public can be ensured. But what would be the consequences of a breach of isolation that might result in exposure of present or future generations? Are there standards of permissible exposure that can be used to guide the design of repositories?

Radiation standards have been promulgated by various national and international organizations for more than 50 years. There has long been essential international agreement as to the radiobiological facts, but there is as yet no general agreement as to a philosophy upon which to base a "permissible dose."

Prudence demands the assumption that there is no threshold, that effects are linearly proportional to dose, and that therefore any radiation exposure will produce harmful effects. Yet humans have evolved in a radioactive world, and variations in natural radioactivity from place to place apparently do not have a major influence on people's decisions as to where they wish to live or work. If a dose of ionizing radiation no greater than the variations normally encountered in nature were considered trivial, one could expect that doses in the range of 10 to perhaps 100 millirems per year would be acceptable to the public.[26] Regulatory agencies in the United States appear to be settling on a dose to the maximally exposed member of the general public of 10 to 25 millirems per year.[27]

If a person were irradiated continuously for 70 years at a rate of 10 millirems per year, the upper limit of estimates of the chance of developing radiation-induced cancer would be 0.02 percent. The risk of dying from "spontaneous" cancer is nearly 20 percent; 10 millirems per year of continuous exposure would therefore increase this risk to an estimated maximum of 20.02 percent.

Ten millirems per year was used here as an illustration of the risk associated with a dose that might be received by a maximally exposed individual. But what might be the consequences of the average exposure applied to the entire population of the United States? The average dose per person from nuclear power production would be very much less than the maximum, but many more people would be exposed.

The Environmental Protection Agency has estimated that the annual average dose per person from the production of nuclear power, including waste processing, may reach 0.4 millirem by the year 2000.[28] If we assume a population of about 300 million, the population dose would be 300 million times 0.0004 rem or 120,000 person-rems per year. The maximum number of excess cancer deaths from radiation exposure can be taken to be approximately 290 deaths per million person-rems.[29] It follows that the excess risk of dying from cancer due to a lifetime dose of 0.4 millirem per year (70 years times 0.0004 rem per year equals 0.028 rem) would be at most about 1 in 100,000. The true risk is probably lower than this estimate, and the possibility of zero risk cannot, as noted earlier, be excluded.

Although the risk to the individual is 1 in 100,000, the maximum number of fatal cancers expected in the lifetime of the entire U.S. population would be 300 million times 1/100,000 or 3,000. These would be in addition to about 60 million cancers that would occur normally in a population of that size assuming no change in the current rate of cancer mortality. Thus, the risk to the individual would be small at most; but the consequences to the population at large would be sufficiently serious to warrant consideration, even though the 3,000 additional cancer cases could not be detected against the higher number due to other factors.

What levels of exposure should be considered permissible for future generations? We might consider ourselves morally justified in taking actions that could expose future generations to risks that the present generation is willing to take. Since people apparently show no concern about levels of natural radioactivity that differ by 10 millirems per year in their everyday actions, this level might be a defensible target for future as well as current generations, particularly if there is agreement that our society has need in the coming decades for the power that can be generated from nuclear reactors.

# CHARACTERISTICS OF RADIOACTIVE WASTES

## Low-Level Wastes

It is important to draw a distinction between "low-level," relatively harmless radioactive wastes and "high-level" wastes that must be effectively isolated from the biosphere for many centuries to prevent them from causing harm. Low-level wastes, disposal of which presents no unusual technical problems, are exemplified by contaminated waste paper, rags, and rubber gloves, animal excreta and carcasses from research laboratories that use radioactive tracers, cleanup materials from nuclear power plants, and radioactive substances used by hospitals for diagnostic or therapeutic purposes. They can be disposed of in a manner similar to nonradioactive wastes. Burial near the surface in landfills is the most common method.

Orderly development of a system of low-level repositories has been seriously impeded by widespread public opposition spurred, in part, by numerous reports of improper packaging of low-level wastes. The disposal of such wastes has been managed to date without harm to the public, yet fears of the dangers of low-level radioactive wastes persist. Sharper definitions of various types of low-level wastes may be needed, since large volumes of material, such as those generated by some medical tests, probably do not require the expensive disposal procedures now used.

Uranium mill tailings, which accumulate in huge volumes, are a special category of low-level wastes. Tailings, the residues that remain after uranium is removed from its ore, contain many long-lived nuclides, including thorium-230 and its daughter, radium-226. This highly toxic nuclide, with a half-life of 1,630 years, is continually produced by thorium-230, which has a half-life of 76,000 years and will produce radium far into the future. It has been known for some years that, because of the presence of radium-226, the concentration of radon is higher than normal in the immediate vicinity of tailing piles. It is also important today that tailings, which have sandlike properties, have been improperly used as a construction material in buildings, resulting in high levels of radon within the atmosphere of the buildings.

Localized increases in radon pollution have been well studied, but little is known about the possibility that radium-226, which is relatively soluble, will be leached by rainfall into drinking water supplies. Great care must be exercised in dealing with this problem in the future, because the quantities of tailings are enormous.

The balance of this chapter is concerned with the management of high-level wastes from the nuclear weapons program and the nuclear power industry. All these wastes, some of which have been accumulating since 1943, must be managed by technologically sophisticated systems that prevent their assimilation into the biosphere.

## High-Level Wastes

Two kinds of high-level waste should be distinguished: "defense" wastes, generated during the past 35 years at government installations primarily as a byproduct of the production of plutonium for weapons; and "commercial" wastes, produced by nuclear power reactors. The two kinds of waste are formed by the same basic nuclear processes, but they differ enough in physical characteristics to require separate discussion.

***Defense Wastes*** Defense wastes result from the chemical treatment, or "reprocessing," of irradiated uranium to separate out the plutonium that was formed by nuclear reactions. In addition, defense wastes include residues from the reprocessing of fuels from nuclear submarines. The reprocessing also separates the remaining uranium from the rods and leaves a waste liquid containing all the other radioactive materials that were produced in the reactor. This highly acid liquid is usually neutralized to make it less corrosive; it is then stored a few meters below the surface of the ground in steel tanks encased in concrete. The principal accumulations of tank-stored waste are at Hanford, Washington, and Savannah River, South Carolina, with lesser amounts at Idaho Falls. Currently, the total volume is about 80 million gallons (300,000 cubic meters), reduced from an original 110 million gallons by evaporation. Neutralization of the liquid produced a sludge that settled to the bottom of the tanks, where, in some of them, it has now attained a nearly rocklike hardness. Evaporation meanwhile has left crystallized "salt cake," with a small amount of highly alkaline liquid, on top of the sludge. When the waste is removed from the tanks for disposal elsewhere, handling the highly radioactive, chemically complex liquid-plus-solid mixture is a complicated engineering operation, as has been demonstrated at the Savannah River plant.

Defense wastes also exist in two other forms. One is an accumulation of metal canisters containing radioactive strontium and cesium, which were partially separated from the Hanford wastes to facilitate evaporation; the other is a powdery solid (a calcine) produced by a different process from high-level liquid waste at Idaho Falls. Finally, radioactive fallout and other forms of radioactive residues are produced by nuclear weapons testing, but these forms are not considered in this chapter.

Only small amounts of defense waste are being continuously generated at present, and much of the existing accumulation is now two or three decades old. Because of aging, the initial high levels of radioactivity and high temperatures have greatly diminished, so that managing the solidified waste is less difficult than it would be for freshly reprocessed material. Careful monitoring of the tanks shows that the waste in its present form poses negligible risk to the public. The risk will remain low as long as current management practices are maintained. More permanent disposal of the waste is clearly desirable, but no decision has yet been reached as to how the disposal should be accomplished. Suggestions include immobilizing the waste in its present tanks, mixing the waste with solidifying agents and injecting it as grout underground, converting the waste to a stable solid and burying it in an underground repository near its present location, and transporting the waste offsite for burial elsewhere. The possibility of disposal onsite in a deep underground chamber is being actively explored at Hanford, and several options are being considered at Savannah River. Thorough study of the various options and careful weighing of costs and benefits seem appropriate courses of action for the next five years.

***Commercial Wastes*** Spent fuel elements from a commercial reactor, consisting of zirconium-clad rods of uranium oxide, are placed immediately after removal in a large basin of water much like a swimming pool. Until recently, it was assumed that the fuel rods would remain in the cooling basins only about six months—by which time their radioactivity and temperature would have considerably decreased—and would then be removed for transportation to a reprocessing center. There the rods would be dissolved and their uranium and plutonium separated, just as in the reprocessing of fuel from military reactors. The uranium would be recycled for isotopic reenrichment, and the plutonium would be fabricated into new fuel. The high-level liquid wastes would then be solidified for eventual emplacement in a geologic repository. If this procedure were adopted, there would be only minor differences between new defense and commercial wastes once both were solidified. Differences would depend upon reactor design and operation to maximize plutonium production in the one case and electrical power in the other.

Because of concern that the separated plutonium might be diverted for military or terrorist purposes, national policy in the United States at present prohibits reprocessing of commercial spent fuel. Thus, the radioactive spent fuel elements themselves, in this country, could be classified as high-level waste for which a method of disposal must be designed. It is by no means certain that the prohibition against reprocessing will be maintained indefinitely, especially since other countries have not followed the U.S. example. This uncertainty leaves the status of spent fuel ambiguous: it is either a waste to be disposed of or a valuable source of electrical energy. In the discussion below, spent fuel will sometimes be treated as a waste form, with no allowance for reprocessing.

Tank for storing radioactive wastes. The steel dome being constructed over an inner steel liner will be layered by a concrete shell, the entire structure then blanketed by several feet of earth. [Battelle Northwest Laboratory.]

The radioactive fuel rods in existing water basins at nuclear power sites are well monitored and pose no risk to the public. The capacity of the present basins is limited, however, and a decision is needed soon as to whether the government should provide space for additional temporary storage in large centralized basins or leave the responsibility for building new basins to each public utility. The decision must rest in part on the length of storage required, and this in turn depends on what method is ultimately adopted for permanent disposal. Storage for long periods has the advantage that radioactivity and heat production are reduced, so that managing the waste becomes easier and the requirements for permanent disposal less stringent and more easily defined. In some other countries, Sweden and Canada, for example, the planned storage period has been deliberately extended for as long as 40 or 50 years. No specific time has been designated in the United States, but delays in planning for ultimate disposal make it evident that the period of temporary storage for most existing waste (both defense and commercial) will of necessity be at least 40 years. The storage period is a design variable that can be adjusted to yield the most effective use. A decision about the period will be needed within the next few years.

The total volume of spent fuel at present is small in comparison with defense waste, although the amounts are expected to be comparable later in this century. The activity in the spent fuel, however, is already greater than the activity in the defense wastes. The volumes of waste, whether or not the spent fuel is reprocessed, are small in comparison with waste generated by other industries. A 1,000-megawatt-electric nuclear power plant, for example, produces about 30 metric tons of unreprocessed spent fuel per year, whereas a coal-fired plant of comparable size generates about 600,000 tons of waste. (However, it should be noted that the nuclear plant would require processing of about 280,000 tons of ore per year, if we assume the uranium content to be 0.15 percent and recovery to be 90 percent.)

Whatever decision is made about managing spent fuel, transportation of fuel elements over considerable distances will be required. Transportation is often cited as a potential source of problems because of the risk of accidents involving radioactive materials. However, since the 1940's there have been 25 to 50 million shipments of radioactive material without injury to the public. Most shipments contained small amounts of radioactivity, but thousands involved amounts of radioactivity comparable with those expected in shipments of high-level waste. The accumulated statistics, which provide a good basis for estimating the frequency and severity of accidents to be expected, show that the risk of public exposure from transportation accidents is very small.[30] It is also important to note that liquid wastes will not be transported, and that the solid waste form will be chosen and packaged with the requirements for safe transportation in mind. The issue is clearly one that will deserve continued surveillance.

For disposal purposes, spent fuel has an advantage over reprocessed waste in that it is already in solid form, whereas reprocessed waste almost certainly will require conversion to a stable solid before it can be safely disposed of. In most other respects, the advantage lies with reprocessed waste. The solid form of spent fuel may actually prove to be a disadvantage, because its composition is fixed, while the composition of the solid formed from reprocessed waste can be adjusted to remain stable in particular disposal environments. A disadvantage of spent fuel is that it contains more longer lived nuclides than does the processed waste. In addition, it generates more total heat than does reprocessed waste and thus would affect repository temperatures for several thousands of years after the earlier fission product decay heat of reprocessed waste had died away. Spent fuel has the further handicap that its volume is markedly greater than the volume of high-level waste left after reprocessing. Finally, the very mobile and long-lasting fission product iodine-129 remains in spent fuel in potentially mobile form and would be subject to later leaching; in reprocessing, this isotope is separated from the waste and can then be incorporated into a waste form specially designed to retain iodine.

## WASTE MANAGEMENT OPTIONS

A number of methods of isolating high-level radioactive wastes have been proposed, including onsite solidification and disposal, use of the marine environment, geologic isolation in mined cavities, insertion in deep holes, injection as a grout in deep rock fissures or mined cavities, and insertion in arctic and antarctic glaciers. It has even been proposed that the wastes be lifted into outer space or into the sun by rocket or transmuted to more rapidly decaying elements; but these options must, for the foreseeable future, be considered infeasible out of considerations of economics, safety, and the current state of technology.

The deep hole concept is based on sinking large shafts to depths of about four kilometers (15,000 feet), but so far experience with the engineering or operating problems that would be encountered at such depths is somewhat inadequate. Thus, this method will probably not be a viable option during the next five years, but further study of it appears to be justified.

A variant of the deep hole concept—rock melting—would use emplacement in deep underground holes in such a manner that the heat of radioactive decay would melt the surrounding rock and the waste itself. Over a period of time, the production of waste heat would diminish, and the rock would refreeze. This method may provide a means of onsite disposal of unsolidified radioactive waste. It may be worthy of further consideration during the next five years, but only to determine whether it should be among concepts to be considered by the end of the century.

Ice sheet disposal is based on the probability that the deep ice of both Antarctica and Greenland is sufficiently stable to permit using it as a repository. However, the great transportation distances involved, the lack of knowledge of the effects of heat on glacial movement, and substantial impediments in international law argue against including ice sheets as an option for the foreseeable future.

Attention during the next five years is likely to be focused on three options: onsite solidification and disposal, use of the marine environment, and geologic isolation in mined cavities.

## Onsite Solidification and Disposal

Because of difficulties in locating waste burial sites, transporting wastes offsite, and removing existing defense wastes from their tanks, increased attention is being given to onsite solidification and emplacement.[31] Considerable experience has already been accumulated at the Oak Ridge National Laboratory and in the Soviet Union with onsite solidification, but only for disposal of low- and intermediate-level wastes. At Oak Ridge, the wastes are mixed with cement to form a grout, which is then injected via wells into rocks that have been hydraulically fractured.[32] Additives in the grout sorb and retard the movement of certain nuclides. In the Soviet Union, intermediate-level wastes were originally pumped into the ground. Recently, the Russians have reported that additives are pumped down with the wastes, and it has been inferred that the wastes are solidified in place.[33]

Recent reviews of defense wastes at Hanford and Savannah River indicate that onsite solidification in its various forms, either in existing tanks or in mined cavities below the surface, may have certain advantages in comparison with offsite disposal—such as less handling and transportation and reduced occupational exposure and cost.[34-35] The advantages and disadvantages must be compared in each case with those of offsite disposal. Onsite solidification is practical for large accumulations of waste, such as those at the major national atomic energy laboratories and production centers.

## The Marine Environment

***Ocean Waters*** The United Kingdom has set a unique example by its use of the marine environment for disposal of low-level radioactive wastes from the Windscale Works.[36-37] Based on studies of the behavior of radioactivity released during experiments conducted about 30 years ago, the quantities released were gradually increased, so that by 1978 the annual discharge included more than 110,000 curies of cesium-137 and more than 1,200 curies of plutonium. Studies of the environmental impact of these practices have demonstrated only minimal human exposure.[38]

More than two decades ago, the National Academy of Sciences began to examine problems that would be encountered if the oceans were used for disposal of radioactive wastes.[39-40] The Academy concluded that huge quantities of radionuclides could be placed in the ocean deeps without hazard. Nevertheless, there has been opposition to using the ocean for waste disposal of any kind (nonradioactive as well as radioactive), and the United States and many other countries have stopped ocean disposal, even for low-level wastes.

Society must be careful that the ecology of the oceans is not damaged by indiscriminate dumping of wastes. However, the carefully managed emplacement of solidified waste in the deep seabed has little in common with "ocean dumping." If waste forms are available that can be deposited so as to avoid ecological or aesthetic injury to the ocean environment, international agreement could be sought that would make it possible for the oceans to accept some share of radioactive wastes.

Enormous quantities of radioactive debris have already been deposited in the oceans as a result of weapons testing. Nuclear explosions are estimated to have produced to date more than four billion curies of tritium and almost six million curies of carbon-14. Almost 200,000 curies of plutonium-239, 7 million curies of strontium-90, and 11 million curies of cesium-137 have already been deposited in the Pacific Ocean.[41] The deposits, furthermore, were on the surface of the ocean, closest to human activity, and were in addition to the natural radioactivity already present in the ocean water. The latter includes 400 billion curies of radioactive potassium, 40 billion curies of radioactive rubidium, and 100 million curies of radium.[42] These quantities are complemented by the vast reservoirs of radionuclides naturally present in ocean sediments. And yet the ocean's contribution to the dose of radioactivity received by humans is so small compared with the dose from foods grown on the land that marine sources have so far been neglected in the description of levels of human exposure compiled by the United Nations.[43]

***Deep Ocean Sediments*** The clay sediments of the deep ocean floor far from plate boundaries are particularly attractive as a containment medium. As noted in a recent DOE analysis, the sediments are soft and pliable at and immediately below the interface of water and sediment, but become increasingly rigid and impermeable with depth.[44] The sediments, which are uniform over considerable areas of the ocean floor, have ion sorptive potential, and there is little movement of interstitial water. High pressures ensure that the sediments remain uniformly saturated and that interstitial water, if heated, cannot boil. The result is an essentially infinite sink for the heat generated by decaying radionuclides.

Although seabed emplacement and containment have not yet been demonstrated for specific geologic and ocean locations, the engineering tools needed to begin working in the deep ocean and seafloor are available. Several emplacement concepts appear feasible, although none has been developed in detail. Most intriguing are injectors or penetrometers, which use gravity for emplacement and would therefore place the fewest technical demands on the vessel used for transport and emplacement. Other methods are drilling and trenching. Criteria for selecting the best method include (1) depth of emplacement necessary to ensure containment of radionuclides, (2) rate and completeness of hole closure after penetration, (3) degree of disturbance of sediment properties, (4) engineering feasibility, and (5) economics.

During the next five years, we can expect a thorough review of the feasibility of using the oceans for radioactive waste disposal. Research and development will probably be expanded, and multinational discussions may be initiated in efforts to reach international consensus.

## Other Methods of Geologic Isolation

Although it has lagged behind slurry injection, as demonstrated at Oak Ridge, in the actual practice of nuclear waste disposal, the concept of isolation of waste in mined cavities in recent years has received more attention than have any of the other concepts proposed. It is therefore the option by which a total waste management system is described in the following section.

Because the most likely way for the radioactivity to reach the biosphere is by water transport, the permeability and flow rates through the host rock and the nearest point of possible transfer from the aquifer to the biosphere are important, as is adsorption on mineral surfaces. If the time for hydrologic transport is sufficiently long, the nuclides will decay before they reach the biosphere.

An alternative to deep geologic isolation is burial at shallow levels in arid regions.[45] This method avoids leaching by ground water, because in parts of the western states the regional water table is 100 to 300 meters below the surface, and transport of water from the surface seldom or never occurs. The thickness of dry material above the water table in such regions is ample for constructing a repository. Active investigation of this option during the next five years is advisable.

Because the technology of isolating of wastes in cavities mined in rock is so advanced, this method will be considered at greater length than the others.

## A WASTE ISOLATION SYSTEM INVOLVING MINED CAVITIES

Waste isolation in mined cavities requires a multibarrier system (see Figure 3) with the following component parts:

1. The solid waste form into which the wastes are processed. In the case of unprocessed spent fuel, the waste form is the fuel itself contained in its original cladding. For processed spent fuel, the radionuclides may be incorporated into a solid glass or ceramic of low solubility or put into a form placed in a matrix of cement, ceramics, or metal. An important characteristic of the waste form is its temperature, which depends in part on its age, in part on the concentration of waste incorporated into the solid matrix, and in part on the thermal properties of the canister's environment.
2. The canister into which the waste form is placed, and possibly an "overpack" surrounding it, which serves to absorb any radionuclides that may escape, as well as to protect the canister.
3. The chemical properties of the repository environment and its response to changes in temperature and pressure.
4. The physical and biological pathways by which the radioactive materials can pass from the repository to humans.

### The Waste Form

The physical form of the waste depends initially on whether or not the spent fuel is reprocessed. If it is not reprocessed, the intact spent fuel

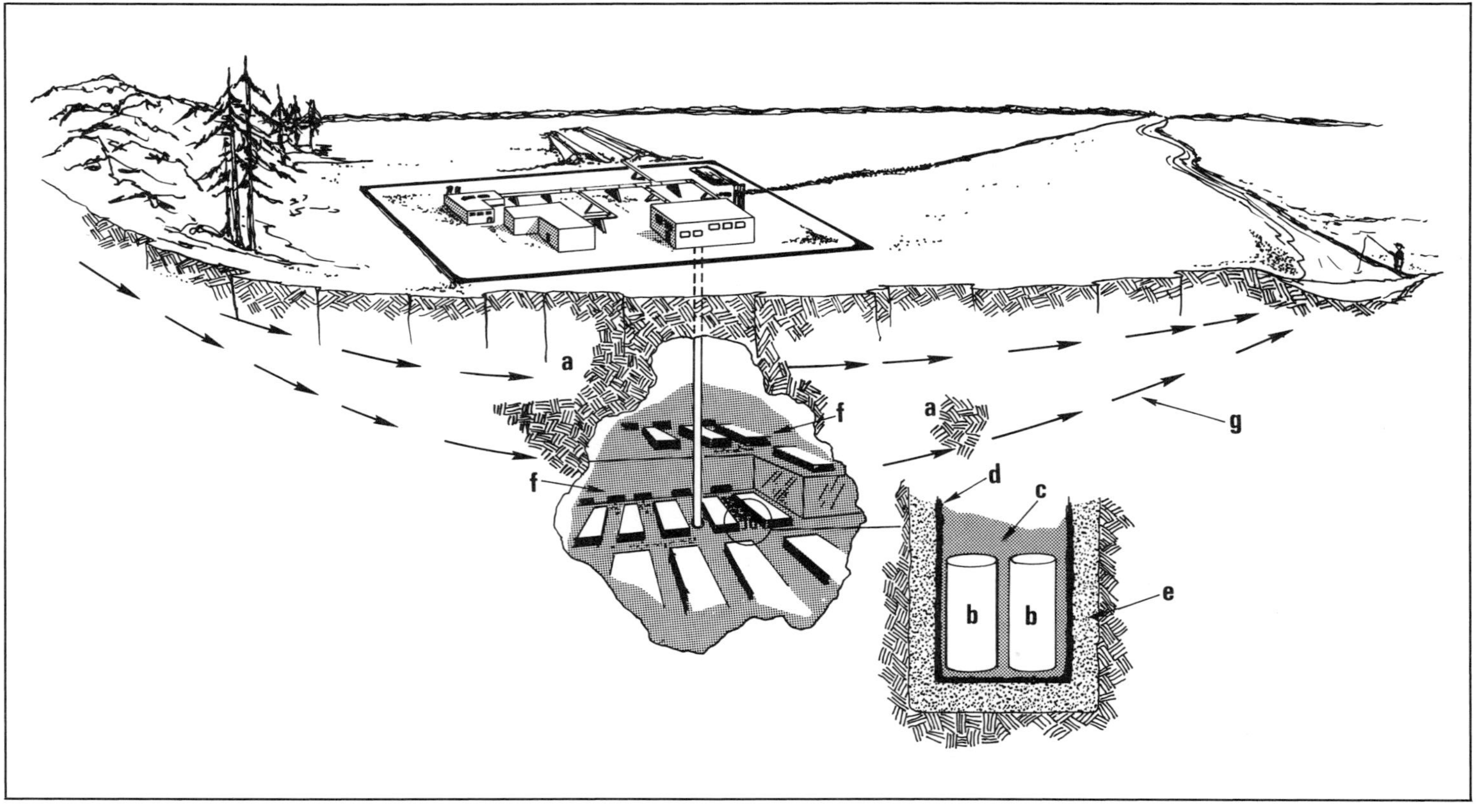
a
a
f
f
d
c
b
b
e
g

elements must be isolated. If reprocessed, they must be cut up and chemically treated to separate the unspent uranium and plutonium from the residues containing the highly radioactive wastes.

The high-level wastes may be incorporated into glass or ceramics, into a concrete matrix, or into synthetic minerals or metals.[46] Research in several countries, especially France, has produced a variety of waste forms that could be mated satisfactorily with other required components of an appropriate system and that could incorporate the liquid wastes of commercial reprocessing plants.[47]

In many programs, borosilicate glass is used as a high-temperature solvent for modest concentrations of fission products. Some glasses can incorporate as much as 25 to 30 percent fission products by volume, but the loadings now being considered are approximately 10 percent. Pilot runs of such glasses containing radioactive wastes have been made in the United States, England, and France.[48-50]

Current research has shown that some other waste forms may be as good as or better than glass, and some of them may be less difficult to manufacture. Certain specific crystalline mineral forms that exist in nature survive over geologic time in a variety of severe natural environments. The concept that ceramic waste forms can be tailored to specific wastes and specific repository environments was developed during the past decade. Several such waste forms have been produced synthetically in laboratories in the United States, and support for their

---

◀ **Figure 3 The mined geologic repository as a system, with examples of natural and manmade protective barriers.** (a) The rock in which the repository is situated provides protection by its relative impermeability to water. It is always possible that some water will penetrate the repository; but if it does, the consequences can be mitigated by the design of the repository system and by the slow movement of the water. (b) The solidified reprocessed waste may consist of a uniform cylinder of glass or ceramic material, or it may take the form of discrete pellets of such material distributed through a resistant matrix. It is also possible to store unprocessed spent fuel in a repository. (c) A filler may or may not be needed. If present, it may be a gas, such as helium, to facilitate heat transfer, or a solid to serve as a corrosion-resistant barrier. (d) The canister is a chemically resistant container for handling the waste. It may also delay access of ground water to the waste. (e) An overpack may be added in the space between waste package and surrounding rock to aid in preventing the intrusion of water and to provide an additional barrier to migration of nuclides. (f) The backfill provides mechanical support against the pressure of the rock, modifies the chemistry of the ground water, and adsorbs selected radionuclides and a limited amount of water. (g) If radionuclides should be leached from the repository, the host rock would provide additional protection by: (1) impeding the flow of water so that radionuclides would not reach the biosphere for thousands or even hundreds of thousands of years; and (2) adsorbing many of the radionuclides on the rock surfaces. Under ideal circumstances, hazardous radionuclides would decay before they could contaminate food consumed by humans.

development has increased recently in this country and in Sweden and Australia.[51-53]

Radioactive waste can also be incorporated into matrices of other materials, such as concrete, in which it does not dissolve but which reduce its mobility. Ceramics and glass containing up to 20 percent fission products have been produced by hot-pressing into quartz and silica-glass.[54] It has been claimed that incorporation of commercial high-level and defense wastes into a cement matrix has resulted in products essentially equal to the typical glasses in leachability and thermal conductivity.[55-56] The potential advantages of incorporation into cement are the simplicity of the process, the lower reaction temperatures, and the possibility of onsite disposal.

On the basis of two years of demonstrated pilot production of borosilicate glass in France and research and development at various other places in the world, it appears feasible to plan for U.S. pilot plant production of one or more waste forms suitable for solid-phase immobilization of commercial wastes.

Ceramics tailored to the specific waste composition, possibly incorporated within a metal matrix, provide a waste form with very low release rates. Two-kilogram ingots of such material have been produced at the Battelle Pacific Northwest Laboratories. The system could be suitable for any host rock, including salt.[57]

## The Canister and the Overpack

The canister is a further barrier in solid-phase immobilization. In its most obvious and typical form, it is a metal or ceramic container for the solidified wastes and any incorporating matrix. Possible metals include ordinary low-carbon steel, stainless steel, copper, and titanium. The proposed Swedish use of copper has received considerable support from geologic and metallurgical scientists and from a recent National Research Council review of the Swedish program.[58] French plans to use ceramic canisters and the Swedish full-scale demonstration of a sapphire canister have been motivated by the long-term stability of these materials.

A principal function of the canister—in addition to making the waste transportable—is to resist corrosion. Thus, any reference to the life of a canister must, to be meaningful, specify the chemical and temperature-pressure environment.

The overpack, recently proposed as a filler material to surround the canister, would have several roles. It could be an adsorbent (and reactant) for nuclides that might escape from the canister, or it could be an adsorbent for ground water and its constituent ions, thereby virtually immobilizing the ground water in the vicinity of the canister. The effect would be to prevent or minimize corrosion of the canister. The addition of ferrous phosphate to the overpack would further reduce corrosion.

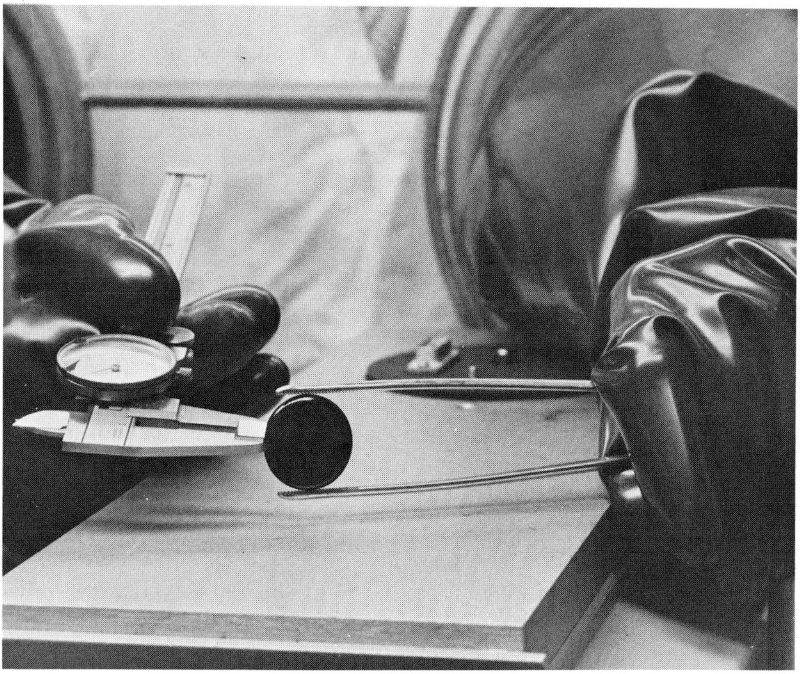

A glass pellet of radioactive wastes "spiked" with radioactive curium to expose it to more radiation than expected with actual radioactive wastes. In two years, this pellet has received the equivalent exposure to radiation of waste glass stored for a half-million years. [Battelle Northwest Laboratory.]

## The Repository Environment

Several quite different rock types and geologic locations offer potentially acceptable sites for repositories. In general, each site must be appraised in terms of certain basic geologic factors. These include tectonic stability, topography, the size of the rock mass being considered, and regional patterns of surface- and ground-water flow. Other important geologic factors relate to rock properties, including heat conductivity, porosity, permeability, water content, chemical reactivity with respect to fluids that may be introduced or generated, response to anticipated temperature increases, state of initial stress, brittleness or plasticity, ease of mining, and value of actual or potential mineral resources.

Ideally, the rock mass should be large, homogeneous, dry, relatively free of fractures, and capable of sorbing or precipitating released radionuclides. Few if any rock bodies can fully meet these ideal requirements, but many possess attributes that in sum qualify them for consideration. Special engineering measures or methods of development can be used to compensate for the drawbacks of certain media. For example, the disadvantage of variable degrees of fracturing in granite (see below) can be taken care of by careful exploration for less fractured zones. The following paragraphs are general comments concerning the suitability of several rock types that are the principal candidates for repositories.

***Salt*** Because of its abundance in many parts of the country, ease of excavation, and the absence of flowing water in existing mines, salt has been a leading candidate for repository siting for more than two decades. Other positive attributes of salt are its high thermal conductivity and its semiplastic behavior under continued stress—fractures in salt tend to reseal themselves. The negative features of salt as a repository medium are its high solubility, its low sorptive power with respect to escaping radionuclides, a tendency for brine inclusions to move toward the source of heat, and its potential value as a mineral resource. In addition, its semiplastic behavior may interfere with retrievability.

***Basalt*** The volcanic flows that underlie the Columbia River plateau of eastern Washington and Oregon are now under serious consideration as a repository at the Hanford Reservation, where much of this country's military waste is stored in tanks close to the surface. The rock is structurally stable and, by virtue of clay minerals along most fracture surfaces, would be moderately sorptive for radionuclides that might escape into ground water. Negative aspects are the relative thinness of some basalt flows and the presence of fractures and interbedded aquifers.

***Shale*** Shale is a common rock, abundant throughout much of the central interior of the United States. Most shales are composed largely of fine-grained quartz and clay minerals. The water content of shale is generally high—on the order of 10 to 30 percent by volume—but low permeability greatly slows the water movement. In fact, shale beds commonly serve to confine interbedded sandstone aquifers. The high capacity of some shales to sorb radionuclides would significantly deter their movement.

However, shale also has a number of negative characteristics. The clay minerals are likely to break down under even moderate increases in temperature, causing release of water and increased porosity and permeability; and the structural weakness of the rock will create problems in maintaining mined openings during the construction and filling of repositories.

***Volcanic Tuff*** Volcanic tuff (compacted volcanic ash) of many different types occurs throughout the western United States. Of particular interest for repository siting, however, is the ash-flow tuff that occurs principally in the Southwest, notably in New Mexico and Nevada. This tuff is found as sheets 100 meters or more thick, some of which have a central glassy "welded" zone surrounded by zones of zeolitized volcanic ash that is highly sorptive. These central zones of welded tuff, with their relatively low permeability and porosity, offer potential sites for waste storage. Negative aspects of tuffs include relatively high water content and possible regional tectonic instability (the deposits are in areas of volcanism and fault movement that have been relatively recent in geologic time) and structural complexity (which may not provide sufficient volume of rock).

***Granite and Related Crystalline Rocks*** Granite and related rocks are common in much of the United States and are the preferred rock type for repository siting in Canada and Sweden. Individual masses extend for as much as hundreds of kilometers. In addition to their abundance, other positive aspects of crystalline rocks are their high thermal conductivity, structural strength, low inherent porosity and permeability, generally low water content, and, for some bodies, a relatively high degree of internal homogeneity. In general, the principal disadvantage is the variable amount of fracturing and the difficulty of predicting the nature of water movement in the fractures.

## Mining Experience

A wealth of experience in underground excavation indicates that the technology is available for excavating repositories needed for the deep disposal of high-level nuclear wastes in any suitable geologic formation.

For example, we know that the design and construction of a successful deep geologic waste repository must take account of the inherent variability in the properties and structure of rocks. Modular design is advantageous because each module can be treated independently—abandoned or isolated, for example—without affecting adversely the adequacy and safety of other modules.

Mining technology is relevant to the development of a waste repository, but the objectives of excavation for mines are different. In a mine, the goal is usually to excavate as much of the ore as practicable, consistent with short-term safety. In a repository, the goal is to disturb the subsurface as little as possible, although, undoubtedly, additional stresses are introduced by the mining operation itself. Minimizing disturbance is important to ensure that the wastes are successfully isolated from the biosphere for sufficient time by the geologic media, and this must be accomplished with the greatest practical degree of safety both in the short and the long terms.

## Geochemical Considerations

Geochemical conditions in bedrock at depths being considered for repository siting will be altered abruptly by excavation of a cavity and the emplacement of heat-generating, chemically heterogeneous radioactive waste packages. In its undisturbed state, the rock mass will be essentially in chemical equilibrium; fluids, whether trapped or moving slowly, will have reacted with the host rock to produce mineral assemblages that are stable in that environment. The wastes will not necessarily be in equilibrium, either internally or with respect to the canister, overpack, or enclosing host rock. Reactions promoted by this lack of equilibrium could, over long periods of time, be responsible for transfer and escape of radionuclides from the repository.

Two idealized situations can be visualized in which transfer of radionuclides by ground water would be virtually impossible. One is essentially complete containment of radionuclides at the source, achieved either by casting the waste in a totally insoluble form or by encapsulating it in a container designed to remain intact for the required period of time. The other is use of a geologic environment in which ground water is either completely absent or, if present, moving sufficiently slowly. Both conditions are approached closely in some natural geologic situations.

More typically, and for the sake of prudence, slow release of radionuclides to the enclosing geologic environment is assumed. Although the geochemistry of this system will be immensely complex, it can be safely concluded that in most rocks chemical reactions and sorption along transport routes will substantially impede the outward movement of most radionuclides.

Undesirable geochemical interactions near the canister can to some extent be countered by the deliberate additions to the waste package and overpack we described earlier.

## Hydrologic Considerations

The hydrologic characteristics of the site and its vicinity must be known if one wishes to predict quantitatively the degree of human exposure that may result from operation of the repository.

The time it takes radionuclides carried by ground water to move from a well-chosen repository to the biosphere may be measured in thousands or even hundreds of thousands of years. In the absence of adequate information about the hydrologic characteristics of the site and its vicinity, conservative estimates must be made of the range of possible transport rates. Because transport of the radionuclides will be slowed by the adsorptive characteristics of the rock, the actual transport times of the radionuclides will depend upon interactions of hydrologic and geochemical factors.

The knowledge required to describe hydrologic transport of radionuclides in potential rock types at the depths of geologic repositories is only beginning to accumulate, but it will be enhanced substantially as we gain experience with actual storage of radioactive wastes. If the flow occurs through fractures in the rock, it would be useful to be able to describe the extent to which undisturbed dense rocks are fractured. The methods should be noninvasive so that the measurements do not themselves result in rock fracturing. Geophysical techniques are needed for estimating the amount of fracturing in large volumes of rocks and for estimating more reliably the ages of ground water in and near a repository site. Such techniques would provide a valuable check on estimates of the rate of ground-water movement.

Prospective rates of radionuclide transport by water must consider possible future fluctuations in flow. Knowledge of the past geologic history of water level fluctuation will be of great help here. Attention must be given to past climatological variations, to fluctuations in water levels in major bodies of water, and to evidence of past dissolution of minerals by ground water and their subsequent precipitation.

Given all the required hydrologic information, it should be possible to provide estimates of the rates of migration over long periods of time. The research required for development of such estimates should be given high priority. For the time being, and until sufficient information is available, a high degree of conservatism must be practiced in estimating transport rates. This applies not only to water flow but to other components of the isolation system as well. In order to establish an upper bound for possible effects of radioactivity mobilized as a result of material transported from a repository, the most conservative assumption for each factor is often selected from the array of reasonable alternatives. However, this procedure can lead to unreasonably high estimates of risk. For example, in the draft environmental impact statement for the proposed Waste Isolation Pilot Plant in New Mexico, the time of transit of radionuclides from the repository to the first point of interaction with the biosphere was estimated to be from 3,500 to 100,000 years.[59] In this case, 3,500 years was selected for purposes of calculation despite the fact that it very probably tended to overestimate the dose. Close examination of that report, as well as of the draft generic environmental impact statement produced by the Department of Energy for the management of commercially generated radioactive waste,[60] reveals many examples of this kind of conservatism. Nevertheless, estimates of exposure to nearby residents, despite the overall conservatism employed, would remain but a small fraction of the dose received from natural sources.

## Summary and Outlook

Of the various options that have been proposed for the management of high-level radioactive waste, emplacement in cavities constructed in deep rock formations is the most likely candidate for a first-generation repository. Either unprocessed spent fuel or wastes from fuel reprocessing could be isolated in this way, and the available evidence supports a substantial degree of confidence that the technical aspects of geologic isolation can be managed in a manner that will protect the public. National policy will probably concentrate on this option during the next five years while recognizing that other options, such as disposal under seabeds, may also become practicable with time.

A major uncertainty still to be resolved is the length of time for which permanent isolation of radioactive wastes must be provided. It has been suggested that after about 600 years the water dilution volume for the reprocessed waste will be comparable with that for the uranium ore from which the fuel originated. However, this has been open to question by recent data suggesting that certain important nuclides behave differently than had been assumed. Careful analysis of the matter in the years immediately ahead should make it possible to determine, with confidence, the length of time for which isolation must be assured.

The civilian nuclear power industry will be the predominant source of high-level wastes in the future, but during the next several years wastes that have accumulated from defense activities will also demand a high priority of attention. These wastes—a legacy of the early years of producing plutonium for defense purposes—exist in various forms, and at various levels of radioactivity, at the major U.S. research and production centers. Monitoring in the vicinity of these storage sites has revealed no evidence that the wastes will present a risk to public health for the foreseeable future.[61-65] Because of previous treatment, defense wastes are quite different chemically from wastes from the nuclear power industry. Only small amounts of defense wastes are being continuously generated at present, and most of the existing accumulation is now two or three decades old.

Several options for dealing with defense wastes have been considered, including immobilization in the tanks in which they are now stored, mixing the wastes with solidifying agents and injecting them as grout underground, converting them to a stable solid for burial near their present locations, and transporting them offsite for burial elsewhere. There is no urgency about reaching a decision as to how these wastes should be managed, but during the next five years careful study of the various options available will contribute to more accurate assessment of the costs and benefits of each. In particular, active investigation can be expected of the possibility that defense wastes can be safely immobilized by onsite solidification and emplacement.

As with defense wastes, there is no technological urgency about immediate selection of a permanent isolation option for high-level wastes from the nuclear power industry. Wastes can be accommodated safely in temporary facilities for many years to allow orderly development of one or more national repositories, using technology adequate to meet protection standards. Nevertheless, because of widespread public concern about the hazards of radioactive wastes, it is in the national interest to proceed expeditiously with plans for their permanent isolation.

Isolation of radioactive wastes in deep marine sediments appears to offer attractive advantages, and the years immediately ahead are likely to see this option fully explored and supported by the necessary research programs. Should the option appear to be viable, it will be necessary to initiate early multilateral discussions aimed at securing international agreement.

As development and demonstration of methods for solidification and immobilization of high-level wastes continue to receive high priority, the ultimate goal will be economical processes by which the wastes can be converted to forms comparable in their solubility to the more resistant minerals found in nature. Such forms would greatly facilitate geologic isolation for the required period of time.

Although national concern about radioactive wastes is focused primarily on high-level wastes, low-level wastes must receive attention as well. In particular, the long-range health implications of accumulations of mill tailings will bear examination. Tailing piles are low-level wastes, but their volume is great enough to justify special attention. The radium contained in the tailings is by far the most hazardous component of the original ore, and because of its relative solubility it may be leached by rain into drinking water supplies. Evolution of radon gas from the tailing piles is reasonably well understood, but less is known about the hazard of potential pollution of surface and ground water by radium.

The task of developing a successful waste isolation system will continue to involve more than scientific and technical considerations: there is a difficult and pervasive educational component. Throughout the critical years ahead, it will be important that information be disseminated about the implications of each major step in the development of an isolation system, and about the interplay of costs and benefits, so that the public can participate—in a timely, effective, and constructive fashion—in the decisionmaking process.

## REFERENCES

1. Robert Cameron Mitchell. *The Public Response to Three Mile Island: A Compilation of Public Opinion Data About Nuclear Energy*. D-58. Washington, D.C.: Resources for the Future, 1979.

2. Ibid.

3. Interagency Review Group on Nuclear Waste Management. *Report to the President*. TID-29442. Washington, D.C.: U.S. Department of Energy, 1979.

4. U.S. Office of the President. *Radioactive Waste Management: Message of February 12, 1980 from the President of the United States Transmitting a Report on His Proposals for a Comprehensive Radioactive Waste Management Program*. Washington, D.C.: U.S. Government Printing Office, 1980.

5. M. Eisenbud. *Environmental Radioactivity*. Second Edition. New York: Academic Press, Inc., 1973.

6. *Natural Radiation Background in the United States*. NDRP Report No. 45. Washington, D.C.: National Council on Radiation Protection and Measurements, 1975.

7. Ibid.

8. Eisenbud, op. cit.

9. *Krypton-85 in the Atmosphere*. NDRP Report No. 44. Washington, D.C.: National Council on Radiation Protection and Measurements, 1975.

10. G. A. Swedjemark. *Radon in Swedish Dwellings*. Summary papers, Symposium on the Natural Radiation Environment III held at the University of Texas, Houston, April 23-28, 1978, p. 188.

11. K. D. Cliff. *Measurements for Radon-222 Concentration in Dwellings in Great Britain*. Summary papers, Symposium on the Natural Radiation Environment III held at the University of Texas, Houston, April 23-28, 1978, p. 191.

12. J. P. McBride et al. "Radiological Impact of Airborne Effluents of Coal and Nuclear Plants," *Science*, Vol. 202(1978), pp. 1045-1050.

13. National Research Council, Committee on Biological Effects of Ionizing Radiations. *The Effects on Populations of Exposure to Low Levels of Ionizing Radiation*. Third Edition. Washington, D.C.: National Academy of Sciences, 1980.

14. Ibid.

15. *Estimates of Ionizing Radiation Doses in the United States, 1960-2000*. ORP/CSD 72-1. Rockville, Md.: U.S. Environmental Protection Agency, 1972.

16. National Research Council, Committee on the Biological Effects of Ionizing Radiations, op. cit.

17. United Nations Scientific Committee on the Effects of Atomic Radiation. *Sources and Effects of Ionizing Radiation, 1977*. New York: United Nations, 1977.

18. National Research Council, Committee on the Biological Effects of Ionizing Radiations, op. cit., p. 118.

19. National Research Council, Committee on the Biological Effects of Ionizing Radiations, op. cit.

20. Ibid.

21. R. D. Evans. "Radium in Man," *Health Physics*, Vol. 27(1974), pp. 497-510.

22. R. D. Albert and B. Altshuler. "Assessment of Environmental Carcinogen Risks in Terms of Life Shortening," *Environmental Health Perspectives*, Vol. 13(1976), pp. 91-94.

23. *Krypton-85 in the Atmosphere*. NDRP Report No. 44. Washington, D.C.: National Council on Radiation Protection and Measurements, 1975.

23a. Annals of the International Commission on Radiological Protection. Report No. 30, Part II. New York: Pergamon Press, 1980.

24. M. Eisenbud et al. *The Mobility of Thorium and Other Elements from the Morro do Ferro*. Progress report through September 30, 1980, to U.S. Department of Energy. Tuxedo, N.Y.: New York University Medical Center, 1980.

25. E. A. Bryant et al. "Oklo, an Experiment in Long-Term Geologic Storage," *Actinides in the Environment*. ACS Symposium Series No. 35(1976), pp. 89-102.

26. M. Eisenbud. *The Concept of de Minimis Dose*. Proceedings of the 1980 Annual Meeting of the National Council on Radiation Protection and Measurements. Washington, D.C.: National Council on Radiation Protection and Measurements, 1980.

27. U.S. Environmental Protection Agency, 40 CFR 190.

28. U.S. Environmental Protection Agency, op. cit.

29. National Research Council, Committee on the Biological Effects of Ionizing Radiations, op. cit.

30. *Final Environmental Statement on the Transportation of Radioactive Materials by Air and Other Modes*. NUREG-0170. Two Volumes. Washington, D.C.: Nuclear Regulatory Commission, 1977.

31. National Research Council, Committee on Radioactive Waste Management. *Solidification of High-Level Radioactive Wastes*. NUREG/CR-0895, FIN#B-1523-7. Washington, D.C.: Nuclear Regulatory Commission, 1978.

32. H. O. Weeren et al. *Waste Disposal by Shale Fracturing at ORNL*. Oak Ridge, Tenn.: Oak Ridge National Laboratory, 1979, p. 257.

33. V. I. Spitsyn and V. D. Balukova. "The Scientific Basis for, and Experience with, Underground Storage of Liquid Radioactive Wastes in the U.S.S.R.," *Scientific Basis for Nuclear Waste Management*. Edited by G. J. McCarthy. New York: Plenum Publishing Corp., 1979, p. 237.

34. National Research Council, Committee on Radioactive Waste Management. *Radioactive Wastes at the Hanford Reservation: A Technical Review*. Washington, D.C.: National Academy of Sciences, 1978.

35. National Research Council, Committee on Radioactive Waste Management. *Radioactive Wastes at the Savannah River Plant*. Washington, D.C.: National Academy of Sciences, in preparation.

36. A. Preston. *The Radiological Consequences of Releases from Nuclear Facilities to the Aquatic Environment.* Proceedings of Impacts of Nuclear Releases into the Aquatic Environment. Vienna: International Atomic Energy Agency, 1975, p. 3.

37. H. Howells. *Discharges of Low-Activity, Radioactive Effluent from the Windscale Works into the Irish Sea.* Proceedings of Disposal of Radioactive Wastes into Seas, Oceans and Surface Waters. Vienna: International Atomic Energy Agency, 1966, p. 769.

38. British Nuclear Fuels Limited. *Annual Report on Radioactive Discharges and Monitoring of the Environment, 1978.* Risley, England: Health and Safety Directorate, 1979.

39. National Research Council, Committee on Oceanography. *Radioactive Waste into Atlantic and Gulf Coastal Waters.* Washington, D.C.: National Academy of Sciences, 1959.

40. National Research Council, Committee on Oceanography. *Disposal of Low-Level Radioactive Waste into Pacific Coastal Waters.* Washington, D.C.: National Academy of Sciences, 1962.

41. M. Eisenbud. "The Status of Radioactive Waste Management: Needs for Reassessment," *Health Physics,* Vol. 40 (1981), pp. 429-437.

42. National Research Council, Committee on Oceanography. *Radioactivity in the Marine Environment.* Washington, D.C.: National Academy of Sciences, 1971.

43. United Nations Scientific Committee on the Effects of Atomic Radiation, op. cit.

44. *Final Environmental Impact Statement: Management of Commercially Generated Radioactive Waste.* Three Volumes. DOE/EIS-0046F. Washington, D.C.: U.S. Department of Energy, 1979, p. 126.

45. I. J. Winograd. "Radioactive Waste Storage in the Arid Zone," *EOS*, Vol. 55(1974), pp. 884-894.

46. R. Roy. "Science Underlying Radioactive Waste Management: Status and Needs," *Scientific Basis for Nuclear Waste Management.* Edited by G. J. McCarthy. New York: Plenum Publishing Corp., 1979, p. 1.

47. National Research Council, Committee on Radioactive Waste Management. *Solidification of High-Level Radioactive Wastes.* NUREG/CR-0895, FIN #B-1523-7. Washington, D.C.: Nuclear Regulatory Commission, 1978.

48. W. F. Bonner et al. "Engineering-Scale Vitrification of Commercial High-Level Waste," *Waste Management '80, The State of Waste Disposal Technology, Mill Tailings, and Risk Analysis Models.* Volume Two. Edited by R. G. Post. Tuscon, Ariz.: University of Arizona Press, 1980, p. 339.

49. K. D. B. Johnson. "The U.K. Program—Glasses and Ceramics for Immobilization of Radioactive Wastes for Disposal," *Ceramics in Nuclear Waste Management.* CONF-790420. Edited by T. D. Chikalla and J. E. Mendel. Washington, D.C.: U.S. Department of Energy, 1979, p. 17.

50. W. Heimerl. "Techniques for High Level Waste Solidification in Europe," *Scientific Basis for Nuclear Waste Management.* Edited by G. J. McCarthy. New York: Plenum Publishing Corp., 1979, p. 21.

51. G. J. McCarthy. "High-Level Waste Ceramics: Materials Considerations, Process Simulation, and Product Characterization," *Nuclear Technology*, Vol. 32(1977), p. 92.

52. A. E. Ringwood et al. "The SYNROC Process: A Geochemical Approach to Nuclear Waste Immobilization," *Geochemical Journal*, Vol. 13(1979), p. 141.

53. S. Forberg et al. "Synthetic Rutile Microencapsulation: A Radioactive Waste Solidification System Resulting in an Extremely Stable Product," *Scientific Basis for Nuclear Waste Management.* Edited by G. J. McCarthy. New York: Plenum Publishing Corp., 1979, p. 201.

54. G. J. McCarthy. "Quartz Matrix Isolation of Radioactive Wastes," *Journal of Materials Science*, Vol. 8(1973), p. 1358.

55. H. S. Godbee et al. "Leach Behaviour of Cementitious Grouts Incorporating Radioactive Wastes," *Bulletin of American Ceramic Society*, Vol. 59(1980), p. 463.

56. H. S. Godbee et al. "Current ORNL Studies on Cementitious Hosts for Radioactive Wastes," *Bulletin of American Ceramic Society*, Vol. 59(1980), p. 392.

57. J. M. Rusin et al. *Multibarrier Waste Forms, Part 1: Development.* No. PNL 2668-1. Richland, Wash.: Battelle Pacific Northwest Laboratory, 1978.

58. National Research Council, Committee on Radioactive Waste Management. *A Review of the Swedish KBS-II Plan for Disposal of Spent Nuclear Fuel.* Washington, D.C.: National Academy of Sciences, 1980.

59. *Draft, Environmental Impact Statement: Waste Isolation Pilot Plant.* Volume One. DOE/EIS-0026-D. Washington, D.C.: U.S. Department of Energy, 1979.

60. *Final Environmental Impact Statement: Management of Commercially Generated Radioactive Waste.* Three Volumes. DOE/EIS-0046F. Washington, D.C.: U.S. Department of Energy, 1979.

61. National Research Council, Committee on Radioactive Waste Management. *Radioactive Wastes at the Hanford Reservation: A Technical Review.* Washington, D.C.: National Academy of Sciences, 1978.

62. National Research Council, Committee on Radioactive Waste Management. *Radioactive Wastes at the Savannah River Plant.* Washington, D.C.: National Academy of Sciences, in preparation.

63. *Final Environmental Impact Statement: Long Term Management of Defense High-Level Radioactive Wastes, Savannah River Plant.* DOE/EIS-0023. Washington, D.C.: U.S. Department of Energy, 1979.

64. *Final Environmental Statement: Waste Management Operations, Hanford Reservation.* Two Volumes. ERDA-1538. Washington, D.C.: U.S. Energy Research and Development Administration, 1975.

65. *Environmental and Other Evaluations of Alternatives for Long-Term Management of Stored INEL Transuranic Waste.* DOE/ET-0081. Washington, D.C.: U.S. Department of Energy, 1979.

# III

# *RESEARCH FRONTIERS*

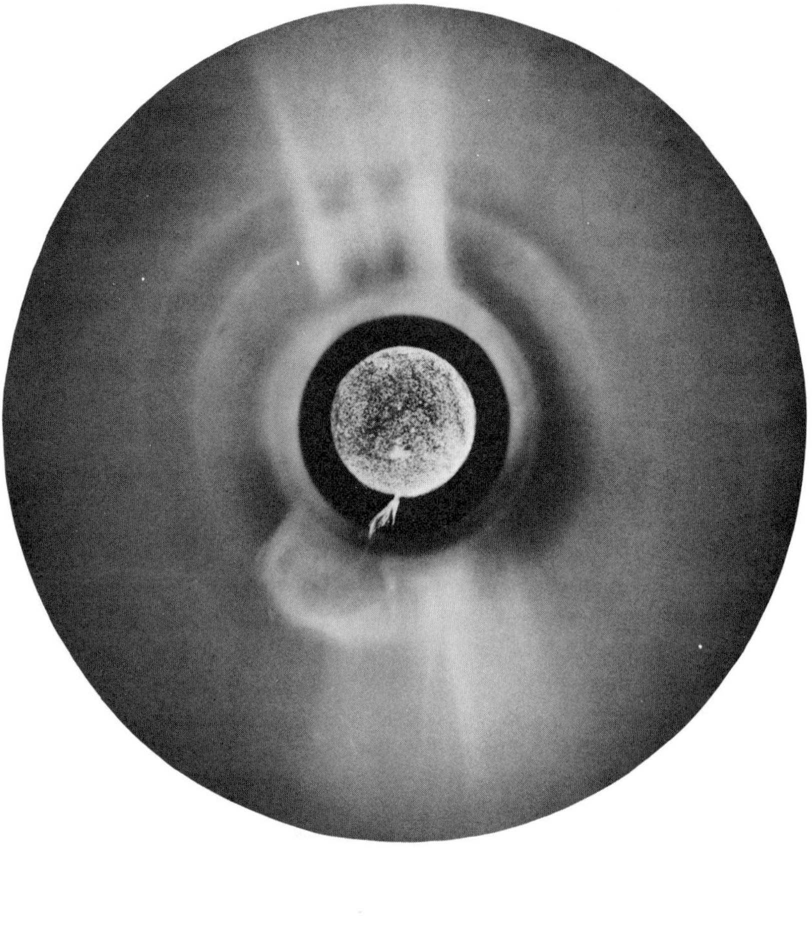

# 9

# *Sun and Earth*

## INTRODUCTION

In past ages, thoughts of sun and earth inspired as much poetry as science among natural scientists. Pierre Gassendi, in the seventeenth century, coined the name *aurora* for the northern lights after the Roman goddess of rosy-fingered dawn. Rainbows were a particularly fascinating scientific curiosity. Sharing an emotion Wordsworth expressed as "My heart leaps up when I behold/A rainbow in the sky," generations of scientists from the time of Newton were enthralled by the subtle beauties of rainbows, as well as by halos about the sun and rings around the moon.

The source of the blue color of the sky was a popular but tantalizing problem in physical optics, solved only recently when rocket measurements provided profiles by altitude of ozone and atmospheric density. Without ozone, the sky overhead would be a grayish green-blue at sunset and yellowish in twilight.

Although scientists were aware of a soft radiant pillar of dust-scattered light stretching from the earth toward the sun in the evening or early morning sky, the prevailing idea until the middle of the twentieth century was that the earth and sun were isolated in a near-perfect vacuum, one almost completely free of gas and dust. In his introduction to the 1954 Liège Symposium entitled "Solid Particles in Stars," Edward O. Hulburt

---

◀ Coronagram of the sun with centered inset of solar surface in ultraviolet light. The coronagraph, by masking the brilliant disk of the sun, reveals coronal streamers and clouds; the inset shows a surge that, in some unknown manner, releases a balloon of gas. [High Altitude Observatory of the National Center for Atmospheric Research and Naval Research Laboratory.]

could still remark humorously, "Why discuss interstellar or interplanetary dust? There is no such thing. The best proof of it is that the wings of the angels are immaculately white!" In the last several decades, we have learned that interplanetary space carries a heavy traffic of particles and fields, as well as invisible ultraviolet radiation and X-rays, which couple the sun and earth in marvelously intricate ways.

During solar eclipses, a faint, pearly white halo—the corona—is seen to crown the sun to heights of several million kilometers. Space-age science has revealed that the gap between the sun and earth is bridged by coronal streamers whose far-reaching fingers brush the outermost fringes of the earth's environment, bringing winds of energetic solar particles that press and eddy against the earth's magnetic field.

From the vantage point of the moon, the Apollo 16 astronauts could view the earth as a large fluorescent screen. Their ultraviolet photographs showed the earth capped by shining auroral ovals encircling its magnetic poles and girdled about the equator by curving belts of high-altitude airglow. And all the earth was bathed in a 50,000-mile-high ultraviolet glow of sunlight scattered from a cloud of atomic hydrogen that envelops the earth—a geocorona.

Looking earthward, the first spacemen photographed our emerald planet, its delicate beauty in vivid contrast to the stark background of jet-black space. That image still conveys a deep sense of loneliness and fragility, a feeling reinforced by our planetary explorations. Studying Venus and Mars has given us an understanding of the narrowness of the ecological zone that permitted a life-giving atmosphere and stable environment to evolve on earth. If the earth were only 5 percent closer to the sun, it is possible that it would have experienced a runaway "greenhouse effect" eons ago, and its surface would be as hot as an oven, like that of Venus. (In the greenhouse effect, solar radiation warms the surface of a planet, and carbon dioxide in the atmosphere traps the infrared heat waves reradiated by the surface.) A one percent shift to a larger orbit might have led to glaciation, and the surface of the earth would resemble the cold and arid deserts of Mars.

Evidence now suggests that tampering with the controls of "Spaceship Earth" may disturb the delicate balance of the ecosystem. We have come to realize with a new sense of urgency that the hazards of natural catastrophes and of human interference with the environment can be rationally assessed only if we have a basic understanding of the vast complex of physical-chemical-biological interactions between solar radiation and every part of space from the surface of the sun to the surface of the earth. Furthermore, as new space-age technologies spawn a host of civilian and military applications, relevant issues of economics and security require a better scientific understanding of the nature of space.

The most direct connection between sun and earth is the flood of sunlight bringing heat to the earth's surface, but there are hosts of subtle solar-terrestrial linkages at every intermediate stage between the sun and the earth. These couplings can be studied on all time scales, from real time to the time of eras stretching back to the origin of the solar system.

The steady power of the sun is about $4 \times 10^{23}$ kilowatts, of which less than a billionth part, about 1.4 kilowatts per square meter (approximately 5 million horsepower per square mile), reaches the outer atmosphere of earth. Only one part in 100,000 of this power is contained in the ionizing X-rays and ultraviolet radiation which produce the ionosphere. Five to ten times as much ionizing energy arrives at the fringe of the earth's magnetic field in the form of fast-moving, electrically charged particles—the solar wind. When the sun is active, its ionizing radiation, both electromagnetic and corpuscular, can increase by several factors of ten. No comparable amounts of energy can be extracted in any form from within the earth and its environment. Although the earth contains $10^{37}$ ergs of rotational kinetic energy, only the tiniest fraction can be released into its environment. The energetic particles trapped in the Van Allen belts are far from sufficient to account for the brightness of auroras. The sun supplies essentially all of the energy of the solar-terrestrial system, reigning supreme over its complex kingdom.

The changing influence of the sun is most dramatic at times of explosive solar flares, when a burst of X-rays can reach deep into the ionosphere to create the conditions that black out shortwave radio communications. In a matter of hours, energetic protons may traverse the interplanetary medium and follow earth's geomagnetic field lines down to the polar caps, where they produce enough ionization to wipe out transpolar communications. A day or so afterward, the full flood of solar-magnetized plasma arrives at the boundary of the earth's magnetic field (the magnetopause), where it creates a shock wave that compresses the magnetic field toward earth. The resulting magnetic storm sets compass needles jiggling and causes the auroral lights to flash and flicker. As the high-latitude storm grows into a worldwide ionospheric disturbance, it garbles radio communications and creates winds that blow through the upper atmosphere. When all of the elements of the sun-earth system have run through their responses, the last residue of a great flare is the faint red airglow that diffuses over the entire night sky from pole to pole. This final gift from the sun to the earth is only the "death rattle" of an enormous explosive outburst on the sun, but it carries the energy of dozens of hydrogen bombs.

## THE SUN AND ITS STRUCTURE

The sun is the engine that drives the circulation of the atmosphere, drawing water from the ocean into the air and creating clouds, wind, rain, and snow. For billions of years, it has shed its steady, benign light and heat on earth, sustaining all terrestrial life.

Over the span of civilized history, no substantial change in the sun's brightness has been detected. Charles Greeley Abbott (1872-1973) devoted half a century to carefully measuring the solar flux and found it constant to better than one percent. Hence, the solar energy delivered to earth is referred to as "the solar constant." Recently, however, evidence has suggested that the solar constant may vary by a few tenths of one percent over a decade. Theoretical and observational studies hint that the sun's thermonuclear furnace may temporarily throttle down significantly for a million years, that the sun's diameter may shrink noticeably over a millennium, and that dramatic cooling trends may be related to a near vanishing of sunspots for time intervals as long as a century. Atmospheric and oceanic buffers must dominate short-term variations of climate, but long-term climatic trends could be keyed to very subtle changes in solar flux. Still, all of the evidence for sun-climate connections is statistically marginal; speculations should be treated with caution.

Because the sun and earth are so delicately coupled, the balance of the ecological system may be precarious. If, for example, the burden of carbon dioxide delivered to the atmosphere by the burning of fossil fuels were to raise the global temperature by 2°C (Celsius) through the greenhouse effect, it could well have ominous repercussions. In the polar regions, the warming would be amplified. The structure of the West Antarctic ice sheet is such that an increase in summer temperature of 2°C could lead to rapid disintegration. Within a hundred years, it might discharge one half to one third of its volume. Complete melting would raise sea level by 5 meters and coastal plains would be inundated. Similarly, a 2°C drop in temperature might set in motion a new glaciation. Doomsday scenarios portray arctic ice sheets slowly pressing southward until Manhattan's skyscrapers crumble in their path. But, again, such predictions should be treated as being highly speculative.

Recent measurements from rockets and balloons show tentative evidence for an increase in solar radiation of about 0.4 percent since 1968. It also appears that smaller variations of the total solar luminosity occur from day to day as sunspots and related optical features move across the solar disk. Climatic models indicate that variations at these levels and time scales should be important to climate dynamics. In the wavelengths shorter than those of visible light—ultraviolet and X-ray radiation—the sun is highly variable. Enhancements of these emissions, however, while representing an increase in the solar luminosity, constitute only a very slight change—less than 0.01 percent of the solar constant. So small a change can have hardly any demonstrable direct effect on climate. However, at ultraviolet wavelengths of around 200 nanometers, variations of several percent in the solar flux have been detected. These changes can be very significant for the photochemistry of trace molecular species in the stratosphere, such as ozone ($O_3$) and nitrogen dioxide ($NO_2$). Since some of these gases absorb visible and infrared radiation, photochemical influences on their concentrations might alter significantly the total energy flux delivered to the lower atmosphere or radiated back to space.

There is some indication that variations in the solar ultraviolet flux affect the atmospheres of the outer planets. Since 1950, the Lowell Observatory has measured planetary and satellite emissions continuously. Changes in the brightness of the visible light reflected by Neptune and Titan, the largest moon of Saturn, seem to be keyed to high solar magnetic activity. (Titan is the only planetary satellite with a substantial atmosphere.) During one magnetic storm, a radiometer aboard the Nimbus 6 meteorological satellite measured no change in the solar constant greater than 0.5 percent; Titan, at the same time, varied in green light by 9 percent. Perhaps solar ultraviolet variations induce photochemical changes in planetary atmospheres which change their reflectivities for visible light.

## Sun and Climate

The climatic record contains wide excursions from "normal climate," as averaged over geological time. Even now, we are in a 10,000-year warming trend following an ice age. The ice-covered regions of the earth are relics of that cooler period. For most of the earth's recorded history, there were no polar ice caps.

The changing seasons of the year result from the tilt of the earth's equatorial plane from its orbital plane. At present, this angle of obliquity is 23.5°, but it varies slowly between 22.1° and 24.5°. In the 1930's, the Serbian scientist Milutin Milankovitch proposed that this gentle nodding of the earth's axis would create a 41,000-year global temperature cycle (see Figure 1). Besides moving back and forth, the earth's axis of rotation also wobbles. This produces a precession of the equinoxes, which slowly varies the relative lengths of winter and summer. According to present theory, this precession produces a 22,000-year temperature cycle. More recent analyses also point toward climatic influence stemming from the changing eccentricity of the earth's orbit. In a 100,000-year cycle, the orbit stretches from almost perfectly circular to slightly elliptical and back again. But the greatest range of this effect on the annual solar flux received at the earth is only about 0.1 percent.

A research project known as CLIMAP (Climate: Long-Range Investigation, Mapping and Prediction), conducted by an international team of scientists in the 1970's, has confirmed the link between climatic change and orbital geometry. The results are based on the use of new radioisotope techniques to analyze core samples of ocean sediments that contain

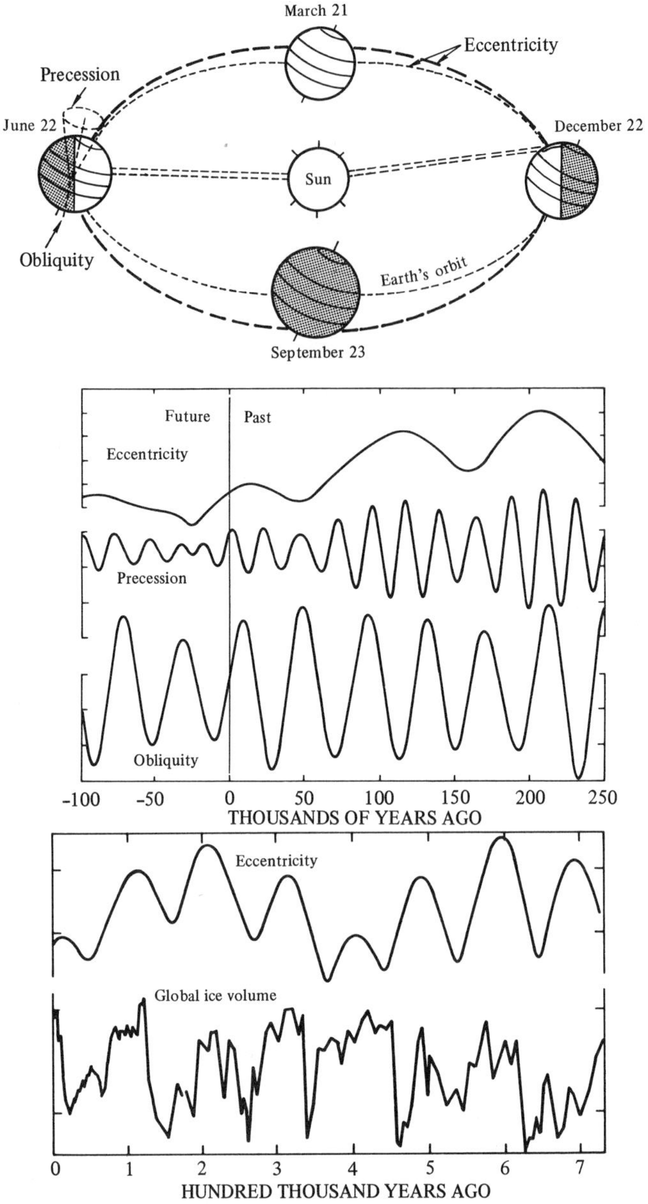


**Figure 1 Orbital variations and their effects on global ice volume.** The geometry of the earth's orbit (top) changes over 22,000-, 41,000-, and 100,000-year cycles (center). The curve for orbital eccentricity tracks with changes in global ice volume over the past 730,000 years (bottom), with the latter determined by the ratio of oxygen-18 to oxygen-16 in fossilized plankton. [SOURCE: *Mosaic,* November/December 1979, p. 5. After J. Imbrie and K. P. Imbrie, "Modeling the Climatic Response to Orbital Variations," *Science,* Vol. 207(1980), p. 944.]

a record of prehistoric temperature variations. The 22,000-year precession cycle, the 41,000-year tilt cycle, and the 100,000-year eccentricity cycle have all been confirmed.

Over the last million years, there is evidence for at least ten major glaciations interspersed with several little ice ages, but the connection between terrestrial solar flux and climate is not simple. Ice sheets do not necessarily develop only in response to sudden large drops in global temperature. For example, a slight trend toward longer winters and a slight fall of summer temperatures can lead to greater accumulations of snow in high latitudes each winter and less melting during the following summers. The coupling of these effects can extend the polar ice cap steadily southward.

Although the Milankovitch theory successfully relates the periodic glaciations of the past few million years to orbital eccentricity (Figure 1), it is very difficult to explain such comparatively large climatic effects from such small changes in incoming radiation. The key perhaps lies not in the total input of radiation but in the seasonal and latitudinal distribution of the solar flux. With an average global change of only 0.1 percent, local variations as high as 10 percent may occur. The effects of small changes can then be amplified by the feedback of ice reflecting radiation back into space.

Only 10 percent of all geological time has been favorable to ice ages, and within an ice age of perhaps two million years only 10 percent of the time was occupied by runaway glacial advances of continental dimensions. The rest of the time consisted of interglacial climates.

***Volcanoes and Climate*** Any comprehensive model of climatic variation must account carefully for the processes of radiation absorption, scattering, and reemission within the atmosphere, as well as for the incoming solar flux. Aerosols and dust in the atmosphere can exert an influence far beyond any variation of the solar flux itself. The evidence that volcanic activity affects climate is highly persuasive. For example, in 1815, an immense volcanic eruption of Mount Tambora in Indonesia spewed vast amounts of fine dust into the atmosphere. Some 100 cubic kilometers of debris were lifted off the mountain, filling the sky with a cloud of ash so dense that it hid the sun. By comparison, the main eruption of Mount St. Helens was a small event; the May 18, 1980, eruption lifted only 2.7 cubic kilometers from its cone. As the dust of Tambora circulated about the earth in the high atmosphere, it was accompanied by profound changes in worldwide climate. In the middle of Europe, the summer of 1816 was the coldest on record in two centuries; in effect, there was no summer. In New England, 3 to 6 inches of snow fell in June and crop-killing frosts appeared through July and August. Incidentally, Benjamin Franklin, during his sojourn in Paris, attributed the exceptionally cold winter of

1783-84 to the eruption of Icelandic volcanoes the previous May and June.

The introduction of particles into the atmosphere through agricultural and industrial processes, although not as dramatic as volcanoes, steadily builds up to very substantial amounts. This effect may be offset, however, as worldwide energy consumption increases. The additional burden of carbon dioxide in the atmosphere from the burning of fossil fuels may enhance the greenhouse effect and produce an overwhelming trend toward higher average global temperatures.

## Solar Features

Astronomers classify the sun as a yellow dwarf, one of the commonest varieties of stars among the 100 billion that inhabit the galaxy. The way solar physicists model the architecture of the sun recalls a character described by Jonathan Swift: "There was a most ingenious architect who had contrived a new method for building houses, by beginning at the roof, and working downwards to the foundation." Astrophysicists, like Swift's architect, start with the roof of the sun, its photosphere. The data they work with are the sun's observed luminosity, its composition, and its size. Without ever seeing deeper into the sun than a few hundred kilometers, it is possible to construct a model of the sun's pressure, temperature, and density all the way to its basement—the thermonuclear core (see Figure 2). At each point in the interior, the inward gravitational pull of all of the sun's mass must match exactly the outward pressure of hot gas and radiation. In such a model, the rates of nuclear reactions are calculated from experimental data obtained with laboratory accelerators.

From 6,000 K (Kelvin) at the surface, the sun's temperature increases inward to 15,000,000 K at its center. Here, the density is 11 times that of lead, but the average density of the sun is roughly equal to that of water. Its chemical composition is about three-fourths hydrogen and one-fourth helium, with all other elements adding up to a trace of about one percent. When four hydrogen nuclei fuse to form one helium nucleus, a small amount of mass is converted to energy according to Einstein's law $E = mc^2$ (energy is equivalent to mass times the square of the speed of light). Nuclear fusion produces 10 million times the energy generated by chemically burning an equal amount of oil or coal. Although 10 trillion atomic fusions provide only enough energy to light a 50-watt bulb for one second, the solar furnace burns at such a ferocious rate that, every second, the equivalent of about 2 trillion trillion trillion hydrogen atoms are annihilated. The resulting flow of energy brings about 200 trillion kilowatts of sunshine to the earth.

Astrophysicists have great confidence in their solar model, but they have been shaken by a failure to observe a predicted flux of neutrinos

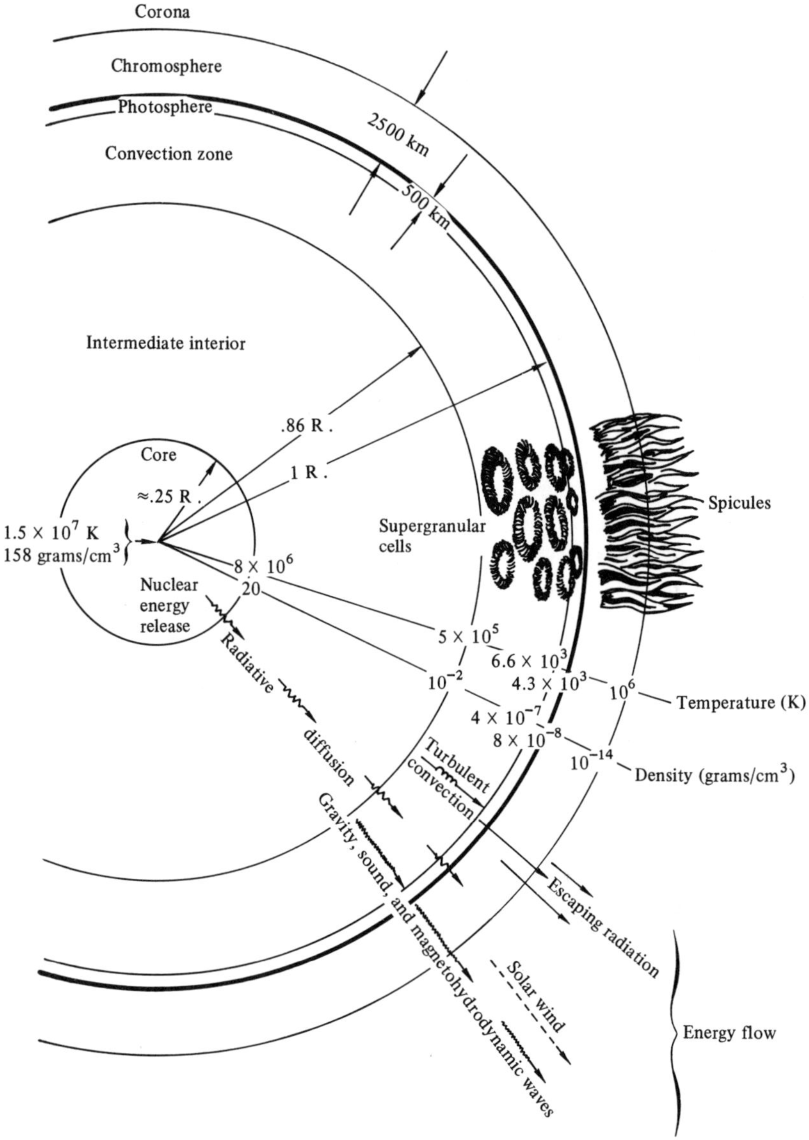


**Figure 2 Solar structures and energy flow.** Models of the sun enable solar scientists to predict its temperature and density from the visible photosphere down to its thermonuclear core. For about 86 percent of the sun's radius, energy is transported solely by radiative diffusion. At that point, turbulent convection sets in, and this convection is thought to account for the large supergranular cells that can be detected on the sun's surface. [SOURCE: Edward G. Gibson. *The Quiet Sun.* Washington, D.C.: U.S. Government Printing Office, 1973.]

from the nuclear reactions at the sun's core. The existence of the neutrino was hypothesized in 1931 to explain some puzzling features of the radioactive beta decay of atomic nuclei, but the particles were not detected experimentally until 1956. According to their postulated nature, neutrinos are ghostlike particles—they have zero rest mass and no electric charge—which are almost impossible to detect. It would take a thickness of thousands of light years of lead to stop a neutrino. They move freely through the universe at the speed of light.

The core of the sun is an enormous neutrino factory. Only one in every ten billion is absorbed in the body of the sun, and only one in every trillion that reach the earth fails to pass right through. Finding such elusive particles requires a truly massive detector. For more than a decade, a neutrino trap consisting of a tank of 610 tons of perchloroethylene (cleaning fluid) has been located deep in a gold mine under the Black Hills of South Dakota. If a neutrino is captured by an atom of the isotope chlorine-37, a radioactive atom of argon-37 is formed which can be detected by an exquisitely sensitive counting technique. In spite of the most painstaking efforts, however, only a few neutrinos per day have been detected—about one third of the theoretical rate predicted for the sun.

Various speculations have been offered to explain the mystery of the missing solar neutrinos. One supposition is that the temperature of the sun's core has decreased by a million degrees without the visible surface being affected yet. Because the radiation generated at the center takes perhaps a hundred thousand years to reach the surface, the change in luminosity may lag that far behind a cooling of the core. An alternative explanation is based on recent experiments that report to have detected slight evidence that neutrinos oscillate between two different identities, electron antineutrinos and tau antineutrinos, for instance. This would have implications as profound for solar physics and cosmology as for elementary particle physics. If neutrinos oscillate, quantum mechanics requires that they have a certain rest mass. Cosmology sets a tight limit on this mass. If it exceeds that limit, neutrinos must exert an overriding constraint on the expansion of the universe. Oscillations also imply that neutrinos may decay spontaneously. This could solve the mystery of the missing neutrinos in solar physics if the electron neutrinos decay in flight from the sun to the earth.

Except in the case of neutrinos, energy is transported from the core through the body of the sun by a slow diffusion of gamma rays and X-rays. A hundred thousand years may elapse before the energy generated in the core reaches the surface. Like a man pushing through a large crowd of people, bumping his way along in zigzag fashion, the high-energy photons released in the core make countless collisions with gas atoms in their random walk outward. Near the center, the average path length between collisions is only about one centimeter, and each collision robs a photon of some energy. Gradually, its wavelength increases, until the radiation that finally emerges from the surface of the sun is mostly visible light and infrared radiation.

The sun's photosphere, or visible surface, is a zone of granulated texture only a few hundred kilometers thick with a temperature of about 6,000 K—too hot for any solid or liquid matter to exist. It may seem puzzling that the sun exhibits a "surface," if it is gaseous. In ordinary air, there is virtually no opacity to visible light, but if the air temperature were 6,000 K, the visibility would be only a few feet. Even though the gas at the sun's surface is very thin, its opacity is very great, and the edge of the sun appears sharp.

Photographs of the sun's surface show that it is coarsely mottled. In a pot of boiling water, bubbles come zigzaging from below, bumping and merging into larger bubbles, to burst at the surface. Similarly, the sun's surface is believed to overlie a convection zone in which hot gas bubbles, seething at temperatures of thousands of degrees, rise in columnar fashion to deliver heat to the surface, after which cooled gas sinks to be heated again. A solar bubble may be 1,300 kilometers across. It is short-lived, lasting only a few minutes. At any moment, the surface is covered with about a million bubbles, which rise at speeds of several thousand kilometers per hour.

Valuable observations made with telescopes show that the surface of the sun "shivers" in a variety of modes. These rhythms are subtle and complex. Most obvious is a motion with a periodicity of about 5 minutes in which the surface heaves as much as 1,000 to 2,000 kilometers, but the motion does not involve the entire photosphere simultaneously. Like the surface of a choppy sea, small local regions bubble independently of each other, and the 5-minute period is a rough average. Recent observations seem to indicate long-period oscillations of small amplitude. Theoretical models predict such movements and relate them to the sun's interior structure, much as earthquake-generated seismic waves tell about the interior of the earth. When all of the vibrational modes of the sun's "heart throbs" are resolved, it may become possible to draw fundamental inferences about amounts of heavy elements in the interior and about how density and temperature vary with depth.

Overlying the photosphere is a reddish chromosphere of spiky forms that rise and fall like fountains of spray. These spicules jet to heights of 5,000 to 20,000 kilometers and fall back again in a matter of seconds. Still farther out, an evanescent corona of faint, pearly light stretches far into space. It becomes visible to observers only when the brilliant solar disk is eclipsed by the moon. Then, wispy streamers of luminous gas can be seen reaching out hundreds of thousands of kilometers. Coronal gas is so thin that stars shine brightly through; yet it radiates strongly in invisible far-ultraviolet and X-ray wavelengths because its temperature is in the million-degree range.

Some solar physicists are beginning to suspect that conventional models of the solar interior may be too simplified. Such models do not include any mixing of the outer layers with the core region of nuclear burning. If such mixing occurs, the solar luminosity might vary on time scales ranging from 1 to 10 million years. Substantial variations in luminosity due to heat transport by convection through the outer layers are also a possibility. A large-scale convection pattern below the photosphere may become unstable and switch suddenly to another mode in which heat transport is lower. The net effect would be an increase in turbulent gas motions. Such heating of the convection zone would cause it to expand against gravity. Time scales of variability could be as short as months. Furthermore, since these convection layers are threaded by strong magnetic tubes of force, the gas motions can twist the field lines until large amounts of energy are wrapped up in coiled fields below the surface. This magnetic field energy might be released explosively years later.

## Sunspots

Aristotle described the sun as a disk of pure fire without blemish. A fleeting glimpse of the sun with the naked eye does indeed give the impression of a uniformly bright disk; but if one looks carefully through a smoked glass, it is possible to detect small dark blotches called sunspots. When Galileo first pointed his telescope toward the sun in 1610 and clearly resolved sunspots, he aroused the anger of theologians of his time, for whom sunspots meant blemishes on the handiwork of God. They refused to look through his telescope for fear that they might be convinced the spots were there. Galileo himself quickly concluded that the spots were attached to the sun, although he first thought they might be clouds above the sun's surface. It was soon recognized that spots move across the face of the sun from left to right in about 14 days, consistent with a rotation period of about 27 days; but the rotation has the peculiar property that spots at the equator rotate faster than do those at the poles.

***Sunspot Statistics*** After the initial excitement of the discovery of sunspots, no systematic study of their variability ensued. It is recorded that some Italian astronomers remarked on possible connections between sunspots and weather; they noted that a "period of unusual drought" accompanied an absence of spots in 1632.

The English astronomer William Herschel suggested in 1801 that sunspots were linked to weather. He found that extreme spottedness of the sun correlated with the price of a bushel of wheat on the London market, which in turn was connected by the law of supply and demand to the weather. His idea predates by half a century the discovery of the sunspot cycle. Jonathan Swift made a similar conjecture as early as 1726.

The discovery of sunspot cycles is attributed to Heinrich Schwabe, a

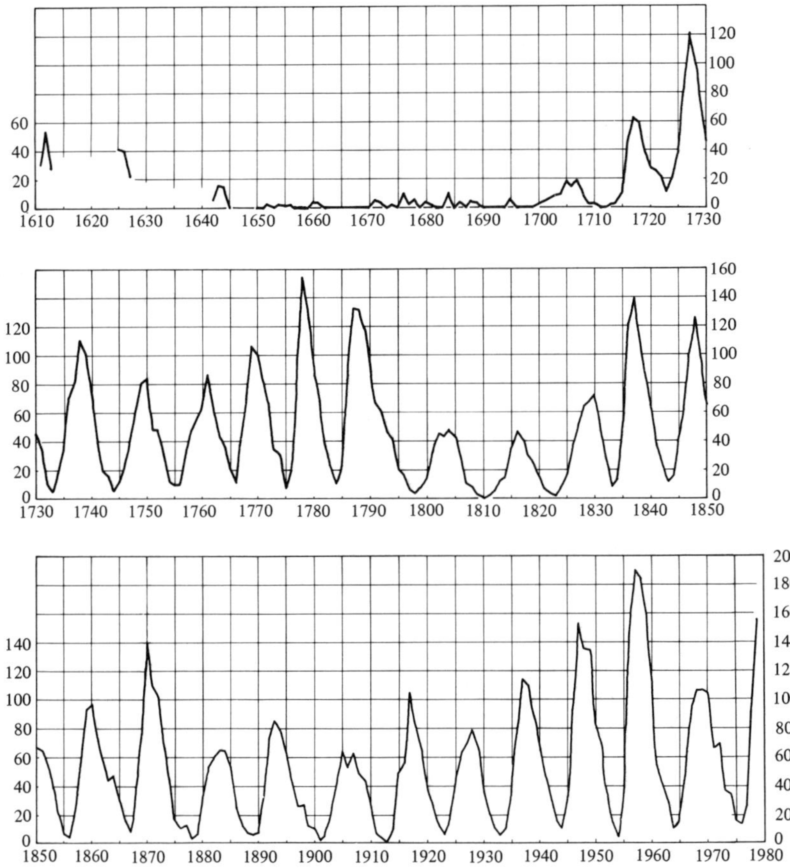


**Figure 3 Average annual numbers of sunspots.** Sunspots have tended to vary in number over an 11-year cycle, but the pattern is not invariable. For a period of approximately 70 years in the seventeenth and eighteenth centuries, almost no sunspots were seen. [SOURCE: John A. Eddy. "The Maunder Minimum," *Science*, Vol. 192(1976), p. 1193. Copyright 1976 by the American Association for the Advancement of Science.]

pharmacist in Dessau, who started to keep careful records in 1826. His objective was to distinguish the transit of a planet across the face of the sun, but, to his great surprise, his records of spots revealed a definite periodicity (see Figure 3). Throughout this century, the number of sunspots has waxed and waned with a rather consistent 11-year rhythm, but, in the longer historical record, it is clear that the 11-year cycle is very erratic. Since Galileo's first telescopic observations, there have been two stretches of time totaling more than 150 years when few spots were noted. From 1645 to 1715, a period known as the Maunder Minimum, spanning the reign of Louis XIV, there is little record of sunspot observations in

Europe and coronal streamers were almost totally missing when eclipses were observed. It was not that seventeenth-century astronomers were uninterested in sunspots. The Astronomer Royal John Flamsteed noted a spot in 1684 and wrote, "These appearances, however frequent in the days of Scheiner and Galileo, have been so rare of late that this is the only one I have seen in his face since December 1676." Maunder himself, in a retrospective study, reported to the Royal Astronomical Society in 1890 that "for a period of about 70 years, ending in 1716, there seems to have been a remarkable interruption of the ordinary course of the sunspot cycle. In several years no spots were seen at all and in 1705 it was recorded as a most remarkable event that two spots were seen on the sun at the same time, for a similar circumstance had scarcely ever been seen during the 60 years previous."

As persuasive as these observations seem to be, there is currently much effort to look into European diary sources, some of which cite sunspots during the Maunder Minimum. In 1975, Chinese astronomers began a general survey of ancient astronomical records. In a search of 8,000 collections of private and public records, they coincidentally discovered 13 reports of sunspots in the Maunder period. It seems that the Maunder Minimum must have been a period of low sunspot numbers, but certainly not spotless.

In recent years, attention has been drawn to the coincidence between the 70 years of low sunspot activity and the "Little Ice Age" of the sixteenth and seventeenth centuries that left Europeans and North Americans numb with cold. Ice skating was popular on the English Channel and the Thames, which rarely freezes now, and Greenland was completely locked in ice. However, current retrospective research casts doubt on the Little Ice Age being more than a comparatively local phenomenon confined to Europe and the northern United States. More complete data for the entire northern hemisphere do not imply a global drop in temperature.

***Tree Rings and Sunspots*** Early this century, the American astronomer A. E. Douglass first tried to link annual tree ring growth with solar activity, using trees from the arid Southwest. He founded the new science of dendrochronology, but he failed to find any link between tree rings and the sun. Measuring the carbon-14 in tree rings to investigate solar activity was first suggested in the early 1960's. The amount of radioactive carbon did indeed vary from year to year by a few percent.

Radioactive carbon-14 is produced by cosmic ray bombardment of the earth's atmosphere. At a maximum of sunspots, a stronger solar magnetic field permeates interplanetary space and deflects cosmic rays entering the solar system. The intensity of cosmic rays and the attendant production of carbon-14 therefore minimize at sunspot maximums. In the process of photosynthesis, trees assimilate atmospheric carbon dioxide containing some carbon-14, so that an individual tree ring carries a natural record of the carbon-14 to carbon-12 ratio in the year it grew. From ring to ring, the carbon-14 content is inversely related to sunspot activity; a 20 percent increase characterizes the years 1634-1715, which encompass the Maunder Minimum of sunspots. The carbon-14 maximum in tree rings and the sunspot minimum match the coldest periods of the Little Ice Age.

In certain bristlecone pines in California, the tree ring record goes back 7,000 years and shows that solar activity has been surprisingly irregular. Similar evidence now comes from analysis of deep-sea sediments obtained by core drilling. The drilling ship *Glomar Challenger* has used a new hydraulic-piston corer in the Gulf of California and the Gulf of Mexico to bring up relatively undisturbed samples of ocean sediments. These samples contain the skeletal remains of sea creatures—plankton—laid down year by year on the ooze of the ocean bed. Different forms of the plankton known as foraminifera, which float near the surface, prefer different water temperatures. By counting the dead shells left by different species in a layer of sediment, it is possible to deduce the water temperature at the time the microscopic creatures were alive. These samples give a temperature record extending millions of years into the past. Colder winters appear as darker bands in sample sections.

The best temperature record is contained in the ratio of oxygen isotopes in the chalky shells of plankton. Atmospheric oxygen is mainly the lighter isotope, oxygen-16. Because water made with oxygen-18 does not evaporate as fast as the lighter molecule made with oxygen-16, the heavier water sticks better on a growing ice sheet, where it remains trapped. Thus, the proportion of oxygen-18 left to be absorbed by foraminifera depends on the amount of ice cover. The ocean bed drillings give us a measure of the variation in oxygen-18 content of the planktonic remains over past history.

There are at least a dozen principal trends in the tree ring data, each lasting from 50 to 200 years. But instead of the record showing mainly high solar activity with occasional minima, it shows mainly low activity. Since the late Bronze Age, there have been only a few brief bursts of strong solar activity. They appear in the records at random, with no underlying periodicity. In the light of this historical record, normal behavior might be an almost spotless condition, with the activity since 1700 being anomalously high. French wine connoisseurs have noted a correspondence between sunspots and good vintages, which inspired one writer to comment that life must have been dull in earlier epochs of low solar activity, with a seemingly endless procession of indifferent burgundy and claret vintages.

Sunspots usually appear in pairs of opposite magnetic polarity. In each solar hemisphere, the leading spot in such pairs nearly always has the same polarity, but this arrangement is reversed in the two hemispheres. The polarities of the leading spots interchange in successive 11-year sunspot cycles so that the true solar cycle is a 22-year magnetic cycle, with similar alternation in the sun's overall magnetic field. The possibility that droughts in the American high plains west of the Mississippi may follow a 22-year cycle has been recognized for some time. In the nineteenth and twentieth centuries, the drought years have been 1823, 1843, 1868, 1890, 1913, 1934, 1954, and 1977. The 1930's dust bowl was especially severe and led to great human suffering. Recently, tree ring data have been analyzed statistically to reconstruct the historical patterns of droughts as far back as A.D. 1600. There appears to have been a tendency for the area affected by drought to pulsate in size with a 22-year cycle over the past several hundred years.

Statistical evidence of a 22-year period also has been found in the ratio of deuterium to hydrogen in two bristlecone pine trees with lives spanning 1,000 years. This ratio has the advantage of not being a local measure. The fractionation of the hydrogen isotopes depends on the mean surface temperature of the Pacific Ocean, from which the water incorporated into the trees evaporated, and on the air temperature where precipitation occurred. The message of all the statistical evidence, however, is not strong enough to allow us to forecast drought cycles with sufficient credibility that action may be taken to reduce their adverse economic impacts.

## Auroral Indicators

Another indicator closely linked to the sunspot cycle is the awe-inspiring natural phenomenon known as the aurora borealis—the northern lights. Ancient literature is so rich in descriptions of auroras that it provides an important complement to the tree ring record.

Some of the earliest references to auroras are found in Aristotle's writing (during the fourth century B.C.). Seneca, in the reign of Tiberius (A.D. 14 to 37), describes how, one night, a blood-red glow was seen in the west. Believing that the seaport of Ostia, at the mouth of the Tiber River, was on fire, Roman soldiers hurried there to fight the flames.

Various biblical references to vivid apparitions in the sky can be attributed to auroras. In 2 Maccabees, written in the second century B.C., there is the following account:

> About this time Antiochus sent his second expedition into Egypt. It then happened that all over the city, for nearly forty days, there appeared horsemen charging in mid-air, clad in garments interwoven with gold—companies fully armed with lances and drawn swords. . . .

During the Middle Ages, frequent literary accounts appear of "fire beams" and "burning spears" in the sky, accompanied by vivid artistic renditions.

After an active solar period in the 1620's, few auroras were seen for a century, and the memory of such events almost vanished. A great aurora in 1716, at the end of the Maunder Minimum, therefore had a startling impact. Edmund Halley gave a thorough account to the Royal Society, describing the arches and rays as they were observed from widely separated places. Halley rejected the popular suggestions that they were sulfurous vapors escaping from the earth's interior and proposed the essentially modern view that they were "magnetic effluvia" constrained to move along the lines of force of the earth's magnetic field.

## Space-Age Observations

The development of rocketry has furthered greatly our ability to observe and understand solar activity. By the time World War II came to an end in Europe, several thousand German V-2 rockets had been launched against targets in England. An enormous underground factory for producing V-2's was located near Nordhausen, in central Germany. Although it was inside the agreed-upon Russian occupation zone, U.S. Army forces got there first and made away with the last remnants of the V-2's in production, enough parts to assemble about 100 rockets at the White Sands Missile Range in New Mexico. While the army studied the rockets' propulsion technology, scientists were offered the opportunity to place instruments in the 2,000-pound-capacity space for warheads. Thus began the systematic conduct of space science. Early work with V-2's was succeeded by experimentation with smaller rockets, such as the Aerobee, Rockoon, and Nike-Deacon, until the era of satellites began in 1957. Skylab, launched in 1973, was the most spectacular approach to solar studies in space.

From the ground, we can observe only the visible light of the sun, a small range of its radio noise, and some of its infrared heat. Above the atmosphere, rocket- and satellite-borne instruments have revealed powerful X-rays and ultraviolet light, which rip electrons loose from atoms to produce the electrified regions of the ionosphere, beginning near a height of 60 kilometers and reaching out to several hundred kilometers. At great heights, the freed electrons act as a mirror to shortwave radio signals. Broadcast transmissions can thus skip around the earth, bouncing between the ground and the ionosphere, over distances far beyond the line of sight.

From the beginning of radio broadcasting, the quality of reception was known to vary with sunspots, but the physical mechanism was not understood until space research revealed the sun's X-rays. A decade of rocket studies showed that X-ray intensity followed sunspot number, and a spectacular campaign of rocket shots, spanning a total eclipse in 1958, finally identified the X-ray emissions very specifically, with the corona immediately overlying sunspots. Fifteen years later, Skylab's X-ray telescopes produced finely detailed images of the X-ray emissions of the sun.

## Solar Flares

In 1859, two English astronomers, R. C. Carrington and R. Hodgson, described a remarkable burst of light within a large group of sunspots. The flash lasted about five minutes and spread rapidly over some 50,000 kilometers. For the next few nights, brilliant auroras stretched to middle latitudes and magnetic compass needles wiggled erratically. Carrington thought he might have witnessed the plunge of a large meteor into the sun; it was a popular notion around the turn of the century that the fall of meteors into the sun was a major source of its energy. We now know that flares are cataclysmic explosions generated by the release of stored magnetic energy.

Superficially, the flash of a solar flare suggests a lightning stroke discharging electricity accumulated in clouds; but a flare releases the power of billions of hydrogen bombs and such energy can be drawn only from the intense magnetic fields in the solar atmosphere. The impulsive phase of a solar flare is generally characterized by hard X-ray and microwave bursts generated by the passage of highly accelerated particles through the corona. A major part of the energy of a flare must be carried by energetic electrons. In the main phase, the radio and X-ray emissions are accompanied by mass motions—surges, eruptive prominences, and expanding clouds of nonthermal particles.

Mass ejection is shown vividly in pictures of the corona obtained from satellites. At the start of the main phase, a shock wave that precedes the cloud of very energetic, ionized particles (plasma) is sometimes observed. Radioheliograph observations reveal the passage of accelerated particles through the corona, but they become trapped in very large magnetic loops. Further studies are needed of their propagation, trapping, and escape into the interplanetary medium.

In the nineteenth century, before the introduction of photography, astronomers had to sketch what they observed. Because an eclipse is so brief—never more than seven minutes—artistic renditions of the corona were set down very rapidly. In 1860, a German astronomer, E. W. L. Tempel, depicted a gigantic whorllike projection of the corona. Most of his contemporaries were skeptical and accused him of artistic license, but, more than a century later, Skylab's astronauts proved him right. They succeeded in photographing 24 transient events of gigantic gas bubbles expanding above the corona at 150 to 500 kilometers per second. From this one-year sample of events, it follows that such transients occur about once every hundred hours on the average; it would take a stroke of luck to duplicate Tempel's experience even once in a hundred years of eclipses. Spacecraft observatories planned for the coming years will televise ultraviolet images of such events, as well as coronal holes, as part of a continuous "weather watch" on the sun.

## Coronal Holes and Solar Wind

Before the days of rocket and satellite astronomy, the corona was pictured as a great cloud of superhot gas bound to the sun by gravitation. Ultraviolet and X-ray photographs from space have now revealed that coronal gas is imprisoned by tightly knit loops of magnetic lines of force rooted in sunspots. When these loops first emerge from within the solar convection zone, they range in size from as large as 300,000 kilometers down to the optical resolution limit of about 1,000 kilometers. The loops eventually disperse over months to years, and the old field lines no longer close in tight arches. Instead, they stretch out from the photosphere as "open" field lines, extending more or less radially into interplanetary space. The ionized coronal gas then flows along these field lines to form long streamers of high-velocity solar wind. As the sun rotates, the streams pass a given point in space every 27 days. Because gas escapes where the field lines are open, the underlying coronal density is much lower, and the X-ray and ultraviolet radiations are correspondingly weak, giving the impression of a dark "coronal hole."

During the year that Skylab's solar telescopes observed the sun, coronal holes covered as much as 20 percent of the surface. Permanent holes exist over the poles of the sun; at lower latitudes, they come and go, with an average life of four to six months, although some last as long as nine. The relatively weak magnetic field that emerges from a coronal hole, almost straight outward, permits gas to pour into space as a high-speed wind. The area of sun covered by coronal holes is closely related to disturbances in the earth's magnetic field and is a very good index of expected auroral activity.

In recent years, new indicators of magnetic activity on the sun have been recognized that may have as much or even more fundamental significance than sunspots. Just prior to Skylab, rockets discovered the phenomenon of "bright points," small pointlike sources of X-ray and ultraviolet emission. At any time, thousands of bright points dot the solar disk, even covering the poles and appearing in coronal holes. Skylab has shown that they sparkle on a time scale of minutes, sometimes flaring to 10 times normal intensity. Of potentially great importance is the magnetic flux that is carried in bright points, which is at least as great and possibly greater than that in sunspots. As sunspots wane, the magnetic flux from bright points increases enough to balance the overall activity of the sun. It almost appears as though sunspots have, until now, distracted astronomers from recognizing more revealing indicators of solar activity.

(a)

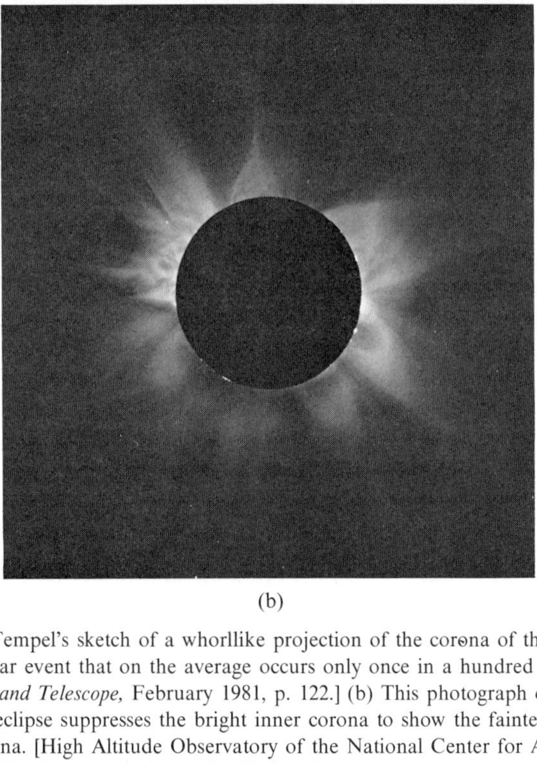

(b)

(a) E. W. L. Tempel's sketch of a whorllike projection of the corona of the 1860 eclipse captured a solar event that on the average occurs only once in a hundred years of eclipses. [*Sky and Telescope,* February 1981, p. 122.] (b) This photograph of the February 16, 1980, eclipse suppresses the bright inner corona to show the fainter streamers of the outer corona. [High Altitude Observatory of the National Center for Atmospheric Research and Southwestern at Memphis College.]

## Solar Research Goals for the 1980's

For the 1980's, solar physicists have compiled a list of fundamental questions that can be attacked with greater observational and theoretical power than ever before. What physical mechanisms produce sunspots? Why does the sunspot number undergo a cyclic variation? What is the deep, underlying solar magnetic cycle? How does the solar magnetic field partition into weak large-scale fields and into concentrated magnetic knots? How do these magnetic concentrations lead to explosive releases of energy in solar flares? How do plasma processes control the flow of energy, momentum, and mass into the corona? How is the solar wind supplied and accelerated? What are the ranges of solar variability in the production of radiation, plasma streams, and energetic particles—on all time scales from minutes to decades? What is the relationship between solar variability and climate—and perhaps weather?

The questions about variability of the solar constant and about spectral changes in the ultraviolet are on their way to being answered. A cavity radiometer on the Solar Maximum Mission Satellite launched by the National Aeronautics and Space Administration (NASA) early in 1980 indicated relative changes in flux with an accuracy of hundredths of a percent, and these variations seem to correlate with sunspot number. Ultraviolet measurements of appropriate accuracy will begin with the second Spacelab mission.

The International Solar Polar Mission is intended to carry instruments over the poles of the sun to give us three-dimensional pictures of the solar wind which will indicate its correlation with coronal holes. Toward the end of the decade, we may see the start of a 1-meter Solar Optical Telescope, launched aboard the Space Shuttle, which would apply resolutions of less than 1 second of arc to studying the microscale structure of the solar atmosphere. On the drawing boards is a solar probe designed to approach the surface of the sun to within as little as 4 solar radii. It would become the first spacecraft to perform *in situ* observations of coronal forms and their dynamics.

# THE INTERPLANETARY MEDIUM AND THE EARTH'S MAGNETOSPHERE

## The Interplanetary Medium

The sun-earth system bears some resemblance to a cathode-ray tube. Thus far, we have discussed in detail the sun, which acts as the gun. It sprays energy toward a target earth that responds as the fluorescent screen. In a cathode-ray tube, the space between gun and target is a high vacuum. Between sun and earth space is an even higher vacuum, but it is far from perfect. Although the energy density of interplanetary space is very small, its volume is so large that the power carried through it by particles and fields is enormous.

Prior to 1957, it was generally believed that the sun influenced the earth's upper atmosphere primarily through photoionization, which created the ionosphere, and through sporadic streams of charged particles, which produced magnetic storms and auroras. Solar magnetic fields appeared to bind the corona tightly to the near vicinity of the photosphere, and the earth's magnetic field served to confine ionized gas to the earth. Interplanetary space was thought to be almost empty, except for a very tenuous extension of a static corona possibly reaching to the earth's orbit and beyond.

Even before the space era brought us *in-situ* measurements, however, evidence that highly variable streams of gas flow from the sun was found in observations of the deflection of comet tails, in the development of geomagnetic storms following solar flares, and in the 11-year modulation of cosmic rays mirroring the sunspot cycle. From cometary data alone, it was possible to estimate that the average wind speed is about 500 kilometers per second, and that it never calms to less than 150 kilometers per second. The wind transports only one millionth as much energy as is transmitted by light and other forms of electromagnetic radiation, but, over the lifetime of the sun, much of its original angular momentum has been carried away by the solar wind, leaving the sun with a low rotation rate today.

The outward expansion of the sun's atmosphere creates the solar wind, which is supersonic throughout the interplanetary medium. Beyond a few solar radii, the rarified atmosphere is nearly collision-free, and electric currents flow with almost negligible resistance. In this manner, magnetic field is frozen into the solar wind and carried into the interplanetary medium. At the same time that solar magnetic field is drawn outward by the wind, solar rotation twists these streams into Archimedes spirals, as seen from above the ecliptic plane.

In the hydrodynamic theory of the solar wind, the flow begins in the lower corona. The observed velocity increases steadily up to about 400 kilometers per second at about 20 solar radii, and the concentration of particles—primarily hydrogen, with 2 to 4 percent helium and a trace of heavy elements—reaches about 8 per cubic centimeter. These figures fluctuate in time and space. Because the theory assumes a spherically symmetric expansion, it offers no detailed model of the long-lived "plasma streams," which have very different velocities and densities.

Large solar wind structures rotate with the sun. They stretch the sun's overall magnetic field roughly in the direction of the ecliptic plane so that the field is divided by a magnetically neutral sheet about the size of the solar system. Above and below this neutral sheet, the magnetic field of the solar wind reverses abruptly. Like a ballerina's skirt, the neutral sheet has a number of folds. Each time that a fold of the sheet passes by, the earth finds itself in a sector of reversed polarity. Since the structure of this sheet may persist for several rotations, successive crossings of field reversals lead to a variety of 27-day recurrent signatures in the magnetic activity of the earth.

It is difficult to trace phenomena within the structure of the interplanetary magnetic field back to large-scale photospheric fields. There may possibly exist an overall solar magnetic pattern that is fundamentally separate from the mechanism responsible for sunspots and small-scale field structures. As was mentioned earlier, the latter are believed to be related to a basic poloidal field, which is stretched into a toroidal field by the differential rotation of the solar atmosphere. When kinks in tubes of magnetic force emerge through the photosphere, they produce magnetic loops rooted in sunspots. However, the general problem of how magnetic flux is intensified in the convection zone and brought to the surface is still only vaguely understood.

## The Magnetosphere

Near the beginning of the seventeenth century, William Gilbert, physician to Queen Elizabeth I, published a lengthy treatise on his two decades of experimentation on magnetism. His writings dispose of such choice popular observations as "garlic causes magnets to lose their virtue," and he was the first to establish the dipolar nature of the earth's global magnetic field. In Gilbert's words, "The terrestrial globe is itself a great magnet."

The source of the earth's magnetic field is not yet fully understood, but we believe that matter flowing in the earth's metallic fluid core must generate electrical currents which induce the magnetic field. Paleomagnetic records reveal that the earth's magnetic field has reversed its polarity about every million years. Indeed, magnetic charts since the seventeenth century show a steady decrease in the strength of the earth's magnetic field, tending toward zero in about 3,000 years.

The distant magnetic field surrounding the earth is a shield that prevents energetic solar and galactic cosmic ray particles from crashing into the atmosphere (see Figure 4). In this protective role, the magnetic field may have had a profound influence on the evolution of life. When energetic particles strike deep in the atmosphere, they produce nitrogen oxides, which in turn react with ozone, decomposing it. The atmosphere then becomes more transparent to ultraviolet light, with harmful consequences to many living organisms. For perhaps a thousand years around the time of a reversal, the magnetic field is close to zero, leaving the earth

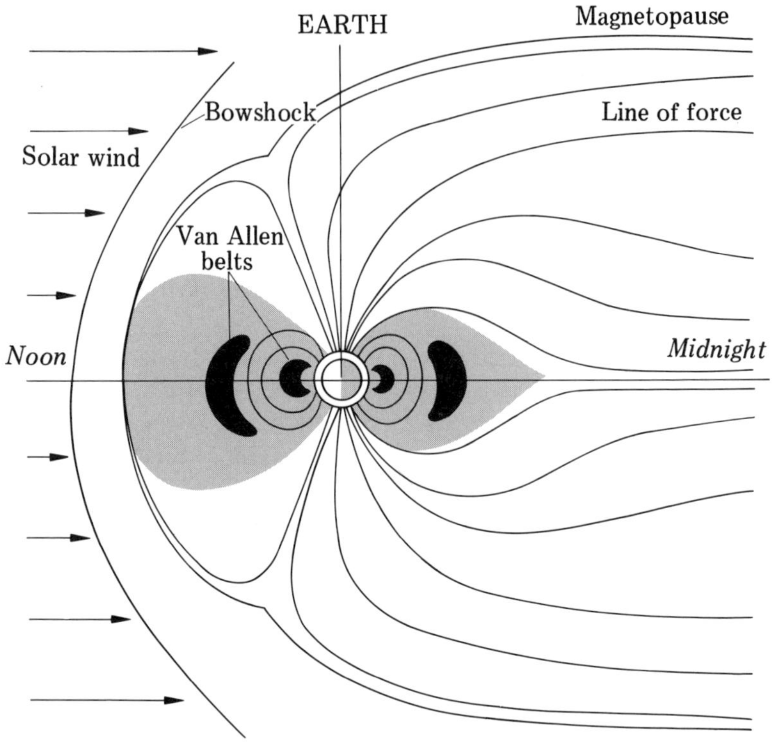


**Figure 4 The earth's magnetosphere.** A cross section of the earth containing the northern and southern magnetic poles and solar wind. The solar wind is deflected to flow around the earth by a shock wave similar to that created by a supersonic airplane. The wind bends back the magnetic lines of force as shown and pumps energy into the particles trapped near the earth to create the Van Allen belts. Auroras are caused by particles leaking along the lines of force near the poles. [SOURCE: G. Field, G. L. Verschuur, and C. Ponnamperuma. *Cosmic Evolution: An Introduction to Astronomy.* Boston: Houghton Mifflin Co., 1978, p. 69. Copyright 1978 Houghton Mifflin Company. Reprinted by permission.]

virtually naked to the impact of the solar wind and cosmic rays. Then, enormous clouds of energetic plasma, released by great solar flares, bombard the terrestrial atmosphere. Biological evidence points to the extinction of simple species of plant and animal life, in particular, protozoa within the order Radiolaria, around the time of reversals.

The present field determines the overall character of the magnetosphere. As the solar wind flows from the sun, it combs the sun's magnetic field lines outward through interplanetary space. Where these hit the outer boundary of the magnetosphere, a standoff bowshock forms. The magnetosphere thus acts like a blunt obstacle to the supersonic solar wind. As the wind flow is deflected around the bowshock, a kind of fric-

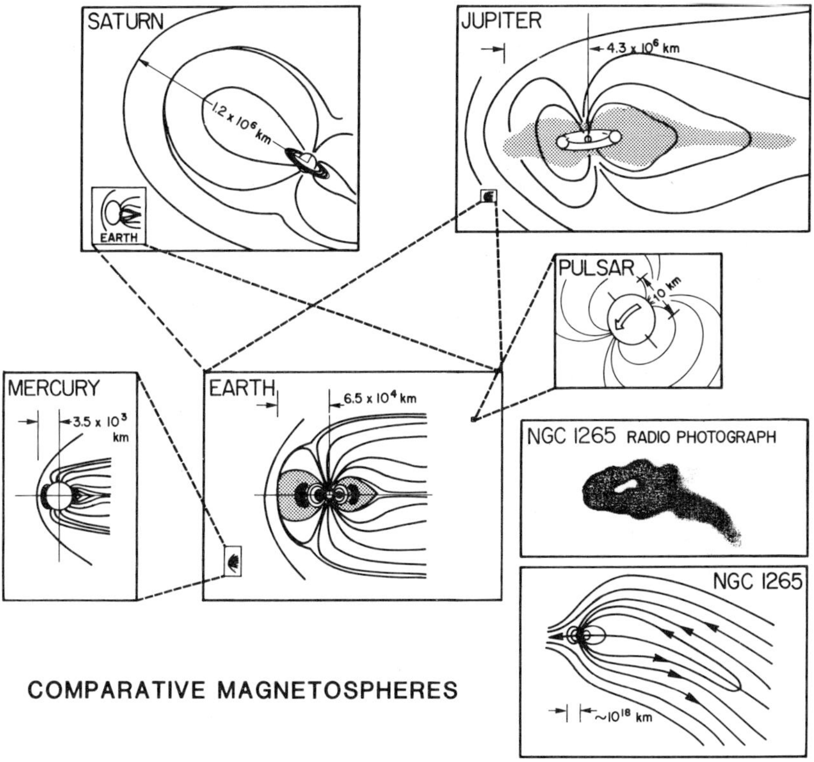


**Figure 5 Magnetospheres of selected planets and objects in the universe.** Fundamental similarities, but on vastly different scales, characterize magnetospheric configurations of planets in the solar system and celestial objects in the universe at large. Within the solar system, Mercury's magnetosphere is tiny compared with the earth's, while Jupiter's is enormous—much larger than a solar diameter. In a typical spinning pulsar, the magnetic field is trillions of times as strong as the earth's. Plasma is locked into the magnetic field until it is spun up to nearly the speed of light. As the radio galaxy NGC 1265 plows through interstellar gas, the ram pressure creates a magnetic tail stretching millions of trillions of kilometers. [SOURCE: National Aeronautics and Space Administration and the Naval Research Laboratory.]

tional drag draws out the magnetosphere into a long tail. (Figure 5 shows the magnetospheres associated with several other planets and objects in the universe.) When this magnetotail wags, geomagnetic responses are generated back at the earth's poles.

The magnetosphere stretches about 65,000 kilometers from the earth to the front of the bowshock. It is a huge bag of plasma containing charged particles with energies extending from the thermal range to hundreds of millions of electron volts. As the pressure of the solar wind varies, the entire bag of plasma quivers in a quasi-periodic mode with characteristic time constants; it fills and empties like a breathing lung. When the bag is hit by the blast wave of a large solar flare, its sudden compression leads to a violent shakeup of the particle population, and energetic particles are dumped into the auroral zones and the ionosphere.

***Magnetic Storms*** The compression caused by a solar flare propagates an increase in geomagnetic field strength all the way to the ground, producing the phenomenon of a "sudden commencement geomagnetic storm." Particles are energized in the Van Allen radiation belts, where they oscillate back and forth in latitude and at the same time drift in longitude—electrons to the east, protons to the west. An equatorial ring current is thus generated at a distance of 3 or 4 earth radii. Its accompanying magnetic field represents the main phase of a magnetic storm.

As particles leave the magnetosphere and find their way into the ionosphere all around the auroral oval, the auroral lights come on. A "polar electrojet" current develops, which produces magnetic substorms at ground level. The particles that shake out of the magnetosphere and enter the ionosphere have energies that greatly exceed the energies of solar wind particles. They are believed to have been stored in the magnetosphere and accelerated to high energy when the acceleration process is triggered by the outburst of particles arriving from the sun. Acceleration may occur in the magnetotail, which stretches for nearly a thousand earth radii away from the sun. Its lines of force return to the polar regions of the earth.

***Whistlers*** Lightning flashes generate radio noise, which propagates in the form of whistlers along geomagnetic field lines back and forth between the northern and southern hemispheres. Whistlers can be clearly heard with a radio receiver as tone bursts of falling pitch. Studies of whistlers first identified a sharp decrease in the density of electrons at about 4 earth radii. This boundary was named the plasmapause. It encloses the toroidal-shaped plasmasphere, a volume of relatively dense, cool hydrogen extending upward from the top of the ionosphere and rotating with the earth. Beyond the plasmapause, the plasma's characteristics change sharply to lower densities and much higher temperatures.

***Micropulsations*** Pulsations in the geomagnetic field have been recognized ever since early studies of the earth's field were made with delicately suspended compass needles. Modern and much faster magnetometers reveal pulsation periods that range from more than a thousand seconds to as short as a fraction of a second. The wavelengths corresponding to these fluctuations can be as large as several earth radii. These magnetic field variations arise from hydromagnetic waves propagating in the magnetosphere. They can be excited by a variety of processes, including instabilities on the surface of the magnetosphere induced by the solar wind, plasma density gradients internal to the magnetosphere, and energetic particle beams propagating in the magnetotail. The magnetic field variations also can penetrate deeply into the earth's crust and oceans and often are used for studies of the earth's conductivity.

## Spacecraft for Exploring the Magnetosphere

The emphasis in solar-terrestrial physics is to understand how all the parts of the sun-earth system work together. To understand the complex of interactive processes and derive predictive capabilities, simultaneous measurements of linked phenomena must be made in widely separated regions of space.

The most recent effort to make spatially and temporally coordinated observations involves three satellites that are a part of the International Sun-Earth Explorer (ISEE) program, a cooperative venture between NASA and the European Space Agency (ESA). In October 1977, NASA's ISEE-1 and ESA's ISEE-2 were launched into nearly identical orbits. As these two satellites chase each other around the magnetosphere, they sense the position and movement of its outer boundary, about 130,000 kilometers above the earth. Where the magnetic field lines carried from the sun by the solar wind merge with those of the earth's magnetic field, the magnetosphere appears to be ripped open. The solar wind's magnetic field merges with the earth's field on the sunward side of the magnetosphere. The solar wind tears back the magnetospheric field and peels it off toward the dark side of the earth, hundreds of thousands of kilometers into the tail of the magnetosphere. As merging occurs and the field lines are sharply bent, particles inside the bend are accelerated as though projected by a slingshot.

In August 1978, NASA launched the third ISEE satellite to a vantage point 1.5 million kilometers above earth; there it monitors the solar wind before it reaches the magnetosphere. Instead of orbiting the earth, the satellite executes small circles in the gravitational well known as the $L_1$ libration point between the sun and the earth. From this outpost, ISEE-3 senses the solar wind early enough to predict the outbreak of magnetic storms and auroras.

For the next round of magnetospheric research, a comprehensive program has been designed—the Origin of Plasmas in the Earth's Neighborhood (OPEN) program, involving a minimum of four spacecraft. Figure 6 depicts four OPEN spacecraft: (1) the Interplanetary Physics Laboratory (IPL); (2) the Polar Plasma Laboratory (PPL); (3) the Equatorial Magnetosphere Laboratory (EML); and (4) the Geomagnetic Tail Laboratory (GTL). The IPL will be placed into a "halo" orbit around the sun-earth $L_1$ libration point. The GTL will arrive at an apogee location in the distant geomagnetic tail by using lunar swing-by maneuvers. It will be

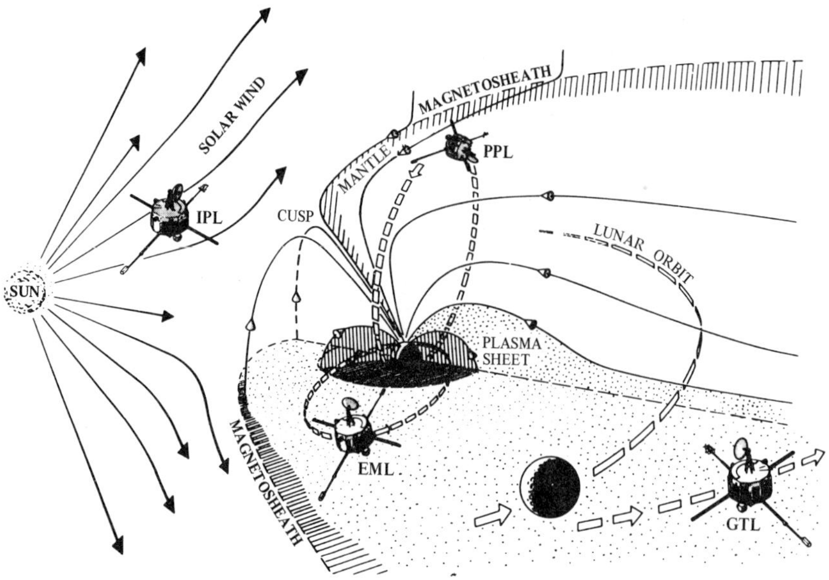


**Figure 6 Spacecraft elements of the *OPEN* program.** Each spacecraft will have propulsion systems providing substantial capacity to change orbits so that a number of different orbital configurations can be used during the course of the program. [SOURCE: Goddard Space Flight Center. *Origin of Plasmas in the Earth's Neighborhood.* Final Report of the Science Definition Working Group. April 1979, p. 18.]

possible to vary the distance of this apogee from 80 to 250 earth radii. The PPL will start out in a polar orbit with an initial apogee at 15 earth radii and will work its way down to 4 earth radii 18 months later. The EML will vary in position from 2 earth radii to 12 earth radii in the magnetotail, while simultaneous coordinated measurements are made with the GTL. With this changing configuration of four spacecraft, fundamental aspects of a great variety of couplings in the magnetospheric system can be explored.

## THE NEUTRAL ATMOSPHERE AND THE IONOSPHERE

### The Neutral Atmosphere

The various regions of the neutral components of the atmosphere are defined by the atmosphere's temperature structure (see Figure 7). The lowest region is the troposphere, where weather is created. Its average height, measured by the tropopause, varies from 8 kilometers over the poles to 16 kilometers over the equator. In middle latitudes, weather patterns from day to day produce large changes in the height of the tropopause, where the temperature reaches a minimum. In the overlying stratosphere, the temperature rises to a maximum at 50 kilometers as a result of solar ultraviolet radiation heating ozone. Above the stratosphere lies the mesosphere (50 to 90 kilometers), where the temperature falls back to another minimum. The height range from the tropopause to the mesopause is now commonly referred to as the "middle atmosphere." It constitutes the principal sink for solar ultraviolet radiation (with wavelengths of 20 to 400 nanometers), galactic cosmic rays, energetic protons from solar flares, and particles accelerated within the magnetosphere.

The thermosphere, above the mesosphere, has the highest temperatures of the neutral atmosphere. Starting from about 200 K at the mesopause, the temperature climbs to about 700 to 1,000 K at 300 kilometers and levels off. In the thermosphere, thermal energy comes from the absorption of solar X-rays and extreme ultraviolet radiation. At greater heights, the atmosphere becomes very rarified; it is nearly collision-free above 500 kilometers, where the exosphere begins and hydrogen escapes to interplanetary space.

Early observations from rockets demonstrated that diffusive equilibrium prevails over turbulent mixing for most constituents of the atmosphere above 95 kilometers. When sodium is released from rockets at twilight and trimethylaluminium is released at night, the resulting vapor trails fluoresce under illumination by the setting sun. Turbulent eddies are beautifully revealed at the upper surface of the mixing region, called the turbopause.

The temperature profile described above is only an average. Ions, electrons, and neutral atoms often have very different temperatures at the same heights, and strong diurnal temperature variations make the sunlit atmosphere 25 to 50 percent hotter than the nighttime atmosphere at high altitudes. At low latitudes in the afternoon, a very distinct density bulge appears. When sunspots are at their height, high-altitude temperatures may be double what they are at sunspot minimums. Hydrogen dissociated from water vapor at altitudes below 100 kilometers diffuses upward to crown the earth with a far-reaching geocorona, stretching to 100,000 kilometers. Helium is a major constituent of the high atmosphere, and, at some phases of the sunspot cycle, high-altitude helium exceeds the concentrations of hydrogen and oxygen.

Before the advent of modern rockets, only the lower 30 kilometers of the atmosphere could be directly sampled by balloons. Pressure varied as expected in this fully mixed, turbulent atmosphere of molecular oxygen and nitrogen, but diffusive separation was expected to control the distribution of atmospheric gases at greater heights. This simplistic picture was challenged by studies of the luminous trails of meteors as they heated to incandescence in the earth's atmosphere near 110 kilometers and evapo-

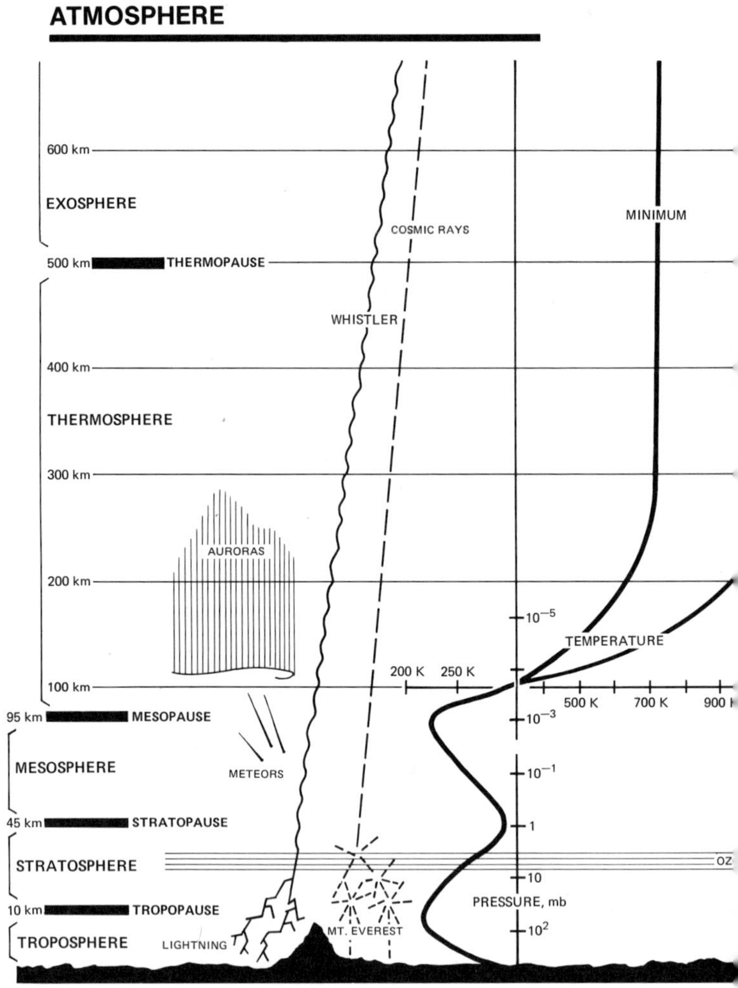


**Figure 7 Structure of the neutral and ionized components of the atmosphere.** Temperature inversions define the tropopause, stratopause, and mesopause. The atmosphere is mixed and the composition of major constituents is essentially constant up to the mesopause. Ozone concentrates in a thin layer in the stratosphere. At higher levels, molecules dissociate and lighter elements separate out by diffusion. Temperatures in the thermosphere maximize at sunspot maximum. Ionized component of the

rated completely before reaching 80 kilometers. These observations required that the air be denser than expected at 100 kilometers if the temperature were the same as observed at 12 kilometers. Accordingly, the temperature at an altitude of 100 kilometers must have risen to that at ground level.

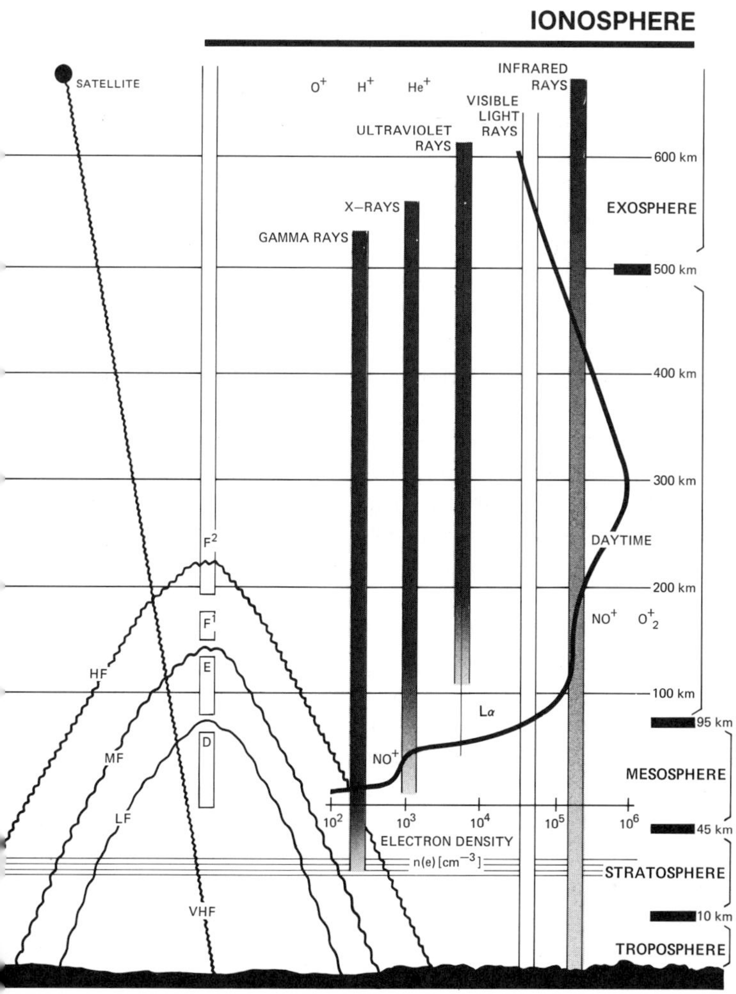


atmosphere is produced by cosmic rays, gamma rays, X-rays, and ultraviolet rays, which have maximum ionization rates in the D, E, $F^1$, and $F^2$ regions. Maximum concentration of electrons occurs in the $F^2$ region, where the ionized component is about 1/1,000 of the neutral-particle density. Shortwave radio signals are reflected from the ionosphere. X-rays from solar flares produce radio blackout at an altitude of 60 to 75 kilometers. Microwaves penetrate the ionosphere for satellite communications.

The overhead sky is not totally black between the stars. On a dark moonless night, far from city lights, the eye can detect a faint glow. Much of this airglow is produced at heights from 60 to 300 kilometers above the ground and can be identified with excited atoms and molecules of oxygen and nitrogen. Far more spectacular are the colored forms of auroral lights seen in rapid motion across the skies in regions surrounding the magnetic poles of the earth. Early triangulation measurements showed that the bottom of these lights was about 100 kilometers above the earth. The auroral light thus provided, before rocketry was developed, a means of learning about the nature of the atmosphere at heights far above the reach of available experimental probes.

Auroral light is composed primarily of the oxygen green line at 557.7 nanometers, the oxygen red line at 630 nanometers, and red emissions from molecular nitrogen. Strong ultraviolet emissions from ionized molecular nitrogen (at 391.4 nanometers) are invisible to the eye. Modern observations reveal intense soft X-rays, radio emission covering the broadcast band from 500 to 1600 kilohertz, and infrasound. All of the evidence points to the existence of a gigantic electrical discharge system capable of driving a thousand billion watts of electrical power into the arctic atmosphere, which responds like a fluorescent screen. This system is clearly coupled to the sun by means of the solar wind of electrified gas, but the details of how the wind sweeps past the earth and channels into the auroral zones still remain to be resolved by future solar-terrestrial research.

Unlike airglow, which arises from the flux of solar radiation on the atmosphere, the aurora is produced by the dumping of energetic charged particles (protons and electrons) out of the magnetosphere. Because charged particles can enter only along magnetic field lines, auroral phenomena are usually confined to well-defined rings, or ovals, surrounding the magnetic poles.

During the International Geophysical Year in 1957-58, more than a hundred "all sky" cameras were deployed for auroral studies. Each routinely photographed the entire sky in a single frame. The ability to observe the aurora from space now opens up the entire electromagnetic spectrum and provides total geographical perspectives very difficult to achieve from the ground. An entire auroral oval can be photographed, and, from sufficiently high altitudes, both the northern and southern auroral ovals can be observed simultaneously. In the ultraviolet range, the aurora can be detected on the sunlit side of the earth because airglow during the day is relatively very weak. Also, no extreme ultraviolet radiation emerges from altitudes below 90 kilometers, and the earth looks nearly black underneath the high-altitude aurora. Most of the incoming particle energy is transformed to soft X-rays, which emerge freely from the atmosphere. Hence, a measurement of the X-ray surface brightness from space can indicate total energy input.

Solar radiation interacts with the stratosphere primarily through the absorption of ultraviolet radiation by ozone. Heating of the ozone layer creates a latitudinal gradient of temperature. The combination of ozone heating and reemission of infrared energy by ozone, carbon dioxide, and water vapor drives a global atmospheric circulation near 60 kilometers altitude which is characterized by strong west-to-east motion in the winter and strong east-to-west motion in the summer.

Because of the difficulty of making *in situ* measurements in the mesosphere, which is too low for satellites, too high for balloons, and traversed too fast by rockets, few basic data for modeling mesospheric processes are available. But the photochemistry and dynamics of the mesosphere are comparatively simpler and better understood than are stratospheric processes. Above 55 kilometers, atomic oxygen is stable in the daytime but converts to ozone at night. Odd hydrogen (H, HO, $HO_2$), produced by the decomposition of water, molecular hydrogen, and methane ($H_2O$, $H_2$, and $CH_4$), is believed to exert strong control over the abundance of odd oxygen ($O_3$ and O). Here is where the reactions occur that release atomic hydrogen from water, molecular hydrogen, and methane. In trying to model the development of the earth's atmosphere in its early history, it is important to estimate the amount of oxygen left behind as hydrogen escaped. One of the major questions about the origins of life is how oxygen production and the development of terrestrial life were coordinated.

## The Ionosphere

The electrically charged component of the upper atmosphere is called the ionosphere. Marconi demonstrated its existence when he transmitted radio signals across the Atlantic Ocean in 1901; its height was measured in 1925 with pulsed radio waves, a forerunner of radar. Shortwave radio communication using the ionosphere is still widely used, in spite of the development of radio relay satellites, and it is of considerable practical importance to understand how ionospheric properties relate to solar ionization and the dynamics of the upper atmosphere.

Free electrons in the thin gas of the upper atmosphere behave like tiny radio relay stations. Radio waves from the ground set the electrons into oscillation so that part of the absorbed radio energy is reradiated toward the ground. When the density of neutral gas is low enough, the electrons suffer very little loss of energy by collisions and constitute an ionospheric mirror for radio waves. In the lower ionosphere, the D region (see Figure 7), the ambient gas density is high enough that the electron oscillations are almost immediately damped out by collisions. The absorbed radio energy is then dissipated as heat in the atmosphere. Ionization in the D region thus creates an absorbing blackout rather than an electron mirror. Higher energy X-rays from solar flares produce the low-lying ionization that is responsible for radio blackout. The accompanying energetic particles that funnel down to the magnetic poles produce deep-seated polar cap absorption.

The distribution of energy in sunlight reaching the ground—from infrared wavelengths through the ozone cutoff in the ultraviolet near 300 nanometers—resembles a 6,000-K blackbody with a maximum near 500 nanometers. At shorter wavelengths, the blackbody emission would be expected to decrease rapidly, and, at X-ray wavelengths, it would be expected to be altogether inconsequential. It was therefore difficult for geophysicists, prior to the use of sounding rockets, to imagine how the ionosphere could exist without short-wavelength ionizing radiation. Within the space of a decade after the end of World War II, though, the solar spectrum was revealed from its X-ray limit throughout the ultraviolet range with instruments carried above the atmosphere on rockets. Every spectral interval was matched with its absorption at a specific height in the ionosphere. All of the ionizing energy required to account for the ionosphere was found to be available at invisible short wavelengths originating in the solar chromosphere and corona, where temperatures range from hundreds of thousands to millions of degrees.

Until comparatively recent years, ionospheric scientists were content to fit grossly averaged data for solar radiation and atmospheric composition to a standard model of solar-terrestrial relationships. That picture was as incomplete as any model of the lower atmosphere without dynamic weather changes would be. With more global data and greater temporal detail, we are coming to recognize that "weather" exists and has a great influence in every level of the ionosphere. Winds, waves, and drifts distort the largest-scale features of static models and produce the fine-scale irregularities that are of such importance to modern communications. Ground observers have known for many years that ionospheric disturbances move from high latitudes toward the equator. It is now clear that auroral heating generates huge high-altitude waves that produce these traveling ionospheric disturbances.

Local irregularities in plasma density spread the return of radio signals from the F region, the highest layer of the ionosphere, into myriad echoes, a phenomenon known as spread F. An early indication of these irregularities came from radio astronomical observations that revealed scintillating signals from point sources such as quasars. This phenomenon is analogous to the optical twinkling of stars that results from turbulence in the lower atmosphere. The F region irregularities produce strong scintillation and fading in high radio frequencies, even in the gigahertz frequencies used by communication satellites, which were once thought to be the answer to trouble-free communications.

The recent series of NASA Atmospheric Explorer satellites disclosed a startling panorama of irregularities in ionospheric structure. Gross variations, as much as two orders of magnitude in plasma density, appear over horizontal distances of only a few kilometers. Fifty percent changes occur over just a few tens of meters. F region irregularities are a constant phenomenon over the polar regions, but also frequently affect the equatorial ionosphere at night, especially near the equinoxes. Satellite observations show typical variations of three orders of magnitude in the amplitude of plasma irregularities over a single polar orbit. In some regions, the plasma is almost perfectly smooth; in others, incredibly rough. At high latitudes, we are undoubtedly seeing the effects of particle precipitation and the small electrical fields associated with auroras. The equatorial behavior as yet has no satisfactory explanation.

At lower altitudes, in the E region of the ionosphere, a frequent irregularity known as sporadic E can reflect high-frequency waves that would normally penetrate the ionosphere. This causes the signals to be received as far away as 2,000 kilometers from the source on a single hop. In summer, it produces severe interference on television broadcasts. Sporadic E has been defined by rocket probes as a sharp stratum of ionization near 100 kilometers. It usually extends over 1,000 to 2,000 kilometers, but it is only 2 or 3 kilometers thick. In middle latitudes, sporadic E is common near midday in the summer. The layer is filled with meteoric ions, but the detailed mechanism of how they concentrate in such sharp layers is not well understood.

While solar control of the high regions of the atmosphere is quite well understood, the bottom of the ionosphere, the D region, contains complex processes governed by trace constituents that are still difficult to untangle. This is the seat of the great disruptions of high-frequency communications that are triggered by solar flares.

The D region is variable on a day-to-day basis as the result of atmospheric factors, as well as the activity of the sun. Perhaps the most striking variation is the winter anomaly at middle and high latitudes, in which the normal electron concentration may multiply tenfold near 80 kilometers. This enhancement appears to be statistically connected with increases in the temperature of the stratosphere near 30 kilometers. Contributing factors may include the effect of large-scale mesospheric circulation on the distribution of ionizable minor constituents and a change in mesospheric temperature sufficient to change the rate coefficient governing the formation of nitric oxide, the principal ionizable constituent. Altogether, the evidence is persuasive that high-altitude meteorological factors have a strong influence on D region variability.

What contributes to D region complexity is the abundance of minor constituents, including $H_2O$, OH, $H_2O_2$, NO, $NO_2$, $N_2O$, CO, $CO_2$, and $CH_4$. It is possible for concentrations of less than a few parts per hundred million or per billion to control ionization rates or to catalyze important chemical processes such as ozone chemistry. Only in recent years have the existence and importance of hydrated ions and complex conglomerates of ions been recognized. It is obvious why the D region has been called the chemical kitchen of the ionosphere.

There is special concern over the disruption of the D region by the debris of nuclear explosions. After such explosions, energetic particles are trapped on magnetic field lines and oscillate back and forth between conjugate points. Some of the particles are dumped into the D region and induce strong high-frequency radiowave absorption. As the electrons circulate about the field lines, they drift eastward. In about one hour, they can blanket the earth with D region absorption. Sometimes, the phenomenon lasts for days as electrons slowly leak out of the geomagnetic trap. If many blasts were to occur, enormous amounts of nitric oxide would be generated in the fireballs, and the ensuing decrease in stratospheric ozone over the entire globe could be very substantial.

### Observational Programs for the 1980's

The programmatic thrust in upper atmosphere research during the 1980's will focus on the stratosphere and the mesosphere. Already approved are (1) Nimbus 7, to investigate nitrogen chemistry and ozone; (2) the Stratospheric Aerosol and Gas Experiment, to measure aerosols and ozone; (3) the Halogen Occultation Experiment, to study chlorine chemistry and its interactions with ozone; and (4) the Solar Mesosphere Explorer, to examine the effects of solar variability on the chemistry of hydrogen and ozone in the mesosphere. Farther ahead in the 1980's will come the Upper Atmosphere Research Satellites, keyed to a comprehensive study of upper atmosphere physics and chemistry.

The upper atmosphere is a chemically active region in which a wide variety of levels, embracing the stratosphere and mesosphere, are coupled by dynamics and transport processes. Vertical motions of the tropopause affect the concentration of ozone and the other constituents of the stratosphere and mesosphere. Possibly, the upper atmosphere has a significant influence on the troposphere through feedback, enough to affect climate and maybe even weather. Variations in the electrical conductivity of the stratosphere and mesosphere are associated with cosmic ray activity. The effect of these variations on the vertical electrical field in the atmosphere may be connected with the development of thunderstorms.

The programs NASA has planned for the 1980's should reveal the kinds of geographical, diurnal, and seasonal dependencies, as well as those of the solar cycle, that link all of the atmospheric and magnetospheric regions to each other and to the sun.

## PRACTICAL IMPACTS OF SOLAR-TERRESTRIAL RELATIONSHIPS

In recent years, a variety of events have brought home to the general public a sense of the practical consequences of solar-terrestrial relationships. Perhaps the most dramatic experience was the plunge to earth of

Skylab. The event had all the necessary ingredients of suspense, complete with prognostications by the media of catastrophe, to frighten a confused public. When the last astronaut left Skylab in 1976, it was thought that the spacecraft was in a safe parking orbit where it could await a visit by an early Space Shuttle flight, which would push it to a higher orbit for safekeeping until it could be refurbished and reactivated. Unfortunately, the plan was frustrated by a delay in the Space Shuttle's schedule and by a rapid rise of solar activity toward sunspot maximum. With high sunspot activity came a hotter and denser atmosphere at Skylab's altitude, which increased the drag and caused the orbit to decay much faster than anticipated. Skylab thus fell victim to sunspots.

## International Programs

Because solar-terrestrial research is intrinsically global in character, it gains substantial benefits from organized international cooperation. Strategic locations for ground-based observation networks often allow less developed countries to participate as important partners. In space, which is the province of the most technically advanced nations, coordinated observations are now regularly conducted by spacecraft from the United States, Western Europe, and the Soviet Union.

About 100 years ago, several nations mounted a joint effort to study arctic weather. The program was named the International Polar Year. Fifty years later, radio science had come of age and provided a new means of studying the electrified regions of the high atmosphere. A second international year was then dedicated to studying the ionosphere. With the end of World War II, American, British, and French research teams undertook to rocket scientific payloads to high altitudes. Until then, direct probes of the upper atmosphere had been limited to meteorological balloons. The early rocket results established the atmosphere's pressure, temperature, and density profiles to altitudes of more than 100 miles; they also detected the ionizing radiation that produces and controls the electrification of the ionosphere.

The idea for a grand campaign of ground-based and, for the first time, space-based observations of the terrestrial environment was so appealing to geophysicists that the International Geophysical Year (IGY) was organized at the highest level of international scientific cooperation ever achieved. In its preparatory year, 1957, 116 rockets were launched and thousands of scientists engaged in coordinated ground-based observations. The orbiting of the Soviet Sputnik in 1957 was followed in 1958 by the launch of U.S. Explorer I, which discovered the Van Allen belts.

By the time the International Geophysical Year was completed in 1958, the power of international scientific cooperation in global research programs had been demonstrated in a most convincing fashion. There was no thought of abandoning the style or organization that the year developed for solar-terrestrial research. Accordingly, it was followed in 1964 by the International Years of the Quiet Sun program and, in 1976-79, by the International Magnetospheric Study. Follow-up programs such as the Middle Atmosphere Program are now well into the planning stage.

## Laboratory Plasmas

An intrinsic similarity exists between much of laboratory plasma physics, space plasma physics, and astrophysics. Plasma instabilities are as critical in controlled fusion experiments with laboratory confinement devices, such as the tokamak, as they are in magnetospheric physics. Perhaps the closest similarity is found in suprathermal solar flare plasmas and the high-temperature, magnetically compressed plasmas of fusion experiments. Figure 8 compares high-resolution spectra of a solar flare with the spectrum of plasma contained in the Princeton Large Tokamak. How remarkably similar they are! The variety of conditions observed from flare to flare in the sun offers as much flexibility in sampling variations of basic parameters as laboratory experiments do. Each area of research has much to learn from the other.

## Ozone

One of the most profound environmental concerns of recent years has centered around the stability of the ozone layer in the stratosphere. Ozone is formed by the chemical combination of an ordinary oxygen molecule ($O_2$) with a free oxygen atom. The free atoms are produced by the sun's ultraviolet rays in the wavelength range 200 to 242 nanometers, which dissociate molecular oxygen. Although ozone is only a trace constituent, amounting to about one part per million of the total atmosphere, it provides a protective blanket against a range of the ultraviolet spectrum that is hazardous to life on earth. The potentially lethal ultraviolet from 280 to 320 nanometers is almost totally absorbed by ozone. Destruction of a substantial percentage of the ozone could raise the specter of skin cancer and various damaging effects on biological organisms, such as plankton in the sea and simple plants on land.

When atmospheric scientists began to suspect that nitric oxide released by supersonic jets could be a catalyst for the destruction of stratospheric ozone, the predicted environmental impact contributed to the decision to terminate the U.S. developmental program for the aircraft. Subsequent studies showed that the chemistry of the stratosphere was much more complex than originally thought; a full analysis of the complete chain of reactions revealed the possibility of a slight increase rather than a de-

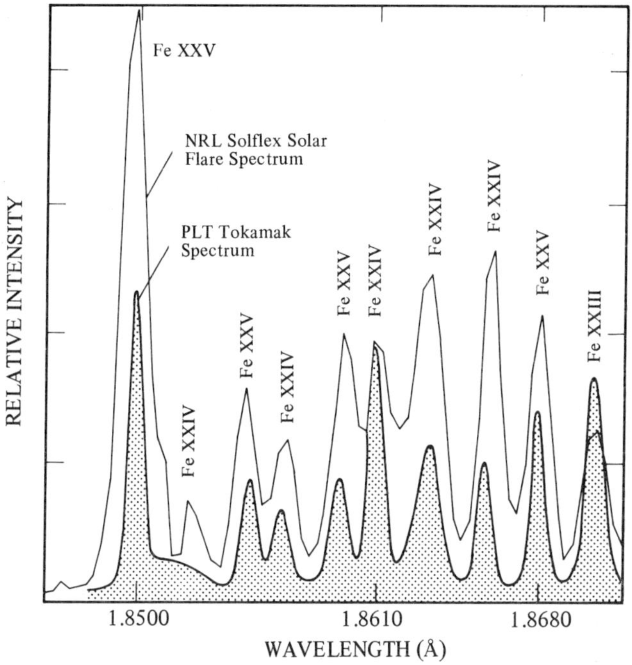


**Figure 8 High-resolution spectra of solar and laboratory plasmas.** X-ray spectrum of a solar flare obtained by a Naval Research Laboratory (NRL) instrument aboard satellite STP-78-1 compared with the spectrum of the plasma in the Princeton Large Tokamak (PLT). The spectral lines are associated with the transitions in highly ionized iron (Fe). To produce these ions requires an electron temperature in excess of $12 \times 10^6$ K. For the spectra shown here, the flare temperature is $17 \times 10^6$ K and the tokamak temperature is $20 \times 10^6$ K. Although the temperatures are almost the same, the spectral line ratios indicate that the flare plasma is close to ionization equilibrium while the tokamak plasma is in the state of rapid heating. [SOURCE: Naval Research Laboratory.]

crease in ozone from the operation of supersonic aircraft. But hardly had the concern over nitric oxide pollution begun to abate when a potentially more serious impact was predicted from chlorofluoromethane (Freon) compounds used in aerosol dispensers and as refrigerants. These chemicals are extremely inert and stable at ground level, but, when they are carried into the stratosphere by atmospheric mixing, solar ultraviolet radiation decomposes them. Free chlorine is released and starts a chain of chemical reactions leading to the destruction of ozone. The process is insidious because the chlorofluoromethanes take a long time to reach the stratosphere after release at ground level. Gas now in the lower atmosphere may not exert its full impact on stratospheric ozone until 50 years hence. The dilemma is to assess the validity of theoretical models when the impact cannot be tested for decades. Prudent steps already have been taken to reduce the use of chlorofluoromethanes wherever environmentally acceptable substitutes are available, such as in spray cans.

Ozone may have a subtle influence on climate. The solar ultraviolet radiation absorbed by ozone heats the stratosphere and generates winds. Although we would not expect the climate of the middle atmosphere to have any substantial direct effect on the troposphere, indirect feedback effects between ozone and vertical motions may influence the development of monsoons, tropical cloud formations, the jet stream, and possibly mid-latitude storm systems. Our interest in stratosphere-troposphere couplings is stimulated by evidence suggesting substantial solar variability in the ultraviolet range where ozone is produced and absorbs energy.

## Radio Communications

Space research has had great success in revealing the nature of solar ultraviolet and X-ray control of a static ionosphere and the impact of the ionosphere on radio propagation. However, the ionosphere is not static; winds, waves, and drifts create a complex dynamics that is very difficult to forecast. At very high frequencies, the effects of ionospheric absorption and reflection tend to disappear. With the advent of communications satellites, which relay radio signals at gigahertz frequencies (i.e., microwave frequencies), it was expected that virtually all of the effects of ionospheric weather on radio communication would be eliminated and that essentially all radio traffic would be handled eventually by satellites. That expectation has been largely fulfilled by communications satellites serving the western world. Surprisingly, however, recent years have actually witnessed the increased use of ionospheric propagation on the lower frequencies, from 10 kilohertz to 30 megahertz, over areas of the earth in the polar regions, which are not easily reached by satellite links at synchronous orbit. Also, in developing countries, ionospheric communications are still used largely for internal purposes, even though these countries have employed INTELSAT for international communications.

Radio communication by ionospheric propagation is presently being used for a variety of purposes, such as long-range navigation, ship-to-shore and ship-to-ship communications, and aircraft communications on transpolar flights. All of these modes of communication suffer from ionospherically induced degradation, such as fading and signal dropout or phase changes in the low and very low frequency signals that are employed for navigation. Major ionospheric irregularities and tilts accompany auroral activity and magnetic storms. At times of solar flares, streams of energetic solar particles travel to earth and funnel down the terrestrial magnetic field lines to the north and south magnetic poles. Both polar caps are then blanketed by polar cap absorption, the result of intense local ionization, and transpolar transmissions are blacked out.

During these events, high-frequency communications are seriously degraded or even wiped out for days at a time down to magnetic latitudes of 60° to 65°.

Accurate forecasting of ionospheric dynamics would have substantial economic benefits, but success thus far has been minimal. Still, even a crude forecasting capability for ionospheric disturbances could lead to more efficient use of frequency bandwidths in an already very crowded spectrum.

Civilian as well as military users of communications systems are concerned with scintillations propagated via the ionosphere that appear at all frequencies. In 1962, Starfish, a high-altitude nuclear burst, produced worldwide scintillation. Video pictures from meteorological satellites are often "snowed" by scintillation. Scintillation correlates with sunspots; as the sunspot number increases, the number of days disturbed by scintillation also increases. Yet the diurnal distribution seems to remain unaffected by sunspot number; scintillations are observed primarily after local ionospheric sunset during the months near equinoxes. Early in the operation of the INTELSAT network (which operates at 4 to 6 gigahertz), problems with scintillation were encountered. The patterns of scintillation indicated that they were produced by irregularities in the F region of the ionosphere. Although the source of F region irregularities and detailed knowledge of their structure are still very poorly understood, enough is now known that, for practical purposes, adequate design margins to ensure satisfactory performance can be included in new satellite communications systems.

## Satellites

Well before the space age, the potential for radio communication by means of orbiting relay satellites was anticipated, but visionary engineers never expected that energetic particle radiation in the then undiscovered Van Allen belts would create a hazardous environment. Walt Disney's 1960's "Voyage to the Moon" featured a space station in a 1,000-mile orbit, right in the heart of the inner radiation belt; what the film failed to acknowledge was the lethal effects of highly penetrating particle radiation.

Particles penetrating the interiors of spacecraft can affect computer memories. Solar cells, although completely covered with protective coating, still lose a few percent of their efficiencies per year. Large-scale integrated semiconductor devices are especially sensitive to radiation damage. An accurate knowledge of the radiation environment is essential to design the most efficient shielding. The problems are exacerbated by the vagaries of the radiation environment in much the same way that the dynamics of terrestrial weather cause problems in ground-level activities.

Synchronous orbit has so many advantages for applications satellites that it has already become loaded with heavy traffic—more than 80 spacecraft of many nations are now in operation there. In the future, the crowding will grow rapidly, as more communications, earth observation, meteorological, and data relay satellites are orbited. One of the surprises found in observations of particle radiation at the altitude of synchronous orbit is the relatively easy access that protons have to this region of space. Proton damage from one major solar flare may exceed that of years of bombardment by trapped electrons. The earth's magnetic shield is not as effective as simple model calculations imply. Detailed calculations must take into account internal and external current systems and hydromagnetic waves in order to predict the observed entry and passage of solar cosmic rays through the inner magnetosphere.

Any isolated probe in the plasmasphere normally will assume a negative potential of a few volts. In plasma cloud regions of the magnetosphere or in the plasma sheet, the potential often reaches 10,000 volts. On one occasion, the ATS-6 satellite in synchronous orbit reached 15,000 volts when the earth eclipsed the sun at the spacecraft. Differential charging between exposed insulators or electrically floating conductors can produce discharges that remove their coatings by sputtering and degrade the thermal control balance. Coated surfaces of optical elements can be destroyed quickly. In several cases, electrostatic discharging has introduced enough noise on spacecraft command lines to produce spurious commands. One Air Force spacecraft is speculated to have failed because of large differential charging in a prolonged and intense geomagnetic substorm. The solutions to these problems may take many forms, but first we must have an understanding of environmental processes in the atmosphere.

## Geomagnetic Storms

A geomagnetic storm is a sharp worldwide decrease in the intensity of the earth's magnetic field at the ground caused by the sudden inward compression of high-altitude magnetospheric plasma. Such storms are accompanied by induced surges in power distribution systems which can have serious consequences. A great geomagnetic storm on March 24, 1940, wiped out 80 percent of all long-distance telephone calls from Minneapolis, Minnesota. A similar storm on February 10, 1958, severely interfered with north Atlantic telegraph cables; voltages as high as 2,650 volts were induced on the Bell System transatlantic cable from Clarenville, Newfoundland, to Oban, Scotland. Voices transmitted in the eastward direction alternately squawked and whispered, yet westbound communication remained near normal.

In a magnetic storm on February 9, 1958, Toronto, Canada, was blacked out by the tripping of circuit breakers in an Ontario transformer station. A great flare in August 1972 caused the failure of a 230-kilovolt power transformer in British Columbia. The cost of such a power transformer failure may be as high as a million dollars. Because the northeastern United States is supplied with electricity by a complex grid that includes southern Canada as well as northeast and north central areas of the United States, transformer failure in this net could lead to a massive power blackout. Fortunately, magnetospheric research in recent years has given power distribution engineers the information needed to design better equipment that minimizes interference.

## Future Technologies

Future applications satellite systems may involve investments of enormous amounts of money. As presently envisaged, a single Space Solar Power Station, rated at 10,000 megawatts, would have a collecting area of about 50 square kilometers and would cost about $15 billion. If better knowledge of space radiation damage could extend the useful life of such a system by only one percent, the savings would exceed a hundred million dollars per station.

In the rapidly growing transmission pipeline industry, control of steel corrosion has become a constant maintenance problem. Corrosion engineers must make survey measurements of the potential between pipeline and earth using direct-current flow measurements accurate to at least $\pm 5$ millivolts and $\pm 100$ milliamps. Induced currents resulting from geomagnetic activity can screen the true steady-state conditions of test measurements easily. Without warning of geomagnetic activity and its telluric effects, corrosion survey work may be scheduled at times when the results will be entirely spurious. In the case of the Alaska pipeline, the problem is exaggerated by the auroral current system, which produces especially large transient earth currents.

Exploration geophysicists carry out surveys with airborne magnetometers, flying more than a million kilometers every year, all over the globe. In a typical survey, the contour lines are flown one kilometer apart and the field is measured to 1 $\gamma$ at 1-second intervals.* Time variations in the geomagnetic field interfere with magnetic contour mapping. If periods of enhanced geomagnetic activity could be predicted, surveying could be scheduled to avoid bad geomagnetic weather. Some success in improved scheduling has been achieved with data from measurements of solar activity, but the forecasting is still hardly more satisfactory than weather forecasting. Again, the problem is aggravated in the aural zone, where periods of excessive magnetic activity lasting several weeks are common.

---

*One $\gamma$ equals $10^{-5}$ gauss. The earth's magnetic field is approximately 0.3 gauss.

## Perturbing the Magnetosphere and Ionosphere

Although we are primarily concerned with the adverse impacts of natural storms in the high atmosphere on technological operations in space, the inverse process also can occur. Human activities on the ground can disturb the natural behavior of the ionosphere and magnetosphere. For example, high harmonics from power grids in North America and Europe are radiated into the magnetosphere, where they appear to be greatly amplified by some unknown physical process. As with natural waves, the manmade waves interact with electrons and perturb their trajectories so that they dump from the magnetosphere to the ionosphere. The process is connected with a phenomenon known as chorus—a birdlike chirping of ascending frequency that occasionally appears on tape recordings. Although this phenomenon has no significant impact on communications, it is scientifically interesting because it reveals a fundamental magnetospheric instability that amplifies waves and precipitates particles into the ionosphere.

One proposal to influence ionospheric perturbations suggests using an intense beam of pulses from a radio transmitter to modulate currents that flow continuously into and out of the auroral ionosphere. The radio source would trigger extremely low frequency (ELF) waves that could communicate with submarines. This system would operate on the free power available from auroral currents, which carry as much as 5 million amperes with a power of 10 thousand million watts. If only a small amount of this natural energy were stimulated to radiate ELF, it would compare very favorably with the transmitter that the Navy's Sanguine test facility planned to stretch across a wide area of Wisconsin, which would have radiated only 1 watt.

## Sun, Weather, and Climate

The greatest enigma in solar-terrestrial relationships is the role of the sun in producing changes in weather and climate. No aspect of solar-terrestrial research has stirred up as much controversy and speculation. Despite tentative evidence for sun-weather relationships, hundreds of years of searching have not turned up a single correlation strong enough to convince the skeptics. Still, we may be lacking a sufficiently refined understanding of solar activity. The obvious cycles of sunspots and the other surface solar features associated with them may not be the key clues. We have recognized the existence of coronal holes and the sector pattern of the solar wind for only a comparatively short time. These phenomena are locked to the 27-day rotation period but are not closely coupled to the sunspot number. Only recently has it been discovered that coronal bright points carry as much magnetic energy as do sunspots, but they appear and disappear out of phase with the rise and fall of sunspots.

Interest in a sun-weather connection has been stimulated by reports of characteristic changes in the size of tropospheric regions of high vorticity in the northern hemisphere, changes correlated with sharp rises in the geomagnetic field strength at nearly the time when the earth sweeps through certain sectors of the interplanetary magnetic field. Several years of observation appear to give a marginally significant correlation. It has been proposed that solar modulations can shift the timing of natural meteorological variations slightly. In this way, a relatively small energy input may trigger much more energetic processes in the lower atmosphere.

## Summary and Outlook

Since the time of Galileo, observations of sunspots have shown a statistically significant correlation with global temperature. Radioactive carbon in tree rings serves as a proxy indicator of sunspots over the past several millennia. The record goes back 7,000 years in the bristlecone pines of California, which reveal an irregular pattern of solar activity. Epochs of missing sunspots appear to coincide with little ice ages of about a century's duration. More immediate has been the history of dust bowls, which exhibit an approximately 22-year cycle, perhaps synchronized with a 22-year solar magnetic cycle.

The main problem now in uncovering correlations on the 100-year time scale between solar activity and climate is with the climate base. Our knowledge of historical climates over past millennia is very sketchy. It seems that, as more and more paleoclimatic information becomes available, the picture of climatic trends becomes more and more complex. Although of less immediate concern to mankind, changes in the sun and in the earth's climate on the time scale of eons are nevertheless important to our understanding of sun-climate connections. Stellar evolution theories tell us that the early sun was only 70 percent as luminous as it is today; modeling the subsequent large growth in solar luminosity with changes in atmospheric composition to fit climatic changes on a time scale of a billion years may help us anticipate the future impact of a large anthropogenic input of carbon dioxide from burning fossil fuels and its accompanying increase in global temperature.

Atmospheric scientists now agree that variations on the order of a tenth of one percent in the solar constant would lead to significant climatic effects. Until recently, observational sensitivity was inadequate to detect such small changes. New radiometers aboard NASA's Solar Maximum Mission are now reporting variations, which correlate with day-to-day variations of sunspot activity, at the level of hundredths of a percent. The challenge for future observations is to monitor the sun continuously with an absolute accuracy of 0.1 percent or better in order to reveal any credible correlation with climate.

Although 97 percent of the solar power that is intercepted by the earth reaches the ground in the form of visible light and heat, the small fraction composed of ultraviolet, X-ray, and energetic particle radiation is absorbed at high altitudes. These ionizing and photochemically potent radiations are highly variable. They are directly responsible for ionospheric and magnetic storms and for the chemical equilibrium of stratospheric ozone. As a consequence of solar activity, sporadic interference with radio and television communications can occur, and the space environment may be made temporarily hazardous to spacecraft systems and astronauts. Although scientific understanding of the basic interactions between solar radiation and the upper atmosphere is well advanced, our ability to predict solar activity and its practical consequences is still only marginal. Newly discovered indicators such as emerging magnetic bright points and coronal holes, from which high-speed solar wind escapes, may lead the way to more useful predictability.

The main goals of future observational programs, based both on the ground and in space, are (1) to achieve a spectral detail in studies of the sun that is an order of magnitude higher than at present, and (2) to deploy spacecraft in strategically spaced arrays within the various elements of the sun-earth system so as to reveal the complex couplings among all its components. Such advances in observational capability will open the way to more useful predictions of solar activity and perhaps even to an understanding of the connections among sun, climate, and weather.

International cooperation, as during the International Geophysical Year, can play an important part in future studies of solar-terrestrial relationships. Moreover, international programs have been an effective means of bringing nations closer together and developing mutual respect. Scientists constantly seek contacts across national boundaries; they share common intellectual values and place high premiums on individual freedom.

From a purely scientific point of view, the sun-earth system is a natural laboratory for plasma physics. Magnetic confinement of high-temperature plasma in the sun and in the earth's magnetosphere can be studied over wide ranges of temperature and density. Much can be learned by comparing natural magnetohydrodynamic instabilities, such as explosive solar flares, with runaway plasma processes in fusion experiments conducted with laboratory devices such as the tokamak.

With respect to the sun itself, the fundamental theory of thermonuclear energy generation can be tested by directly observing solar neutrinos. Thus far, failure to detect solar neutrinos has been viewed as a catastrophe for stellar theory. Now the mystery of the missing solar neutrinos may yield to new evidence from elementary particle physics that neutrinos oscillate between two states, one of which the solar neutrino experiments of the past decade could not detect. Our understandings of elementary particle physics and of solar energy generation are thus intrinsically connected.

## BIBLIOGRAPHY

Syun-Ichi Akasofu and Sidney Chapman. *Solar-Terrestrial Physics*. Oxford: Oxford University Press, 1972.

John A. Eddy. *A New Sun—The Solar Results from Skylab*. SP-402. Washington, D.C.: U.S. Government Printing Office, 1979.

Herbert Friedman. "The Sun," *National Geographic*, Vol. 128(November 1965), pp. 713–742.

George Gamow. *A Star Called the Sun*. New York: Viking Press, 1964.

H. S. W. Massey and R. L. F. Boyd. *The Upper Atmosphere*. New York: Philosophical Library, 1959.

Donald H. Menzel. *Our Sun*. Cambridge, Mass.: Harvard University Press, 1959.

*Research in Geophysics. Volume One: Sun, Atmosphere and Space*. Edited by Hugh Odishaw. Cambridge, Mass.: M.I.T. Press, 1964.

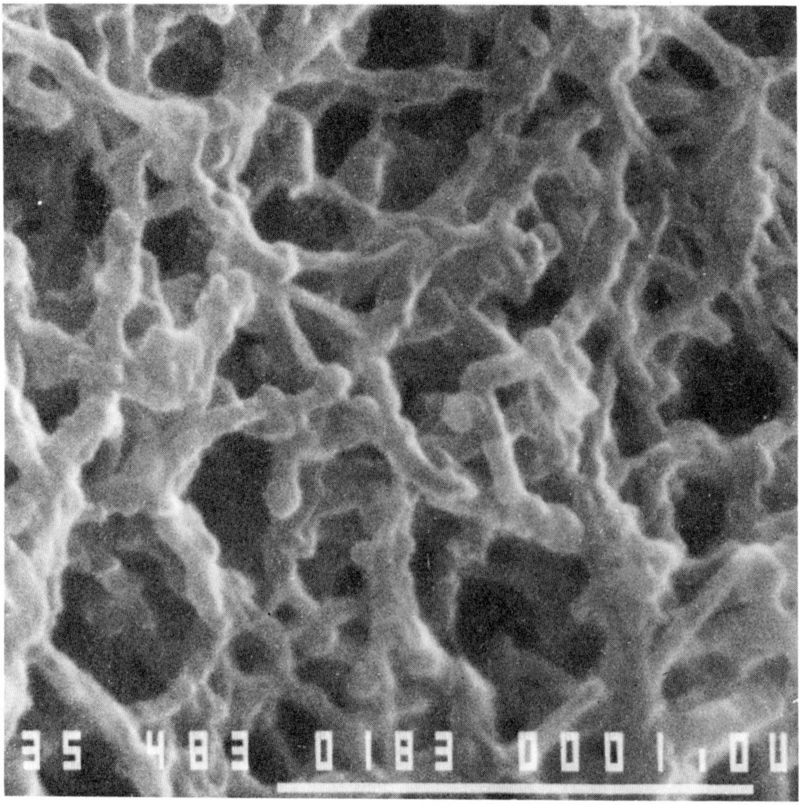
35 483 0183 0001,0U

# 10

# The Science of Macromolecules

## INTRODUCTION

Substances composed of very large molecules, or macromolecules, abound in man's environment. Since time immemorial, they have provided humans with starches and proteins for food, with cotton, wool, and silk for clothing, with hides and leather, with rubber, and with wood for shelter and fuel. The structural materials of all forms of life consist of macromolecules. The muscles, tendons, ligaments, and skins of mammals, the chitin of insect skeletons, animal hair, and bird feathers, the cellulose and lignin of trees, all are composed of macromolecules of diverse kinds. Enzymes and the intricate genetic substances that govern living processes are macromolecules. Macromolecules are not merely essential to all forms of living matter, they are the fabric of life itself.

The macromolecular materials, or polymers, at man's disposal have been greatly increased in modern times, both in number and variety, through chemical synthesis. Synthetic rubbers, plastics, and fibers with properties surpassing those of their natural counterparts are among the fruits of these achievements. New synthetic polymers continue to enrich the material resources available to us.

The diverse properties of commonplace substances such as air, water, metals, textiles, synthetic polymers, and the constituents of living organisms arise from the combinations of atoms that form these substances.

---

◀ Scanning electron micrograph of polyacetylene doped with iodine. The magnification is 100,000; the white bar represents 1 micron. The addition of iodine converts the polymer to a metallic-type conductor. [H. Gibson, Xerox, Webster Research Center.]

These combinations depend on chemical bonds that connect one atom with another. (See "Chemical Terms and Conventions" at the end of Chapter 11.) In the language of the chemist, chemical bonds determine the molecular structure that characterizes a given substance and accounts for properties peculiar to it.

The molecules of the principal constituents of air, nitrogen and oxygen, consist of only two atoms joined by a chemical bond. The forces between these small molecules are too weak to overcome their rapid, incessant motions at ordinary temperatures. They consequently occur as gases. The hydrocarbon molecules that make up gasoline and kerosene are larger, they contain from 12 to 50 or so atoms. The atoms in inorganic minerals, salts, and metals occur in vast regular arrays like stacks of oranges, each atom being affiliated with one or more of its immediate neighbors either by a chemical bond or through electrostatic attraction.

Atoms of many commonly occurring elements have the capacity to enter into liaisons with two or more other atoms; that is, they possess valences enabling them to form chemical bonds with several other atoms. Some atoms, notably carbon (C) and sulfur (S), readily form stable bonds with other atoms of the same kind; these atoms in turn may be bonded to others, and so on. Thus, when sulfur is heated, its atoms form long-chain macromolecules comprising many hundreds of sulfur atoms:

—S—S—S—S—S—S—

Upon cooling, the long chains of sulfur atoms reapportion themselves in cyclic molecules of various sizes, the most prominent one being

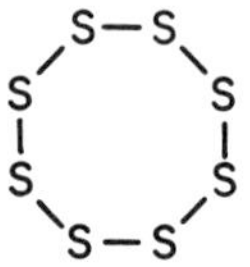

These cyclic $S_8$ molecules aggregate in ordered array to form a hard crystalline solid, the common yellow form of sulfur. In contrast, the long chains of "elastic" sulfur impart plastic and rubberlike properties to the material.

The carbon atom, with a valence of four, is especially prone to join repetitively with other carbon atoms. This propensity is illustrated by the series of linear-chain hydrocarbons found in natural gas, liquid fuels, and paraffin wax. Several members of this series are shown on the next page. The first three are gases at ordinary temperatures; octane is a liquid that boils at 126°C (Celsius).

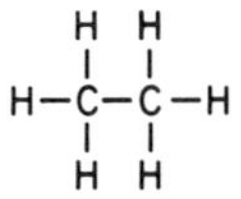 or $CH_3-CH_3$

ETHANE

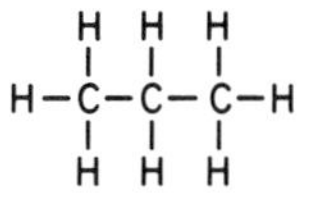 or $CH_3-CH_2-CH_3$

PROPANE

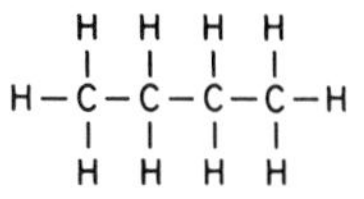 or $CH_3-CH_2-CH_2-CH_3$

BUTANE

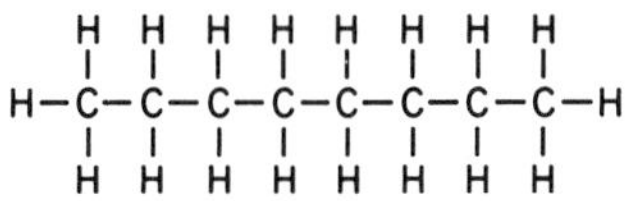 or $CH_3-(CH_2)_6-CH_3$

OCTANE

The concatenation of carbon atoms can continue indefinitely, as in linear polyethylene, a synthetic polymer consisting of tens of thousands of methylene groups ($CH_2$) consecutively connected by carbon-carbon bonds. In other instances, the methylene groups are joined cyclically like the links of a necklace. Cyclohexane ($C_6H_{12}$), consisting of six methylene groups connected in the form of a ring, is an example. In other molecules, some of the carbon atoms are joined directly to three or four other carbon atoms, instead of to just two of them. Nonlinear "branched" molecules of more complicated design are thus constructed.

Silicon (Si) and oxygen atoms readily combine chemically in alternating succession

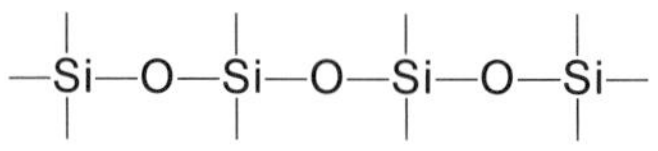

as manifested in silicate minerals and in silicone polymers. Phosphorus (P) and nitrogen (N) likewise lend themselves to alternating combinations that yield cyclic or chainlike molecules.

The multivalent character of atoms and their capacity to enter into consecutive chemical bonds with other atoms permit the existence of molecular structures consisting of many atoms. The connections may be linearly sequential, as in the examples already cited, or they may be reticular with each atom bonding to three or more other atoms. Graphite, for example, consists of carbon atoms in a sheetlike array resembling that of hexagonal floor tiles; each carbon atom occurs at a vertex and is chemically bonded to three neighboring atoms in the plane of the array. Certain silicate minerals contain silicon and oxygen atoms similarly bonded in sheetlike patterns. Mica is an example. In other instances, atoms may be joined by bonds extending in three dimensions. In diamond, for example, each carbon atom is bonded to four other carbon atoms in a symmetric tetrahedral array.

The linear macromolecule consisting of many concatenated atoms provides the basis for a class of materials with an unmatched diversity of properties. Its potentialities have been exploited extensively by nature and by man. Structure and function in all living organisms are founded on substances with this molecular motif. Textile fibers, plastics, natural and synthetic rubbers, cellulose and starch from plants, the innumerable proteins from living organisms, and the polynucleotides (DNA and RNA—deoxyribonucleic acid and ribonucleic acid) that store and transfer genetic information are examples. Tens of thousands of atoms may be joined by chemical bonds in a single macromolecule of one of these substances. The dominant structural pattern in each instance is a chain of atoms connected consecutively by chemical bonds, which may number upward of a thousand.

Macromolecules are large, or macro, only in the sense that they consist of thousands of atoms. Their lengths typically range from 100 to 100,000 nanometers (a nanometer is $10^{-9}$ meter). In breadth, however, they are no larger than the small molecules dwelt upon in traditional chemistry textbooks—in other words, no larger than about 1 nanometer. They are too thin, therefore, to be seen in a microscope, or even to be satisfactorily resolved in an electron microscope.

Polyethylene, with its comparatively simple chemical structure, (see the first substance in Figure 1), and the silicone polymer polydimethylsiloxane (see the second substance), are illustrative of synthetic macromolecular substances. The "mer," or repeating unit, of polyethylene is the methylene group; in the silicone chain, it is $Si(CH_3)_2O$.

The third and fourth substances in Figure 1 are "natural" polymers that occur widely in living organisms. The respective repeating units are NH—CHR—CO and $C_6H_{10}O_5$, the latter given here in a condensed notation. The principal structural materials of animals are proteins (the third substance in Figure 1), including most notably elastin in ligaments, blood vessels, and aortic tissue; collagen in tendons and bone; myosin and actin in muscle; and keratin in hair and nails. Polysaccharides, of which cellulose (the fourth substance) is the foremost example, impart form and strength to plants. Other polymers produced by living organisms include natural rubber, lignins, starch, and the polynucleotides, the latter two represented by the last two examples in Figure 1. The properties of polymeric materials that qualify them for their many uses include strength, durability, flexibility, and high elastic deformability; this latter

1. Polyethylene

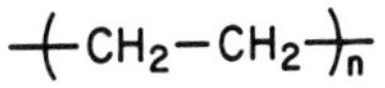

2. Polydimethylsiloxane (a silicone polymer)

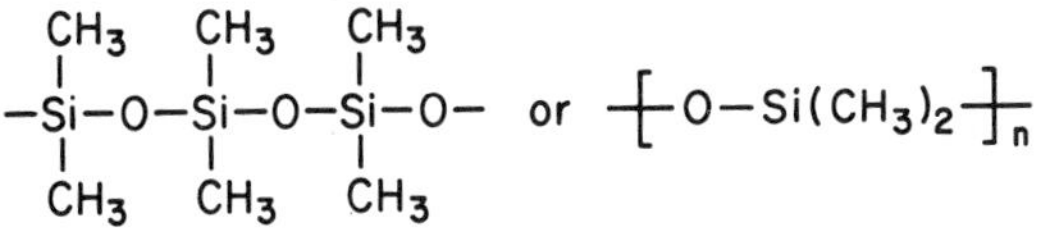

3. Polypeptide (the primary structure of proteins)

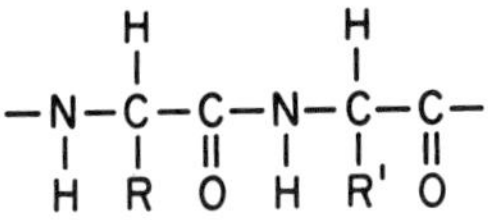

4. Cellulose

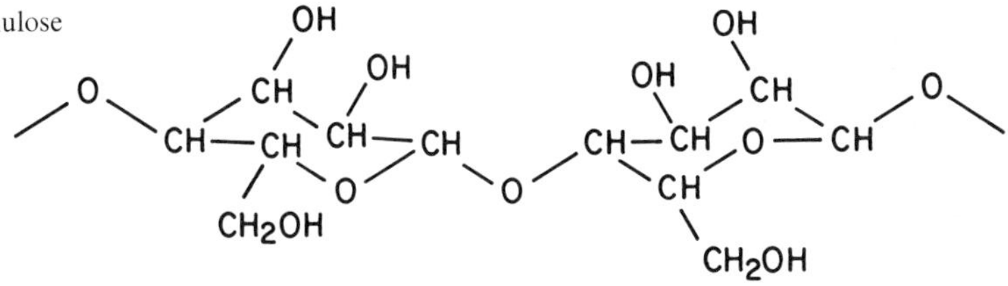

5. Amylose (a form of starch)

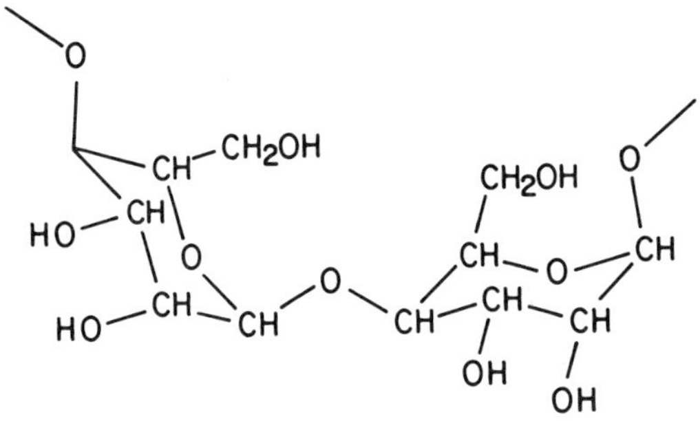

6. Polynucleotide (the primary structure of nucleic acids DNA and RNA)

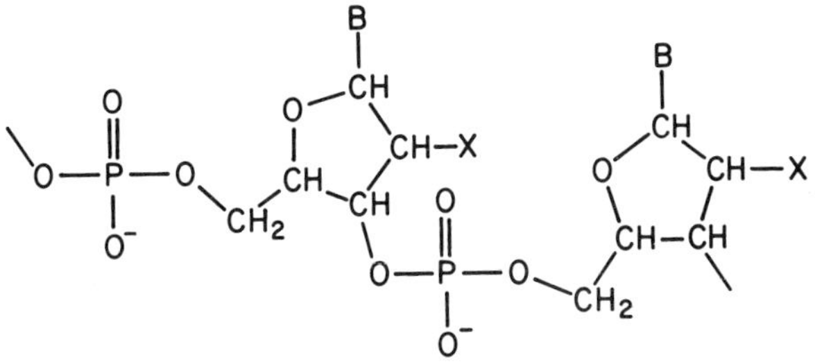

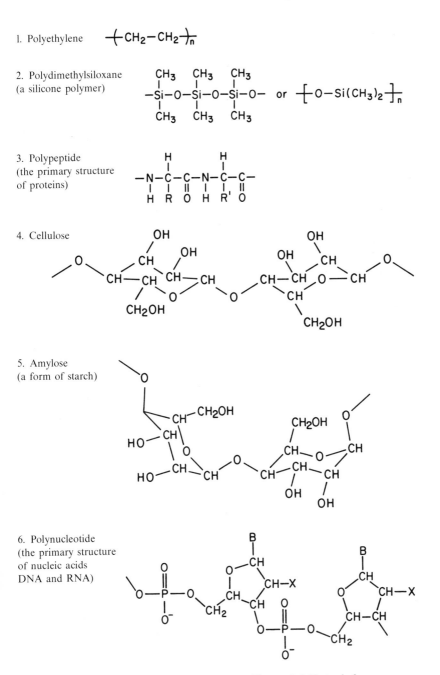

**Figure 1 Structural formulas of selected polymers.** The symbol X stands for a hydrogen atom (H) in DNA, and B stands for any of four base groups (the free bases contain an additional hydrogen atom in place of the bond): adenine ($N_5C_5H_4$), cytosine ($N_3C_4H_4O$), guanine ($N_5C_5H_4O$), or thymine ($N_2C_5H_5O_2$). In RNA, X stands for a hydroxyl group (OH).

property, rubberlike elasticity, is unique to substances consisting of long-chain macromolecules.

A remarkable feature of nearly all naturally occurring polymers is the strict adherence of their molecular structures to exact specifications. This is nowhere better illustrated than in the family of proteins, which includes not only those mentioned above but also numerous enzymes, each of which performs a unique catalytic or regulatory function. The various proteins are differentiated according to the chemical constitution of variable groups (R) in the successive NH—CHR—CO units in protein molecules. The units are selected from some 20 different nutrients, the amino acids, which likewise are distinguished by the chemical nature of their variable groups. For a given protein, a particular kind of unit is required at each location along the chain. The chain may be likened to a very long word written from an alphabet of 20-odd letters. The designated succession of units, or spelling of the word, is replicated precisely throughout the entire length of each protein macromolecule, which may comprise from a hundred to as many as several thousand units.

## SYNTHESIS OF MACROMOLECULES

All macromolecular chains are formed from small molecules, or monomers. Their formation is accomplished through chemical reactions that link the monomer in linear succession to generate a long chain. Different kinds of chemical reactions are required for the various types of monomers, as illustrated in the following sections. Once a macromolecular chain has been synthesized, its chemical constitution may be modified by subsequent chemical reactions.

### Condensation Polymerization

Certain complementary pairs of molecules have the capacity to join together and simultaneously release a small molecule such as water. A molecule bearing a carboxyl group (COOH), for example, may condense with another molecule bearing either a hydroxyl (OH) or an amino ($NH_2$) group. Examples illustrating these reactions are

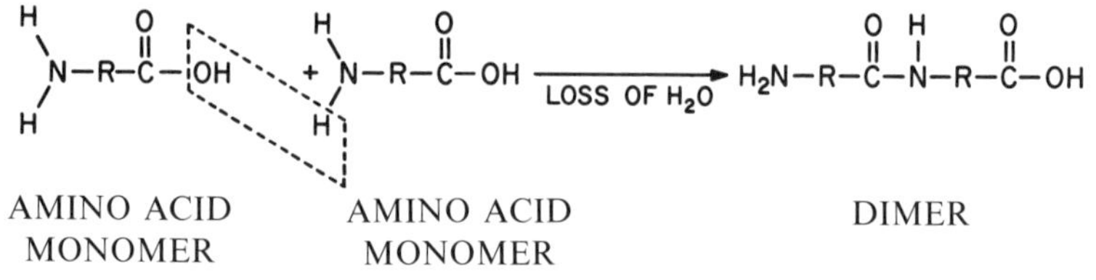


where R is a divalent group or substituent, and

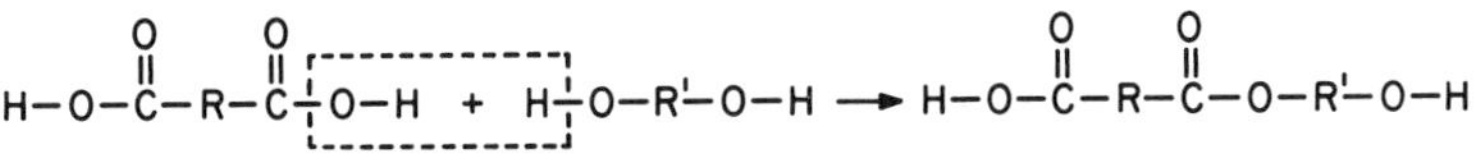

DIBASIC ACID MONOMER | DIHYDROXY MONOMER | DIMER

where R and R′ denote divalent groups. Continuation of these processes leads to the formation of long-chain polyamides or polyesters

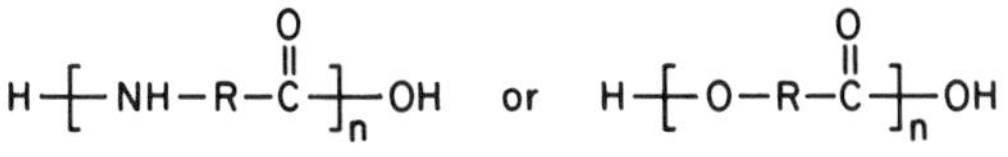

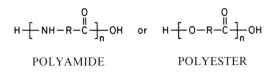

POLYAMIDE | POLYESTER

where the number, n, of mers may be several hundred. Polymerization proceeds because the dimer formed in the first step can condense with a monomer, with another dimer, with a trimer, or with a higher polymer of any size. As long as a reactive carboxyl, hydroxyl, or amino group exists at both ends of the chain, the molecule may condense with another molecule to yield a larger polymer.

A wide diversity of condensation reactions are known, ranging from those that take place in living cells to yield proteins, starch, cellulose, or the polynucleotides DNA and RNA to those used for manufacturing polymers such as nylons, polyesters, and polycarbonates. Polycondensations in living organisms take place under the influence of enzymes, which are themselves protein macromolecules. Manmade condensation polymers are not as complex and precisely regulated in structure as biological ones, but a greater variety of monomers can be used in their formation.

It is important to recognize that biologically generated protein and polynucleotide macromolecules generally are tailored to a definite chain length, with the several monomers involved entering the chain in a precisely determined sequence. For example, all the molecules of an enzyme derived from a particular species of animal or plant have identical units, or residues formed from amino acids, arranged in the same linear succession (see the third substance in Figure 1). By contrast, synthetic condensation polymers contain chains of diverse lengths, and, when two or more monomers are used to synthesize the chain, they become incorporated therein more or less at random. This randomness and the statistically variable chain length are universal features of most synthetic macromolecules. Although techniques are now available for controlling the sequence of units in certain condensation polymers, these processes are laborious and inefficient and are not amenable to practical application.

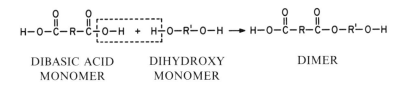

Linear chains are formed if each monomer molecule is difunctional, in other words, if just two reactive sites are present on the same molecule, as in the examples above. However, if one or more of the participating monomer molecules possess three or more reactive sites, nonlinear branched structures will be formed. For example, glycerol, which has the formula

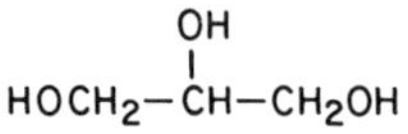

and contains three hydroxyl groups, can generate a branch point if it is incorporated in a polyester chain—for example, if it replaces part or all of the dihydroxy monomer in the second of the two examples above. Alkyd resins are so formed with glycerol and phthalic acid (a dibasic acid) as the principal ingredients. A glycerol monomer with each of its three hydroxyl groups esterified, or reacted, introduces a branch into an otherwise linear chain that can be represented by:

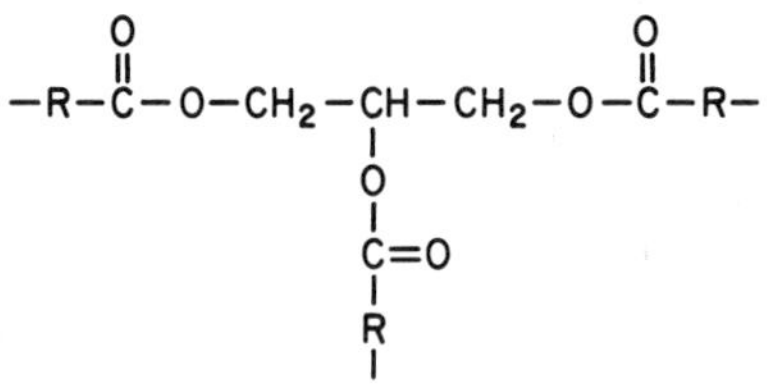

If the number of such fully reacted glycerol units exceeds one third of the number of unreacted chain ends (—OH plus —COOH), then a continuous labyrinthian *network* structure will be formed, as represented schematically in Figure 2.

**Figure 2 Polymer network or gel.** Diagram of a portion of a polymer network or gel. Cross-links are represented by dots; the chains connecting them, by lines.

This network, or gel, extends throughout the polymer (although some of the polymer may not be a part of it). It is bounded only by the surface of the polymer or by the walls of the container. It may imbibe a liquid solvent, thereby becoming swollen to a degree that depends on the density of junctions in the network, but because its chains are interconnected it cannot dissolve. Networks may also be formed by introducing cross-links between different linear chains so that they become permanently connected one to the other at the site of the cross-link. Gelatin gels, vulcanized rubber, and elastin, the elastic protein of ligaments, blood vessels, and skin, are familiar examples of cross-linked polymer networks. The junctions holding the chains together differ in each instance, but the structural patterns resemble in many respects the one illustrated.

The synthesis of condensation polymers containing rings of atoms in the macromolecular chain is of special interest. The most common ring system used is the so-called aromatic ring derived from benzene ($C_6H_6$). This hexagonal ring of six carbon atoms is unsaturated, meaning that it is deficient in hydrogen atoms compared with the full number (12) that the valence of its carbon atoms would allow. Consequently, the ring possesses an excess of six electrons above the number needed to form six carbon-carbon bonds. It is one of the most stable entities known in organic chemistry. The benzene ring is depicted in chemical formulas by a hexagon with three double bonds denoting the six mobile electrons. A hydrogen atom is understood to be bonded at each corner except where another atom or group of atoms is indicated. The rings can be incorporated into polymers in various ways, some of which are shown below.

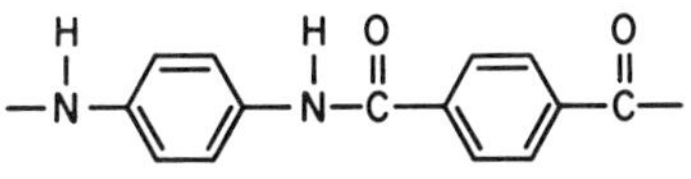
AROMATIC POLYAMIDE

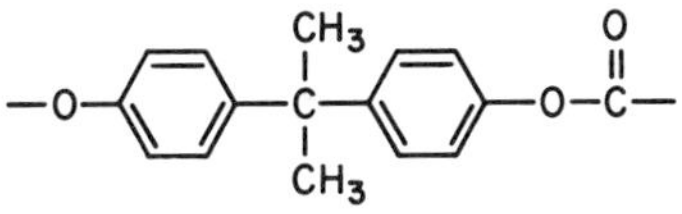
AROMATIC POLYCARBONATE

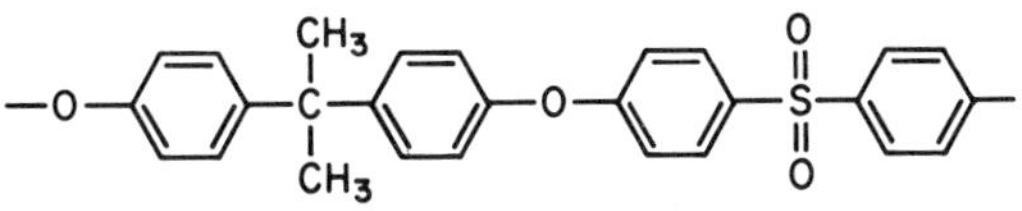
AROMATIC POLYSULFONE

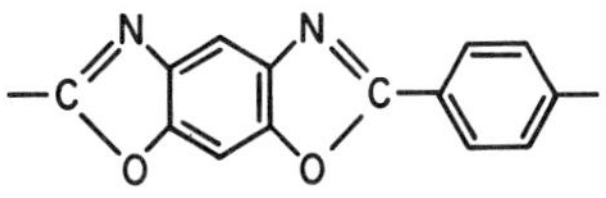
POLYBENZOXAZOLE

One reason for the current interest in macromolecules of this type is that the benzene rings stiffen the chains against coiling (see the section "Spatial Configurations of Macromolecules"). The geometric arrangement of bonds in the first and fourth of these examples is such as to confer an extended, rodlike shape on the macromolecules. Polymers with these structural characteristics have high melting points and are stable up to comparatively high temperatures. An important characteristic of extended-chain polymers is their tendency to form liquid-crystalline solutions, in which the macromolecules are partially oriented parallel to one another, as in a crystal, but are otherwise disordered, as in a liquid (see the section "The Fibrous State").

## Ring-Opening Polymerization

A second method for the synthesis of macromolecules involves the opening of molecular rings. As shown below, the silicone polymer polydimethylsiloxane (see Figure 1) can be prepared by heating a cyclic analog:

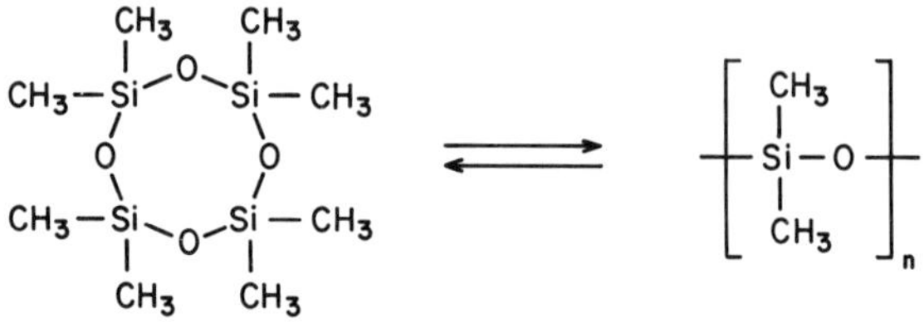

Note that the ring and the polymer contain exactly the same repeating unit, $Si(CH_3)_2O$. Hence, the conversion of the ring to the polymer does not involve the release of a byproduct. The interconversion between rings and chains occurs readily in the presence of acids or bases. These "ring-chain equilibria" play an important part both in the formation of many macromolecules and in their breakdown at high temperatures.

Polysiloxanes are members of a class of macromolecules known as inorganic-backbone polymers, polymers in which carbon atoms do not occur as members of the chain backbone. They are related to the silicate minerals found in glass, brick, concrete, asbestos, ana mica. The mineralogical materials lack molecular flexibility because of their complex sheet-like or three-dimensional matrix structure. Polyorganosiloxanes are derivatives of silicate minerals in the sense that the cross-linking units of the mineral have been replaced by organic methyl groups. The difference between the properties of a three-dimensional silicate rock and a single-strand silicone is striking. Silicones are among the most flexible polymers known; they remain flexible, or elastomeric, even when cooled to the temperature of dry ice (−78°C).

Another ring-opening polymerization is exemplified by the phosphorus-nitrogen system shown on the next page. These compounds are called phosphazenes. When the cyclic trimer is heated to 250°C, it yields a rubbery polyphosphazene with some 15,000 repeating units linked end to end. If the polymer is heated to around 350°C, it "depolymerizes," i.e., it reverts to small cyclic molecules. Thus, the conversion of rings to polymer and polymer to rings can be cycled by changing the temperature.

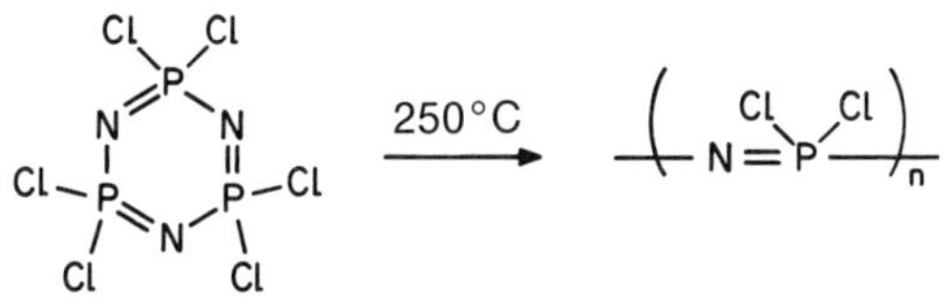

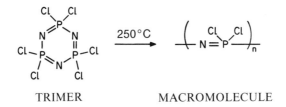

TRIMER MACROMOLECULE

Much current interest also revolves around the ring-opening polymerization of the cyclic, four-atom sulfur nitride ($S_2N_2$) ring:

This solid crystalline compound polymerizes to yield the gold-colored "metal" known as poly(sulfur nitride). This polymer looks like a metal and conducts electricity.

Several well-known carbon-based polymers also are prepared by ring-opening reactions, as illustrated in the diagrams below. One form of nylon is obtained from the cyclic organic compound known as caprolactam. Polyoxymethylene can be formed from the cyclic molecule trioxane, which is itself obtained from formaldehyde.

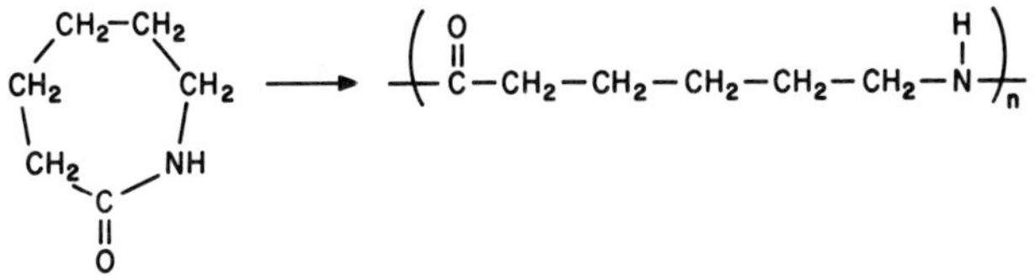

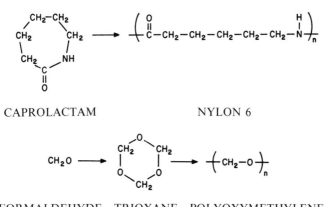

CAPROLACTAM NYLON 6

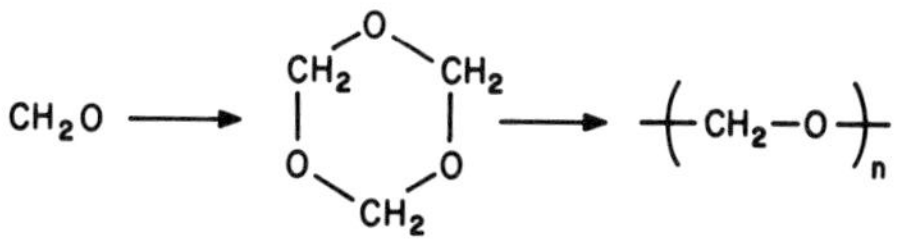

FORMALDEHYDE TRIOXANE POLYOXYMETHYLENE

## Vinyl Polymerizations

Unsaturated molecules containing a carbon-carbon double bond possess the latency for polymerization. So-called vinyl monomers (see next page)

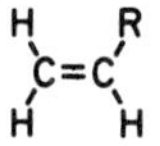

such as ethylene (in which R is a hydrogen atom), styrene (in which R is a benzene ring), and vinyl chloride (in which R is a chlorine atom) are examples. When attacked by a free radical or by other reactive agents, the double bond may "open," thereby providing two valences that can link with other monomers.

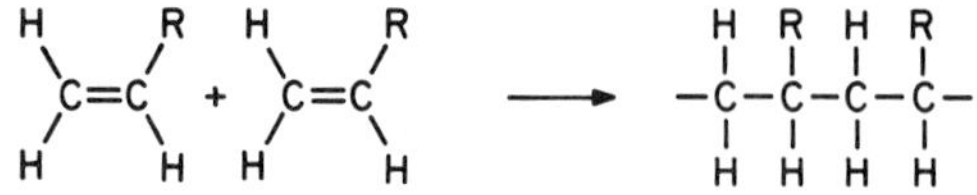

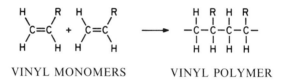

VINYL MONOMERS VINYL POLYMER

Some of the earliest synthetic polymers known were made inadvertently by research chemists who were attempting to prepare unsaturated monomeric compounds. The conversion of comparatively reactive double bonds to stable single bonds provides a powerful driving force for the polymerization. Polymerization reactions of this kind are rarely encountered in biological systems.

Most vinyl polymerizations are chain reactions. This means that a very small number of molecules in the system become "activated" by heating, by exposure to light, or by traces of catalytic substances that serve as initiators. An activated molecule can acquire another monomer to form an activated dimer, which then adds a third molecule to form a trimer, and so on, as shown below.

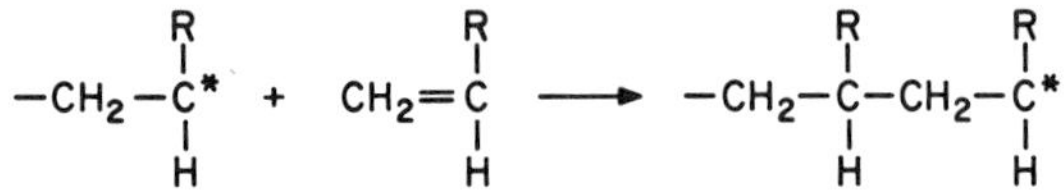

In this process, the active site, denoted by an asterisk, may be a free radical (characterized by an odd electron—one not paired with another, as in a chemical bond) or a positively or negatively charged ion. It is successively transferred from one terminal unit to the newly acquired monomer as the chain grows. Once started, chain growth can proceed very rapidly.

The polymerization of many small unsaturated molecules to form one large molecule is an amplification process. One initial event, the start of chain growth, may induce polymerization of 10,000 monomer molecules. The large polymers produced by the amplification of this single event have properties that differ greatly from those of the small monomers from which they are derived. The original monomer molecules may be volatile, but polymers are nonvolatile. The monomers dissolve quickly in many ordinary liquids, but polymers dissolve slowly or not at all.

Further amplification occurs if individual macromolecular chains are joined together by cross-links. The original polymer molecules may dissolve (albeit slowly) in liquids, but cross-linked polymers cannot do so. An average of less than one-half cross-link per macromolecule of many thousands of units suffices to form an insoluble network. Cross-linking can be induced by the action of light, as well as by ordinary chemical means. These amplification processes are reminiscent of those responsible for silver halide photography. In fact, light-induced polymerization and cross-linking are being used increasingly in nonsilver imaging processes.

Vinyl polymers of countless varieties have been synthesized, and their properties vary widely. This diversity stems from three factors. First, the chemical nature of the side group R is susceptible to almost limitless variations. Thus, it is relatively easy to prepare a wide range of polymers, each with the same backbone structure but containing a different side group. As with proteins, the side group determines the chemical and physical properties of macromolecules of this class. For example, if this group contains fluorine atoms, the polymer will be water-repellent. If it is a hydroxyl group, the polymer dissolves in water. When the side group is simply a chlorine atom, the polymer is the well-known material poly(vinyl chloride), and when it is the cyano group (CN) the polymer is polyacrylonitrile, a fiber-forming material. Some side groups provide sites for chemically linking individual chains together—an important feature in synthetic elastomers. The range of choice of the substituent group R has enabled chemists to prepare a wide variety of materials (see Table 1).

A second dimension of variation in vinyl polymers is afforded by the fact that two or more different monomers (i.e., monomers having different R groups) can be copolymerized in the same chain. The resulting polymers exhibit properties that may be varied according to the proportions of the several groups present and their sequence along the chains. Here, an analogy with the diversity of protein structures is apparent. The sequence of the R groups in synthetic polymers cannot yet be controlled, however, with the precision that is the rule in proteins.

A third avenue for affecting the chemical structures and properties of vinyl polymers depends on arranging the side groups in specific, controlled orientations, as shown below.

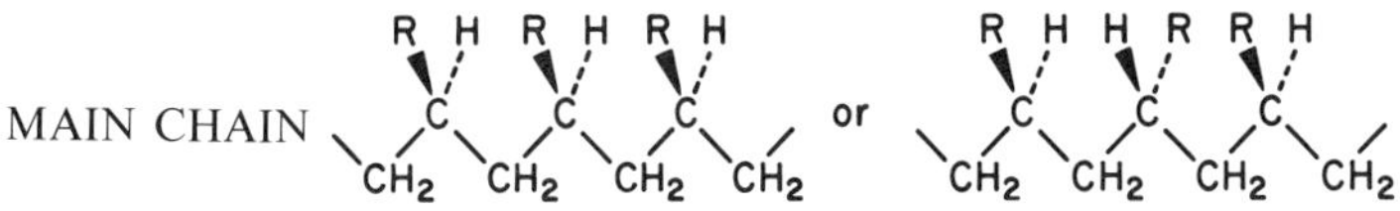

**Table 1** Changes in characteristics of vinyl polymers as side groups are altered

| Type of side group (R) | Name and properties of polymers with the structure 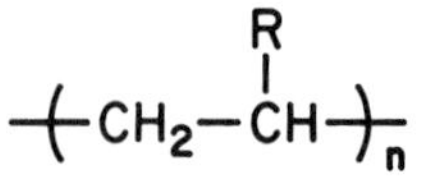 |
|---|---|
| $-H$ | Polyethylene. A colorless, translucent material that forms tough films, fibers, etc. Good electrical insulator. Insoluble in water. |
| $-OH$ | Poly(vinyl alcohol). Soluble in water. |
| $-Cl$ | Poly(vinyl chloride) (PVC). A hard, tough polymer used widely in technology. Often softened by the addition of "plasticizers." |
| $-CN$ | Polyacrylonitrile. A fiber-forming polymer. |
| $-CH_3$ | Polypropylene. A tough polymer, used widely for fibers, films, and molded plastics. Insoluble in water. |
| $-COOH$ | Poly(acrylic acid). Soluble in water. A polyelectrolyte. |
| 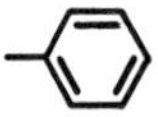 | Polystyrene. A hard, transparent, glassy polymer. Insoluble in water. |
| $-OC_2H_5$ | Poly(ethyl vinyl ether). A rubbery material. |
| 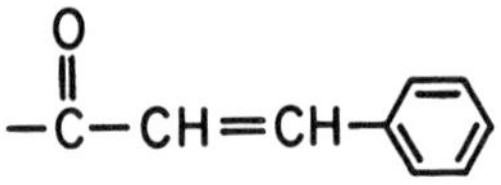 | Poly(vinyl cinnamate). A photosensitive polymer used for photo-cross-linking. |

In the first instance, all the side groups are on one side of the chain; in the second, they alternate from one side to the other (distinguished by wedge-shaped and dashed bonds). These are called stereoregular polymers, and their properties are strikingly different from those of polymers in which the side groups occur at random. The familiar polypropylene of widespread commercial importance is a stereoregular polymer of the first kind represented above.

The synthesis of stereoregular polymers is generally accomplished by choosing a suitable catalyst to control the addition of each monomer unit at the active site. Some of the most effective catalysts are organometallic compounds in which the metal is titanium, vanadium, or aluminum. These catalysts orient the incoming monomer molecule in the precise manner needed to generate stereoregularity. This is perhaps the closest that chemists have yet come to emulating the selectivity of enzymes. However, the manner in which these catalysts function is not fully understood. Research to improve our understanding of these reactions is of foremost importance if we are to learn how to control macromolecular structure more effectively.

Alternative ways that do not require the use of special catalysts for the synthesis of stereoregular polymers are being investigated. One approach that has been demonstrated in the laboratory involves packing vinyl monomer molecules into submicroscopic tunnels that penetrate solid crystalline matrices. The molecules line up in single file in each tunnel like railroad box cars in a marshaling yard. Irradiation with gamma rays then initiates linkage to form a long stereoregular chain.

A subject of growing importance is the polymerization of acetylenes. Acetylene itself ($C_2H_2$) can now be converted to a stereoregular polyacetylene, which conducts electricity as a semiconductor or, when "doped" by adding an ionic impurity, as a metallic-type conductor.

## Synthesis of Macromolecules by Living Organisms

As mentioned earlier, the synthesis of biological macromolecules in living organisms is carried out according to exact specifications. These specifications are encoded in DNA chains comprising the genes. They are written in a four-letter alphabet consisting of the four bases (represented by B in the sixth example of Figure 1) that distinguish the four kinds of units in the polynucleotide chains. DNA serves as the repository for all information that governs the function, synthesis, and reproduction of the organism.

Cell reproduction necessitates replication of the DNA chains and their stores of essential information. This is accomplished by the process, widely publicized, whereby the DNA double helix at first unwinds. Each of the two complementary strands thus released serves as a template for the synthesis of its complement. Building blocks for DNA are nucleotide monomers formed by condensation of a molecule of deoxyribose, a sugar, with phosphoric acid and one of the four nitrogen-containing, flat-shaped bases denoted by B in Figure 1. The four kinds of monomers are joined precisely in the sequence specified by the template chain through a chemical condensation catalyzed by an enzyme called DNA polymerase. The chain grows at a speed of about 1,000 monomers per minute. The resulting DNA chain consists of an alternating succession of sugar and phosphate residues, as shown in Figure 1, with the appropriate base attached to each sugar residue. This giant macromolecule, comprising hundreds of thousands of units, encrypts a staggering quantity of information. Template control is the key to accurate replication. This is the mark of biosynthetic processes. It distinguishes them from present methods for synthesizing manmade polymers.

Recognizing the quintessential role of DNA, biochemists have developed sophisticated methods for isolating the genes it contains, for determining the sequences of bases therein, and for synthesizing genes and the corresponding proteins. Progress has been rapid, yet much remains to be accomplished. These biochemical techniques, involving the use of enzymes, have been applied also to the copying of genes and to their transfer from one location to another in the string of genes in a DNA chain, or even from the DNA of an animal to that of a bacterium. The potentialities of these methods have been widely discussed. They are already exploited to produce insulin and soon may be used to produce interferons. (Chapter 17, "Prospects for New Technologies", expands further on the scientific and commercial possibilities of this technique.)

The sequence of amino acids that uniquely characterizes the polypeptide molecule of a particular protein is inscribed in the DNA chain of the gene for that polypeptide. Three consecutive bases identify 1 of the 20 amino acids. The succession of triads of bases in a gene specifies the sequence of amino acids in one of the many proteins essential to the organism. This information is transferred to the site where the protein is synthesized and amino acids are combined in the prescribed order by an elaborate series of processes. Briefly, the DNA chain serves as the template for synthesis of a complementary "messenger" RNA macromolecule made from monomers in which the sugar residue is ribose instead of deoxyribose (see Figure 1). Polymerization is effected by an enzyme, RNA polymerase. The sequence of bases in the parent DNA is accurately transcribed in the complementary bases of the RNA chain.

The messenger RNA migrates to a ribosome, a complex aggregate of specialized protein and RNA macromolecules that collectively carry out the synthesis of proteins. Also present are many kinds of smaller "transfer" RNA molecules. Each bears at one end a triad of bases that "reads" the code on the messenger RNA signifying a particular amino acid. At the other end, the transfer RNA is equipped to bind this and only this amino acid. The transfer RNA molecule faithfully collects the amino acid from the store of nutrients in the cellular fluid and carries it to one of the ribosomes, where it is delivered at the position designated by the messenger RNA to the protein molecule under construction. The constitution and vital chemical action of ribosomes are imperfectly understood. They may be likened to automated factories in which recorded instructions organize the work routines of enzymes for the manufacture of products of specified design. Enzymes that direct and catalyze the chemical combination of amino acids to form the polypeptide chains are highly specialized proteins which likewise are synthesized by the sequence of events briefly indicated. Elaborate "feedback loops" monitor the process of protein synthesis and maintain "quality control" by correcting errors that may occur in replication and synthesis.

Polysaccharides, such as starch or cellulose (see Figure 1), are among the simpler biological polymers in the sense that only one or two types of monomer molecules are incorporated in the polymer chain. The spatial orientation, or stereochemistry, of each monomer unit is uniform throughout the length of the macromolecule. Polysaccharides are generated in the cell from small molecules, such as glucose, in the presence of phosphoric acid derivatives and nitrogen-containing bases. Linkage of the glucose units into a long chain is accomplished under the control of special enzymes.

Starch differs from cellulose in the geometry of the bonds connecting the glucose units, as shown by the structural formulas in Figure 1. These bonds are so joined in cellulose chains as to favor an extended conformation; in amylose, they impose a bend at each junction between units. Cellulose consequently is readily amenable to parallel aggregation in the form of fibers of high strength and stiffness (see "The Fibrous State"), which serve as the structural material of plants. The macromolecules of amylose and other forms of starch are not suited to this purpose; they are relegated principally to the role of storing sugars for fuel and sustenance.

Amylopectin, the major component of starch, differs from amylose through the presence of branch points at frequent intervals along the polysaccharide chains. Glucose units joined to three, instead of two, other units provide the branches. They are introduced by branching enzymes. So far as is known, polysaccharides are not constructed in contact with a template or "memory" macromolecule analogous to DNA. The kind of chemical linkage between sugar units that distinguishes one of the many kinds of polysaccharides from another apparently is determined at the instant of monomer addition under the influence of the particular enzyme responsible for synthesis. In this respect, these processes resemble more closely the stereospecific vinyl polymerizations mentioned earlier than they do the synthesis of polynucleotides and of proteins.

Revelation of the elaborate array of interrelated biosynthetic processes, together with identification of the participating macromolecules, must surely stand as one of the most impressive achievements of science in this century. In sheer complexity, these processes overwhelmingly surpass synthetic procedures devised in the laboratory. Yet our knowledge of them is largely descriptive; fundamental comprehension of them within the framework of molecular science is lacking. An adequate understanding of the chemical processes involved in biosynthesis would not only redound to the benefit of biology and medicine; it would also almost surely point the way to more versatile and powerful methods for synthesizing artificial macromolecules.

## Modification of Preformed Macromolecules

Thousands of different macromolecular structures have been synthesized in the laboratory from simpler constituents. Yet the molecular architecture can be diversified by another approach. This involves the chemical modification of preformed macromolecules.

Since the beginning of civilization, naturally occurring polymers have been chemically modified. The cross-linking of proteins in the tanning of leather is an example. During the late nineteenth century, it was found that the properties of cellulose can be modified drastically by chemically altering its hydroxyl side groups. Derivatives adapted to a wide range of uses thus came to be commercialized. The solubilization and reconstitution of wood cellulose as rayon or cellophane are achieved through chemical reactions of these hydroxyl groups. Likewise, the conversion of cellulose to cellulose nitrate (gun cotton) or to cellulose acetate depends on the substitution of nitrate or acetate groups for the hydroxyl groups, respectively.

The polyphosphazenes are readily susceptible to modifications by chemical reactions of their side groups as illustrated below.

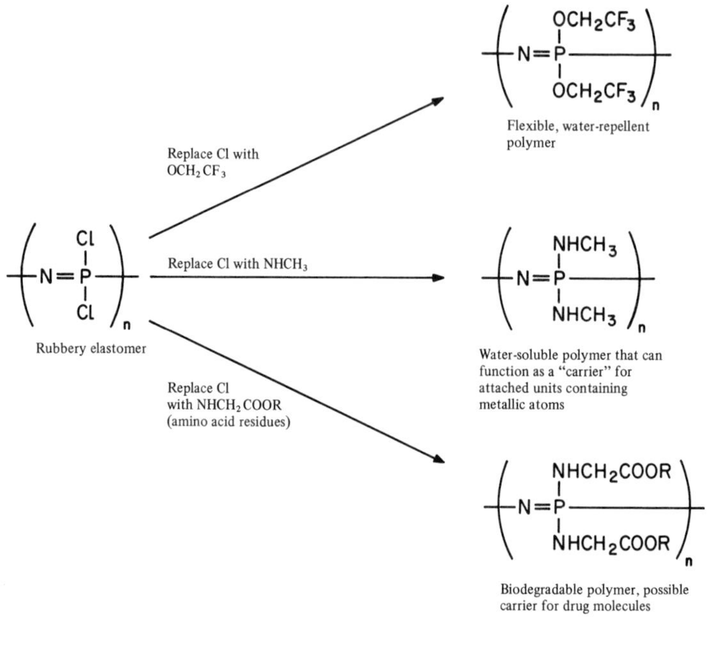

The two chlorine atoms attached to each phosphorus atom furnish sites of chemical reactivity; they can be replaced by a wide variety of other groups, such as $OCH_3$, $OCH_2CF_3$, benzene rings, biologically active groups, or metal atoms that function as catalysts.

The influence of a long polymer chain on the chemical reactivities of groups attached to it is an important subject of current interest. The heme sites in hemoglobin (see "Globular Conformations: Enzymes"), which perform the oxygen-carrying function in blood, and the catalytically active metal atoms strategically situated in certain enzymes are profoundly affected by the neighboring units of the macromolecules of which they are a part. Much research is committed to an understanding of these influences.

Chemotherapeutic drug molecules can be attached chemically to polymer chains. Investigating the effects of the polymer chain on the activity of the drug so bound is a research area of lively interest at present. Through judicious choice of the polymer, the drug may conceivably be delivered selectively to a specific organ or to certain kinds of cells. The absorption of the polymer-drug complex by the cells may be affected too. The design of macromolecules that can decompose slowly in the body to harmless small molecules and, at the same time, release a pharmacologically active agent is an obvious extension of these ideas that may come to fruition in the near future.

In a related way, polymers can be used to immobilize either enzymes or living cells. The objective is to attach a water-insoluble "leash" to the enzyme or cell in order to restrict its mobility. This allows the biologically active agents to be handled and recovered simply by manipulation of the polymeric support. Using this technique, the biochemical reactions that normally take place in living organisms can be carried out in laboratory equipment or in a continuous-flow reactor, often on a large scale. The enzyme or cell may be chemically linked to the polymer or may be physically trapped within the interstices of a swollen cross-linked polymer network. In either case, the biochemical reactions that take place on the support are almost identical to those that occur within a living organism. It is also possible in principle to immobilize a variety of different enzymes on the same polymeric support in order to allow a sequence of reactions to take place, mimicking the complex cascades of reactions that occur in living cells.

Synthetic macromolecules may similarly be used to anchor metallic catalysts and surround them with the twisting coils of polymer chains. The catalyst would then be protected from contact with other catalyst molecules in the system. Its activity may thereby be enhanced. Only small-molecule reactants and products would be able to enter and leave the protective macromolecular cage. Research in this little-explored area is especially promising of novel and useful results.

In a more speculative vein, one may envision the prospect of sequentially coding the side groups in long-chain macromolecules in a manner reminiscent of the ordered sequences in biological macromolecules. Such coded macromolecules may eventually be employed as templates for the synthesis of other macromolecules of definite constitution, in imitation of the biosynthesis of proteins. Alternatively, they could provide a means for storing archival information at the molecular level. One of the most promising routes to such systems is through the chemical modification of preformed polymers. These possibilities will not be realized until the subtlety of our synthetic procedures has been improved considerably.

## SPATIAL CONFIGURATIONS OF MACROMOLECULES

### Chemical Bonds and Molecular Geometry

The chemical formulas in Figure 1 accurately describe the patterns of chemical bonds and the atoms connected by them. They fail, however, to depict the geometric relations of these atoms and the spatial forms of the molecules so represented. They may be likened to wiring diagrams that delineate electrical circuits but not the locations of the wires and of the components they connect.

Actually, chemical bonds impose definite geometric relations between atoms. Bond lengths specifying the distances between centers of the atoms joined by chemical bonds are quite precisely fixed. The angles between pairs of bonds incident at a given atom likewise are well defined. In particular, the successive carbon-carbon bonds in the hydrocarbons shown in the introduction are not collinear, as might be inferred from their formulas as there expressed. The same caveat applies to the bonds that form the backbone of polyethylene and other macromolecules. The polyethylene chain is more realistically represented in Figure 3. In this diagram, the bonds of the carbon skeleton are arrayed in the plane of the paper. Adherence to the proper angles between successive carbon-carbon bonds is achieved by ordering the bonds in zigzag sequence.

Detailed information on the lengths of chemical bonds and on the angles between pairs of bonds has been gathered through extensive research on structural chemistry conducted in laboratories all over the world, mainly during the past half century. Since atoms and molecules are too small to be seen, even with the most powerful microscopes, indirect methods must be used. The principal experimental methods used for this purpose are the diffraction of X-rays by crystalline materials, the diffraction of a beam of electrons by gases, and infrared spectra. Remarkably precise values for the lengths of various chemical bonds and for

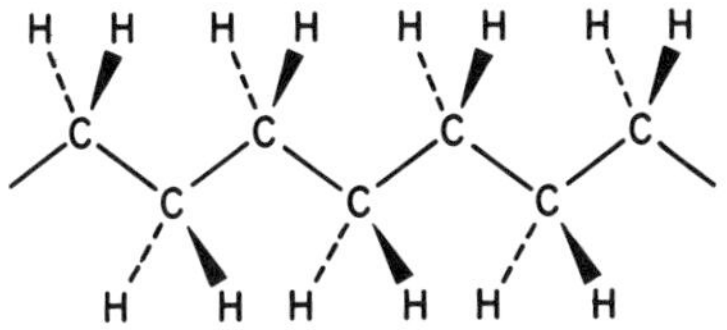

**Figure 3 Structural representation of a hydrocarbon chain in its maximally extended planar form.** Carbon atoms occur in the plane of the paper; hydrogen atoms are situated in front of and behind the plane of the paper, as indicated by wedge-shaped bonds and dashed bonds, respectively. The length of the carbon-carbon bonds is 0.153 nanometers; the angle between them is 112°.

bond angles have been acquired through skillful exploitation of these methods.

Structural data of this kind enable one to proceed from the abstract notation of a chemical formula on a sheet of paper to the conception of a real molecule in three-dimensional space, replete with all its atoms accurately situated relative to one another. Comprehension of the spatial form of the molecules, or macromolecules, constituting a given substance is essential if one is to grasp the connections between the physical and chemical properties of that substance and its chemical constitution at the molecular level.

Representation of a macromolecule in the manner shown in Figure 3, with all of the atoms of its backbone in a plane, does not suffice to convey adequately the spatial forms that a long-chain polymer may assume. Specifically, it fails to take account of the rotations that are permitted around chemical bonds. If one or more of the carbon-carbon bonds in Figure 3 are twisted, while the angles between bonds are preserved, the structure of the backbone will no longer be coplanar. Rotations about the many bonds of the chain can lead to a multitude of different configurations in three-dimensional space. This basic fact points toward what is perhaps the most important attribute of the macromolecule, namely, the diversity of spatial forms it may assume. This attribute is largely responsible for the unique mechanical and physical properties of polymeric substances.

Investigations of the rotational propensities of bonds in comparatively simple molecules show that certain angles of rotation provide more harmonious arrangements of bonds and atoms than do others. Three options for rotations about the carbon-carbon bonds in hydrocarbons are typically preferred overwhelmingly above all others. These are shown in Figure 4. In the form shown in Figure 4a, the four carbon atoms are coplanar, and the first and third carbon-carbon bonds are parallel; the rotation angle $\phi$ is zero. This form corresponds to the planar zigzag configuration of polyethylene shown in Figure 3, all bonds of which are so conformed. Rotations of $+120°$ or $-120°$ about the central carbon-carbon bond

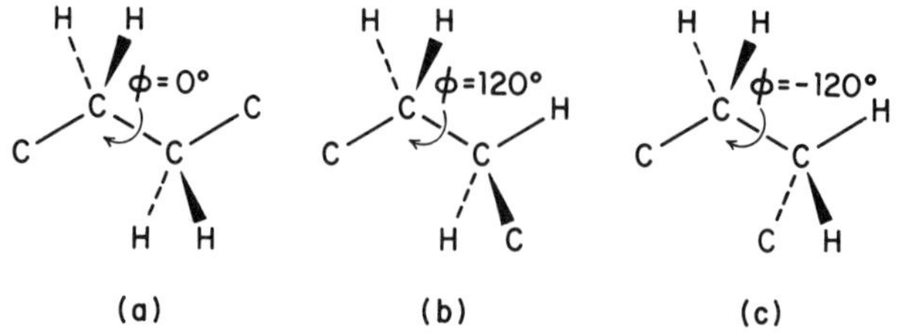


**Figure 4 Rotations about the carbon-carbon bond show planar (a) and twisted forms (b) and (c).** In (b) and (c), a carbon-hydrogen bond replaces the third carbon-carbon bond as the one parallel to the first carbon-carbon bond. In (b), the fourth carbon atom lies in front of the plane of the first three carbons; in (c), it is behind the plane.

generate the forms shown in Figures 4b and 4c, respectively, in which the third bond is redirected above or below the plane occupied by the first two bonds, in this case the plane of the paper. It is a very satisfactory approximation to consider the bond rotation to be confined essentially to one of these three discrete values; all other possibilities may be disregarded without making significant errors.

Of the three forms shown in Figure 4, the planar one (a) is generally preferred over either of the other two in a ratio of about two to one, owing to its lower energy. If one of the bonds in the chain is nonplanar, in other words, if it is (b) or (c), then the opposite form, (c) or (b), is strongly disfavored for the bonds next to it because the combinations (b)(c) and (c)(b) cause the preceding and following $CH_2$ groups to overlap one another. "Interferences" of this kind are intensified considerably if, as in many other macromolecules, some of the hydrogen atoms are replaced by larger atoms or groups. These complications may readily be taken into account by thorough analysis of the molecular geometry associated with various bond rotations and combinations of rotations.

## Trajectories of Randomly Configured Macromolecules

Figure 5a shows a computer-generated random configuration for a hydrocarbon chain of 50 bonds. The rotation about each bond was assigned one of three possible angles: 0°, +115°, or −115°, these being the optimum values. Selections were executed by assessing the probabilities of occurrence of the immediate alternatives, these probabilities being dependent on the rotation assigned to the preceding bond. The selection process is equivalent to spinning a roulette wheel with three ranges apportioned according to the probabilities incident at each step. Bond lengths and angles between successive pairs of bonds were assigned their actual values, as established by exhaustive structural studies on linear hydrocarbons.

Included for comparison in Figure 5a are four chains in the planar zigzag configuration represented in Figure 3, each of the same length as the random chain. The second and fourth of these fully extended chains are rotated by 90° about their long axes relative to the first and third chains. This alternation in angle about the length of the extended chain replicates the molecular arrangement in crystalline polyethylene and its hydrocarbon analogs.

The irregular configuration shown in Figure 5a is of course one of many that may occur. Repetition of this computer experiment would yield another statistical structure that would almost certainly differ drastically in detail, but not in general form, from the one shown. From many trials of this kind, one may determine average characteristics of the configurations of a chain of 50 bonds.

Configurations generated in the same way for chains of 500 and 3,000 carbon-carbon bonds are shown in Figures 5b and 5c, respectively. The scales of measurement are reduced in order to accommodate the larger chains. Typically, polyethylene chains may range from 3,000 to 50,000 bonds or more.

The spheres shown in these figures represent the molecules if by some means they could be induced to coalesce into droplets such that all of the spheres would be filled by the molecule's atoms. The volume of the dense sphere in Figure 5c is about one eightieth of the domain of the random chain of 3,000 bonds. This randomly configured macromolecule therefore occupies only a little over one percent of the space it pervades. The shorter chains depicted in Figures 5a and 5b occupy approximately 10 and 3 percent of the space of their respective domains. The fully extended chains are not shown in Figures 5b and 5c; their lengths, 63 and 630 nanometers, respectively, would overrun the scales of these diagrams.

The tortuosity and irregularity of the random chains are well illustrated in these figures. Other kinds of macromolecules differ in the spatial geometry of their bonds and in their preferences for various torsional rotations. The silicone polymers (Figure 1), poly(vinyl chloride), polypropylene, and polystyrene (Table 1), and natural rubber differ notably from polyethylene in their structural detail and in the bond rotations they prefer, but the trajectories of their chains are superficially similar to those of polyethylene, when viewed at the level of resolution of the longer chains shown in Figures 5b and 5c. The aromatic polyamides cited previously (p. 379) are much more highly extended owing to the stiffness imparted by the benzene rings. They may be likened to long rods with only gentle curvature at any point.

Polypeptides and proteins, when the native configurations of the latter are disrupted (i.e., denatured) by dissolving them in suitable solvents, are somewhat less tortuous and more extended than polyethylene. The difference is one of degree, however, and the overall pattern remains similar. Cellulose (the fourth substance in Figure 1) and its derivatives are

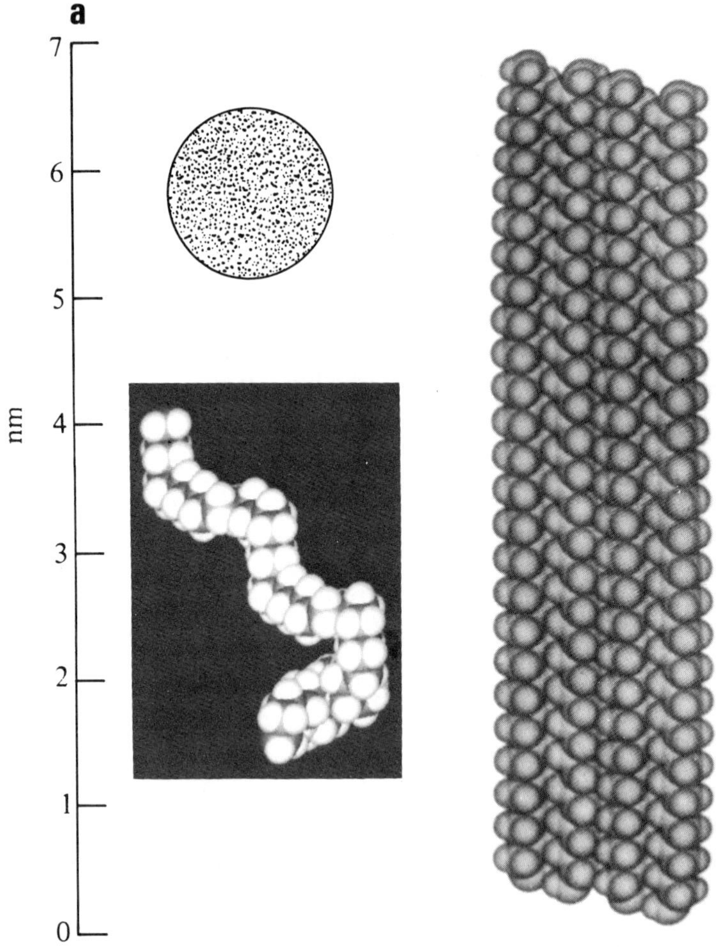


**Figure 5 Computer-generated configuration of hydrocarbon chain molecules.** Computer-generated random configurations of hydrocarbon chain molecules consisting of (a) 50, (b) 500, and (c) 3,000 carbon-carbon bonds. Indicated for comparison with each configuration are spheres representing how much space would be occupied by the atoms of the molecule if the molecules were fully compacted. Also shown in (a) are fully extended planar chains as they are arranged in a crystal. [SOURCE: National Research Council. From data provided by Richard J. Feldmann, National Institutes of Health; Paul J. Flory, Stanford University; and Do Yeung Yoon, IBM Research Laboratory, San Jose.]

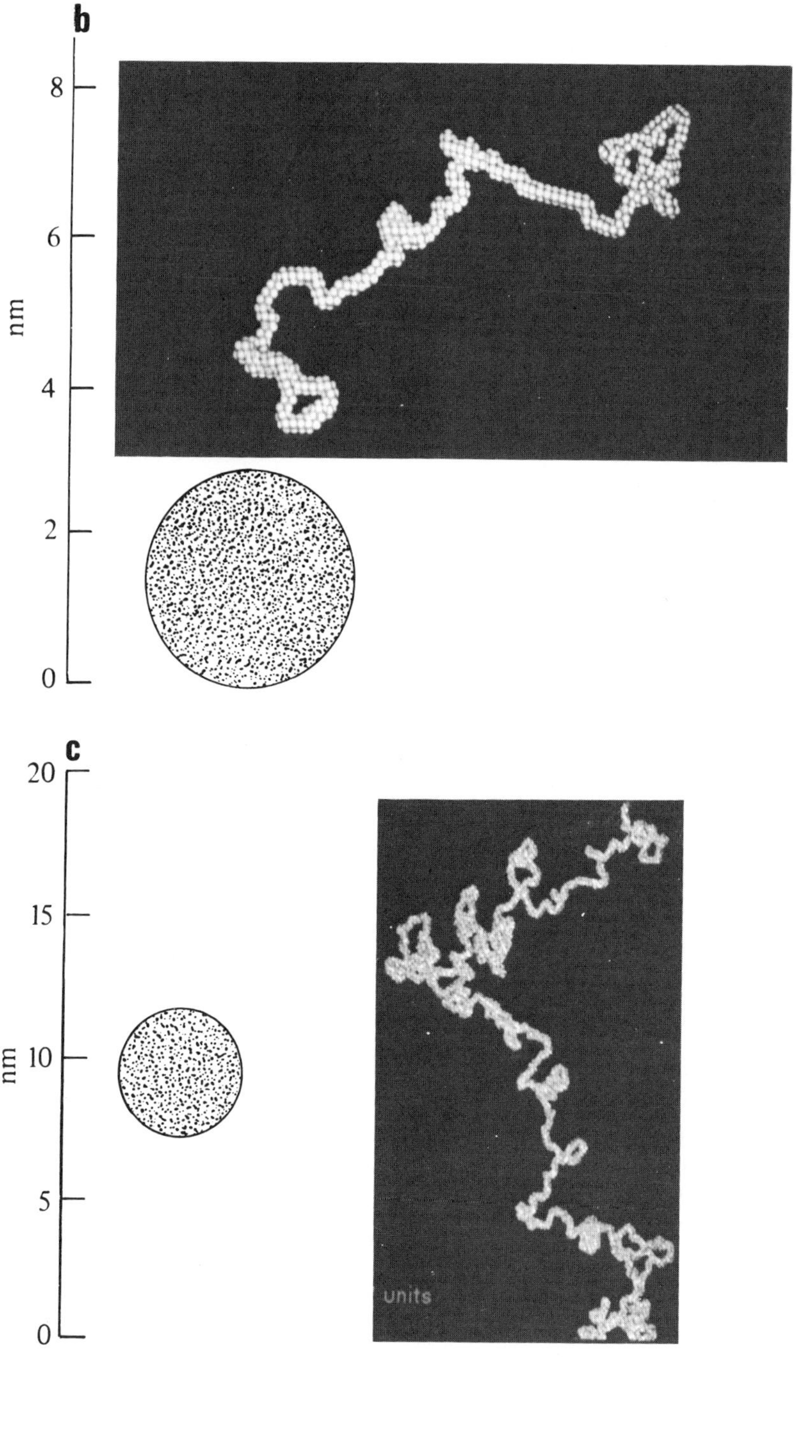
b
8
6
4
2
0
nm
c
20
15
10
5
0
nm
units

more highly extended. Amylose from starch (the fifth substance in Figure 1), which differs in chemical structure from cellulose only in the direction of one of its interunit bonds, is characterized by a much less extended trajectory. Over short distances, it winds its way along a path that is approximately helical, but it frequently deviates from strict adherence to this pattern. Polynucleotides also are randomly configured when their native structure, the double helix comprising two complementary chains wound together, is disrupted.

Especially to be noted is the well-established generalization that synthetic macromolecules and those of natural origin display similar characteristics when disordered. In each category, some are flexible and tortuous, others are stiff and extended.

## Configurations and Molecular Statistics

The configurations shown in Figure 5 are mere samples from the many that may occur for polyethylene chains. For a macromolecule consisting of many skeletal bonds, the number of configurations is prodigious. Consider, for example, a polyethylene chain of a mere 1,000 bonds. With three rotational possibilities available to each bond, the total number of configurations for the macromolecule as a whole is the product of 1,000 threes, or $3^{1,000}$. This is a number with 477 digits—a number that literally defies comprehension.

Confronted with so many spatial forms, one might well conclude that molecules of such complexity must defy rigorous scientific analysis. On these grounds, it would seem that inquiries into the behavior of large macromolecules must perforce be limited to qualitative observations and generalizations. On the contrary, the supposed difficulties turn out to be illusory rather than real. As so often happens in science, an approach from an alternative point of view leads to insights and methods that circumvent difficulties which otherwise would appear to be insurmountable.

The immensity of the number of configurations accessible to the macromolecule presents circumstances ideally suited to statistical methods. Whereas this number is far too great for a full count, the vast population can be sampled by randomly selecting a manageable set of representative configurations. In this way, results of any desired accuracy, in principle, may be determined. One may evaluate, for example, the average distance (r) between the ends of a chain by generating a sufficient number of configurations like those shown in Figure 5. The value of r for each may be determined and the average taken. The average radius of the region occupied by a random macromolecule, or any other characteristic or property that depends on the spatial configuration, may be determined similarly. Several thousand chains may have to be generated in order to achieve the accuracy desired. The task is easily within the capabilities of present computer technology. (Actually, much more efficient mathematical methods are available which circumvent the tedious task of sampling individual configurations.)

The connection between the statistical methods described above and the actual macromolecules that occur in real materials may be established by the following line of reasoning. A sample of a polymeric material large enough for experimentation will necessarily consist of very many molecules. A computer-generated set of configurations may be considered to be representative of these molecules at a given instant in time. It follows that the average size of the domain, for example, pervaded by a configuration as determined by computation, should correspond to the mean size averaged over all molecules of the real sample. This correspondence may be confirmed, in fact, by experimental measurements. Appropriate techniques for this purpose involve sensitive measurements of the scattering of light by the macromolecules. Similar information can be gleaned from the scattering of X-rays, or from measuring the viscosity of a dilute solution of the macromolecules.

Alternatively, one may focus attention on a single macromolecule and consider the configurations it assumes at successive intervals of time. Being subject to incessant bombardment by neighboring molecules moving about under the influence of thermal agitation, the macromolecule writhes from one to another of its virtually endless repertoire of configurations. These changes occur with great rapidity. In effect, then, in the course of its temporal history the macromolecule samples the many forms accessible to it. The configurations occurring at successive intervals of time may be assumed to correspond to a representative set of the myriads of possible configurations. We can assert, therefore, that the average of the configurations assumed by a given macromolecule during a period of time commensurate with or greater than that required to perform an actual experiment on a real sample in the laboratory is equivalent to the instantaneous average over the many macromolecules in a sample large enough to be subject to experimental measurements.

This assertion is a premise of statistical mechanics, a science whose foundations were formulated with profound depth of insight by J. Willard Gibbs, professor of mathematical physics at Yale University from 1871 to 1903. His legacy to science has few parallels.

The foregoing discourse on the configurations of macromolecules is a paradigm of the principles underlying Gibbsian statistical mechanics. This branch of science was created for the express purpose of comprehending materials consisting of hosts of small molecules. Analyzing the positions in space and detailed movements of the billion trillion molecules in a drop of water is obviously impossible. But by sampling the "complexions" (i.e., configurations) of many identically constituted drops, or, equivalently, by sampling the single drop at successive intervals of time in its ceaseless reorganization due to thermal motions, one might conceivably gain an understanding of the connections between the molecules of water and the properties of this substance.

Having mentioned a droplet of a common liquid by way of illustration, the additional point should be made that whereas the methods of statistical mechanics are indubitably applicable in principle to such a system, in practice their application to small molecules irregularly compacted one against the other is fraught with difficulties. Each molecule is under the influence of molecules impinging on it from all directions. Its immediate neighbors are affected by their neighbors, and so on. The successive "generations" of neighbors proliferate in ever increasing numbers. The associated difficulties require adopting approximation methods. Results obtained are inaccurate and often unreliable. Progress in understanding systems of small molecules in condensed states, namely, solids and liquids, has therefore been slow, especially in the latter case.

Long-chain polymers, on the other hand, consist of members (mers) that are linearly connected. Each unit or bond is succeeded by only one adjoining unit or bond; this neighbor is in turn succeeded by just one more, and so on. The succession does not proliferate, as in the case of a molecule surrounded on all sides by a number of other molecules. In the language of the mathematician, the system is one-dimensional. It has long been known that the abstract formalisms of statistical mechanics can be solved exactly when they are applied to molecules or other entities in a hypothetical world of one dimension. Macromolecules exist in the real world of three dimensions, but they are amenable to the mathematics of imaginary one-dimensional systems. Hence, rigorous analysis of the configurations and related properties of macromolecules is readily feasible, their superficial complexities notwithstanding. Appreciation of this fact is comparatively recent, and it has yet to be fully exploited for the purpose of gaining a fuller understanding of the relationships between chemical structure and the properties of these important substances.

## Configurations of Macromolecules in Polymeric Materials

The foregoing discussion addresses implicitly the configuration of a single macromolecular chain existing in isolation from all others. It applies, for example, to the macromolecules in a solution so dilute that they are separated from one another. The discussion of Figure 5 emphasized that only a small portion of the space pervaded by a long macromolecule in a random configuration is actually occupied by it. For a polyethylene consisting of 3,000 bonds (see Figure 5c), the occupied portion of the pervaded space is about one percent. The remainder may be filled by molecules of solvent.

What happens if the solvent is removed? Intermolecular forces require that the space vacated somehow be filled. Possibilities that come to mind are the following: (1) the macromolecules may collapse to smaller, dense configurations resembling the spheres depicted in Figure 5, each remaining discrete from the others; (2) they may associate in bundles in which neighboring chains are parallel to one another; or (3) they may remain randomly configured with each of them interpenetrating the domains pervaded by other nearby chains, the macromolecules becoming entangled like long strands of limp spaghetti. The first possibility is realized only in globular proteins (see "Globular Conformations: Enzymes") with highly specialized constitutions that predispose them to adopt compact conformations. The second occurs in crystalline polymers. The third possibility is manifested under suitable conditions in nearly all polymers, of both natural and synthetic origin. It is universally prevalent in the molten and rubbery states, and also in amorphous and glassy plastics (e.g., polystyrenes and polyacrylates). Even crystalline polymers contain a substantial residuum of amorphous material in which the chains are disorganized.

Precisely how the irregularly configured molecular chains are arranged in the molten or amorphous state has long been in contention. According to one body of opinion, they form clumps or clusters in which sections of the chains are organized in bundles possibly resembling crystallites but less perfectly ordered. The opposing view holds that the long chains assume configurations equivalent to those of isolated random chains, each threading its way tortuously through the maze of surrounding chains without, however, intruding on the space occupied by the atoms constituting them. Although the forces between neighboring molecules are large, they should not favor one configuration over another, according to this body of opinion. Hence, the configuration of the long chains should be affected little, if at all, by their mutual sharing of the space at their disposal.

Much indirect evidence indicated that the latter view is the correct one. Experiments on the elastic properties of amorphous polymers, on their absorption of solvents, and on the conversion of linear chains to cyclic rings and vice versa strongly supported this view. It remained, however, for sophisticated experiments involving the use of neutrons and elaborate equipment to provide the ultimately compelling evidence on this important issue. These experiments, brought to fruition within the past decade, merit brief description.

Neutrons are uncharged subatomic particles of great penetrating power. A stream of them may nevertheless be scattered (i.e., deflected) by collisions with atomic nuclei. In particular, nuclei of the heavy hydrogen isotope deuterium (D) scatter neutrons much more than do the nuclei of ordinary hydrogen. A sample of the polymer to be investigated is prepared, therefore, in which the hydrogen atoms are replaced by deuterium. A small proportion of this D-polymer is then thoroughly mixed with the ordinary H-polymer so that molecules of the two kinds are intimately interspersed. Substitution of deuterium for hydrogen does not alter the properties of the polymer appreciably, apart from enhancing its ability to scatter neutrons. Hence, the D- and H-polymers mix indiscriminately without affecting each other.

The mixture is exposed to a beam of neutrons of the desired (low) velocity. The numbers of neutrons scattered in different directions, particularly at small angles from the incident beam, are detected by counters and recorded. The situation is altogether analogous to "seeing" the light scattered at various angles by small particles (the D-polymer) dispersed in a relatively transparent medium of different refractive index (the H-polymer). By precisely determining the "profile" of the scattering at various angles, one may deduce an accurate measure of the size of the domain pervaded by the macromolecule as it is configured in the condensed state (molten or amorphous solid).

Suitable equipment for performing the desired neutron scattering experiments was developed in the early 1970's, first in Grenoble, France, and in Jülich in the Federal Republic of Germany. Facilities for performing such experiments recently became available at Oak Ridge National Laboratory and are being established elsewhere in the United States.

Within the past several years, neutron scattering experiments have been undertaken by many groups of investigators, principally in Europe, on various polymers: poly(methyl methacrylate), polystyrene, molten polyethylene, molten polypropylene, and molten poly(ethylene oxide). In every instance so far investigated, the average size of the random configuration of the guest molecules labeled with deuterium equals, within a few percent, that of the isolated chain as it occurs when dissolved in a large quantity of an indifferent solvent. Thus, the average size of the macromolecule, and therefore its configuration, is virtually unaffected by the necessity that it coexist with other chemically similar long-chain molecules. This assertion applies quite universally provided that the polymer is not crystalline (see "Physical and Mechanical Properties of Polymers" below). It has been confirmed for hard glassy polymers such as the first two examples cited, for viscous molten polymer liquids, and for rubbery solids. Figure 5 therefore gives a realistic picture of the spatial irregularity of macromolecules in diverse situations.

The foregoing generalization, which has gained acceptance only within the past few years, promises to be of foremost importance in guiding future research. In consequence thereof, results of studies on macromolecules in virtual isolation from one another are directly applicable to melts, to plastic and glassy polymers, to rubbers, and indirectly to fibers and films. The configurational characteristics of a macromolecule in the actual environment of its practical use or function may now be deduced from the chemical formula specifying its atoms and the bonds connecting them. Only a few of the many macromolecular substances of everyday importance, both in living systems and amongst the artifacts of commerce, have been investigated from this standpoint. However, the time is ripe for research that should enable us to understand more clearly why, for example, polymers of styrene and methyl methacrylate suffer from embrittlement, how these deficiencies may best be alleviated by incorporating "blocks" of rubbery units (such as butadiene or isoprene) in their chains, and why some rubbers recover more efficiently from deformation (stretching, bending, or twisting) than do others. We may thus gain a deeper understanding of those properties of macromolecular substances that commend them for an ever increasing range of uses.

## Rodlike Conformations

Random configurations like those shown in Figure 5, which have been discussed above at length, are not universal among all polymers under all conditions. Stiff chains, such as those of aromatic polyamide and polybenzoxazole, shown earlier under "Condensation Polymerization", are better represented as long rods. Even macromolecules whose bonds allow rotations that engender random configurations may organize themselves in parallel, crystalline arrays when packed densely together, in a film or molded block of polyethylene, for example. A rodlike form is obligatory when chain molecules are packed in parallel array. The fully extended planar zigzag conformation shown in Figures 3 and 5a is illustrative. In other macromolecules, a helix generated by uniformly twisting the skeletal bonds frequently provides a preferable conformation. The external shape of the helix is rodlike and hence acceptable for parallel arrangement with other identically conformed helixes.

In occasional instances, the atoms and groups attached to the chain skeleton may occur in singularly favorable situations when the chain is wound in the form of a particular helix. This conformation may then be preferred over all other conformations, including the host of random coils. The $\alpha$-helix of the polypeptide chain, shown in Figure 6, is illustrative. It owes much of its stability to hydrogen bonds, shown by dashed lines (see "Chemical Terms and Conventions," following Chapter 11). $\alpha$-Helices are stable even when they are separated from other molecules by dissolving them in a large quantity of a solvent that does not disrupt the hydrogen bonds within the helix. Independently stable helices are especially prominent among biological macromolecules. The familiar double-stranded DNA helix is the foremost example. Collagen, the primary component of tendons, consists of three protein chains wound together in a slow-turning helix. It provides the fibrillar material that reinforces skin and other tissue, as well as the mortar that binds the bricks (hydroxyapatite) of bone; it is also the precursor of gelatin.

The helical polypeptides and polynucleotides are converted to random

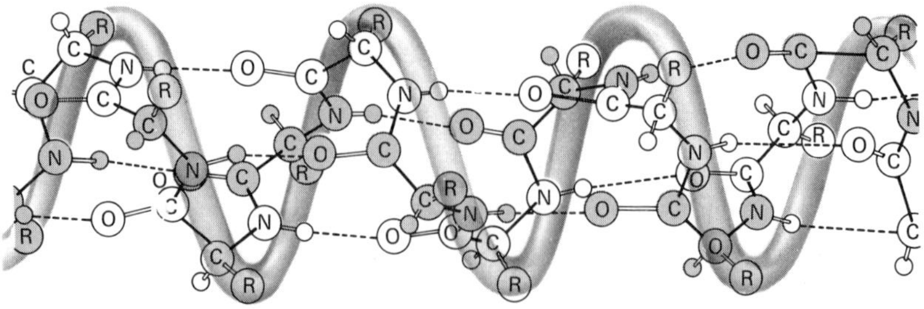

**Figure 6 L-Polypeptide chain.** L-Polypeptide chain in the form of an $\alpha$-helix. [SOURCE: C. Cohen. "The Protein Switch of Muscle Contraction," *Scientific American*, Vol. 233(November 1975), p. 36. Copyright © 1975 by Scientific American, Inc. All rights reserved.]

coils either by a suitable change of solvent or by raising the temperature. As random coils, they exhibit a set of properties that corresponds in every respect to that of synthetic macromolecules.

## Globular Conformations: Enzymes

A long macromolecular chain having sufficient flexibility may be conformed in yet another fashion. As briefly noted earlier, it may be compactly coiled so that it completely fills the space it occupies, leaving no room for other molecules within its spatial domain. The spheres depicted in Figure 5 are illustrative. They are much smaller than the irregularly articulated random coils.

Most macromolecules, and especially those prepared by *in vitro* synthesis, are not easily induced to conform in this manner. One might expect that adding a poor solvent (a precipitant) to a dilute solution of the polymer should cause the random configuration to collapse as the macromolecule compacts itself to minimize contact with its inhospitable environment. Before this happens, however, the macromolecules coalesce with one another to form agglomerates which precipitate from the solution.

Nature has overcome this tendency to agglomerate in the case of globular proteins by producing polypeptides that include both hydrophobic (water-repelling) and hydrophilic (water-attracting) units in the same chain. By judiciously situating the respective kinds of units along the chain, evolutionary processes have so constituted the protein as to enable it to "fold" itself into a compact globule with the hydrophobic units in the interior, out of contact with the surrounding aqueous environment. The hydrophilic groups, at the exterior, serve to render the globule compatible with the environment and thus discourage agglomeration with other globules.

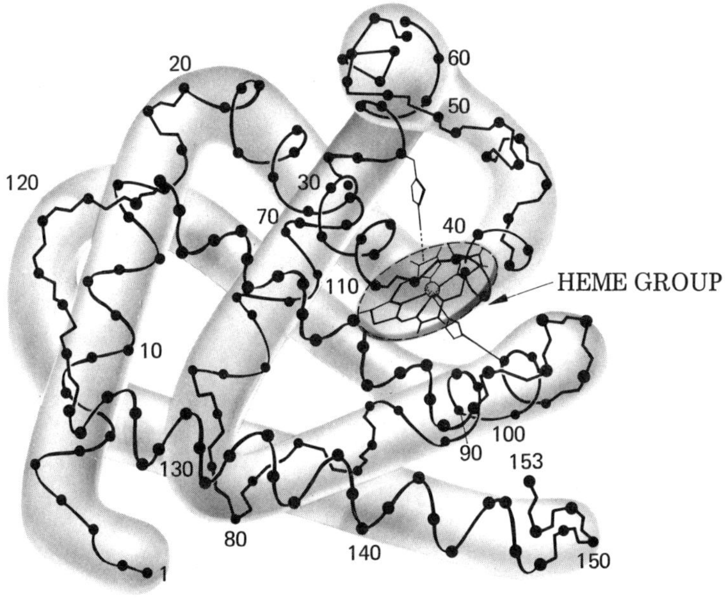


**Figure 7 Myoglobin in its native form.** Schematic representation of the globular protein myosin. Dots denote amino acid units, with every tenth unit numbered serially from the amino terminus (1) to the carboxylic acid end (153). The space occupied by the chain backbone is delineated approximately by the shaded regions. The disk-shaped heme group is also identified. [SOURCE: M. F. Perutz. "The Hemoglobin Molecule," *Scientific American*, Vol. 211(November 1964), p. 71. Copyright © 1964 by Scientific American, Inc. All rights reserved.]

The spatial arrangement of the chain backbone in myoglobin, a typical globular protein, is illustrated in Figure 7. It consists of a unique sequence of 153 amino acid units, determined by the DNA template that controls its synthesis. Myoglobin is one of the smaller globular proteins; others may comprise as many as several hundred units. Each of them is a jigsaw puzzle assembled from pieces (amino acid units) of some 20 different shapes and sizes strung into a chain. The pieces can be fitted together satisfactorily in only one way; that is, there is a unique pattern for the given protein. In fact, only by ingenious design has nature succeeded in producing polypeptide chains that can be compactly folded with hydrophobic units in the interior and other units positioned precisely in the manner required for the particular globular protein to function. Replacement of but one of the amino acid units by another amino acid selected at random will usually preclude folding of the chain in a globular conformation that is at once compact, stable, and capable of performing its intended biological function.

Myoglobin contains a heme group (see Figure 7) bearing an atom of iron that acts as a repository for oxygen in muscle. If the heme group were exposed to the surrounding water, its iron atom would be oxidized and the oxygen-binding capacity of the complex would be destroyed. The active site is exquisitely situated in a locus of the globular structure where it is protected against destruction but can still easily receive and deliver oxygen.

Hemoglobin consists of two pairs of subunits that resemble myoglobin. The four subunits fit together to form the hemoglobin complex, which carries oxygen in the blood stream from the respiratory system to cells throughout the organism. Because of their compact size and shape, globular proteins may move about with a minimum of friction. They are well designed therefore to transport various substances essential to life.

Enzymes are globular proteins that regulate or catalyze vital chemical reactions in living organisms. They number in the thousands. Each bears an active site in the form of a cleft where the geometric arrangement and conjunction of chemical groups promotes the selective binding of certain molecular species and their chemical transformation.

Enzymes are intricate chemical devices. Through innumerable mutations put to the test during eons of evolution, the sequence of units assembled in each of them has been optimized, enabling enzymes to induce specific chemical reactions with remarkable selectivity and high efficiency. This functional adaptation of enzymes far surpasses the considerable advances in synthetic methods achieved in chemical laboratories during the past century. Through differentiation and elaboration of only a few chemical types, nature has succeeded in evolving a plethora of macromolecules, each designed to serve a specific purpose or function essential to a living organism and its reproduction. The mystery surrounding these achievements evokes profound awe and wonderment. Yet there is no evidence to suggest that nature invokes principles or laws beyond those discernible through man's scientific inquiry, including his experimentation in the laboratory.

## PHYSICAL AND MECHANICAL PROPERTIES OF POLYMERS

Polymers may occur as viscous liquids, rubbery solids, glasses, or semicrystalline solids. Polymers of the same chemical composition may occur in any of these states, depending on the temperature and their chain length. Thus, polystyrene and polymers of methyl methacrylate (Lucite and Plexiglas) are hard brittle glasses at ordinary temperatures, where the density of packing of the molecular chains and the forces of attraction between them combine to arrest rearrangement of their configurations. But above 100°C, their molecular torpor is overcome and they exhibit properties ranging from those of viscous liquids to those of elastic solids, depending upon the lengths of their chains. Similarly, commercial polypropylene and polyamides (e.g., molding-grade nylon) are hard, tough, semicrystalline solids, but when heated above 200°-265°C they melt to form viscous liquids.

## Viscous Flow

The flow of liquids involves the movement of molecules past one another. In a polymeric fluid, such relative motion requires a given macromolecule to extricate itself from the many neighboring molecules with which it is entangled and to reentangle itself with other passers-by. This tedious process must be repeated continuously, entirely under the impetus of undirected thermal motions that are alone responsible for random changes in configuration. The force acting on the material and causing it to flow exerts only a gentle bias on the incessant thermal movements, to and fro, occurring at the molecular level. It is small wonder, then, that liquid polymers flow with extreme sluggishness—that they are very viscous. The fact that the viscosity increases markedly with chain length is likewise comprehensible. The viscosity depends also, but to a lesser degree, on the temperature (inversely) and on the strength of intermolecular forces.

The consistency of a liquid polymer may vary from syrupy to semisolid. For chain lengths greater than 10,000 bonds, liquid flow becomes scarcely perceptible; elastic rubberlike deformation then becomes dominant.

## High Elasticity

The molecular displacement and diffusion required for flow in the manner of a liquid may be suppressed entirely by connecting the long chains to one another through chemical bonds that function as cross-links. Deformation of the resulting network (see Figure 2) is elastic rather than viscous or plastic. Thus, if stretched or otherwise deformed, the material recovers its original shape and dimensions when the deforming force is removed.

If the cross-links are sparse, occurring only at intervals of 100 to 1,000 bonds along a given macromolecule, then the material may exhibit the high elasticity of typical rubberlike substances: it is highly deformable at temperatures where it is neither glassy nor crystalline. It may be stretched to several times its initial length without rupturing, and the material recovers its original dimensions when the stress is removed (in contrast to the flow of a liquid).

Figure 2 offers a schematic representation of a rubberlike network. The chains between cross-links, consisting of several hundred bonds, are much more irregularly configured than those shown in the figure; they may resemble in this respect the configuration of the 500-bond chain shown in Figure 5b. When the network polymer is stretched, the cross-links defining the ends of a chain must move apart in the direction of elongation. The many alternative configurations accessible to the chain enable it to comply readily with the redeployment of its ends; rotations about a few of its bonds suffice to meet the demands dictated by drastic changes of the material's shape. The capacity of a polymer to accommodate large deformations without sustaining rupture or other permanent changes finds immediate explanation in the diversity of configurations that its chains may adopt.

The origin of the force tending to restore the material to its initial dimensions also has its explanation in the configurational versatility of macromolecular chains. It may be understood by considering the number of configurations accessible to the chain between cross-links (see Figures 2 and 5b). As one might expect, this number diminishes as the distance separating the ends of the chains is increased. It is a universal rule that molecules subject to thermal agitation gravitate toward circumstances that offer them the greatest number of options. In the language of statistical mechanics, they tend toward the state of greatest probability. This tendency manifests itself in a force pulling on the ends of the chain. A detailed analysis of the configurations a polymer chain can assume leads one, by way of a somewhat abstruse statistical argument, to the conclusion that the average force exerted by the chain is directly proportional to the distance between its ends. This deduction from the statistical theory of chain configurations provides the fundamental basis for the theory of rubber elasticity.

Recent investigations have affirmed that the retractive force of a stretched elastic network (i.e., a vulcanized rubber) is due in its entirety to these forces exerted by its individual chains. When rubber is stretched, the distances spanned by most of its chains are lengthened; the effect on a given chain depends on its orientation in relation to the stretching direction. The force measured in an experiment is the sum of the responses of the individual chains to the changes they undergo. It is proportional to the number of chains in the network, and hence, for a given strain, it increases proportionately to the degree of cross-linking.

High elastic deformability is a property found exclusively in substances consisting of long macromolecular chains. Moreover, nearly all of the many varieties of polymers are potentially capable of exhibiting this property. The only requirements are that the chains be joined by occasional cross-links or other durable connections that impart a network structure, and that the substance be neither glassy nor crystalline. Thus, cross-linked polystyrene, a hard glass at ordinary temperature, becomes rubberlike when heated above its softening temperature. Nylon fibers and polyethylene film, both partially crystalline, melt to amorphous bodies that are rubberlike if they have been cross-linked.

Elastin, the principal elastic protein of ligaments, aortic tissue, blood vessels, and skin, consists of long polypeptide chains cross-linked at intervals of 50 to 100 units through enzymic action on amino acid residues specially equipped for the purpose. It exhibits all the properties of typical rubberlike polymers. Experiments confirm that, like the latter, it has randomly configured chains, with the restoring force that is induced by deformation originating in alterations of the configurations accessible to these chains.

Other proteins, such as collagen in tendons and bone, fibroin in silk, myosin in muscle, and keratin in wool, hair, and fingernails, become rubberlike when denatured—that is, when their native configurations, which may be helical or extended, are disrupted. In fact, rubber elasticity plays a key role in the retraction of muscle *in vivo* (see "Muscle and Motility in Biological Organisms" below). Motility in living organisms in general, including that of individual cells, may similarly depend on the changes of shape that macromolecules are uniquely able to undergo.

Physical science offers few instances where the connection between a mechanical property amenable to quantitative measurement can be related so definitively to the molecular character of a material. High elastic deformability can be elicited from most polymeric substances under suitable conditions; it is by no means a property limited to natural rubber and a restricted class of its synthetic analogs. Even among nonrubbery polymers, changes in molecular configuration akin to those involved in rubber elasticity are involved when the material undergoes elastic deformation. In biological systems, the role of rubber elasticity may be far more pervasive than has been realized.

## Mechanical Strength and Crystallinity

Polymers are weak in the high-elastic state. Although they sustain high degrees of stretching, the force required to stretch them is low. Their strengths at rupture may be only 20 to 50 kilograms per square centimeter, unless, as in the case of natural rubber, stretching induces transformation to a fibrous state reinforced by crystallization (see below). The mechanical fragility of typical rubbers is a consequence of the chaotic molecular disarray that is essential for rubberlike elasticity. At high elongations, the load is borne unequally by the chains of the network; some of them, by virtue of their situations, are more susceptible to overextension than are others. When these chains are broken, others in turn are vulnerable, and so on to the eventual catastrophe: rupture of the specimen.

If the filamentary molecular chains could be arranged in parallel bundles like the strands of a rope, they should then respond in unison to a force applied in the direction of their alignment. Their strength should be much greater than that of a jumbled mass of filaments that can be broken individually. This kind of organization can be achieved most readily through crystallization of the macromolecules. Alternatively, it may emerge quite spontaneously if the macromolecules are inherently stiff and prefer rodlike structures. These two avenues to materials of extraordinary strength are considered in the following paragraphs.

The forces of attraction between macromolecular chains may induce them to pack together efficiently and in perfect array, like the tiles on a roof. Precise molecular arrangement is the mark of the crystalline state. All chains must be identically conformed if they are to be assembled in an ordered array. Moreover, an extended configuration (see Figure 5a) is the only form that allows the chains to be regularly packed together. At temperatures below the melting point, the intermolecular forces of attraction prevail over the advantages of greater freedom offered by the random molecular chaos prevalent in the liquid state. The converse holds above the melting point.

When crystallization takes place in a molten polymer cooled below its melting point, the difficulties of extricating a long macromolecule from entanglements with its neighbors in the melt, and of arranging it in extended form, preclude the growth of large crystals of perfect order, such as those readily obtained in substances consisting of small molecules. Instead, thin lamellar, or sheetlike, crystals develop. Additionally, the transformation is incomplete; a substantial fraction, typically 20 to 50 percent, remains amorphous, that is, disordered. The amorphous material is interspersed between the crystalline lamellae, the alternating layers being stacked, as illustrated in Figure 8.

The structure in the crystalline lamellae can be precisely determined by analyzing their X-ray diffraction pattern. It has thus been established that the chains run through the crystalline lamellae in a direction transverse to the surfaces of the lamellae, as shown in Figure 8. The crystalline layer measures only 10 to 50 nanometers in thickness; its breadth is much greater. The arrangement of chains in the amorphous interlayers cannot be so readily determined.

From the fact that a sequence of 100 to 300 bonds suffices to span crystalline lamellae from one face to the other, whereas an entire chain contains on the order of 10,000 bonds, one may infer that a given macromolecule must either pass repeatedly through the same lamella or must be engaged in a number of them. Which of these alternatives prevails has long been in contention. Recent research employing neutron scattering techniques has revealed that trajectories described by individual molecules retain the expanse prevalent in the liquid melt from which the crystalline array is formed (see "Configurations of Macromolecules in Polymeric Materials," p. 398). Evidently, the chains exercise the privilege

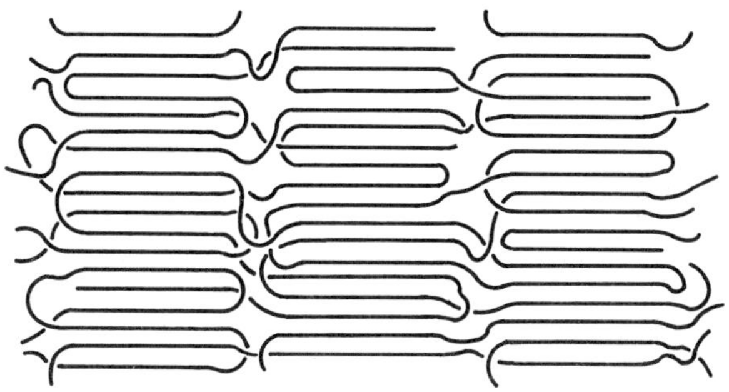

**Figure 8 Semicrystalline polymer.** A section of a semicrystalline polymer, showing three crystalline lamellae separated by amorphous (random) interlayers.

of free passage through the amorphous layer separating one lamella from the next. A macromolecule therefore becomes engaged in several lamellae. At the same time, its path must frequently return to the same lamella. A more refined analysis of the neutron scattering results shows that the point of return seldom is adjacent to the point of the preceding exit, as had long been assumed.

These recent seminal findings concerning the nature of the amorphous interlayer enable one to discern the essential aspects of the morphology of semicrystalline polymers. The resulting concepts, surveyed in the following paragraphs, are destined to advance profoundly our understanding of the properties of semicrystalline polymers—properties that are of foremost importance in the enlarging range of applications of polymeric materials.

Growth of the crystalline lamellae must proceed from their edges through accretion of chain sequences in the required extended configuration. It is a rapid process. The small thickness of lamellae is a direct consequence of the impediments in the melt, from which crystal growth takes place, to disentanglement and rearrangement of chains. In the brief time interval available for deposition, only a limited section of the participating chain can undergo the required rearrangements. Hence, the crystal must be formed from short sections. Entanglements present in the melt are excluded from the crystalline regions and hence are relegated to the intervening amorphous layers. They inhibit acquisition of material from the interlayers, blocking the further growth of the crystalline lamellae.

The crystalline regions are themselves quite rigid. Entanglements crowded into the amorphous layers suppress the kind of mobility that accounts for ease of deformation in the rubbery state. Consequently, typical semicrystalline polymers like polyethylene, polypropylene, nylon, and polyesters (Dacron, Terylene, and Mylar) are relatively hard and tough but compliant nonetheless. This combination of properties recommends them for many uses.

## The Fibrous State

If a polymer is stretched rapidly while crystallizing, or if it is "drawn" extensively after crystallization (a process that involves recrystallization), the chains in crystalline regions become oriented parallel to the stretching direction. The lamellar morphology shown in Figure 8 is supplanted by crystallites of narrow dimensions, as depicted in Figure 9. The alignment of the chains in the crystalline regions enables them to respond collectively to a force tending to stretch the fiber. Both stiffness and tensile strength consequently are high. For current commercial synthetic fibers, the strength may reach 10,000 kilograms per square centimeter, which is superior to the strength of naturally occurring fibers and nearly half that of high-strength steel.

Nevertheless, the stiffnesses and strengths of nylon, polyester, and other conventional synthetic fibers are inferior to the levels theoretically attainable for perfect molecular alignment. Amorphous regions, persisting in substantial amount, have been implicated as sources of compliance lowering the fibers' stiffness and as the defects responsible for fracture. These conjectures are supported by recent achievements in several laboratories in the Netherlands, the United Kingdom, and the United States. By ingenious extrusion and drawing techniques applied to polyethylene of unusually high molecular weight (300,000 bonds per macromolecule), fibers have been prepared with exceptional order. Their strengths range upward of 40,000 kilograms per square centimeter. This is twice the strength of steel. If the nearly eightfold difference in densities is taken into account, then the polyethylene fibers are 15 times stronger than an equal weight of the best steel!

For practical applications, polyethylene fibers are inherently deficient because of their comparatively low melting point, about 140°C, which precludes their use at temperatures much above the boiling point of water. Moreover, the elaborate drawing techniques used in the experiments cited probably would be impracticable commercially. The foregoing achievement is significant mainly in showing that macromolecular substances, if properly constituted, offer strengths unsurpassed by other materials.

The deleterious effects of entanglements may be diminished or suppressed entirely by forming fibers from macromolecules that are predisposed to assume highly extended configurations. This circumstance may be realized by choosing a polymer whose chemical structure requires the bonds of its chain skeleton to be aligned, for example, through enchainment of benzene rings (see "Condensation Polymerization," p. 376). A

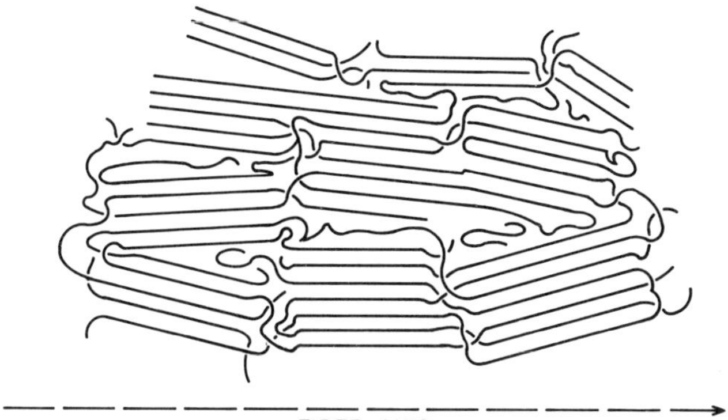


**Figure 9 Morphology of a drawn fiber, showing crystallites and amorphous inclusions.**

polymer can also be selected that can be induced, under suitable conditions, to adopt a helical rodlike form in advance of its assemblage into a fiber. Examples of the latter kind are especially prominent among biological macromolecules. With macromolecules rigidly fixed in rodlike form, forces of attraction between them are not essential to produce an ordered array. The molecular shape alone dictates a fibrous state of aggregation in which the long rods are assembled in parallel bundles.

If rodlike macromolecules of one of the kinds mentioned are dispersed in a solvent, and their concentration is then increased, a point will be reached beyond which the individual particles are no longer free to choose orientations independently of one another. Like long logs floating on a pond, they organize themselves in semiparallel arrays (Figure 10). The resulting order is dictated by molecular shape. This state, called "liquid crystalline," differs from solid crystalline in that the alignment of molecules is imperfect and their lengthwise positions are in disarray; the logs in the analogy above are not neatly stacked with their ends adjacent.

The liquid crystal retains the mobility of a liquid. Consequently, it can easily be spun to a fiber in which the rigid molecules are well aligned in parallel. Most important, the fibers formed in this manner are free of the disordered intercrystalline regions so difficult to suppress in fibers spun from flexible, randomly entangled polymers. Fibers of aramid (aromatic polyamide) thus prepared and now commercialized exhibit strengths much greater than that of steel, weight for weight, and more than half that quoted above for fibers specially prepared from polyethylene. Moreover, aramid fibers retain their strength up to temperatures of 250° to 300°C.

The study of liquid-crystalline polymers is a new field presenting a variety of challenges in synthesis, morphology, properties, and theory.

**Figure 10 Liquid crystal, consisting of long, rodlike molecules.**

Progress in this area gives promise of novel materials of exceptional strength, rigidity, and durability. The advantage of low density in comparison with metals and other conventional materials especially recommends polymers of this class where reduction of weight is important, as in transportation.

The fashioning of macromolecules in shapes best suited to their function, prior to their assembly into larger structures, is a practice that is apparently widely exploited in nature. The genesis of fibrous proteins is illustrative. The collagen triple helix is the precursor of the microfibrils that make up tendons. Myosin, the principal protein of muscle, adopts the form of an $\alpha$-helix before it is assembled in myofibrils. Plant cellulose is a relatively stiff, highly extended macromolecule readily amenable to fibrous aggregation and free of the entanglements present in most synthetic and artificial fibers available at present. In other instances, liquid crystallinity may mediate the combination of biological macromolecules in a manner specifically required for structure and function. It offers a particularly elegant and simple route to self-organization, whereby molecular constitution and shape are translated into the intricate supermolecular entities that are observed in cells and that control cell behavior. The possible importance of liquid-crystalline states in biology is only beginning to be appreciated.

## Muscle and Motility in Biological Organisms

Muscles of animals are highly elaborated devices for converting chemical energy to mechanical work. The intricacy of their structure is apparent from Figure 11. This diagram summarizes the combined results of many years of painstaking efforts in a number of laboratories in various countries. Advanced techniques of electron microscopy and X-ray diffraction have provided most of the information supporting present understanding of the fine structure of this elegant mechanochemical device.

Basically, the vertebrate skeletal muscle fiber pictured in Figure 11 consists of repeating sets of tiny filaments separated by membranes recurring at distances of about 3 micrometers. Attached on either side of the membrane are thin filaments, about 7 nanometers in diameter, consisting of three different proteins, of which actin is the major one. These thin

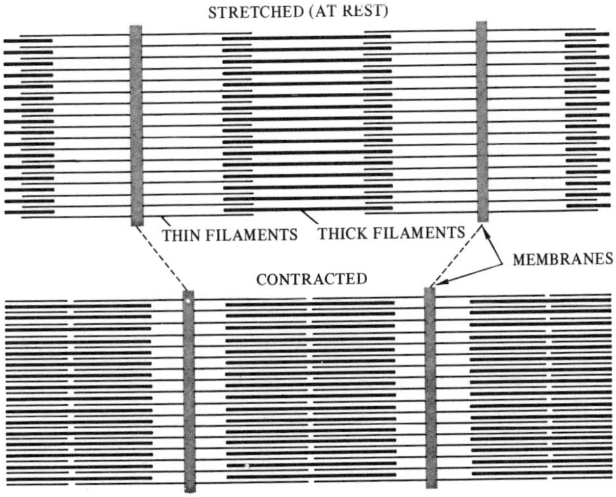


**Figure 11 Diagram of muscle sarcomere.** [SOURCE: J. M. Murray and A. Weber. "The Cooperative Action of Muscle Proteins," *Scientific American*, Vol. 230(February 1974), p. 59. Copyright © 1974 by Scientific American, Inc. All rights reserved.]

filaments terminate in the midportion between membranes. In this region, they are interdigitated with thicker filaments (15 nanometers in diameter) made of myosin, the principal protein of muscle. The myosin molecule is a hybrid consisting of a globular head and a long tail. Like most globular proteins, the head functions as an enzyme. It binds energy-rich ATP (adenosine triphosphate), the fuel for the muscular engine, at its active site. It has the capability also to bind to a molecule of actin in the thin filament when the muscle is activated by events following a neural impulse. In combination with actin and under the influence of calcium ions released by the neural command, the globular head catalyzes the hydrolysis of ATP (by reacting with water), whereby energy is released to power the contraction of the muscle system.

The myosin tail is an $\alpha$-helix (see Figure 6) measuring about 1 nanometer in diameter and 150 nanometers in length. Two such helices gently twisted together constitute the primary contractile element of muscle. Several hundred of these two-stranded cables make up the thick filament shown in Figure 11.

At a higher level of magnification, the myosin heads are found to protrude from the myosin filament and to reach toward the thin (actin) filaments, as shown in Figure 12. This figure is an enlargement of the midsection of the muscle fiber in its resting (a) and contracted (b) states, shown in Figure 11. Many of the myosin heads are attached to the adjacent actin thin filament in the latter state.

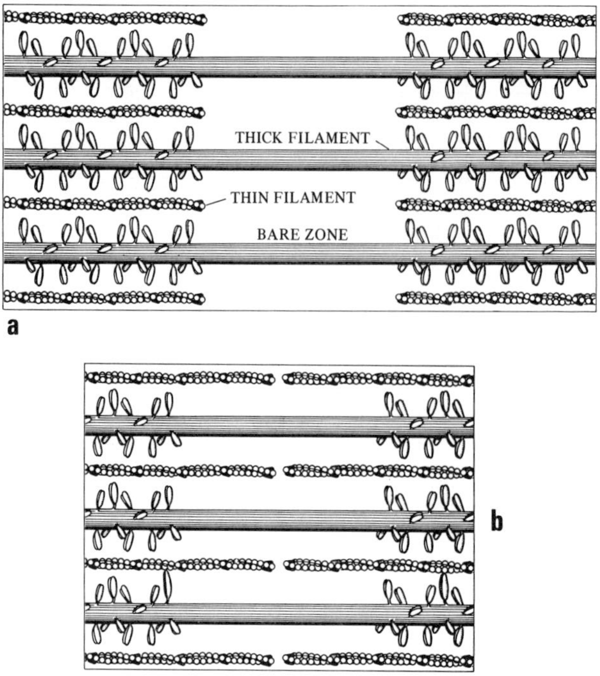


**Figure 12 Central portion of the sarcomere.** Enlarged sketch of central portion of the sarcomere: (a) in resting muscle; (b) in contraction. [SOURCE: J. M. Murray and A. Weber. "The Cooperative Action of Muscle Proteins," *Scientific American*, Vol. 230 (February 1974), p. 59. Copyright © 1974 by Scientific American, Inc. All rights reserved.]

According to recent studies that have yet to gain widespread acceptance, the contraction cycle comprises the three stages represented in Figure 13. In the resting state (a), the myosin head, laden with a molecule of ATP, is attached firmly to the thick filament surface. Activation of the sarcomere, or unit of muscle between membranes, by a nerve impulse and the ensuing release of calcium ions cause the tip of the myosin head to attach itself to an actin molecule in one of the neighboring thin filaments. Hydrolysis products, ADP (adenosine diphosphate), and phosphoric acid ($H_3PO_4$), are discharged. Probably as a result of the local change in acidity, the head and the adjoining section of the two-stranded $\alpha$-helix tail are released from the thick filament, as shown in Figure 13b. With the myosin tail freed from the stabilizing environment at the surface of the thick filament, hydrogen bonds of the $\alpha$-helix in the transducing segment (shown hatched in Figure 13) are ruptured, and partial melting occurs as indicated in Figure 13c. The resulting randomly coiled polypeptide chains exert a contractile force exactly as if they were chains of stretched rubber. This force pulls the thin filament toward the center of

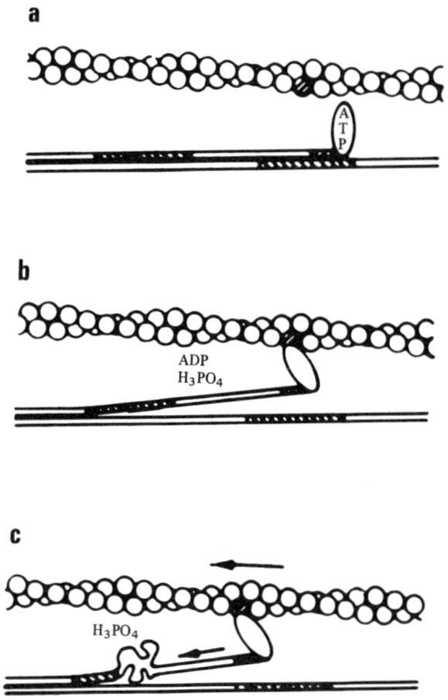


**Figure 13 Myosin macromolecules.** Action of myosin macromolecules during stimulation of muscle: (a) two-stranded pair of myosin molecules near the surface of the thick filament in the resting (elongated) state; (b) attachment of the myosin head to actin in the adjacent thin filament following stimulation; (c) "melting" of the transducing section (hatched) helices to random coils with consequent contraction of the myosin pair and displacement of the actin thin filament. [SOURCE: Adapted from W. F. Harrington et al. "Rapid Helix-Coil Transitions in the S-2 Region of Myosin," *Proceedings of the National Academy of Sciences,* Vol. 76(1979), p. 1112. See also Chapter 3 of *The Proteins,* Volume Two. New York: Academic Press, Inc., 1979.]

the sarcomere assembly. Concurrent action in the opposite direction on the set of thin filaments that reaches inward from the adjoining membrane (see Figure 12) brings the membranes closer together.

Having performed its function, the head of the myosin macromolecule acquires another molecule of ATP and immediately disengages itself from the actin site. Once the pulse of phosphoric acid from the ATP hydrolysis has been dissipated, the helical form of the transducing element is restored and it returns to the surface of the thick filament, along with the pendant myosin head. The previous process is repeated, with the head attaching itself to a site farther along the translated actin thin filament. Through sequences of these events, the complex machinery of the muscle fiber transduces chemical energy into mechanical work.

Recent studies have revealed the presence of muscle proteins in many other cells. They appear to be responsible for the motility of cells in general, and indeed for motions within cells. Motility is essential for living processes, and the muscle proteins may be largely responsible for this aspect of all forms of life.

As the foregoing description demonstrates, the primary transducing element in muscle turns out to be remarkably simple. It consists of macromolecules arranged in ordered (helical) configurations which, under stimulus, are transformed to random coils. The random coil prefers a shorter length than does the helical segment from which it is formed; hence, it delivers a force that brings about contraction. This force, in fact, is calculable from the same concepts and theory that apply to a stretched rubber. No principle is invoked beyond the most general precepts of macromolecular science.

The contractile element of myosin plays a role corresponding to that of the cylinder in an internal combustion engine. The principles of chemistry and physics underlying the combustion process and its action on the piston are well known, and certainly are not unique to power plants in motor vehicles. But if the valves, spark plugs, distributor, carburetor, cooling system, crankshaft, and the rest of an engine's appurtenances are left out of the account, the internal combustion engine and its operation are quite incomprehensible. So it is, *a fortiori*, with the muscle machine. The intricacies of the macromolecules constituting muscle, their assembly into a highly refined supermolecular organization, and the articulation of their mutually related functions stagger the imagination and dwarf the performance of manmade machines and accomplishments in laboratories.

Yet even a glimmering of the workings of muscle at the level of macromolecules is highly rewarding. It lends encouragement, challenge, and incentive to the quest for deeper understanding of the fascinating processes that are so elegantly performed and orchestrated in living organisms.

## EPILOGUE

Production of polymers and their effective application depend on high technology. A correspondingly high level of science pertaining to macromolecules is therefore imperative. The provision of a background of knowledge that will secure the foundations for modern biology places further demands on this area of science. In all branches of science, the desire to know for the sake of knowing is an ever present driving force. In this area, however, societal needs add a special urgency to advancement of science through research.

Notwithstanding the manifest importance of macromolecules both in the realm of synthetic polymers and in all living organisms, cultivation of the fundamental aspects of knowledge concerning macromolecules and their behavior has fallen gravely into neglect, especially in the United

States. Less than 5 percent of the research efforts in chemistry and physics departments in leading American universities is devoted to this important area. The subject is generally not included in the undergraduate curricula of colleges and universities. The education of graduate students in polymer science is slighted correspondingly. This contrasts with advanced education and research commitments in Japan, the U.S.S.R., and the Federal Republic of Germany, where macromolecular science receives much greater attention.

# Summary and Outlook

Macromolecules consisting of thousands of atoms are the basic ingredients of all living matter. Additionally, they comprise many of the materials, such as fibers, leather, wood, and rubber, used in the artifacts of man since primitive times. Also called polymers, meaning many members, these diverse substances that occur in profusion both in living organisms and in familiar objects and materials share a common motif in their chemical design: the macromolecule typically is a long chain consisting of a hundred or more units, or monomers, joined consecutively by chemical bonds. The capacity of the polymer chain to assume various geometric forms endows these materials with properties not found in substances consisting of smaller molecules. Thus, rubber elasticity is a property peculiar to polymers with randomly configured chains. Alternatively, long macromolecular chains may assume extended, rodlike forms, often helical, which pack together in regular array. Such crystalline arrangements impart high strength and toughness to the material. By perfecting the molecular order, strengths far exceeding that of steel have recently been achieved. Polymeric materials are especially advantageous where low weight is important, as in vehicles for transport.

Through chemical synthesis, the range of polymers available for myriads of applications has been greatly increased in this century. Various synthetic fibers and synthetic rubbers have largely displaced their natural counterparts for many purposes. The use of plastics continues to increase. In fact, the total volume of polymeric materials produced synthetically or processed from natural sources exceeds that of all metals combined. The diversity of properties that can be elicited from polymers through endless variations in their molecular constitution and architecture ensures that the current growth in their use, and in the scope of their applications, is destined to continue.

The versatility of the macromolecular theme is nowhere so well illustrated as in the complex variety of biological polymers. Included are the proteins and carbohydrates (e.g., cellulose) that provide the structural material of animals and plants, respectively, the globular proteins (enzymes) that function as catalysts mediating living processes, and the polynucleotides (DNA and RNA) that serve as repositories for all essential information and as agents for applying that information in biosynthesis and replication. Biological macromolecules are tailored to precise specifications; the sequence of amino acid units in a protein chain is exactly specified, the same in all macromolecules of a given protein. This is a feat that cannot be duplicated by present artificial methods of synthesis of macromolecules. The achievements of nature in this regard present an extraordinary challenge to polymer science, calling for more sophisticated methods of synthesis.

In recent years, we have witnessed breathtaking advances in knowledge concerning intricate biological structures (e.g., the complex array of fibrous proteins in muscle) and biological processes (e.g., protein synthesis). Yet our knowledge in this vast area is largely descriptive. Fundamental understanding in terms of basic principles governing molecular behavior and the properties of molecules is largely lacking. This goal can only be achieved by advancing the frontiers of knowledge concerning the properties and behavior of macromolecules, which is the quintessence of macromolecular science.

On the technological front, major advances can be foreseen that will depend on design and synthesis of macromolecules for ever more demanding applications. We are at the threshold of promising opportunities for use of polymeric materials in the fabrication of more elaborate and efficient microelectronic devices, for example. The unique role of polynucleotides in living organisms demonstrates the potentialities of long-chain macromolecules as media for storage of information. Biomedical applications of polymers require new candidate materials for artificial organs. Delivery of drugs to the target organ or to pathological cells presents a further challenge to macromolecular science and the ingenuity of those engaged in its pursuit. Ever stronger, more durable materials for innumerable applications are needed with increasing urgency as nonrenewable resources dwindle.

## ADDITIONAL READINGS

H. R. Allcock and F. W. Lampe. *Contemporary Polymer Chemistry*. Englewood Cliffs, N.J.: Prentice-Hall, Inc., 1981.

F. W. Billmeyer, Jr. *Textbook of Polymer Science*. New York: John Wiley & Sons, Inc., 1971.

*Macromolecules: An Introduction to Polymer Science*. Edited by F. A. Bovey and F. H. Winslow. New York: Academic Press, Inc., 1979.

P. J. Flory. "Spatial Configuration of Macromolecular Chains," *Science*, Vol. 188(1975), pp. 1268-1276.

L. Mandelkern. *An Introduction to Macromolecules*. Heidelberg Science Library. New York: Springer-Verlag, Inc., 1972.

D. R. Uhlmann and A. G. Kolbeck. "The Microstructure of Polymeric Materials," *Scientific American*, Vol. 233(December 1975), pp. 96-106.

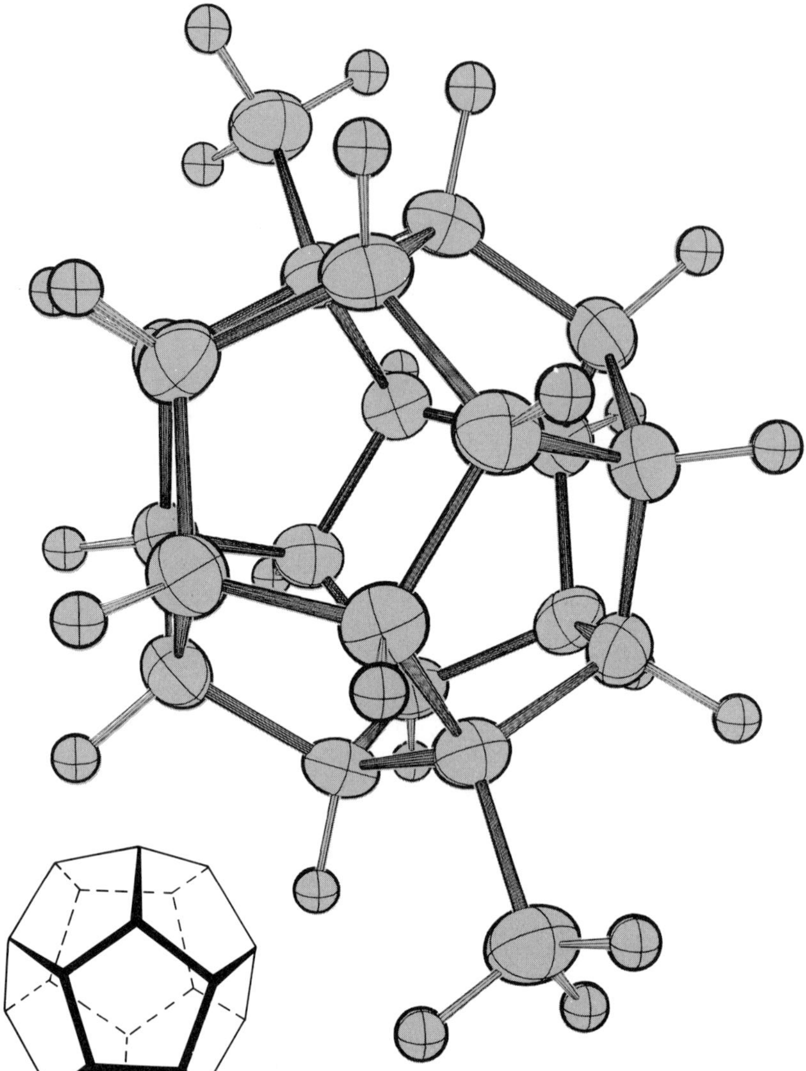

# 11

# *Chemical Synthesis of New Materials*

## INTRODUCTION

Chemical synthesis is the creation of chemical substances, or compounds, from combinations of atoms. In a practical sense, atoms themselves are used only infrequently in synthesis; much more common is the combination, or reaction, of two or more preexisting natural or synthetic compounds to produce another compound.

Chemical synthesis supplies mankind with a wealth of materials needed for our well-being, for our health, and even for our survival. Early in civilization, man learned empirically how to carry out the chemical processes required to convert natural materials into bronze, iron, glass, and ceramics. Today, the science of chemical synthesis supplies us with a myriad of essential products. Many of those reading these lines would not be alive were it not for synthetic and semisynthetic drugs, or without the surgery that depends upon the use of synthetic anesthetics and antiseptics. Without synthetic fertilizers, hundreds of millions throughout the world would starve; our food supply is also augmented by the use of synthetic agents that control plant pests and diseases. Chemical synthesis supplies us with high-octane gasoline and high-performance lubricants; with elastomers (synthetic rubber) and construction materials such as fiberglass and plastic laminates; with fibers for clothing, films for packag-

◂ Computer-generated view of the first molecule to be synthesized that contains a dodecahedral framework of atoms—the first hydrocarbon analog of Plato's highly symmetrical perfect solid, the dodecahedron (inset). Carbon atoms form the 20 vertices for the 12 faces, each of which is a regular pentagon. [Leo A. Paquette, Gary G. Christoph, and Bob Hummels, Ohio State University.]

ing, alloys for high-performance magnets, and inorganic solids of precisely controlled composition for transistors and other electronic devices; with dyes, detergents, photographic materials, and a host of other essential products.

Our future, too, depends on chemical synthesis. Bacteria and insects undergo mutations and acquire resistance to the drugs and insecticides now in use; chemical science is in a race with nature to maintain or improve our precarious and only partial mastery of bacterial and insect-borne diseases. Synthesis is providing safer and more selective chemicals to increase food supply: weed-control agents that translocate from foliage to root to kill the weed; systemic fungicides that control fungus on food crops and then decompose in a short time to harmless products; and insect sex attractants (pheromones) that lure only one species of insect to a trap. Viral diseases and cancer present challenges to both chemistry and biology. America is headed toward the construction of a huge industry for the synthesis of liquid fuels, an industry that will be based on catalysts and catalytic reactions discovered in synthesis research. Not of least importance, research on chemical synthesis develops facts that lead to continual refinement of theories of chemical bonding and reactivity; these theories, in turn, suggest new areas for chemical exploration.

The challenges that confront chemical synthesis are to learn how to create more useful substances with minimum energy expenditure, to continue to teach users of chemicals how to apply these substances effectively and safely, and to handle and dispose of the wastes and byproducts of chemical synthesis in safe and ecologically acceptable ways.

While chemical synthesis is the heart of chemistry, it is critically dependent on the other branches of chemistry. Theoretical chemistry suggests objectives for synthesis, physical chemistry suggests chemical routes to achieve objectives and ways to improve syntheses, and analytical chemistry provides essential information about the identity and structure of the products of synthesis.

Chemistry interacts strongly with other sciences. Just as some of the major advances in physics have depended on new materials from chemistry, so chemistry profits from new methods of analysis and of structural determination that have their origins in basic physics. Discoveries in the biological sciences stimulate the synthetic chemist to attempt to duplicate or simulate biological compounds and processes. Mathematics provides tools for unraveling the intricacies of chemical structures and reactions, and computer science is beginning to play a role in the design of synthesis procedures.

This chapter summarizes some of the newer areas of exploratory chemical synthesis, as well as some potential implications of this science.*

---

*Some chemical terms and conventions for drawing structural formulas are explained in "Chemical Terms and Conventions" at the end of this chapter.

## NEW CATALYSTS AND REAGENTS FOR CHEMICAL SYNTHESIS

The success of chemical synthesis depends on the ability of chemists to devise and use catalysts and reagents that will effect reactions selectively, both as to the specific atom attacked and as to the stereochemical outcome of the process.* Although chemists have accumulated a vast array of catalysts and reagents, and can use them to synthesize a wide variety of molecules, these syntheses are clumsy and imprecise compared with those that are brought about in nature by enzymes; the contrast between the efforts of chemists and the ease and precision of biosyntheses provides a measure of the present challenge. Although most new catalysts and reagents provide only small advances, the sum of many such advances has altered, and will continue to alter, synthetic chemistry. A few new catalysts and reagents are discussed below; together with many others, they contribute significantly to our ability to prepare the host of synthetic products that distinguish modern technology.

### Catalysts

Catalysts promote chemical syntheses by increasing the rate at which specific chemicals react. If the chemicals can undergo more than one reaction (and this is usually the case), a catalyst generally accelerates only one of these reactions, and thereby promotes it at the expense of the others. A catalyst usually performs these roles without being incorporated into the reaction products or being permanently altered. Thus, a small amount of catalyst can repeat its function many times to transform large quantities of reactants into desired products.

Chemists recognize two kinds of catalysts in addition to the natural enzymes. Heterogeneous catalysts are finely powdered or highly porous solids, often metals or metal oxides, that present a large surface area on which the reaction takes place. Homogeneous catalysts include acids, bases, and metal compounds that are soluble in the reaction medium and that derive their activity from the characteristics of a particular metal atom that is bonded to one or more inorganic or organic ions or molecules—such a combination is called a metal complex.

Heterogeneous catalysis was recognized almost 150 years ago by the Swedish chemist J. J. Berzelius, who coined the term "catalysis." One of the most widely used reactions brought about by a heterogeneous catalyst is the addition of hydrogen to carbon-carbon double bonds, which can be promoted by finely powdered metals such as nickel. This type of reaction is used today in, for example, the hydrogenation of corn oil to margarine.

---

*Stereochemical refers to the spatial arrangement of atoms in a molecule; see "Chemical Terms and Conventions."

Heterogeneous catalysts are used in such large-scale processes as the cracking and reforming of petroleum, the synthesis of methanol from carbon monoxide and hydrogen (this combination, called syngas, is available from coal and steam), and the synthesis of ammonia from nitrogen and hydrogen.

Modern physical techniques are making it possible to study the nature of catalyst surfaces and, to some degree, to determine the details of the reactions that occur on them. This knowledge is helping chemists to synthesize catalysts of ever increasing specificity for catalyzing selectively only the desired reaction. For example, the Fischer-Tropsch process that was developed in Germany during World War II to make synthetic fuel uses a heterogeneous catalyst to convert syngas to a complicated mixture of aromatic and aliphatic hydrocarbons, alcohols, ketones, and other compounds. Recently, chemists have synthesized complex silicate catalysts, related to the natural minerals known as zeolites, in physical forms that contain micropore channels of carefully controlled and uniform size, a few thousand nanometers in diameter (a nanometer is $10^{-9}$ meter). With these catalysts, methanol (from syngas) can be converted selectively to either aromatic compounds or olefins, depending on the particular catalyst used. This advance in selectivity opens a route to convert coal into hydrocarbons that can be used in high-performance fuels or in feedstocks for industrial chemical synthesis.

The development of homogeneous catalysis has been paced by the advances in knowledge of the characteristics of heavy metal atoms and of their capabilities to bond with ions and molecules to form relatively stable, soluble complexes. A soluble rhodium complex, formed in methanol from a rhodium salt, an iodide, and carbon monoxide, catalyzes the combination of carbon monoxide with methanol under modest pressure with 99 percent selectivity to form acetic acid, widely used as a solvent and chemical intermediate. Methyl acetate, derived from methanol and acetic acid, can be combined with carbon monoxide in the presence of a catalyst to give acetic anhydride, a widely used industrial chemical. The same reactants with a different catalyst give an intermediate compound that can be thermally cracked to produce vinyl acetate, an important raw material for the synthesis of polymers. Ethylene glycol, a polymer intermediate and antifreeze, has been made directly from carbon monoxide and hydrogen by using another soluble rhodium complex catalyst at high pressure. The number of basic chemicals that can be made from syngas, and hence from coal, by using selective catalysts will continue to grow.

All of the 20 natural $\alpha$-amino acids that are the building blocks of proteins have the general formula

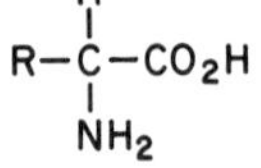

where R is the symbol for an unspecified group. They are called $\alpha$-amino acids because the amino group ($NH_2$) is attached to the $\alpha$, or first, carbon from the acid group ($CO_2H$). In all of these except glycine, in which R is simply a hydrogen atom, the $\alpha$-carbon is a chiral center, and the amino acid can exist in two stereoisomeric forms.* However, those amino acids that make up proteins, which are synthesized in nature by enzyme-mediated chemistry, have only one of the two possible forms. On the other hand, laboratory synthesis of an $\alpha$-amino acid produces a mixture of both forms in equal amounts. Because these two isomers have the same physical and chemical, but not biological, properties, they cannot be separated except by chemical manipulation involving another compound with a chiral center that is in only one of its two possible forms. Moreover, if only one form of the amino acid is desired, half of the laboratory preparation must be discarded.

It has recently been found that a catalyst prepared from rhodium and a chiral phosphorus compound promotes the selective addition of hydrogen to a double bond to form only one or the other form of an $\alpha$-amino acid, a result comparable with that achieved by an enzyme in a living system.[1]

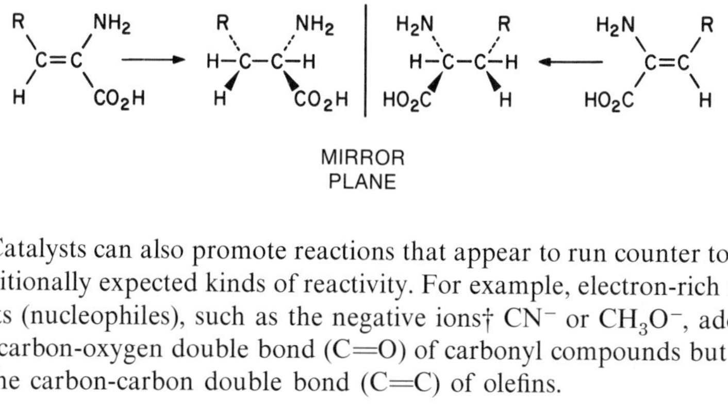


Catalysts can also promote reactions that appear to run counter to the traditionally expected kinds of reactivity. For example, electron-rich reagents (nucleophiles), such as the negative ions† $CN^-$ or $CH_3O^-$, add to the carbon-oxygen double bond (C=O) of carbonyl compounds but not to the carbon-carbon double bond (C=C) of olefins.

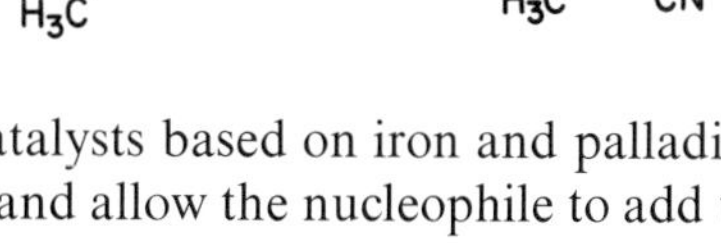

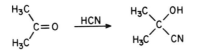


However, soluble catalysts based on iron and palladium coordinate with the C=C of olefins and allow the nucleophile to add to this double bond,

* Most molecules are three-dimensional rather than planar as represented on paper. "Chiral" and "stereoisomeric" refer to spatial orientations of atoms in a molecule; see "Chemical Terms and Conventions."

† These ions are electron-rich and usually react with electron-deficient centers, or nuclei, and thus are called nucleophiles. Conversely, the electron-deficient centers are called electrophiles.

making it possible to synthesize, directly from readily available olefins, some compounds that formerly had to be made in more than one step from more costly materials.

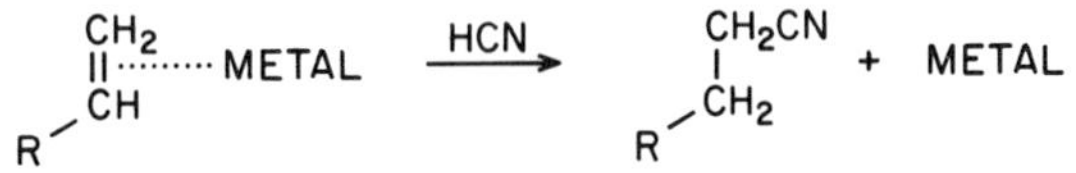


A similar effect has been achieved in an important class of carbon-carbon bond-forming reactions called displacement reactions. The group displaced by the reaction of a nucleophile can now be controlled with a soluble palladium catalyst.[2] The broken arrows below point to the new carbon-carbon bond. The ability to direct the courses of these nucleophilic reactions is a significant advance in chemical synthesis.

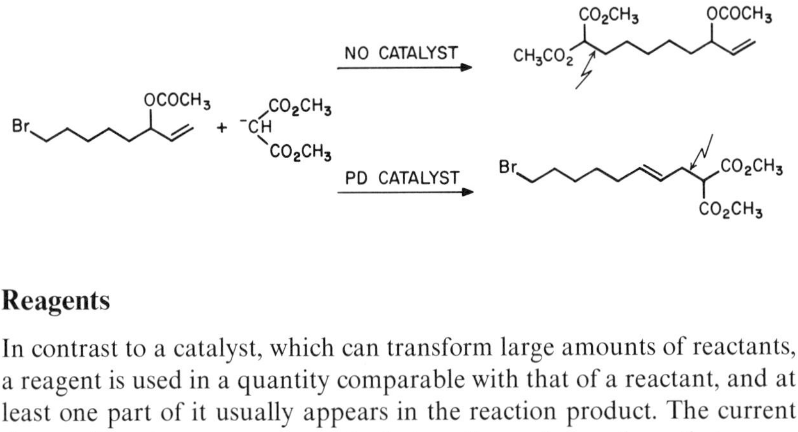


## Reagents

In contrast to a catalyst, which can transform large amounts of reactants, a reagent is used in a quantity comparable with that of a reactant, and at least one part of it usually appears in the reaction product. The current thrust of work on new reagents derives from observations that many organo-inorganic compounds can insert their organic groups in unconventional ways, leading to new synthetic routes. (Organo-inorganic refers to an organic, or carbon-containing, compound in which a carbon atom is bonded to an atom of an inorganic element, such as boron [B], magnesium [Mg], copper [Cu], lithium [Li], zirconium [Zr], etc.)

One of the most versatile synthetic reactions for forming a new C-C bond is the addition of the organomagnesium bromide reagent, RMgBr (the Grignard reagent, named for its discoverer, the French chemist Victor Grignard), to a ketone as shown in path a.

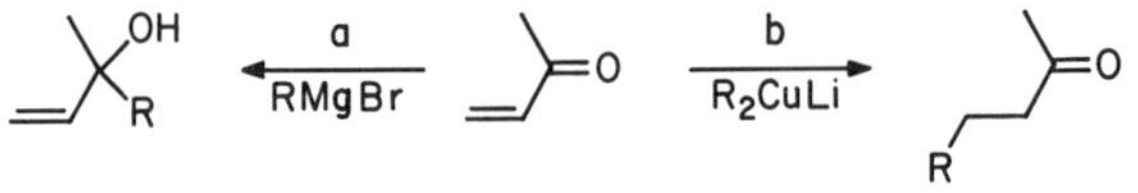

However, studies of the occasional erratic behavior of thc Grignard reagent revealed that trace metal contaminants in the magnesium used to prepare the reagent profoundly altered its behavior. The finding that copper was one such contaminant led to investigation of organocopper reagents, which were found to add to the $\alpha,\beta$-unsaturated ketone in a totally different manner (path b).* This discovery has provided a route to synthesis of the prostaglandin hormones and their analogs. The observation of this dramatic difference between organomagnesium and organocopper reagents stimulated a search for useful reactions of other organometallic reagents. An example is the discovery that organozirconium complexes add to terminal carbon-carbon double bonds to give intermediates that can be hydrolyzed to primary alcohols—just the opposite of the traditional acid-catalyzed hydration.[3] (A primary alcohol has the OH group attached to a $CH_2$; the compound on the left, with the OH attached to CH, is a secondary alcohol.)

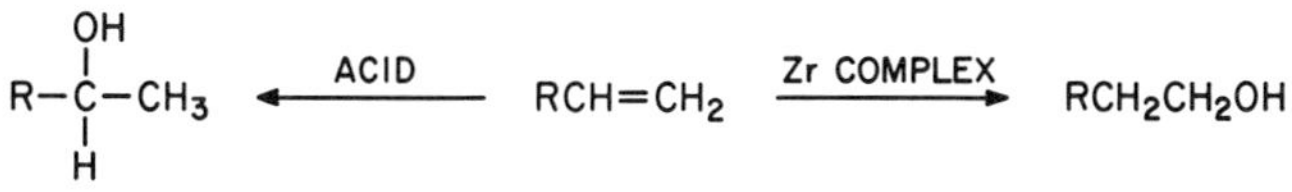


Reagents based on sulfur, phosphorus, and selenium are enabling the chemist to perform unusual and selective reactions that have been the sole province of living systems. For example, the carbon of a carbonyl group normally reacts only with a nucleophile. In nature, vitamin $B_1$, a carbon-sulfur compound, can invert this reactivity and allow the carbonyl carbon to react with an electron-deficient group (an electrophile). A related inversion has been achieved in the laboratory using sulfur-containing reagents. (When different unspecified groups are to be indicated, they are designed R, R′, and R″.)

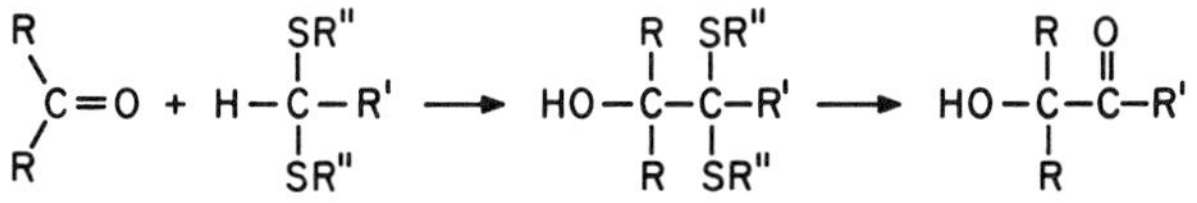


One of the most versatile reactions for forming carbon-carbon bonds is the aldol condensation. For example, two carbonyl compounds can be condensed in the presence of a basic catalyst:

---

*Double and triple bonds often are called "unsaturated." "$\alpha,\beta$-unsaturated" refers to the double bond (unsaturation) between the first ($\alpha$) and second ($\beta$) carbon atoms from the carbonyl (C=O) group.

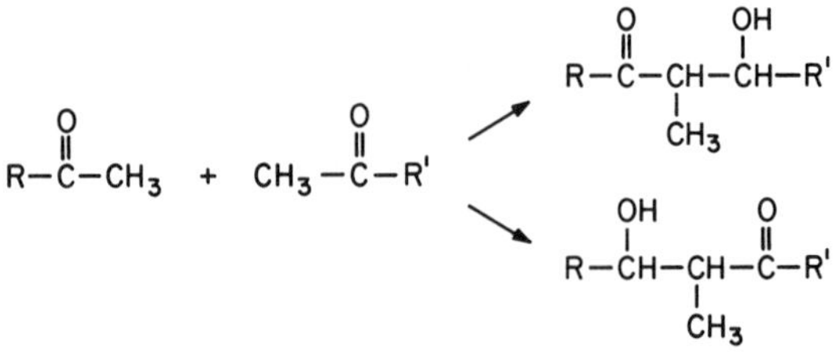

The utility of this reaction in synthesis is limited because, when it is carried out with two different carbonyl compounds, a mixture of the two isomers shown above is usually formed. The limitation has been overcome by use of a boron derivative of one of the carbonyl compounds, which controls the course of the reaction so that only one product is formed.

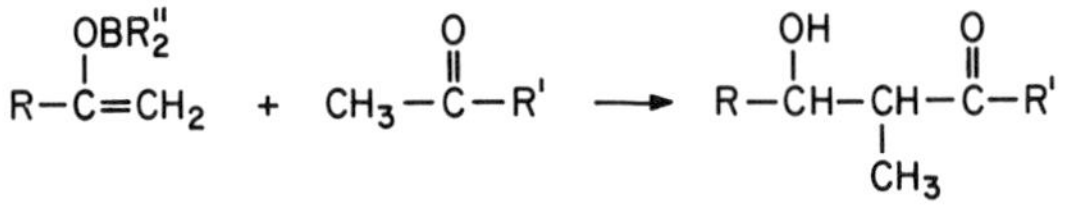

In some aldol condensations, two chiral centers are formed, implying that four stereoisomers can exist, although only two of them need be considered. Here again, use of a boron derivative of the ketone in the condensation leads to the formation of predominantly one or the other of these isomers, depending on the nature of R and R′.[4]

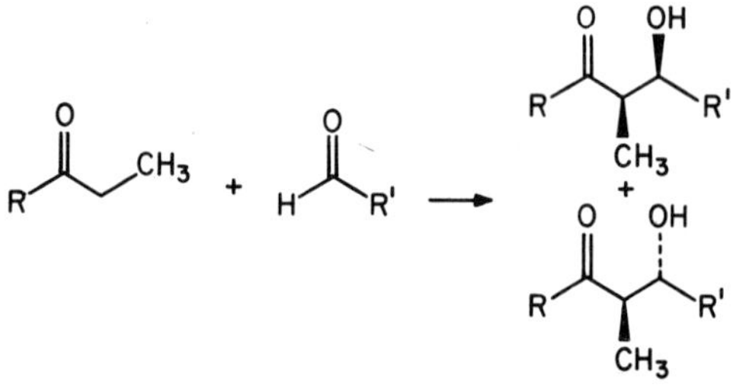

Nature uses the aldol condensation in many ways to synthesize parts of large complex molecules such as carbohydrates, steroids, and fatty acids. In nature, the course of the reaction is controlled by enzymes so that only a single product is formed. The ability of chemists to exercise similar control of this versatile reaction in the laboratory has important implications for the synthesis of natural products and their analogs.

Development of more selective catalysts and reagents will lead to cleaner processes with fewer byproducts and will increase the number of commodity chemicals that can be made from the most available raw materials. The recent advent of sophisticated techniques for studying the chemistry of surfaces will provide a new rationale for devising solid catalysts. The surge of progress in organo-inorganic chemistry will be a basis for the development of novel soluble catalysts and new selective reagents.

## NATURAL PRODUCT CHEMISTRY[5]

A significant force in the development of chemistry is the synthesis of the myriad types of compounds that nature has learned over millennia to create for her own purposes. In some cases, the total synthesis of a natural compound (e.g., camphor), or its partial synthesis from an abundant natural material (e.g., aldosterone, a mammalian hormone, from plant steroids), has made available in practical quantities a compound that would be more expensive, or even impractical, to obtain from natural sources. Moreover, surmounting the challenge of synthesizing complex natural compounds in the laboratory has led to the development of many new and ingenious synthetic procedures that have become broadly useful.

The structure of vitamin $B_{12}$ [1] was shown 25 years ago to be a cobalt complex with the formula $C_{63}H_{88}CoN_{14}O_{14}P$ and with 15 chiral carbon centers (equivalent to $2^{15}$ or 32,768 possible stereoisomers, of which only one occurs naturally). The synthesis of a molecule of this complexity,

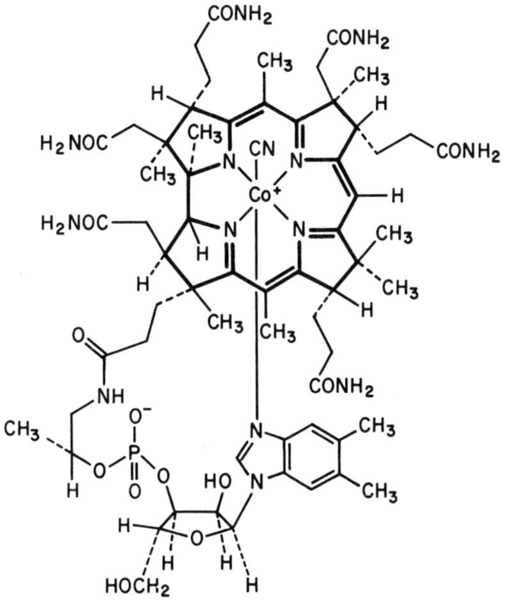


[1]

which involves many stereochemical considerations, difficulties in controlling desired reactions so that they occur at only one of several possible reactive sites as the molecule is assembled, and determining a workable sequence of the many synthetic steps, would have seemed impossible only a few years ago. Yet through the collaborative research of the late R. B. Woodward (Harvard University) and A. Eschenmoser (Federal Institute of Technology in Zürich), this molecule has been totally assembled in the correct stereochemical configuration from simple precursors. The accomplishment stands as a landmark in the field of organic synthesis.

It appears that chemical reaction technology has become sufficiently advanced that almost any synthesis of a stable substance is possible, given sufficient time and manpower. However, in selecting synthesis targets an important question must be the ultimate practicality of the venture. For example, in recent years intensive effort has been directed toward finding substances that would be effective therapeutic agents against various forms of cancer. One of the more promising naturally occurring substances yet tested is maytansine [2]. This compound is found

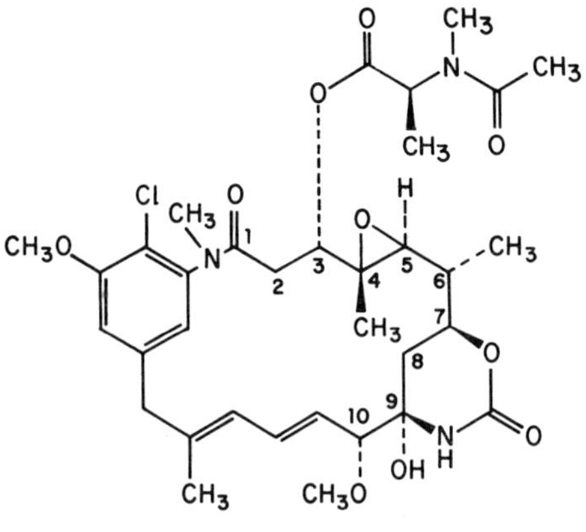


[2]

in an East African plant, *Maytenus serrata*, but only 6 grams of the compound have so far been isolated from 10 tons of dry plant wood and bark. For such a rare but potentially valuable chemical, the economic advantage of undertaking the synthesis is in favor of the chemist. However, the synthesis poses a number of challenging questions. How does one construct the basic large ring system? How does one control the stereochemical relationships at the seven chiral centers at carbon atoms 3, 4, 5, 6, 7, 9, and 10 in the molecule? The efforts of two research teams headed by A. I. Myers (Colorado State University) and E. J. Corey (Harvard University) have recently culminated in two independently conceived and executed syntheses of this antitumor agent.

The hormones called prostaglandins constitute a class of potentially medically important substances found in mammalian tissue and in some marine plants. Minute quantities of these compounds have exceedingly high activity in regulating a number of physiological functions, including aspects of reproduction, blood pressure, inflammation, and blood platelet aggregation. The prostaglandins are derived biogenetically from arachidonic acid [3]; from this and other building blocks, chemists have made a

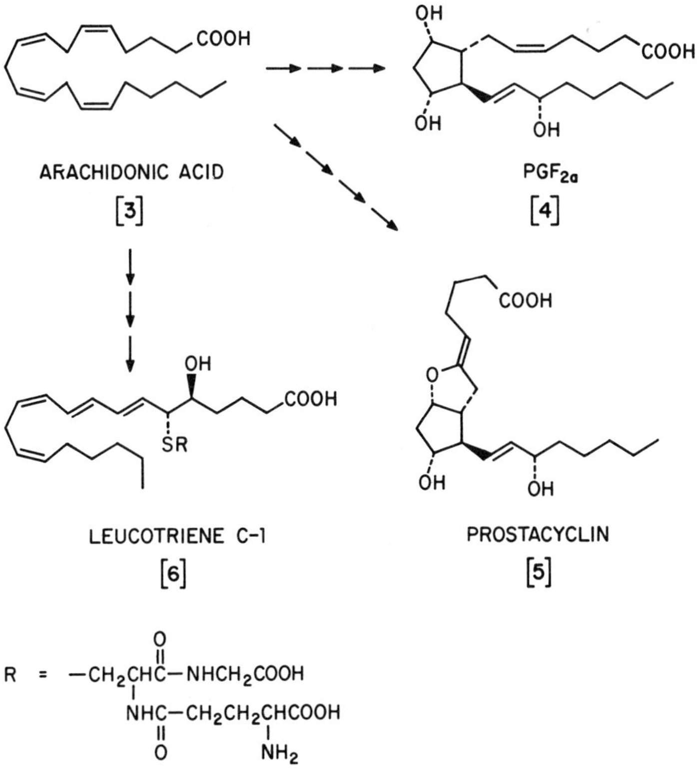


host of natural prostaglandins as well as many analogs of the natural structures. Prostaglandin $F_{2\alpha}$ [4] and prostacyclin [5] (see Chapter 2, "On Some Major Human Diseases") illustrate the basic structural features of these hormones. A recent addition to the family is leucotriene C-1 [6], or SRS (slow reacting substance), which affects allergenic responses. In fact, the recent synthesis of leucotriene C-1 provided the rigorous proof of its structure. The amount of natural leucotriene C-1 available was so minute that its structure could not be determined by conventional techniques. A set of plausible structures was deduced from spectroscopic data, and each of these was synthesized by an unambiguous route. Comparison of the synthetic isomers with the natural compound permitted the correct isomeric structure to be identified. Because the hormone SRS is involved in allergenic responses, a search is under way for compounds that will inhibit its biosynthesis and thus possibly alleviate certain forms of asthma.

The new methodology for synthesis that has evolved from research on the prostaglandins is impressive. New techniques have been developed for stereospecific synthesis of unsaturated compounds (in which the stereochemistry at a chiral center is controlled to produce only one stereoisomer), for the construction of five-membered rings of carbon atoms, and for the protection of certain reactive sites in a molecule while performing reactions at others.

The last few years have seen tremendous advances in stereospecific synthesis and in chemical manipulation of compounds with several reactive sites. These advances, paced by the advent of new selective catalysts and reagents, and the use of computer technology to select promising reaction pathways, will enable the chemist to achieve syntheses of complex natural molecules with a degree of sophistication that could hardly have been foreseen 15 years ago.

## SYNTHETIC APPROACHES TO ENZYME ANALOGS[6]

Enzymes are natural catalysts, the substances in living organisms that perform the chemical transformations of life. They are proteins, ranging in size from a few hundred to several thousand $\alpha$-amino acid units. An enzyme catalyzes a biochemical reaction by speeding up only one of the several reactions that are usually possible, thus producing only a single product because the other possible reactions are hopelessly slow in comparison. The reactions of life processes may proceed as much as $10^{10}$ times faster under the catalytic influence of an enzyme than they would without it. Thus, processes that in a living organism take one second would take more than 300 years without the catalyst.

Natural enzymes are used in industrial manufacturing. A familiar example is the use of the enzymes in yeast to convert carbohydrates to ethanol. There are drawbacks: compared with other routes to ethanol, the enzymic process is slow, requires large reaction vessels, and can form only a dilute aqueous solution of ethanol that must then be concentrated. Nevertheless, the fermentation process remains an important commercial route to ethanol. In a very different type of synthesis, advantage is taken of enzyme specificity in the use of another enzyme system to introduce oxygen to a particular position of a complicated steroidal structure in the synthesis of cortisone, an antiinflammatory hormone. Here the enzyme provides a single desired product, in contrast to the mixtures produced by standard oxidation procedures.

Enzymes are selective in three ways. First, there is selectivity as to the type of chemical reaction performed. Second, enzymes are selective with respect to the site of the molecule on which the reaction is performed. Enzymes bind the reacting materials in a cavity, where they are held together in a specific spatial relationship of one to the other. Thus, reaction occurs only at one site in a molecule, even when other possible reaction sites are present. Third, because enzymes are themselves chiral substances, they are generally highly stereoselective catalysts. That is, enzymes not only direct the reaction to occur at one particular atom of a substrate, they even direct it to one side or the other of the atom. (A substrate is a compound whose reaction is mediated by another compound, often an enzyme.) The result is to produce only one of the two possible stereoisomers, whereas most standard chemical procedures produce both. For many biologically important compounds, particularly those of natural origin, it is generally true that only one of the two stereoisomers is biologically active, while the other is either inactive or may even have adverse biological properties.

There are powerful incentives for synthesizing catalysts that would have the superb selectivity of natural enzymes without some of their limitations. Natural enzymes function only in dilute aqueous solution and in a narrow range of temperature and acidity. They catalyze only the reactions that are of interest to living systems, which do not encompass many of the reactions that the chemist may wish to accomplish.

Most approaches to synthetic enzymelike substances are aimed at imitating two characteristics of the natural catalysts. First, enzyme molecules have a spatial cavity that accommodates the reactive portions of the substrates. Second, within that cavity are chemical groups that hold the reactive portions of the substrates in close juxtaposition. These characteristics, which have been revealed in detail only through recent developments in sophisticated instrumentation, have led to study of synthetic molecules that have spatial cavities like those in natural enzymes. Doughnut-shaped molecules such as the cyclodextrins (compounds with six, seven, or eight sugar units combined into a ring) and related compounds (with benzene rings instead of the sugar units) have been shown to bind other molecules into the cavity of the doughnut and to have the enzymelike property of directing a reaction to occur at only one site of a molecule. Thus, chlorination of anisole [7] that is bound to a cyclodextrin produces only *p*-chloroanisole [8], whereas chlorination of unbound anisole produces a mixture of two isomers.

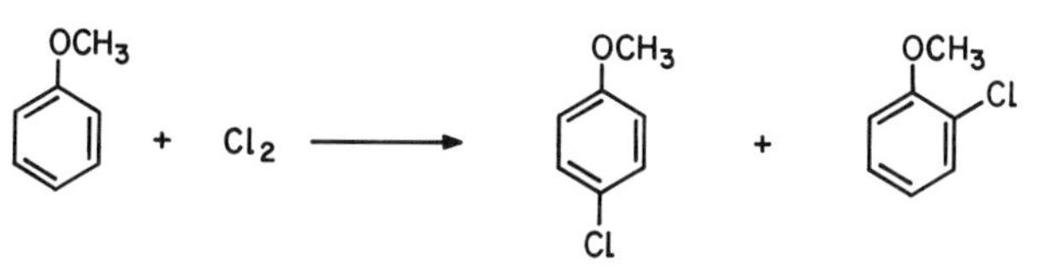


[7] [8]

The second step in simulating a natural enzyme is to attach reactive groups to the cavity molecule. For example, a cyclodextrin molecule, shown schematically below, that has two imidazole rings attached to it is

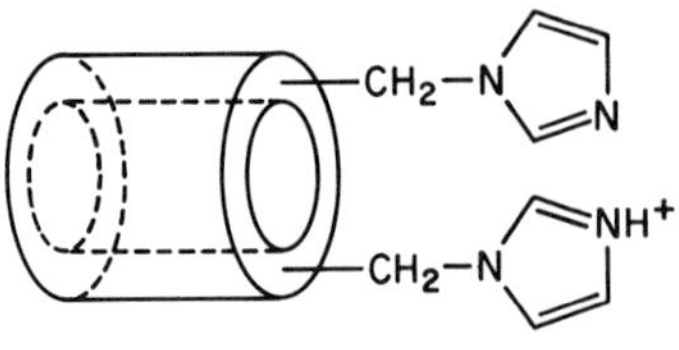


related to the natural enzyme ribonuclease, which hydrolyzes ribonucleic acid (RNA). This synthetic material has the chemical selectivity of a natural enzyme. In the reaction shown below, usual hydrolysis procedures give [9] and [10] in a ratio of 60:40, whereas when the hydrolysis is catalyzed by the synthetic enzyme only product [10] is obtained. Recently, a synthetic enzyme analog has been devised to perform the selective introduction of an oxygen atom in the synthesis of cortisone.

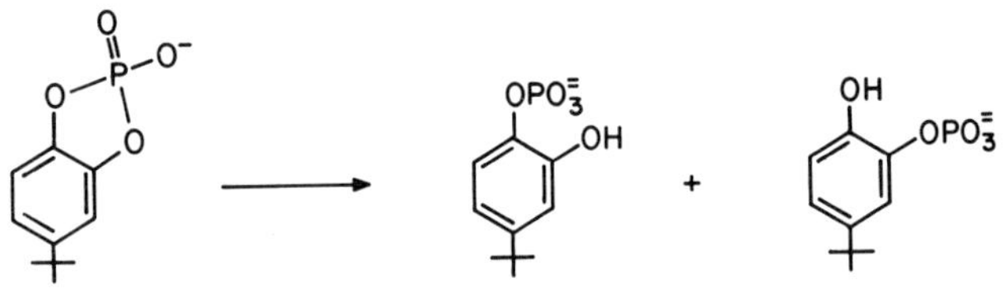


[9] [10]

The goal of producing synthetic catalysts with the great selectivity of natural enzymes is an important one, and only the first steps have been taken toward achieving it. Progress will be made as chemists develop more intimate knowledge of enzyme structures and harness the techniques that are emerging from work on synthesis of natural products.

## SYNTHESIS OF NEW PHARMACEUTICALS

Historically, pharmaceuticals have come primarily from three sources. Many have been obtained directly as natural products from plant and animal tissues—for example, quinine from the bark of the cinchona tree, morphine from the opium poppy, and heparin (a blood anticoagulant) from swine liver. Second, many natural pharmaceuticals have been found among the incredibly diverse structures that are formed as metabolic products by microorganisms; examples include antibiotics such as penicillin, erythromycin, and the cephalosporins. The third principal source has been chemical synthesis, which has led to the development of many drugs, such as the sulfanilamide antibiotics, hydralazine (an antihypertensive), and indomethacin (an antiarthritic).

Chemical synthesis also plays an important role in creating useful drugs from natural products. Many natural product drugs have limited utility because they may be too quickly deactivated by chemical modification in the body, or may be effective by only one route of administration (e.g., oral or injection), or may have too narrow a range between an effective dose and a dose that produces undesirable side effects. Such deficiencies can often be overcome by synthesizing an analog of the natural compound: morphine is an effective analgesic by injection, but is much less active orally, whereas the semisynthetic analog oxycodone is more active by the oral route—an important distinction when continual pain is being treated. Natural antibiotics sometimes induce tolerance in some strains of their target microorganisms or induce sensitivity in some patients; here, chemical modification of the natural product to form a semisynthetic drug often overcomes these problems.

Natural products are far from exhausted as sources of new pharmaceuticals. Microorganisms, including over 50,000 defined species that occur widely in soils, offer a virtually unlimited reservoir of fermentation products for study. Most of the pharmaceuticals that have emerged from this source are antibiotics, and intense activity on making semisynthetic modifications of these structures, such as the penicillins, cephalosporins, and cephamycins, will continue. In addition, products of microorganism fermentation are now beginning to be studied for other kinds of pharmacological activity.

Elucidation of the structures of pharmacologically active natural products will continue to be an important objective. This information can be used by biochemists and pharmacologists to define biological mechanisms and to probe the causes of certain diseases such as arthritis and schizophrenia.

## Enzyme Inhibitors

Advances in the knowledge of biological mechanisms are providing not only insight into the ways in which some drugs work but also a rational basis for synthesizing drugs that are targeted for a specific biological mechanism that may be involved in a disease.

A number of drugs have been shown to act by inhibiting specific enzymes. One example is allopurinol, which is used in treating gout and which inhibits the action of xanthine oxidase, an enzyme that is involved in the production of uric acid in the body. Another is indomethacin, which interferes with an enzyme responsible for the body's synthesis of the prostaglandin hormones. The synthetic antibacterial drug trimethoprim and the natural antibiotics cycloserine, penicillin, and fosfomycin block the action of enzymes that are essential to bacterial growth. Most of these products were developed from empirical observations of biological activity, their mechanisms of action being worked out later. In recent years, there have been increasing efforts to design drugs that will inhibit the functioning of specific enzymes. An example of this approach is the new antihypertensive compound captopril. Its synthesis was based on knowledge of the role of the polypeptide* hormone angiotensin II in the regulation of blood pressure and on detailed information about the enzymes involved in the biosynthesis of angiotensin II.

The trend toward designing drugs that can modulate biosynthetic or metabolic pathways is sustained by the growing base of knowledge about the structure and functioning of enzymes. In some cases, the enzyme's own mechanism is used to inhibit its action; drugs that do this are called suicide inactivators. In other cases, the inhibitor is a stable analog of an intermediate in the normal enzymatic reaction that, because it cannot go on to a final product, remains bound to the enzyme and thus blocks the enzyme's normal function.

Although enzyme inhibitors are only one approach to new drug development, the precision with which potent, selective inhibitors can be designed promises increasing use of them. They will be used as new drug candidates wherever the clinical significance of the target has been defined. Where it has not, they will be used in animals and in the clinic to help unravel biosynthetic pathways in disease states, thereby expanding the basic understanding that is required for medical breakthroughs.

## Receptors

The concept of receptors was originated by Paul Ehrlich in the early 1900's, when he postulated that drugs exert their pharmacological activity by interacting with specific chemical groups—receptors—in body organs. This postulate of specific binding sites was later extended to account for the action of hormones and of neurotransmitters (natural chemicals that transmit electrical impulses between nerve segments).

In the last two decades, the concept of physiological receptors for hormones and neurotransmitters has evolved from an empirical definition into the field of molecular pharmacology. The existence of specific receptors implies that these sites can recognize and bind other compounds, called ligands, that share in one way or another specific structural features of the receptors: molecular configuration, volume, electrical charge—variables that are subject to rational manipulation and are thus amenable to drug design.

---

*The term "polypeptide" describes a polymer composed of units of $\alpha$-amino acids. Proteins are high molecular weight polypeptides.

Prior to the late 1960's, information about receptor sites could only be inferred from measurements of biological response, a limitation that handicapped attempts to answer basic questions about receptor geometry and mechanisms of information transmittal. The situation changed dramatically with the introduction of radioligands (ligands tagged with radioactive isotopes) as probes to identify and study receptor sites. This technique has contributed much to the understanding of events at receptor sites and has extended the scope of the field beyond hormone interactions into the area of binding sites for neuromodulators (natural chemicals that control nerve impulses) and for drugs themselves.

The possibilities suggested by the concept of receptors can now be refined by radioligand techniques to guide drug design in much the same manner that measurements of enzyme inhibition by drugs have been used in the past. The effects of variations in the structure of drug candidates on their activity and specificity can be determined without complications from the variables of absorption, transport, and metabolism. With the establishment of a data base founded on relative ligand affinities, structural hypotheses of receptor geometry can be developed and tested.

Examples of successful application of this strategy include propranolol, an antihypertensive drug that blocks a receptor site for compounds that tend to increase cardiac output, and cimetidine, an antiulcer drug that blocks a histamine receptor site and thus inhibits the secretion of gastric acid.

Discoveries in the area of biologically active peptides have expanded the opportunities for creating new receptor-specific pharmaceuticals. Peptides, like proteins, are built primarily from the 20 nutritionally essential $\alpha$-amino acids; however, they are smaller than proteins, containing from a few to some tens of amino acid units, in contrast to the hundreds in proteins. A few peptide pharmaceuticals have been used for years—insulin, oxytocin (to stimulate uterine contractions in childbirth), and vasopressin (an antidiuretic).

Biochemists have long wondered why morphine, a plant product, is capable of producing analgesia in mammals. They reasoned that there must be receptor sites in the brain to which morphine can bind, but wondered why such sites would ever have evolved, unless to bind compounds that occur naturally in the brain, which morphine certainly does not. The answer was found recently in some peptides, called enkephalins, that were isolated in very small amounts from mammalian brains. These peptides produce deep analgesia when injected into the brains of rats. Furthermore, even though their chemical compositions are quite different from that of morphine, both they and morphine can assume very similar spatial conformations. It appears that mammals have evolved these internal analgesics that can on occasion be called upon to alleviate severe pain, and that morphine happens to be able to fit into the same receptor sites as these peptides. The discovery of the enkephalins has opened a new avenue in the search for pharmaceuticals modeled on biologically active peptides, as well as offering the possibility of gaining new insight into opiate drug addiction.

The ultimate development of new peptidelike pharmaceuticals will undoubtedly depend on the parallel development of recombinant DNA technology. Biologically active peptides are present in only very small amounts in natural sources, and they are difficult to synthesize by current laboratory techniques. However, recombinant DNA technology is tailor-made for the assembly of $\alpha$-amino acid units into peptides with a desired, predetermined sequence (see Chapter 17, "Prospects for New Technologies), and it will surely become increasingly important as useful biologically active peptides are identified and needed for pharmaceutical purposes. Insulin prepared by this technology is now nearing commercialization.

## Molecular Modeling and Computer Aids in Drug Design

Medicinal chemistry seeks understanding of the relationship between molecular structure and biological response. Molecules are three-dimensional, and the geometric arrangement of functional groups affects their chemical properties as well as their ability to interact with receptors. These are the interactions that the medicinal chemist is exploring as he synthesizes analogs to optimize a biological effect; thus, his thinking must have a three-dimensional orientation. Molecular models have long been an important aid to three-dimensional visualization, but now the computer is emerging as a more powerful tool.

Interactive computer graphics has led to the development of systems to aid in modeling the three-dimensional structure of potential new drugs and of drug-enzyme and receptor-agonist interactions. (An agonist is a drug that binds to a receptor and elicits a response; an antagonist is a drug that binds to a receptor but does not elicit a response.) When these systems have evolved further, they will play a significant role in the development of new therapies.

As the medicinal chemist and biologist design drugs of greater specificity, the complexity of the target molecules will increase, and greater geometric control and chemical sophistication will have to be exercised in constructing new drugs. The development of new catalysts and reagents will be invaluable in this task, as will the ability to control economically the stereospecificity at chiral sites. Syntheses performed by man or microorganism rarely produce only one product, hence the need for more selective and energy-efficient separation methods. Selective adsorbents, extractants, and semipermeable membranes will continue to be developed for this purpose.

## UNUSUAL ORGANIC STRUCTURES

There are many unusual organic molecules for which structures can be drawn but whose properties cannot be predicted with certainty from current theories of chemical bonding. In many cases, it is not even possible to be certain whether the molecule would be stable under any conditions. Attempts to synthesize such molecules are important in order to learn what structures are possible and thus to refine theories of chemical bonding. In the past, the choice of molecules to synthesize was dictated largely by simple qualitative theories. However, as more experimental data have become available, quantitative theories of bonding are evolving to provide better understanding and better predictions of the relationships among structure, reactivity, and properties. An outstanding contribution to the theory of chemical bonding has been the enunciation of a principle known as the Woodward-Hoffmann rules for conservation of orbital symmetry, which was an outgrowth of the solutions of the many problems encountered in the synthesis of vitamin $B_{12}$. This principle has led to a basic understanding of why certain reactions that involve the formation of, or the opening of, rings of carbon atoms proceed readily while others do not, and has also led to the successful prediction of new reactions. The development of this theory has engendered a very productive collaboration between chemical theorists and experimentalists.

It is implicit in the phrase "unusual organic structures" that these preparations present difficult challenges in either methodology or conception. Furthermore, to obtain sufficient quantities of a compound for study of its physical and chemical properties, the synthesis scheme should be efficient and reasonably short. Accordingly, syntheses of unusual organic structures often lead to the development of new and widely useful synthetic methods.

The carbon atom in most organic compounds forms four covalent bonds with other atoms and is thus said to be tetracovalent (see "Chemical Terms and Conventions"). In such compounds as open-chain saturated (i.e., with only single bonds) hydrocarbons, these bonds are directed spatially toward the corners of a regular tetrahedron, with the angles between each pair of bonds approximating 109°. Compounds with tetracovalent carbon present in small rings or in certain caged structures, with consequent large deviations from the ideal bond angle, are much more reactive because of this strain, and their carbon-carbon bonds are broken more easily than are bonds that are near the tetrahedral angle.

Accordingly, there has been interest in exploring the synthesis and properties of saturated hydrocarbons in which the carbon-carbon bond angles are forced as far as possible from the ideal 109°. One way of doing so is to explore Platonic hydrocarbons—molecules that correspond to perfect solids of antiquity: the tetrahedron, the cube, and the dodecahedron. In the simplest compound modeled on the first of these, tetrahedrane, all of the carbon-carbon bond angles are deformed from 109° to 60°, and in the compound modeled on the second, cubane, to 90°. For many years, it was believed that these molecules would be too strained to exist as stable compounds. In principle, cubane [11] could, by reorganizing its bonds, rearrange to the thermodynamically more stable isomer [12], containing two double bonds. However, cubane has been synthesized and has been found to be a stable compound that does not isomerize except in the presence of a metal catalyst. The Woodward-Hoffmann rules provide a theoretical basis for understanding why this rearrangement does not occur spontaneously.

[11] [12]

The tetrahedrane ring system is even more strained, and theoretical chemists predicted that it might be isolated only at low temperatures, that at higher temperatures it would spontaneously convert to cyclobutadiene and then to acetylene. Although tetrahedrane has not yet been made, its tetra-*t*-butyl derivative [13] has, and this molecule is stable up to 130°C (Celsius), where it does isomerize to tetra-*t*-butylcyclobutadiene [14].*

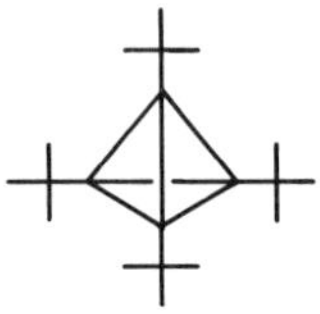

[13]

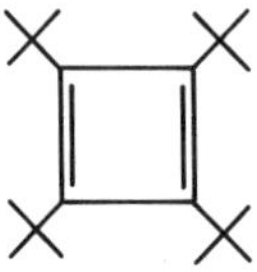

[14]

The molecule modeled on another of the regular polyhedrons, dodecahedrane, has 12 five-sided faces, should have normal carbon-carbon bond

---

*The symbol *t* denotes "tertiary." The tertiary butyl, or *t*-butyl, group is

$$CH_3-\overset{\overset{\large CH_3}{|}}{\underset{\underset{\large CH_3}{|}}{C}}- \quad \text{or}$$

angles, and is of interest because of its very high symmetry. The long-standing challenge of synthesizing this 20-carbon ring system has recently been met by L. A. Paquette (Ohio State University), who has made 1,16-dimethyldodecahedrane. The high symmetry of the ring system is manifested by the fact that the hydrocarbon does not melt, but begins to discolor, above 350°C. X-ray crystallographic analysis has confirmed the high degree of symmetry and the anticipated normal carbon-carbon bond angles. Research is in progress to determine other properties of this novel type of compound.

An example of extreme strain is provided by a [2.2.2]propellane, such as [15], in which the cage structure forces all of the bonds of the two carbons at the juncture of the rings to be directed toward the same side of the molecule. Although this compound has been isolated at low temperatures, it rearranges at room temperature with a half-life of 28 minutes to its open-ring isomer [16].

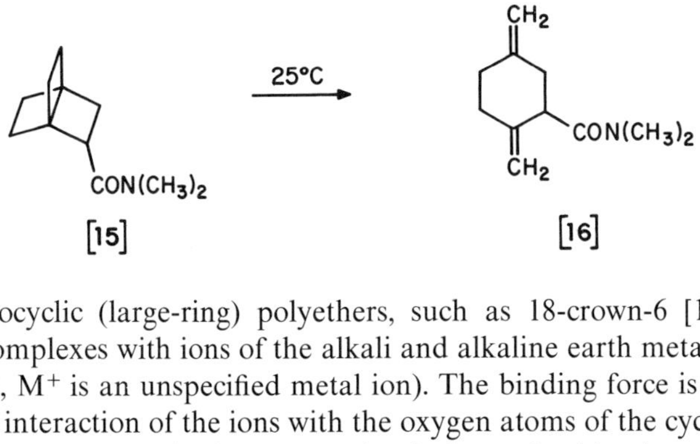


[15] [16]

Macrocyclic (large-ring) polyethers, such as 18-crown-6 [17], form stable complexes with ions of the alkali and alkaline earth metals (in the diagram, $M^+$ is an unspecified metal ion). The binding force is the electrostatic interaction of the ions with the oxygen atoms of the cyclic ether. The size of the cavity in the crown ether is determined by the number of $—O—CH_2—CH_2—$ units that make up the ring. Because various ions have different sizes, crown ethers of different ring sizes can complex certain ions preferentially, making it possible to separate some from others. This principle has been extended to three-dimensional cryptates [18],

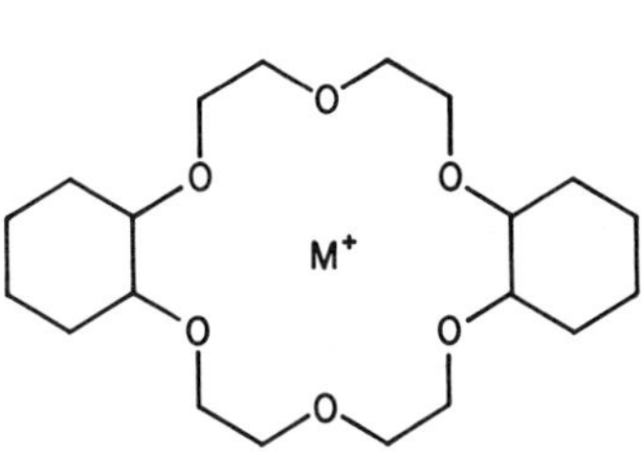


[17]

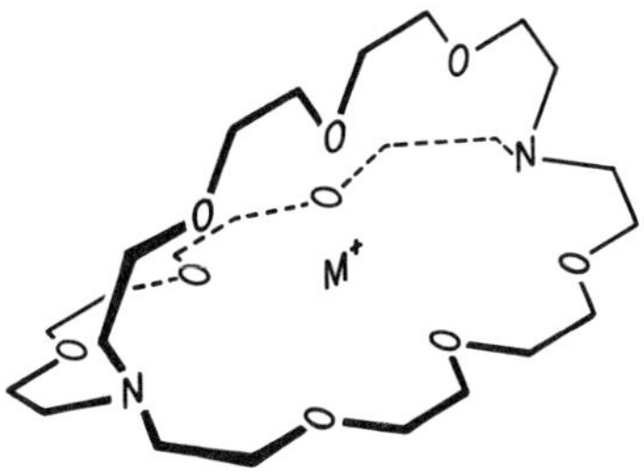


[18]

which are even stronger binding agents. This selective ion-binding property has been exploited by synthesizing crown ethers that have a chiral center and using them in an automatic machine to separate racemic mixtures of stereoisomers such as ionic amino acid derivatives. This achievement is a significant improvement over the classical method of preparing the stereoisomers of such compounds.

The chemistry of macrocyclic ethers is a very active field because of the possibilities that these compounds offer for separation of ions and because of the roles they can play in influencing the activity of some antibiotics and in transporting ions across biological membranes.

Many years ago, E. Hückel of the University of Marburg deduced that the aromatic stability of benzene and similar compounds should be a general property of planar, conjugated rings that have $4n + 2$ $\pi$-electrons, where n is an integer. (Conjugated rings have double or triple bonds in every other position.) The implications of Hückel's prediction were of particular interest for larger rings. Early syntheses of large conjugated rings gave products that were not sufficiently planar to test the hypothesis. More recently, syntheses of the dihydropyrenes [19], the methanoannulenes [20], and the acetylene-cumulenes [21] have provided molecules that have benzenelike stability and other properties that fully confirm Hückel's predictions.

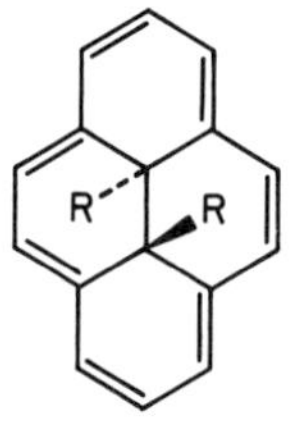


[19]

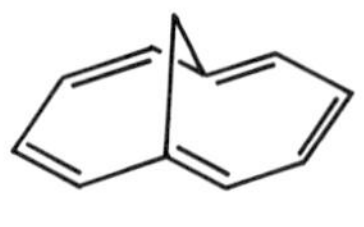

[20]

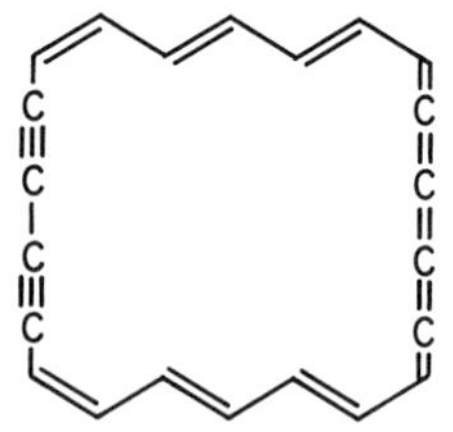


[21]

A further prediction from extensions of Hückel's theory is that, as the number of atoms in the ring becomes approximately 30 or more, aromatic stability should disappear. However, a 30-membered analog of [21] has been made and shows aromatic stability, so the challenge remains to prepare still larger rings to test the theory.

The synthesis of cyclophanes was undertaken to test questions of bonding, strain, and the nature of $\pi$-electron interactions. In graphite, the planar sheets of six-membered aromatic rings of carbon atoms are separated by a distance of 0.34 nanometer. In [2.2]-*p*-cyclophane [22], the *para* bridges constrain the benzene rings to a separation of 0.309 nanometer. A more extreme example is provided by superphane [23], in which the separation is only 0.262 nanometer. (The term *para*, abbreviated *p*, refers to two positions at opposite apexes of the benzene ring hexagon.) The interaction of the aromatic rings in these compounds leads to a further delocalization, or spreading, of the $\pi$-electrons, so that the molecule as a whole behaves as if it has only one $\pi$-electron system. The same effect is also observed in multilayered cyclophanes, such as [24]. An outstanding challenge is to synthesize a cyclophane polymer. If the delocalization of the $\pi$-electron system in fact occurs over the entire polymer molecule, such a polymer should be an electrical conductor.

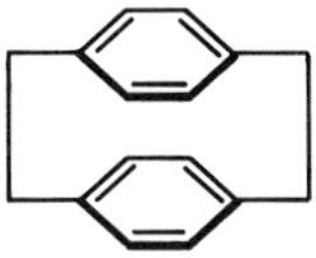

[22]

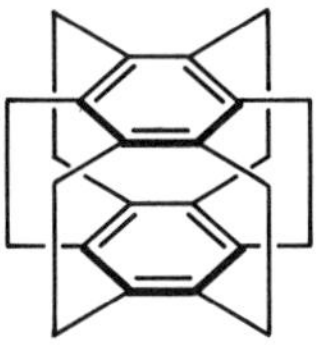

[23]

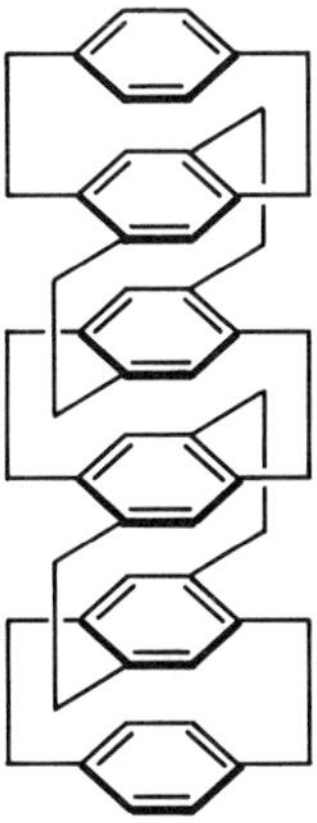

[24]

The synthesis of unusual organic structures, many of which defy classical theories of chemical bonding and stability, is opening new lines of thought about structural chemistry and offers the promise of organic and organo-inorganic compounds with potentially useful properties not traditionally associated with classical organic molecules.

## PHOTOCHEMICAL SYNTHESIS[7-8]

Photochemical reactions differ from conventional thermal reactions in a fundamental way that allows the preparation of compounds that are otherwise difficult or impossible to make. When a molecule absorbs a photon (a quantum of light), it becomes electronically and vibrationally excited, that is, it possesses a large excess of energy. Absorption of a photon can excite a molecule to a reactive state at or below room temperature and can induce kinds of reactions that do not occur at higher temperatures. The excited molecule can change its geometry and yield structures that are too unstable to be formed by simple heating. These unusual configurations are often preserved when the molecule loses its photochemical excitation by forming a new bond or bonds, and an unexcited molecule with a highly strained structure may result. In effect, some of the energy of the absorbed photon is stored in the new structure, much as potential energy is stored in a coiled spring.

Structures with strained three- or four-membered rings are commonly produced as a result of photoexcitation of structures that possess either no rings or unstrained six- or higher-membered rings. For example, benzene [25] is a very stable compound with a low-energy six-membered ring structure. Other configurations of six carbons each bearing one hydrogen atom are conceptually possible, but they have highly strained three- and/or four-membered rings and therefore possess much more energy than does benzene. Until recently, it was believed that ring isomers of benzene would be impossible to prepare because their high energy content would trigger their immediate rearrangement to the much more stable benzene structure. Nevertheless, the remarkable selectivity of photochemical excitation has permitted chemists to prepare three of these benzene isomers. For example, photochemical excitation of benzene itself with light of 200-nanometer wavelength generates Dewar benzene [26] and benzvalene [27]. Of greater synthetic importance, photolysis of the polycyclic nitrogen compound azaprismane [28] produces three benzene isomers, including prismane [29], in useful yields. Now, publications on the chemistry of some of these benzene isomers are beginning to appear.

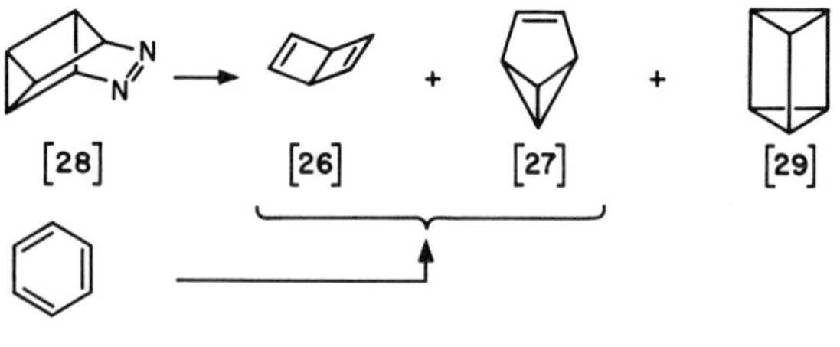


[28] [26] [27] [29] [25]

Photochemical excitation provides a way of making chemical species that are so reactive that they have only been obtained by trapping them in an inert environment, such as a matrix of frozen argon at about −200°C. This technique has made possible the study of a number of theoretically important chemical species such as unsaturated small-ring compounds (e.g., cyclobutadiene), biradicals (e.g., trimethylenemethane and tetramethyleneethane), and molecular fragments (e.g., carbenes).

The last two decades have seen the development of a wide variety of lasers that produce monochromatic (i.e., only one wavelength) light ranging from the ultraviolet to the infrared. The very narrow frequency bandwidth of lasers allows selective excitation of very narrow absorption bands in molecules. This property of lasers is not of general use in solution photochemistry because most molecules in solution absorb a broad range of wavelengths of light. However, in the gas phase, small molecules often absorb only very narrow wavelength ranges. This property can be used to separate different isotopes of an element, which have almost identical chemical and physical properties. An example is a potential process for separating the two principal isotopes of natural uranium ($^{235}U$, natural abundance about 0.72 percent, and $^{238}U$, natural abundance about 99.27 percent). The uranium is converted to its volatile hexafluoride ($UF_6$) and is irradiated with laser light of a wavelength that is selectively absorbed by $^{235}UF_6$. The photochemically excited molecules undergo a chemical reaction to give a product that has properties different from those of $^{238}UF_6$, permitting it to be separated from the mixture.

The high intensity of laser light also provides a new avenue for synthesis because it can generate high concentrations of reactive species. As a result, reactions between reactive species, rather than between such species and the solvent, become significant. Materials that are unattainable with ordinary excitation sources may be produced with laser excitation. For example, laser photolysis, or decomposition, of diphenyldiazomethane [30] yields totally different products than does lamp photolysis.

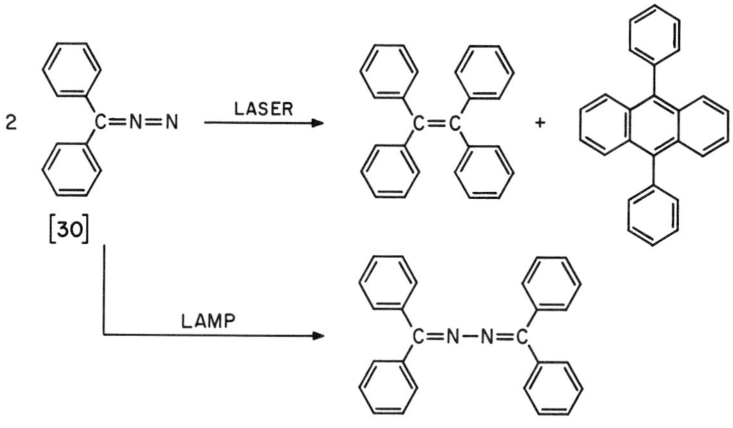


Photosynthesis, by which green plants convert the energy of light into energy stored in chemical compounds, is the basis for all life. Plants use the energy in sunlight to convert compounds of low energy content, carbon dioxide and water, into compounds of higher energy content, such as carbohydrates, liberating oxygen formed by the oxidation of water. This process is a complicated series of chemical reactions that involves, among many mediators, chlorophyll, nucleotide phosphates, and enzymes. The overall path of the carbon of carbon dioxide in photosynthesis has been determined, but chemists have not yet been able to devise practical ways of duplicating important features of photosynthesis. The key to doing so may lie in the development of synthetic enzymelike catalysts.

## COMPUTER-GUIDED SYNTHESIS

A chemist setting out to synthesize an organic compound must devise a plan, or more often a set of contingency plans, that encompasses such considerations as potential starting materials, what reaction steps can be used and in what sequence, the necessity for protecting some reactive sites in an intermediate compound while conducting a reaction at another site, how and at what stage to remove site-protecting groups, and, in many cases, how to achieve a desired stereospecificity at chiral centers. The complexity of such plans increases exponentially with the number of reaction steps required for the synthesis. In drawing up such plans, the chemist calls upon an accumulation of knowledge of reactions from the literature and his own experience and then assembles what appear to be feasible reaction sequences by a mental go/no-go process that is strongly biased by his own experience and intuition. The elements of this process are conceptually amenable to modern computer technology. To examine how computer technology is applied in such planning, we will first examine the options open to the chemist in designing a synthesis.

The first and most basic technique for designing a synthesis involves working backward from the structure of the target molecule. This target is examined to find compounds that can be transformed into it by a single chemical step. Each of these precursors is then similarly analyzed until satisfactory starting materials are identified. This method is called retrosynthesis.

A second approach that the chemist may take in designing a synthesis is to examine the target molecule for particular structural features that can be formed by either very restrictive or extremely general routes. For example, six-membered rings can be produced easily by many reactions, whereas the number of known practical ways to make rings of more than ten atoms is quite small, thereby constraining the flexibility of the synthetic route.

The third possible line of attack in designing a synthesis is based on identifying a structural resemblance of the target with common starting materials. This approach has potential drawbacks in that it may restrict unreasonably the choice of pathways, but it is quite useful if a fairly complex starting material, not many steps removed from the target, is available. This mode of design is particularly attractive when the production of a family of related compounds is desired. In such cases, it is economical of both time and effort to find some intermediate that can be used in the synthesis of all the desired compounds.

Substantial progress has been made in devising computer programs for the first of these approaches, retrosynthetic analysis. This kind of analysis can, in theory, be applied to a large number of possible precursors of the target molecule and then, in turn, to comparably large numbers of the preceding intermediate structures, producing a vast number of possible synthetic routes. As retrosynthetic analysis is practiced by most chemists, only one, or at most a few, complete routes are generated; for any synthesis that involves more than a very few steps, it is not feasible for the chemist to lay out and assess more than a few of the many possible synthetic routes. This fact reflects not only the limit to the number of structures a chemist can consider but also the amount of time he can spend on the task and the number of reactions in his information base (analogous to the computer's memory). A digital computer is a more tireless device than the human brain for performing this kind of repetitive operation.

A key element in the use of computer technology in planning synthetic routes is computer graphics: the capability to display on a cathode-ray tube not only the basic structure of a molecule but also its various possible spatial configurations. In performing a retrosynthetic analysis, the chemist "draws" the structure of the target molecule on an electronic data tablet, using the same conventions that he would use on paper. The structure appears on the cathode-ray tube as it is drawn.

When the structure has been drawn, the computer programs decipher the chemical information it contains, such as rings, reactive sites, and aromatic character. The chemist now calls on a stored data base to explore types of compounds and reactions one step back from the target structure, and then selects those that appear most promising. Successive iterations of this process enable the chemist to work back to reasonable sets of starting materials and to develop promising routes to the target molecule.

Since the heart of the computer program is the data base comprising the reactions that the program "knows about," building these data bases is a critical activity. Chemists must pore over the literature to extract examples of any reaction that they wish to include, as well as to identify those aspects of the molecular environment that make the reaction go well or poorly. A program can use only the chemistry that is programmed into it, and seemingly obvious reactions will be ignored if they are not in the data base. The more reactions, and the more information about them, the better. More efficient ways to create the data base must be found to make these programs widely useful. There is also a need for a way of representing reactions so that the programs can "learn"—be given a large number of reactions from the literature and be able to identify similarities and differences that affect their relative efficacies.

The quality of the results obtained with the interactive programs for synthesis planning is a direct function of the knowledge of the chemist-user, because he guides the programs along directions that his experience tells him are likely to be fruitful. Accordingly, these programs give him frequent opportunity to interact by specifying the types of chemistry that are to be considered. For example, if the target molecule has a large number of functional groups, he may choose a strategy that reduces the number of such groups early in the retrosynthetic plan, corresponding to adding the groups in the later stages of the synthesis. Likewise, if there is a particular class of reactions that he especially wants to use, he can specify that the program select routes that use them.

An illustration of a retrosynthetic analysis produced by this procedure is the following identification of a series of feasible intermediate compounds and synthetic reactions leading from a target molecule [33] back to the available starting materials [31] and [32]. In this diagram, the open arrows indicate the direction of the backward retrosynthetic analysis performed by the chemist and the computer; the actual synthesis would be performed in the opposite direction.

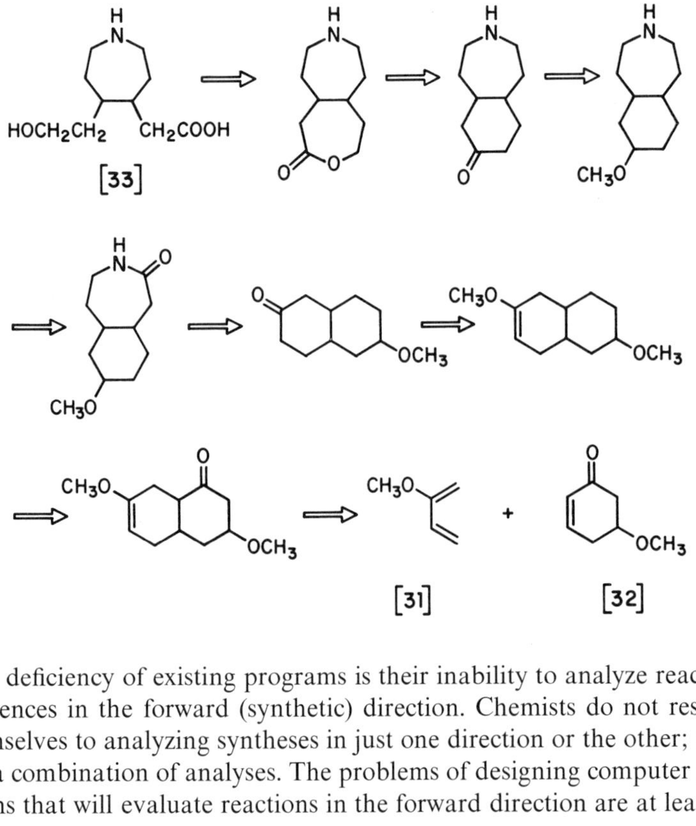


A deficiency of existing programs is their inability to analyze reaction sequences in the forward (synthetic) direction. Chemists do not restrict themselves to analyzing syntheses in just one direction or the other; they use a combination of analyses. The problems of designing computer programs that will evaluate reactions in the forward direction are at least as challenging as those for the backward direction. Ideally, programs for designing synthesis should have modules for both kinds of analysis that operate in synchrony. This would be especially valuable for identifying practical common intermediates when designing syntheses of related compounds.

Computer design of syntheses should be considered a valuable aid to, but not a replacement for, the chemist. The computer is able to search quickly and tirelessly for reaction possibilities, within the scope of its programmed memory; and this capability will surely increase as more information is programmed into computers. What the computer cannot do, and the chemist can, is to inject into synthesis planning (1) an instinctive feeling that a certain reaction may work in a specific circumstance while another probably will not, and (2) the "flash of inspiration," for example, that a reaction he read about years ago might be modified by a technique that he read about elsewhere, to solve a problem in a particular synthesis.

Computer graphics technology is beginning to be used in teaching chemistry. For example, a student can be given on a cathode-ray tube screen a target molecule and be asked to lay out reactions and intermediates for synthesizing it. He proceeds to do so by "drawing" on the cathode-ray tube successive intermediates and the reactions that he thinks should lead to them. Should he select a reaction that would not produce a desired intermediate, the computer tells him that the reaction will not work, and why. Obviously, the teaching process in this technique relies heavily on the number and types of reactions that are programmed into the computer data base beforehand. A virtue of this technique is that the student has the opportunity to "learn by doing" more rapidly than he could in trial-and-error laboratory work. However, the technique must be considered an adjunct to, not a replacement for, learning in the laboratory; experimental technique must be learned in the laboratory, and the opportunity for some face-to-face contact with a teacher must not be lost.

The current costs of purchasing and operating the equipment for computer-assisted chemistry are a limitation on its widespread adoption. Nevertheless, the technology is evolving rapidly through cooperative efforts of users and equipment manufacturers. It is anticipated that these costs will come down significantly and that this powerful aid to the planning of chemical synthesis and the teaching of chemistry will become broadly used.

## SOLID-STATE SYNTHESIS

The synthesis of solid materials with novel electronic properties has been at the heart of the electronics revolution that has so profoundly altered our society. The basic preparative techniques used for making semiconductor compounds—techniques like zone refining, float-zone crystal growth, crystal pulling, and liquid-phase epitaxy*-have provided the materials foundation for such developments as the transistor, solid-state lasers, light-emitting diodes (LED's), and magnetic bubble technology.

## Superconductors and Conductors

Superconductivity, the flow of an electric current without resistance (see Chapter 17, "Prospects for New Technologies"), has so far been observed only in materials at very low temperatures, not far above absolute zero. Widespread commercial use of superconductors awaits the discovery of materials that have critical temperatures ($T_c$, the highest temperature at which they remain superconducting) somewhat closer to room temperature. Superconductors with the highest $T_c$'s yet found all possess a crystal structure called the A-15 phase—a type of cubic crystal structure with very high symmetry. It has been shown recently that the $T_c$'s of superconductors in the A-15 phase are very dependent on the degree of perfection of the A-15 crystal structure—the fewer defects in the crystal structure, the higher the $T_c$.† Synthetic techniques for making superconductors with a high degree of crystal structure perfection have been essential to the manufacture of high-field superconducting magnets.

Trends of properties in the periodic table of elements imply that compounds of niobium (Nb) with Group III and Group IV elements such as aluminum (Al), gallium (Ga), silicon (Si), germanium (Ge), and tin (Sn), in the A-15 phase, may have $T_c$ values significantly higher than the highest value for any element (9.25 K for niobium).‡ Some of these compounds, for example, $Nb_3Sn$ and $Nb_3Al$, can be prepared in the A-15 structure by direct melting of the components and have $T_c$ values only about 2° below the boiling point of hydrogen (20.3 K). Others, such as $Nb_3Ge$, $Nb_3Ga$, and $Nb_3Si$, have metastable A-15 structures that can be obtained only by special techniques such as sputtering, rapid quenching, or explosive compaction. Samples so prepared have shown $T_c$ values around the boiling point of hydrogen. There is reason to believe that still higher $T_c$ values can be obtained with these compounds as synthetic techniques are improved to increase the crystal perfection of the A-15 phase.

Organic materials with high electrical conductivity, and with hints of superconductivity, have been synthesized recently. One example is a

---

* Epitaxy is the deposition of a thin layer from a liquid or gas onto the surface of a host crystal, where the deposited layer assumes the crystal structure of the host crystal rather than its normal structure.

† Actual crystals are almost never the perfect, regular arrays of atoms that their representations on paper would indicate. They usually contain defects, such as occasional missing atoms, impurity atoms, or lattice discontinuities analogous to geological faults.

‡ K is the unit of the Kelvin scale and is used without the degree symbol. Zero on the Kelvin scale is about −273.1°C.

"doped" polymer of acetylene, $(CH)_x$. Polyacetylene has a metallic appearance; while it is not a good conductor as formed, doping it by the introduction of small amounts of impurities, such as arsenic pentafluoride ($AsF_5$) or halogens, increases the conductivity dramatically up to 1,000 mhos* per centimeter at room temperature, comparable with that of mercury. It is believed that this conductivity is associated with the transport of $\pi$-electrons along the carbon backbone of the polymer. Another example of a synthetic conductor is the polymer $(SN)_x$, prepared by solid-state polymerization of crystalline $(SN)_2$, sulfur nitride. The golden fibrous crystals of $(SN)_x$ have a room temperature conductivity twice as high as doped $(CH)_x$, 2,000 mhos per centimeter, parallel to the chain axis. $(SN)_x$ becomes superconducting at 0.5 K.

Another recent class of organic conductors is illustrated by the charge-transfer salts, such as $TTF^+TCNQ^-$ and $(TMTSF)_2{}^+PF_6{}^-$ (these are acronyms for very long chemical names). These salts are ionized in the solid state, and electron transport occurs along stacks of individual molecules. They are good conductors at room temperature and have a phase transition at a specific low temperature, below which they are poor conductors. The low-temperature phase transition of $(TMTSF)_2{}^+PF_6{}^-$ can be suppressed by pressure, and this complex becomes superconducting below 0.9 K under 12 kilobars pressure.

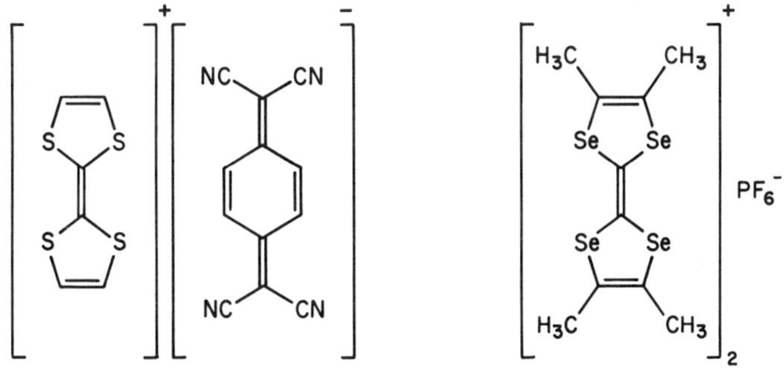


## Sheet Structures and Intercalation Compounds[9-10]

Some materials crystallize in such a way that their atoms are bound to one another in a planar network, resulting in a rather rigid, flat sheet on both the molecular and macroscopic scales. Familiar examples of such structures are graphite, molybdenum disulfide, and talc. The well-known

*The mho is a unit of conductance and is the reciprocal of the ohm, a unit of resistance.

lubricating properties of these solids reflect the ease with which one sheet can slide across another and the relative weakness of the interaction between the sheets.

Because of this weak interaction between molecular sheets, it is often possible to insert other atoms or molecules (guests) between the sheets of the sheet-structure material (the host). In many cases, the insertion can be performed by putting the guest in contact with the host at room temperature or with heating. Insertions can also be performed electrolytically at a conducting host electrode. The resulting material is called an intercalation compound ("intercalate" means "to interpose"). These compounds reflect specific relationships in the ratio between host and guest, as indicated by such formulas as $C_8K$, $C_{10}K$, $C_{24}K$, etc., for combinations of graphite with potassium (K) metal.

When a sheet structure is converted to an intercalation compound, the host sheets move apart to accommodate the guest, resulting in expansion of the original material perpendicular to the sheet plane; expansions as great as eightfold have been observed. However, there is essentially no effect on the strength and rigidity of the host sheet, nor is there any dimensional change in the directions of the sheet plane.

In most intercalation compounds, the host and guest atoms exchange electrons, often with significant effects on the electrical properties of the compounds. The sheet structure titanium disulfide ($TiS_2$), used as an electrode, can serve as a host and accept lithium (Li) guest atoms to form the intercalate $LiTiS_2$ without significant change in the physical structure of the $TiS_2$. The process is electrically reversible, and the cycle can be repeated many times with no apparent change in the $TiS_2$. This system has been proposed as the basis for an electrical storage battery.

Intercalation can raise the superconducting critical temperatures of sheet structures. For example, tantalum disulfide, with a $T_c$ of 0.8 K, has been intercalated with potassium hydroxide to give a material with a $T_c$ of 5.3 K. Efforts to find superconducting intercalates with a $T_c$ near or above the boiling point of hydrogen have not yet succeeded.

Intercalation compounds of graphite with alkali metals, halogens, and inorganic anions have been known for many years. However, only with the recent development of a special form of pyrolytic graphite, in which the graphite sheets are highly ordered and parallel, have the unusual electrical properties of these intercalates begun to attract attention. For example, intercalation of such anions as $NO_3^-$, $HSO_4^-$, or $ClO_4^-$ with this form of graphite increases the electrical conductivity in the sheet plane greater than 20 times, to better than that of aluminum; the conductivity perpendicular to the sheet plane remains low.

It has been reported that intercalation of this form of graphite with arsenic pentafluoride ($AsF_5$) gas gives a compound with an in-plane electrical conductivity comparable with that of silver. It is not yet certain, because of preparative and measurement problems, that the conductivity of this compound, $C_8AsF_5$, is that high, but it is certainly at least as high as that of aluminum. The reason for the dramatic increase in conductivity is not known, but it has been established that there is electron withdrawal from the graphite because at least part of the $AsF_5$ is changed to hexafluoroarsenate anion ($AsF_6^-$) and the neutral molecule $AsF_3$. Graphite can also be fully intercalated by $AsF_6^-$ ions to give the product $C_8^+AsF_6^-$. Curiously, this intercalate has an electrical conductivity an order of magnitude lower than the fully intercalated relative $C_8AsF_5$, despite the fact that the electron removal from graphite is greater in the former than in the latter. This decrease in conductivity with an increase in graphite sheet charge is not yet understood.

The chemistry of intercalation and the properties of these products are only beginning to be understood. The development of practical uses for materials that are highly conductive in two directions and not in the third must await the reproducible preparation of larger and more structurally perfect samples. Nevertheless, it is believed that continued study of intercalation chemistry will lead to unusual materials.

## Optical Communications[11]

Optical fibers have potential information-carrying capacities (bandwidths) several orders of magnitude greater than those of microwave radio systems, and are already being used in some communications systems. These fibers are prepared by a synthesis called modified chemical vapor deposition that involves reacting very pure silicon and germanium tetrachlorides ($SiCl_4$ and $GeCl_4$) and other reagents with oxygen in a glass tube to make $(SiGe)O_2$. The reaction is controlled to provide a radial gradient in composition, and hence in refractive index, so that light is guided down the fiber with low loss. When the reaction is complete, the tube is collapsed to eliminate the center hole and then hot-drawn to give a long, very fine fiber.

This process has provided fibers with losses as low as 2.5 decibels (44 percent) per kilometer, using light with a wavelength of 830 nanometers. These fibers permit a transmission bandwidth of 0.4 gigahertz per kilometer, and installations with these characteristics have carried more than 600 two-way telephone conversations per fiber over repeaterless spans 10 kilometers long. Losses as low as 0.5 decibel (11 percent) per kilometer have been achieved experimentally with longer wavelength light (1,300 nanometers) and will permit substantially greater transmission bandwidths, more than 2 gigahertz per kilometer.

The LED or laser used to produce the monochromatic 830-nanometer light employs gallium aluminum arsenide ([GaAl]As), prepared by a liquid- or vapor-phase reaction on a single-crystal gallium arsenide substrate. For transmission at wavelengths equal to or greater than 1,100 nanometers, where losses are lower, transmitters are made by depositing single crystal layers of (GaIn)(AsP) on substrates of indium phosphide (InP). The InP is prepared by pulling a crystal from an InP melt that is covered with molten boron oxide ($B_2O_3$). Methods for growing high-perfection crystals of doped InP by this technique have been developed, giving transmitters with improved light output and lifetimes. The extension of these sophisticated techniques will be key to future advances in optical communications.

## Magnetic Materials[12]

Amorphous (noncrystalline) magnetic materials, which first appeared in the 1970's, continue to be a source of new magnetic phenomena. They offer a combination of properties superior to those of polycrystalline magnetic materials because they are free from the losses and other problems that arise at crystal grain boundaries. Magnetic materials like to crystallize, so amorphous phases must be made by an array of synthetic techniques that includes splat cooling (propelling a liquid with an air blast onto a cold surface), roller quenching (quenching a liquid between chilled rollers), and spinning (cooling a liquid against a cold, spinning wheel). With all of these methods, cooling rates greater than 100,000° per second provide amorphous forms of materials that are otherwise obtainable only in crystalline forms.

Two important classes of amorphous ferromagnets are made by these techniques: the transition metal-metalloid (TM-M) alloys and the rare earth-transition metal (RE-TM) alloys. The TM-M alloys contain about 80 percent iron, cobalt, or nickel, with the balance being boron, carbon, silicon, phosphorus, or aluminum. They are ferromagnetic at room temperature and have low losses in power applications, and some exhibit useful low magnetostriction (change in dimensions with change in magnetic field) and low thermal expansion. One amorphous ferromagnet is being used as a transformer core because its low loss results in less wasted energy—one percent versus the five percent of a polycrystalline core.

# ADVANCED INSTRUMENTATION IN CHEMICAL SYNTHESIS

An instrumental revolution in chemistry took off in the 1950's as chemists began to realize the power of infrared, X-ray, and the newly invented nuclear magnetic resonance spectroscopies to provide quickly information about chemical structures that had previously been obtainable only by laborious chemical study and intuition, or not at all. Since then, advances in electronics, optics, analytical techniques, and data processing have provided the chemist with an array of sophisticated instruments that have added new dimensions to his ability to unravel complex structures, to gain intimate understanding of reaction mechanisms, and to design synthesis procedures. Much of the recent progress in chemical synthesis has been strongly dependent on the kind of information that is obtained with modern analytical instrumentation. Mathematicians have made an important contribution to this advance by discovering methods for analyzing electronic signals so as to filter out background noise. Micro- and minicomputer processing of data with these methods, which are called Fourier transform methods, has provided substantial improvement in the speed and sensitivity of spectroscopic techniques.

## Infrared Spectroscopy

The energy of infrared radiation matches that required to distort many interatomic bonds in compounds. For example, alcohols absorb infrared radiation of frequencies that correspond to the energy required to stretch the oxygen-hydrogen bond and the carbon-oxygen bond. For this reason, infrared (IR) spectroscopy has become a standard technique for the detection of many kinds of functional groups in organic molecules. The introduction of Fourier transform methods to IR has improved its sensitivity and has reduced the time required to obtain a spectrum from between 10 and 30 minutes to less than 1 minute, an important advantage when many samples must be run. Excellent quality IR spectra are now obtained routinely with less than 1 microgram of sample, whereas not long ago a fraction of a gram was required. The use of Fourier transform IR is expected to increase substantially.

## Raman Spectroscopy

Raman spectroscopy, a useful adjunct to IR spectroscopy, has been revitalized by the availability of coherent monochromatic light from lasers. The method involves passing a beam of light through a sample and analyzing with an optical spectrometer the light that is scattered at 90° from the incident beam. Raman spectroscopy measures many of the same molecular properties as IR spectroscopy, but is useful because it can reveal some structural features that IR does not and because of the high resolution that can be obtained with a very small sample. For example, it has been possible to obtain a good Raman spectrum of a pesticide, and thus identify it, inside a single plant cell.

## Nuclear Magnetic Resonance Spectroscopy

Nuclear magnetic resonance (NMR) spectroscopy is based on the fact that the nuclei of certain elements, when placed in a strong, uniform magnetic field, absorb radiofrequency energy at frequencies that are determined both by their own atomic characteristics and by the effects of the types and positions of nearby atoms in a compound. NMR spectroscopy is applicable to naturally occurring isotopes of most of the elements that occur in organic compounds, including $^{1}H$, $^{13}C$, $^{2}H$, $^{19}F$, $^{31}P$, $^{15}N$, and $^{11}B$,* the first two being the most important to the organic chemist. NMR provides much more detailed information about structures than does IR.

The well-established and relatively inexpensive 60-megahertz proton sweep NMR continues to be the workhorse instrument of the organic laboratory, giving information about the numbers of hydrogen atoms bonded to various other atoms in the molecule. However, instruments with more powerful magnets, operating at 200 to 500 megahertz, can provide much more detailed information about the hydrogen atoms by spreading the peaks of absorbed frequencies farther apart. NMR spectra of the naturally occurring isotope $^{13}C$ give especially valuable information about the basic structural framework of organic molecules. Thus, the $^{13}C$ spectrum of even a very complex molecule shows the number of carbon atoms in the molecule, the number of hydrogen atoms bonded to each carbon, whether a carbon atom is singly, doubly, or triply bonded to another carbon, and whether a carbon atom is singly or doubly bonded to an oxygen or nitrogen atom. Synthesis chemists use NMR to monitor the course of reactions by carrying them out in the sample tube of an NMR spectrometer and following the changes in the $^{1}H$ or $^{13}C$ spectra.

The magnetic field in an NMR spectrometer must be extremely uniform over the sample volume and extremely stable in time; variations of the magnetic field in both of these respects can now be controlled within one part in $10^9$ or less. Superconducting solenoids, which operate without power loss at the temperature of liquid helium, are being used increasingly because the magnetic fields obtained are two to four times greater than are those that can be obtained with ordinary iron-core electromagnets. An important advance has been the introduction of spectrometers that provide much greater sensitivity by using radiofrequency pulses to excite the nuclei. Although the spectra so obtained are usually very complex, they can be computer-analyzed by the Fourier transform technique.

Although NMR has been applicable only to liquid samples, the advent of the superconducting solenoid and pulse technology, and the technique of "magic angle" spinning of the sample, have now made it possible to obtain NMR spectra of solid samples. (The "magic angle" technique involves spinning the sample tube with its axis oriented at a specific angle to the direction of the magnetic field, thus narrowing the normally broad,

---

*These symbols designate specific isotopes of these elements. Thus, $^{1}H$ denotes the principal isotope of hydrogen, with an atomic mass of 1, as distinguished from its isotope $^{2}H$ (deuterium), the latter constituting about 0.015 percent of natural hydrogen. $^{13}C$ denotes the less common isotope (an abundance of about 1.1 percent) of natural carbon, which is principally $^{12}C$.

diffuse spectrum of a solid into identifiable peaks.) These new NMR methods are being used to study the composition of insoluble polymers and natural and treated coals. Application of NMR to solid materials should grow at a rapid pace.

## X-Ray Crystallography

X-ray crystallography has been used for many years to determine the spatial relationships of atoms in inorganic and organic crystals. The power of this technique was shown dramatically in its use to determine the spatial arrangement of the atoms in DNA.

The development of theoretical tools for analyzing the measured intensities of scattered X-rays, and the availability of automated diffractometers and minicomputers to perform the necessary Fourier transform of the measured data, have made it possible to determine the complete structure of a relatively simple organic compound in a few days or weeks. X-ray analysis can, in one operation, provide the molecular formula of an unknown substance as well as its conformation. This technique is particularly useful for ascertaining the structures of highly complex biological molecules, such as enzymes and other proteins (although substantial time is required for these analyses), if they or their derivatives can be obtained in crystalline form. It is also invaluable for determining the structures of intercalation compounds and many other types of inorganic compositions. Some commercial laboratories will now carry out X-ray structural determinations of low molecular weight organic compounds for as little as $1,000.

## Mass Spectroscopy

In mass spectroscopy (MS), the molecules of a compound are converted to gaseous positive (or more rarely, negative) ions, usually by electron impact. The ions are then separated and analyzed according to their ratios of mass to charge. Some of the ions may come from the complete molecule by simple loss of an electron, and such an ion indicates the molecular weight of the compound. High-resolution mass spectrometers can separate and distinguish between ions that differ in mass by only a small fraction of one mass unit and thus can provide the molecular weight of the compound with sufficient accuracy that the number and kinds of atoms in the molecule can be calculated. For example, 2-methylpiperazine ($C_5H_{12}N_2$) in such a spectrometer shows a mass of $100.100 \pm .001$, and methyl 2-butenoate ($C_5H_8O_2$) shows a mass of $100.052 \pm .001$. These masses are readily distinguished in the high-resolution mass spectrometer. Furthermore, if one enters the mass number $100.100 \pm .001$ into a properly programmed computer, it will identify

2-methylpiperazine, and possibly one other compound, as having that mass. Although the number of possible structures that the computer identifies increases somewhat as the mass number increases beyond 100, it is usually possible to eliminate all but one or two by using information from other analytical techniques, thus often enabling the chemist to infer the correct structure of an unknown compound. Similar information can also be obtained from ions that are derived from fragments of the original molecule, in some cases allowing inference of the complete molecular structure.

## Chromatography

Chromatography has become an invaluable technique for separating the components of a reaction mixture or a biological sample and for monitoring synthetic reactions.

Gas chromatography (GC) involves passing a volatilized mixture with an inert carrier gas such as hydrogen or helium through a long thin tube that contains a stationary phase, i.e., a solid adsorbent or a liquid coated on an inert solid. The components of the mixture are separated because of differences in their adsorption or solution by the stationary phase. A recent development is a flexible, fused-silica capillary column (30 meters long and 2 millimeters inside diameter) that requires only 0.05 nanoliter of sample and that provides 4 times the speed and 50 times the resolution of older columns (2 meters long and 2 millimeters inside diameter). This supermicro technique permits direct analysis of compounds that formerly had first to be converted to chemical derivatives for GC analysis. An especially powerful technique for analyzing mixtures involves using a mass spectrometer as a detector to analyze the various components as they emerge from the GC tube.

In high-performance liquid chromatography (HPLC), a solution of the mixture is passed under pressure through a tube filled with a solid adsorbent that retards passage of the components to different extents. Thus, the mixture is separated into its components, which can be eluted selectively from the adsorbent with a different solvent. The combination of HPLC with MS is more difficult than the combination of GC and MS, but it is equally powerful.

Both HPLC and GC can be used for preparative purposes on a larger scale to isolate from a mixture a few hundred milligrams to several hundred grams of a compound in a day.

Thin-layer chromatography is a valuable technique for monitoring the course of a reaction and for rapid detection of the components of a mixture. One form of thin-layer chromatography involves placing a small drop of a solution of a mixture near one edge of a square plate that is coated with a solid adsorbent, such as silica. The solvent is evaporated, and the plate is then "developed" by holding it vertically and just immersing the edge with the spot in a different solvent. As this solvent migrates up the adsorbent on the plate, the components of the mixture are carried upward in a thin line, with each component concentrating at a spot, the location of which depends on the rate of adsorption of the component by the silica. The technique is usually performed by placing drops of several solutions along the edge of the plate—one of the mixture under test, and others of pure compounds that are suspected components of this mixture. Comparison of the locations of the developed spots of these latter with those from the mixture indicates which, if any, of the pure controls are present in the mixture.

Many promising techniques are yet to come, and the improvement in analytical instrumentation will continue. One exciting new technique, which is not yet available commercially, is field flow fractionation. This is a technique by which molecules or particles of high molecular weight polymers or inorganic solids, which are not amenable to liquid chromatography, can be separated into fractions of various molecular weight ranges or particle sizes. It involves passing solutions or colloidal dispersions of such materials between closely spaced parallel plates in a strong force field, such as a strong gravitational field created in a centrifuge. This technique is expected to become widely used as it becomes more fully developed.

## Summary and Outlook

Chemical synthesis is the science of combining atoms or molecules to form new materials. These substances, together with similar ones found in nature, are the basis for all the materials used by modern civilization, such as fertilizers, drugs, fibers, semiconductors, and magnetic materials. Synthesis has made great progress in recent years and will certainly lead in the future to other advances that will affect all sectors of society.

Catalysts and reagents are the keys to the discovery of new reactions that proceed more efficiently or more selectively or that permit preparation of molecular structures not previously available. The field is burgeoning and offers much promise for additional progress. Studies of natural products and of ways to synthesize them have led to much new chemistry and often to means of preparing useful natural products that are difficult to obtain from natural sources—particularly pharmaceuticals. The first steps have been taken in the preparation of synthetic analogs of enzymes, the catalysts used by nature to carry out many synthetic reactions, with the aim of mimicking and even improving on the incredibly efficient way in which enzymes control biological reactions. The knowledge so obtained should make it possible to synthesize molecules that will catalyze reactions that natural enzymes cannot perform. Much is yet to be done to discover new pharmaceutical agents to control diseases that respond only poorly or not at all to chemotherapy, and the increasing understanding of basic biology is providing a more sophisticated approach to the synthesis of such agents. New knowledge concerning receptors in the brain has led to the discovery of small peptide hormones; this discovery opens new directions in the search for pharmaceuticals. Further study of such materials will involve close collaboration between chemical synthesis and molecular biology.

Synthesis aimed at testing chemical structural theory has led to many new compounds that some years ago were thought incapable of existence. Important objectives remain—structures have been postulated that have not yet been prepared, and remarkable new properties should be realized when successful syntheses have been devised. Progress in this field will be aided by recent theoretical and experimental advances in photochemical synthesis, a procedure that permits the preparation of substances of high energy content.

Computer technology has advanced to a point where substantial guidance in planning synthetic routes can be obtained. This area will progress rapidly as both hardware and software are adapted to the objective.

Synthesis of solid-state materials has led to impressive changes in the electrical, computer, and communications industries. New synthetic magnetic materials are beginning to be used to reduce energy losses in the generation and transmission of electrical power. Superconductors are being used in specialized applications, and, when new ones are found that remain superconducting at higher temperatures, a host of new uses, including even power transmission, will emerge. Much interest centers on sheet structures containing added molecules between the sheets (intercalation compounds), since great changes in properties, including electrical properties, occur as a consequence of the insertion of the added molecules. Optical fibers made of a glass of precisely controlled composition are just beginning to be used in communications. They carry information in a light beam with dramatically higher density than is possible with electrical cables or microwave radio transmission systems.

Modern instruments are advancing in sensitivity and speed and are key to building up the effectiveness of research in chemical synthesis. It is now possible to analyze minute amounts of materials with a sensitivity that is orders of magnitude greater than was possible a few years ago, and to separate complex mixtures quickly and effectively. Structures of crystalline solids can now be determined with computer-controlled instruments that provide automatic data acquisition and analysis. Continued improvements in laboratory instruments are expected that will make chemical synthesis of new materials even more productive. The very high cost of these advanced instruments may be a constraint on realizing the full potential of scientists working in chemical synthesis.

Not only the progress but even the maintenance of our civilization is dependent on the materials that have been and will continue to be created by chemical synthesis. Advances in the science of synthesis in the last 20 years have made it possible to assemble structures that it had been thought too complicated to make or even not capable of existing, and we are confident that today's "rules" of chemical combination and structure will not long persist without change. The challenge is to create the new materials, to learn how to make them efficiently, and to learn how to use them effectively and safely. Much progress is being made in meeting this challenge, and much more is to come.

## REFERENCES

The following review articles provide additional material on the subjects indicated by the corresponding reference number in the text.

1. D. Valentine, Jr. and J. W. Scott. *Synthesis*, Vol. 5(1978), p. 329.
2. B. M. Trost. *Tetrahedron*, Vol. 33(1977), p. 2615.
3. J. Schwartz and J. A. Labinger. *Angewandte Chemie, International Edition in English*, Vol. 15(1976), p. 333.
4. Z. G. Hajos. *Carbon-Carbon Bond Formation*. Edited by L. R. Augustine. New York: Marcel Dekker, Inc., 1979.
5. *Stereoselective Synthesis of Natural Products.* Edited by W. Bartmann and E. Winterfeldt. New York: Elsevier-North Holland Pub. Co., 1979.
6. *Biomimetic Chemistry*. Advances in Chemistry Series, No. 191. Washington, D.C.: American Chemical Society, 1980.
7. N. J. Turro. *Modern Molecular Photochemistry*. Menlo Park, Calif.: Benjamin/Cummings Publishing Co., 1979.
8. *Photochemistry, Specialists Periodical Report*. Volume Ten. London: Chemical Society, 1979.
9. F. R. Gamble and T. H. Geballe. "Inclusion Compounds," *Treatise on Solid State Chemistry*. Volume Three. Edited by N. B. Hannay. New York: Plenum Publishing Corp., 1976.
10. L. E. Ebert. *Annual Review of Materials Science*, Vol. 6(1976), p. 181.
11. S. E. Miller and A. G. Chynoweth. *Optical Fibers.* New York: Academic Press, Inc., 1979.
12. P. H. Gaskell. *Journal of Physics C: Solid State Physics*. C12, 4337. 1979.

## CHEMICAL TERMS AND CONVENTIONS

Chemical compounds are often represented by *empirical formulas* that indicate the kinds and numbers of atoms they contain. These formulas are made up of the *symbols of elements,* which usually are derived from one or two letters in the element's name. Those used most frequently in these chapters are H (hydrogen), C (carbon), O (oxygen), N (nitrogen), Cl (chlorine), Br (bromine), S (sulfur), and P (phosphorus). Others are identified where they are used.

The unit of a compound identified by the *empirical formula* is a *molecule*. The empirical formulas for some simple molecules are $H_2O$ for water, $CH_4$ for methane, and $C_2H_6$ for ethane. The architecture of a molecule is shown by a *structural formula*, which describes the bonding pattern of the atoms. In these formulas, the lines that connect atomic symbols represent *covalent bonds*, which are chemical bonds formed by the sharing of a pair of electrons by two atoms.

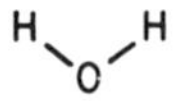
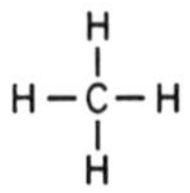
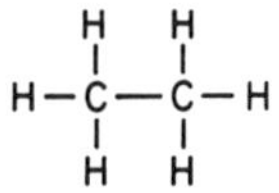

WATER, $H_2O$ METHANE, $CH_4$ ETHANE, $C_2H_6$

Each atom in a compound is said to have a *valence*, which corresponds to the number of bonds it forms with other atoms. Thus, the valences of hydrogen, oxygen, and carbon, as shown in the formulas for water, methane, and ethane, are one, two, and four, respectively.

In some molecules, two atoms may be bonded by sharing two or more pairs of electrons. Such multiple bonds are represented by two or more lines connecting the atomic symbols.

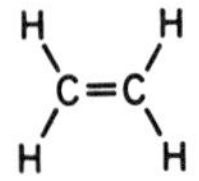

ETHYLENE, $C_2H_4$ ACETYLENE, $C_2H_2$

These compounds are called *unsaturated,* because they can be chemically combined with hydrogen to form the *saturated* compound ethane, which has only single bonds.

The pair of electrons in a single bond, and one of the pairs in a double or triple bond, are called *σ-electrons*, and the single bond is sometimes called a *σ-bond*. Additional pairs of electrons in double and triple bonds are called *π-electrons*, and the bonds, *π-bonds*. *π-Bonds* are sites at which many chemical reactions occur.

Many compounds are made up of, or contain, rings of atoms. For example, benzene, $C_6H_6$, is a ring of six carbon atoms, each bonded to one hydrogen and to each other by alternating single and double bonds. Rings with such alternating single and double bonds are called *aromatic*, and are generally more stable and less reactive than compounds with an isolated double or triple bond. Their structures are usually drawn without showing the hydrogen atoms.

BENZENE, $C_6H_6$

A given combination of atoms can often be arranged in more than one structural form. For example, two different structures can be constructed for butane, $C_4H_{10}$. Actually, both of these compounds exist, each with its

$$CH_3-CH_2-CH_2-CH_3$$

normal-, or *n*-BUTANE

$$CH_3-\overset{\overset{\displaystyle CH_3}{|}}{\underset{\underset{\displaystyle H}{|}}{C}}-CH_3$$

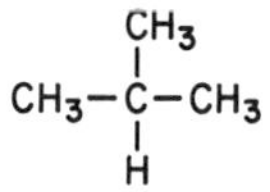

iso-, or *i*-BUTANE

own physical and chemical properties. They are called *isomers*—compounds with the same composition but different structures. The number of possible isomers of the saturated open-chain hydrocarbons increases rapidly with the number of carbon atoms; there are 75 possible isomers of decane, $C_{10}H_{22}$. Another example of two isomeric substances is provided by the composition $C_2H_6O$:

$$CH_3-O-CH_3$$

DIMETHYL ETHER

$$CH_3-CH_2-OH$$

ETHANOL

A different kind of isomerism occurs because two carbon atoms joined by a double bond are not free to rotate about the double bond. If each of these carbon atoms bears dissimilar groups, two isomeric structures are possible. Thus, 2-butene has two isomeric forms, one with both methyl groups on the same side of the double bond (*cis*-2-butene) and the other with the methyl groups on opposite sides of the double bond (*trans*-2-butene). As with the two isomers of butane, both of these compounds exist, each with its own physical and chemical properties. The *cis* and *trans* isomers of biologically active compounds often have significantly different biological properties.

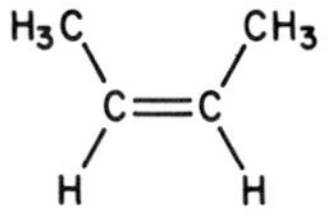

*cis*-2-BUTENE          *trans*-2-BUTENE

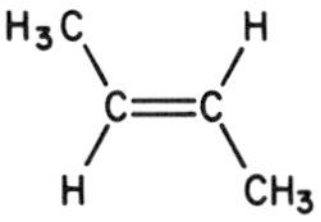

Still another kind of isomerism can occur when four different groups are bonded to a carbon atom in a compound, but it is not revealed by the kinds of plane-projection structural formulas shown above. The *α-amino acid* leucine, a building block of proteins, can be represented by either of the following plane-projection structural formulas:

$$\underset{\displaystyle CH_3}{\overset{\displaystyle CH_3}{\diagdown}}CH-CH_2-\underset{\underset{\displaystyle NH_2}{|}}{\overset{\overset{\displaystyle H}{|}}{C}}-CO_2H \quad \text{or simply} \quad i\text{ - }C_4H_9-\underset{\underset{\displaystyle NH_2}{|}}{\overset{\overset{\displaystyle H}{|}}{C}}-CO_2H$$

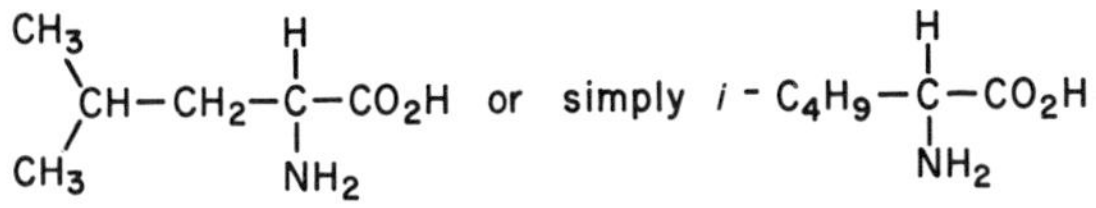

Leucine is called an α-amino acid because the amino ($NH_2$) group is bonded to the α, or first, carbon atom from the acid ($CO_2H$) group. Note that there are four different groups bonded to this α-carbon atom. The bonds to a carbon atom normally extend not in four directions in a plane but rather in space toward the corners of a regular tetrahedron with the carbon atom at its center. If we visualize the α-carbon atom at the center of a regular tetrahedron and put the four groups bonded to it at the corners, we find that we can arrange them in two, and only two, ways that cannot be made identical by rotations of the tetrahedra. These two ar-

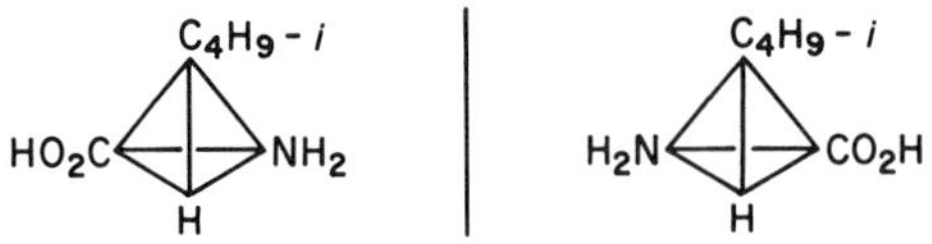


MIRROR
PLANE

rangements are mirror images of each other and cannot be superimposed, just like our right and left hands; they are called *stereoisomers* (from the Greek "stereo," meaning three-dimensional). Such isomers have almost identical physical and chemical properties but usually not the same biological properties; for example, only one of the two stereoisomers of leucine is a constituent of natural proteins. The carbon atom with the four different groups attached, which is responsible for this kind of isomerism, is called a *chiral* atom or center; "chiral" means having the property of handedness.

When we depict the spatial characteristics of molecules, we often use a projection of the three-dimensional form onto the plane of the paper, omitting most or all of the hydrogen atoms. Such a representation of *n*-butyl alcohol is shown on the right, where the unsubstituted end and each point at an angle is understood to be a carbon atom with hydrogen atoms attached as required to satisfy the carbon atom's valence of four.

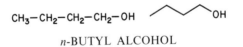

$CH_3{-}CH_2{-}CH_2{-}CH_2{-}OH$ 

*n*-BUTYL ALCOHOL

When the *stereochemistry*—the spatial arrangement about a chiral carbon atom—is to be shown, a wedge-shaped line is used to indicate a position above the plane of the paper, and a dotted line is used for a position below the plane. Using these conventions, the two stereoisomers of leucine can be drawn as shown below. The carbon atom bearing the wedged and dotted lines is the chiral center.

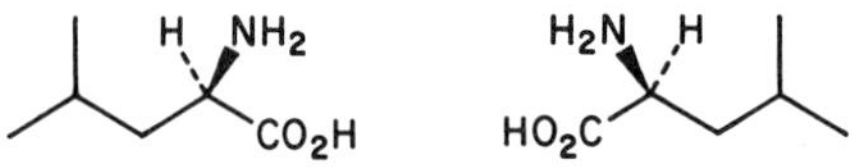

A very important class of molecules is called *polymers*. These are large molecules (*macromolecules*) with a repetitive structure. In *addition polymers*, the repeating segment has the same composition as that of a small molecule, called the *monomer*. For example, ethylene, $CH_2{=}CH_2$, can be polymerized to polyethylene, where n may have any of a range of statisti-

$$-\!\!\left(CH_2-CH_2\right)\!\!-_n$$

POLYETHYLENE $(C_2H_4)_{\underline{n}}$

cally distributed large values—of the order 500 to many thousands. The term *copolymer* denotes a polymer formed from two or more different monomers. Thus, ethylene can be copolymerized with vinyl acetate, $CH_2{=}CHOCOCH_3$, to give ethylene/vinyl acetate copolymer, where m has the same characteristics as n, and there is usually a random distribution of the two monomer units in the polymer chains.

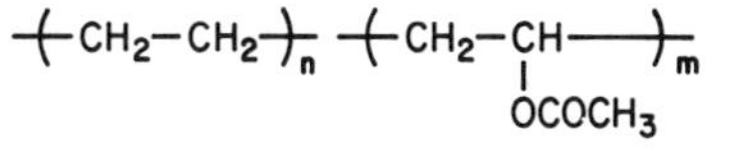

Polymers occur widely in nature. Proteins are polymers derived from several α-amino acids such as leucine. Cellulose is a polymer derived from a class of compounds called sugars; table sugar (sucrose) consists of two molecules of simpler sugars—glucose and fructose.

The properties of many molecules are strongly influenced by a relatively weak kind of bonding—the *hydrogen bond*. Covalently bonded hydrogen has the property of being able to attract electrons of some other atoms, such as nitrogen, oxygen, or fluorine. Water molecules participate in such bonding in liquid water and in ice, the hydrogen bond being shown by the dotted line.

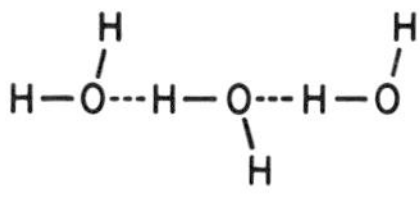

Although hydrogen bonds are some 10 to 20 times weaker than normal covalent bonds, they have important effects on the properties of compounds in which they occur. For example, water might be expected to boil below its actual freezing point (O°C) were it not for the energy required to break the hydrogen bonds in liquid water. Numerous hydrogen bonds recurring at regular intervals on certain long-chain polymeric molecules have profound effects on their molecular behavior and structure, as in the cases of many proteins and of deoxyribonucleic acid (DNA).

# 12
# On Some Recent Developments in Mathematics

## INTRODUCTION

Among intellectual disciplines, mathematics occupies a unique position. It is, with the possible exception of astronomy, the oldest. It has been taught continuously as an integral part of school curricula for over two thousand years. It is in many respects an art, but it is also the language of science. It is pursued, in the words of the nineteenth-century mathematician C. G. J. Jacobi, "pour la gloire de l'esprit humain," while at the same time it serves the practical needs of business and technology.

After World War II came a steadily growing incursion of mathematics into more and more areas of human activity. Even sports have not escaped: regression analysis and related statistical techniques aided by the availibility of high-speed computers are increasingly used in analyzing athletic performance and in planning football strategy. That overbooking on airlines leads only to relatively rare inconveniences is not a matter of blind luck but of careful statistical analysis. Similarly, queuing theory—a sophisticated branch of probability theory—has been used in designing telephone exchanges so as to prevent costly and irritating backups of calls.

The staggering number and bewildering variety of the applications of mathematics at one end of the spectrum, combined with its abstractness at the other, render the task of presenting a popular yet undistorted picture of today's mathematics extremely difficult. To make matters worse,

◀ Production numbers from *Oklahoma!* In general, all problems of allocation, including scheduling rehearsal times for musicals, can be restated as problems of coloring graphs (see p. 483). [Charles A. Goodrum. *Treasures of the Library of Congress.* New York: Harry A. Abrams, 1980. By permission of Mrs. Oscar Hammerstein and William Hammerstein.]

mathematical research, unlike research in some of the scientific disciplines with an empirical background, is focused on a variety of themes and problems; mathematics has no fundamental agreed-upon problem at any given time. Its course seems to the outside observer to proceed on seemingly unrelated fronts, leaving an impression of division and fragmentation. Nevertheless, it is both surprising and illuminating that ideas and techniques from apparently unrelated fragments of the mathematical enterprise have come together in recent mathematical developments with remarkable power and effect.

Although much of contemporary mathematics is abstract and not easily accessible to nonspecialists, the real conceptual and technical advances that have been achieved have serious applications to many problems in science and human life. This chapter offers a few glimpses into the problems and preoccupations of today's mathematics and attempts to give the reader some sense of the liveliness and excitement of the subject.

## Scaling an Intellectual Mountain

The forces that push mathematics forward are many, and the most elemental of these is akin to the drive which for centuries made men climb mountains because they were there.

The mathematical analogs of the Mont Blancs and Mount Everests are problems that for long periods of time defied solution. When at long last such a peak is scaled, there is jubilation, mixed in some cases with soul searching as to whether the achievement was worth the effort.

One of the sections in this chapter is devoted to the conquest of the four-color problem, an example of the "mountain climbing" side of mathematics. This problem, which concerns proving that a map on a plane, or on a sphere, can always be colored using just four colors so that no two countries with a common border are of the same color, is over one hundred years old. It is easy to show that three colors are not enough and that five will suffice. But the answer to whether four colors will also suffice turned out to be stubbornly elusive. For maps drawn on a torus—a surface like that of an inner tube—the required number of colors is seven. This underscores the topological nature of the problem, since a torus and a sphere are topologically distinct, neither being deformable without breaking or pasting into the other.

The problem attracted the attention of many mathematicians. In fact, a method introduced as early as 1913 led to its recent solution. In the last stages of this conquest, the help of the computer proved decisive, but it also raised a perplexing philosophical question: whether a proof which cannot be checked step by step by humans, because of the prohibitively long time that would be required, is entirely acceptable.

Now to the soul searching. Isn't the four-color problem "merely a puzzle"? Should not one direct one's energy, ingenuity, and resources to dealing with "serious problems"? Well, the truth of the matter is that at any given time we do not know what are "merely puzzles" and what may turn out to be "serious problems." Trisecting an angle by means of ruler and compass seemed a mere puzzle, but attempts to prove that such a construction is impossible ultimately led to much of modern algebra and were responsible, at least in part, for introducing the concept of a group, surely one of the most fundamental and applicable concepts of all mathematics.

To take another example, in 1917 an Austrian mathematician, Johann Radon, showed that it is possible to reconstruct a function f(x,y) of two variables by knowing the integrals along all of the lines in the plane. A puzzle? Not quite, for, although the inventors were unaware of it, the extension of Radon's theorem to functions of three variables (reconstructed from its integrals over all planes in three-dimensional space) is the mathematical basis for the X-ray mapping technique known as computerized X-ray transmission tomographic scanning (CAT), for which the Nobel Prize in Medicine or Physiology was awarded in 1979. It is perhaps worth mentioning that Radon's discovery was foreshadowed as early as 1906 by Hendrik Lorentz in the course of lectures on the theory of waves—a subject quite different from that which motivated Radon.

## Forging New Links to Physics

The scaling of another peak is described in the section of this chapter on the classification of simple groups. While the four-color problem is somewhat peripheral to the mainstream of mathematics, the classification of simple groups is unquestionably in the center of an active and lively field of modern algebra. It, too, has had a long history, and it, too, attracted the attention of many mathematicians. It is much more technical, and, in contrast to the four-color problem, the methods used in carrying it out are mathematically more intricate. It is much less likely to be considered a mere puzzle, largely because it has deep connections with seemingly unrelated parts of mathematics. The connection with a separate field of mathematics called the theory of automorphic functions, which is briefly mentioned at the end of the section on groups, is especially fascinating and is likely to attract much further work.

Conquests of peaks are relatively rare, and it is pure happenstance of history that two of them can be included in this Five-Year Outlook report. Most of the time, progress in mathematics is more like exploring a territory with uncharted, and even undetected, areas. The developments that will probably be adjudged the most exciting of the past five years fall into this category.

Although much of mathematics, especially the crucial problems, can be traced to influences coming from external sources, the bulk of its creativity is motivated internally. Unhampered by empirical facts of the external world, the mathematician can let his imagination soar with no checks save those of logical consistency and some sense of historical continuity. Abstractions are followed by abstractions, and generalizations yield more generalizations.

And then once in a miraculous while the result of such a process makes contact with physical reality. Einstein's theory of general relativity is a striking example. The story, or rather one of two stories, begins with the work of Carl Gauss on curved surfaces in ordinary three-dimensional space. Gauss made the pivotal discovery that there is a measure of curvature (called in his honor Gaussian curvature) that is intrinsic to a surface; for example, a flat creature living on a surface and unaware of the fact that the surface was imbedded in three-dimensional space could by appropriate measurements calculate how much the surface curves at a given point.

Following Gauss, G. F. B. Riemann sketched a program for generalizing the theory of curved surfaces to a theory of curved spaces of arbitrarily many dimensions, thus laying the foundations of the branch of differential geometry that we now call Riemannian geometry. Riemann's ideas were taken up and developed by Italian mathematicians, and it turned out that in higher dimensions curvature cannot be described by a number (a scalar); one needs instead a much more complicated construct called a tensor. These Italian mathematicians then laid the foundation of tensor calculus, which became the principal tool of Riemannian geometry.

Then came Einstein, who about 1907 began struggling with the theory of gravitation and was searching for an appropriate mathematical framework for his ideas. When his friend, the mathematician Marcel Grossmann, called Einstein's attention to Riemannian geometry, the search was over. The celebrated Einstein equations of general relativity express the equality of two tensors, one of which is related to the curvature of space-time.*

## Nothing More Practical Than Good Theories

In the past few years, mathematicians and physicists have witnessed a similarly miraculous confluence of ideas. On one end were physically

---

* In general relativity, there is an added complication, insofar as the square of the differential distance ds is given (in appropriately chosen coordinates) by the formula

$$ds^2 = (dx_1)^2 + (dx_2)^2 + (dx_3)^2 - (dx_4)^2$$

rather than by the formula

$$ds^2 = (dx_1)^2 + (dx_2)^2 + (dx_3)^2 + (dx_4)^2,$$

which is a generalization to four dimensions of the familiar Pythagorean theorem. The minus sign in the first formula reflects the special role of the fourth coordinate—time.

motivated theorics called gauge fields that were developed as frameworks for dealing with weak and strong interactions; on the other was an inspired generalization of Riemannian geometry that was conceived in the kind of intellectual solitude that tends to characterize abstract mathematics. The section of the chapter on gauge theories provides some details of this remarkable encounter, but it may be useful here to give an imperfect snapshot of the mathematical side of the picture.

Riemannian geometry, rooted as it was in the theory of curved surfaces, made essential use of tangent spaces, analogs of tangent lines to curves and tangent planes to surfaces. It also defined the all-important concept of parallel displacement, in analogy with earlier theory. To explain further, the underlying skeleton of Riemannian geometry is a space manifold. This manifold is the analog of a curved surface M which has a tangent plane $T_x$ attached to each point x of M and a rule that specifies how to perform parallel displacements of elements. In Riemannian geometry, keeping in mind the analogy with curved surfaces, parallel displacements are defined in terms of measurable properties of the underlying space M.

Now, one can replace tangent spaces by more complicated structures—groups of matrices, for example—that are not necessarily related to the space M to which they are attached. One can also define analogs of parallel displacement, loosening the intimacy that existed between the tangent spaces $T_x$ and the underlying manifold M. To be sure, some restrictions have to be imposed; but the freedom gained by departing from the strict confines of Riemannian geometry is great—so great, in fact, that Maxwell's equations of the electromagnetic field can be couched in geometrical terms, as can the much more complicated fields that have been proposed by physicists to deal with weak and strong interactions.

These and related developments in differential geometry tended to focus on its global aspects, not on its local ones, which at least as far as Riemannian geometry is concerned had been largely worked out by the 1920's. To understand what is at stake, it is sufficient to consider smooth curved surfaces again. On each such surface, a small neighborhood of a point is indistinguishable from a correspondingly small piece of the tangent plane at this point. This is why our ancestors, before the days of long-distance travel, thought that the earth was flat. Thus, all small neighborhoods of smooth surfaces look alike, namely, flat. Only when all of the tangent planes are put together does an idea of the overall or global appearance of the surface emerge.

A question arises at once as to what local information for all points of the surface is needed to reconstruct the surface, and, analogously, to reconstruct curved spaces of higher dimensions. Of special interest are global questions in general relativity, for these are intimately connected with the most exciting problems of astrophysics, such as the existence of black holes.

There are many questions in general relativity and cosmology that can be attacked using recent developments in the global aspects of differential geometry, and some of the successes are mentioned in the last section of this chapter.

The problem of reconstructing surfaces or curved spaces from local information requires solving differential equations. But unlike most differential equations in mathematical physics, which are linear and in general describe small changes and effects, these equations are nonlinear. Such nonlinear equations are for the most part terra incognita even to mathematicians. The oldest and best known set of nonlinear differential equations are those of fluid flow—the equations of hydrodynamics.

In spite of their having been intensively studied for over a century, and in spite of their great practical importance, for example, in airplane design, relatively little is known about nonlinear equations. For example, there is still no satisfactory mathematical theory of turbulence. Mathematicians do know that nonlinear equations give rise to unusual phenomena, the best known example being the shock wave, which was predicted by Riemann before it was actually observed.

In recent years, progress has been made in understanding nonlinear equations, including two notable successes. The first concerns the discovery and classification of a class of important solutions to a set of equations proposed by Chen-Ning Yang and Robert Mills to describe weak and strong interactions in particle physics. This remarkable feat was accomplished with tools adapted from the "purest" parts of mathematics, namely, topology and algebraic geometry, underscoring once again that one never knows what in a given situation may turn out to be "useful."

The second success is a remarkable story in itself, concerning a novel method for solving a whole class of nonlinear differential equations. It started with a little-known approximate differential equation that was derived late in the last century to explain a striking phenomenon observed in narrow water channels. The phenomenon is that of solitary waves (solitons). These were first observed in 1847 by an Englishman named John Scott Russell. He saw a single standing wave propagating along a canal leading to the river Thames. An equation was derived by two Dutch scientists that approximates this behavior, in the sense that the equation has solutions of a form consistent with the shape of the observed solitary wave. The matter then rested for over 70 years, until four American scientists discovered that solving the equation could be reduced to solving a problem—the inverse scattering problem—related to a linear equation that in this case, remarkably enough, turns out to be the Schrödinger equation familiar from quantum mechanics. This method, which became known as the inverse scattering method, was then extended to many other nonlinear equations, and the game is still being played all over the world.

It should also be pointed out that in the search for periodic solutions of the nonlinear equation describing solitons, one encounters "hyperelliptic functions," first introduced by the ubiquitous Riemann in his treatment of a certain class of algebraic curves. Other aspects of this work have had an impact on algebraic geometry, providing yet another illustration of the remarkable ways in which mathematics develops and grows.

The first section of this chapter deals with problems that owe their being to the computer. That does not mean that they could not have arisen if the art of electronics had not kept pace with the ideas of Alan Turing and John von Neumann that form the basis for today's computing machines. Questions as to whether and how easily something can be computed belong naturally to logic and have preceded the electronic assemblies that actually perform the computing. But the computer gave many of these problems an actuality that they otherwise would have lacked, and it is not overgenerous to credit the computer with spawning a challenging new branch of mathematical research.

An illustration is in order, specifically, a well-known and widely discussed problem, that of the traveling salesman. Suppose that a salesman is to visit n cities, 1, 2, . . . , n, and that the cost of going from city i to city j is known. The question is how to plan a route through all the cities that will have the lowest total cost.

This problem is of a type encountered in economics under the general heading of allocation of resources. One has to consider all possible itineraries—a finite number—and compare their costs. But if the number n of cities is large, e.g., 50, the number of all possible routings is enormous, and from a purely practical point of view the problem may be insoluble. It is therefore of considerable interest to have an estimate of how long it would take a computer to solve the problem. This is not only a difficult mathematical problem; it also has close and profound connections with many sorts of questions in pure mathematics, such as in the theory of polyhedra in n-dimensional spaces.

This brief introduction and the sections that follow cannot even come close to transmitting to the reader the full scope of mathematical activity taking place today. This chapter can hardly touch upon algebra, which is, as it has been for many decades, an active field of mathematical research. It can only allude to topology and algebraic geometry. It has to leave out probability theory, a mathematical field of great activity which is unique in that it is an integral part of the whole of scientific methodology. Neither is there space to deal with the rapidly growing field of mathematical physics, which is coming to grips with some of the thorniest mathematical difficulties posed by quantum field theories and related subjects.

Still, it is hoped that the reader will be left with some impression of the remarkable edifice called mathematics, which under one roof can house in harmony and unity subjects ranging from the traveling salesman problem to differential geometry, from the classification of finite groups to solitary waves.

## THEORETICAL COMPUTER SCIENCE

The front page of the *New York Times* of November 7, 1979, displayed the headline "A Soviet Discovery Rocks World of Mathematics." Although the headline was a gross exaggeration and the article itself contained serious misinformation, the fact that a mathematical result was considered so newsworthy accurately reflects the importance of the active field of research called algorithmic analysis, which is part of theoretical computer science. This field, developed in large part by American researchers, uses mathematical methods to predict the efficiency of procedures, or algorithms, as they are called, for use in solving problems by computer.

Some of the basic issues involved in algorithmic analysis can be appreciated from a simple example: searching an ordered list. Imagine a given list of items stored in a computer in a definite order, such as a telephone directory. Since a computer that is being used to determine whether or not some particular name is in the phone book cannot flip the pages in the way a person would, it could instead be programmed to compare the name being sought with each successive name in the stored list until the name is found or until the entire list has been unsuccessfully searched. This crude procedure, which does not even make use of the fact that a phone book is arranged in alphabetical order, will require at worst as many comparisons as there are items in the list (when the item being sought is not in the list).

A much more efficient algorithm, called the binary search procedure, would compare the item being sought with an item midway through the list. This one comparison enables the computer, because the list is ordered, to eliminate from consideration half of the list. This process would then be continued, the computer looking halfway down the items that remain at each stage. Searching a list of 128 items by the first crude method may require as many as 128 comparisons; the binary search procedure would require only 7 comparisons. The reduction becomes even more dramatic for larger lists.

This example shows how an adroit choice of algorithm can greatly reduce the number of steps required to solve a problem. Algorithmic analysis is concerned with issues involving the number of steps that specific algorithms require to solve a problem, and, in particular, how many steps are required in the "worst" case.

## Yes-or-No Algorithms

Many of the most important problems studied in algorithmic analysis are called decision problems; these are problems to which the solution is a simple yes or no. The list-searching problem just discussed is a decision problem; either the item being sought is or is not in the list. An algorithm for the solution of a decision problem is a general procedure that will work, at least in principle, on any particular example to which it is applied. Thus, the binary search procedure for the list-searching problem can be applied to any item together with any ordered list to produce either the answer "yes—the item is in the list" or "no—the item is not in the list." How many steps an algorithm will require depends, of course, on the size of the particular example to which it is applied. The particular example is called the input to the algorithm, and its size is conveniently measured by the number of symbols required to write it down.

Researchers in algorithmic analysis distinguish between tractable and intractable algorithms. A tractable algorithm is one that will produce answers in a "reasonable" amount of time for inputs of "reasonable" size. An algorithm that always works in principle, but is not tractable, is naturally called intractable.

### GLOSSARY

**Algorithm** In computer science, a procedure consisting of discrete steps that seeks an answer to properly formulated problems.

**Decision problem** A problem that can be answered with either a yes or a no.

**Polynomial time algorithm** An algorithm that can solve a decision problem in less than a certain number of steps. This number is determined by a subsidiary procedure applied to the number of symbols in the input and consisting of some number of additions and multiplications that is fixed in advance.

**Nondeterministic algorithm** A pseudo algorithm that yields an answer to a decision problem by knowing, at each step in the process, one of the various possible correct steps to take.

**NP-complete problem** A decision problem that has a nondeterministic polynomial time algorithm and is equivalent, via a polynomial time algorithm, to every other decision problem with a nondeterministic polynomial time algorithm.

**Oracle** An imaginary device that furnishes the correct answer to a decision problem without using any algorithm.

Since tractability is not a precise notion because of the vagueness of the word "reasonable," work has centered on polynomial time algorithms for decision problems. What makes an algorithm be polynomial time is the availability of a subsidiary complexity-predicting procedure that employs only some fixed-in-advance number of additions and multiplications. (The number of additions and multiplications is fixed in advance of being presented with the actual data. For example, to calculate $n^3$ takes two multiplications, but to calculate $3^n$ requires a number of multiplications that cannot be fixed in advance of knowing n.) This subsidiary procedure, when applied to the size of an input, will compute a number that is guaranteed to exceed the number of steps the algorithm will take in determining whether the correct answer for the input is yes or no.

In practice, algorithms regarded as tractable are also found to be polynomial time algorithms, and algorithms regarded as intractable tend not to be. A typical intractable algorithm may require a number of steps that is an exponential function of the size of the input. The intractability is then an instance of the familiar explosiveness of exponential growth. Computing an exponential function of the size of an input requires a number of multiplications that depends on that size, and hence this number cannot be fixed in advance, as is required of polynomial time algorithms.

The discovery mentioned at the beginning of this section, which was made by the Soviet mathematician L. G. Khachian, was that there is a polynomial time algorithm for a specific important decision problem: the general linear programming problem. A typical example, or input, for the general linear programming problem is as follows. Let there be warehouses in Detroit, Syracuse, and Denver for storing shoes with capacities of 17,654, 28,231, and 19,158 pairs, respectively; and let there be wholesale distribution centers for shoes in Chicago, New York, and Los Angeles with capacities of 5,215, 46,183, and 13,645 pairs, respectively. Let the cost of shipping one pair from a warehouse to a distribution center be given by the following.

| Origins | Destinations | | |
|---|---|---|---|
| | Chicago | New York | Los Angeles |
| Detroit | 5¢ | 9¢ | 17¢ |
| Syracuse | 6¢ | 4¢ | 18¢ |
| Denver | 10¢ | 12¢ | 11¢ |

Can all of the pairs of shoes be moved from the warehouses to the destinations for a cost of less than $6,000?

Applications of linear programming are ubiquitous; they occur in business, economics, and engineering. Because of the usefulness of linear programming, there has been a great deal of work on suitable algorithms for solving the general linear programming problem. But, until Khachian's work, no polynomial time algorithm for this problem was known.*

## Does P = NP?

It is possible that future refinements of Khachian's methods will lead to improved linear programming algorithms, but this is not the aspect of the matter that most excites researchers in algorithmic analysis. Rather, the existence of a polynomial time algorithm for the general linear programming problem is so provocative because it sheds new and surprising light on an ambitious effort by theoretical computer scientists to classify decision problems according to the least number of steps required by algorithms for these problems.

The observation that originally initiated this was that, for a large class of decision problems for which only intractable algorithms were known, polynomial time procedures could be found if an element of nondeterminism was permitted. Such nondeterministic procedures would not be genuine algorithms. For an input with a correct answer of yes, such a procedure would be guaranteed to reach this answer only if at each stage in the algorithm the correct next steps were chosen from the alternatives. It is as though a computer were equipped with a magic roulette wheel which it could consult about its next operation and which would always advise it correctly.

The class of all decision problems that possess polynomial time procedures of this type is called NP, for nondeterministic polynomial; by contrast, the class of problems with genuine polynomial time algorithms is called P. Clearly, the class P is included in NP. Evidence suggests that many tractable decision problems are in P, and that many apparently intractable problems are in the class NP. These observations have strongly suggested to mathematicians that the classes must be different, that P $\neq$ NP. However, despite the fact that the question "Is P = NP?" has been regarded as the central problem in theoretical computer science for a decade, it has so far resisted all efforts to settle it.

The focus of much of the research on NP has been on what are called

---

*The popular simplex method, developed by George Dantzig of Stanford University, is not a polynomial time algorithm. Although Dantzig's algorithm works rather well in practice, it is known to be exponential in the worst case. The real importance of Khachian's discovery is not that it will lead at once to improved practical algorithms for linear programming. In the present state of knowledge, the simplex method works far better on examples likely to be met in practice than would algorithms derived from Khachian's work.

NP-complete problems. A decision problem A is called NP-complete if A is in the class NP and if every problem in the class NP can be translated into A by a procedure which works "rapidly" relative to the size of the input. Stephen A. Cook of the University of Toronto introduced the notion of NP-completeness in a ground-breaking paper in 1971 and proved that a number of important decision problems are indeed NP-complete. A year later, Richard Karp of the University of California at Berkeley proved that a number of other problems are also NP-complete. Among these is the traveling salesman problem: given various cities (the distance between them being known) and a number D, does there exist a journey of no more than D miles that passes through all of the given cities?

Since NP-complete problems are all polynomial time reducible to one another, they are, in a sense, all equivalent. In particular, by using translation procedures, a polynomial time algorithm for any NP-complete problem can yield a polynomial time algorithm for every other NP-complete problem. Thus, if even one NP-complete problem turned out to be in P, it would follow that P = NP.

The prevailing judgment among computer scientists is that it is most unlikely that P = NP and that, therefore, no NP-complete problem is in P. Because of the tendency to regard membership in the class P as equivalent to tractability, these considerations suggest that proving a problem to be NP-complete is strong evidence of its intractability. This has led to efforts to develop effective techniques for deriving such proofs, and hundreds of decision problems, many of them of considerable importance, have been shown to be NP-complete over the past decade, mostly by American computer scientists.

In his 1972 paper referred to above, Richard Karp discussed the linear programming problem, among others. It is very easy to see that the linear programming problem belongs to NP, but Karp was unable to show it to be NP-complete. Although no one could settle the matter, it was widely believed that the linear programming problem was NP-intermediate, that is, neither in P nor NP-complete. Khachian's proof that the problem is actually in P of course ended the matter. But the rebuff to the intuition of experts sent a shock wave through the algorithmic analysis community and led at least some to wonder whether polynomial time algorithms might not also exist for NP-complete problems, that is, whether P might not actually be equal to NP.

## The Easy Versus the Hard Way

Why has the question of whether P = NP proved so difficult to settle? Some recent results obtained by computer scientists at the University of California at Berkeley and at Cornell University can be interpreted as implying that there is no "easy" way. These results involve an "oracle," a fruitful notion first introduced in 1940 by the great English logician and computer scientist Alan Turing. (Turing, incidentally, was part of a team largely responsible for the spectacularly successful effort by the English to decipher German military communications during World War II—Project ULTRA.) The idea is to imagine a computer furnished with a magic device that instantaneously and at no cost furnishes the yes or no answer for some decision problem when given an input. Naturally, a computer enhanced by such an oracle can accomplish more than one without. Most proofs in theoretical computer science remain valid when they are "relativized" by introduction of an oracle. Hence, a proof that P = NP should imply a proof that more can be accomplished in polynomial time nondeterministically than deterministically, even when an oracle is allowed. However, in 1976, it was shown that while this is the case for some oracles, it is false for others. In the same year, it was shown that when a suitable oracle is chosen, the question cannot be resolved even from a system of axioms powerful enough to derive all of ordinary mathematics.

These indications that the P = NP problem is not likely to be settled by ordinary methods suggest that some of the more exotic parts of mathematical logic have a role to play in computer science. This connection is especially interesting because of the part that mathematical logic played in formulating the basic concepts of computer science. The first successful model of computation was supplied in 1936 by Turing and, independently, by the American logician Emil Post. Turing's work was a theoretical prediction that an all-purpose computer was feasible. This so-called Turing machine remains the basis of much work in theoretical computer science. In fact, some of the concepts discussed here are formulated more precisely in terms of Turing machines. For example, in speaking of the number of "steps" in a computation, as we have been doing, it is necessary to be precise about just what constitutes a step, and Turing's model permits us to do so unambiguously.

Turing's work was based on the work of the great logician Kurt Gödel and was parallel to, but independent of, the work of the American logician Alonzo Church, who used a mathematical system called the lambda calculus, which has been of considerable recent interest to computer scientists. Polynomial time reducibility is a refinement of a notion introduced in 1944 by Emil Post.

The problems of theoretical computer science arise from computing practice, and the concepts developed by theoretical computer scientists in turn provide a necessary conceptual framework for those working in the more applied parts of the field. The next five years will certainly see much new and interesting work on the tractability of decision problems.

## THE FOUR-COLOR THEOREM

How many different colors does a mapmaker need? English cartographers of the nineteenth century noticed that it was never necessary to use more than four colors to distinguish the countries of any map they had ever seen. Is this an accident of the particular political history of Europe, or is there a more general principle at work here? Is it true that all maps that could possibly ever exist can be colored in four or fewer colors, and, if so, can this be proven? The answer to both questions is yes—four colors are enough, and this can be proven.

The four-color theorem is a curious problem. Unlike those classically associated with mathematics, the objects being considered are neither numbers nor functions but ordinary geopolitical maps. The operation on these objects is not addition, multiplication, or differentiation; it is the application of colors, the activity a child undertakes with a coloring book. Even so, it is still a sophisticated mathematical problem to show that there cannot exist a map requiring five or more colors. Such proofs of the nonexistence of a specific object are often among the most difficult and rewarding in mathematics.

This problem, although it is so simple to state, proved intractable to many mathematicians during more than a century of frustrating effort. The solutions that we have today depend heavily on enormous calculations performed by computers. They are not the clear and simple explanations of why the result is true that mathematical proofs ideally should be. For this reason, investigators of this topic continue to seek a fuller understanding of the problem and the principles behind it. It is also the nature of the proofs we have today that has led to philosophical problems unique to our age.

## The Challenge Begins

The question of whether it is possible to draw a map, no matter how wildly convoluted, that requires more than four colors was first asked in 1850 by Francis Guthrie, who was then a student at Edinburgh. Shortly thereafter, the problem was communicated to Augustus de Morgan, a noted mathematician of the time, who publicized it to the mathematical world.

A purported solution was published in 1879 by the English barrister A. B. Kempe. His proof was faulty, but it contained many interesting ideas, and its flaws were not exposed for over ten years.

An entire branch of mathematics called graph theory has evolved over the last 100 years in part to solve this very problem. The development of graph theory illustrates the dual nature of mathematical progress. Sometimes theoretical results pursued for their own aesthetic nature are later discovered to have practical value; and sometimes work on specific practical problems gives rise to general theoretical discoveries. This is especially true of the relationship that has developed between graph theory and computer science.

To return to the statement of the four-color theorem, let us consider a map of the 48 contiguous states. We wish to color each state in such a way that any two states sharing a common border receive different colors. By the phrase "a common border," we mean more than just one point. Utah and New Mexico touch at only one point, and so we shall allow them to be colored with the same color, just as the squares of a checkerboard can be colored with the two colors red and black. One way of coloring the United States is to use 48 different colors, each for a different state. This guarantees that no pair of neighboring states receives the same color. To be more efficient, we ask the question "what is the least number of colors with which we can color the map?"

We can see immediately that it is not necessary to use 48 colors. The state of Maine borders only on New Hampshire. We can color the 47 states other than Maine with 47 different colors, and then color Maine with any one of these colors other than the one used for New Hampshire. It is also easy to demonstrate that we need at least four different colors for our map. The state of Nevada borders on five states: Oregon, Idaho, Utah, Arizona, and California. It must receive a color different from any of the colors of these five states. If we tried to color the whole map in three colors, then the ring of five states around Nevada must be colorable in two colors—$C_1$ and $C_2$. If Oregon is colored $C_1$, then bordering Idaho must be colored $C_2$. Utah borders Idaho and so must be colored $C_1$. Arizona borders Utah and so must be colored $C_2$. Now California borders Arizona, which was colored $C_2$, but it also borders Oregon which was colored $C_1$. So California cannot be colored. The ring of five states cannot be colored in two colors; at least three different colors are needed for them. And Nevada, which borders on all five of the other states, must then be colored with a fourth color.

So far, we have shown that to color the United States requires at least 4 and at most 47 different colors. Surprisingly, the final answer to the question is that four colors will suffice to color the entire map. Even more surprising is that even if the borders between the states are redrawn in any possible fashion and any number of new states are created, we can still color the map in four colors. This is the four-color theorem.

In the usual statement of this theorem, we refer to the areas to be colored as countries (not states). We require that the countries be in one solid piece and not be separated like Michigan. We further require that all the countries be contiguous. With these stipulations, the four-color theorem states that "the countries of any map can be colored in four or fewer colors such that no two countries which share a common border receive the same color."

## The Duality of Map and Graph

The concepts of country and coloring may seem very far from the usual considerations of science and technology, but the four-color theorem is

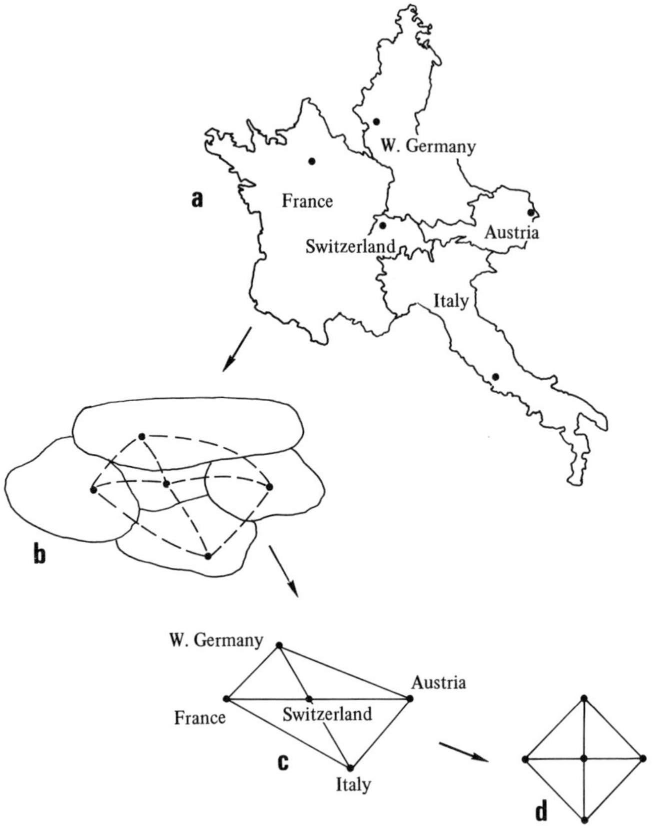


**Figure 1**

not as remote from application as it may appear. Consider a specific map of countries as in Figure 1a. Inside each country designate a single capital city by a point. Draw a line between two cities if their respective countries share a common border. The lines between capitals can be drawn without intersecting, as in Figure 1b.

We shall call the points *vertices* and the lines connecting them *edges*. The resultant diagram is the dual graph of the map (see Figure 1c). The dual graph, an abstraction of the map, carries with it all the information about which country borders which, but it loses all specific information about size and shape of the countries. This lost information is irrelevant to the consideration of coloring.

The graphs that can be obtained in this way are called planar graphs, because they can be drawn on a plane in such a way that the edges do not meet anywhere but at a vertex (Figure 1d).

The problem of four-coloring maps then becomes the problem of four-coloring the vertices of planar graphs. The four-color theorem states that we can label the vertices of any planar graph with a set of four or fewer labels (a label is in effect a color) such that no pair of vertices connected by an edge receive the same label. It is in this form that the four-color theorem has its greatest applicability.

For example, suppose that we are producing a Broadway musical that contains several production numbers. The casts involved in these musical numbers are not disjoint, in other words, the same singer may be needed in more than one song. We wish to establish a rehearsal schedule such that as many songs as possible can be practiced at the same time. Of course, no two songs can be rehearsed simultaneously if they require the presence of the same singer, who can only be in one room at any one time. This is a problem of graph coloring in the following way.

Let each musical number be represented by a vertex. Connect two vertices with an edge if the corresponding musical numbers have at least one singer in common and cannot be rehearsed at the same time. The result is a graph. The problem of assigning rehearsal slots is the problem of labeling the vertices of this graph with rehearsal times, instead of colors, in such a way that no pair of vertices connected by an edge is assigned the same time. The question of determining the fewest separate rehearsal times required is equivalent to the problem of coloring this graph in the least number of colors.

Consider the following concrete example. There are seven musical numbers with casts as given below, where A, B, C, D, E, F, and G are singers.

#1: A, D, F
#2: A, B
#3: A, B, C, D, E
#4: B, C
#5: D, E, F, G
#6: F, G
#7: C, E, G

The graph representing these seven musical numbers is shown below.

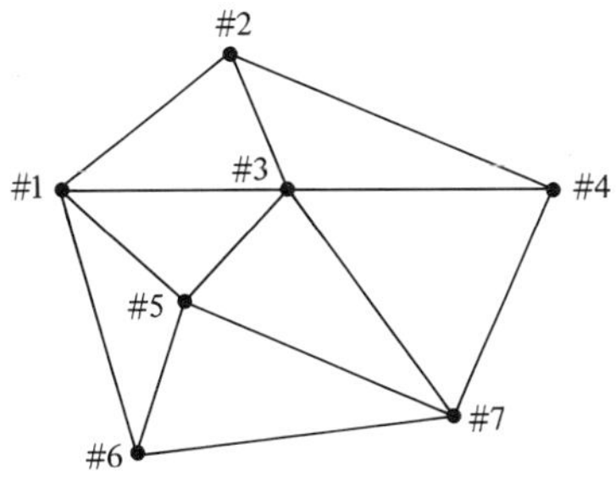

An edge connects any two vertices whose corresponding musical numbers have a singer in common. Because they have singer B in common, #2 and #4 are connected; on the other hand, #3 and #6 are not connected. This graph is a planar graph (its edges do not intersect), so the four-color theorem says that this graph can be colored in four or fewer colors, as below. Notice that if we drop song #7, the graph that results can be colored in three colors.

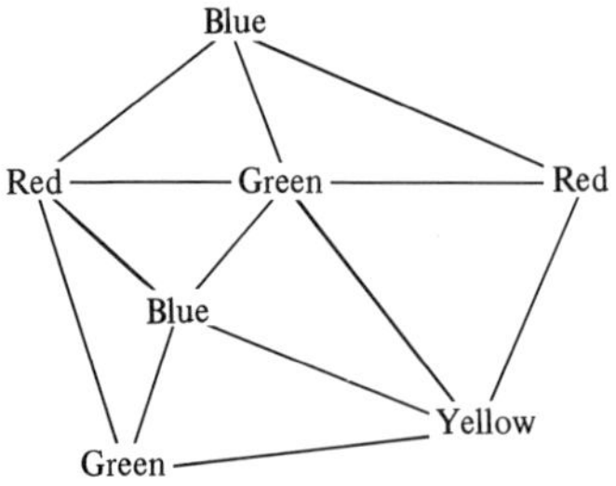


The two musical numbers colored blue can meet at one o'clock; the two musical numbers colored red can meet at two o'clock; the two musical numbers colored green can meet at three o'clock; and #7, colored yellow, can meet at four o'clock. This particular set of rehearsals cannot fit into fewer than four different time slots.

This method applies equally well to scheduling congressional committee meetings—the number of time slots for meetings should be kept to a minimum, but two committees cannot be assigned to meet at the same time if they have a member in common.

One of the most frequent uses of graph coloring is found in the operation of a computer. As a computer executes the instructions in its program, certain pieces of information, such as the results of arithmetic calculations, must be stored throughout the whole run of the program; other information is only needed up to a point and can then be forgotten. The information that will have to be referred to in the future is stored in index registers, which at any one time can contain only one piece of information each. If one piece of information can be forgotten before another piece is created, then it is possible that both may be kept in the same index register at different times. Two pieces of information whose useful life spans overlap cannot be kept in the same index register.

Let all the pieces of information used in a specific program be equated with vertices. Connect two vertices by an edge if the two pieces of information cannot be assigned to the same index register. We have now constructed a graph. The question "what is the least number of index registers needed by this program?" is the same as "what is the least number of colors with which we can color this graph?" All the vertices assigned the same color can be allocated to the same index register. Since this graph need not be planar, the four-color theorem may not apply, and we may have to use many index registers. This is not a frivolous example. The problem of assigning index registers is one that computers must solve millions of times a day, and the approach used is entirely analogous to that of graph coloring.

These examples illustrate that in general all problems of allocation can be restated as problems of coloring graphs, and therefore the study of chromatic graph theory (graph coloring) should have an important bearing on issues that can be stated as allocation problems. One of the fastest growing fields of information science is the development of mathematical models that use the structures of graphs and colorings to describe and analyze sets of data.

## Critical Maps and Unavoidable Sets

The structure of the proof of the four-olor theorem, as it has evolved over the past century, is in the form of a *reductio ad absurdum* argument. This means that we begin by stipulating that the theorem is false, because there is some map (call it M) that requires more than four colors to be properly colored. We then logically examine the conclusions that we can deduce about M; and when we have discovered enough information, we find that M must possess conflicting properties and so cannot exist. The fact that no counterexample can exist suffices to prove the theorem.

Suppose now that some maps did exist requiring more than four colors. Most of them would be very large, by which we mean that they would contain many countries. Let us call a map which can be colored in four or fewer colors a *colorable* map. Let us start with some uncolorable map $M_1$. A map is called smaller than $M_1$ if it has fewer countries than $M_1$. Let us look through all the maps smaller than $M_1$ to see if there is one that is also uncolorable. If there is, then let that map be called $M_2$; and let us look through all the maps smaller than it to see if any of them are also uncolorable. We then call any such map $M_3$ and look for a still smaller uncolorable map.

This process cannot go on forever, since sooner or later we will be down to maps of four or fewer countries, all of which are colorable. Let this process stop at the map M. This map has two properties: (1) it is uncolorable, and (2) all maps smaller than it are colorable. There may be other maps that also have these two properties, but they will necessarily have the same number of countries as M. All maps with these two properties are called critical.

Either the four-color theorem is true or there exist critical maps. It is possible to determine some detailed facts about critical maps.

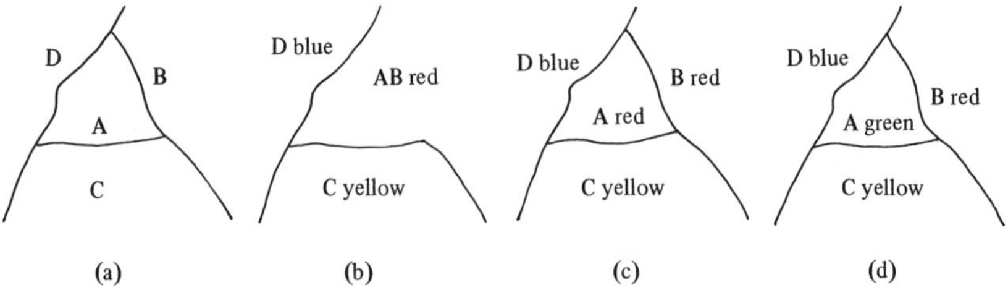


**Figure 2**

Theorem: No critical map can contain a country with only three neighbors.

Proof: Let us suppose, to the contrary, that the critical map M contains a country, A, with three neighbors, as shown in Figure 2a. Then suppose that countries A and B become united so that the boundary between them is erased, as in 2b. The map resulting from this has one fewer country than M, so it must be colorable. Let us color the whole new map with the colors red, blue, yellow, and green. After this coloring takes place, suppose that countries A and B become disunited again and the border between them reappears, as in 2c.

We have now returned to the original map M, and it is *almost* properly colored. Every country has a color, and no two countries of the same color are neighbors, except for A and B. However, this conflict can be resolved. We can reassign to country A a new color so that it will not conflict with the colors of any of its neighbors. This is shown in 2d. This recoloring can

## GLOSSARY

**Dual graph** A representation of a map in which each country is indicated by a dot and lines connect neighboring countries.

**Critical map** A map (if there are any) which cannot be colored in four colors and such that all maps with fewer countries are colorable in four colors.

**Configuration** A grouping of countries inside a map.

**Reducible** A configuration is reducible if it can be shown that it is never part of any critical map.

**Unavoidable set** A set of configurations such that *any* map must contain at least one of them.

**Discharging** A method of producing unavoidable sets.

always be performed, because country A has only three neighbors and it is always possible to choose a color different from those assigned to all three.

In this way we have shown that map M is colorable. But this contradicts the hypothesis that M is critical. Therefore, no critical map M could exist which had a country with three neighbors.

What we have shown above is that some configurations, such as the one in Figure 2a, simply cannot exist in any critical map. Any such configuration we call reducible.

In 1913, George David Birkhoff of Harvard University developed a powerful method of proving that certain configurations, such as that shown below, are reducible. Many more configurations were found to be reducible in the ensuing years.

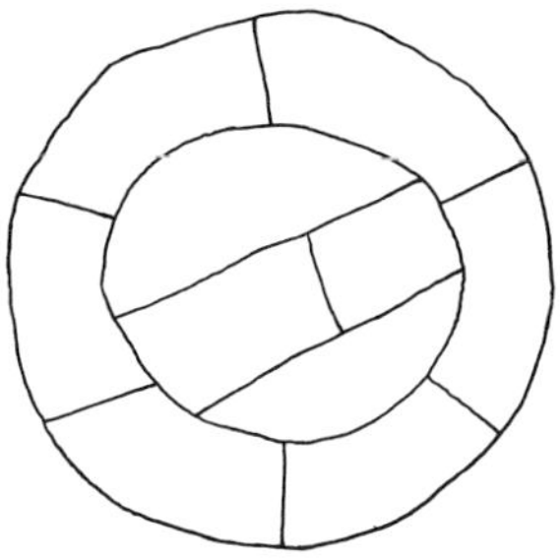

Reducible configurations are what *must not* be found in critical maps. But there is another side to the problem—discovering what *must* be found in all critical maps. Using a formula discovered by both René Descartes and Leonhard Euler, Henri Lebesgue proved in 1940 that every critical map that exists must contain at least one of the seven configurations shown in the figure below. The configurations are so complicated that they are given in their dual graph forms, where a digit has been used to indicate the number of edges coming out of each vertex.

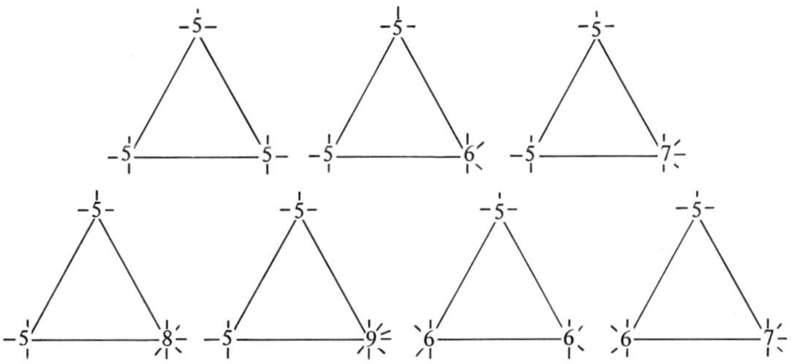

A set of configurations with the property that all critical maps that exist must contain at least one of the configurations is called an unavoidable set.

If someone were to be able to show that the seven configurations found by Lebesgue were each reducible, then the four-color theorem would be solved by *reductio ad absurdum*. Every critical map M would have to contain one of the configurations, since the set is unavoidable, but could not contain any of the configurations, since they are all reducible. This contradiction would show that no critical map M could exist.

To date, no one has been able to show that Lebesgue's configurations are reducible, but the theorem was finally solved along much the same lines.

## Progress in the 1970's

Oystein Ore of Yale University published an influential book in 1967 devoted to the four-color theorem, in which he collected all the bits of research known at that time. The next significant research came from H. Heesch in 1969, who had considerable experience in determining whether configurations were reducible. He had attacked many large configurations and had discovered that configurations which he could prove to be reducible had certain specific characteristics; he called such configurations Z-positive. He did not prove a theorem that all Z-positive configurations were reducible, but he gave massive empirical evidence for this fact. He then modified Lebesgue's procedure for finding unavoidable sets and looked for an unavoidable set all of whose configurations were Z-positive. This he found by using a method he called discharging. Heesch produced an unavoidable set of some 9,000 configurations, all of which were Z-positive, and so all of which he believed could be shown to be reducible.

In the abstract, this appears to be the solution to the four-color problem except for one catch. The classical method for demonstrating that a given configuration is reducible requires an immense amount of calculation, and the complexity of this calculation grows enormously as the size of the configuration increases. Heesch not only had a large number of configurations to test, but many of the configurations were themselves gigantic, so much so that the computation required was well beyond the scope of the fastest and largest computers. As these machines get more and more powerful, we may see at some future time that Heesch was a mere computer run away from proving the four-color theorem, but the result of his work at the time was inconclusive.

Heesch had developed a procedure for producing an unavoidable set with certain properties which indicated that the configurations were likely to be reducible. Walter Stromquist of Harvard modified the discharging procedure in 1975 to produce a smaller and more manageable unavoidable set with even more desirable properties. But this set was still too big to handle.

Then, in 1976, Kenneth Appel and Wolfgang Haken of the University of Illinois at Champaign-Urbana announced that they had discovered a modification of discharging that produced an unavoidable set of some 1,900 configurations, all of which they were able to show were reducible.

This meant that the four-color theorem was in fact proven. The calculation had been reduced to one small enough to be performed by a computer, albeit in hundreds of hours of machine time. However, there was a great amount of unhappiness about the solution, because the calculation was so large that a person could not check it by hand. That is, there was evidence that the solution existed, but not one mathematician knew of the solution's details, not even Appel and Haken.

This situation has never before occurred in mathematics. It presents definite philosophical problems. There is a distinct possibility that a computer could make an error in so vast a calculation. There is also the possibility of an error being made in the interfacing between man and machine, invalidating the results. It is conceivable, too, that a computer program does not actually do what people believe it does. And there are many more qualms and fears which can be expressed.

Some, however, hailed this proof as the dawning of a new era in which mathematics will become a quasi-experimental science. They look at a computer run as a laboratory experiment giving inductive evidence of a nature totally opposed to the classical conception of mathematics as a purely deductive subject. To date, this is still a minority opinion and scorned in most mathematical circles.

There is one other confounding difficulty with this proof. In order to duplicate the experiment, a mathematician must write independent programs to generate the unavoidable set and test its reducibility and must then spend the enormous amount of time running these programs. This no one has done.

Verification of the truth of the theorem, though not of Appel and Haken's work, came from Frank Allaire at the University of Waterloo in 1979. He used a different discharging procedure, which generated an even larger unavoidable set. However, using mathematical theorems he was able to demonstrate the reducibility of most of the configurations in his set, leaving only a few hundred to be checked by computer. The result is thus proven to the satisfaction of everyone who will accept as definitive a machine calculation that cannot be checked by people.

## Recent Work and Future Directions

Much evidence indicates that no breakthrough is going to produce an unavoidable set that hand calculations could show to be reducible, at least using the Birkhoff method of demonstrating reducibility. Thus there are two possible directions for new research. Either a proof may be found along entirely different lines, or an improvement must be made in the method of demonstrating reducibility. The latter has been done.

In 1975, Daniel I. A. Cohen of Harvard University proposed a method of determining reducibility which converts the problem into one of solving a system of equations. This method was shown to dominate the classical method in the sense that any configuration which can be shown to be reducible in the old way can also be shown to be reducible in the new way. This new technique, called the method of block-count consistency, has one further advantage—after a computer has demonstrated that a specific configuration is reducible, it can print out a proof of this fact, which can be independently checked by a human. A computer is necessary for finding the proof but is not required to check it. The proof generated by this method is admittedly stupendously long (each case is several hundred pages), but it can be made complete and can be examined.

In 1980, Victor Miller of the IBM Watson Research Center at Yorktown Heights implemented this procedure for a computer. His goal is not to check the unavoidable sets of Appel and Haken and of Allaire (which were constructed with the classical reducibility method in mind) but to create a smaller unavoidable set that will yield to this method of showing reducibility.

Looking beyond the inevitable resolution of the four-color problem, which someday, one hopes, will have a short analytical proof, there are deeper and more immediate questions raised by the computer's invasion of mathematical sanctity. There are already other examples of computer-generated results that are also analytically uncertain. One such concerns the search for large primes used in making "unbreakable" codes. In this instance, a computer calculation determines that a number is "most likely" a prime, giving a probability estimate as its result; such a process is difficult for a classical mathematician to swallow.

Historically, the great mathematicians often relied on extensive hand calculations to guide them in their theoretical work. The development of new theorems has often followed the model of theoretical physics, from hand experiments to conjectures to proofs. Mathematicians had expected that computers would play an important role in mathematics only in the calculation and speculation stage. Few believed that computers could help to discover actual proofs, even though some limited but interesting work was done on this subject very early by Hao Wang and others.

The curriculum in mathematics has always been responsive to the computational needs of society and has matured accordingly. Methods for extracting square roots by hand and extending the accuracy of logarithm tables by interpolation have been dropped from the syllabus. It may not be long before calculus courses reflect the fact that symbolic integration is best performed by machine algorithms totally unlike those found in standard texts. Exactly how mathematicians will respond to the challenge presented by the four-color problem and the possibility of computer-assisted (if not dominated) proofs is yet to be seen.

## CLASSIFICATION OF FINITE SIMPLE GROUPS

The concept of a group is one of the most useful and important ideas discovered in mathematics in the last two centuries. It arose from considerations of symmetry in mathematical systems, but has also been applied to symmetries in many areas of science. Groups have been one of the great unifying mathematical ideas connecting many different fields of study. In mathematics, the concept is of great importance in topology, differential geometry, number theory, and harmonic analysis; in science, it has been extensively employed in spectroscopy, crystallography, atomic physics, and particle physics.

Groups are certain mathematical objects—like geometrical shapes or numbers—which are very easy to understand. As is common in most of science, one of the challenging tasks for mathematicians is to catalog and describe completely the objects being studied. Often, this is accomplished by breaking down the objects into smaller, more fundamental objects which are the constituents of all the objects and which can be listed. An analogous breakdown in chemistry is the periodic table of elements; in physics, it is the search for the fundamental particles. In the study of groups, the basic building blocks from which all groups are constructed are the so-called simple groups, and the determination of all simple groups has been a major theme in the history of the subject.

Researchers working to complete a unified group theory have recently finished the classification of all finite simple groups. That is to say, they have now drawn up a list of all the fundamental groups, as chemists drew up the periodic table. This undertaking, done on a scale not seen before in the history of mathematics, has been carried on for the last three decades by hundreds of mathematicians in many countries around the world. The complete results, taken together, will occupy about 10,000 pages of highly condensed notation in mathematical journals.

The next few years promise the consolidation and checking of this project and the beginning of a whole new era for group theory. One of the important tasks for the near future will be to understand a startling discovery made at the end of the classification process: a totally unexpected connection with the theory of numbers. Another vast project is the representation theory of finite groups, which explores the way groups arise in many applications. (Representing a group amounts to showing a specific example of it, as in the manipulation of a particular geometrical object.) An exciting period is under way.

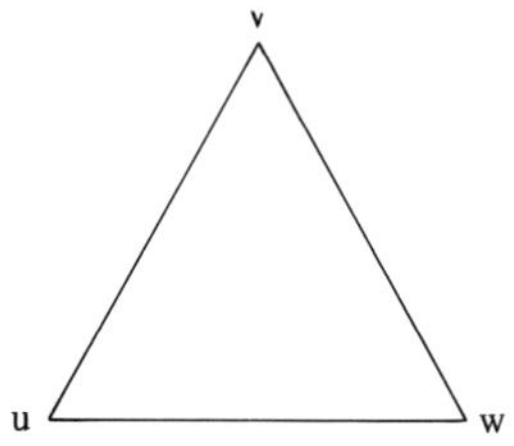


**Figure 3**

## Hidden Complexities

The idea of a group is deceptively easy. This is because the applications of the idea are not immediately apparent; a quick understanding of the basic idea does not lead directly to an appreciation of its relevance.

Consider the symmetries of a familiar system, and one of the simplest mathematical objects, the equilateral triangle (Figure 3). A symmetry of this triangle is a way of moving the triangle so that the result leaves the triangle looking exactly as it did before. For example, suppose we rotate the triangle 120° in a counterclockwise direction, so that the vertex labeled u is carried over to the vertex labeled w, the vertex labeled w goes to the one labeled v, and v goes to u. We have simply rotated the triangle about its midpoint, leaving it in the plane of the page. Call this symmetry operation by the name R. We can also rotate the triangle in the clockwise direction, again by 120°, carrying the vertex labeled u to the one labeled v and so forth. Call this operation C.

It is important to note that by the symmetry R we mean the result of the operation and not the manner in which it is performed. If the triangle is picked up out of the plane and rotated before being put back in the plane, this is still the symmetry operation R, provided the net switching of vertices is the same as we just described. The same convention applies to any symmetry operation in mathematics.

There are also some symmetry operations involved with turning the triangle over. In fact, we can flip the triangle, leaving the vertex u in place and interchanging the vertices v and w. To visualize this, imagine a line drawn from u to a point in the middle of the edge opposite it, the edge joining v and w, and rotate the triangle around this line until it has turned over. We call this operation U. Similarly, we can turn the triangle over leaving v alone and interchanging u and w, and we call this V. Finally, the operation leaving w alone and turning the triangle over is called W.

There is also one more symmetry, one which is so easy it can be missed, namely, the operation of leaving the triangle completely alone. This is called I for "identity." The six operations R, C, U, V, W, and I are the totality of all possible symmetry operations of the equilateral triangle.

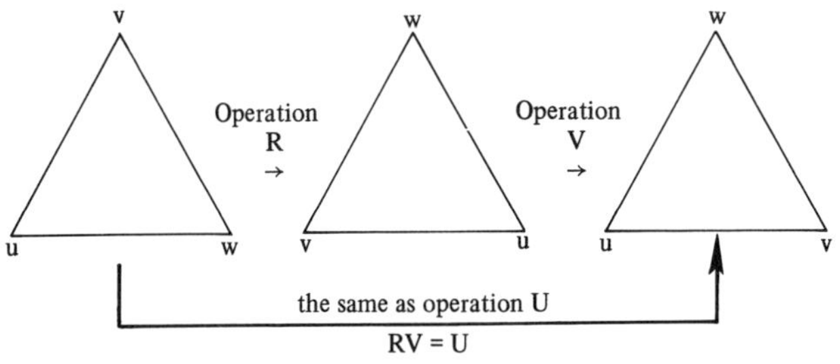


**Figure 4**

We are now at the threshold of a complete definition of the concept of a group. The key idea is understanding what happens when we combine operations. For example, suppose that we first perform the operation R and follow it by the operation V. What is the result? We have moved the triangle twice; each time the result leaves the outline looking the same, so the result of these two operations must be one of our six symmetry operations. Let us trace through the steps and see what happens (Figure 4). The operation R carries u over to w and then V sends it back to u, as V interchanges u and w. Next, R carries v to u and then V carries it to w. Finally, R takes w to v where V leaves it. Hence, the combination of R followed by V carries u to itself and interchanges v and w. This is exactly what U does. In fact, the result is just U. We write this in the following manner: RV = U (see Figure 4).

In general, when we "multiply" two operations and write a product like RV or UW, we mean the operation that results from combining the two operations in the order given. For example, if we perform U and then follow it by W, we will obtain C. This should not be too surprising, because the result of performing U and then W turns the triangle over twice so that the result must be one of I, R, or C. The complete table of the results of multiplying out all the products of two symmetries is given in Figure 5. To see what the result of VC is, look in the left-hand column where the V is listed and follow the row until you are under the C listed at the top. There you will find U, and VC = U is the result.

We now have an example of a group, namely, the entity consisting of the six operations together with the rule of combination. A group is then a collection of operations and a rule for combining them to yield other operations in the group, as in the situation above.

There are many examples of groups. As an illustration, we chose an especially easy one; it had only six operations. Such finite groups are among the most important, but many other groups of fundamental importance are not finite. In general, mathematical systems can be finite or infinite. Besides the usual number systems we all learn about in school—

| | I | R | C | U | V | W |
|---|---|---|---|---|---|---|
| I | I | R | C | U | V | W |
| R | R | C | I | W | U | V |
| C | C | I | R | V | W | U |
| U | U | V | W | I | R | C |
| V | V | W | U | C | I | R |
| W | W | U | V | R | C | I |

**Figure 5**

which are certainly infinite—there are very important finite number systems, such as the numbers on a clock dial.

## Simple Is Complex

Most groups when broken up are seen to be constructed of smaller groups, as molecules are constructed of atoms. The building blocks of group theory, the groups from which other groups are made, are called simple groups. This is a great misnomer, as the simple groups are far from easy to find and study. To explain this notion, we shall describe its opposite—what it means for a group to be nonsimple. In other words, we shall describe what it means to break up a group into smaller ones; the groups that cannot be so broken up are the simple groups.

To return to the example given above, the symmetry operations of the triangle can be divided into two types: operations that flip the triangle over, and operations that do not. That the elements of the group can be partitioned in this way shows that the group is not simple. We have divided the operations into two classes, the first class consisting of I, R, and C, the second, of U, V, and W. Note that the first class forms a group, but that the second class does not because UV is R, which is not in the class.

An important observation is that if we choose a member from the second class, U, V, or W, and a member from the first class, I, R, or C, then the product always lies in only one class, in this case the second. Similar statements can be made for other combinations of the two classes. We can see this in Figure 5. The 36 entries of the table fall into four squares, each three by three and each containing entries from only a single class. (For an example, see the figure at the top of the following page, which is the same as the lower right-hand corner of Figure 5.)

Let us denote the first class by E and the second by X. The above observation now gives us a way to multiply the classes. The product of X

| | U | V | W |
|---|---|---|---|
| U | I | R | C |
| V | C | I | R |
| W | R | C | I |

with X is E, because if we take any member of X and multiply it by any member of X then the result always lies in E. A table of the products of the classes is given in Figure 6. The entity consisting of the classes E and X together with the multiplication just described forms a group! We have constructed a group by breaking down the structure of our original group. The three operations I, R, and C also form a simple group by themselves, according to the multiplication table already given for them; this is also displayed in Figure 6.

| | E | X |
|---|---|---|
| E | E | X |
| X | X | E |

| | I | R | C |
|---|---|---|---|
| I | I | R | C |
| R | R | C | I |
| C | C | R | I |

**Figure 6**

We have just given an example of how our original group has been broken down into a group with three elements and a group with two elements. These two groups cannot be further broken down in this manner, and so they are examples of finite simple groups.

As of August 1980, mathematicians know what all the finite simple groups are. The epic task of finding these building blocks is complete, and the results of this program can be described. The finite simple groups are of two kinds. The first kind consists of groups falling into one of 18 families. These are groups associated with classical types of geometries, built up from number systems that are finite. Simple groups of the second kind are called sporadic, and classification theorems show there are only a finite number of these—26, to be exact. Table 1 gives the 18 families, denoted by their mathematical symbols.

**Table 1** Families of finite simple groups

| |
|---|
| Nineteenth century |
| $Z_p$ |
| $A_n$, $n > 4$ |
| $A_n(q)$ |
| $B_n(q)$, $n > 1$ |
| $C_n(q)$, $n > 2$ |
| $D_n(q)$, $n > 3$ |
| $^2A_n(q)$ |
| $^2D_n(q)$, $n > 3$ |
| Early twentieth century |
| $G_2(q)$ |
| $E_6(q)$ |
| 1950's |
| $F_4(q)$ |
| $E_7(q)$ |
| $E_8(q)$ |
| 1960's |
| $^3D_4(q)$ |
| $^2E_6(q)$ |
| $^2B_2(2^{2m+1})$ |
| $^2G_2(3^{2m+1})$ |
| $^2F_4(2^{2m+1})$ |

*Note:* In this table, p is a prime, m and n are positive integers, and q is a prime power.

Table 2 lists the 26 sporadic groups and also gives the order of each, that is, the number of operations that each contains. The smallest of these, $M_{11}$, has 7,920 operations, so it is considerably bigger than our one easy example. The largest, $F_1$, called the "monster" because of its size, has about $10^{54}$ operations in it.

The first five of the sporadic groups, $M_{11}$, $M_{12}$, $M_{22}$, $M_{23}$, and $M_{24}$, were discovered in the 1860's, but the rest were found only in the last 15 years. The largest one, $F_1$, was just recently constructed, although its existence was predicted in 1974. It consists of rotations in a space of 196,883 dimensions.

The key to much of the whole classification process which led to the complete listing of the finite simple groups is the notion of a centralizer. It can be illustrated with the six symmetries of the triangle described earlier. As noted, the order in which multiplication is done affects the answer. For example, RV = U while VR = W. However, RC = I and CR = I. In this latter case, we say that R and C commute, as numbers commute in arithmetic. The totality of elements of groups that commute with R is called the centralizer of R.

In general, if X is an element of a group, then the totality of all elements Y that commute with X (that is, all Y such that XY = YX) is called

**Table 2** Sporadic finite simple groups

| Group | Number of elements |
|---|---|
| $M_{11}$ | $2^4 \times 3^2 \times 5 \times 11$ |
| $M_{12}$ | $2^6 \times 3^3 \times 5 \times 11$ |
| $M_{22}$ | $2^7 \times 3^2 \times 5 \times 7 \times 11$ |
| $M_{23}$ | $2^7 \times 3^2 \times 5 \times 7 \times 11 \times 24$ |
| $M_{24}$ | $2^{10} \times 3^3 \times 5 \times 7 \times 11 \times 24$ |
| $J_2$ | $2^7 \times 3^3 \times 5^2 \times 7$ |
| HS | $2^9 \times 3^2 \times 5^3 \times 7 \times 11$ |
| MC | $2^7 \times 3^6 \times 5^3 \times 7 \times 11$ |
| Sz | $2^{13} \times 3^7 \times 5^2 \times 7 \times 11 \times 13$ |
| $C_1$ | $2^{21} \times 3^9 \times 5^4 \times 7^2 \times 11 \times 13 \times 23$ |
| $C_2$ | $2^{18} \times 3^6 \times 5^3 \times 7 \times 11 \times 23$ |
| $C_3$ | $2^{10} \times 3^7 \times 5^3 \times 7 \times 11 \times 23$ |
| He | $2^{10} \times 3^3 \times 5^2 \times 7^3 \times 17$ |
| $F_{22}$ | $2^{17} \times 3^9 \times 5^2 \times 7 \times 11 \times 13$ |
| $F_{23}$ | $2^{18} \times 3^{13} \times 5^2 \times 7 \times 11 \times 13 \times 17 \times 23$ |
| $F_{24}$ | $2^{21} \times 3^{16} \times 5^2 \times 7^3 \times 11 \times 13 \times 17 \times 23 \times 29$ |
| $F_5$ | $2^{14} \times 3^6 \times 5^6 \times 7 \times 11 \times 19$ |
| $F_3$ | $2^{15} \times 3^{10} \times 5^3 \times 7^2 \times 13 \times 19 \times 31$ |
| $F_2$ | $2^{41} \times 3^{13} \times 5^6 \times 7^2 \times 11 \times 13 \times 17 \times 19 \times 23 \times 31 \times 47$ |
| $F_1$* | $2^{46} \times 3^{20} \times 5^9 \times 7^6 \times 11^2 \times 13^3 \times 17 \times 19 \times 23$ |
| | $29 \times 31 \times 41 \times 47 \times 59 \times 71$ |
| $J_1$ | $2^3 \times 3 \times 5 \times 7 \times 11 \times 19$ |
| $J_3$ | $2^7 \times 3^5 \times 5 \times 17 \times 19$ |
| 0 | $2^9 \times 3^4 \times 5 \times 7^3 \times 11 \times 19 \times 31$ |
| R | $2^{14} \times 3^3 \times 5^3 \times 7 \times 13 \times 29$ |
| $J_4$ | $2^{21} \times 3^3 \times 5 \times 7 \times 11^3 \times 23 \times 29 \times 31 \times 37 \times 43$ |
| Ly | $2^8 \times 3^7 \times 5^6 \times 7 \times 11 \times 31 \times 37 \times 67$ |

* *Note:* The first 20 are to be found inside $F_1$; of the last six, the first four may be inside $F_1$, but the last two are definitely not.

the centralizer of X. This collection of elements always forms a group in its own right. In our example, the centralizer of R consists of I, R, and C.

This idea is used in the study of simple groups in the following manner. If G is a finite simple group, then it is possible to prove that its centralizers are groups with special structural properties, and it is possible to determine the structure of the whole simple group from its centralizers. This idea was also the key to the discovery of most of the sporadic groups. First, a group that looked likely to be a centralizer in a simple group was chosen, and, second, the nature of the simple group that contained it was uncovered in a step-by-step procedure until enough was known to actually build the group. This approach has enabled theorists to prove, once and for all, that the list of finite simple groups is now complete.

The value of having completed this project is twofold. First, finite simple groups are of such fundamental importance that their complete cataloging was inevitable and mandatory for aesthetic reasons internal to mathematics. Second, finite groups are also crucial to the study of many topics in applied mathematics and science, such as discrete structures for computer science, enumeration of combinatorial designs, contemporary theoretical physics, defining stereoisomers in organic chemistry, and cataloging structures in crystallography.

## Future Directions in Mathematics

Mathematics is full of startling and unexpected connections among its different branches. These relations are a great force for unification. The largest of the sporadic groups, $F_1$, the monster, has led to an amazing connection between simple groups and discoveries in the analytical theory of numbers that date from the nineteenth century. A real understanding of this connection now eludes mathematicians. The coming years will see attention focused here and should lead to new ideas and further exciting discoveries.

Among other topics, number theory studies automorphic functions and forms, which are mathematical operations and objects that are invariant under certain groups of transformations. For example, some of the most important automorphic functions are those that are invariant under the substitution of $q + 1$ for $q$. The elliptic modular function $J(q)$ is one of these of exceptional interest, and it is this function that is strongly linked to the group $F_1$. There is a formula

$$J(q) = q^{-1} + 744 + 196{,}884q + \ldots$$

where the dots denote terms in higher powers of q. Remember that $F_1$ is a group of rotations in a space of 196,883 dimensions. This differs by one from the coefficient of q in the expansion of $J(q)$, which is no coincidence at all. In fact, all the coefficients of the powers of q in the above formula are positive integers, and all are the dimensions of spaces in which the group $F_1$ appears as a group of transformations.

This is just the beginning. Other numbers that can be derived from the group $F_1$, called character values, are the coefficients of other very important automorphic functions that appear in number theory but are beyond the scope of this chapter. In fact, the functions that arise in this way are themselves interesting objects to be explained. The impact of these revelations has been felt in other areas. Automorphic functions also arise in the study of Lie algebras, and connections parallel to those mentioned above have been found by mathematicians looking for just such analogies.

A second future direction will be a better understanding of the whole classification program. This is a project that will receive much attention from researchers in the coming years. The enormous length and complexity of the classification, achieved over a long period of time, suggest that there are ideas to be discovered which will lead to drastic clarification of several aspects of group theory. The fact that most of the finite simple groups are in the 18 families of groups, which are all of a similar geometrical nature, suggests that there should be methods of studying simple groups which bring in geometrical ideas early in the classification, something that is not done at the present. Already a number of researchers are involved in more studies of the structure of groups in geometrical spaces not necessarily limited to three dimensions. This attention should expand in the near future.

A third future direction is representation theory, the study of the way groups relate to vector spaces and more general objects in n dimensions. This has long been established as a central part of the theory of groups and is now coming to the forefront of research. A number of related topics will also be of great interest. Block theory studies the relation between representations of groups on vector spaces over different number systems. This connects to the whole area of representations of algebras, a subject that has experienced a renaissance and several breakthroughs in the last decade. Homological algebra, which mostly developed from roots in topology, is having a major impact on representation theory and is an area that is opening up. The study of the representations of the simple groups will continue to expand and make great strides. This is directly related to work on the representations of algebraic groups and brings many sorts of geometrical ideas to bear on problems in representation theory.

## GAUGE THEORIES, RELATIVITY, AND GEOMETRY

In his book *Introduction to the Calculus*, William Osgood wrote, "It is customary in physics to take geometry for granted as if it were a branch of mathematics. But in substance geometry is the noblest branch of physics." Perhaps his statement is a deep truth in the sense of a statement attributed to the great Danish physicist Niels Bohr: "A deep truth is one who's opposite is also a deep truth." That geometry is useful in physics is not news. It permeates mechanics, electromagnetism, and of course relativity. What is noteworthy is the applicability of modern global geometry to problems in elementary particle physics and relativity. It is the aim of this section to explain what global geometry is, how in the form of fiber bundles it applies to gauge theories in high-energy physics, and how global methods have recently settled old problems in relativity. The concepts here are sophisticated and unintuitive to many readers; nevertheless, we hope their patience will be rewarded.

The physical forces known to us are four in number. The most familiar are gravity and electromagnetism, for we feel or see their effects. The two remaining forces are the weak nuclear force, which accounts for radioactivity, and the strong force, which holds the nucleus of the atom together.

The notion of gauge theories (see "Differential Geometry and Physics" below) has been introduced into theoretical physics as a unifying principle. It is expected to be the key to understanding the strong force and is the most likely candidate for the unification of all known forces. It has been used successfully to unify the weak and electromagnetic fields.

Albert Einstein, whose theory of relativity made gravity a gauge theory, strove in his later years to find the geometry and symmetry principles needed to unify gravity and electromagnetism. He failed; what he may have been looking for is the geometry at the basis of gauge theories, that of fiber bundles.

It came as a revelation to many mathematicians and physicists in the mid-1970's that gauge theories involved connections or vector potentials on fiber bundles. One of the principal architects of gauge theories, Chen-Ning Yang, wrote, "I found it amazing that gauge fields are exactly connections on fiber bundles, which the mathematicians developed *without reference to the physical world.*"

"Fiber bundles" is an accurate name. It *is* a bundle of fibers. The fibers can be complicated high-dimensional surfaces, and the "bundle"—how the fibers fit together—can be just as complicated. We shall soon give some simple examples. But first a review of classical differential geometry is in order. It will help us understand the emergence of the fiber bundle concept and its role in modern global geometry.

## Differential Geometry and Fiber Bundles

The foundation of differential geometry was laid by Gauss and Riemann in the nineteenth century. In Gauss' study of curved surfaces, the primary information used is the distance between neighboring points on the surface. If this theory is applied to the surface of the earth, one would work exclusively with the data available from teams of surveyors—lengths of straight paths and the angles between them. A triangle drawn on a flat plane has three angles which sum to exactly 180°. But the angles of a triangle drawn on a sphere sum to more than 180°. In this way, the surveyors can determine how much the earth is curving by measuring triangles. This inspired Gauss to explain by similar means how to measure curvature of a general surface. Comparing measurements, such as those of triangles in a small region, with the results one would obtain in a flat plane, one can precisely define and compute how that region deviates from being perfectly flat. This important notion of Gaussian curvature is an example of a "local" concept in that it depends only on the information contained in a small region. It does not require circumnavigation of the globe or photographs from outer space.

The total surface, its global shape, is the concern of a branch of mathematics called topology. Spheres, toruses, and planes are different, even when each is distorted slightly.

Global geometry tries to get at the fundamental differences in shape using local geometric invariants. The most stunning example in the theory of surfaces is the celebrated theorem of Gauss and Bonnet, which says in effect that by knowing only the curvature of each small region on the surface of the earth, one can determine that the planet is a ball, as distinct from a torus or a multiholed torus. For a sphere, no matter how it is distorted, the integral of the Gaussian curvature is $4\pi$. For the torus, it is 0, while for the object below the integral is $-4\pi$.

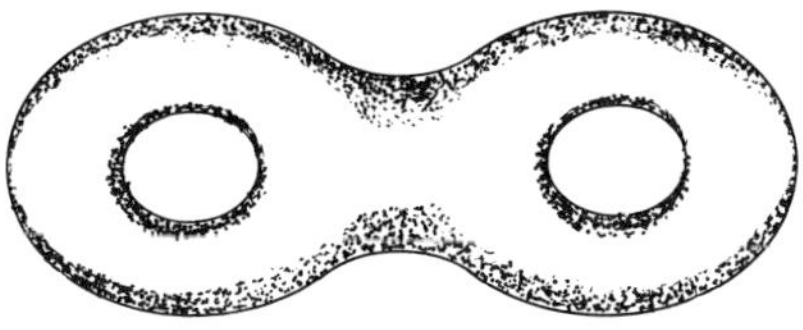

The generalization of the basic concepts of surfaces to the differential geometry of spaces of arbitrarily many dimensions is historically associated with Riemann. Sixty years before Einstein, Riemann suggested that a curved space might give the most natural description of the universe. Einstein's gravitational field equation gives an explicit relationship between the energy or mass and the curvature of a four-dimensional space-time. In this way, Einstein identified gravitation with the curvature of the universe. This brought new applications and directions for the old techniques of differential geometry. Global geometry becomes applicable when we try to obtain information about the shape of our universe by accessible measurements without stepping out of our universe.

Auxilliary spaces are useful in studying ordinary surfaces and their higher dimensional analogs. The simplest is the space of tangents to a surface. The totality of these tangents is the tangent bundle. It is a fiber bundle of planes; each fiber is the plane tangent to a point of the surface (see Figure 7a). We can also make a bundle of circles by taking each fiber to be the circle of unit radius that is tanget to a point of a surface.

A different bundle is the normal bundle of a surface embedded in a surrounding space. In this case, the fibers are the directions normal to the surface at each point of the surface (see Figure 7b). The dimension of these fibers depends on the dimension of the surrounding space. For example, for a surface in four-dimensional space, the normal bundle is a bundle of (normal) planes. The bundle of normal directions of unit length is another circle bundle.

Until the 1930's, differential geometry was mainly concerned with tan-

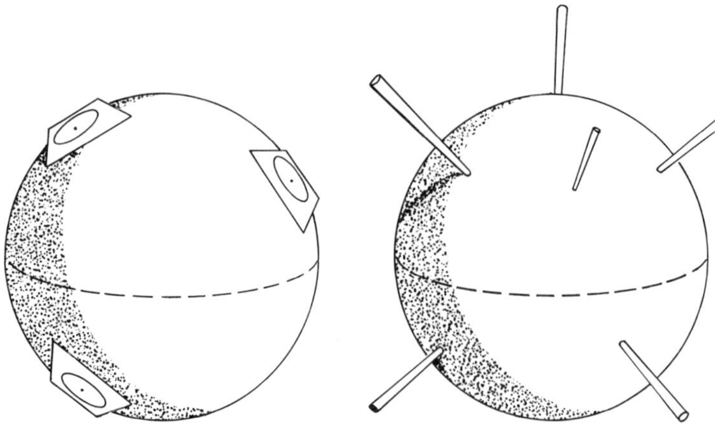

a b

**Figure 7**

gent bundles and their associated tensor bundles. An important conceptual breakthrough occurred at that time. In studying how surfaces sit in higher spaces, Hassler Whitney was led to the study of normal bundles and then to a consideration of more general bundles completely unrelated to the tangent bundle.

We now examine some simple fiber bundles. A cylinder is a fiber bundle whose fibers are line segments, one segment for each point of a circle (see Figure 8a). It can be constructed by taking a piece of tape and pasting the ends together. The Möbius band is another fiber bundle whose fibers are line segments, with one for each point of a circle. It can be constructed from tape by twisting the tape once before pasting the ends together (see Figure 8b).

A similar construction gives an interesting bundle of circles for the two-sphere, the surface of the earth. At each point of the earth, place a circle. Think of the two-sphere as the pasting of the upper hemisphere with the lower hemisphere along the equator, the equator measured by its meridian angle $\theta$. Paste the circles along the equator by identifying the circle from the upper hemisphere at equatorial point $\theta$ with the circle from the lower hemisphere by rotating through angle $\theta$.

Other circle bundles could be constructed by rotation through angle $k\theta$ for any whole number k. All the circle bundles can be made this way. For example, the bundle of tangent circles for the two-sphere corresponds to $k = 2$. It is remarkable that Paul Dirac, while studying magnetic monopoles in 1932, found this classification of circle bundles on the two-sphere outside of the stream of mathematical development.

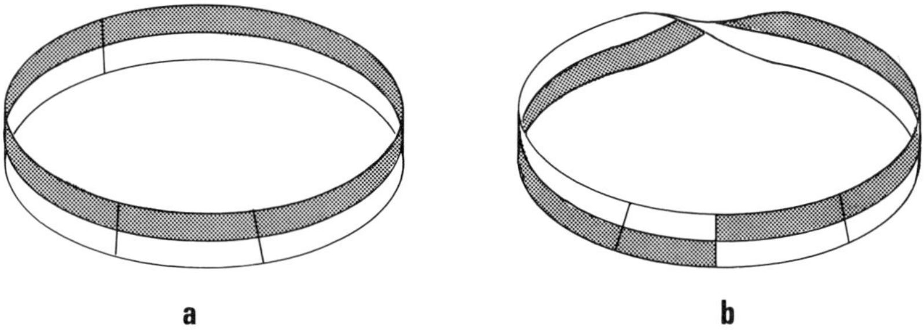


Figure 8

## Differential Geometry and Physics

The classification of bundles and their invariants, known as characteristic classes, proceeded in the 1930's and 1940's with the work of Whitney, Eduard Stiefel, L. S. Pontrjagin, and Shiing Chern. Among other things, they found interesting integral formulas for the invariants of bundles, generalizing the Gauss-Bonnet formula mentioned above.

In the 1970's, these global invariants cropped up in physics. To explain how, we need to explore the idea of curvature a bit further. An observer carrying a frame of reference or clock around a closed curve can compare his frame with that of an observer left behind. How much they differ measures the average curvature over a surface bounded by the curve. This curvature for space-time, according to Einstein, is the gravitational field.

The electromagnetic field can also be thought of as a curvature, one associated with a circle bundle. The observer carries a circle along with him and records the angle of rotation when he returns home, thus measuring the average field in a surface bounded by the curve. (A component of the electromagnetic field $\mu\nu$ is the infinitesimal rotation the observer experiences if he traverses an infinitesimal square in the $\mu\nu$ direction).

It is believed that the weak force and the strong force can be interpreted in a similar manner. However, the circle, the group of rotations in the plane, is replaced by more complicated symmetry groups, groups of rotations in higher dimensional planes. The circle bundle is then replaced by a bundle of these groups.

A gauge theory studies the ways of transporting a complicated symmetry group along curves in space-time and examines their associated curvature (or force) fields. Gauge theories break away from space-time in much the same way that fiber bundles break away from the tangent bundle. Space-time has global symmetries, the Poincaré group. But electromagnetism and the weak and strong forces have additional local internal symmetries giving extra degrees of freedom—the fibers in the bundle of local symmetries. It is natural, then, that fiber bundles are an appropriate framework for gauge theories.

The specific applications come about in the following way. Unfortunately, we must now become more technical. Physicists try to infer the quantum mechanical behavior of a system from a classical Lagrangian density. For pure gauge theories, the Lagrangian action functional in geometric terms is the average of the field strength squared.* The gauge fields A divide into topologically distinct classes determined by an integer k, often called the topological charge.† Heretofore in physics, the discreteness in quantum mechanics, represented by quantum numbers, came from eigenvalues of operators and ultimately from symmetries of groups. There is now the intriguing possibility that some quantum numbers and conservation laws may be topological in nature. In any case, the physicist, in calculating physical quantities, integrates over all configurations of the gauge field. The fact that these fields split into nontrivial topological classes may give interesting physical effects.

If perturbative methods are used to evaluate the integral, it is essential to find the critical, or stationary, points of the Lagrangian density. Special solutions of these nonlinear global partial differential equations on $R^4$, Euclidean four-dimensional space, giving local minima were found. These solutions have singularities at a single point and are called instantons or pseudoparticles. The singularities can be "gauged away"; that is, the solutions can be made free of singularities if we replace $R^4$ by the four-sphere (the unit vectors in $R^5$, or $R^4$ with a point at infinity added) and the ordinary bundle by a general plane bundle. More generally, it has been found that any solution with finite action and simple (meaning isolated) singularities really lives on a high-dimensional plane bundle over the four-sphere. If fiber bundles had not been invented ear-

---

*Specifically, for the knowledgeable reader, a Yang-Mills field is a vector potential $A_\mu$ with a value in a Lie algebra (say su(N), the $N \times N$ matrices $m_{ij}$ with $\sum_i m_{ii} = 0$ and $\overline{m}_{ij} = -m_{ji}$), so that $A_\mu = A_\mu^a T_a$, where $T_a$ are generators for the Lie algebra. The field strength is

$$F_{\mu\nu} = \partial_\mu A_\nu - \partial_\nu A_\mu + [A_\mu, A_\nu]$$

The action functional is

$$S(A) = -\int_{R^4} trF_{\mu\nu}F_{\mu\nu}dx^4$$

A stationary point will satisfy the equations of motion

$$0 = D \cdot F = \partial_\mu F_{\mu\nu} + [A_\mu, F_{\mu\nu}]$$

†The formula for k in terms of the field is

$$-\frac{1}{8\pi^2}\int_{R^4} tr(\epsilon^{\mu\nu\alpha\beta}F_{\mu\nu}F_{\alpha\beta})$$

In a manner similar to plane bundles for the two-sphere, the integer k, called the Pontrjagin index or second Chern class, classifies higher dimensional plane bundles for the four-sphere.

lier, they would have been, in the 1970's, in order to describe these instanton solutions (just as Dirac found the circle bundles on the two-sphere).

The solutions minimizing the action are called self dual solutions, and their degrees of freedom can be calculated. It turns out that there are more self dual solutions than the instanton count. What do they look like? It was observed that the problem of finding self dual solutions could be transformed into a problem in algebraic geometry. That problem amounts to finding all algebraic plane bundles over $CP^3$, complex projective three-space, i.e., the complex lines in complex four-space. This startling observation is based on the Penrose twistor program,* a way of looking at massless particles in physics that is very different from the traditional view. Happily enough, algebraic geometers had independently been studying the classification problem of algebraic plane bundles over $CP^3$, and all the self dual solutions have been found.

It is always striking when developments in a branch of pure mathematics (in this case algebraic geometry, seemingly far removed from practical matters) give the solution to a problem posed in some other field of science. Yet it happens often enough so as not to be surprising. But one must pause and contemplate "the unreasonable effectiveness of mathematics in the natural sciences," as Eugene Wigner put it. What is gratifying in this particular case is that the solutions can be put in the form of an "ansatz," or rule, that is usable and easily understood without the language of algebraic geometry. One can check directly that the ansatz gives solutions with the right number of degrees of freedom. Now the physicist need appeal to the mathematics only to ensure that there are no other solutions than the ones displayed.

We have sketched above how global differential geometry and algebraic geometry have helped find the self dual solutions, on the classical level, of the Lagrangian equation of motion for a nonabelian gauge theory. Quantum field theory begins and builds on these classical solutions. How they and their configurations, particularly the new solutions, contribute to the nonperturbative properties of gauge theories is still unsettled. In fact, how to compute in continuous nonperturbative gauge theory is unknown. Because the geometry and topology of fiber bundles are so intimately connected with gauge theories, it may turn out that mathematical insights will help provide the key to the nonperturbative theory.†

---

*The twistor program has been successful in finding global solutions to other partial differential equations, giving new nonabelian magnetic monopoles and new Einstein manifolds.

†They have been useful in present-day attempts to find a grand unified field theory. Earlier in this chapter is a discussion of the classification of simple finite groups. The simple continuous groups were classified by Elie Cartan early in this century. Physicists seek the symmetry group of a grand unified gauge theory from among these. It is amusing that a prime candidate is one of the exceptional groups. It is a pity that lack of space disallows an exposition of the rich theory of continuous groups, their representations, and their applications to mathematics and physics.

## The Positive Mass Conjecture

Returning to relativity, global methods in geometry have been useful in settling what is called the positive mass conjecture. Consider an isolated physical system that consists, for example, of a group of planets floating around in empty space under no forces but their own gravity. According to the classical mechanics of Newton, each object has its mass and the total mass of the system is the sum of these separate masses. In the setting of general relativity, however, this apparently straightforward concept raises delicate and significant problems.

In relativity, mass is identified with energy, according to the formula $E = mc^2$. Gravitational energy must then also contribute to the natural definition of total mass. Moreover, relativity acknowledges the possibility of negative mass, which can cancel at least part of the positive mass of matter. The assertion that the total mass of an isolated system, as defined by Einstein in 1918, is still positive is known as the positive mass conjecture. It was shown to be true in 1979 by Richard Schoen and Shing-Tung Yau.

The geometry of space-time is somewhat peculiar. In order to sort out the difference between the three dimensions of space and the single one of time, some paths have positive length and some have negative length. The paths which light travels along have zero length. From the early days of relativity, a rather involved study was required just to understand the local theory and to write down some explicit examples in special cases, such as a simplified universe with no matter except for a single symmetric star. However, in the last 40 years, the techniques of global differential geometry, including the venerable Gauss-Bonnet theorem itself, have been used very efficiently to deal with problems of general relativity in the large. This is particularly true of the singularity theorems of Stephen Hawking and Roger Penrose. These theorems demonstrate, among other things, the existence of regions of unbounded density under physically reasonable conditions. An example is a black hole, a star that has collapsed so that even light may not escape its gravitational attraction. In addition, the occurrence of a singularity (or unique event) in the universe's past is implied, which is commonly referred to as the big-bang theory of the origin of the universe.

The question of positive mass arises as a result of trying to retain the vital conservation laws of energy and momentum in the model of general relativity. The contribution of gravity to mass is exactly defined to obtain these laws. It is assumed that in every small region, the positive mass of matter locally dominates the contribution of gravity. It must then be determined that the gravitational effect of distant matter does not globally dominate, since this would in some circumstances give a negative total mass.

In systems with negligible gravitation, relativity reduces to Newtonian mechanics. As mass is certainly positive in Newtonian mechanics, there are abundant physical reasons to believe that total mass is positive under the assumption that the gravitational force is weak. Even in this limited case, however, a rigorous proof was furnished only a few years ago.

One reason for believing the conjecture without the assumption of negligible gravitation is that, as gravitation increases, the gravitational force causes the formation of black holes with positive mass before the negative potential energy exceeds the positive energy contributed from the matter. This type of phenomenon does indeed show up in the proof of Schoen and Yau.

We now describe more precisely the positive mass conjecture and its difficulties. As in any field theory, we have the local conservation law for the energy-momentum tensor, $T^{ab}$, which can be written

$$T^{ab}_{;b} = 0$$

The fact that space is curved prevents this equation from being integrated to give global conservation laws. Physically, this means that $T^{ab}$ represents matter fields only and does not include a contribution from the gravitational field. The difficulty was first resolved by Einstein in 1918. He proposed that for each given set of coordinates $x^a$ one introduces a pseudotensor

$$t^b_a$$

to represent the gravitational energy and momentum. It satisfies the equation

$$\frac{\partial}{\partial x^b}(T^b_a + t^b_a) = 0$$

These could be integrated to give global conservation laws. Einstein showed that if the metric approached the Minkowski flat metric in a suitable way at spatial infinity, the integrated energy and momentum would be invariant under local coordinate change and under global Lorentz change of coordinates at infinity. One can show that this total energy and momentum, as measured at infinity, could be defined simply from the asymptotic form of the metric on an asymptotically flat, space-like initial hypersurface. As in all other physical theories, one wants the energy-momentum four-vector to be a future timelike vector. This is an intuitive way to phrase the positive mass conjecture. It can be formulated precisely and becomes a theorem in global geometry. For example, a special case says that a three-dimensional torus with nonnegative scalar curvature is flat.

# Summary and Outlook

From sports statistics to economic forecasts, from travel reservations to the scheduling of committee meetings, the increasing use of mathematics has altered the way in which people go about their lives and think about the world. Yet research at the frontiers of mathematics is carried out at a level of abstraction that to the untrained observer may seem light-years removed from possible applications. Relying heavily on internal sources for inspiration, mathematicians have pressed their explorations into distant regions of abstract thought on a variety of fronts. One of the wonders of this pursuit is when a breakthrough at the very edge of research reveals interconnections not only into other branches of mathematics but into the other sciences and society as well.

One area of active recent research is the field of theoretical computer science, a branch of mathematics that has developed rapidly since the pioneering work of Alan Turing and John von Neumann in the 1930's. Current investigations are improving an algorithm that solves the general linear programming problem, which is a common tool in economics, engineering, and many other sciences. Computer programs that solve these programs are becoming faster and less complex, and mathematicians now wonder if there may not be similar algorithms for a wide variety of other such problems. Because of the many links between the theoretical and applied aspects of computer science, future research in algorithmic analysis could prominently affect the evolution of one of modern society's most important machines.

The computer has also played a prominent role in a recent mathematical triumph, in the process raising philosophical problems that are likely to intensify in the years ahead. This is the proof of the four-color theorem, which in its simplest form states that it is possible to color any map using only four colors so that no two adjoining countries are given the same color. The roots of this problem go back more than a century. It was first solved in 1976, and even then only with a great deal of help from a computer. But the years of searching for an answer were hardly barren—attempts to solve the four-color problem helped give rise to an entire branch of mathematics, graph theory, which independently has been found to be of fundamental importance to twentieth-century electronics, especially to computers. The particular subject of map coloring can be applied to all problems of allocation, from economics to the storage of information in a computer.

The philosophical problems arising from the proof of the four-color theorem are based on the fact that the original proof involved calculations which were too long to be checked by humans. Many mathematicians question whether a problem solved by such calculation can truly be said to be solved. Underlying this qualm is the much larger issue of how computers will be used in future mathematical research—simply to generate deductive evidence, or as an active partner in producing proofs.

Another historic achievement that has occurred in mathematics in the past few years has been the final classification of all of the finite simple groups, which are the building blocks from which all larger groups are made. This process has been a massive undertaking, involving hundreds of researchers and volumes of results, but now that the classification is complete the field is about to enter a new, productive phase. The final stages of the classification revealed an unexpected relationship between group theory and number theory, and exploring the byways of this surprising connection should produce new insights into both fields.

The development of mathematics does not, however, depend solely on solving specific difficult and challenging problems. There is an aspect of mathematical growth that is more akin to pushing the field's frontiers into uncharted territory, where one sometimes finds new and dazzlingly beautiful worlds. Such was the way in which geometry, with its roots in antiquity, grew through the centuries. The worlds it conquered and annexed often seemed to be purely creations of the human mind. Then, perhaps miraculously, the new worlds created "out of nothing" (in the words of Janos Bolyai, one of the founders of non-Euclidean geometries) gave Einstein the clue to the understanding of the global aspects of the universe.

Recently, the frontiers of geometry have expanded still more, and there is now real hope that even the subatomic world of elementary particles will become part of the new world that mathematicians have created "out of nothing."

The sense of beauty that accompanies the discovery of interconnections within mathematics or between mathematics and other sciences has been a powerful driving force throughout the discipline's history. As mathematics expands in disparate directions, new associations are found that link together even the remotest branches of the field. These new relationships increase the applicability of mathematics so that it continues to serve mankind as high art and society's practical servant.

## ADDITIONAL READINGS

Modern differential geometry, and especially its connections with topology through the theory of fiber bundles, is frontier mathematics and has not yet settled into a state inviting popularization. Thus, the section "Gauge Theories, Relativity, and Geometry" may well be the first attempt at such popularization.

Kenneth Appel and Wolfgang Haken. "The Solution of the Four-Color-Map Problem," *Scientific American,* Vol. 237(October 1977), pp. 108-121.

Herbert J. Bernstein and Anthony O. Phillips. "Fiber Bundles and Quantum Theory," *Scientific American,* Vol. 245(July 1981), pp. 122-137.

Norman L. Biggs, Keith Lloyd, and Robin J. Wilson. *Graph Theory: 1736-1936.* Oxford: Clarendon Press, 1976.

*Complexity of Computer Computations.* Edited by Raymond E. Miller and James W. Thatcher. New York: Plenum Publishing Corp., 1972.

Robert L. Forward. "Spinning New Realities: Theorist Roger Penrose Gives Einstein's Universe a New Twist," *Science 80,* Vol. 1(December 1980), pp. 40-49.

H. Georgi and S. L. Glashow. "Unified Theory of Elementary-Particle Forces," *Physics Today,* Vol. 33(September 1980), pp. 33-39.

Sheldon Lee Glashow. "Quarks with Color and Flavor," *Scientific American,* Vol. 233(October 1975), pp. 38-50.

Frank Harary. "On the History of the Theory of Graphs," *New Directions in the Theory of Graphs.* Edited by Frank Harary. New York: Academic Press, Inc., 1973.

Gerard 't Hooft. "Gauge Theories of the Forces Between Elementary Particles," *Scientific American,* Vol. 242(June 1980), pp. 104-138.

Donald E. Knuth. "Computer Science and Its Relation to Mathematics," *The American Mathematical Monthly*, Vol. 81(April 1974), pp. 323-343.

A. I. Mal'cev. "Groups and Other Algebraic Systems," *Mathematics—Its Content, Methods and Meaning.* Edited by A. D. Aleksandrov et al. Volume Three. Cambridge, Mass.: M.I.T. Press, 1963, pp. 263-351.

Oystein Ore. *The Four-Color Problem.* New York: Academic Press, Inc., 1967.

Arthur L. Robinson. "Unified Gauge Theories: An Atomic Fly in the Ointment," *Science*, Vol. 198(December 1977), pp. 908-909.

B. A. Trakhtenbrot. "Algorithms," *Perspectives on the Computer Revolution.* Edited by Z. W. Pylyshyn. New York: Prentice-Hall, Inc., 1970, pp. 69-86.

Steven Weinberg. "Unified Theories of Elementary Particle Interaction," *Scientific American*, Vol. 231(July 1974), pp. 50-59.

Chen-Ning Yang. "Einstein's Impact on Theoretical Physics," *Physics Today*, Vol. 33(June 1980), pp. 42-49.

# IV

# RESEARCH SYSTEMS

# 13

# *Research in Europe and the United States**

## INTRODUCTION

### Major Concerns in the U.S. Research System

Since World War II, America has been a world leader in science and technology. The partnership between the scientific community and the federal government has been instrumental in achieving this high level of scientific excellence. Our present system for supporting research has served us well for 30 years: first, in maintaining world leadership in basic research; second, in providing the educated people needed at advanced levels to meet national objectives in science and technology; and third, in providing a foundation for the applied research in industry that leads to improved productivity. A great strength of the system has been that it has successfully linked basic research in the universities with industrial and agricultural innovation in a way that few other national systems have been able to match. The land-grant tradition of American universities is

* Although primary responsibility for this chapter rests with the chapter coordinator and the Committee on Science and Public Policy of the National Academy of Sciences, the field work, research, and interviews upon which the chapter is based, and preparation of drafts, were performed by Drs. Bruce L. R. Smith and Charles V. Kidd.

◀ "Cyana," a French manned submersible able to dive to 3,000 meters. "France has been able to continue the construction of a few large and very expensive domestic facilities, directly administered by the government." [Centre National pour l'Exploration des Océans.]

a unique asset that has played an important part in providing service to the nation, as intended by the Morrill Act of 1863, in "the agricultural and mechanical arts." Some important sectors of American industry have a long tradition of carrying on their own research, some of it fundamental, that has enabled them to take advantage of basic research done at universities and also to make full and effective use of the graduates trained through the university research system.

Over the past decade, Americans have won 41 Nobel Prizes. Researchers in the United States continue to publish a major, although in some areas slightly declining, portion of research papers in a large number of scientific fields. A predominant share of the important basic research in the country continues to be conducted in our major universities. In comparison with most other nations, including the European countries reviewed in this chapter, the United States looks more to the universities for the conduct of basic research. Britain is much closer to U.S. practice in this respect than are France and Germany, both of which rely heavily on national laboratories or independent research institutions.

Approximately half of the basic research in U.S. universities is carried out by faculty members working closely with small numbers of graduate students. Supported largely by federal funds, academic researchers have worked in relatively small groups made up of one or a few professors plus graduate students, and the research has typically been closely coupled to teaching. Much of this research is supported as individual projects, the ideas for which are generated by individual scientists who seek outside funds to carry out their ideas. The remaining share of research is done by formally organized research laboratories, institutes, and centers associated with universities.*

Funding has come increasingly from federal sources, but no single agency dominates the allocation of funds. The share of federal support for university research by the National Science Foundation has remained at approximately 20 percent over the past decade, and the "mission" agencies—those with specified functional responsibilities such as providing for the nation's defense, health, energy, and other needs—fund the rest of the federally supported research done in the universities.

Research activity is widely dispersed in the university system, although a relatively small number of major research universities perform a large share of the research and receive a large share of the federal research funds. There has been considerable mobility as scientists have moved easily from one university to another. Well-equipped laboratories exist in all regions of the country, and research in nearly all fields of science has

*Approximately 50 percent of federally funded university research is for projects of $150,000 and under; 60 percent for projects of $250,000 and under; the remainder is for larger grants or contracts to institutes, centers, or laboratories.

originated in academic departments in all regions. A decentralized structure also prevails within most universities, with authority for decisions resting with individuals and small subunits rather than with central administrations.

While reliance on the universities as research performers is heavy, it is by no means exclusive. Nearly half of our basic research and much more than half of our applied research are conducted in industrial laboratories, non-profit research institutions, and in-house governmental or other national laboratories. Very large facilities, such as the great accelerator laboratories at Batavia, Stanford, and Brookhaven, and the large observatories like Kitt Peak and Mount Palomar, have carried on scientific activities requiring expensive instrumentation beyond the means of separate universities. Such facilities on the whole, however, have operated less as independent institutes than as facilities for university researchers. Organized research centers, institutes, and laboratories exist in increasing numbers within many universities, where science is pursued on a larger scale than in individual projects undertaken by a professor with a few graduate students. Such laboratories are usually supported by mission agencies, which see useful applications in the development of particular fields of science. The work of the laboratories thus reflects a combination of mission-oriented and basic scientific priorities.

## Emerging Problems in the U.S. Research System

This system, which in recent years has been the envy of many other nations around the world, is facing increasingly severe stresses and structural problems. Major recent surveys of the American research system, including that by Bruce L. R. Smith and Joseph J. Karlesky, *The State of Academic Science*[1] (1977), and the reports of the National Commission on Research[2] (1980), have warned that, while the nation's research system remains strong, significant emerging problems could pose a threat to the long-term health of the system.

In light of these concerns, the Committee on Science and Public Policy of the National Academy of Sciences decided to undertake a review of research systems in Britain, France, and Germany, comparing them with the American system. This review has not been comprehensive, but has focused on aspects of the other systems that seem particularly relevant to the American experience. The Committee has sent representatives on site visits to each country, reviewed relevant national statistics, and discussed its findings with officials and other observers in the countries reviewed. This chapter summarizes the principal findings in keeping with the Five-Year Outlook's purpose of identifying significant trends and emerging issues for science-policymakers in Congress, the Executive Branch, and the wider concerned public. The chapter is written from the perspective of an American viewing the research systems of Europe. The approach is to discuss each nation, then compare the salient elements of the European systems with ours, and finally assess the potential implications for the United States.

The major stresses found to be developing in the American situation include:

- Resource constraints, in part the result of inflation's impact on research budgets.
- Increasing costs of instrumentation for more advanced research activity.
- Growth of pressures for short-term returns on research investments.
- The impact of demographic changes affecting the university research system.

Despite efforts to maintain and even to increase federal support for basic research in recent years, inflation threatens to erode the financial base supporting university research. Income from fees, tuition, and endowments, and from state governments and other sources is part of the research-support system. Any serious weakening in these areas of support, which are also threatened by economic and demographic forces, could have far-reaching effects. The increasing administrative complexity of our system of research support has also led to frictions and burdensome accounting requirements that divert substantial time and effort from research to the search for funds. Our system of research support, despite its many advantages, often provides for research in forms that create uncertainty for the scientist and administrative burdens for both the recipient and the funding agency. In the future, tightened budgets and resource constraints could lead to a conservative attitude among researchers, fearful of jeopardizing their funding if they introduce new and venturesome lines of inquiry.

A related strain on the system is the increasing scale, cost, and complexity in many areas of research not normally considered a part of "big science," that is, high-energy physics, astronomy, and space physics. In fields where large-scale national facilities exist to further research, the U.S. research system enables university researchers to share access to large machines. Now, in addition to big science fields, more areas of science have become highly instrumented in recent years and costs have risen sharply, although they are still below the threshold of a large national or regional facility. How can the nation maintain a broad base of well-instrumented laboratories in these increasingly expensive areas of inquiry? Will there be a trend away from the very small research group and toward the organized research unit, toward cooperative use of equipment, or to regional facilities? Is our research system capable of maintaining reasonable balance between fields of science, between large- and small-scale efforts, and between organized and individual research? The increasing scale and complexity of scientific research will lead to difficult choices about where facilities should be located, which universities and which regions should remain active in the research areas requiring expensive instrumentation, and many related issues.

The growth of pressures for short-term results is another serious strain on the nation's research system. There is heightened concern with productivity and with the competitiveness of the U.S. economy. Many observers appear to believe that the research system has not contributed sufficiently to the solution of those pressing issues and that the urgency of the nation's economic problems requires more effective application of knowledge specifically to those problems. The solution of nearly every major economic and social problem facing the nation undoubtedly depends in part upon science and technology. How this powerful force can be harnessed most effectively to contribute to the solution of such problems as the development of new energy sources, conservation of energy, protection of the environment, defense, and industrial innovation is a major challenge to government, industry, and the universities. The need for balance between basic and applied research efforts also remains imperative. Free-ranging scientific inquiry guided only by the discipline-oriented creativity of individual scientists must be maintained. The search for current solutions should not overwhelm free-ranging inquiry because of the revolutionary advances that may flow from such inquiry during decades to come. How to assure the most productive balance between these competing claims on resources and on the attention of scientists will be an acute problem in the future.

Finally, demographic changes may produce lowered science enrollments and fewer teaching positions, causing a major future concern for academic science. Decreases in the college-age population have triggered this complicated set of problems for the nation's universities. The size of faculties has been linked closely to enrollments and, as enrollments increased rapidly in the post-World-War-II period, many university faculties expanded substantially. Graduate programs increased in size and supplied the scientists who filled the new teaching positions created in the overall expansion of the university system. If the university system reaches a steady state or even decreases slightly in size, fewer new teaching positions will be available. Many academic departments have already found themselves overexpanded with heavily tenured ranks. This has meant diminished career opportunities for young academic scientists and aging faculties in many universities—a trend that could impede the flow of adequate numbers and quality of young scientists into research careers. Moreover, the historic close relationship between research and graduate education in the United States may change in consequence of these demographic changes. The coupling of research and teaching becomes less firm as the capacity for growth in American universities becomes more limited and as research activities eventually come to have a rationale apart from the education function. Growing numbers of nonacademic research staff members in many universities play little or no role in the educational functions today. New forms of support for research teams, centers, institutes, and laboratories loosely attached to universities, in which the majority of the researchers are not involved in undergraduate teaching, or separate from universities, may emerge in the future. Whether such changes are either feasible or desirable in our circumstances is unclear, but it seems likely that such issues will become matters of growing public debate, especially as the nation strives for a pattern of research support that does not rest exclusively on national or regional trends in university enrollments.

The dynamic forces affecting the research system add to the concern for the continued originality, vitality, and leadership of U.S. basic research. Looking at other countries in the light of our problems, we may see more clearly the strong aspects of our system that we must particularly strive to preserve. We also may see options in the experiences of other industrialized nations that, with suitable modifications, might be adopted here.

Before World War II, Europe played a leading role in the fundamental discoveries that have shaped modern science. The war dealt a severe blow to European science and, for many years after the war, the major European nations were faced with the task of rebuilding their scientific resources. They looked to the United States, in a reversal of the earlier role, for the fundamental discoveries and for the leadership role in basic science that we had now acquired in addition to our traditional reputation for industrial and technological strength. In the postwar period, European energies concentrated initially on economic recovery, but the economic recovery made possible—and now in turn has been stimulated by—the recovery and flourishing growth of its science and technology. Clearly, European science has experienced a substantial recovery in recent years. In quantity and quality, with variations from field to field and country to country, Western Europe has regained much of the scientific eminence that it lost during the upheavals surrounding the war and has broken new ground in a number of fields.

Today, total research and development expenditures in science and engineering in the United States are still much higher than those in individual European countries, but important changes have been taking place. In Germany, there has been nearly a decade of rapid growth in expenditures for research and for higher education—a period of expansion reminiscent of the early 1960's in the United States and Britain.

**Table 1** Comparative national R&D expenditures, 1978-79 (in billions of dollars)

| Expenditures | U.S.[3] | % | U.K.[4] | % | France[5] | % | West Germany[6] | % |
|---|---|---|---|---|---|---|---|---|
| Total national | 52.0 | 100 | 7.6 | 100 | 10.0 | 100 | 15.2 | 100 |
| Total governmental | 26.0 | 50 | 4.2 | 55 | 6.0 | 60 | 7.5 | 49 |
| Total non-governmental | 26.0 | 50 | 3.4 | 45 | 4.0 | 40 | 7.7 | 51 |
| Defense R&D (% of total governmental[7]) | | 51 | | 46 | | 30 | | 12 |

Recently, it has begun to moderate. The period of austerity in Britain has forced a more selective approach to research support and generally very tight research budgets. France has been in a position between Germany and Britain in terms of total national investment in research and development (R&D). In the aggregate, total expenditures on R&D by the major European countries moved from less than half that of the United States in 1965 to about two thirds of total U.S. R&D expenditures in 1978-79. (The comparative figures are shown in Table 1.) Defense R&D constitutes much of the U.S. margin.

The situation is different for basic research exclusively. By 1979, the combined basic research expenditures of Britain, France, and Germany almost equaled those of the United States. While these are general figures, they reflect the resurgence of science in Europe and the shift away from postwar dominance in basic research by the United States.

## Purpose of the Comparative Study

It is not our aim in this chapter either to predict or to recommend new policies and arrangements. We seek to alert the Congress to significant developments in other countries where vigorous research activities are in progress. This knowledge can potentially widen the range of options available to us as we appraise our own research system, seeking to preserve its strengths and to adjust to changing circumstances. Each of the countries reviewed in this chapter has its own system based on its traditions and needs. We cannot, of course, simply import institutional practices and arrangements from other nations. But, if we understand more fully the similarities and differences, the strengths and weaknesses of different national research systems, we may be in a better position to cope with the coming strains on our system. Most importantly, a knowledge of the full range of options available is essential as policymakers confront the emerging problems in the changing circumstances that surround the nation's science effort.

In the surveys of the European nations, it became clear that many of the distinctive features of our system either do not exist or are much less prominent in those nations. Examples include:

- Heavy reliance on the universities as performers of basic research.
- A university system composed of both public and private universities.
- Support provided to individual scientists in the form of relatively small project grants and contracts rather than block funding of centers or institutes.
- Provision of support through several mission agencies instead of through central science agencies.
- Opportunities for independent work for the scientist at a relatively early stage in a scientific career.
- Substantial mobility for the scientific workforce between universities and across different institutional sectors.

In these respects, the European nations, while exhibiting significant differences among themselves, differ substantially from the United States. The differences will become evident in greater detail in the discussion of each country. Fixed points of reference familiar in the American scene will not be present. At the same time, it is clear that Britain, France, and Germany are deeply involved with the same kinds of problems that face our own research system. It is their responses to these common problems that make this comparative experience of considerable interest.

## UNITED KINGDOM

### Major Features of the Research System

The basic research system in the United Kingdom, diverse but oriented around the university effort, shares many of the strengths and weaknesses of the American system. In Britain, however, the effects of resource constraints on the research system are more evident than in the United States, France, or Germany. Britain has been able to avert the worst consequences of austerity by a policy of selectivity and stability in research support. This policy makes it likely that the quality of British scientists—as demonstrated by the fact that no other country has produced so many Nobel Prize winners per million inhabitants—will continue to be high. However, if present trends continue, the narrowing base of scientific excellence, with declining investments in instrumentation and in research support, will erode the quality of British science in the future.

The British system is like our own in its heavy reliance on the universities as key basic research performers; however, British universities do not generally manage large-scale facilities. The universities are financed through a dual system. Institutional support, used in part to cover the operating costs of university laboratories, is provided through the University Grants Committee. Then, in addition, funding through small-scale project grants to individual scientists is provided by five Research Councils. In recent years, this dual support system has come under increasing strains because the institutional grant has increased only slightly while costs have escalated. As a consequence, some modification of the working assumptions of the system seems likely in the near future. Britain deserves credit for making the transition from the era of rapid growth in expenditures for scientific research to one of leveling off or even contraction over a substantial period without major disruption of the research system. But the problems that the United Kingdom has faced—in manpower, scientific equipment, shrinking resources, and other common concerns—have tended to be more severe than in other major Western nations. They have posed great challenges to both government and the scientific community.

Research in the United Kingdom is carried out in a variety of settings—in the laboratories of private concerns, of nationalized industries, of industrial research associations, of military and civil government departments and ministries of research councils, and of universities. The role of universities in basic research is of critical importance, setting the tone and the direction of the overall effort and lending distinction and international prestige in many fields of modern science. Separate publicly financed research units, primarily associated with medical schools and hospitals, also play a role, but a much less prominent one than in France. Large-scale physical facilities, funded and operated by the Research Councils for academic scientists who are the principal users of the facilities, are a significant feature of the British system. However, in the United Kingdom, as in Germany and France, such facilities are not managed by universities or university consortia, as they often are in the United States.

Despite financial austerity, research of the highest quality continues to be performed in many British laboratories, especially those of the leading universities. With exceptions in some fields in some periods, the scientific community in the United Kingdom has not been rich in material resources. But the country has produced outstanding investigators in many fields, many of whom have worked in theoretical areas of their disciplines not requiring large experimental machines. The areas of notable achievement in British science include radio astronomy, high-energy and solid-state physics, neurophysiology, and molecular biology. The Astronomy, Space, and Radio Board of the Science Research Council reported in 1979 that

> . . . in spite of the decrease in its budgets, which have been declining steeply for the last four years, the Board can report again this year substantial scientific success in each of its main fields. It is not intended . . . to dwell on . . . individual successes, simply to report that they do not yet show any fall from the high standard of recent years.[8]

Part of the reason the United Kingdom has been able to withstand the effects of austerity, to date at least, is reliance on selectivity in research support. This has been a major theme of British science policy in recent years. There is relatively little pressure in the United Kingdom to distribute research funds on the basis of criteria other than scientific excellence. Competition for research funds and for recognition among geographic regions and among different types of universities is less intense in Britain than in the United States, and cultural attitudes allow for deference to recognized elite groups more readily in Britain than in the United States. Moreover, funding levels have been predictable. There has been little growth; neither have there been many sharp cutbacks or extreme fluctuations.

The many components of the research system in the United Kingdom operate with a high degree of independence, with no strong central coordinating body. Some of the major sources of governmental support for civil R&D are the Department of Industry, the Department of Energy, the Department of the Environment, and the Department of Agriculture, Fisheries, and Food. While this dispersion, at first glance, is reminiscent of the American system, Britain relies less on the operating agencies and more on central support agencies to fund basic research. In common with the other countries reviewed, Britain is giving increasing attention to the practical application of research findings. This is reflected, for example, in the establishment of the Department of the Environment in 1973 and the Department of Energy in 1974. There also have been important changes within departments. One of the most significant of these is the establishment of Requirements Boards in the Department of Industry, whose task is to identify and finance R&D of high importance to industrial productivity and innovation and to other urgent national requirements. The National Research and Development Corporation, a quasi nongovernment organization, was created to bring inventions to the point of industrial application. The Advisory Council for Applied Research and Development (ACARD), a group of eminent scientists and engineers appointed to advise the Secretary of State for Education and Science, has been influential in directing attention to specific emerging areas of high potential commercial significance, such as microelectronics, biotechnology, and information technology.

Turning to the dual system that supports university research, one must note that both segments of that system are publicly financed. The first element of the system is general support for all functions of universities, with the most important segment of this general support being the annual (formerly quinquennial) recurrent block grant, at present amounting to about $1.6 billion and administered by the University Grants Committee (UGC).*

An estimate by the Advisory Board for the Research Councils indicates that about a quarter—$400 million—of the recurrent grant can be reasonably attributed to the support of research. The UGC also provides capital and equipment grants to universities for teaching and research. Finally, fees are an increasingly significant part of the income of universities in the United Kingdom and form the remaining segment of the general support part of the dual system.

The universities have traditionally enjoyed wide latitude in the use of the recurrent grants, but there is a general agreement between the universities and the UGC as to how the money will be spent. On rare occasions, the Committee has earmarked part of a grant for specific purposes and has suggested how the universities might use the funds most effectively. The long-standing principle is that the recurrent UGC grant will cover faculty salaries, building maintenance and services, general university costs generated by research, and the "basic costs of well-found laboratories." A well-found laboratory is one equipped and staffed for the research investigations traditionally undertaken by the faculty using the laboratory. In the past, the goal has been to provide such a base for research in all major disciplines at all universities.

The second element of the dual support system is funds earmarked for particular university research activities. Most of this money is provided in the form of research grants administered by the five Research Councils—the Science Research Council (SRC), the Medical Research Council (MRC), the Natural Environment Research Council (NERC), the Agricultural Research Council (ARC), and the Social Science Research Council (SSRC). Additional funds for direct support of university research are provided by government departments and private foundations.

The research grants provided by the five Research Councils currently amount to about $140 million per year, about a third of that portion of the recurrent grant devoted to research. These grants for specific research projects cover such costs as special equipment, supplies, costs of time on large experimental machines, costs of data manipulation, and, to an in-

* R&D expenditures are cited in dollars, with pounds converted to dollars at the rate of 1 pound = $2.00.

creasing degree, salaries of technicians. They supplement the funds available from the recurrent grants. Moreover, they provide a means for individual scientists and teams to secure support in direct national competition, and a means of emphasizing special fields of science and problems.

As with all elements of the R&D system in the United Kingdom, the councils operate with a substantial degree of independence. However, they are linked by an informal committee consisting of the heads of the Research Councils. In addition, the influential Advisory Board for the Research Councils (ABRC) advises the Secretary of State for Education and Science on the division of funds among the councils. (Its advice in this respect is virtually always taken.) The ABRC also provides advice to the Secretary on general policy issues related to basic research.

The scientific community in the United Kingdom has numerous informal and influential ties with the government departments and ministries. Scientists play an advisory role that usually involves a *de facto* power of decision in connection with the standard, relatively small, basic research projects. In that context, the system strongly resembles the peer review process used extensively in the United States. The scientific judgments are tempered and sometimes overridden by economic and political judgments when the proposed undertaking is very expensive—as, for example, a proposal to build a particle accelerator. In policy matters, scientists participate in a variety of formal advisory boards and councils, including the Advisory Board for the Research Councils and the Advisory Council for Applied Research and Development. Hence, decisionmaking on science and technology reflects wide participation. The Committee of Vice Chancellors and Principals acts, moreover, as a spokesman for the universities on matters of mutual interest affecting them. This Committee is entirely nongovernmental but, like the West German Rectors' Conference, interacts actively with public officials and carries weight when it takes a strong position on issues.

The scientific community, like the rest of society, is subject to the overriding priorities of the government and, for the foreseeable future, these priorities reflect the conditions of austerity and economic difficulty facing the country. However, within the broad limits on expenditure set by general economic and political forces, the scientific community has important influence on scientific priorities and on the structure and membership of the advisory system. This influence is particularly strong during periods when, as is now the case, the level of R&D support is not expanding but is virtually constant.

The decisionmaking process differs in notable respects from that in the United States. Much depends upon custom and conventions that do not fit any simple scheme. For example, the University Grants Committee, which functions as a buffer organization between government and the universities, has wide discretion on the allocation of funds to universities, but it has no statutory base to legitimate its actions or to guide its decisions. In the parliamentary form of government, decisions by the cabinet and by ministries are not subject to detailed review and reversal by Parliament, as is the case with Congress in the United States. Nevertheless, parliamentary committees hold hearings and their reports are of growing significance. For example, the Select Committee on Science, Education, and the Arts, formed in 1979, has issued an influential report on university fee policy for foreign students.

## Structural Arrangements for Basic Research

A quarter of all R&D in the United Kingdom is performed in government-owned and -operated laboratories. These laboratories are of two kinds. The first is the large set of laboratories—typified by the National Physical Laboratory—managed and staffed by the various ministries. Academic scientists play little part in their activities. The second set consists of large basic research facilities operated by the Research Councils—primarily the Science Research Council—for use mainly by academic scientists.

Half of the total R&D in the United Kingdom is performed in industrial laboratories (as compared with 70 percent in the United States). Of total R&D expenditures, 20 percent goes to university laboratories, as contrasted with 13 percent in the United States. As for basic research, half of that in the United Kingdom is performed in the universities, as is also the case in the United States. The university research is highly concentrated, with half of all government funds for university research going to the 13 largest research universities, of which 12 percent goes to the two largest research universities—Oxford and Cambridge. The total R&D effort of the United States is about seven times that of the United Kingdom—about $54 billion as compared with $7.6 billion in 1979.

This different scale of research activity has resulted in several distinctive features. First, with a smaller resource base than Germany, France, or the United States, the British have concentrated their large-scale basic research efforts in relatively few nonuniversity laboratories and in areas with a record of high scientific productivity. These large-scale research activities are rarely attached to and managed by universities. On the other hand, the relatively small project grants to individuals and to small groups for basic research in universities continue to be important. The project funds are largely spent for small scientific efforts at universities, but some project grants provide funds for experiments involving large equipment. The only other substantial support specifically for research performed at universities is aid to research units of the Medical Research Council attached to medical schools and hospitals. These units are similar to the associated laboratories of the French Centre National de la Recherche Scientifique (described in the section on France) and to research centers attached to universities in the United States.

The United Kingdom, like France and Germany, has relied increasingly on cooperative Europe-wide research efforts, and its contributions to such efforts have been important expressions of its research policy. While selected large-scale facilities for some fields of physics have been constructed recently in the United Kingdom (notably the Synchrotron Radiation Source at Daresbury), all national facilities for particle physics have been closed down. British scientists in this field have carried on their work in Europe, most notably at CERN (European Organization for Nuclear Research), outside Geneva.

Subscriptions to international organizations, administered by the Science Research Council, include contributions to the European Organization for Nuclear Research, the European Space Agency, various North Atlantic Treaty Organization science programs, the Institut Laue-Langevin, the European Science Foundation, and the European Molecular Biology Organization. These absorb 28 percent of the Science Research Council's budget and 16 percent of the combined budgets of all five Research Councils. The British have decided to remain active in the most expensive areas of modern science through collaborative international efforts.

Although the mission agencies—especially the Ministries of Defence, Industry, Energy, Agriculture, Health, and Environment—are substantial supporters of R&D, their role differs from that of the mission departments in the United States. Ministries are predominantly concerned with applied research—exploratory, advanced, and engineering development work intended to contribute directly and over the short run to the solution of practical problems—leaving the research of scientific interest but less immediate applicability to the five Research Councils. In contrast, in the United States in 1979, the Departments of Agriculture, Defense, Energy, Health, Education and Welfare, the Interior, and the National Aeronautics and Space Administration financed 80 percent of all federally funded basic research as well as most applied research and nearly all exploratory and advanced development work.

The decision in the United Kingdom to establish Research Councils independent from the operating departments of government was based on the philosophy of the 1918 Haldane Report which concluded that, if responsibility for basic research were placed mainly in the operating departments, the current problems of the departments would exert pressures leading to an unhealthy emphasis on research with short-term objectives and to an inadequate level of basic research with long-term payoff. The Haldane Report led directly to the establishment of new independent Medical and Agricultural Research Councils.

The Science Research Council provides external support for a broad range of the basic sciences, finances participation in international research facilities, provides advanced fellowships, encourages cooperation between industry and universities, administers fellowships, and operates research facilities. The Social Science Research Council provides for the support of the social sciences and operates independently of the Science Research Council. The other three Research Councils support fundamental research in areas loosely related to the missions of their counterpart operating departments—Health, Agriculture, and Environment. In addition, they promote certain areas of applied research of potential use to the operating departments.

During the early 1970's in the United Kingdom, the impression developed that the Medical, Agriculture, and Natural Environment Research Councils as well as the in-house department laboratories were not adequately responsive to the short-term needs of the various government ministries and departments. The 1972 report, *Framework for Government Research and Development*, popularly known as the Rothschild report, crystallized these concerns, calling for strengthening the customer-contractor relationship across the board and for the transfer of funds from the councils to the user departments to finance contracts with the councils (or other contractors) for relevant research. The report also called for the appointment of a chief scientist for each department and ministry with substantial research responsibilities. It also recommended the formation of an Advisory Board for the Research Councils to provide general advice on science policy to the Secretary of State for Education and Science, to give more specialized advice on the allocation of funds among the councils, and to promote close liaison between the councils and the user departments.

There is a consensus in Britain that the appointment of chief scientists in the departments, as well as a chief scientist in the influential Central Policy Review Staff of the Cabinet Office, has improved both in-house research programs and interdepartmental coordination. The Advisory Board for the Research Councils is believed to have performed well. On the other hand, problems have developed in administering the customer-contractor principle. In the case of the Medical Research Council and the Department of Health and Social Security, the new system worked poorly, as evidenced by the fact that almost all of the transferred funds have in effect been spent on research that the council would have done anyway. Accordingly, the arrangements between the department and the MRC have been modified so that the department sets out a range of problems and the MRC suggests how biomedical research might be applied to them. For the other councils and departments, the Advisory Board for the Research Councils has suggested that the departments demonstrate greater regard for the long-term health of the councils by placing longer and broader commissions with them and by simplifying administrative procedures. Furthermore, the Advisory Committee on Applied Research and Development has urged the councils to be alert to critically important emerging areas of technology, which may be overlooked when there is no obvious single customer and when new technologies will appear only several years in the future. A case in point is the failure of departments to identify biotechnology as a promising field.

The urgent national need to increase industrial productivity has moved the Science Research Council toward greater emphasis on applied research, on research and training collaboration between industry and academic institutions (universities and polytechnics), and on a large number of special problems. Among the areas selected for emphasis are applications of numerical control to manufacturing, coal technology, marine technology, polymer engineering, advanced ground transport, and energy in buildings. In total, these measures represent a substantial modification of the priorities of the Science Research Council.

British industries, both nationalized and private, do extensive research in their own laboratories and contract for research activities with profit-making firms and university laboratories. Industrial research tends to be oriented toward applied and developmental work, to be concentrated in large firms, and to be centered in relatively few sectors of the economy—notably electronics, fine chemicals, energy and petrochemicals, pharmaceuticals, aviation, communications, and optics.

To an increasing extent, British university science and engineering departments are undertaking research under contract with industrial firms, and consulting with industrial firms by faculty members is becoming more prevalent. The Science Research Council has initiated a Teaching Company Scheme which is providing continuing links between industrial firms and academic departments. On the whole, however, these relationships are newer and less extensive than those between industry and the Grandes Écoles in France, technical universities such as Aachen and Munich in Germany, and the large research universities in the United States.

The research role of nationalized industries—such as the U.K. Atomic Energy Authority, the Central Electricity Generating Board, the National Coal Board, British Gas, and British Steel—is significant. In 1978, the nationalized industries expended about $125 million on research and development.

## Forms of Support and Financing of Research

Total national expenditures on R&D in the United Kingdom approximated $7.6 billion in 1979. Of this amount, $3.4 billion (45 percent) was expended by industry (both private and nationalized) and $4.2 billion (55 percent) was expended by government. Defense R&D outlays of $2 billion accounted for half of the public expenditures, and the Departments of Industry and Energy each provided another $260 million (11 percent) of the total monies spent on public R&D. The complete science budget, totaling $800 million in 1980, represents the regular outlays plus funds for commissioned research for the five Research Councils (including also relatively small sums for the Royal Society and the National History Museum). The science budget accounts for less than a fifth of the total government R&D expenditures, but the actions of the councils are particularly important because they concentrate on fundamental research. The council budgets, along with a portion of the recurrent capital and equipment grants of the University Grants Committee and some assistance from the mission agencies, provide the principal support for basic research in the United Kingdom. The Science Research Council, the largest of the councils, spent $360 million for basic science in 1980. The Medical Research Council is the next largest, providing $140 million. The Agricultural and Natural Environment Research Councils each spent about $120 million and the Social Science Research Council spent about $30 million.

The Research Councils operate primarily through five forms of support: project grants to individual university scientists (20 percent of the total); research units (mostly medical) attached to universities and medical schools (10 percent); training awards and fellowships (15 percent); operation of laboratories for university user groups (15 percent) and for in-house staff (mostly ARC, MRC, and NERC) (25 percent); and subscriptions to international organizations (15 percent).

There are clear trends in the relative importance of the various types of support. The rising costs and scientific utility of advanced instrumentation have involved the councils—particularly the Science Research Council—increasingly in providing costly central instrument facilities. Manpower also has become an increasingly important problem, and training awards and fellowships have consumed a greater proportion of total resources. Activities designed to make research contribute more effectively to industrial innovation and productivity are becoming increasingly important. Contributions to the support of the large international machines, such as those at CERN, are high and rising as the minimum charge for remaining active in the forefront of particle physics rises. These developments have reduced the proportion of council funds available for grants to individual investigators, now 20 percent of the total council funds, and decreased the proportion of high quality applications that can be funded.

A set of forces is undermining the dual support system. The recurrent grant of the UGC for 1979–80 was $1.6 billion and the combined equipment and capital grants amounted to about $150 million. This support, although a substantial increase over that of a decade ago, represents virtually a level rate of expenditure when allowance is made for high inflation rates, and is a decline in resources per student. Continuing high rates of inflation and a policy of strict expenditure control by the government suggest that the universities will face a period of financial stringency.

Because of the pressure on the recurrent grant, the UGC is unable to provide the basic support for "well-found laboratories" for the system as a whole. There has been a continuing shift in financial responsibility to the councils for such direct research expenses as basic laboratory equipment and the salaries of technicians. The assumption of these costs by the councils has reduced the support available for individual research projects.

The basic roles of the UGC and the councils have been called into question by trends of recent years. The possibility that the UGC will formally abandon the idea of providing "well-found laboratories" in all major disciplines in all universities is under serious discussion by a joint committee of the ABRC and the UGC. The UGC probably will become more selective in support for academic science by adjusting the recurrent grant to provide relatively greater assistance to outstanding individual teams and laboratories, and for areas of science and problems selected for priority support by the Research Councils. Suggestions by the UGC to the universities may well be made more often and with somewhat greater urgency. At the moment, this is occurring only in specialized areas. An influential minority view is that this degree of selectivity will be inadequate, and that some programs or even entire unversities must close down. This probably will not occur unless enrollment declines sharply a few years from now. Such greater selectivity in the allocation of the recurring grant would involve an important change in basic philosophy and in administrative style and relationships.

The role of the Research Councils probably will change also, with the prospect that the councils will assume greater responsibilities for some direct research costs—such as salaries of technicians, utility costs, and costs of laboratory equipment—previously borne by the UGC. However, there is little prospect that any portion of general university costs not generated entirely by research will be shifted to the budgets of the councils.

## Trends in Scientific Manpower and in Supporting Resources for Research

The United Kingdom faces two particularly important and interrelated manpower policy issues. The first is the appropriate volume of support for graduate students—called postgraduate students in the United Kingdom. The second is the proper response to the small number of new positions available over the next few years in government laboratories and universities.

The development of higher education in the United Kingdom has been strongly guided by a central principle of the 1963 Robbins Report: higher education should be available for all who are qualified and willing to pursue it. The principle has been implemented in particular at the undergraduate level, where all students of United Kingdom origin who are accepted by universities for a first degree course are eligible for mandatory, though means-tested, maintenance grants and for full reimbursement of tuition fees.

Now the system is under pressure from economic forces. British graduate students are increasingly reluctant to accept some types of stipends when the additional study will lead only to a series of short-term positions. Under one category of advanced fellowship awards, for example, the Science Research Council awarded 33 stipends in 1977 and only 19 in 1979. For the Cooperative Awards in Science and Engineering—a special SRC doctorate program involving in-service training in industry—the number of eligible candidates was substantially less than the number of available awards in 1979. In general, the council in recent years has found that the number of suitable candidates has been inadequate for a number of its various special programs; this is reminiscent of the reluctance of some outstanding younger German scientists to accept five-year Heisenberg awards. Students have increasingly tended to seek employment in industry after the initial degree in science or engineering rather than to follow a course of advanced study in the face of uncertain university career prospects after completion of their studies.

There are growing concerns over the continuation of a large-scale graduate studentships program during a period of financial stringency in the face of the prospective low demand for scientists and engineers in university teaching positions and in government laboratories. These challenges may lead eventually to a reduction in government support for graduate study, as has happened in the United States.

Britain shares with other countries the prospect of few new academic positions because of a decline in research expenditures and restricted general university budgets. Several modest programs have been designed to address the problem. About 20 fellowships a year are awarded for up to five years for carefully selected scientists under the age of 35 who do not hold tenure positions yet—the Advanced Fellowship Program. In addition, a small number of senior research scientists with tenure positions in universities will be given the opportunity to leave their teaching duties in order to engage in full-time research. The vacated teaching positions will provide tenure openings for younger scientists—the Special Replacement

Fellowship Program. While these are regarded as useful measures, they are widely viewed as modest in scope and unlikely to have a decisive effect.

The average age of researchers in the Research Council laboratories and the government laboratories, as in the universities, is rising. A large proportion of the research staffs will remain with their laboratories until the normal retirement age of 65. Since the laboratories are not growing, they will be able to offer few permanent jobs to younger scientists. Further, there is very little movement of staff from nonuniversity research laboratories to universities because of the diminished academic opportunities; previously, a natural outlet into university employment had promoted a greater degree of staff turnover and career mobility. Thus the system as a whole appears to face both depressed long-range demand for highly trained people and decreasing mobility among the segments.

As a consequence of the virtual absence of new tenured positions in universities, some university research programs are suffering, as are the Research Council laboratories. Short-term contract positions are offered but it has become increasingly difficult to recruit young research assistants on these terms. Many who have accepted such positions are unhappy. Added to these problems are difficulties in recruiting and retaining technical support staff, largely because salaries are lower than those in industry.

All in all, the manpower problems are becoming as severe a constraint as budgetary considerations. However, if more of the brightest students in science and engineering seek industrial employment, the innovative capacity and competitiveness of the economy could be stimulated.

In addition to scientific-manpower concerns, the issue of supporting resources for academic science, especially laboratory equipment and facilities of various kinds, has been debated intensely in Britain. Standard small-scale laboratory equipment in British universities has been deteriorating rapidly in recent years. The standard working equipment in university laboratories was installed in volume about ten years ago when the expansion of British higher education took place. Much of this equipment is now reaching the end of its useful life. Neither the University Grants Committee nor the Research Councils have been able to divert enough funds from other high-priority programs to refurbish the laboratories generally. The outlook is for careful selection of relatively few laboratories to be modernized and for a much more extensive sharing of equipment.

The situation with respect to large-scale instruments and facilities has been more favorable. The extensive contribution of the Science Research Council to the operation of European facilities—such as the European Organization for Nuclear Research and the Institut Laue-Langevin—accounted for 24 percent of the council's budget in 1978-79, an amount exceeding its total domestic research grants for the year. But there also has been an effort to create large-scale facilities domestically in selected areas—for example, the Synchrotron Radiation Source at Daresbury and the laser laboratory and neutron-beams facility at the Rutherford Laboratory. Other domestic facilities, however, have been shut down. These include the Nimrod proton accelerator at the Rutherford Laboratory and the NINA electron synchrotron machines at Daresbury.

## FRANCE

### Major Features of the Research System

The French experience shows that it is possible to have a strong and flourishing basic research enterprise that places heavy emphasis on a system of laboratories parallel to and associated with universities and the Grandes Écoles, rather than relying primarily upon the universities and Grandes Écoles. Yet the universities provide the institutional base for a large fraction of the nation's most brilliant and productive scientists, and the advanced training of scientists is begun in the universities. There is increasing recognition that the long-run strength of the system of parallel laboratories depends on the strength of the universities and Grandes Écoles. Nevertheless, the steps taken thus far to ensure their health have been weaker than those taken to ensure the strength of the parallel laboratories; this is due partly to the fact that many of the universities have faced intractable difficulties. An important issue for France is the extent to which achievement of such goals as increasing scientific career mobility, enhancing the capacity of industry to use research results, and rapid development of priority research fields may be frustrated by weaknesses in the institutions of higher education—particularly the universities. How the obstacles to the development of the universities' research capabilities can be overcome is a many-sided question. This basic problem, in a different institutional and political context, also is faced in the United States, where the federal government finances research and associated costs in universities on a large scale but provides no funds for general university support.

Direction of the research system of France is more centralized than those of the United States, the United Kingdom, and Germany. This basic feature of centralization is consistent with the structure of French government and society, and it is reinforced by the habits and traditions of scientists and administrators. At the same time, central planning is indicative rather than overriding, and many strong national sources of R&D support operate with a substantial degree of independence.

A Secretary of State for Research reports directly to the Prime Minister. His duties include preparation of the R&D component of the five-year general economic plan and coordination of the work of the central R&D organizations with that of the ministries and large state enterprises. The Secretary of State for Research receives staff support from the DGRST (Délégation Général à la Recherche Scientifique et Technique), a small organization with both a staff and an operating role. The DGRST manages a limited number of high-priority basic R&D programs, called concerted-action programs (actions concertées), contracts with public and private groups, and maintains a reserve fund for special programs. In addition, it provides staff support to the Secretary of State for Research.

The most significant single organization responsible for the promotion of basic research is the National Center for Scientific Research (Centre National de la Recherche Scientifique—CNRS) under the trusteeship ("sous la tutelle") of the Ministry of the Universities. It is a large organization, covering most fields of modern science. For 1979, it had a budget of about $800 million and employed approximately 8,500 scientists and 14,000 supporting personnel in its own laboratories or in laboratories associated with universities.*

At the level of national policy, R&D priorities are developed as part of the successive five-year plans, which set general indicative goals. Even though these goals are not firmly established as targets, they represent a national effort to make the allocations of R&D resources more coherent and comprehensive than in the United States, the United Kingdom, or Germany. To illustrate, priorities for the next (the eighth) five-year plan include steps to enhance the value of selected mature fields (information, space technology, and microbiology), interdisciplinary and interregional cooperation (materials and extension of the techniques of physics to other disciplines), and research on complex interacting systems (atmospheric sciences, oceanography, biosocial interactions, and ecology).

Administratively, many critical decisions relating to the operation of government laboratories are made centrally but with the help of the scientific community. To illustrate, the size of each CNRS laboratory, the total budget, and the number of new positions are determined centrally. However, the laboratory director has broad authority to shift operating funds from one category to another (but money cannot be shifted from the operating to the personnel categories) and to recommend staff promotions. All university faculty members are civil servants. They are paid under a salary schedule that is identical for all universities and they receive their paychecks directly from the Ministry of the Universities, not from their respective universities. Decisions on which teaching and research units in universities may offer advanced degrees are made centrally.

---

* R&D expenditures are cited in dollars, with francs converted to dollars at the rate of 1 franc = $0.25.

The system is centralized not only in the sense that decisions tend to be made in Paris but also because a large portion of the physical plant and human talent for research are concentrated there.

Yet the system is not centralized in every aspect. First, regionalization designed to reduce the dominance of Paris has emerged as an important objective of national policy. Some measure of success has been achieved with the growth of strong research centers in the Nancy-Strasbourg, Lyon-Grenoble, Marseille-Nice, and Toulouse-Montpellier-Bordeaux areas. The CNRS recently has set up regional centers that have decentralized many kinds of administrative decisions. A number of large government R&D agencies have strong bases independent of the central science agency. The Ministry of Defense, responsible for about 33 percent of all government financing of R&D, relies heavily on its own research establishments but also has an extensive network of contracts with industry and with universities. It essentially pursues its own research priorities. The Ministry of Industry supports important conglomerate R&D activities in the atomic energy, space, data-processing, oceanography, and various energy programs, including solar, geothermal, and hydrocarbon. In addition, the Ministry of Industry has primary responsibility for stimulating technological advance in industry, a major goal of the seventh five-year plan. The National Institute for Health and Medical Research (INSERM) is responsible for medically related research and research in the life sciences. In addition, there are such significant agencies as the National Institute for Agricultural Research (INRA), the National Center for Space Studies (CNES), the National Center for Telecommunications Studies (CNET) and the National Center for Exploitation of the Oceans (CNEXO). Other officials with extensive research responsibilities, such as the Secretary of State for the Postal Service and Telecommunications, also operate with considerable independence both from each other and from the Secretary of State for Research.

Both professionals and nonprofessionals engaged in research tend to place long-term job security above other career goals in the French research system. This powerful tendency is reflected in two particularly important ways—unionization and regulations protecting job rights. The organization of both professional and supporting staff promotes, among other goals, protection of job rights, establishment of equitable terms and conditions of employment, and elevation of pay and perquisites. National laws and regulations state in detail the qualifications required for each level of position, criteria for attainment of permanent status, job performance requirements, seniority rules and dismissal procedures, eligibility for pensions, and the like. These laws and regulations are not uniform among the major sectors of research performers. The desire for security also accounts for the legal requirement that workers hired for more than two years must be given permanent positions (which causes a rapid turnover in research support staff). The priorities given to security and to equity place high barriers in the path of efforts to sustain flexibility in research organizations and mobility among them.

The close formal linkage between national goals and the research system, whether the earlier focus on national prestige or the more recent concern with economic competitiveness, is an important feature of the system. The research system in France is tied closely to national priorities and the most visible spokesmen for science are full-time officials of the national government. The Secretary of State for Research is the most prominent public spokesman on scientific matters, along with the Minister of Industry on matters with a major technological component and the Minister of the Universities on matters related to academic sciences. This link between national goals and the research system generates strains between government and parts of the academic community; the latter see the system as one increasingly dominated by political rather than scientific considerations.

Working scientists, however, are represented at a number of levels in the system through participation in advisory groups. The most highly placed of these is the CCRST (Comité Consultatif pour la Recherche, la Science et la Technologie). This group of 16 scientists—sometimes called the Committee of Sages—advises the Secretary of State for Research on a wide range of scientific issues, but it has no influence with the technical ministries. A special research commission of scientists, with numerous subgroups, participated in the preparation of the seventh five-year plan, and another group is preparing background papers on R&D for the eighth plan and for the ten-year outlook.

Research in France has traditionally been strong, and remains strong, in a number of basic fields, including, for example, mathematics, particle physics, and nuclear physics. In a number of fields of modern physics with important practical applications, France has moved rapidly upward internationally; these include optics, fluid mechanics, information and automation sciences, semiconductor properties, and the physics of surfaces and interfaces. On the other hand, some areas of physics—for example, molecular physics and material sciences—have not developed strongly. There are a number of points of strength in chemistry—inorganic catalysis, the chemistry of natural substances, and organic chemistry; but photochemistry, thermodynamics, and electrochemistry are not as strong.

Research in France in many areas of the life sciences is of outstanding international quality. These include cellular and molecular biology, immunology, intercellular information (hormones and neurobiology), developmental biology and the biology of reproduction, biology of interactions between living organisms and their environments, and basic bioenergy (for example, thermodynamics of biochemical reactions and photosynthesis). In contrast, both nutrition and fundamental and applied bacteriology are weak fields.

## The Performance of Basic Research

The major performers of basic research in France include: university scientists who frequently work at laboratories associated with the CNRS, INSERM, and other government agencies; the in-house government laboratories of CNRS and INSERM, or other laboratories administered by the Ministries of Defense and Industry and other ministries; a limited number of private institutes, such as the Pasteur Institute; and a number of collaborative European research centers located on French soil.

A major role is played by the CNRS, whose laboratories resemble Germany's Max Planck Institutes in some respects. All of them are devoted to coherent areas of basic research, led by outstanding scientists, funded by block grants for the work of the laboratory as a whole rather than by smaller project grants, and staffed by scientists with secure tenure. They differ notably, however, from the Max Planck Institutes and from the free-standing research institutes of the United States in that they are run by the government. CNRS staff members are state employees under a special personnel system, and not employees of a self-governing society, as are those in the Max Planck Institutes. Moreover, the CNRS devotes about half of its resources to external grants to investigators in laboratories associated with universities, whereas the Max Planck Institutes and the institutes in the United States support no external research.

At the operating level, the CNRS has a National Science Research Committee, with 41 disciplinary panels, to which hundreds of working scientists contribute time. These panels have been responsible for recommendations, almost invariably followed, on the programs, funding, and personnel actions of every CNRS laboratory, including the directly operated CNRS laboratories as well as those associated with universities. Half of these advisers are elected by two groups of electors, the senior staff and the junior staff of the universities and of CNRS. The remaining half are appointed by the Minister of the Universities after receiving recommendations from the Director General of CNRS, the DGRST, various ministers, and politically influential groups. The ministerial designations tend to establish an equilibrium among fields of specialization, background (academic or industrial), regional origin, age, and philosophical and political views. But, to some degree, the advisers represent constituencies and points of view. In this respect, the advisers are more than scientific peers and this gives the advisory system in France a political coloration that is more visible and pronounced than in the other countries.

The associated laboratories of CNRS (laboratoires associés) are physically located on the campuses of universities and Grandes Écoles. The largest concentration is in the major research universities, including parts of the University of Paris (such as Paris XI [Orsay], Paris VI, and Paris VII), the University of Strasbourg, and the University of Grenoble. In recent years, additional points of strength have emerged in the universities at Lyon, Lille, Montpellier (medicine particularly), Rennes, and Toulouse. The associated laboratories, drawing upon, augmenting, and orienting the basic support provided by the universities, provide the conditions under which the most capable faculty scientists can conduct research while retaining faculty status but largely free of the rules governing research in the universities. In addition to providing a source of support, association with CNRS continues to carry great prestige for a university scientist and most serious academic researchers seek CNRS sponsorship for their work. Conversely, many outstanding CNRS investigators seek professorial status because of the prestige and the benefits, such as superior pensions, associated with it. The universities frequently provide substantial resources for the associated laboratories.

Approximately 30 percent of faculty in the "exact and natural sciences" in French universities have some association with the CNRS, and the CNRS support ranges from symbolic to critical. The CNRS retains direction of the associated laboratories as well as its own direct scientific operations but, in practice, there is sometimes virtually complete cross-assimilation among faculty affiliates and CNRS employees at a laboratory. In a few major research universities where research is being done on a massive scale across the broad range of disciplines in the physical, biological, and social sciences, the CNRS employees enjoy faculty privileges and some of them teach. In turn, the faculty associated with CNRS activities participate fully and may assume leading roles in the direction of CNRS laboratories.

In addition to supporting laboratories as a whole, CNRS supports the research of relatively small but seasoned teams (équipes) of university scientists and of groups of younger university investigators. Under this mode of support, the research of the entire group, rather than projects of individual investigators, is aided.

University investigators may apply, either as individuals or as collaborators, for support of specific projects initiated by them under the relatively small concerted actions (actions concertées) program of the DGRST (budgeted at $36 million in 1979) and the selected themes program (actions thématiques programmées) of the CNRS (budgeted at $16 million in 1979). The National Institute of Health and Medical Research also has a program of selected themes for support of individual investigators, budgeted at $8 million in 1979. Overall, however, the total system operates primarily to support laboratories and groups, with much less attention to support of projects proposed by individual investigators.

France, like Germany and Britain, has a dual support system for funding universities, but support for research in universities is not as important an element as in the other two countries. Support for the general functions of universities, most importantly for the functioning of the research system, covers the salaries of faculty members engaged in research. Some funds specifically for research are channeled through the university itself, as is the money provided by the Research Mission of the Ministry of the Universities. However, most research support is provided through channels that bypass the university as an institution. For example, the important associated laboratories of CNRS, which support the research of a high proportion of the most able university scientists, are administratively distinct from the universities. Similarly, other research support provided by government agencies and industry goes directly to individuals and teams.

Steps have been taken to increase the capacity of the universities to serve as the institutional base for research. For example, beginning in 1972, the universities of Strasbourg, Paris XI (Orsay), and Grenoble began to build research offices to define institutional policies to help investigators and to encourage collaboration across faculty lines. In 1977, the Ministry of the Universities established the Research Mission (Mission de Recherche) to improve the research infrastructure of universities, to strengthen the disciplines most handicapped by extreme scattering of faculty among the universities, to ensure the optimum distribution of research equipment, and to encourage interdisciplinary research. The funds available to the Ministry of the Universities for the purpose of strengthening research in the universities totaled $140 million in 1979 (as compared with the $500 million allocated to associated university laboratories by the CNRS in that year). The funds available to the Research Mission have been, in general, widely distributed in relatively small amounts. It is too early to assess the success of these grants in strengthening the research capacities of universities.

The special French institution, the Grandes Écoles, trains a small number of carefully selected students for technical and managerial leadership in government, state-operated enterprises, and private industry. They have not, with some notable exceptions, had strong research programs in the past. However, beginning about a decade ago, the Grandes Écoles have developed extensive research programs of high quality.

## Forms of Support and Financing of Research

For a number of years, especially from 1958 to 1968, R&D activities in France grew spectacularly. R&D expenditures increased at a rate of 15 percent per year in real terms and the percentage of GNP devoted to R&D almost doubled, to a level of about 1.8 percent. The increase in spending reflected concentration on enhancing national prestige as well as on advancing science and technology. These efforts included the design and construction of the Concorde airplane, the atomic bomb, an impressive space program, and large solar and nuclear energy efforts. The decade 1968-78 was one of slow growth in R&D expenditures, reduction in the scope and scale of new undertakings, and fewer new R&D job positions. In spite of the problems facing the French R&D system, it has continued to grow. Some increase in new positions in government laboratories (but not universities) has occurred, large amounts of money have been provided for instrumentation, and new research areas have been opened up. No major research programs have had to be curtailed or abandoned in France. New priorities have emerged; in particular, there have been less attention to national prestige and an increased emphasis upon industrial innovation, encouragement of new fields such as biotechnology, a greater degree of energy self-sufficiency, and the international competitiveness of French industry.

Total national expenditures for R&D in France in 1979 amounted to approximately $10 billion. Industry provided nearly $4 billion (40 percent) of the total national investment in research. Of the remaining $6 billion, about 95 percent came from government, with the nonprofit sector and higher education providing the remaining 5 percent.

While industry provides 40 percent of the funds, it performs 60 percent of the total R&D in France. Government support for the R&D in industry takes such forms as subventions, research contracts, or reimbursable advances to firms in electronic industries for the design and construction of high-technology military systems and the testing of prototype civilian aircraft. The other 40 percent of the national R&D effort is produced in government laboratories, universities, and nonprofit institutions. The largest share of the work is done in government laboratories.

Government financing of R&D in France falls into two categories: funds "within the research envelope" and "outside the research envelope." The half of government R&D outlays that are outside the research envelope are estimated expenditures for R&D activities not separately identified in agency budgets. This group of R&D activities is dominated by defense—nearly 70 percent of outlays outside the envelope. Estimated university salary costs attributed to research costs in all fields comprise about 20 percent of expenditures outside the research envelope. The remaining 10 percent outside the envelope is distributed between aeronautics and telecommunications.

Funds within the research envelope are explicitly for research and identified as such in advance in the budget. These funds are dominated by the large R&D enterprises under the Ministry of Industry, especially atomic energy, space, and oceanography. This ministry provided about $1.4 billion (42 percent) of the total funds within the research envelope in 1979. The two agencies under the Ministry of the Universities—CNRS and the Mission de Recherche—provided $800 million and $140 million respectively in 1979 (28 percent of the total funds within the envelope). The DGRST with $160 million (5 percent), the Ministry of Health with $150 million (5 percent), the Ministry of Agriculture with $200 million (6 percent), and the Ministry of Foreign Affairs with $200 million (6 percent) are the major remaining sources of funds within the research envelope. While direct comparisons are difficult because of such factors as different amounts spent on staff salaries, the combined budgets of the CNRS and the Mission de Recherche are about three quarters of the budget of the National Science Foundation.

The Ministry of Foreign Affairs is a supporter of research within the research envelope by virtue of the fact that the French contributions to CERN and the Institut Laue-Langevin are included in its budget, since they are international cooperative ventures. Until a few years ago, British international scientific contributions were part of the foreign affairs budget, but they have been transferred to the Science Research Council.

The forms of support provided for basic research by CNRS and INSERM in their external grant programs differ from American practices. Support tends to be in larger aggregates than in the United States, going mostly to teams of scientists or entire laboratories rather than to individual investigators. The length of support is usually for a four-year period, with opportunity for renewal. Funding is sometimes reduced, or support for a program may be withdrawn altogether, if the scientific results are judged to be disappointing. Review panels of the CNRS play an important role in decisions relating to research priorities and directions, and to funding levels in the associated laboratories.

## Manpower and Supporting Resources

Manpower problems in the French research and higher education systems are cause for continuing concern. Exposure of undergraduate science and engineering students to active research has not been a strong feature of the system. Recruitment of scientists and engineers for high-priority tasks has been difficult because barriers to the easy movement of scientists between universities, industry, and the government research enterprises are substantial. But the manpower issue of greatest immediate concern is the reduction in the rate of growth of the research system, resulting in a scarcity of new positions for young scientists and apprehension that too few of the best minds will be attracted to basic science careers.

Investigators in the government laboratories were recruited in large numbers in the 1960's. Today, the average age of CNRS scientists is 40 years and a high proportion of them will remain in their positions for many years. The age pyramid is such that the CNRS envisages substantial blocking of promotions until 1990; thus, the research productivity of individual scientists almost inevitably will be less significant than seniority as a criterion for promotion. The government has responded by adopting in 1979 a policy of increasing the number of permanent positions, primarily in CNRS and INSERM laboratories, by 3 percent each year. This step was taken even though it could mean the diversion of funds from other purposes at a time when budgets were not increasing rapidly. Successive national budgets have implemented the stated policy and have specified the number of new positions available to each agency. This commitment to an annual increase of 3 percent, and not merely to a one-time or transitional action, is a stronger measure to provide continued career opportunities for young scientists in basic research than the substantial German Heisenberg program or the modest equivalent British programs.

In 1980, additional steps were taken to speed the entry of younger scientists into the government laboratories and to encourage mobility between government laboratories, industry, and universities. A young scientist formerly served a long (six to eight years) probationary period as "attaché de recherche" before entering the lowest grade in a government laboratory—the "chargé de recherche." Senior scientists were not required to follow uniform rules with respect to the length of the probationary period served by junior scientists, and many of them felt that the long and flexible probationary period resulted in inequities. To improve the system of recruitment, the government has promulgated rules stating that scientists will be eligible for appointment at the first level—attaché de recherche—not later than age 27 (with some exceptions and with a phasing-in period). They then serve a fixed four-year probationary period at that level. About 80 percent have typically been promoted to the next level—chargé de recherche. Those who are not promoted must leave the laboratory and cannot be retained in a short-term or indefinite status in any CNRS laboratory.

The 1980 budget also provided additional positions at the chargé-de-recherche and maître-de-recherche levels. The government has created incentives to mobility by giving preference in promotion to people willing to move geographically and from CNRS to university or semipublic employment, and to switch scientific fields. These changes—both the new age limits and the considerations of mobility—are controversial. Some scientists believe that mobility should be encouraged not by decree but by providing incentives to attract good people. The upper limit of 27 years for appointment to the attaché grade is suspected of being a hidden device to move out of the system many of the younger people who have been hoping for permanent positions. The universities have complained that, while the number of new positions has been increased in both industry and the government laboratories, they have had to face a situation of virtually no growth in new permanent posts. Spokesmen for higher education have sought an extension of the 3-percent principle to both universities and Grandes Écoles, especially in the light of the ostensible commitment to enhancing their capacities as research performers. The

Ministry of Finances has resisted proposals for new positions in the higher education sector.

In 1979, the Ministry of the Universities, under extreme pressure from faculty organizations, transformed 2,900 middle- and senior-level faculty grades into higher-level grades in order to promote those then in the lower positions. This satisfied some grievances of faculty members but, by filling posts whose number is unlikely to be increased in the foreseeable future and by adding to the total cost of the education budget, the possibility of adding positions for young teacher-investigators in the future was decreased further.

## Instrumentation

France has been able to continue the construction of a few large and very expensive domestic facilities, directly administered by the government. They include the ion accelerator GANIL at Caen, the research reactor Saturne III at Saclay, a superconducting tokamak (TORE III), and oceanographic vessels. A large investment has been made in computers. But the peak of investment in large national machines has passed. They do not dominate the nation's total investment in instrumentation and facilities for scientific research nor do they play a dominant role in the country's research strategy.

The increasing need for international cooperation in scientific activities, mostly but not entirely occasioned by the need to cooperate in building and operating large machines, is recognized in France as in all European countries. In 1979, France contributed a total of $90 million to 11 international research facilities; a $70 million annual contribution to CERN dominated this total.

France has not neglected standard laboratory equipment. In 1979, about $650 million (or 15 percent) of the total government civilian R&D expenditures of $4.4 billion were spent on scientific equipment. Over a third of those funds were for the general purpose civilian R&D agencies primarily engaged in basic research—CNRS and DGRST. Most of the equipment funds were for standard laboratory equipment, including some relatively large items such as electron microscopes and nuclear magnetic resonance machines. These are concentrated in CNRS or other government laboratories and in the major universities and Grandes Écoles. Modernization of equipment has remained a priority concern even at some cost to current research support in times of budget stringency. A large proportion of the instruments used by French scientists has been imported in the past, and a strong effort is being made to stimulate a domestic instrumentation industry.

Collaborative use of scientific equipment has been widely discussed in France, in part because of recognized inefficiencies in current use. A review of instrumentation at the University of Strasbourg a few years ago revealed that there were 32 electron microscopes in the university, all purchased, operated, and maintained independently. In 1977, a DGRST report prepared in connection with the seventh plan noted that "Ownership and control of equipment by a team or a laboratory is giving way progressively to cooperative use at all levels. . . ." Extensive consultation has taken place within the scientific community on the technical and administrative aspects of collaborative uses. There has been an effort to set out the rights and responsibilities of all parties using equipment collaboratively and to develop generally acceptable principles of cost sharing.

The state of equipment in France and the prospects for maintaining a broad base of modern scientific instrumentation appear to be more favorable than in the United Kingdom but not more so than in Germany. In all three countries, there is a realization that new forms of facilities management are emerging and that the way of life of many scientists may be changing correspondingly.

## Strengths and Weaknesses of the French System

The French research system has the strength that derives from having a concrete set of proposed policies and actions. Explicit policy choices are suggested through the process of preparing the five-year R&D plan and other priority programs. But centralization has weaknesses as well as strengths. It tends to inhibit quick and easy reorientation of research efforts in response to unanticipated opportunities, to delay decisions, and to remove the power of decision from the operating level. These faults have been recognized and the CNRS recently has set up decentralized offices with authority to make decisions involving substantial expenditures.

The French research system provides for greater stability in support than exists in the United States. Support is given largely to teams of scientists and to whole laboratories rather than to individual investigators.

The kind of detailed accounting for expenditures of funds required of investigators and universities in the United States is not required in France because most of the funds are spent in government establishments. Nor do the complexities relating to the payment of indirect costs to universities in the United States exist in France. CNRS and the other large national research organizations pay all of the costs of research conducted by associated faculty members—supplies, equipment, supporting personnel, laboratory animals, shop services, and the like.

Training in the uses and conduct of research is not a prominent part of the undergraduate education of many of the nation's most gifted science and engineering students, including those who have attended the Grandes Écoles. This is now seen by French observers as a reason French industry does not have a strong tradition of basic research, and why in some areas of advanced technology France is not in a strong competitive position internationally. The recent expansion of research in the Grandes Écoles is intended to cope with these problems.

The United States can learn from the relatively advanced efforts at collaborative uses of scientific instruments that France has pursued for several years. The French have placed emphasis on collaborative ventures at all levels—locally within large laboratories, regionally, nationally, and internationally.

The rigidities in the French system are viewed in France as a serious problem. To deal with it, the French government has sought to encourage mobility of scientists and has taken more far-reaching steps than other European nations to address the problem of diminished career opportunities in basic science for young investigators.

## THE FEDERAL REPUBLIC OF GERMANY

### Major Features of the Research System

The Federal Republic of Germany has a pluralistic and decentralized system of research performance and support. In the breadth and diversity of the sources of support, the German system resembles that of the United States. Funds are provided by the federal government, the state governments (the Länder), private foundations, and industry. The major research performers are also diversified, including universities, non-university institutions like the Max Planck Institutes, large-scale or big science research establishments, industrial laboratories, and in-house government laboratories tied to mission agencies. The principal performers of basic research are the universities, the Max Planck Institutes, and some of the big science establishments. The large synchrotron accelerator DESY, for example, is one of the 12 installations belonging to the Association of Big Science Establishments (AGF); the Association also includes installations that support the German space, nuclear energy, and other large development programs.

In general, German science has made a substantial recovery from the dramatic upheavals that shook the nation in the 1930's and 1940's. From 1933 to 1939, 35 to 40 percent of all German scientists fled the country in the wake of the Nazi attack on Jewish scientists—an unprecedented break in the scientific tradition of an industrialized nation. The war continued the disruption of German science. But, since the war, German science has regained some of the preeminence it enjoyed in the early part of this century. Germans have won 13 Nobel prizes since World War II, principally in chemistry, the life sciences, and experimental physical science. The outstanding fields of German science in the contemporary period include biochemistry (steroid and enzyme chemistry, biophysical chemistry, natural products chemistry), physical chemistry, metallurgy, low-temperature physics, engineering sciences (mechanical and electrical), and certain biomedical clinical fields (opthalmology, eardrum surgery, neural physics). Good research also is being done in particle physics, molecular biology, nuclear-structure physics, solid-state physics, certain areas of mathematics, and some fields of humanistic scholarship.

Despite this substantial recovery, German science has not regained fully the preeminent position it once held in world science. The appraisal of serious observers in the nation is that German basic science is good in many fields and outstanding in a few, but that there is not enough excellence and pace-setting work. The student disturbances in the late 1960's and the rapid expansion of the higher education system in the 1970's have played some part in affecting the climate for basic research in the universities, leading to worries about a possible loss of momentum in the general recovery of German science or even a possible decline in the quality of scientific research. The nonuniversity research institutions and German participation in Europe-wide scientific activity have remained largely unaffected by university developments in the short run, but nearly all facets of the German basic science effort would be affected if the university research climate were to deteriorate.

Research activities in the Federal Republic, as in other Western nations, are influenced by short-run pressures. But, to a significant degree, the German basic research system is self-governing and relatively immune from short-run pressures. This self-governance is assisted by structural arrangements and operating procedures whereby research support is provided for areas of inquiry largely determined by the scientific community itself. An underlying associated factor that also helps to account for the substantial autonomy of the basic science system is the memory of the damage done by government intervention in the Nazi period.

At the level of the federal government, major support for basic research is provided through agencies whose purpose is to promote research rather than such functional missions as defense, health, transportation, and the like. The Ministry for Research and Technology (BMFT) and the Ministry for Education and Science (BMBW) in 1978 provided more than three fourths of the $450 million* in federal research funds going to universities, with the mission agencies providing the rest. The

*R&D expenditures are cited in dollars, with marks converted to dollars at the rate of 1 mark = $0.50.

principal organization engaged in the implementation of research support programs with the universities is a society (the Deutsche Forschungsgemeinschaft, or DFG) that operates outside the formal framework of the government. It awards research grants to university scientists with funds provided by the federal and state governments. The initiative for research proposals comes largely from the scientific community. The proposals are evaluated by scientific peers, and priorities in resource allocation are largely set by the flow of quality proposals and by scientific judgments as to the readiness of particular fields for rapid advance. Priority programs and special collaborative programs are selected for emphasis, but these tend to be assessed in terms of scientific opportunity rather than of social priorities. Nonetheless, many of these projects prove to be of practical benefit to the nation (for example, the early analysis and identification of pollution sources in Lake Constance and the analysis of toxic materials in the workplace).

With respect to applied research programs, and especially in the innovation-promoting ventures with industry, federal officials have taken a more active role in shaping research priorities. The principal federal agency whose mission includes support of applied research is the Ministry for Research and Technology (BMFT). The BMFT was created in 1972, when the Ministry for Scientific Research was split into two new ministries—the BMFT and the BMBW (the Ministry for Education and Science). Since 1972, the BMFT has been the major federal research-support agency funding the federal government's share of the big science establishments, the Max Planck Society, applied research, and the nuclear, space, energy, and other development programs; it also is a coordinating agency for all federal R&D expenditures.

The state governments have primary responsibility for the system of higher education and a major role in supporting research in the universities. Of the total $2.65 billion spent on research in the higher education sector in 1978, the state governments provided 80 percent, the federal government 17 percent, and private industry 3 percent.* Another $5.5 billion are spent on instructional activities; the bulk of these funds is provided by the states. These figures contrast with corresponding figures in the United States, where only about 12 percent of total higher education expenditures, instead of nearly a third, as in Germany, is devoted to research support. State governments contribute as well to the support of the nonuniversity research institutions, a smaller percentage to the big science establishments (10 percent), and a 50/50 share of funds to the Max Planck Institutes. The state governments, besides their contributions

*This 80 percent, however, is an estimated figure of that portion—approximately a third—of total state government expenditures on higher education (including faculty salaries) that can be reasonably attributed to the promotion of research.

to joint federal-state research programs and their support for universities, also engage in a wide variety of research-support programs of their own, including efforts to promote industrial innovation and the competitiveness of high-technology firms located within their jurisdictions.

The leading figures in the Deutsche Forschungsgemeinschaft (DFG), the Max Planck Society, the Science Council (Wissenschaftsrat), and the Association of Big Science Establishments (AGF) are the principal spokesmen for science with the public and with other government agencies. The DFG is a research-supporting entity, the Max Planck Society and the AGF are research-performing organizations, and the Science Council is an advisory body that serves the government, the DFG, and the Max Planck Society in a variety of advisory tasks. In 1980, for example, the Science Council conducted a study that recommended the creation of a new Max Planck Institute for Materials Sciences. The West German Rectors' Conference, an advocacy organization that speaks for the universities, is another group that occasionally plays an influential role in the public debate on university-related research issues.

None of these organizations is formally a government organization, although all except the West German Rectors' Conference are supported by public funds. This pluralist science-policy leadership has the task of explaining to the public its needs, problems, and possible contributions. Within broad limits, it can help chart general directions and can support the impulses for new research priorities arising from within the decentralized scientific community. But, when science and education seek major new resources from society, or a research role of a novel character, the leadership must persuade the political leadership of the country. That political leadership includes the Bund-Länder Commission for Educational Planning and Research Promotion, officials of the Ministry of Science and Technology, the finance ministers of the state governments and of the federal government, the Ministers of Culture of the state governments, and the state and federal legislatures. While all research systems are ultimately accountable for performance and for fiscal propriety, the research system in the Federal Republic of Germany reflects a high degree of self-governance and insulation from short-run social pressures.

## Institutional Structures for Basic Research

As in the United States and the United Kingdom, the universities in Germany are the major performers of basic research in volume of effort and number of research personnel, and they perform applied research as well. Approximately 78,000 scientists perform research, both basic and applied, in all fields in German universities, representing an investment of $2.65 billion in 1978. The big science establishments, which perform basic research requiring expensive instrumentation and capital costs, form a second important part of the system.

There is a third major element: the Max Planck Society, a system of research institutes set up outside the universities that is significant to the nation's basic science effort. The Society, whose 50 institutes are the principal nonuniversity research organizations engaged in basic research, in 1978 had a budget of approximately $360 million and employed about 2,100 scientists. The role of the Max Planck Society is more significant than the figures alone suggest, since the leading work in certain fields is performed in Max Planck Institutes and the tone of the overall national effort is influenced importantly by Max Planck scientists.

Other significant nonuniversity organizations that perform a mixture of basic and applied research are the approximately 46 joint federal-state research institutes—institutes of the "blue list." The list was set up in 1975 under an umbrella agreement between the federal and the state governments, pursuant to the authority of Article 91B adopted in 1969 as an amendment to the federal constitution. These institutes spent some $165 million on R&D in 1975 and employed 1,400 scientists and technicians. The Fraunhofer Society has 26 research institutes, a number of which perform basic research (although the society's primary mission is applied research). Some 50 independent social science research institutes also perform some basic research as a part of their contract research activities, which are predominantly applied in character.

Certain institutes attached to universities in the United States, such as the Biodynamics Laboratory at Berkeley and the National Magnet Laboratory at the Massachusetts Institute of Technology, resemble Max Planck Institutes in many respects. But the differences in philosophy, origin, funding, and mode of operation are still substantial, and distinguish our modes of operation from the German practice. Significantly, these institutes in the United States reflect the priorities of separate mission agencies and philanthropic donors rather than a central strategy for the development of science, as is reflected in the Max Planck Society.

The Max Planck Institutes, despite their importance, are not designed to be general performers of basic research in all fields. Thus, they do not have quite the same role as the CNRS laboratories in the French research system. A Max Planck Institute is created for special research reasons: to support outstanding scientists in fields that are deemed ready for significant research activity. Consideration is given to whether the research already exists in sufficient magnitude and quality in the universities. Unlike the German universities, there is no general presumption that the Max Planck Societies should be active across the whole spectrum of the sciences, nor is it assumed that a Max Planck Institute should have a permanent existence. For peak periods and chosen outstanding scientists, certain fields are selected for attention, but the Planck Institutes are not intended to replace or to substitute for the normal research efforts of the universities.

German industry performs a substantial amount of research, some part of which may be classed as basic. The German chemical, optics, machinery, aerospace, and electronics industries, in particular, rely heavily upon research; the research in optics and machinery, however, does not match the magnitude of the effort in the chemical, electronics, and aerospace industries. German industry, apart from these fields of endeavor, is a performer of basic research only to a limited degree. The style of industrial R&D in Germany differs from that in the United States in the sense that, with respect to the leading industrial laboratories, there is less pioneering work in new areas and more emphasis on solidity and thoroughness in traditional areas. The bulk of the R&D effort in German industry appears to be product-oriented, developmental, and applied. As in the United States, the industrial R&D effort is concentrated heavily in the larger firms. Some smaller firms, however, such as Steigenwald in electron-beam technology, have been built around the successful exploitation of new technologies. A clearer division of labor, more sharply defined than in the United States, generally exists between the universities and nonuniversity societies as basic research performers and industry as an applied research performer.

For certain fields, cooperation in Europe-wide organizations is a major element of the German basic research effort. The European Molecular Biology Organization (EMBO) in Heidelberg, the German contribution to CERN (the president of CERN is a German), the Institut Laue-Langevin in Grenoble, and the European Space Sciences Organization are important examples. Even facilities that are national, such as DESY, are being used increasingly by international teams. The cooperative ventures attract leading scientists and provide a highly stimulating environment for significant work.

## Forms of Support and Promotion of Research

The total R&D expenditures in the Federal Republic amounted to $15.2 billion in 1978. Of this total, only 12 percent was devoted to defense R&D, in contrast to 51 percent in the United States. Public sources—federal and state—of the funds amounted to approximately half, and private industry spent an equal amount on research, development, and technical activities of its own. Of the $7.5 billion in public expenditure, 37 percent was spent in universities, 38 percent in research support outside the universities, and the remaining 25 percent went to research and development activities in private industry.

All universities are state institutions in Germany. The German Länder, or state governments, support university research in a fashion that is somewhat similar to support for public universities in the United States, but the proportion of research support that they provide is much higher. Significantly, in Germany, each professor has a certain minimum level of basic support from the university so that the investigator typically can perform some level of research without outside grant support. The salary of the principal investigator, the basic equipment and laboratory instrumentation, operating costs, certain support personnel, and a certain minimum research support are borne by the university. This practice provides a measure of continuity in research support that does not exist for most scientists in either public or private universities in the United States. Indirect costs are not a factor since these are borne by the university and are not part of a grant.

Scientists seek such additional support as may be needed to carry out particular research projects. The extra support goes for additional equipment, the salaries of scientific coworkers (usually graduate students or other young scientists), and other costs associated with the research projects or programs. Most active university researchers seek outside research support. The proportion of outside support varies greatly. Research efforts of a significant size normally require the university scientist to seek substantial additional funding and, in recent years, researchers have relied upon outside support for a growing share of total support. The relative decline in the universities' share of support for research in recent years results from higher instructional costs arising from greatly increased enrollments and from the expanded body of university faculty now seeking support.

The principal source of outside funding for basic research in the universities is the Deutsche Forschungsgemeinschaft (DFG). The DFG grew out of the old "Emergency Association of German Science" formed in 1920; it was reestablished in 1949 after the war and has remained an autonomous organization since then. It has the legal status of a nonprofit corporation under civil law. The resources for the DFG are contributed on a 50/50 basis from the federal government and the Länder for the Normalverfahren and priority programs, and on a 75/25 basis for the special collaborative programs. The DFG's resources, like the total of public funds spent on research and development, have grown over the past decade, from roughly $96 million in 1968 to $350 million in 1978. The Federal Republic has shown the most rapid and sustained growth in public support of research of any of the European nations. But the growth of resources for scientific research has begun to slow and, in the next few years, tighter R&D budgets are generally expected. Approval rates for proposals submitted to the DFG in the first quarter of 1980 were slightly below those in the same period a year earlier; the size of average awards also decreased slightly.

Research support from the DFG is provided in various ways. The largest part is in the form of small individual project grants of one to three years' duration (called the normal procedure or Normalverfahren). In 1978, the project-grant system comprised 42 percent ($153 million) of the DFG's total budget of $350 million. Approximately 5,000 project-grant proposals are submitted to the DFG each year. They are reviewed by elected peers in the specialty areas and, normally, the whole review process takes place within a four-to-five-month period. This figure reflects a slippage in recent years from the earlier norm of three months for the review of proposals. The support provided under the project-grant system goes largely to pay for extra staff associated with the research project, principally younger scientists. In 1978, money from the DFG in all categories of support paid for approximately 6,300 scientists, mostly in the younger age brackets, or about 10 percent of the total of all scientists working in universities.

The next largest category of DFG grants is the special collaborative research programs (or Sonderforschungsbereiche). Examples are "Photochemistry with Lasers," "Stochastic Mathematical Models," and "Energy Transfer During Atomic and Molecular Collisions." In 1978, there were 105 such programs and they received $113 million (or 32 percent) of the DFG budget. This category of assistance provides long-term and stable, but not permanent, support for groups of researchers at a single university or nearby institutions working on common areas. Typically, support is provided for a ten-year period, with provision for evaluation at three-year intervals. Some elements in the U.S. system are analogous to the special collaborative research programs. The Materials Laboratories, started by the Advanced Research Projects Agency in 1962 at 12 universities, were based originally on a minimum of five-years' advance funding. The Joint Services Electronics Program, sponsoring such units as the Research Laboratory of Electronics at the Massachusetts Institute of Technology, also featured block funding on a stable basis for several years into the future. But these programs in the United States have suffered some erosion of support, falling back toward short-term funding. The research group (Forschergruppe) programs provide support to teams of fewer than ten scientists working in related areas at a single institution. This category of support is similar to the special collaborative programs but is smaller, amounting to only $4 million for 1978. Support for the research groups is typically for a five-year period, longer than the individual project but shorter than the special collaborative program.

Another large category of support from the DFG is the priority program (or Schwerpunktprogramm). In 1978, this program provided $57 million, or approximately 25 percent, of total DFG expenditures. The priority programs differ from the special collaborative programs in that they support investigators from different universities around the country who are linked loosely together around a common disciplinary theme.

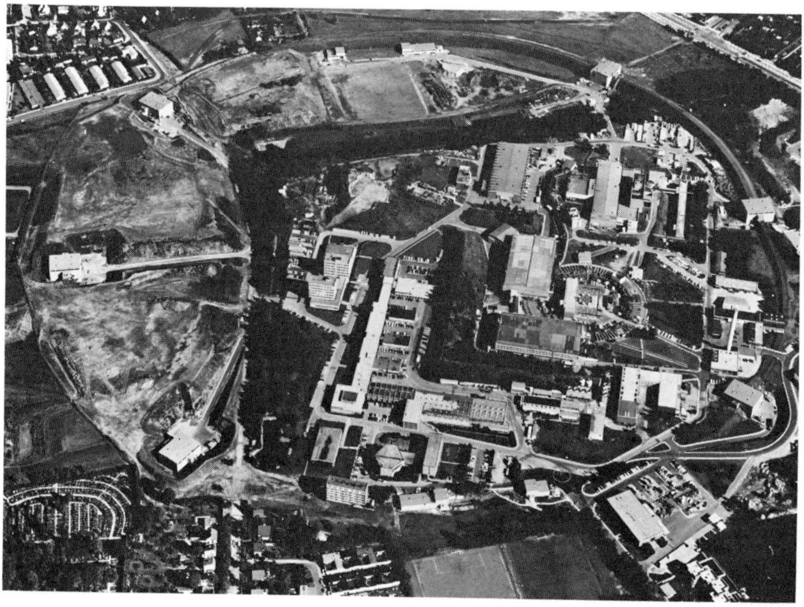

The German Electron Synchrotron (DESY) at Hamburg. [German Information Center.]

This is in contrast to having investigators at a single institution or neighboring institutions work on a common research project. Like the collaborative programs, the priority programs provide stable and long-term (between five and ten years), but not permanent, support to a substantial group of researchers. The 15 new priority programs initiated in 1980 (making a total of over 100 programs with nearly 1,500 individual projects) included such subjects as "psychological ecology," "remote sensing," and "process technology in foundry work."

German university research laboratories are plentifully equipped with modern instruments. The heavy wave of investment in instrumentation has been so recent that obsolescence is not a major problem yet. However, in three to five years, the present equipment will be out of date. Germany then may be in the same situation that currently faces France, the United Kingdom, and the United States—serious deterioration of the national stock of scientific instruments. Observers of the economy and the course of R&D investment are far from certain that funds will be forthcoming to refurbish the laboratories on the scale and in the style to which investigators have become accustomed.

Apart from the laboratory equipment, the array of German big science installations is impressive. In this category are a number of nuclear research establishments, including the German Electron Synchrotron (DESY), the German Cancer Research Center, the Mathematics and

Data Processing Research Corporation, the Nuclear Research Center in Karlsruhe, and the Nuclear Research Installation in Jülich.

Finally, private foundations are important in supporting basic research in Germany. Approximately 80 percent of the funds distributed by the major German foundations are to support research, contrasting with 15 percent of the funds given by the major U.S. foundations. The U.S. foundations no longer play the important role in research support that they played before World War II, when foundations were a major source of funds for university research; however, they still provide greater amounts of funding in absolute terms than do the smaller German foundations. The Donors Association for German Science (Stifterverband für die Deutsche Wissenschaft), a major foundation funded by contributions from German industry, provided \$30 million for research support in 1978 and administered some 80 other foundation funding programs in the amount of \$70 million for research. The Volkswagen Foundation provided \$55.2 million for research assistance in 1978, and in the same year, the Fritz Thyssen Foundation provided over \$5 million for research. The Humboldt Foundation plays a more specialized role in providing for younger scientists from other countries and in giving certain mid-career scientists an opportunity to spend a year of study and research at a German laboratory of their own choosing. The foundations are of considerable importance to the basic research effort because they typically can provide funds more flexibly than the public sources of support and can adapt more quickly to developing opportunities for research advances.

The role played by the mission agencies in supporting research in the universities contrasts sharply with the significant role played by such agencies in the United States. Aside from the Ministry of Education and Science (which provides the funds for the DFG and also supports some university research directly) and the Ministry of Research and Technology, mission agencies have only a very limited role as supporters of university research. Of the \$450 million in federal research funds spent in universities in 1978, the DFG provided about two thirds and the BMFG nearly a quarter. The mission-oriented agencies, most notably the Ministry of Defense, provided the remainder. The mission agencies in the Federal Republic devote substantial sums to research, but most of their money is used for mission-oriented research (Ressortforschung), which is performed in industry and in in-house federal laboratories.

## Scientific Personnel and Manpower Issues

The Federal Republic has experienced the problem of a potentially unbalanced age structure in its scientific work force. The problem consists of both an aging workforce in key research sectors and diminished opportunities for young scientists. University faculties are heavily populated by a large group of relatively young faculty members who will not retire for many years, a situation similar to that in the United States, but more acute than in America because of the rapid buildup in the late 1960's and early 1970's. In many nonuniversity research institutes, the scientific staffs have roughly the same age distribution because they were recruited at the same time, and there has been little opportunity in recent years to add younger scientists. The goal is to avoid sharp swings in the supply of and demand for scientific skills; in particular, the Germans seek to smooth the transition for the current group of young scientists completing advanced degrees from an era of expansion in the research system to a post-expansion era of limited growth.

Three factors complicate the manpower issues in the Federal Republic. First, the expansion in German higher education and in the research system took place later and occurred more rapidly than in the United States. It is expected, therefore, that in Germany the effects of rapid expansion will be felt for a longer period. Second, Germans tend to be less mobile both geographically and occupationally than Americans, and adjustments through normal turnover are not as likely to occur. Third, a very sharp drop in enrollments is expected beginning in the mid-1980's.

Access to higher education in Germany increased from about 5 percent of the age cohort to nearly three times that amount over the two decades from 1960 to 1980.* New positions in higher education, which began to increase in the early 1960's, decreased and practically disappeared in 1973-74. The costs of the higher education system have risen so dramatically that finance ministers in the various state governments are reluctant to authorize new positions at nearly all levels. But the number of students attending the universities has continued to increase, and projections indicate that enrollments will not drop until the mid-1980's, when a very sharp decline will occur. Although the number of students matriculating in the universities has increased, the number of graduates, especially at the advanced degree levels, has not risen correspondingly. In some fields, the number of graduates at the advanced degree levels has even declined.

The problems of adjusting to the rapid expansion in the higher education system have been substantial and have left significant administrative difficulties for the universities. The federal government, as a part of its assertion of wider responsibilities in education and research, adopted a University Framework Law in January 1976, but it has not been fully implemented yet through appropriate enacting measures in the states. The climate of uncertainty has been a factor in discouraging some stu-

---

* Student enrollment in higher education totaled 180,000 in 1960 and increased to 800,000 in 1980, but only 550,000 of this number were enrolled in universities and technical universities, narrowly defined.

dents from seeking careers in university teaching and in causing some switches in fields and interruption of studies to take jobs at an earlier stage than was the case a few years ago.

The prevailing strong desire for job security affects the career patterns of young scientists in Germany. There is less mobility in career patterns in German society generally than in the United States. In the context of the research system, this means less movement from one university to another, from a university to a nonuniversity institute, or from a university to an industrial firm. German science policymakers voice a strong need for greater mobility in career patterns. There are some signs of change. The increasing movement of students in the physical sciences into industrial positions in the face of diminished academic opportunities is a noteworthy trend. The movement of able young scientists into industrial positions and applied science careers is welcomed for its hoped-for contribution to productivity and economic growth. Students with advanced degrees have the lowest rate of unemployment of any group in the society. But there is fear that gifted students may prematurely abandon basic research careers. The flow of young scientists into certain fields, such as physics and clinical medicine, has begun to be interrupted already. Although students in these disciplines have increasingly sought employment rather than finishing advanced degrees, no general decline in the number of graduate students is evident across all fields of science.

The Federal Republic has developed policy initiatives to help keep young scientists actively involved in research careers. In response to an initiative of the important scientific organizations, in 1977, the federal and Länder governments created the Heisenberg program to provide about 150 new fellowships each year for outstanding young scientists to devote themselves full time to research of their choosing for a five-year period. The fellowship stipend is set at a generous level. Experience with the program is still limited and no full evaluation can be made yet, but the initial results have fallen short of expectations. Too few outstanding young scientists have sought places in the program, and most of those who have accepted appointment have cautiously remained at their home universities in the hope that permanent positions will develop. Fellowships are also available in limited numbers from the DFG for young scientists engaged in "habilitation" research. (Habilitation is a degree awarded to scientists who, typically, have a Ph.D., are in their early thirties, have completed a major research project, and now are ready to seek permanent teaching or research positions.) The Deutscher Akademischer Austauschdienst, or German Academic Exchange Service, also provides fellowships for overseas study and research to young scientists at the habilitation stage. In 1980, the Humboldt Foundation started a new program, assisted by funds from the BMFT, to provide for research abroad by young German scientists.

Relatively well-paid permanent positions may still be found at various levels, from scientific coworker in a laboratory to a junior-level professorship. Since the universities are still in a phase of compliance with the University Framework Law, there are incentives for young academic scientists to remain in their current positions, which they hope may be reclassified as permanent civil service positions when the states adopt their laws implementing the University Framework Law.

A related initiative for young German scientists is a pending plan to add approximately 700 new positions to the Association of Big Science Establishments (AGF), to be apportioned among the 12 big science institutes on a formula basis. These positions would be used for hiring young scientists; in return for this infusion of new blood, the AGF will give up an equal number of positions as retirements come due over the next decade.

The West German Rectors' Conference has announced its support for a further major initiative, a proposal for a public commitment to the principle of increases in faculty size proportional to enrollment increases in the universities. This plan has been controversial within the scientific community. Some university spokesmen feel that acceptance of this principle would undermine an important strength, namely, the belief that a university chair carries research as well as teaching obligations and a reciprocal obligation by the Land government to provide research support.

Support for the young scientist in Germany typically is not provided through an individual fellowship. Rather, support in the early stages of the scientific career comes through participation in a professor's research project or through employment as a scientific coworker. Limited amounts of fellowship money are available at the habilitation stage. The young scientist also may seek a research grant from an outside source, often through the DFG project grant or normal procedure (Normalverfahren). Later, after the habilitation stage, the young scientist will turn routinely to the DFG, a foundation, industry, or elsewhere for support.

Students at all levels, undergraduate or graduate, typically pay no tuition or other fees. Two states, Hesse and Bavaria, recently have imposed nominal registration fees. Students derive some nontrivial financial benefits from their status, such as cheaper bus and train tickets, discount theater prices, noncontributory health care, and other benefits. Financial burdens arise for the advanced student mainly from the very long periods of study (and consequent foregone earnings) required in many disciplines.

## Strengths and Weaknesses of the German System

A fundamental strength of Germany's research system is the enduring popular respect for learning—in the humanities and the social sciences as well as in the exact sciences. The scholar continues to be respected, and the pursuit of scholarship benefits from stable, material support, a high degree of freedom in the choice and conduct of research activities, and in high social prestige.

Another strength of the German system is its solid financial base. Research support is provided through a multiplicity of sources, including the federal government, the state governments, industry, and private foundations. The latter are more important sources of funds for science than in France or the United Kingdom and, in relative terms, more foundation funds are devoted to research than in the United States (although our foundations still provide larger absolute amounts to research). Greater continuity of support is provided for researchers in the Federal Republic than in the United States. The basic support from the university provides a degree of stability that generally is lacking for most American scientists, even those of the highest distinction. The project-grant system plays a role in Germany, but is a smaller part of the total pattern of research support. The collaborative and priority programs of the DFG provide funding for groups of researchers for up to ten years. The Max Planck Institutes serve as an additional important element in a stable long-term pattern of basic research support. Some of the nation's outstanding research takes place in these Institutes, although the largest share of the basic research effort as a whole continues to be in the universities.

In the eyes of many German observers, our system generates a measure of wasted motion, but it also fosters competition and intense activity. There are fewer centers of research excellence in Germany, and many German observers would like to see greater diversity and selectivity and less uniformity in their research system. They believe that our system stimulates the rapid development of new fields. It is less clear to them that our system provides for purposeful exploitation of areas that are currently not among the exciting frontier subjects but which may be important for the health of science and the economy in the long run.

The weaknesses of the German system are essentially the reverse of its strengths. There is a trade-off between continuity of support and the resources available competitively for new lines of inquiry. Support for research teams rather than for individual projects may produce less strict standards of quality control for the work of each scientist. Although the U.S. system is perhaps quicker to exploit new basic research opportunities than the German system, over a period of time the latter responds to new lines of inquiry and builds securely for future work in a new area. The differences in the two systems lie mainly in the greater use of research institutes and centers in Germany.

Attractive pay and job security contribute to a lack of mobility in the system as individuals cling to comfortable positions at all levels in the career ladder. The overall costs of the research system lead finance ministers to be wary of new initiatives, particularly with respect to new positions, since an already large professional group of scientists will remain for years in secure civil service positions. Autonomy can lead, in some instances, to the isolation of the scientific community and to its failure to participate effectively in affairs of legitimate public concern and in decisions that affect its own future.

The Nobel prizes won by German scientists in the physical sciences since World War II have been mainly for work done in the period before the rapid expansion of the research system and the increased levels of public support of scientific research. As social goals, such as access to higher education, have swelled enrollments in the nation's universities, the climate for research has appeared to suffer. The rapid expansion in the German higher education system may have occurred at the expense of quality. The presence of larger numbers of relatively young professors with tenure limits the opportunities of what may be in some cases more talented younger scientists now coming up through the system. Even though nonteaching institutes play a major role, any weakening of the basic research effort in universities would seriously impede the full recovery of German science.

## COMPARISON OF THE U.S. AND EUROPEAN RESEARCH SYSTEMS

### Organization of Research

The European nations differ among themselves as well as with the United States in the organization and functioning of their research systems. There are also common features in the research systems of Europe and the United States, and each of the nations reviewed in this chapter faces many of the same or similar problems—inflation and rising costs, tighter budgets for research, declining opportunities for young scientists, the increasing difficulty of maintaining well-equipped laboratories, short-run pressures on the research system, and others. This section seeks to highlight the major similarities and differences between and among the European countries visited and the United States in the organization, functioning, and financing of their research systems.

In France, Germany, Britain, and the United States, a variety of organizations perform basic research. First and foremost are the universities. Elements of universities and units attached to universities—schools, departments, institutes, centers, and laboratories—play a role in every country considered in this chapter. Then there are organizations performing basic research that are associated with but not integral parts of the universities. In each country, government laboratories perform basic research into some degree and, in each country, there are private, nonprofit, free-standing organizations that perform basic research. Finally, the laboratories of industrial firms (including those of nationalized industries in France and the United Kingdom) play a substantial role in basic research.

However, although these major types of research institutions exist in each country, the mix of types differs widely from country to country, as does the importance of each in the total national effort. The United States relies more on the universities as research performers or managers of large research facilities than do any of the other three countries. France and Germany supplement the universities with parallel systems of research laboratories. In France, this includes the government laboratories of the CNRS and other national research laboratories. In Germany, the Max Planck Institutes play a highly important role in many of the frontier areas of basic science. These institutes have no direct counterpart in the U.S. experience, although some free-standing laboratories such as the Carnegie Institution of Washington, various medical research institutes (Salk and Worcester, for example), and some units attached to universities, such as the Biodynamics Laboratory at Berkeley and the National Magnet Laboratory at the Massachusetts Institute of Technology, closely resemble them. The British and U.S. research systems are much more alike with respect to the university's role but, in Britain, there is less reliance on universities for the operation of large facilities and major laboratories than in the United States.

Many large-scale installations in the United States are managed by universities or consortia of universities. This mode of operating large facilities (observatories, particle accelerators, oceanographic research vessels, etc.), for which the government has provided both construction and operating funds, is not used in Europe. The prevailing European form of university participation is through scientific advisory groups from the universities serving the government installations.

The European nations in general have relied more heavily than the United States on international collaborative research efforts and jointly sponsored facilities. They have merged their national scientific efforts or they have augmented their separate national efforts with institutions supported by contributions from the participating nations. Examples include the European Organization for Nuclear Research, The Institut Laue-Langevin, the European Molecular Biology Organization, and the Joint European Torus. Within the countries, most notably in France, there is increasing emphasis on the cooperative use of expensive, middle-level scientific instruments. Some of the Europe-wide scientific efforts have displayed a vitality, a level of intellectual ferment, and a degree of support that have surpassed national efforts. For the three European countries, cooperation offers a promising means of maintaining quality of

DELCO (Direct Electron Counter), an experiment at the Stanford Linear Accelerator Center (SLAC), with participation by physicists from the California Institute of Technology, Stanford University, and SLAC. "Many large-scale installations in the United States are managed by universities or consortia of universities. This mode of operating large facilities, for which the government has provided both construction and operating funds, is not used in Europe." [Stanford Linear Accelerator Center.]

effort at a time when national resources for science are no longer growing rapidly. The Europeans have collaborated in a number of technological ventures as well, the most successful being the air bus, which has made important gains in the world market for aircraft. Other ventures in technological cooperation have been much less successful. As the cost of experimental basic science rises, the United States may have to engage more extensively both domestically and internationally in the kind of collaborative scientific activities that the Europeans—with their smaller-scale national efforts and resources—already have found necessary.

## Funding of Research and Resource Constraints

The three European nations present important differences with respect to their levels of funding for research. The resources devoted to research vary from the relatively high level of support per investigator in Germany to the austerity evident in the United Kingdom. In Germany, there has been nearly a decade of rapid growth in expenditures for higher education and for research generally—a period of expansion that is only now beginning to lose momentum. In contrast, Britain has faced a protracted period of austerity that has sharply limited the growth of research budgets, but budgets have remained relatively stable and predictable during this period of austerity. France's rapid growth in R&D expenditures occurred a decade ago and the past decade has been one of much slower growth, putting France in a middle position between Germany and the United Kingdom in total national R&D expenditures but somewhat lower than both of those countries in the percentage of GNP devoted to R&D. In the aggregate, total expenditures on research and development by the major European countries have moved from a level of less than half that of the United States in 1965 to about two thirds of the U.S. total for 1975. But France, Britain, and Germany together spend on basic research each year an amount that is about equal to what the United States spends. One feature of interest in the funding patterns is the wide difference in the amount spent on defense R&D within the overall total: the United States is the highest, spending about 50 percent of its total R&D investment in the defense area; Britain is quite similar to the United States, with 46 percent of its total R&D investment going to defense-related activity; 30 percent of France's total R&D investment is in the defense area; and Germany spends only 12 percent of its total R&D investment in national defense.

The United States also differs from the three other countries in the relative prominence given to the biomedical sciences and the life sciences within its basic research effort. In the United States, the National Institutes of Health have a total budget roughly three times that of the National Science Foundation and support much research that in the three other nations is supported through the general science funding agency. In France, the budget of the National Institute for Health and Medical Research (INSERM) is only 20 percent of the budget of the National Center for Scientific Research (CNRS). Support of most basic research in the life sciences is the responsibility of the Science Research Council in the United Kingdom and the Medical Research Council budget is only 40 percent of the Science Research Council budget. In Germany, there is no separate agency with responsibility for funding basic research in the life sciences; the life sciences are treated equally with other fields of basic science by the agencies responsible for the general support of basic science.

An important difference in the support policies for university-based research between the three European nations and the United States is represented by the "dual support" systems in the European countries. The first element of such support for German and British universities is general institutional support provided by central government to the universities for "well-found" laboratories. It is understood that this support should provide a certain minimum level of funding for research as well as cover educational costs. (In France, by contrast, there is little provision for research through the general support mechanism.) Salaries of supporting staff, supporting services, building maintenance, operating costs, and the like are paid for through this general government support. In none of the three countries are the salaries of principal investigators, except in very rare circumstances, covered under research grants; hence, the issue of the proper accounting for faculty time allocated to teaching and to research does not arise. In Britain, general support is provided by recurrent grants of the University Grants Committee, while in France the Ministry of the Universities provides institutional support from the national government directly to the universities. In Germany, general institutional support is provided not by the federal government but by the states (Länder), much as the states support public universities in the United States. The German states, however, typically devote a greater proportion of their general institutional grants to research support than do American state governments. (In contrast, in the United States the national government provides virtually no funds for the general support of universities.)

The second element of the dual support system is additional public (and sometimes private) funds specifically earmarked for research activities either in the universities or in the laboratories affiliated with universities. Only this second element—support for specified research activities in universities—is present in the U.S. federal system of financing research. Of course, there is no counterpart in European countries to the important private American universities (except that various colleges of Oxford and Cambridge in the United Kingdom have some private funds). The differences between the U.S. and the other systems of supporting universities are therefore fundamental, although in recent years some of the differences have grown less sharp.

The European practice of providing universities partial support for research from general institutional support mechanisms, assisted by a system of supplementary funding for particular research projects, has encountered strains as the cost and complexity of scientific research have increased. The "well-found" laboratory no longer can be supported in all

British universities through the University Grants Committee, and the Research Councils have had to assume a larger share of direct research costs, such as instrumentation and salaries of technicians. Similarly, the base funding for research associated with the university chair in Germany has been held at very low levels in recent years, and scientific investigators have relied on external sources for an increasing share of their total research support. In brief, there has been a degree of convergence between the support systems for basic research in the three European countries and those in the United States as scientists in those nations have had to rely to a greater degree on extramural research support. The Europeans have recognized that the increasing costs of experimental research cannot be handled primarily through the general support for the universities as institutions. The growth of external support for research performed in universities in Germany, France, and the United Kingdom has generated a situation that is, in some respects, similar to the "indirect costs" problem in the United States. The similarity is that, when substantial funds earmarked for research are made available from sources outside the universities, especially for large and expensive research programs, the performance of this research frequently generates additional costs that must be funded from university budgets.

The three countries reviewed provide support for longer periods of time and for more aggregated research efforts than does the United States. The American system relies more heavily upon support of relatively small discrete research projects, generally under $150,000 per year, proposed by individual scientists and funded for relatively short periods (usually one to three years). In the United States, 60 percent of all support for university-based research is in the form of the individual project grant. This support accounts for half of the federal funds provided to universities for research. Germany, France, and the United Kingdom, on the other hand, give a larger share of the overall research budget to teams of scientists, research groups, institutes, and entire laboratories. The European nations do continue to use project support in varying degrees. In Germany, the central research-funding agency provides 40 percent of its funds for small-scale projects, and the five Research Councils in Britain devote 30 percent of their total budgets to small individual projects. In France, research projects of individual investigators are financed by a number of agencies, but such funding amounts to only about 5 percent of the funds supplied to teams and laboratories. Only the funds of the Research Mission of the Ministry of the Universities in France, however, provide support specifically for small individual projects—and this constitutes only 5 percent of the total R&D support to universities. The American system, with its greater reliance on project grants, tends to place power in the hands of individual scientists as contrasted with laboratory or institute directors, and to provide the individual investigator who succeeds in competition for funds with control over research resources at a relatively early age. On the other hand, the support is less predictable and the system forces scientists of all ages to spend great amounts of time on administrative work.

In France, Germany, and Britain, a relatively high proportion of national funds for basic research conducted by university scientists is administered by a central science agency—the CNRS in France, the DFG in Germany, and the SRC in Britain. Some "mission" ministries, such as the Ministry of Industry in France, the Ministry of Research and Technology in Germany, and the Department of Energy in Britain, support basic research in universities. That support is not, however, on the scale of that of the mission departments in the United States, where nearly 80 percent of the federal funds for research in universities still is provided by mission agencies—Health and Human Services, Agriculture, Defense, Energy, Commerce, Transportation—and not by the central science agency, the National Science Foundation. The American system for funding most of the nation's publicly supported basic research through mission agencies sets this nation apart from the three major European countries.

## Peer Review

The system of peer review of research proposals as known in the United States operates differently in the three European nations. The fact that the Europeans award funds for longer periods—and place a heavier reliance on research teams, institutes, and entire laboratories—leads to less individual competition and a generally less elaborate administrative process than is required here for the evaluation of proposals and award of a large number of small grants. In Europe, the research communities are also generally smaller and it is thus more difficult to provide neutrality and anonymity in peer evaluations. In Germany, peer reviewers are elected by the entire body of scientific investigators in university and nonuniversity centers, and they serve fixed terms with full public recognition of their roles. Some peer reviewers in France are also elected, but by narrower constituencies. Yet, in each of the nations reviewed, various mechanisms of peer evaluation—site visits, periodic appraisals, and other reviews by qualified scientists—form an integral part of the process of administering public support for scientific research. The systems differ in various respects, including the degree of openness, the significance of political considerations surrounding the process, the frequency of review, and the role played by government as opposed to outside scientists, but each system strives for quality control through mechanisms of evaluating scientific work by scientists.

## Impact of Demographic Changes

In the three European countries, as in the United States, there have been growing concerns over the diminishing academic career opportunities for young investigators. Expansion in the higher education and research systems of the United Kingdom, France, and Germany typically took place later and occurred more rapidly than was the case in the postwar growth in American universities, and the rates of growth have declined more sharply than in this country. The result has been a more severe problem of declining opportunities for young scientists than we have faced in this country, but the three European countries have taken more extensive steps to address the problem. France has taken the most significant action by providing a 3-percent increase in new permanent positions in CNRS and INSERM laboratories each year, principally for the purpose of hiring younger scientists. Germany also has taken some substantial steps, including the creation of special long-term fellowships designed to hold some of the most able young scientists in research for a transitional period. The United Kingdom has created special fellowship programs also, but on a more modest scale. The United States alone among the four countries has taken no government action to address the problem.

The shortage of new academic jobs is causing substantial numbers of young scientists and engineers in Germany, France, and the United Kingdom to shorten their periods of training and to seek early employment in industry and government. This increasing interest in industrial rather than academic employment is seen as desirable by European observers in view of its potential contribution to the economy and to strengthening the research base in industry. But there are also fears that shortages may develop in the numbers and quality of advanced students in certain fields of science and engineering.

## Short-Term Pressures on the Research System

In each of the European countries reviewed in this chapter, as in the United States, there are increasing pressures to apply science and technology to urgent problems such as inflation, declining productivity, soaring energy costs, raw materials shortages, environmental issues, and regional development. This effort to bring the fruits of research to bear more effectively on such problems has created strains in the research systems of the three countries, as it has in our own. New opportunities for service also have emerged. Some fields of science, and especially certain key areas of technological development such as energy and biotechnology, have been singled out for greatly increased government support. The European experiences reflect rather different assumptions in policy approaches than does the U.S. experience in this area. The differences in approach relate mainly to the more self-conscious effort by the Europeans (by France in particular, but also by Germany and, to a lesser extent, by Britain) to set the pace and direction of technological change through government action. France, Germany, and Britain have all experimented more extensively than the United States with a range of programs designed to promote industrial innovation. The objective in particular instances is to stimulate exports but, more generally, the efforts seek an escape from the current economic situation of low growth, inflation, and declining economic competitiveness. The programs include direct and indirect subsidies, rapid depreciation allowances, special assistance to small firms, exploitation of patents, subsidies for the hiring of technical personnel in certain sectors, direct support for key technologies, and a miscellany of other programs too numerous and varied to mention in detail. There are also various schemes in each country to strengthen ties between universities and industry, including direct support for applied research in industry and in universities, participation of university researchers in efforts to develop cooperative generic technologies, and exchanges of scientists between industry and the universities. In total, these programs constitute a more varied and intensive effort than exists in the United States.

In general, the European nations reviewed here have gone farther in the direction of intervention into the normal workings of the marketplace in the interest of promoting innovation than has the United States. France has gone the farthest in this direction, with Germany next and with Britain involved in certain industrial sectors but keeping aloof from direct government intervention in other sectors. Despite this growing governmental activity, it remains difficult to assess whether the initiatives have accomplished their stated purposes. At times, the rationale for government action to promote industrial innovation in these countries is to compensate for failures in their systems to relate innovative technical activity to the marketplace. With notable exceptions, the university systems in Europe and the nonuniversity basic research institutes have been less characterized by applied work, public service roles, and close links between knowledge production and practical applications than have the American universities, especially those with land-grant traditions. The link between universities and the community in the United States, most strongly exemplified by the land-grant tradition, is a unique American strength. The Grandes Écoles and Établissements of France, the technical universities of Germany, and the civic and technological universities of Britain, in contrast to their traditional university systems, have had close ties with industry and have functioned extensively in a national service role.

United States leadership is still evident across a wide spectrum of scientific activity. But European science has clearly rebuilt much of the eminence that it held in the pre-World-War-II period. After a period of rapid expansion, the research systems of the three European countries have begun to experience a slowdown in the growth of support, with related stresses and strains that generate a number of concerns. A strong effort to understand how others have dealt with these common concerns and to learn from the experiences of other industrialized nations is appropriate.

# Summary and Outlook

This Summary and Outlook points to those aspects of the recent experiences of the three European countries studied that seem to be of particular interest and relevance to Americans concerned with the support of science.

The intention is not to suggest that particular institutional arrangements, policy initiatives, or procedures can be or should be incorporated directly into the U.S. system of supporting scientific research. The elements of a research support system are, of course, interrelated and derive their significance from a unique national context. The aim here is to bring attention to some alternative approaches that seem to have worked well in other nations and may be useful for stimulating thought about our own system. Many observers in Europe view our system with admiration. The American research system, as those observers see it, has qualities of competitiveness, wider opportunities for scientists at a younger age, great dynamism, and high capacity to exploit new research opportunities; they do not find those qualities in equal measure in their own systems. They seem less convinced that our system provides as well as theirs for the orderly development of designated scientific fields, particularly those not currently on the most exciting frontiers but that may have great long-term potential for science and for economic development. In the spirit of understanding more fully the strengths and weaknesses of our own system, the following summary observations are offered.

## UNIVERSITY AND NONUNIVERSITY RESEARCH INSTITUTES

Research institutes—associated with universities or free-standing nonprofit institutions—are a prominent part of the U.S. research structure. Those associated with universities include gigantic research establishments financed by the federal government at a level of about $7.5 billion per year. But they also include smaller specialized research groups, often centered around a special facility and generally financed by federal mission-oriented agencies. The National Institutes of Health alone finance some 600 centers at universities at a cost of $340 million (approximately the budget of the Max Planck Institutes of Germany) in the area of the life sciences. These university-based research institutes are similar to the university-associated laboratories of the French

CNRS in that they provide a stable environment for research centered around a broadly defined scientific area. They differ from the CNRS laboratories in that they are a part of the structure of individual universities.

Nonprofit independent research organizations are the second type of research institute. At one end of the spectrum, there are five large research and development centers funded by federal agencies at a total level of $300 million per year. In the middle range are 16 large, independent, nonprofit institutions with federal funding at a level of more than $600 million per year. Their missions are wide ranging—biomedical sciences, basic research, urban research, strategic advice to the armed forces, and industrial consultation. At the extreme end of the spectrum are about 300 small, specialized research institutes quite heavily concentrated in the biomedical sciences. These types of independent institutions have some noteworthy characteristics. They are nonprofit organizations. They are part of neither universities, industry, nor government. In total, their functions are wide ranging, but the individual organizations tend to have coherent central missions. A high proportion of the institutions receives funds from industry and many—particularly in the biomedical sciences—are supported heavily by private giving. They are characterized by a high degree of voluntarism, independent local initiative, decentralized decisions, and organizational innovation.

In contrast, such centers in France and Germany tend to be administered within a single organizational framework and to be financed almost totally by government. These characteristics generate strengths and weaknesses. On the one hand, relationships between centers, movement of personnel, and communication are facilitated. This provides opportunities for the orderly development of fields based on scientific rather than mission-oriented priorities, for a greater degree of overall direction in starting up and closing down research installations, for coordination of effort and avoidance of duplication. On the other hand, the variety and flexibility, responsiveness to varied needs, local initiative, citizen participation, and industrial involvement that characterize the system of independent nonprofit research institutions in the United States are much less prominent in France and Germany.

## ALTERNATIVE FORMS OF SUPPORT FOR RESEARCH

In Britain, France, and Germany, research support is less exclusively tied to small projects for short periods of time. Each of those countries has found it useful to supplement project funds with other forms of support—for example, priority programs for groups of scientists, special collaborative efforts, and teams of investigators working on common themes. The distractions and discontinuities that can and often do accompany the continual pursuit of funds in the United States are thus not a problem for scientists in those countries. The United States might benefit from the development of alternative support mechanisms to provide greater continuity of research support in selected fields of science and for selected groups of workers.

## POLICY INITIATIVES TO COPE WITH DIMINISHED OPPORTUNITIES FOR YOUNG SCIENTISTS

The French, Germans, and British all have felt the need to establish programs specifically to provide opportunities for young researchers to launch careers as investigators. As noted above, the problem is more severe in the three European countries than it has been for the United States because of the rapid recent buildup in their systems of higher education and research. The problems facing young scientists in this country today are also serious, however. The future productivity of the nation's research system cannot be assured when career opportunities are as sharply diminished for young investigators as they now are in a number of scientific and engineering fields. In the long run, the strength of the nation's economy requires graduate as well as undergraduate teaching programs and a flourishing research effort in the universities. The variety of programs undertaken in Europe in response to this problem—fellowship programs, arrangements in which senior professors devote full time to research, coupled with the hiring of younger scientists in teaching positions, new positions created in national laboratories, and other plans—seem to be strong practical alternatives.

## COOPERATION IN RESEARCH ACTIVITY

Another aspect of the three European nations' experiences that may offer useful lessons is their extensive use of cooperative scientific ventures. Because of limited resources, the Europeans have relied more extensively than has the United States on cooperative programs at various levels—within individual laboratories, regionally within their own countries, and internationally with each other. We should take careful note of these experiences with a view to learning how the Europeans have coped with the practical difficulties of cooperative research at a time when the changing scale, cost, and complexity of scientific activity seem to suggest that steps in the same direction may be appropriate for us. With fewer laboratories able to afford full instrumentation in a number of fields, it will be important that public officials and the scientific community give serious attention to the problems of collaboration in research on a regional level and even locally within large research universities. The user-group for large facilities possibly could work as effectively in this country as it does in Europe. It is the experience of the Europeans with respect to the fields of "small science" which are beginning to require more sophisticated and expensive instrumentation that is of particular interest to American scientists.

## NATIONAL POLICIES TO STIMULATE INDUSTRIAL INNOVATION

In the European nations reviewed in this chapter, there are increasing pressures to apply science and technology to urgent problems of society. The European perception of this problem reflects different assumptions and approaches with regard to the appropriate degree of government intervention in supporting and guiding technological developments. The three nations reviewed share, with differences, more activist views of the role of government in stimulating industrial innovation than has been characteristic of the United States. Although the traditional view in the United States has been that technological developments in the civilian sector are best left to the private marketplace, there are emerging pressures for more government intervention in coupling research to industrial innovation as a response to current economic conditions. In exploring these unfamiliar waters, it will be useful for the United States to be fully aware of what has and has not worked well in Europe. Areas in which the European experience may be particularly helpful include the following: innovative activity with respect to the technologies of small firms; university-industry relationships; labor-management cooperation in rationalizing production technology; government policies to encourage exports; and tax treatment of innovative activity in industry. Perhaps most important is the example of the cooperative, nonadversarial dialogue between government and business in France and Germany.

The European nations, while striving explicitly to mobilize their technical resources to meet national needs, have provided substantial stability and, in varying degrees, a measure of insulation to the basic research elements of their systems from the increasing short-run pressures on the total research systems. They have achieved this through methods stemming from habit and tradition but also by funding practices that provide stable support for free-ranging scientific inquiry by basic researchers. As we seek to achieve a reasonable balance between basic and applied research, and between short-term and longer-range inquiry, the mix of support mechanisms and the practice of the science support agencies in the three countries of Europe merit our close attention.

## REFERENCES


1. Bruce L. R. Smith and Joseph J. Karlesky. *The State of Academic Science: The Role of the Universities in the Nation's R&D Effort,* Volume One. *The State of Academic Science: Background Papers,* Volume Two. New Rochelle, N.Y.: Change Magazine, 1977.
2. (Five separate reports.) *Accountability: Restoring the Quality of Partnership.* March 1980. *Funding Mechanisms: Balancing Objectives and Resources in University Research.* May 1980. *Industry and the Universities: Developing Cooperative Relationships in the National Interest.* August 1980. *Research Personnel: An Essay on Policy.* August 1980. *Review Processes: Assessing the Quality of Research Proposals.* May 1980. Washington, D.C.: National Commission on Research.
3. "National R&D Spending Expected to Exceed $57 Billion in 1980," *Science Resources Studies Highlights.* Pub. No. 79-309. Washington, D.C.: National Science Foundation, Division of Science Resources Studies, May 8, 1979, p. 309.
4. *An Official Handbook.* London: Her Majesty's Stationery Office, 1980, p. 379. Economic Trends No. 321. *R&D Expenditures and Employment.* London: Her Majesty's Stationery Office, July 1980, p. 99.
5. Projet de la Loi de Finances pour 1980. Document Annex. *La Recherche Scientifique et Technique en France en 1979.*
6. Reihe. Berichte und Dokumentationen, Band 4. *Bundesbericht Forschung VI.* P. 73.

7. *Science Indicators 1978, Report of the National Science Board 1979.* Washington, D.C.: National Science Board, National Science Foundation, pp. 146-147.

8. *Report of the Science Research Council for the Year 1979-80.* London: Her Majesty's Stationery Office, November 5, 1980.

## BIBLIOGRAPHY

Robert Bock et al. "Academic Research in the 1980's: Pressure and Response," *The National Research Council/1980.* Washington, D.C.: National Academy of Sciences, 1981, pp. 89–109.

*Continued Viability of Universities as Centers for Basic Research.* National Science Foundation Advisory Council, October 20, 1978.

Délégation Générale à la Recherche Scientifique et Technique. *Le Progrès Scientifique.* (Series.) Nos. 196, 199, 202, 204, 205. Paris: Documentation Française.

Reinhold Geimer and Hildegarde Geimer. *Science in the Federal Republic of Germany: Organization and Promotion.* Fourth Edition. Deutscher Akademischer Austauschdienst, 1978.

Charles V. Kidd. "Academic Positions: Outlook in Western Europe, Canada, and the United States," *Science,* Vol. 212 (April 1981), pp. 293–298.

S. E. Luria. "On Research Styles and Allied Matters," *Daedalus,* Spring 1973.

*Research Excellence Through the Year 2000: The Importance of Maintaining a Flow of Faculty into Academic Research.* Washington, D.C.: National Academy of Sciences, 1979.

*Third Report of the Advisory Board for the Research Councils 1976-78.* London: Her Majesty's Stationery Office, February 1979.

C. H. Townes. "Differentiation and Competition Between University and Other Research Laboratories in the United States," *Daedalus,* Spring 1973.

Dael Wolfle. *The Home of Science.* New York: McGraw-Hill Book Co., 1972.

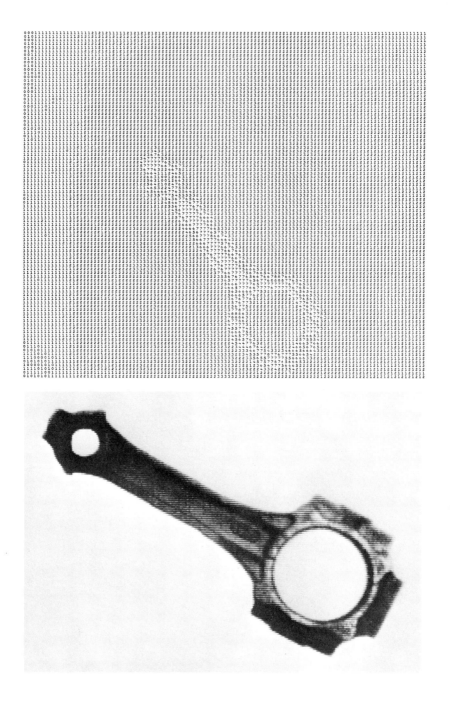

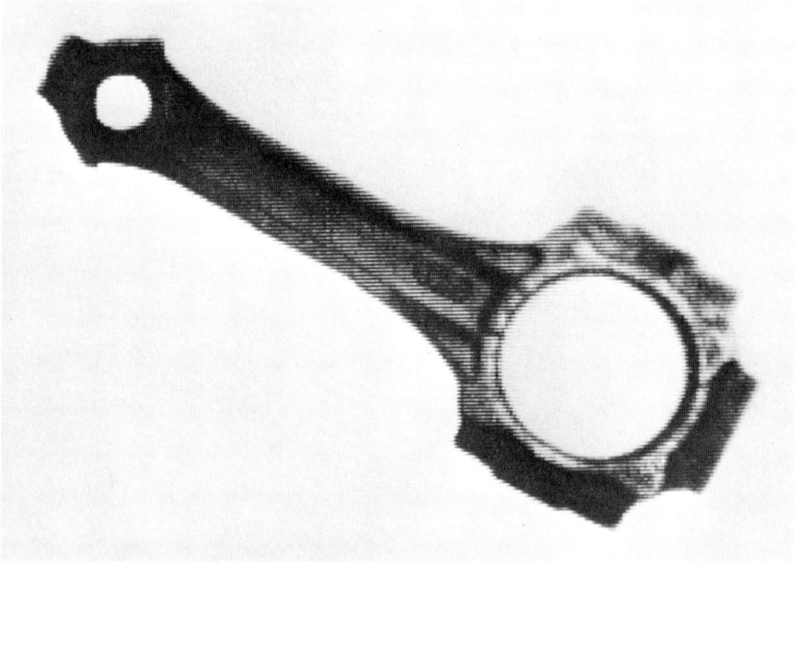

# 14

# *Research in Industry*

## INTRODUCTION

American industry has been characterized by a great ability to find new and better ways of meeting man's needs, of increasing his productivity, and pioneering innovative ways of solving old problems. This quality, combined with abundant resources and an energetic population, has given this nation one of the highest standards of living in the world. Much of this success has been achieved as a result of a multitude of new ideas, a philosophy that encourages the development of these ideas, and the financial means to carry them out.

During the past 15 to 20 years, however, for a variety of reasons, there appears to have been a decline in the industrial preeminence of the United States. This is best seen in the decline in the share of the international market formerly dominated by this country. This chapter looks at the current level of industrial research and development (R&D) in the United States and compares this level with those in other developed nations. Perhaps it should be stated at the outset that, even though this chapter has been motivated by a concern for our industrial competitiveness, it should not be assumed that this country's relative loss in competitive position is primarily the result of changes in the attention which has been given to research. Other factors, such as the overall quality of management and the cost of regulations, as well as high interest rates and the

◄ Digital and television pictures of a connecting rod, from research on the use of computers to enable industrial robots to "see." In the manufacture of cars, "computer-controlled robots that possess vision and other senses will come into use for assembly and inspection." [General Motors Corporation.]

attitudes and composition of the labor force, have played important roles. Here, we try to indicate the research which some major industries will be undertaking over the next five years and the pressures and factors that can be expected to affect the direction of that research.

It is difficult to make comparisons of international R&D efforts, partly because of slightly different reporting bases from country to country, partly because of the variable influences of inflation and exchange rates, and partly because, although many large international companies carry on R&D in only one country, the benefits may be shared by many countries. Still further, a substantial portion of the U.S. national budget for R&D is devoted to military and space research which, although it does have a significant impact on this country's economic vitality, provides an input that it is difficult to quantify. Nonetheless, reasonable comparisons can be made by noting the ratio of R&D expenditures to the gross national product (GNP) in different countries.

Table 1 shows this ratio for five of the leading industrial countries belonging to the Organisation for Economic Co-operation and Development (OECD).[1] The numbers are revealing, for they show that over a 15-year period the U.S. ratio has declined steadily while the ratios for three of the four other countries have gained substantially.When space and defense expenditures are deleted, there is an upward trend of private expenditures for R&D in the United States, but the trends in West Germany, France, and Japan are for increases at a substantially greater rate.

Although there has been a decline in R&D as a percentage of GNP in the United States during the past 15 years, the division of expenditures among basic research, applied research, and development has been almost constant, at 13 percent, 22 percent, and 65 percent, respectively.[2]

For many years, about 70 percent of the annual funding for basic research came from federal sources. During the period from 1963 to 1978, however, federal support for industrial basic research dropped from 22 percent to 15 percent of the total, while funding for universities/colleges increased from 6 percent to 10 percent.[3] In terms of who has been performing the research, there also have been equivalent changes in basic research by sector. The university/college sector increased from 41 percent to 52 percent of the total, while the industrial sector declined from 27 percent to 16 percent. The work carried on by other sectors has been about constant.[4]

This trend has been confirmed by a survey of over 100 firms that together account for over one half of all industrial R&D expenditures in this country. The survey disclosed that the proportion of company-financed R&D expenditures devoted to basic research declined significantly between 1967 and 1977. In four fifths of the industries, the proportion of funds devoted to relatively risky projects also declined. In some industries, there was a substantial decline in the proportion of R&D expenditures devoted to relatively long-term programs.[5] Thus, in aggregate,

**Table 1** National expenditures for research and development as a percentage of gross national product (1963-78)

| | 1963 | 1968 | 1973 | 1978 |
|---|---|---|---|---|
| United States | 2.87 | 2.83 | 2.34 | 2.25 |
| West Germany | 1.41 | 1.97 | 2.32 | 2.28 |
| France | 1.58 | 2.08 | 1.77 | 1.79[a] |
| United Kingdom | 2.30[a] | 2.29 | 2.06[a] | 2.05[a] |
| Japan | 1.44 | 1.61 | 1.89 | 1.92[a] |

[a] Estimated values

SOURCE: Adapted from National Science Foundation. *Science Indicators 1978.* Washington, D.C.: National Science Foundation, 1979, Table 1.1, p. 140.

much of the responsibility for the development of basic scientific knowledge upon which future development will depend is passing from industry to the universities and colleges.

Some comment on the interrelationship of technology and productivity may be pertinent to understanding the impact of R&D expenditures on the national economy. Although in absolute terms the United States has the largest value of real gross domestic product per civilian employee of all of the major industrialized countries (as of 1979, this value in France was 88.8 percent that of the United States', West Germany's was 87.9 percent, Japan's was 66.4 percent, and the United Kingdom's was 59.5 percent), the rate of growth in productivity in those other countries has been substantially faster.[6] If the current trend continues, most of these countries will surpass the United States by the mid-1980's.

Concern about this trend was reflected in the Carter administration's support of research and development and in its recognition that industry has an important role to play in advancing the technical capability of the United States. An indication of this conviction is that the 1980 budget increased the outlay for federal R&D from 1.17 percent of GNP to 1.21 percent. The original budget showed this trend continuing to 1.25 percent, but budget constraints reduced this figure to 1.20 percent. The outlay for basic research has been almost constant, at 0.16 percent of GNP for 1979-81 revised.[7]

## Economic Significance of Research and Development

***R&D Expenditures and Productivity Increases*** One way to explore the relationship of R&D to the U.S. economy and what returns can be expected from it is to measure the amount spent on R&D by an industry or firm and its rate of productivity increase. During the past 20 years, a number of such studies for specific industries have shown that R&D seems to have had a significant effect on the rate of productivity increase. In the chemical industry, for example, a study at the University of Chicago concluded that a firm's rate of productivity increase was directly related to the size of its expenditures on R&D. In the period under study, the marginal rate of return—that is, the discounted rate of return from an additional dollar spent—was about 50 percent for R&D in chemicals.[8]

Another study at the University of Pennsylvania concluded that the marginal rate of return from R&D was about 40 percent or more in the petroleum industry, and about 30 percent in the chemical industry if technical change in the form of new equipment or plant was considered (but much less if it was not).[9] Data for almost 900 manufacturing firms showed a private rate of return from R&D of about 17 percent. It seemed to be much higher than this in the chemicals and petroleum industries and much lower in the aircraft and electrical equipment industries.[10]

In a study for the Committee for Economic Development, Edward Denison attempted to include factors that were omitted in other studies—for example, changes in labor quality associated with increases in schooling—and found that technological change was responsible for about 40 percent of the total increase in national income per person employed during 1929-57 in the United States.[11]

***Technological Change and Economic Growth*** Case studies designed to shed light on how new technology—such as electronic innovations—affects the economy indicate that these technologies play a major role in determining the size, viability, and profitability of particular industries and firms, as well as their competitiveness in international trade. For example, the significant contribution of R&D-intensive products in the U.S. trade balance can be seen in Figure 1. The technical and economic benefits of such advances are self-evident, but it is not easy to obtain a quantitative summary of their contribution to the rate of economic growth. For one thing, it is difficult to separate the effects of technological change from those of investment in physical capital, since new technology frequently must be embodied in physical capital—new machines and plants. Nor can the effects of technological change be easily separated from those of education, since the social returns from increased education are enhanced by technological change, and the rate of technological change is influenced by the extent and nature of society's investment in education.

***Social and Private Rates of Return from Innovations*** By a social rate of return, economists mean the rate of return to society as a whole. (This includes the diseconomies involved in the early obsolescence of older technology and the displacement of labor.) Although it has long been

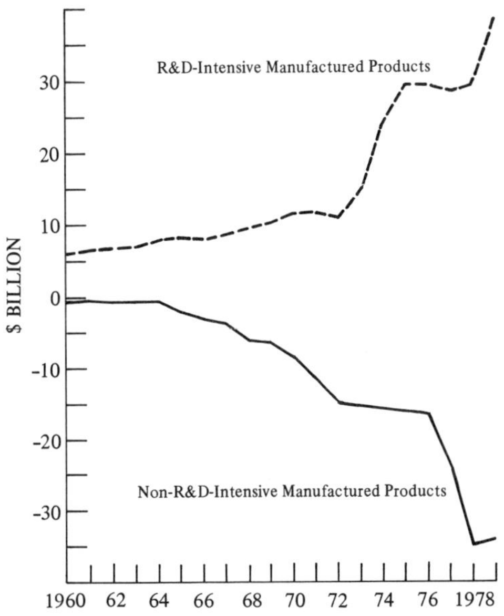


**Figure 1 U.S. trade balance (exports less imports) in R&D-intensive and non-R&D-intensive manufactured product groups (1960-78).** [SOURCE: Adapted from National Science Foundation, *Science Indicators 1978.* Appendix, Table 1.21, page 161.

recognized by economists that estimates of such rates of return are crucial in formulating public policy toward civilian technology, no estimates of this sort were made for individual industrial innovations until 1977, when detailed estimates were made at the University of Pennsylvania for a sample of 17 innovations.[12] Although the sample cannot be regarded as randomly chosen, there is no obvious indication that it is biased toward relatively profitable innovations (socially or privately) or toward relatively unprofitable ones. In large part, it contains undramatic, run-of-the-mill improvements in products and processes, such as a new type of thread or an improved machine tool. As many studies indicate, this is the type of work that accounts for most industrial R&D.

The findings are quite striking. The median social rate of return was about 55 percent, which indicates that the investments in these new technologies paid off handsomely from society's point of view. To check this result, the same methodology was used to estimate, for one of the nation's largest firms, a lower bound on the social rate of return from its investment in new process technology in 1960-72. This result also was about 50 percent. Of course, the sample, although large relative to what was previously available, is too small to support definitive conclusions, but the results certainly suggest that, even taking into account the riskiness of innovative activity, the rate of return from investments in new technology has tended to be high and the social rate of return is much higher than the private rates.[13] This is quite consistent with other econometric studies that have relied on more indirect methods and have used more highly aggregated data.[14]

## Research and Development in U.S. Industries

It is apparent that there are indeed social and economic returns related to expenditures on R&D. While, as we have indicated, a declining R&D trend is evident, it is not clear whether all industries are affected equally by changes in funding, nor is it clear how different industries are affected by research challenges from abroad. The task of looking at U.S. industry as a whole is far too large to be undertaken here. Instead, we have chosen to review five major industries that are among the top dollar investors in R&D in this country—chemicals, pharmaceuticals, aircraft, automobiles, and electronics—and one sector—energy.

Although each industry has its own characteristics and pattern of spending, they all share some common problems. In a practical sense, competition limits the fraction of a corporation's income that can be spent on research and development. Traditionally, internal decisions have had to be made to determine the funds spent for support of existing products and the fraction that could be devoted to new and perhaps more risky programs. However, a new factor has been introduced in recent years, and that is the necessity of spending some of these scarce funds to meet society's expectations of industrial responsibilities in the form of pollution control and health and safety regulations. While available capital funds always have been a constraint for many industries, the declining ability to generate new capital as a result of lessening profits, rising costs, inflation, and increasing socially required investments has limited the exploitation of all except the most promising and safest (but not necessarily the most innovative) products of the R&D process.

Obviously, these factors affect different industries in different ways. Those that are capital intensive, such as the automotive and chemical industries, have been heavily affected. Industries such as the computer complex, which from a manufacturing standpoint have much greater flexibility, are far less affected.

By looking at these industries, we hope to shed some light on what is happening to technology in the United States—where it is headed, and what problems it faces. We should note here that the terms "basic" and "applied" research do not mean the same things to all people. In the automotive industry, for instance, engineers and scientists feel that their research is basic, even though it results in commercial applications. A general rule of thumb might be that, in industry, basic research is long range and applied research is short range. In examining the health of research in the United States, one also must differentiate between the R&D process and the ability to put that process into practice. Funds for R&D are ill spent if there is no capital available to implement the results of the research.

## THE ELECTRONICS INDUSTRY

The market for U.S. electronic equipment in 1979 is estimated to have been over $70 billion at factory prices. Computers, including peripherals, and communications equipment each accounted for about 30 percent of the total. The consumer and federal markets are estimated at about 15 percent each, and other industrial and commercial electronics equipment at about 10 percent. Although recent growth has been approximately 14 percent per year in current dollars, with economic slowdown and inflation, few sections of the electronics industry will grow in real terms in the near future. Micro- and minicomputers, data communications equipment, and integrated circuits will be among the exceptions. This section will deal primarily with these items.

### Industrial Research and Development

The history of the electronics industry shows that research, development, and design of electronic components and equipment have been conducted mainly by the industrial laboratories. The industry is highly research intensive, employing about one quarter of all of the scientists and engineers in U.S. industrial laboratories. Basic research averages about 3.5 percent of all electronics R&D, somewhat above the average of 3.1 percent for all industries, but some individual laboratories devote up to 10 percent of their R&D funds to basic research. However, the industry includes many small equipment design and assembly firms with no R&D beyond specific product design. Federal funds have supported about one third of all electronics R&D; this is twice the federal portion of the electronics market, due to the special requirements and relatively small manufacturing runs of the national defense market.

There has been a steady growth in the level of electronics research and development, mainly applied R&D, performed in U.S. industrial laboratories. Expressed as a percentage of net sales for manufacturing companies, R&D investment, including federal funding, was fairly constant in the 1970's for computing and accounting machines (6-12 percent), for electronic components (6.5 percent), and for communications (7.5 percent). R&D in radio and television receivers dropped from 2.5 percent to 1.3 percent. Continuing R&D investment and large investment of capital in the communications, computer, and semiconductor component segments of the industry have led to rapid price decreases. Because of these unit price decreases, the penetration of these products into the economy has made them the major growth segment of the industry in terms of dollar values of sales.

The electronics industry is likely to continue to invest in applied research and development since this has been seen as the principal factor in corporate growth, whether in the early years of General Electric, Westinghouse, or Bell Laboratories, or in the more recent experience of Hewlett-Packard, Fairchild, or Intel. On the other hand, basic research requires building a large community of scientists in each area of interest and investing in their efforts over many years, with little assurance of when and how benefits will be realized. This investment can only be made (without undue burden to current operations) by large firms with broad areas of interest. When ultimately applied, investments in basic research benefit the company, the industry as a whole, and the nation.

## Key Technologies

Three technologies seem to be central to continued progress in electronics R&D—integrated circuits, to allow low-cost, reliable systems of increasing capability; software, to direct the achievement and operation of these systems; and interface devices, to couple systems in different parts of the world.

***Integrated Circuits*** Continued growth in the capability of silicon integrated circuits, with commensurate improvements in cost and reliability, has come to be expected. This progress has been made possible through improvements in photolithography. Conventional optical systems of essentially ideal performance are being used and yet are beginning to be inadequate. Systems using ultraviolet and X-ray wavelengths, synchrotron radiation, and electron beams (this last derived in part from electron microscopes) are in various stages of development. New processes of lithography probably will be required by these more exotic technologies. The desired features, defined by this lithography, have to be created by deposition, removal, implantation, and other modifications of material in layers that may be less than one micron in thickness.

The dimensions of circuit elements are being reduced so far that our existing knowledge of materials properties is inadequate and new basic research is required. The future of very large scale integration is based on the continued interplay between materials, processing, interconnection, testing, device structure, architecture, software, and design automation from simulation through test word generation. The potential for interdisciplinary trade-offs of these technologies represents one of the most challenging opportunities ever faced by the electronics industry. Other basic research programs will be needed and the new microelectronics centers such as those at Cornell, Stanford, the University of California at Berkeley, and the California Institute of Technology may be important models for future programs. However, these programs undoubtedly will require costly equipment. The coupling of academic and industrial research will become more important, with new device concepts, better understanding of extremely thin surface phenomena, computer aids to design, and software science being among the promising areas.

The increase in the capability of integrated circuits has been most rapid in logic and memory circuits. Integrated circuit manufacturers now can put as many of these digital circuits into a single chip of silicon as were used in total to build the early vacuum tube computers. This has led to the increasing use of digital electronics in communications systems and to increasing overlap between computer and communications technologies.

As continued improvements in lithography and processing allow the fabrication of more complex integrated circuits, new challenges appear. The foundation of the integrated circuit industry is the ability to make and sell many thousands or millions of devices of one design. But there is the question of what functions (other than memory) can be put into larger and larger integrated circuits without making the product so specialized as to destroy this mass market. Assuming that this challenge can be met, new computer design procedures still are needed to lay out and subsequently test the million or more transistors that each circuit will use. Automatic fabrication facilities also will be needed to avoid operator errors. Finally, all of these changes in the integrated circuit industry involve increasing capital needs. New ventures are possible only with massive capital from outside sources, and the ability of nationally supported Japanese companies to challenge the U.S. merchant suppliers is a matter of great concern.

***Software*** The use of software aids in design and testing already has been noted. These programs and application programs are becoming more complex, but programming productivity—the number of lines of code per person per unit time—has not changed appreciably. Some effective progress has been made in that lower cost memory and faster logic circuits permit the economical use of higher level languages and, thereby, offer more logical power from each line of code. The problem is being reduced somewhat by the use of "structured programming" to facilitate debugging of programs and by the design of modular programs, where sections of code are reusable in other applications and portable between different machines. Program modules for common applications and language compiler modules to bring the native language of machines closer to human forms will be offered on read-only memory chips by integrated circuit manufacturers. However, more fundamental understanding—

software science—is needed if we are to achieve dramatic improvements in the capability and quality of software. Again, this suggests a closer coupling between academic and industrial research.

We can hope to achieve a real synergy between software and silicon. Within this decade, we can look for research on totally new computer and program architectures. Up to now, we have used improvements in electronic components to make bigger and faster versions of the first computer designs; in this decade, we may see the conception of new architectures uniquely deploying the capabilities and the richness of interconnection available from silicon integrated circuits. New software concepts will be integral to the development and deployment of these architectures.

***Interface Devices*** While integrated circuits and software dominate the intelligence of electronic systems, other technologies are needed to couple this intelligence to people and to machines. Display devices, ranging from glow discharge tubes (Nixie tubes) and mechanical flap displays to cathode-ray tubes, gas discharge panels, light-emitting diodes, and liquid crystal displays, all have been pursued. The recent trend has been toward technologies offering opportunities for miniaturization and providing inexpensive mass fabrication, i.e., toward approaches requiring large capital investment and extensive development. Optical communication systems are a new product of several lines of applied research and technology development. Again, large investments are essential; by the end of the 1980's, much of the growth in broadband transmission will be in this new technology, and optical fiberguide links may well be carrying wideband services into the home.

As progress in electronics offers improvements in capability, cost, and reliability, new applications in home appliances and automobiles are being realized. These involve new technologies and new environments, both physical and institutional. Automotive applications will be most significant in engine control, where requirements of fuel economy, pollution control, and adequate performance can only be met by more sophisticated intelligence. Sensors for atmospheric and manifold pressures, shaft positions, fluid temperatures, etc., and a similar range of actuators, are being developed in a number of large and small companies vying for the very large automotive market.

The incursion of electronics into the home appliance control market also offers high potential volumes, but it is limited now by the fact that the sensor, display, and power-switching elements are more costly than the microprocessor intelligence to which they interface. Yet electronic controllers, optoelectronic displays, and calculator-type keyboards have found their way into top-of-the-line appliances. Penetration into lower cost models will occur as volumes rise and costs fall.

Scanning electron micrograph of a portion of a programmable logic array with locations for up to 4,002 logic elements. [International Business Machines Corporation.]

## Trends

Research and development challenges in electronics over the next five years are likely to be governed by four main factors—decreasing costs per function in electronics equipment, rising energy costs, foreign competition, and the increasing overlap between computer and communications technologies.

Decreasing costs will continue to make new products feasible and will lead to replacement of electromechanical and mechanical elements in existing products. Rising energy costs are becoming a significant factor in design selection and are creating new markets such as engine controls. These costs also may cause the increased use of communications services as a substitute for business travel and even for travel to work. The United States still appears to have a substantial lead over other countries, but Japan has established a national program to accelerate its activities in this area. It appears to be quite successful, with the Japanese-made 16K-bit memory chips now taking about 35-40 percent of the worldwide market. They also appear to have an important position in the next generation of such memory chips. The work in this nationally supported effort appears to have gone far beyond anything done in the United States outside IBM and Bell Laboratories.

Finally, continuing research and development in digital electronics is merging computer and communications technologies. However, the traditional separation between the regulated communications industry and the open-market computer industry in the United States has not accommodated the increasing overlap between the two technologies. The Federal Communications Commission's decision in its Second Computer Inquiry (April 1980) is a first step in reshaping regulations to recognize the importance of distributed information services. The degree to which the path that is taken by regulators and legislators accommodates the evolution of the technology may be the largest external factor determining the ability of U.S. electronics research and development to serve the nation.

## THE CHEMICAL INDUSTRY

Historically, the chemical industry has been research intensive, accounting for about 8 percent of all R&D expenditures in the United States during the 1970's. It also has emphasized basic research more strongly than have most other industries, to the extent of performing nearly 40 percent of all industrial basic research. Another characteristic has been internal funding of a high percentage of its R&D work, with consequently low dependence on government funds.

Since World War II, this high level of research activity has resulted in a plethora of new products, in many cases providing the foundation for major new industries and societal changes. The replacement of natural fibers such as cotton, silk, and wool by synthetic fibers is an example of such a development. The overall result during this period has been a chemical industry growth rate nearly double that of the gross national product.

In the mid-1970's, however, the world began to undergo some basic changes, and nowhere were the effects felt more than in the chemical industry. The rapid postwar growth occurred in a period characterized by plentiful and low-cost energy and hydrocarbon feedstocks. As we begin the 1980's, this situation no longer obtains. The cost of manufacture for those chemical products derived from hydrocarbons, particularly when manufactured in energy-intensive processes, is increasing rapidly. Thus the potential displacement of other materials by chemical products is governed by new and still-changing economic relationships. Along with drastic economic changes, the chemical industry also is facing new challenges in the control of toxic substances. These problems range from the disposal of solid wastes to the control of the atmosphere in the chemical plant itself. This environmental challenge has been sharpened by the enhanced capability to measure and detect toxic substances, as well as by new regulatory controls that often specify the method of control, rather than the desired result. Another change has occurred in the competitive environment; now, the chemical industry in the United States must face the reality of worldwide competition. Large, well-financed competitors in western Europe and Japan are supporting major R&D programs, as well as competing aggressively across the board in established businesses. The major European companies are spending higher percentages of their sales dollars on R&D than are American companies and are focusing on the same areas of research. While they have the same problems of rising fuelstock costs, they nevertheless appear to be building a technological base that could well make them a dominant force in world markets by the late 1980's.

Superimposed on all of these changes is the persistent inflation in the U.S. economy. This has radically altered the traditional relationships governing earnings and new investment; it raises troubling questions about the availability of capital for continued industry growth.

## General Trends

The world changes of the 1970's are affecting all chemical industry research in several important ways. First, the basic priorities of chemical research are shifting. The rapid growth of the industry since World War II was fueled by a constant stream of innovative products—which in turn had been the objective of most of the industry's research effort. New product research in the next five years certainly will continue to be performed, but it will no longer occupy the dominant role it assumed in the past. Instead, an increasing share of the industry's R&D will be directed toward improved process economics, reflecting the higher share of total manufacturing cost attributable to the cost of raw materials and energy use.

A second major trend will be in the growing importance of productivity increases. The difficulty of generating enough capital to replace and expand the industry's productive capacity will call for an enhanced research effort to maximize the productivity of facilities already in place. In the past, productivity increases often could be obtained by constructing new plants on an ever larger scale. These economies of scale, however, are rapidly diminishing as the industry matures, providing more incentive for research emphasis on productivity.

A third general response to the changing environment will occur in the further move to make use of process automation and control. On the one hand, the improved materials yields to be gained by better process control will be an important source of productivity increase; on the other, the labor savings attendant upon automation will help substantially to combat the international competitive challenge. Much of the industry's research will be devoted to making use of more sophisticated computers, process instrumentation, and industrial robots.

Generally, these trends will be apparent across the entire chemical industry. The specific emphasis, however, will differ according to the nature of the products, markets, and competitive situations. It is logical to comment briefly on anticipated research trends under four major headings: basic chemicals, polymer products, agrichemicals, and photographic products.

***Basic Chemicals*** Basic chemicals include not only large-volume inorganic chemicals such as sulfuric acid, chlorine, and titanium dioxide but also the large group of organic materials that are used as solvents and raw materials in the manufacture of downstream products. Research in this area will be heavily weighted to changing the raw material bases of the organic products from natural gas and petroleum to domestically produced raw materials (e.g., coal, shale) with more assured availability and more stable prices. This will feature efforts to shift from the chemistry of ethylene and other olefins derived largely from petroleum and natural gas liquids to the chemistry of synthetic gas, derivable from many carbonaceous raw materials. Alternative routes to olefins from shale oil, coal, and renewable resources also will be actively pursued.

In view of the rapidly escalating cost of energy, emphasis also will be placed on the reduction of process energy requirements. In some cases, this will involve replacing energy-intensive processes with those having lower energy requirements. In most cases, however, it will involve such things as improved insulation, recovery and use of waste heat, and substitution of cheaper, more available fuels.

Research and development in basic chemicals also will be increasingly concerned with environmental matters. The public concern for safe disposal of solid wastes will fall with particular heaviness on this segment of the industry. Research will concentrate on process changes that minimize the amount of solid waste generated, on specific treatments that could convert the waste to a less harmful form, and on conversion of solid wastes to usable products.

Many manufacturing operations for basic chemicals traditionally have been difficult to control in terms of water and air pollution requirements. More research will be concentrated on determining the precise composition of liquid waste streams and gaseous effluents; this will be accompanied by process research to minimize the toxic components and seek more environmentally benign processes.

***Polymer Products*** Polymer products (synthetic fibers, plastic resins and films, elastomers, and finishes) were most responsible for the explosive postwar growth of the chemical industry. Generally manufactured from hydrocarbon feedstocks, they benefited from the availability of cheap feedstocks and energy. It is in this area, of course, that the changes of the 1970's have had their most profound effects.

Even in a greatly changed environment, much polymer research in the 1980's will continue to be directed at developing new products. Many societal changes have created demands for polymer products. For instance, the downsizing of automobiles to increase mileage and limit emissions has resulted in new opportunities for high-performance plastics to replace metals. Another example is the replacement of glass containers by plastics; the latter consume less energy on an overall basis. Substantial research efforts will continue to be aimed at developing such products and applications. As polymer science matures, it will be increasingly less likely that new polymers with high-volume applications will be developed. The trend, instead, will be toward the synthesis of new polymers with unusual properties and performance capabilities, particularly in high-stress environments. In addition, considerable emphasis will be placed on the development of polymeric composites, in which the versatility of the composite structure can be used to tailor products to varied applications. (For further details, see Chapter 10.)

Although product research in polymers will continue to be important, it is clear that the overall emphasis will shift to performing more process research than in the past. The industry's emphasis on increased productivity, improved materials yields, reduced energy consumption, and increased automation certainly will apply to the polymer segment, and a major part of the research effort in this area will be committed to those objectives.

The polymer industries will have their share of environmental challenges. The finishes industry will devote a major effort to decreasing the solvent content of finishes. High-solids enamels, water-based enamels, and powder finishes will be the principal new products coming from this research. Work also will be concentrated on reducing or eliminating suspected toxic substances from polymer product formulations.

***Agrichemicals*** In spite of the increased demand for land for nonagricultural uses, the more advanced countries have met their needs for food and fiber with the aid of mechanization, fertilizers, and crop protection chemicals. These chemicals contribute to higher crop yields, better food quality, and enhanced food safety. Many agricultural industries are sustained only through the use of chemicals that control potentially devastating diseases. While these products are of great benefit to society, their widespread use has caused public concern about their effect on the environment and human health. Undesirable features of some products, such as concentration in the food chain, have led to curtailment or outright banning. These contrasting factors will continue to provide the backdrop for agrichemicals research in the 1980's.

All plant growth, development, reproduction, and fruiting are initiated by compounds produced and circulated within the plant. Some chemicals that regulate plant growth are being sold now, but substantial research in this area should lead to many useful new products in the next decade. Growth promoters, defoliants to aid harvesting, fruit ripeners, and fruit looseners are only a few of the new kinds of plant growth regulants under study.

The general toxicity of insecticides is a major environmental concern. Most commercial insecticides are nerve poisons and to some extent they eliminate beneficial as well as destructive insects. Many of these potent compounds have harmful effects on humans, and it is obviously undesirable for applications of these chemicals to spread into populated areas. One important research objective continues to be the discovery and development of insecticides that are nonnerve poisons.

A promising new development for controlling plant disease is the use of biological agents, which are now becoming commercially available.

Pheromones, the natural chemicals that insects use to attract partners for mating, offer a potentially valuable tool for safer insect control. Research will continue to focus on developing practical ways to use these potent but currently expensive compounds. Better understanding of insect biochemistry will lead to the discovery and development of new insect control chemicals.

Despite improvements in pesticide safety, society will continue to press for reduced use of pesticides and for pesticides that do not persist in the environment. Industrial research toward these objectives will continue to be important.

***Photographic Products*** Fluctuations in the price of silver are leading most photographic manufacturers to intensify research efforts to reduce or eliminate the use of silver halide as the light-sensitive element in films. However, high-speed photographic films will continue to use silver halide crystals since no substitute has been found that has equal sensitivity to light and equivalent image quality. Efforts to reduce the silver halide content of current films will accelerate.

Where high light sensitivity is not required, research will be aimed at nonsilver systems, or at films with very low silver content, including diazo, photopolymer, and photochromic systems. In addition, films will be designed that use silver halide crystals in small amounts to capture the images but use some lower cost material, such as carbon, to provide optical density.

Photopolymer technology will be expanded to provide films for use with projection and with laser exposure. Extension to cameras and electronic exposure systems should be a reality by 1985.

The trend in research on imaging applications will be toward electronic systems. For example, digital radiography is being developed to achieve image quality in X-ray applications through the use of sophisticated image enhancement methods. In printing applications, both electrostatic and magnetic methods will be developed that will eliminate the pollution and costs associated with the use of ink. Electronic manipulation and storage of color picture information in digital form will benefit from the accelerating developments in microcomputers. There will be an increased use of lasers as essential components of new systems in readout and imaging steps.

## THE PHARMACEUTICAL INDUSTRY

The research-intensive portion of the pharmaceutical industry is directed toward advances in therapy and prevention of diseases, with an average of 9 percent of corporate sales invested in R&D activities. There are several factors that influence how these funds are spent.

First, the cost and time needed to develop a new drug have escalated dramatically during the past 20 years, with the average cost increasing from $1.3 million to approximately $50 million and the average time increasing fourfold—from two to eight years. Economically, a company cannot support such an investment of its resources without assurance of an adequate return. This means that drug candidates must have a strong probability of demonstrating clear advantages over existing compounds or of meeting significant unfulfilled medical needs.

Traditionally, drug discovery has depended heavily on screening thousands of compounds for biological activity. The active compounds were then chemically modified until specific pharmacological properties were attained. Refined through the years, this process produced large numbers of therapeutically significant drugs, even though scientists often did not know how or why the drugs worked.

In recent years, biochemical processes within the human cell, the structure and function of the cell surface and its receptor molecules, the body's immune response systems, and other biological mechanisms have been elucidated, allowing researchers to design drug molecules to perform specific functions with minimal side effects.

The explosion of knowledge in biochemistry, molecular and cell biology, immunology, and neurobiology should result in a new era of drug discovery and development—one in which drugs are targeted to interrupt a specific disease process instead of simply treating its signs and symptoms. The tools of the researcher also are much improved. Better instrumentation allows more accurate measurements, more sophisticated analysis of the molecular structure of complex biochemical substances, and faster, more accurate, and less expensive performance of repetitive, routine tests. This new environment is an intensely interdisciplinary one in which the intuition and skill of the chemist are combined with the knowledge of biochemists, pharmacologists, and computer scientists as well as other researchers in universities, industry, and the government.

## Research Trends

Major advances can be anticipated in a number of fields, although not all of them will be realized during the next five years. One of the clearest trends is that the biological sciences will play a pivotal role in drug discovery and development. The following paragraphs highlight some areas undergoing rapid expansion, but space limitations require that other important fields be omitted.

Developments in biochemistry have encouraged greater exploitation of enzymes as specific targets for drugs. The recent success in synthesizing angiotensin-converting enzyme inhibitors as antihypertensives is one of the first examples of drug design based upon structural and functional information on enzymes. Efforts to identify, isolate, characterize, and develop inhibitors (or activators) for enzymes crucially involved in metabolism or its regulation will increase. Conventional enzymological studies frequently have dealt with soluble, abundant, extracellular enzymes; new efforts will deal increasingly with those enzymes that are bound in membranes, are present in very low concentrations, or occur in relatively exotic life forms.

The use of radioactive ligands has permitted better biological characterization of another class of drug targets—hormone and neurotransmitter receptors. (The word "receptor" refers to the specific chemical groupings to which the ligands bind in order to exert their pharmacological effect.) The identification and characterization of such receptors will permit increased understanding of how hormones and neurotransmitters act and will permit the design and construction of specific, potent antagonists (blockers) and agonists (mimics). For example, little is known about postreceptor events (those occurring after a substance is bound to its receptor) associated with receptors for insulin, growth hormone, prolactin, parathyroid hormone, and somatostatin; thus, the pathophysiology of their receptors remains unelucidated. Such work almost certainly will be associated with studies of the mechanism of transport processes. Alteration in membrane permeabilities for sodium, potassium, and calcium ions may well underlie postreceptor events associated with one or more of the hormones indicated above as well as with important physiological processes in the cardiovascular, gastrointestinal, renal, and nervous systems.

It is anticipated that immunology will suggest novel approaches to the diagnosis and treatment of important animal and human health problems. Advances in the next ten years will fall into two broad categories:

1. Technical advances which translate into better diagnosis and treatment.

2. Conceptual advances which reveal novel approaches to inhibit or stimulate immunological pathways, that is, to control individual cellular events in the generation and expression of immune responses.

One example of the first is the combined use of biological engineering and the chemical synthesis of defined antigens, which allows the creation of immortal hybrid cell lines that secrete a monoclonal antibody of precise specificity. Such antibodies can be used:

1. For the diagnosis of infectious, endocrinological, and neoplastic diseases.
2. As reagents for the isolation and identification of unique structures from natural sources.
3. As therapeutic agents for the directed killing of tumor cells, for example.

In the second category, an effort will be made to understand the regulation of humoral and cellular agents that control the level and specificity of the immune response. Such knowledge should provide the means to control the signs and symptoms of acute and chronic inflammatory processes resulting from this response, such as asthma, psoriasis, and rheumatoid arthritis. Since an immunological parameter has been implicated in a majority of chronic degenerative diseases such as rheumatoid arthritis, modulators of the immune response might disrupt the progression of the disease rather than simply ameliorate symptoms.

Metabolic pathways will continue to occupy the attention of scientists in the pharmaceutical industry. As new compounds that are involved in important physiological processes are isolated, it becomes crucial to determine the routes of synthesis and degradation as well as to identify associated enzymes. For example, the recent elucidation in the structure of SRS-A (slow-reacting substance in anaphylaxis) should help to unravel the biochemical events in asthma, and novel drugs for treating asthma may result from compounds that interfere with the synthesis, release, or degradation of SRS-A. The study of metabolic pathways also may reveal unique differences in the metabolisms of a parasite and its host and suggest new targets in using drug therapy for parasitic diseases.

Neurobiology should produce some breakthroughs in therapy during the next decade. The role of biologically active neuropeptides is clearly important, and the list of such peptides is increasing rapidly. The potential therapeutic implications of synthetic peptides will grow substantially as additional active peptides are isolated and as peptide analogs resistant to proteolysis are synthesized. The naturally occurring opiates (such as morphine) provide an excellent example of the potential therapeutic relevance of nonpeptide versions of biologically active peptides—in this case, the enkephalin-endorphin group of molecules. Molecules based upon the enkephalins may lead to a family of potent, nonaddicting analgesics. In addition to neuropeptides, neurotransmitter substances, their agonists and antagonists, and their roles in both peripheral tissue and the central nervous system will continue to provide novel targets for therapy.

Molecular biology and gene technology have shown dramatic progress, with recombinant DNA technology, in particular, being one of the most exciting and potentially productive areas in biomedical research. (See pp. 710–715 of Chapter 17, "Prospects for New Technologies.") Pharmaceutical applications of this technology include:

1. The production of polypeptides and proteins such as human insulin, human growth hormone, and interferon.
2. The production of antigens for vaccine development.
3. Increased yields of natural products such as antibiotics.
4. Basic research in the regulation of gene expression.

Although initial concern over the potential biological hazards has diminished, guidelines still need to be developed for production-scale processes. The first commercial product probably will be human insulin.

There has been a great deal of excitement associated with the testing of interferon, an antiviral protein that confers resistance to infection by a number of viruses. It is being tested in certain viral infections, such as chronic active hepatitis B and herpetic infections of the eye and the central nervous system. In addition, preliminary trials against several forms of cancer have demonstrated some limited activity; however, it is unlikely that interferon will be a "magic bullet" for curing cancer. The most tantalizing fact, which is of the greatest scientific significance, is that interferon is active at all. It appears to possess multiple biological activities, including effects on cell multiplication, immunological defenses, and cancer cells, as well as antiviral activity. Basic research on how it works may greatly expand our knowledge of biological processes while providing new models for drug design in these critical areas.

## Development Trends

***Testing*** The identification of an active compound is only the first step in the long process required before a drug is marketed. Extensive effort is expended in examining the drug under *in vivo* conditions, determining the duration of its activity, identifying and evaluating its metabolites, studying its disposition in tissues, assessing its safety, designing an economic process to produce it, and demonstrating safety and efficacy in the target species. Efforts are being made to develop new animal tests and to create new animal test models that will help predict potential therapeutic action in man.

Clearly, one of the major challenges facing the ethical drug industry is establishing that its products are both effective and safe. The subtlest aspects of safety evaluation lie in the areas of mutagenicity, carcinogenicity, and teratogenicity (that is, how a drug affects fetal development). Because these subjects are little understood and cannot be studied in man, a great deal of research is directed toward their elucidation in tissue culture and animal models. Mutagenicity may be the best understood of these subjects, because *in vitro* tests have improved our ability to evaluate mutagenic potential.

For carcinogenicity, the problems are more complex and safety evaluation is less certain, mainly because the mechanisms of carcinogenesis are unknown. At present, the carcinogenic potential of drugs is tested in long-term studies in mice and rats, which adds about three years to the time required for safety assessment. However, the results of these studies are not always conclusive as predictors of carcinogenic potential in man since there are metabolic differences between rodents and man. In addition, the incidence of spontaneous tumors and other pathological changes that are not related to drugs may obscure drug-related results.

Teratogenic research has many of the same problems as carcinogenicity. Since teratogenicity results from subtle perturbations in embryonic development, the biological/biochemical mechanisms remain obscure. Most current industrial research on teratogenicity tests compounds in animals and extrapolates these data to humans.

Very few compounds that possess therapeutic benefits present no safety risk. The quantitative risk/benefit analysis of therapeutic compounds will continue to be emphasized in basic research and in the development of improved methodologies for evaluating risks.

***Metabolic Studies and Delivery Systems*** In the area of drug metabolism, requests by regulatory agencies for basic information on the toxicological significance of metabolites and of biotransformation reactions undoubtedly will increase. Studies will be required that will reveal interactions between new chemical entities and known agents. It is anticipated that studies on mechanisms of action, identification of reactive intermediates, and characterization of the enzymatic processes for metabolizing new drugs will be needed.

In the future, the dosing of drugs will be done more precisely than at present. As the processes of absorption, distribution, metabolism, and excretion for pharmacological compounds are better understood, the dosage regimen will be based upon sound pharmacokinetic principles. To meet these specific regimens, controlled dosage systems, capable of more precise delivery of drugs to the body, are being developed. For orally administered dosage forms, control of the transit time of the drug in the gastrointestinal tract will be prerequisite to maintaining the minimal drug levels for prolonged periods, particularly beyond 24 hours. Also, methods for targeting drugs to specific tissues are being explored.

An area of increasing importance for the pharmaceutical industry is that of devising economic and environmentally safe processes for synthesizing compounds. The challenges facing process development have intensified dramatically during the past ten years and this trend is expected to continue over the next decade. Molecular structures of new products are more complex, raw material and energy costs are increasing, and manufacturing plant costs are growing steadily; all of these factors combine to increase manufacturing and capital costs.

***Role of Computers*** One of the most obvious methods of enhancing productivity and lowering costs is the effective use of computer resources. In basic research, this includes molecular modeling of new drugs; pattern recognition; storage, retrieval, and analysis of data; laboratory automation; interactive graphics; and personal computing. In development, computers have been used in safety assessment, in process control; in managing the vast information generated in preclinical and clinical studies; and in establishing a system to monitor adverse drug reactions. The future trend will be toward distributed processing and the gradual establishment of automated storage and retrieval of drug information.

## Major Issues

***Government Regulations*** Proposed congressional revisions to the Food, Drug and Cosmetic Act, as well as the rewriting of the Investigational New Drug/New Drug Application (IND/NDA) guidelines by the Food and Drug Administration, will increase the burden of the pharmaceutical industry in complying with regulations. Although the full impact cannot be assessed at this time, some proposals have the potential for making research more expensive, for decreasing the incentive to innovate, for lengthening the new drug approval process while easing the entry of generic (untrademarked) products, and for driving more research overseas. The absence of consistent legislative and regulatory requirements throughout the world increases the need for harmonizing regulatory standards.

***Multinational Research*** A number of foreign countries are stimulating the local development of high technology through accelerated depreciation for capital expenditures, additional tax deductions for increased R&D expenditures, and pricing advantages. As a result of regulatory and pricing incentives, tax advantages, and market needs, U.S. companies are establishing laboratories and conducting more clinical research in those countries where they need to market their products. This trend is encouraged by the disincentives of U.S. tax policy, which limit the credits for foreign taxes paid by multinational companies.

An analysis of the relative standing of the U.S. pharmaceutical industry reveals that there are no significant differences in sales performance, market share, and scientific achievements among the top ten multinational companies, of which five are based in the United States. During 1975-78, cumulative worldwide sales of the top ten companies increased by 8 to 13 percent per year; during the same period, 9 of them lost an average of 10 percent of their share of the world market.

A study of the top 49 multinational pharmaceutical companies reveals what may be an emerging trend—namely, the increasing importance of the Japanese pharmaceutical industry. The breakdown for the top 49 multinational drug companies, with worldwide sales in 1978 ranging from $250 million to $1,000 million and with individual market shares of 0.3 to 1.9 percent, is as follows: 22 United States, 10 Japanese, 7 West German, 4 British, 3 Swiss, 2 French, and 1 Italian. During the years 1975-78, worldwide sales of the ten Japanese companies increased from 30 to 60 percent per year per company. Over the same period, their cumulative increase in the world market varied from 25 to 100 percent. Clearly, a continuation of this trend could have a significant impact on the future of the U.S.-based pharmaceutical industry.

***Inflation and Profitability*** Inflation has become a more significant factor than in the past. For example, the cost of specialized chemicals, animals, and supplies for R&D is increasing at a faster rate than the Consumer Price Index. Petrochemicals, which are the starting substances for the production of many drugs, have increased in price by 30 percent over the past twelve months alone. Laboratory animals command the highest price the market will bear; in the past year, the prices for some species increased 20 percent or more. The prices of specialized laboratory glassware and other supply items also have increased faster than the Consumer Price Index. In addition, the rapidly rising cost of purchased medical services, including those provided by hospitals and physicians to support the clinical development of new products, has had a major impact. These factors can be countered only through cost control and improved productivity.

## Prospects

The present scientific outlook for the pharmaceutical industry is optimistic. Scientists throughout the world have made great strides in unraveling biological mechanisms, and application of this basic knowledge to the control of disease processes should occur during the next decade. Moreover, industrial research is well equipped to perform the long, complex, and expensive steps required in translating a new drug discovery into a commercial product.

It is less clear, however, whether pharmaceutical research in the United States will continue in the long term to operate at the leading edge of science. That will depend, in part, on the financial risks and rewards associated with such R&D. Costs and time for drug development have increased substantially and public policy has encouraged competition, as opposed to innovation. Two things are required if a substantial long-term investment in research is to be sustained. The first is not so much incentives as the absence of disincentives. The second is reinforcement of the right to own inventions for a limited period of time. Clearly, if government and public policy are supportive in these respects, dramatic achievements can be anticipated.

## THE AIRCRAFT INDUSTRY

### Research Trends

The aircraft industry's research involves a complex intermeshing of many disciplines. Its requirements are such that expensive and lengthy technological refinements in adaptation and in systems integration are necessary to minimize any possible dangers or risks that are anticipated in unproved new technology. Similarly, aircraft manufacturing processes and control systems place heavy emphasis on improvements in productivity to minimize recurring costs and program investment risks. It is the fallout from both the requirements of aircraft systems and the efforts to minimize risks that, in part, makes the industry valuable as a market for high-technology innovation and a source of adaptive innovation. For instance, computer-controlled manufacturing using aircraft wire harnesses has been adapted for automotive production, as have powered controls and, in recent years, the lighter weight and more energy-efficient structural materials used in aircraft. Similarly, nondestructive test and inspection procedures used by the aircraft industry have found widespread industrial applications.

***Commercial Technology*** Today, the most significant needs and opportunities involving commercial aircraft research are to improve the fuel efficiency of the basic airplane and its operating environment. Research is aimed toward achieving, through improved aerodynamics, a potential 25 to 40 percent increase in fuel efficiency. This involves changes in the basic configuration, wings, and airfoils of aircraft, increases in surface smoothness, and other major developments, such as advanced boundary layer control and fanprops.

"Today, the most significant needs and opportunities involving commercial aircraft research are to improve the fuel efficiency of the basic airplane and its operating environment." [Boeing Commercial Airplane Company.]

Improvements of the aircraft's structural efficiency through thin-wall castings and increased use of lighter weight materials (composites and new alloys) also offer sizable gains in fuel efficiency. The research thrust over the next five years will be to validate for commercial applications the structural integrity of potential innovations. The greatest opportunities (and the greatest risks) for reduction of structural weight lie with the nonremovable primary structures of the aircraft.

The technology required for optimizing designs using composites is so incomplete, however, that limited use of its potential by U.S. manufacturers can be expected over the next five years. The Europeans, in contrast, have launched bold research programs to accelerate composite applications.

Manufacturers of airframes and engines are conducting programs to improve propulsion efficiency by reducing the weight of the turbofan engine and its specific fuel consumption through the use of lighter weight materials and optimized designs of inlet and exhaust systems, nacelles, and engine placement. The potential gains may be offset, in part, by regulatory requirements for noise and emission levels and by increased maintenance costs.

Considerable fuel savings may be realized by minimizing ground burn and shortening the amount of time spent in holding patterns. This requires incorporating improvements into the air traffic control system, scheduling, and ground operations. The technology to avoid nonproductive burning of fuel is largely available to the air transportation system today; however, its incorporation into the air and ground-side elements of the air traffic control system is expensive and time consuming.

Research on alternative fuels for aircraft is receiving little attention compared with extensive industry and government efforts to maximize the fuel efficiency of conventionally fueled aircraft. Incorporation of cryogenic fuels (such as liquefied or solid methane and hydrogen) into the air transportation system would present formidable development costs in aircraft conversion and ground-side logistics. At this time, a more desirable route may be to produce synthetic jet fuel from alternative hydrocarbons. Extensive research is necessary, however, to determine the long-term effects such fuels may have on the engine and the environment.

Solid-state technology is receiving considerable attention, with the latest transport designs including digital electronic systems for flight management and navigation and situation displays provided by cathode-ray tubes. These advances simplify cockpit instrumentation and improve safety, as well as fuel efficiency. Their application to controlling load distribution will reduce structural weight and improve aerodynamics.

Finally, the industry has incorporated computer-based advances that integrate its technologies from research and development with its manufacturing technologies and systems. Such integration offers a revolutionary improvement in the quality and costs of aircraft design, as well as in the recurring costs associated with manufacture.

***Military Research*** Excluding its fighters, the U.S. military inventory contains, for the most part, aircraft whose airframe development and initial production date back to the 1950's and 1960's or earlier. The nation's only strategic bomber, the B-52, was developed in the late 1940's, and the last delivery took place in 1962. The C-130 entered military service in the early 1950's, and the C-141 and C-5A in the 1960's. With a few exceptions, a similar situation exists for the current inventory of helicopters, which are largely derived from technology of the 1960's.

Later production runs improved some types of aircraft, and a few have undergone costly modification programs to extend airframe life or keep pace with changing mission requirements. Nonetheless, the technology of their era represents an often serious compromise with the more advanced technology of the military environment in which they operate today. This technological gap may be even more serious as the ramifications of rapid deployment forces become better defined.

The aircraft industry's research outlook, therefore, places heavy emphasis on improving the military efficiency and the service life of the plane and its propulsive system. For a number of types of aircraft, these improvements will bring about efficiencies in aerodynamics, structures, and propulsion similar to those previously described for commercial aircraft.

A second area of focus will involve a greater refinement and integration of the military avionics subsystems. These can improve crew efficiency in flight management and boost the effectiveness of an aircraft's weapons and defensive subsystems. There will also be increased research on the "observables" of an aircraft's stealth characteristics—its radar cross section and heat emissions, for example—in order to inhibit detection of the craft by enemy sensors.

## Economic Trends

Advances in technology are based on the premise of economic return and the availability of investment capital. Advances in commercial aircraft in the United States are feasible only if they provide a suitable return on investment to the manufacturer and customer alike. The opportunities for profit and sustained research lie in maintaining a competitive edge throughout a long production program that may include incorporating continuous incremental advances into the product and initiating major improvements to existing models, both of which require sizable reinvestments. Over the lifetime of a successful commercial program, the reinvestment factor may, in fact, approximate the initial cost for development. It should be added here that the long production and testing times required for changes in aircraft mean that, as in the automobile industry, federally mandated changes are not instituted easily. An approach that emphasizes iterative government proposals and industry responses, such as that followed for regulations on aircraft noise, are likely to be more successful in achieving industry compliance than regulations that are formulated without any consideration of technical feasibility.

Market timing for new or improved products is critical and depends on the readiness and confidence to apply potential new technology. Without supportive federal policy and possibly some federal funding, the technology needed to hold the U.S. share of the market cannot be applied in a timely manner. Thus far, the industry has maintained a competitive advantage in the world market by producing aircraft that are superior in cost, efficiency, and quality. In current dollars, the aircraft industry leads all other manufacturing industries in its positive contributions to the U.S. trade balance. The industry's world position, however, may become uncertain as it faces serious challenges from abroad, economic decline at home, and a lack of cohesive or consistent government policy to deal with either.

***Foreign Competition*** As the research infrastructures of other nations began to emphasize aerospace activity, the United States pursued policies of contraction, a situation which essentially has doubled the competitive momentum of foreign aircraft research over the past decade. The full impact of these counter directions will be seen in the years ahead.

The role of an aerospace industry (particularly the aircraft sector) in stimulating other high-technology activities is now quite clear in Japan and the countries of western Europe. In the late 1960's, the European Economic Community assessed its aeronautical future, recognizing that a unique strength of the American industry was its ability to maintain production runs of sufficient volume and length to provide the necessary economic base and stability for expansion (for instance, it is not unusual for a manufacturer to require over 15 years of production in an airplane program before reaching the break-even point). The Europeans acquired their fundamental research and elements of expansion through government stimulation, joint ventures, and consortia. Although several years behind the Europeans, Japan recently has earmarked the aircraft industry as the leader of its knowledge-intensive industries for the future.

These countries have linked national growth objectives to their aircraft industries and research infrastructures and now represent a serious threat to U.S. aircraft markets. In 1977, foreign manufacturers had 4 percent of the world commercial aircraft market. By the end of 1979, they had 30 percent and U.S. manufacturers produced only 200 military aircraft for U.S. and allied armed forces. The foreign share of the market for wide-bodied aircraft, which contain the latest technology, increased from 16 percent in 1977 to 36 percent in 1979.

Associated with the vigorous challenge from abroad is a growing U.S. dependence upon foreign technology for state-of-the-art aircraft components such as electronics and optics. This is a further indication of the declining competitive state of U.S. aircraft industry and its productive capacity as a whole. The scarcity of domestic supplies of critical parts and components and the long lead time for their acquisition have affected production capabilities already and threaten to stall the industry should a vigorous rearmament program be instituted.

***Research Funding*** The aircraft industry generally builds on corporate knowledge and the research programs of the National Aeronautics and Space Administration, the Department of Defense, the Department of Transportation, other government agencies, universities, colleges, research institutes, and elements of the supplier community. This research provides a major portion of the theoretical seeds which the industry refines and translates into specific applications. The continued instability and uncertainty of research funding for these institutions may well impair the long-term technological superiority of U.S. aircraft if, in fact, this has not occurred already. The industry needs viable alternatives for energy sources that can be adapted for aircraft, as well as alternatives for its strategic materials. Basic research is an essential first step to obtaining them.

## THE AUTOMOBILE INDUSTRY

The automobile industry has undergone substantial changes in the past decade in response to government mandates and changing consumer needs. Because today's automobile is made up of over 15,000 individual parts, many of which interact with each other at very high speeds, close tolerances, and high temperatures, changes in one area require modifications elsewhere. Making even small changes in automotive systems requires massive investments in technical manpower and facilities. It is not surprising then that innovations have tended to be evolutionary in nature. For the next five years, research will concentrate on such areas as changes to meet environmental regulations, improvements in fuel economy, more efficient internal combustion engines, aerodynamics, advanced and new propulsion systems, safety, productivity, and the search for new materials to meet these needs.

### Research Trends

***Environmental Concerns*** Research and development spending to meet safety, emission, and fuel economy standards has grown until it accounts for close to half the R&D budgets of most American automobile manufacturers. Research on gaseous engine exhaust emissions will continue, with emphasis on after-treatment systems that are more durable and efficient. Closed-loop computerized engine controls will play a major role in this regard. Control of diesel particulates appears to be very difficult, however, and much work remains to be done. At present, the most promising approach appears to be the use of traps or filters that can be regenerated by burning the trapped particulates at appropriate intervals. Progress is needed on trapping efficiency and durability in the regeneration cycle.

Emissions from manufacturing operations will continue to receive attention in the 1980's. The most promising avenue appears to be the development of paints and coatings that use either very small amounts of hydrocarbon solvents (powder paints) or that use water as a solvent. Much progress has been made to date. Solid waste disposal research will focus on paint sludge and other byproducts that make poor or unacceptable landfill materials.

Noise is also an environmental problem and research will be continued to make quieter engines, improve noise attenuation in exhaust systems, and produce quieter tire tread patterns (while preserving tire traction). Aerodynamic noise, generated by fans, grills, window moldings, and side-view mirrors, is being studied and will be reduced.

***Fuel Efficiency*** Our nation has had access to cheap energy since the beginning of the Industrial Revolution. During most of this time, gasoline has sold in the marketplace in the United States at far below its real costs. This is the environment in which the U.S. automobile evolved and its patterns of use were established. As conventional petroleum resources are depleted, these patterns must change unless other suitable sources of economical energy can be developed. Liquid fuels from coal or oil shale cannot be expected to supplant a large fraction of our petroleum needs in the next decade, even though these distillates seem to be economically nearly competitive with today's world prices for petroleum products.

When synthetic fuels, including ethanol, do become available, the technology to make use of them will exist already, and continuing research will improve the efficiency and adaptability of engines using such fuels. Ideally, fuel refinement and engine design will be optimized as a system to produce the maximum overall efficiency. There may be additional restraints, however, on some of these synthetic fuels, as mentioned in the chapter on transportation.

The factors to which the fuel economy of a vehicle are most sensitive are vehicle weight, the combined efficiency of the engine, transmission, and axle (the powertrain efficiency), and aerodynamics. Vehicle weight reduction has been used as the most effective route to fuel conservation. This can be done by making the vehicle smaller, by making it from lighter weight materials, by more efficient structural design, or by some combination of these. A dramatic shift in the size of vehicles in the American market has taken place already and, from the appearance of the market trend, this shift will continue.

More than half of the engine power of a typical American car traveling at 50 miles per hour is used to overcome aerodynamic drag. This factor, however, is very sensitive to the speed of the vehicle. Improved aerodynamics may permit fuel economy improvements of up to 10 percent in the 1980's.

A significant change in the automobile in the 1970's was a rapid growth in the use of electronics and integrated circuit devices. This was brought about by the need for better engine controls to improve fuel economy and control emissions. The 1980's certainly will see this continue even more rapidly as closed-loop fuel controls and computer-controlled exhaust treatment systems are employed. The introduction of electronic or computer control systems, with sophisticated integrated circuit designs, may well be the most dramatic innovation in automotive technology for the 1980's.

Research on more efficient transmissions, lubricants, fans, tires, air conditioning, and other accessories will produce still further gains in fuel economy in the next decade. Regenerative braking will receive greater attention. Research and development of a variety of other alternative engines and power systems will continue but, at this time, it appears that the diesel engine offers the best fuel economy.

Not to be overlooked are the fuel economy gains to be achieved through better road and street layout, better traffic control, and an energy-conscious driving public. In the short run, we have no choice but to use our petroleum products with greater efficiency.

***Materials*** More research and development are needed to accelerate the use of alternative lightweight materials. Automobiles are sold at \$2-\$3 a pound, they are made at rates of tens of thousands per day, and they must last for more than a decade. These facts establish the three most important criteria for the materials used in their manufacture. These materials must be economical, they must be easy to form, machine, and assemble at high speeds, and they must be durable. Much work has been diverted to the search for materials with such characteristics, and it will continue through the 1980's and beyond. High-strength, low-alloy steels, aluminum, and magnesium are already in use, replacing mild steel in metal bars, sheets, and castings. Polymers are used extensively, and the use of reinforced plastics will grow as economics and production cycle times improve.

Materials scarcity in the form of interrupted supplies may characterize the 1980's and beyond. Recycling capability is a must. Most metals are readily recycled with current technology, but much work is needed to develop methods for the segregation, recycling, and use of plastics and polymers from scrapped automobiles.

Other materials find their way into the automobile in small amounts.

Chromium, nickel, antimony, and tin are examples. Platinum, rhodium, and palladium serve as exhaust catalysts. The worldwide availability of most of these materials is very uncertain. In the 1980's, research will continue on the question of nonnoble, metal catalysts. Efforts also will continue to find better catalyst support materials, corrosion-resistant coatings, more durable paints, high-temperature structural ceramics, and a lighter replacement for window glass. In short, materials will be a dominant area of research in the automobile industry in the 1980's. Electronics, computers, and manufacturing methods also will receive intense attention.

***Safety*** Much progress has been made in the 1970's in improving the crash protection provided by the automobile. A better understanding of the behavior of the human body in trauma situations has been most important in this regard. Research on the dynamics of the human body will continue, as will research on more protective automobile structures, particularly in side impact accidents. The search goes on for more convenient and effective restraint systems suitable for passengers of all ages and configurations. Although all new large and small automobiles meet the federal motor vehicle safety standards, which include withstanding a 30 mph front impact into a fixed collision barrier, field data indicate that, when small vehicles are involved in accidents, serious injury and/or fatalities are more likely to occur to their occupants than those in large cars. This increased risk is primarily a result of momentum transfer because there is less impact-absorbing material in the front of small cars.

Safe highway and road design has made progress, but more research is necessary. Improvements in highway durability and better repair techniques are required. Vehicle handling and drive controls will improve. Better driver training, particularly in developing skills that take full advantage of a vehicle's capabilities in avoiding accidents, would improve the safety record. The problem of the drinking driver also needs more attention.

***Productivity*** The next decade will see many developments directed at improving the productivity of manufacturing operations and the quality of the product. Automated systems to improve product quality and manufacturing productivity will receive much attention. Computer-controlled robots that possess vision and other senses will come into use for assembly and inspection as they are developed and improved. Computer control and scheduling of worldwide materials and component sourcing, assembly, and distribution will develop in the 1980's. With these developments will come an increasing challenge to work toward unifying standards and regulations on a global basis in order to reduce unnecessary impediments to worldwide commerce.

# ENERGY

## Trends in Energy Research and Development

Energy supplies and costs in the near-term future will affect not only the day-to-day operations of U.S. industries but also the nature and direction of the research these industries undertake, as well as the overall state of the economy. For these reasons, the following discussion reviews briefly the current state of energy research and development in the United States. This is admittedly a vast subject, and there is space here for only an indication of some general trends. While the federal government has consolidated into a Department of Energy (DOE) much of the research that a decade ago was scattered among such agencies as the National Science Foundation, the Bureau of Mines, and the Atomic Energy Commission, research in the private sector is still characterized by wide-ranging, diverse, and sometimes competitive approaches. Traditionally, the manufacturing and natural resource industries have funded and performed most of the R&D associated with energy production, but the vast increases in overall national funding levels for energy R&D, estimated as a 3.4-fold increase in real level of effort in the 1970's, have attracted other participants, notably aerospace companies. These have been joined by small, newly established firms building solar equipment and other forms of energy supply and conversion equipment.

Some integrated energy companies have begun to emerge recently, with oil companies, for example, diversifying into everything from coal mining to electric motor controls and nuclear fuel fabrication; on the whole, however, the energy industries are not well defined. This makes it difficult to discuss common trends in energy research in industry. For instance, petroleum geologists, developers of solar cells, and builders of electric generating equipment scarcely speak the same technical language and, until recently, they certainly did not consider themselves colleagues working together with a common objective. In a time when the funding of energy R&D has escalated almost as rapidly as the price of oil, a wide spectrum of technical talent has been hurriedly mobilized to confront a major national dilemma. It is not surprising that no consensus has emerged yet about the roles of the various technical approaches.

Historically, the position of the federal government has been to call for R&D on all available energy options. To some extent, this was a tenable position if the problem were viewed as sufficiently long range that R&D, rather than rapid implementation, was the appropriate strategy. (For example, in DOE's present \$3.1 billion R&D budget, funds for nuclear fission are comparable to funds for fossil energy, and solar, conservation, and environmental research all have large budgets.) Until now, it has proved quite possible to satisfy both nuclear advocates and solar advocates by continuing the funding for breeder reactor development while starting new large programs in such areas as solar thermal power plants, photovoltaics, and ocean thermal energy conversion. The long-range nature of these technologies was not particularly disturbing in the years from 1974 through 1978, when energy prices were largely stable and energy shortages seemed short term.

In 1979 and 1980, a major change occurred. The Iranian crisis brought back energy shortages in the form of gasoline lines. All energy prices soared, highlighted by an almost 60 percent increase in the price of gasoline at the pump in a one-year period. Finally, inexorable inflation resulted in calls for budgetary restraint and, in March 1980, this was translated into a proposed cut of 14 percent in the DOE budget. A dramatic change in the nation's perception of its energy problem means that we may have to narrow our options and make some tough choices.

Many advocates of specific technological approaches still consider the evolution from experiment to pilot plant to demonstration plant as the usual course of events for their projects, but the next five years probably will be filled with controversy and change for energy science and technology. Some programs must be discontinued; already, the continuance of such a politically popular program as the Solar Thermal Central Receiver Pilot Plant at Barstow, California, is in doubt. The sorting out of technical options in energy during the next five years will not always seem sensible, since social, political, and economic pressures may sometimes prove stronger than technological considerations.

There has been a shift toward shorter term R&D already, and an R&D emphasis itself has been overlaid with an emphasis on nontechnological policy options such as building up oil reserves or inducing price changes. However, science and technology will play an important role in implementing or evaluating the consequences of such policies. For example, implementation of marginal pricing schemes for electricity has involved not only experiments on customers' reactions to new tariffs but R&D on new communication and metering hardware. A decade ago, electric and gas utilities were not research intensive; in 1969, their funding of R&D amounted to only \$50 million, about 0.2 percent of their revenues. It was accepted practice for the supplier to carry out the necessary and often extensive R&D on new equipment. This situation changed in the early 1970's and was accelerated in 1972 with the formation of the Electric Power Research Institute, followed by the Gas Research Institute five years later. The utility industry as a whole now spends \$700-\$800 million per year on R&D.

The energy demand growth rate has changed rapidly from the 4.2 percent per year of the 1960's to the 1.5 percent per year of the 1970's; orders for nuclear reactors fell within a few years from about 10,000 megawatts of orders per year to about the same number of cancellations;

government energy R&D expenditures have soared by as much as 60 percent in a single year. This sort of volatility in objectives and driving forces will continue to characterize U.S. energy, even though the structure and products of the energy industry will change relatively slowly.

## Topics in Industrial Energy Research and Development

A large number and variety of R&D programs are being pursued by U.S. energy industries. They include:

- Oil and gas exploration, drilling, fracturing, and enhanced recovery; transport; oil refining.
- Coal mining equipment and systems, coal cleaning, coal transport, fluidized bed combustion.
- Shale oil, oil sands, heavy oils, underground coal gasification, unconventional natural gas.
- Systems to generate electricity from coal, including combined-cycle integrated gasification, fuel cells, magnetohydrodynamics; flue gas cleaning; gasified coal.
- Direct and indirect liquefaction and gasification of coal.
- Nuclear reactor safety and reliability, nuclear fuel technology, breeder reactors, uranium enrichment, reactor instrumentation and operation, nuclear waste processing and storage.
- Automotive engine technology, steam and gas turbines, electric vehicles.
- Electricity transmission and distribution, energy storage.
- Heating and ventilation technology, gas and electric appliances.
- Solar heating, solar electric generation, photovoltaic technology, wind energy, passive solar building design, biomass energy.
- Geothermal energy.
- Magnetically confined fusion, inertial fusion.
- Hydrogen as a fuel.
- Environmental considerations associated with energy, including acid rain effects, carbon dioxide buildup, health effects of ionizing radiations.

This listing includes only defined energy production and conversion technologies, the framework of most industry and DOE programs. More research on such subjects as fuel chemistry, high-temperature materials, the physics and chemistry of the atmosphere, biochemical processes, heat transfer, and corrosion and wear is needed to reinforce the largely applied emphasis of much of the work under way. Beyond this, continuing strong programs in plasma physics, in geophysics, in solid-state physics, and in other disciplines are needed for new energy options for the future.

There are many difficult, unresolved questions in energy technology. A typical question involves the relative merits of gas-cooled nuclear reactors and light water reactors. In the United States, the gas reactor has not been able to overtake the massive commercial lead of the light-water reactor, although it still has advocates who are convinced that it is superior in performance and safety potential. In the United Kingdom, debate between the two reactor types rages fiercely; in the past, nuclear development in that country has been somewhat impeded as a result.

Examples of unresolved questions of a similar nature are:

- Direct versus indirect coal liquefaction.
- Fixed, fluidized, entrained, and molten bath coal gasification.
- The appropriate trade-off between energy conservation in buildings and indoor air pollution (from carbon monoxide, radon, etc.).

Beyond these difficult technical questions, there are questions about the best ways to obtain the answers. Is it necessary, for example, to build and test, at the demonstration plant level, all of the approaches to coal liquefaction, or can some judgment be made on the basis of pilot plant experience plus analysis? This question is being debated currently in regard to solvent refining. Present plans are for two demonstration plants to be built but, with other technical questions, this could prove an unacceptably expensive answer.

## Legislative and Regulatory Factors

The progress of energy science and technology in the United States has been, and will continue to be, profoundly influenced by legislative and regulatory initiatives. Several technological options, such as gasohol, solar water heating, and minihydro, are being stimulated by taxes and other incentives now built into legislation. The case of gasohol is especially significant, since several studies in 1979 cast doubts upon the ability of alcohol from grains to produce more energy than was consumed in the production process. Nevertheless, in 1979, it was a popular and fast-growing program, with much congressional support, since it addressed a potential grain surplus, offered decentralized energy production to a minor extent, seemed to provide an additional gasoline supply visible to consumers at the pump, and provided an avenue for farm self-sufficiency in fuels.

There were five significant legislative and regulatory initiatives in 1978-80 that have consequences for energy science and technology in the next five years:

1. Funding of $20 billion was authorized for synthetic fuels; of this, almost $18 billion have been appropriated, with authorization to apply for $68 billion in 1984.
2. The Energy Mobilization Board was an attempt to expedite the construction of priority energy projects of all kinds, but it has been deferred indefinitely.
3. The windfall profits tax will remove funds from the oil companies, some of which would have been spent for advancing petroleum science and technology, and give them to the federal government, where it now appears that much of it may be spent for other purposes.
4. Energy performance standards for new buildings, if they are approved, will call for considerable reductions in the energy consumption of new residential and commercial buildings, with consequences for construction, heating, ventilation, and air-conditioning technology.
5. The National Energy Conservation Policy Act, by making conservation an option that must be considered on the same basis as additions to capacity by electric utilities, will slow growth and hence may slow the adoption of new generation technologies. At present, some utilities are making interest-free loans to consumers for conservation projects; they are, in effect, in the strange position of paying their customers not to buy their product, even though it is abundantly available and its price is not rising faster than inflation.

## A World Perspective

No other country has the resources to maintain as widespread an energy R&D program as the United States. As a result, there has been more specialization overseas on technological options considered appropriate to specific national situations, such as the breeder reactor in France, the United Kingdom, the U.S.S.R., and Japan; solar and geothermal energy in Japan; and coal conversion in West Germany and South Africa. Multinational collaboration via the establishment of specialized R&D facilities, such as CERN (European Organization for Nuclear Research) in Switzerland for fusion and high-energy physics, is also a feature.

As a result, technological development in some energy areas is more advanced outside the United States than inside. The most conspicuous

A 75-megawatt test reactor at the Jülich nuclear research center in the Federal Republic of Germany. "There has been more specialization overseas on technological options considered appropriate to specific national situations." [German Information Center.]

example is breeder reactor technology, exemplified by the 1,200-megawatt Super Phenix reactor well advanced in construction in France and the 600-megawatt breeder reactor that has just started operation in the U.S.S.R. Nuclear fuel reprocessing, which has been under a cloud in the United States because of a national concern for proliferation, is being pursued actively in the United Kingdom, France, Japan, and elsewhere. The U.S.S.R. has proceeded with the development of magnetohydrodynamics and high-voltage electrical transmission more aggressively than has the United States; the United Kingdom had, until recently, an ambitious program of wave energy development; Germany and the United Kingdom are ahead of the United States in automating underground coal mining and developing new mining equipment such as longwall miners.

Some of these technologies may not be appropriate for the United States; in fact, some, like the wave power initiative, seem not to be appropriate anywhere. However, U.S. participation in multinational energy R&D projects undoubtedly will increase in the next five years. Both Japan and Germany are participating in the development of solvent-refined coal in the United States, while this country, via the International

Energy Agency, is participating in a demonstration of pressurized, fluidized bed combustion of coal in the United Kingdom. The United States, West Germany, and the United Kingdom also have worked together in a methanization project in Scotland.

Many U.S. energy industries are multinational corporations and represent a conduit by which overseas energy developments reach the United States. For example, the South African experience in producing synthetic fuels from coal could become available in the United States, subject to suitable licensing, because a U.S. engineering company designed and built the latest plants there.

On the implementation side, another trend that may accelerate in the next five years is the formation of multinational consortia or new companies to handle major energy construction projects. Large hydroelectric projects developed in Africa and South America in the past decade have been handled in this way, and similar arrangements may be needed for nuclear projects to ensure a viable world nuclear reactor construction industry.

## Conclusion

While the national picture of energy R&D has shown some encouraging signs, it is still characterized for the most part by uncoordinated approaches in the private and public sectors. Of necessity, the next decade in energy research and development will be one of hard choices between competing technological options. Some choices have been made already: research on some coal liquefaction and gasification processes has stopped, for example, and some of the less plausible nuclear reactor alternatives are no longer being explored. While selection is essential, a hurried and premature assessment of performance and economics may lead to choosing a seemingly optimum system that in the end no one will be willing to make or sell.

There are also dangers in the push for rapid commercialization. In the few years that a large federal energy effort has existed, there has been a marked shift away from long-term research toward short-term results. The consequence has been a decline in radically new ideas, and even a decline in the applied research essential to ensure the success of existing projects. The next decade will see a continuous fight to preserve valuable, long-term exploratory research in energy. One criterion that might be useful in setting policy is determining the degree to which the research could bring completely new energy resources into existence.

One problem that has yet to be faced in the selection and commercialization of energy research options is the problem of low growth. For example, there are some interesting innovations in oil refining technology but, if no new U.S. refineries are needed, these innovations will be slow to appear in practice. The automobile manufacturers and the building industry will not be much interested in investing in new technology for improving energy efficiencies at a time when they are trying to stave off bankruptcy. The electric utilities accepted new technology eagerly while electricity use was growing at 7 percent per year but, with peak loads now growing very little, they must deal with a reduction in demand when their planning was based on increased growth.

Choices in energy science and technology will be profoundly influenced by the availability of capital. All new energy supply options are very capital intensive and, as a new energy project like the proposed Alaskan gas pipeline proves, putting together a great deal of capital can be a long, uncertain process. Partnerships and consortia likely will become the main route for sharing the capital burden—not just partnerships to build new energy facilities, but partnerships for R&D—and not just industrial consortia, but partnerships involving universities, governments, and overseas interests as well.

In the short space of a decade, energy has become of central concern to U.S. society. There will be no relaxation in the next decade. Defining the best research options, collecting technical teams of appropriate size and experience, and finding the proper institutions and incentives for commercialization are all challenges for the 1980's.

## Summary and Outlook

Although there are some marked differences among the industries reviewed here, there are also some common trends and problems.

Basic research has constituted almost a constant fraction of the nation's R&D effort for the past 20 years, but the focus of effort has shifted away from industry toward universities and colleges. In the long run, this shift may be beneficial because the product of academic research will be available to a broader spectrum of industry than would industry-generated proprietary research. Adequate funding is essential to carry on this important activity.

However, industry cannot forgo all basic research. In many cases, academic research may not relate to an industry's specific field of interest, or the cost of such research might far exceed the financial capacity of an academic institution. Finally, there can be an important interplay between research activities in the university and those in industry that could be beneficial to both.

Rising energy costs will affect almost all industries, not least the energy-producing industries themselves. The automotive and aircraft industries already have modified their products to increase fuel efficiency, as well as to reduce energy used in production. The choice of processes in the chemical industry is influenced by energy costs. In contrast, rising energy prices have created a new market in the electronics industry—that of engine controls. Indications are that research in this area will continue.

Increased competition from abroad has affected U.S. industry already and will continue to do so. The success with which this challenge is met will depend not only on the economic and political climates in the United States but on the strength and vitality of the U.S. R&D effort and on the success with which U.S. industries use technology developed abroad. In the energy sector, for instance, it is expected that U.S. participation in multinational energy R&D will increase in the next five years. The stimulation by foreign governments of their aircraft industries, combined with U.S. policies leading to contraction, has resulted in a serious threat to U.S. aircraft markets. The auto industry has felt this competition for some time, as has the electronics industry in the area of consumer products. The chemical industry expects to encounter increasing competition from western Europe by the end of the decade. It is interesting to note that, at a time when several of the other industrialized nations are setting up direct R&D support for industry, at the recent Tokyo round of trade negotiations the United States committed itself not to subsidize industrial R&D.

Federally mandated changes affect all industries. In some, meeting such changes not only requires an increasing share of the R&D budget but it also is resulting in a concentration on short-term rather than long-term research. In industries that are both capital intensive and produce a highly complicated product, such as an automobile or airplane, changes are not made easily. They involve redesign and testing of more than one part and often major capital expenditures. In many cases, the funds for such changes are not readily available and must be obtained from budgets that are already constrained. Standards that specify an end without considering the means often fail to achieve compliance. On the other hand, in industries such as the chemical industry, where a variety of means are available, regulations often specify the control technology rather than the result. There is a real need for an increased government-industry dialogue to consider both the results desired and the technological options available before final standards are set.

Productivity increases will become more important in both capital-intensive and labor-intensive industries. But such increases inevitably involve increased capital spending since they call for the replacement of existing facilities in the first case and the installation of labor-saving equipment in the second. As mentioned previously, such funds are in short supply. It is worth noting that recent increases in the productivity rates of West Germany, France, and Japan far exceed those of the United States. The overall productivities of these countries may surpass that of the United States by the end of the decade.

The government has recognized the returns generated through R&D by supporting an increased national R&D effort. However, in the most recent revision of the 1981 budget, the proposed funding, when regarded as a fraction of GNP, has reverted to about 1980 levels. A number of proposed initiatives for industrial innovation have been severely curtailed. While there seems to be a commitment in principle to keeping U.S. industry at the leading edge of R&D, such cuts, combined with the problems outlined above, do not promote an optimistic view. In contrast, R&D investment by other countries is becoming an increasing share of their gross national products. Finally, industrial effort is shifting away from long-term research and from basic and applied research toward development. These trends seem to indicate that little progress has been made in reversing the downward drift of the R&D effort in the United States, which is declining in relation to the efforts of the other major developed nations.

## REFERENCES

1. National Science Foundation. Derived from *Science Indicators 1978.* Washington, D.C.: National Science Foundation, 1979, Table 1.1, p. 140.
2. Ibid. Derived from Table 2.11, p. 178.
3. Ibid. Derived from Table 2.12, p. 179.
4. Ibid. Derived from Table 2.18, p. 185.
5. Edwin Mansfield. "Basic Research and Productivity Increase in Manufacturing," *American Economic Review,* December 1980.
6. National Science Foundation. Preliminary figure for *Science Indicators 1980.*
7. American Association for the Advancement of Science. *Research and Development, Report 5.* Washington, D.C.
8. T. Jora Minisian. "Research and Development Productivity Functions and Rates of Return," *American Economic Review*, May 1969.
9. Edwin Mansfield. *Industrial Research and Technological Innovation*. New York: W. W. Norton & Co., Inc., 1968.
10. Zvi Griliches. "Returns to Research and Development in the Private Sector," *New Developments in Productivity Measurement and Analysis.* Chicago: University of Chicago Press, 1980.
11. Edward Denison. *The Sources of Economic Growth in the United States.* New York: Committee for Economic Development, 1962.
12. Edwin Mansfield et al. *The Production and Application of New Industrial Technology.* New York: W. W. Norton & Co., Inc., 1977.
13. Mansfield et al. Op. cit.
14. Edwin Mansfield. "Federal Support of R and D Activities in the Private Sector," *Priorities and Efficiency in Federal Research and Development.* Washington, D.C.: Joint Economic Committee of Congress, October 29, 1976.

Other sections of this chapter contain proprietary figures from reliable industrial sources.

# V

# NEW TECHNOLOGIES

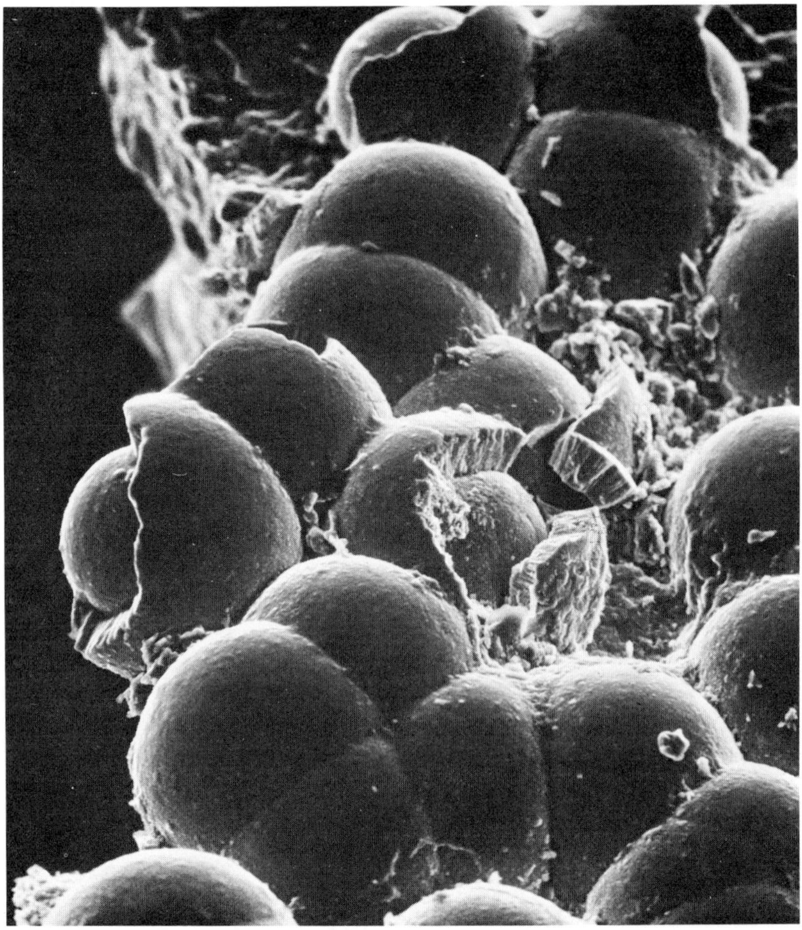

# 15

# *Fuel Science and Technology*

## INTRODUCTION

Extensive use of fuel is an integral part of our industrial civilization. In the last century, the sources of fuels and the technologies for using them have undergone great changes, determined by the available fuel supplies, the pressures of population growth, demands for higher living standards, and advances in science and technology. Equally great changes are foreseen for the future.

Probably the most important such changes are those implied by the expected decline in the availability of petroleum and natural gas, our dominant fuels. The obvious replacements for these fuels are this country's huge coal and oil shale deposits, as well as biomass in the forms of agricultural and forestry wastes and special energy crops. However, because so many of the potential applications require liquid or gaseous fuels, these solid fuels will need to be converted to synthetic liquid and gaseous fuels on larger and larger scales as time goes on. This chapter describes the state of the technology for doing so, indicates some important areas for scientific research, and discusses some economic, environmental, and social implications.

Some historical perspective can be gained from Table 1. In 1850, wood accounted for more than 90 percent of the nation's fuel; coal provided most of the rest. Except for small amounts of whale oil and other animal and vegetable oils used for illumination, all fuel was in solid form, with its

---

◀ Anthracite coal, magnified some 600 times. "The general chemical structure of coal is still a matter of debate." [U.S. Geological Survey, U.S. Department of the Interior.]

**Table 1** Annual consumption of fuels in the United States, in quads ($10^{15}$ Btu)

| Year | Wood | Coal | Natural gas and oil | Total energy consumption (includes hydro, nuclear, etc.) | Total energy supplied by solid fuels (percent) |
|---|---|---|---|---|---|
| 1850 | 2.1 | 0.2 | 0.0 | 2.3 | 100 |
| 1875 | 2.9 | 1.4 | 0.0 | 4.3 | 100 |
| 1900 | 2.0 | 7.0 | 0.5 | 9.9 | 91 |
| 1925 | 2.0 | 15.0 | 5.7 | 23.0 | 72 |
| 1950 | 1.2 | 13.0 | 19.5 | 35.0 | 57 |
| 1975 | 0.8[b] | 16.0[a] | 56.0[a] | 77.0[a] | 22 |
| 1979 | 1.0[c] | 15.1[c] | 57.0[c] | 78.2 | 21 |

[a] *U.S. Energy Supply Prospects to 2010.* Washington, D.C.: National Academy of Sciences, 1979.
[b] "Peat Draws Attention as Hydrocarbon Source," *Chemical and Engineering News,* November 7, 1977.
[c] *Monthly Energy Review.* Washington D.C.: U.S. Department of Energy, Energy Information Administration, May 1980.

SOURCE: Adapted from H. C. Hottel and J. B. Howard, *New Energy Technology.* Cambridge, Mass.: M.I.T. Press, 1971, p. 4.

inconvenience and high air pollution levels and the thermal inefficiency attendant upon combustion equipment of the day. By 1900, petroleum, natural gas, and hydroelectric power were making small contributions to the national fuel supply, but about 90 percent of the fuel used was still in the solid form.

After 1900, the proportion of solid fuels in the fuel supply declined rapidly. By 1975, they contributed only 22 percent and were used mainly for generating electricity and industrial steam. Oil and gas supplied 73 percent of the total. This predominance reflected in part the abundance and low cost of petroleum and natural gas. Of equal importance, however, were their ease of handling and the low cost and high performance of home heating units, airplanes, automobiles, and other end-use equipment.

Some perspective on the implications of these consumption statistics for future fuel supplies is afforded by one current estimate of recoverable fossil fuel reserves and resources, shown in Table 2.

U.S. oil and gas consumption in 1979 totaled 57 quadrillion (57,000,000,000,000,000) British thermal units (Btu) per year. Since the estimated ultimately recoverable resources of natural gas and oil amount to only about 17 years' supply, the need for importation, conservation, and substitution of other resources is apparent. (The United States, in fact, imports about half its oil today, largely from regions deemed to be unreliable suppliers for the long term.)

Some estimates of ultimately recoverable resources can be viewed as underestimates. Large quantities of natural gas, for example, are believed

**Table 2** Fossil fuel reserves and resources in the United States

| Mineral | Heat of combustion, in quads: Recoverable reserves[a] | Heat of combustion, in quads: Ultimately recoverable resources[b] |
|---|---|---|
| Natural gas | 230 | 500 |
| Oil | 200 | 470 |
| Oil shale | [c] | 13,000 |
| Peat | 700[d] | 1,400[d] |
| Coal | 6,000 | 79,000 |
| Tar sands | 30[e] | 100[e] |

[a] Known resources, economically extractable with prevailing technologies.
[b] Reserves and all other deposits known or believed to be economically extractable.
[c] No reserve estimates are given for oil shale because commercial exploitation has not yet taken place.
[d] "Peat Draws Attention as Hydrocarbon Source," *Chemical and Engineering News*, November 7, 1977.
[e] References vary; entries represent average values.

SOURCE: Adapted from *U.S. Energy Supply Prospects to 2010*. Washington, D.C.: National Academy of Sciences, 1979, p. 4.

to be present in the geopressured brines of the Gulf coast and in certain low-permeability formations. Large-scale extraction of these deposits may be economically attractive in the future; however, they must compete with synthetic gas from coal. And, similarly, the oil shale resource estimates of Table 2 omit the very large deposits of Devonian oil shale in the eastern and midwestern United States, because extracting oil from these shales will be difficult and costly. In the long term, however, they will probably become important resources. Tar sands resources in the United States are smaller and more difficult to process than those in Canada; however, they could be of future importance.

Biomass, or plant matter, might also make important contributions to future fuel supplies. Estimates of the magnitude of the contribution vary widely, depending to a considerable extent on judgments about the amount of agricultural and forest land that can be dedicated to fuel crops. Table 3 gives one estimate, a "national commitment" scenario for the year 2010, of 8.8 quads. Since total energy consumption estimates[1] for that year range from 70 to 100 quads or more, this contribution might be 10 percent of the total. Here again, the starting materials are solids that would need to be converted to gas or liquid fuels for many uses.

Although there is much judgment and guesswork involved in such estimates, they do illustrate the need to shift emphasis from petroleum and natural gas to other resources.

The world's ultimately recoverable oil resources are roughly estimated at about 10,000 quads, or 20 times the amount of U.S. oil resources.[2] However, growth in world oil demand and the increasing difficulty of

**Table 3** Biomass as a source of fuel in the United States, estimated for the year 2010

| Source | Quads per year |
|---|---|
| Municipal waste | 1.9 |
| Agricultural waste | 3.5 |
| Energy crops | 3.4 |
| TOTAL | 8.8 |

SOURCE: *U.S. Energy Supply Prospects to 2010.* Washington, D.C.: National Academy of Sciences, 1979, p. 172.

adding to reserves have resulted in the rate of oil production exceeding the rate of discovery of new reserves since 1970. While increases in world oil prices have brought about increases in drilling activity, the drawdown of reserves will probably continue irreversibly. This will, of course, limit the supply of oil available for importation and accelerate the development of alternative energy sources throughout the world.

Importation of gas in the United States and in the rest of the industrialized world is expected to grow. While the prospects for future gas discoveries are better than those for oil discoveries, much of the new gas will be found in areas remote from major industrial markets. The lack of transportation facilities will present a major barrier to the development of those resources. Even if the transportation system is expanded widely, resource limitations probably will limit output of natural gas in the not distant future.

Coal is expected to become more important worldwide. The United States has about 25 percent of the world's estimated coal resources and reserves. Approximately 90 percent of the world's coal is held by the United States, the Soviet Union, China, and Australia. Resource limitations will not present a serious problem for many decades, and coal presents a major opportunity for world trade.

The world's largest oil shale deposits are believed to be in the United States, Brazil, and Australia. Exploitation of this resource will help reduce oil imports, although exploiting it more than is necessary to serve domestic needs is unlikely.

Rapidly increasing oil and gas prices and decreasing world resources will result in slow but major changes in fuel use through substitution and conservation. In electricity generation in the United States, the use of oil and gas (now 23 percent of total energy input) is expected to decline drastically (to around 5 percent of the total), with coal and nuclear energy achieving the major growth during this century. Oil and natural gas supply about 75 percent of industrial fuel. Substitution of coal or coal-generated gas is expected to reduce this share to less than 40 percent by the year 2000.[3] In residential and commercial buildings, heating and air conditioning are expected to change to greater use of electricity and gas. Growth in energy demand by both the industry and buildings sectors is expected to continue.

Demand for transportation fuels is expected to decrease during the 1980's as more people turn to smaller and more efficient automobiles and light trucks. In the slightly more distant future, that decline will probably be counterbalanced by increases in truck and airplane traffic; thus, transportation fuel consumption may grow slightly in the 1990's.

If there were no environmental, industrial, or financial constraints, the nation could produce enough synthetic fuel to eliminate almost all oil imports and at the same time slow the consumption of domestic oil. There are such constraints, however, and science and technology will have to provide means for reducing them.

The urgency with which the Congress and policymakers view the fuel situation is indicated by the Energy Security Act of 1980 (Public Law 96-294). The act, among its provisions, establishes the federal Synthetic Fuels Corporation and authorizes it to provide up to about $20 billion in financial assistance to synthetic fuel projects between 1980 and 1984. The act also includes special tax incentives and financial assistance provisions for those who produce alcohol fuels from biomass. These provisions will probably help establish commercial production of synthetic liquid fuels over the next decade; however, some coal gasification projects are already proposed without special federal assistance.

## FUEL OPTIONS

In determining which synthetic fuels should be produced, the major considerations involve ease of storage and handling, combustion-generated pollution potential, cost, and efficiency of conversion from the solid raw material. Table 4 summarizes the relative ratings of some major synthetic fuel options in terms of these criteria.

Energy use and fuel supplies vary with time. Fuel storage is therefore necessary. Liquid fuels receive by far the best rating in terms of ease of storage and handling for use in combustion equipment. Solid fuels are readily stockpiled, but they are relatively cumbersome to handle and transport. Gaseous fuels are simple to transfer and burn, but are relatively difficult for consumers to store.

Combustion oxidizes all fuel constituents, releasing them as gases (water, sulfur oxides, nitrogen oxides, and carbon dioxide) or as solids (soot, polynuclear aromatics, and inorganic particulates). Except for water vapor, all of these emissions can be harmful and, with the current exception of carbon dioxide, are regulated to minimize damage to human health and the environment. Control technology at the point of combus-

***Bertram shoveled some coal into the dragon's middle mouth***

Even exotic energy solutions need a source of energy. [Paul T. Gilbert. *Bertram and His Fabulous Animals.* New York: Rand McNally, 1937. Pictures by Minnie H. Rousseff and Barbara Maynard, Library of Congress.]

tion can be complex and expensive. For large-scale users such as electric power stations, it is cost-effective to burn such unrefined fuels as coal and petroleum residuum and to remove pollutants from the stack gases. For small-scale users, however, it is generally most effective to remove polluting elements such as ash, sulfur, and nitrogen from the fuel before combustion.

The cost and efficiency of converting one fuel to another are interrelated. In general, the purer the product, the more expensive and energy-consuming the conversion process. The cost rating for each heating value in Table 4 is given in relative terms, with the cost of producing gasoline from coal by direct liquefaction assigned a value of 1. (Cost ratings are necessarily approximate at this time.) The efficiency is the heating value of the product divided by the heating value of the original coal or shale in a self-sufficient plant. Efficiencies will change with further development of the processes.

**Table 4** Comparison of major forms of refined synthetic fuels

| Fuel[a] | Storage and handling | Ease of pollution control | Approximate cost rating[b] | Efficiency[c] | Major end use |
|---|---|---|---|---|---|
| Mixture of carbon monoxide and hydrogen (intermediate-Btu gas) | Difficult | Good | 0.7 | 0.65 | Industrial fuel and intermediate product for other fuels |
| Methane (high-Btu gas) | Difficult | Good | 0.8 | 0.60 | Home heating |
| Methanol and ethanol | Base case | Best liquid | 1. | 0.55 | Internal combustion engines |
| Gasoline | Base case | Nitrogen oxides a problem | 1. | 0.58 | Internal combustion engines |
| Jet and diesel fuel, and home heating oil | Best | Nitrogen oxides and soot are a problem | 0.7–0.8 (shale)<br>1.0 | [d]<br>0.55 | General transportation |

[a] Assumed raw material coal, unless noted. [b] Cost ratings are relative; value of 1 = the cost of producing gasoline from coal by direct liquefaction. [c] Efficiency = product's heating value divided by heating value of original coal or shale in self-sufficient plant. [d] Shale efficiency not comparable.

*Note:* The accuracy of estimates only justifies one significant figure. The cost difference between methane and intermediate-Btu gas is probably closer to 0.05 than to the 0.1 indicated.

Intermediate-Btu gas is a mixture of carbon monoxide and hydrogen formed by reacting water and oxygen with solid fuels. Its low cost, versatility as an intermediate in the production of fuels and chemicals, and clean combustion may result in large-scale production for industrial and power generation use. Its toxicity and low volumetric heating value make it incompatible with the existing natural gas distribution system, requiring that it be used near the site of production, usually within 100 miles.

Synthetic natural gas, or methane, will be manufactured to maintain gas supplies for residential and commercial heating and to replace natural gas in industrial and power generation use.

The alcohols methanol and ethanol are chemically the simplest of the liquid fuels and, in terms of combustion, the best. Methanol does not form soot and polynuclear aromatics when burned, and it allows the high-temperature thermal fixation of nitrogen (forming nitric oxide) to be much more easily controlled (by water dilution) than do hydrocarbons. Methanol's combination of high octane number and lean inflammability limit, and the possibility of using exhaust heat for methanol vaporization and decomposition, can with further research and development give efficiencies in the spark-ignition automobile engine matching those of the diesel engine, but with much cleaner combustion. (Methanol-gasoline blends, however, lose these advantages.) Ethanol is established as a gasoline blending agent, but has less potential than methanol for improving engine efficiency and for mass production. Both alcohols clearly have great potential and can be manufactured by available technology. The use of methanol, however, may be limited because of the fuel's toxicity and low volumetric heating value (half that of gasoline).

Gasoline, diesel fuel, jet fuel, heating oil, and industrial fuel oils can also be manufactured from solid fuels. Pollution control with these substances is more difficult than with alcohols. However, it is expected that continued advances in the science and technology of combustion and in the storage and operating advantages of these fuels will maintain them as the major synthetic liquid fuels for the foreseeable future.

The specifications of fuels, which have been optimized in the past for petroleum as a raw material and for current combustion systems, will not remain constant. Raw materials, manufacturing processes, and end-use equipment will all change drastically with passing time, and fuel properties will evolve with them.

## Conversion Science and Technology

The need for synthetic fuels has been felt in the past, and there is a substantial base of science and technology for the much larger scale production planned for the future. Production of coal gas by reaction with steam became important in Europe and the United States after 1861, when the Siemens brothers first used this technique with their invention, the open hearth steel furnace. The gas (a mixture mainly of hydrogen and carbon monoxide with varying amounts of nitrogen) was used industrially when an ash-free fuel was required, as in glass, lime, and steel manufacturing, and was also used for lighting, heating, and cooking in cities. By the middle 1920's, about 11,000 gas producers were in use in this country. They were displaced by natural gas in the United States at about the time of World War II, but coal gas remained in limited use in Europe and South Africa. Coal gasifiers made by the German companies Lurgi, Winkler, and Koppers-Totzek represent the current commercial gasifier technology.

Nor is coal liquefaction new. Germany, with abundant coal and limited native petroleum, built a number of coal liquefaction plants before World War II. These provided about 80,000 barrels per day of liquid products and helped fuel the German war effort. This quantity is small compared with the current U.S. consumption of about 18 million barrels per day. The Bergius process combined coal with hydrogen at pressures of 4,000 to 10,000 pounds per square inch. Because of its high costs, the process did not survive the war, but it was the ancestor of the "direct" liquefaction process now being developed in the United States. Another liquefaction process, Fischer-Tropsch, reacts coal gas over a catalyst to make a variety of liquid fuels. Development and commercialization of this "indirect" process have continued in South Africa; the SASOL facilities there represent the state of the art in commercial coal liquefaction.

In the United States, future petroleum supplies have appeared inadequate enough at certain times to spur interest in synthetic substitutes. During the 1930's, small-scale feasibility studies of the German technology were carried out, but the discovery of the East Texas oil field ended this work. In 1944, petroleum supplies again appeared inadequate; Congress enacted a R&D program, and interest in coal gasification and liquefaction was revived in industrial and federal laboratories. These programs expired with the discovery of the large Middle Eastern oil deposits. Government and industry funded some small-scale work in the 1960's, but the current, greatly accelerated program began with the oil embargo in 1973.

The petroleum refining and petrochemical industries, over the years, have also produced science, technology, and commercial bases of considerable value to a synthetic fuels industry. Methanol, for example, can be produced efficiently from coal by existing commercial technology, and catalytic processes for petroleum refining are excellent starting points for purification of coal tars and shale liquids.

Fermentation of grains and sugars to produce ethanol is, of course, the oldest process of current interest for fuels production. While improvements in efficiency and extension to cellulose and other nonfood raw materials are needed, this old technology offers an important point of departure.

All of the natural solid fuels, when heated, decompose to char, liquid, and gas. This reaction is known as pyrolysis. The proportion of the solid fuel's heating value remaining in the solid char varies with the raw material, averaging 60 to 90 percent for coal and lignite, less than 20 percent for the organic fraction of Colorado oil shale, and less than 5 percent for cellulose. Development of technology for converting coal to liquid and gaseous fuel therefore concentrates on attaining higher conversions to more desirable products than can be obtained by simple pyrolysis. For Colorado oil shale and biomass, pyrolysis followed by refining of the crude liquid and gas product is promising. The char produced need not be wasted; it can be burned to produce heat for pyrolysis.

Figure 1 presents a generalized diagram of the modules used to produce clean fuels from solids. The manner in which these six modules are combined depends on the raw feed materials available and the products desired. For each module, there are many and varied competing processes, each with its own name and its own stage of development.

A common characteristic of the primary production modules (gasification, gas conversion, pyrolysis, and liquefaction) is the production of multiple products. Liquefaction produces gas, and gasification, liquids. If a plant is designed to produce only gas or only liquid products, it is necessary to convert gas to liquid or vice versa, with a corresponding added cost and energy consumption. Since both types of products are needed, this is wasteful. However, industry and government, for reasons that will be discussed later, tend to foster single-product synthetic fuel projects.

## GASIFICATION

An element common to most synthetic fuel plants is gasification and gas cleanup. The gas produced can be used as low-Btu gas (mostly carbon monoxide, hydrogen, and small amounts of methane, diluted with nitrogen) or intermediate-Btu gas (the same mixture without the nitrogen). It can be used as a source of hydrogen, which is required for liquefaction and refining, and it can be converted directly to end products, such as high-Btu gas (methane, equivalent to natural gas) or liquid fuels like alcohols, gasoline, and diesel fuel.

More than 35 gasification processes are listed in the 1977 National Research Council report on coal gasification[4]; all depend on the same set of stoichiometric reactions:

| | |
|---|---|
| $CH_x + H_2O \longrightarrow CO + (1 + x/2)\,H_2$ | Water gas reaction |
| $C + 2H_2 \longrightarrow CH_4$ | Methane synthesis |
| $CO + H_2 \longrightarrow CO_2 + H_2$ | Water gas shift reaction |
| $C + O_2 \longrightarrow CO_2$ | Oxidation reaction (to supply heat to the water gas reaction) |

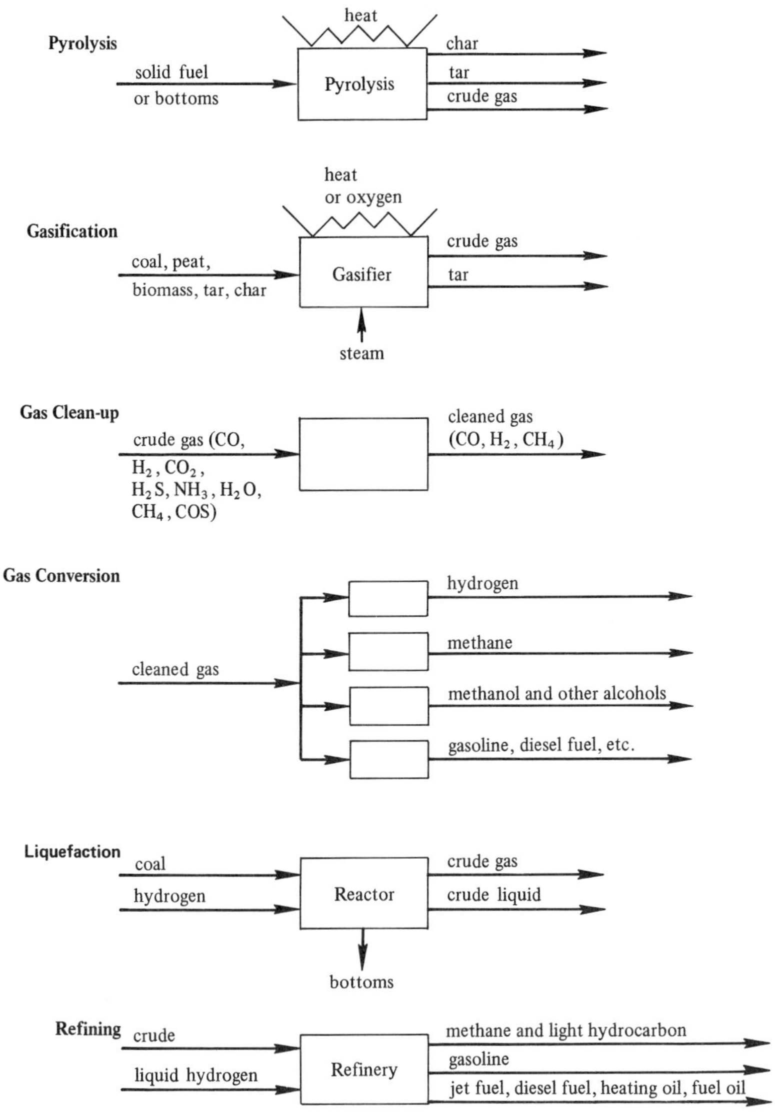


**Figure 1 Processes to produce clean fuels from solids.**

Large amounts of heat are consumed by the water gas reaction, which is the primary mode of gasification. The remaining reactions liberate heat; however, oxidation is required to liberate enough heat to balance the highly endothermic water gas reaction. The oxidation can take place externally to heat char, which then is used to heat fresh coal or char, or (more commonly) the oxidation can take place in the same reactor with the other reactions.

If this latter course is used and air supplies the oxygen, the product is diluted with nitrogen and thus becomes low-Btu gas. If pure oxygen is used, the product is intermediate-Btu gas, sometimes known as synthesis gas or syngas. Because of the high cost per Btu of storing or transporting it, low-Btu gas is normally used as fuel in facilities within 20 to 30 miles of the gasifier. Intermediate-Btu gas is more versatile, since it can be economically pipelined up to 100 miles and can serve as a starting point for the production of many fuels and chemicals.

At very high temperatures (above 2,400°F), methane and tars are destroyed, so carbon monoxide and hydrogen are the only important fuel products. Coal ash is molten at very high temperatures, and can be removed as a liquid. At low temperatures, the ash is a solid, and methane and tar appear as products. In some cases, tar is an undesirable product, but it can be a valuable fuel. As temperatures fall below 1,400°F, significant quantities of methane are formed, increasing the heating value of the gas; also, reaction rates slow, and the use of catalysts such as potassium salts becomes necessary.

The variables that determine reaction rates are important in the design of gasification systems over this entire temperature range. Knowledge of this factor should improve over the next five years. An understanding of catalysis in these reactions is especially important.

## Gasifier Types

The manner in which solids are brought into contact with the reacting gases is critical to the design of gasifiers, and the techniques employed serve to classify the many proposed gasifiers. The major categories are illustrated in Figure 2.

***Fixed-Bed Gasifiers*** In fixed-bed gasifiers like those produced by the German Lurgi Corporation, coal is fed in at the top and gaseous reactants enter at the bottom of the reactor. The coal moves downward as a solid plug at such a rate that ash leaves the bottom of the bed. The major reaction zone is partway up the reactor. The bottom section preheats the incoming gases, and the top section preheats the coal. The very efficient internal heat exchange yields a high-temperature reaction zone with minimal combustion. If pure oxygen is used, this temperature can easily exceed the melting point of the ash and, if "dry" ash is desired, an excess of steam must be used for temperature control. This type of reactor is subject to plugging, and mechanical stirrers are used in both the top and bottom. Even so, the commercial Lurgi system is limited to noncaking coals such as lignites and western bituminous coals.

Slagging fixed-bed gasifiers, in which temperatures are high enough to keep the ash molten, can process three or four times as much coal in a

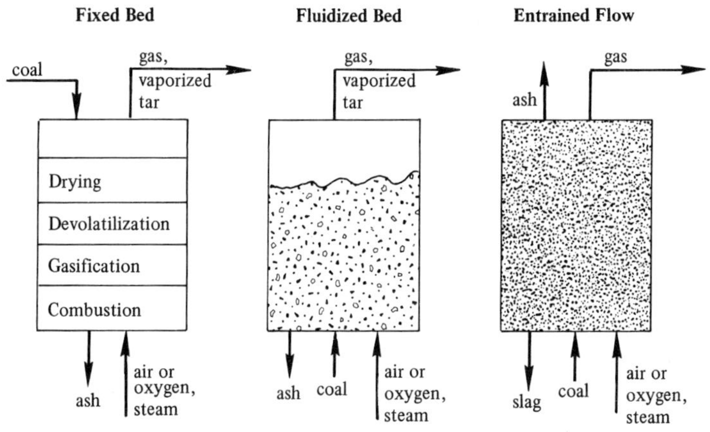


| | Fixed Bed | Fluidized Bed | Entrained Flow |
|---|---|---|---|
| **Commercial:** | medium pressure Lurgi | low pressure Winkler | atmospheric pressure Koppers-Totzek |
| **Second Generation:** | medium pressure British Gas/Lurgi slagging gasifier | low pressure COGAS | medium pressure Shell-Koppers, |
| | | high pressure HYGAS | medium-high pressure Texaco |

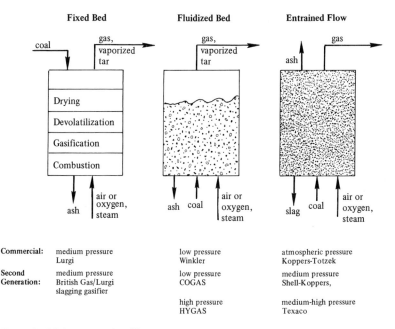

**Figure 2 Major types of gasifiers.**

unit of a given size because of the higher reaction temperature and lower steam throughput. An additional advantage is a lower steam requirement, which improves thermal efficiency by about 5 percent. While not yet commercially proven, slagging gasifiers will very probably be available in the next decade as second-generation systems in competition with the newer fluidized-bed and entrained-flow systems (described below).

Underground gasification can be considered a special kind of fixed-bed gasification. Air rather than oxygen is normally used; the result is low-Btu gas, which must be used locally. Considerable progress is being made in technical analysis, and demonstrations have been carried out in western coal seams. Because of the need to avoid ground subsidence and interaction with aquifers, and because a suitable coal seam is required, this approach is less generally applicable than mining and surface gasification. However, technical progress in the near future may allow this technology to take its place among the available gasification techniques.

***Fluidized-Bed Gasifiers*** If the gas velocity in a gasifier is increased, at some point the coal particles in the reactor begin to move relative to each other. If most of them remain in the bed, the reactor is known as a fluidized-bed gasifier. (If the gas velocity is high enough to draw the particles along with it, the reactor is called an entrained-flow gasifier.

This technology is described later.) In fluidized-bed systems, the residence time of solids in the bed can be very long compared with the residence time of gas. This allows operating at relatively low temperature despite the lower char gasification rates. The heat of reaction can be supplied by circulating char heated in a separate vessel by combustion with air, which obviates the need for a special plant to separate oxygen from the air.

Dealing with caking coals is more difficult in fluidized-bed gasifiers than in entrained-flow reactors, and several approaches to avoiding agglomeration of ash are under development. These include preheating the coal, staged pyrolysis, rapid dilution with large amounts of char, and the addition of inorganic materials such as lime and potassium carbonate, which inhibit agglomeration and catalyze gasification. By raising the temperature to just below ash fusion temperature, the ash in a fluidized-bed gasifier can be induced to form balls, which can be separated from the finer coal particles fed to the bed. This allows more efficient use of the bed volume and is therefore of current developmental interest.

In principle, fluidized-bed systems offer the advantage of allowing pyrolysis tar and gas to be recovered, and they scale up to commercial sizes more easily than do fixed-bed systems. The fluidized-bed Winkler process is used commercially to a limited extent, and more recent development efforts have produced several candidates for commercialization.

***Entrained-Flow Gasifiers*** The gas velocity in a fluidized-bed gasifier can be further increased to produce entrained flow of the coal particles, as in the commercial Koppers-Totzek system. This process can handle a wide variety of coals. Because oxygen is used, the temperature is high (2,400°F to 2,800°F at the exit), and ash is molten. The high temperature gives high reaction rates, minimizing the output of tar and methane. Commercial entrained-flow units are limited to low-pressure operation, but the potential for cost and efficiency improvements through higher pressure operation, at 350 to 1,000 pounds per square inch, is recognized. A high-pressure version (Shell-Koppers) is in the advanced development stage. Texaco has developed a commercial entrained-flow gasifier for use with heavy oil and is currently adapting it to use coal. Commercialization plans have been announced for both of these second-generation entrained-flow systems.

## The Choice of Processes

The choice of gasification processes will largely depend on local coal resources, the products desired (e.g., tar and gas), and the end use (e.g., power generation, industrial heating, or chemical feedstock), as well as on the states of development and the costs of particular gasification technol-

**Table 5** Cost breakdown for a Lurgi gasification plant

| Component | Percent of total plant investment |
|---|---|
| Gasification | 16 |
| Steam and oxygen production | 28 |
| Gas purification | 22 |
| Effluent treatment | 12 |
| Miscellaneous | 22 |
| TOTAL | 100 |

SOURCE: *Assessment of Long-Term Research Needs for Coal Gasification Technologies.* MITRE Technical Report, MTR-79W00160. McLean, Va.: The MITRE Corporation, 1979, p. 8.

ogies. It seems clear that current Lurgi technology for noncaking coals will be a major starting point, and that slagging fixed-bed and high-pressure entrained-flow systems will follow over the next decade. For the third wave of developments, the possibilities are much more varied. Underground and advanced fluidized-bed systems are considered likely to become important as the industry develops.

## Gasification Costs

Table 5 provides a breakdown of investment for the components of one gasification facility. (Capital expenditures will account for somewhat more than half of product costs; the precise proportion will depend on raw material costs and financing terms.) Investment in gasification units themselves amounts to only 16 percent of the total. However, the other, larger investments are to a considerable extent set by the gasifier's characteristics. Oxygen production, for example, is a major item that can be eliminated if air is used. But the use of air produces a dilute, low-Btu gas of lower heating value and can increase gas cleanup costs. For intermediate-Btu gas, production using air requires a special gasifier that provides gasification heat by circulating char heated in a separate vessel or transfer line.

Much of the cost of coal gasification, as in other synthetic fuel processes, results from the discharge of waste heat. More than 30 percent of the heat of combustion of the coal is generally lost. (The rest is recovered in the products.) The development of gasification technology will be aimed largely at improving thermal efficiency by reclaiming waste heat and by other means, all of which imply investments in equipment. Each process will have an economic optimum. Typically, a 10 percent improvement in efficiency reduces the product cost by 15 percent.

As fuel costs increase relative to construction costs, the incentive for efficiency grows also. It can be anticipated that energy losses, with continued development, can be decreased to 15 to 20 percent of the feed heating value, with a corresponding decrease in product cost, resource consumption, and discharge of wastes.

## Gas Cleanup

Raw gas leaving the gasifier contains impurities, such as tar, ammonia, hydrogen sulfide, and carbon dioxide, that must be removed. If the gas is to be converted to other fuels or chemical products, its impurity content must be reduced drastically. Fortunately, gas processing in petroleum refining requires the same separations and is a highly developed commercial technology. However, some separations, such as particulate removal and tar and phenol recovery, are more severe with coal.

Substantial energy losses occur during cleanup, and efforts to find more efficient processes are under way. If the gas is to be used in a nearby power plant or furnace, for example, a great deal of energy could be saved if the gas could be cleaned at the gasification temperature, which is generally very high. The hot gases could then be used directly in advanced, combined-cycle power systems. The use of lime to capture sulfur during gasification is one approach being studied. Dust removal at high temperatures cannot depend on the electrostatic precipitators and bag filters of normal industrial practice. High-temperature cyclones provide one option.

## Gas Conversion

A large fraction of intermediate-Btu gas will be converted to other products. The cost of conversion is generally much less than the cost of producing the original gas. For example, the additional facilities required for converting the gas to methane add only about 10 to 20 percent to the plant investment. This step, however, decreases the thermal efficiency of the total process slightly, because of the heat released when carbon monoxide and hydrogen react to form methane and water.

***Hydrogen*** Hydrogen production will be one of the major activities in coal conversion processes. The manufacture and use of hydrogen accounts for a major fraction of the cost and energy loss in coal liquefaction and in refining coal and shale liquids. Manufactured from natural gas and petroleum residuum, hydrogen is at present produced in large quantities for ammonia manufacture and, to an increasing extent, for petroleum refining. In the future, the increased costs and limited supplies of natural gas and petroleum in this country will cause a shift to gasified coal and char as hydrogen sources.

The carbon monoxide-hydrogen mixture is reacted with steam in the presence of a metallic iron catalyst, producing the water gas shift reaction

$$CO + H_2O \longrightarrow CO_2 + H_2$$

This mixture is cooled to remove the carbon dioxide; the remaining carbon monoxide is removed by catalytically combining it with hydrogen to form methane. This methanation reaction consumes hydrogen, dilutes the resulting gas with methane, and generates heat. The result of these temperature and composition changes is to reject heat to the biosphere and to reduce hydrogen yield. While this technology is highly developed, additional heat conservation and catalyst improvements can be expected because of the very large use of hydrogen throughout the world.

Because of hydrogen's key role in synthetic fuels, the search for better sources should be emphasized. In the long term, heat from nuclear or solar energy and electricity could supply hydrogen directly from water without the use of carbon. The continued search for major improvements in gasification technology specifically designed for hydrogen production is important.

***Methane Production*** High-Btu gas (methane) production uses the same sequence of reactions as does hydrogen production. The water gas shift is still needed, since methane synthesis requires a hydrogen to carbon monoxide ratio higher than that of the gas from high-temperature gasifiers. The methanation reaction liberates a large amount of heat, and very large reactors that can better deal with this heat release are needed. The heat of combustion of the methane produced is less than that of the carbon monoxide-hydrogen mixture used in its manufacture. This loss of useful energy accounts for the high cost and low efficiency of high-Btu gas manufacture relative to low- or intermediate-Btu gas production. The loss, however, can be reduced by maximizing methane production during gasification. Here, the heat released by methanation counterbalances the heat required for the steam-carbon reaction. As the gasification temperature is lowered, increasing amounts of methane are formed; while high-temperature slagging gasifiers might be chosen for hydrogen production, the low-temperature dry-ash systems are most appropriate for methane manufacture.

Unfortunately, as the temperature is lowered, reaction rates decrease. Use of catalysts such as potassium carbonate can permit lower temperatures and increased methane yields. This is a subject of considerable R&D interest.

***Synthesis of Liquid Fuels*** While methane production from carbon monoxide-hydrogen mixtures is chemically the most favored synthesis reaction, special catalysts and process conditions can produce a wide variety of useful liquid products. Methanol, for example, is now manufactured in large quantities from synthesis gas derived from natural gas. Suppression of methane formation has been a major goal in catalyst research and process optimization efforts. The need for suppressing methane output is obvious when methane in the form of natural gas is the starting point. However, production of methane as a byproduct of coal liquefaction is more acceptable. Acceptance of methane as a valuable byproduct can lead to simpler and more economical methanol production. Production of methanol offers an efficiency advantage over hydrocarbon production; less heat is liberated in the synthesis step. Because of methanol's great potential as a fuel, important advances in its production from solid fuels can be anticipated.

Liquid hydrocarbon fuels can also be produced from synthesis gas. This catalytic process was discovered in 1925 by Franz Fischer and Hans Tropsch in Germany; it was used for part of that nation's synthetic fuel production during World War II. A 7,000-barrel-per-day American version was built in Brownsville, Texas, by a consortium of companies (Carthage Hydrocol, Inc.) using natural gas as a source of synthesis gas. Increases in U.S. gas prices and importation of low-cost petroleum from the Middle East brought this operation to an end. Little research on the process was done in this country during the era of cheap imported petroleum, but the process, using coal, has been further developed and used in South Africa. South African production of liquid hydrocarbon fuels is being expanded to supply a major part of that nation's gasoline requirements. The increasing cost of imported petroleum has revived research and commercial interest in the United States also.

While the Fischer-Tropsch process can be criticized for low overall thermal efficiency and for producing a very wide range of products—from methane to hard wax—it appears to offer considerable potential for improvement. Its broad product distribution is exemplified in Table 6. Since the major objective of the operation illustrated in the table was gasoline production, the large outputs of both lighter and heavier products demonstrate the unselective nature of the process.

The gasoline product tends to have a low octane number, and maximizing gasoline output results in a large output of lighter hydrocarbons. The Fischer-Tropsch process may be better adapted to the production of high-hydrogen-content diesel and jet fuels. In the United States, fuel demand projections indicate a growing need for these mid-distillates and a decreasing demand for gasoline. Reoptimization of this process for mid-distillate production calls for advances in catalysis. Both mid-distillates and gasoline production are the subject of active current synthetic fuels research, and important advances can be anticipated. Of special

**Table 6** Product yield of Fischer-Tropsch process

| Product | Percent by weight |
|---|---|
| Methane | 11 |
| Liquefied petroleum gas | 23 |
| Gasoline | 40 |
| Diesel fuel | 11 |
| High-boiling fractions | 4 |
| Oxygenated compounds | 11 |
| TOTAL | 100 |

SOURCE: Adapted from J. C. Hoogendoorn, "Experience with Fischer-Tropsch Synthesis at Sasol," *Clean Fuels from Coal Symposium*—1. Chicago: Institute of Gas Technology, 1973, pp. 353-365.

interest is the search for catalysts that allow more selectivity in the choice of products. Progress toward this goal is being made in laboratory studies.

Synthesis gas conversion allows selective production of the single-carbon species, methane and methanol. By first producing methanol, and then catalytically converting it to hydrocarbons over a molecular sieve catalyst recently discovered by Mobil, it is possible to obtain a high yield of high-octane gasoline. The first commercial application of this process is being planned in New Zealand, where natural gas will be converted to gasoline. It is anticipated that the use of methanol as a starting point for fuels or for fuel components, such as ethers and esters, will supplement methanol's direct use as a fuel.

## LIQUEFACTION

The problem of producing good-quality liquid fuels from solids is primarily one of increasing the hydrogen to carbon ratios of the products. As Figure 3 shows, the atomic ratios of hydrogen to carbon for the important transportation fuels are between 1.8 and 2.2, while those of coals fall in the range of 0.5 to 1.1. Shale oil and tars are in the 1.3 to 1.7 range.

The hydrogen content of the fuel can be increased by the rejection of carbon, as in pyrolysis, or by the addition of hydrogen. Both approaches are important. Pyrolysis involves heating the solid fuel to a temperature at which the molecules rearrange to concentrate the hydrogen in gas and tar, leaving behind a char containing relatively little hydrogen. For most coals, a yield of 30 to 40 percent of the heating value of the coal can be recovered in tar and gas, with 60 to 70 percent of the heating value remaining as char. For Colorado oil shale, about 20 percent of the heating value remains as char; this amount is small enough to be consumed as fuel in the pyrolysis process. If the char from coal pyrolysis is gasified and

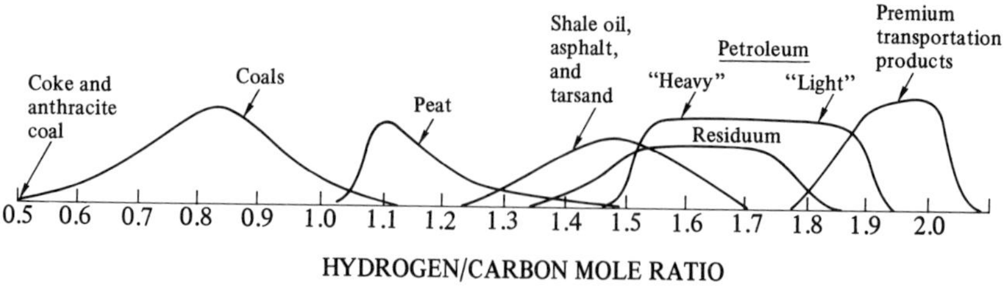


**Figure 3 Hydrogen to carbon ratios for various hydrocarbon sources and end products.** [SOURCE: D. D. Whitehurst. "A Primer on Chemistry and Constitution of Coal," *Organic Chemistry of Coal.* ACS Symposium Series No. 71. Edited by John W. Larsen. Washington, D.C.: American Chemical Society, 1978, p. 13.]

converted to liquids, it can yield 2 to 2.5 barrels of liquid per ton of coal. Adding hydrogen directly to coal produces somewhat higher yields.

## Coal Liquefaction

If the char produced by pyrolysis can be used without cooling, liberation of liquid and gas is efficient, consuming little more heat than such alternatives as hydrogen addition or synthesis from mixtures of hydrogen and carbon monoxide. Since transporting and marketing char do not appear attractive, pyrolysis probably will find its place as an adjunct to gasification or immediate char combustion.

The coproduction of pyrolysis liquids in gasification processes, such as the commercial Lurgi process, is often considered a disadvantage if gas is the only desired product and the liquids produced are sent back to the reactor to be gasified. Similarly, in liquefaction by hydrogenation or by Fischer-Tropsch synthesis, the inevitable production of gas is often considered undesirable, and the gas is burned to supply process heat or is broken down to carbon monoxide and hydrogen for reuse. Since gasification and liquefaction plants will both be needed, it appears more efficient to retain gas and liquids once they are burned.

While economic and environmental considerations may limit production of byproducts, marketing and institutional considerations are also important. A gas company, for example, may not have a ready market for crude pyrolysis liquids, which are difficult to handle and refine; a refiner producing liquid products may wish to avoid the regulatory problems of interstate gas marketing and may lack a suitable local market for gas. These problems are analogous to the problems encountered in industrial cogeneration of heat and electricity. In the case of liquid and gas manufacture, success in dealing with these problems would favor the use of pyrolysis prior to char combustion or gasification.

Liquefaction by hydrogen addition was practiced by the Germans be-

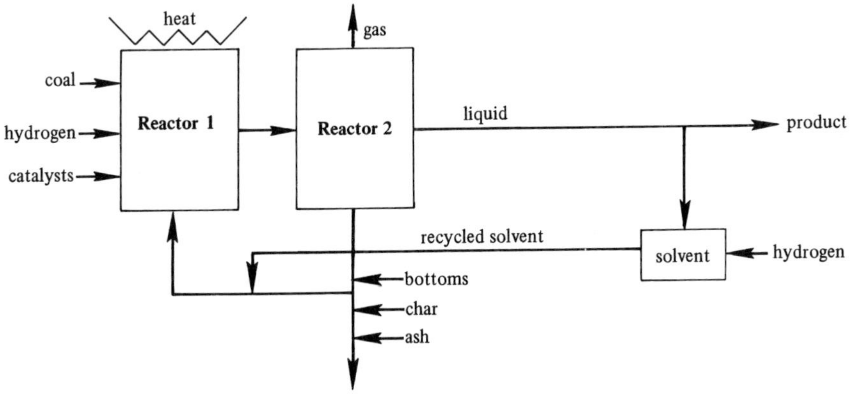


**Figure 4 Major components of coal liquefaction processes.**

fore and during World War II. The very high pressure process used was expensive and consumed large amounts of hydrogen. None of the modern, lower pressure approaches has yet been commercialized, but plans for demonstration plants are being formulated, and many competitive processes are being studied. These processes differ mainly in their approaches to hydrogen management.

Some molecular groups in coal have been found to be able to donate hydrogen to other coal molecules. This occurs during pyrolysis; but if the coal is slowly heated in the presence of a solvent, generally derived from the coal liquids produced, less char and more liquid are recovered. Formation of liquids can be augmented by adding more hydrogen.

If hydrogen is dissolved in coal liquids under high pressure and at a temperature of 800°F to 900°F, some of the coal liquid molecules may, in the presence of a catalyst, react with hydrogen and then donate this hydrogen to the coal to form more liquid. This catalytic solvent hydrogenation can occur either in the coal reactor or externally. Direct reaction of hydrogen with some of the coal molecules is also believed to be important.

Figure 4 is a diagram of the major components of coal liquefaction processes. Pyrolysis is represented by Reactor 1 without the solvent recycle stream and without the addition of hydrogen.

The solvent-refined coal process SRC-I corresponds to Reactor 1 without the hydrogenated solvent recycle stream or the addition of catalysts. This process produces a tar, solid at room temperature, from which most of the mineral matter has been removed. The tar is suitable as a boiler fuel and can be converted to lighter products by additional hydrogenation.

The SRC-II process also recycles the bottoms from the separation process to yield a distillable product. The process depends on the natural catalytic activity of the coal ash, and results vary with the ash's composition.

Addition of catalysts can increase the rate of hydrogenation. The H-Coal process, for example, uses a standard hydrotreating catalyst from the petroleum refining industry. Research on catalysts offers major process improvements and is being carried on very actively in a number of organizations. Another approach is to hydrogenate separately a fraction of the liquid product, which can then be sent to the reactor as a hydrogen donor. The Exxon Donor Solvent (EDS) process uses this approach.

The process developments discussed above are perhaps the most advanced, but others are being pursued. The optimum combination of the steps shown in Figure 4 will depend on the coal used and on the product goals selected. Competitive development coupled with continued advances in coal and catalytic science will ultimately determine the best systems. During the next five years, several of the competing approaches will be ready for commercial plant design and construction.

## Shale Oil Extraction

Shale oil, because of its great potential importance, deserves separate discussion. A desirable feature of Green River oil shale (found in a small area at the juncture of Colorado, Utah, and Wyoming) is that simple pyrolysis can recover as liquid and gas over 80 percent of its organic matter. The Devonian shales found east of the Mississippi form much larger quantities of coke on pyrolysis; as with coal, higher liquid and gas yields from these shales require heating with hydrogen at elevated pressure. Since large coal reserves are available in the area where the Devonian shales occur, and since mining, treating, and disposing of the large amounts of stone involved in shale oil extraction there would be costly, Devonian shale is less likely than Green River shale to be commercially important in the near future. However, the ease of obtaining a liquid relatively high in hydrogen content from Green River oil shale makes it an attractive option for the first generation of synthetic fuels.

Table 7 shows a typical cost distribution for producing from oil shale a synthetic crude with the sulfur, nitrogen, and arsenic removed. Retorting and upgrading (mild refining) account for 62 percent of the cost in this case. (Room-and-pillar mining of a rich seam and above-ground retorting are assumed.) Mining and solids disposal are major costs and are also sources of environmental problems. Requirements for increased resources recovery, lower local water consumption, and further reductions in environmental impact might well increase these costs. These factors are discussed later in this chapter.

While room-and-pillar mining followed by surface retorting can be

**Table 7** Cost distribution for shale oil (syncrude)

| Operation | Percent of total cost |
|---|---|
| Solids preparation and disposal plus lease of resource | 8 |
| Mining and crushing | 22 |
| Disposal and revegetation | 8 |
| SUBTOTAL | 38 |
| Retorting | 41 |
| Upgrading | 21 |
| TOTAL | 100 |

SOURCE: Adapted from M. A. Weiss et al. "Shale Oil: Potential Economies of Large-Scale Production, Preliminary Phase." M.I.T. Energy Laboratory Working Paper No. MIT-EL 79-012WP, revised June 1979.

considered the standard technique for early use of shale resources, *in-situ* (underground) retorting as well as surface mining plus retorting are major variations.

Surface mining is a major option for a substantial fraction of the Green River formation, which consists largely of multiple layers of rich shale interspersed with layers of lean stone. Surface mining can recover all of the rich stone and, unlike first-generation room-and-pillar mining or *in-situ* retorting, does not leave behind supporting pillars or walls. Second-generation techniques for using the stone left behind in pillars or barrier walls can increase the resource recovery of underground mines, but surface mining will retain a substantial advantage in this regard. While recovery is very much a function of local conditions, surface mining might be expected to recover around 79 percent of the mined oil content of rock in place. *In-situ* retorting and underground mining might generally fall in the 10-40 percent range, with *in-situ* retorting averaging lower than mining.

***Surface Retorting*** Shale oil retorting has a 600-year history, but modern continuous retorts date back only to about 1950. Since that time, shale oil production in the United States has periodically been considered on the verge of commercialization, and a number of retorts have been developed by various combinations of government and industry. While none has been tested on a commercial scale, several are considered ready for design and construction of commercial plants. The major classes of retorts are illustrated in Figure 5.

The simplest system—the Paraho direct heating system—uses vertical lime kiln technology. Stone moves through the vertical kiln countercurrently to air; heat is generated by combustion of the char remaining after pyrolysis. The oil and gas leave together at the opposite end of the

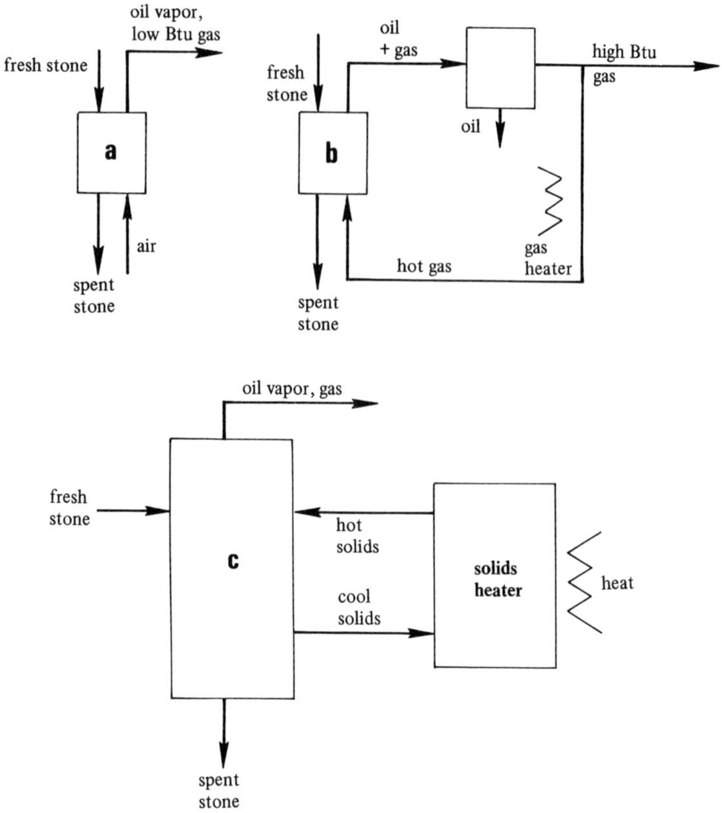


**Figure 5 Major classes of shale retort.** [SOURCE: Adapted from H. C. Hottel and J. B. Howard. *New Energy Technology.* Cambridge, Mass: M.I.T. Press, 1971, p. 196.]

kiln (Figure 5a). This system has the disadvantages of producing low-Btu gas and of giving somewhat lower oil yields than do competing systems. It has been tested at approximately 300 barrels per day and has produced more than 100,000 barrels of oil for use by the Departments of Defense and Energy and the Electric Power Research Institute in refining and end-use studies. Construction of a full-size (7,000-barrel-per-day) module is proposed.

An "indirect heating" variation of the system, also tested by Paraho, recycles the heated product gas (Figure 5b). This produces a high-Btu gas and avoids burning part of the oil in the reactor. A similar system has been developed in Brazil and is scheduled for commercialization there. However, that system does not use the residual char to produce heat; thus, it consumes a substantial amount of the product gas for heating. The Union Oil Company has been developing a version of the vertical kiln in which the stone is forced upward by a "rock pump," and products are withdrawn from the bottom. Superior Oil has proposed construction of an 11,000-barrel-per-day moving-grate system that can recover shale liquids and low-Btu gas, as well as other minerals, such as sodium bicarbonate, soda ash, and alumina. The remaining char is burned to supply heat to recirculated inert gas.

Heat can also be supplied by hot solids, as Figure 5c shows. The TOSCO II unit of the TOSCO Corporation recirculates hot ceramic balls which, after separation from the spent stone, are reheated by burned product gases. This unit uses a rotating, inclined cement kiln and can give yields greater than 100 percent Fischer assay.* A 300-barrel-per-day unit has been used to produce 180,000 barrels of shale oil. The TOSCO Colony project proposes construction of 8,000-barrel-per-day units for a pioneer commercial operation. Since the typical commercial proposal is for a production rate of 50,000 barrels per day, many such units would be required.

An ideal retort would have an oil recovery rate of more than 100 percent Fischer assay, would recover all the gas as marketable high-Btu gas, would make use of char combustion heat, could be scaled to larger capacities, would process fine shale particles, and would produce an easily handled spent stone. None of the first-generation systems meets all of these criteria; substantial improvement in second- or third-generation retorts seems possible. The use of more finely divided fluidized solids, which has been studied on a small scale, offers an interesting approach. While the first-generation mining and retorting systems could be improved considerably, they can probably be scaled to commercial operations by conservative design practices.

***In-Situ Retorting*** There are several approaches to *in-situ* retorting. The modified *in-situ* technique, considered nearest to commercial readiness, involves mining approximately 20 to 40 percent of the stone from the retort zone. For the other 60 to 80 percent, explosives break up stone to fill the zone with a vertical bed of rubble through which air and product gases can pass. The stone removed is retorted and disposed of aboveground. A substantial fraction of the seam being exploited must be left as walls to contain the *in-situ* retort. After preparation of the retort, air is passed downward through the broken rock, the top of the zone is ignited, and a combustion wave moves slowly downward, heating the rock and consuming the char left after pyrolysis as well as the oxygen in the air. The product gas is low-Btu, useful for local steam and electricity genera-

*The Fischer assay test method is a carefully controlled pyrolysis procedure used to rate oil shale retorting processes.

Construction at a modified *in-situ* oil shale tract in northwestern Colorado. [Occidental Petroleum Corporation.]

tion. The liquids produced are collected at the bottom of the retort. These liquid yields are lower than those of surface retorts, where there is more control of retorting conditions; they may approach 60 percent of the retorted stone's oil content. This process has been brought to the point of

**Table 8** Sample compositions of various crudes and typical transportation fuels

| Characteristic | SRC-I | SRC-II | Paraho shale oil | Light Arabian petroleum | Transportation fuels |
|---|---|---|---|---|---|
| Atomic hydrogen to carbon ratio | 0.85 | 1.3 | 1.65 | 1.8 | 1.9–2.1 |
| Percent by weight | | | | | |
| Sulfur | 0.9 | 0.3 | 0.5 | 1.7 | 0.01–0.2 |
| Nitrogen | 2.0 | 0.85 | 2.1 | 0.08 | 0.01 |
| Inorganics | 0.2 | 0.004 | 0.01 | 0.0 | 0.0 |
| Percent boiling above 650°F | 95.0+ | 10.0 | 75.0 | 50.0 | 0.0 |

SOURCE: *Refining Synthetic Liquids from Coal and Shale.* Washington, D.C.: National Academy Press, 1980.

consideration for commercial use by the Occidental Petroleum Company. A second *in-situ* technique being researched involves fracturing a block of oil shale without partial mining. The fractured block is then ignited as above.

## Refining Synthetic Liquid Fuels

The crude liquids from coal and shale must be refined into products acceptable for consumer use. Refining involves changing the boiling range of the fuel (as in converting tar to gasoline), removing unwanted elements such as nitrogen and sulfur, and adjusting hydrogen contents to improve combustion characteristics.

Table 8 compares the characteristics of some refinery feeds and products. The range of atomic hydrogen to carbon ratios in the table illustrates the widely varying need for hydrogen additions in refining. Use of heavy fuel oils is expected to decrease and the major liquid products of the future will be transportation fuels (gasoline, diesel, and jet fuel) with hydrogen to carbon ratios of 1.9 to 2.1. In the mid-distillate fuels, this high hydrogen content is needed to reduce smoke and emissions of polycyclic aromatic hydrocarbons and to improve ignition characteristics. Petroleum crudes generally require little or no additional hydrogen. Coal liquids, on the other hand, are quite deficient in hydrogen. Coal liquefaction processes vary considerably in the amounts of hydrogen added during liquefaction. The SRC-I solvent-refined coal process adds the smallest amount and produces a low-hydrogen tar. The SRC-II process, by recycling this tar, adds more hydrogen to reach, for the particular example, a hydrogen to carbon ratio of 1.3, still far below current product requirements. Other processes can produce a lower or higher hydrogen content depending on the product objectives.

Transportation fuels vaporize at temperatures below 650°F for mid-distillates and 430°F for gasoline, as set by freezing-point and combustion requirements. All refinery feeds require some reductions in the sizes of the larger molecules to meet this requirement. This reduction can be accomplished in coal liquefaction largely by recycling the components with higher boiling points and by increasing temperatures and hydrogen pressures. In the refinery, this conversion can be accomplished by passing the feed over a cracking catalyst (catalytic cracking) or by a combination of hydrogenation and cracking (hydrocracking). Catalytic cracking is the major process in petroleum refining and will probably be important in shale oil conversion.

Crude petroleum contains compounds of nickel, vanadium, and iron in greatly varying amounts. These metals tend to remain in the higher boiling fractions and are removed by boiling range conversion or hydrogenation. Undistilled coal liquids also contain titanium, in addition to the metals found in petroleum, while shale oil has an unusually high content of arsenic (20 parts per million), which must be removed in refining.

In petroleum refining, metals removal has often been avoided by marketing the high boiling point residual oil, which contains the metals, for power generation and industrial heating. Pyrolysis of these residual oils is also practiced, in which case the metals remain in the coke formed. To an increasing extent, these practices will be replaced by special catalytic hydrocracking processes in which the metals are deposited on the catalyst and at the same time the boiling range is reduced and some nitrogen and sulfur are removed. Such processes could be adapted to the conversion of coal liquefaction products.

Arsenic removal from shale oil is a special case. Arsenic poisons the catalysts used in refining and must be removed early in the refining process. First-generation processes that involve depositing the arsenic on a low-cost catalyst have been announced.

Nitrogen and sulfur can both be removed by catalytic hydrotreating of the oil. Sulfur removal is a highly developed technology; however, the removal of nitrogen with current technology consumes large amounts of hydrogen, because the conditions required for nitrogen removal result in hydrogenation also of the low-hydrogen (aromatic) components of the fuel. While this results in high-quality mid-distillate products, much of this hydrogen must be later removed in gasoline manufacture to increase octane ratings. Research is needed to improve this process.

The hydrogen contents of mid-distillate fuels are set by the abilities of combustion devices to burn the fuels without forming soot and polycyclic aromatic hydrocarbons. One goal of combustion R&D is clean combustion of low-hydrogen fuels. Similarly, combustion systems capable of

**Table 9** Conversion of biomass to synfuels, with coal conversion for comparison

| Feed | Process | Product | Ratio of product heating value[a] to dry fuel value of raw material | Holding time |
|---|---|---|---|---|
| Corn | Aerobic fermentation | Ethanol | 0.5-0.6 | 2 days |
| Coal or biomass | Gasification, liquefaction | Methane or methanol | 1.0 | 1-2 hours |
| Wood | Anaerobic fermentation | Methane | 0.5 | 2 weeks |
| Coal or biomass | Gasification, methanation | Methane | 1.0 | 1-2 hours |
| Wood | Pyrolysis | Tar and gas | 0.7-0.8 | 1 minute |
| Coal | Pyrolysis | Tar and gas | 0.2-0.4 | 1 minute |

[a] Energy requirement of processes not counted.

burning high-nitrogen fuels with acceptable nitrogen oxide emissions will greatly reduce hydrogen consumption. Strong interaction between fuel qualities and the changing requirements of fuel-using devices is important in minimizing the overall cost and energy consumption of fuel manufacture and use.

## BIOMASS CONVERSION BY FERMENTATION

Solid plant materials can be gasified or liquefied by much the same methods as those used with coal. The additional alternatives of producing ethanol or methane by fermentation have been known and practiced for many years.

Table 9 compares several conversion options for biomass. Fermentation of sugars and carbohydrates by yeast to make ethanol is an established commercial process of high selectivity for the fermentable components of corn, and 50 to 60 percent of the heating value of the corn can be converted to ethanol with a fermentation time of around two days. The solid residue from the fermentation of some grains is useful as a livestock feed supplement.

Woody materials must first be hydrolyzed to produce fermentable substances. Lignin and some other components leave a residue that must be dried for burning or other use. The process can be compared with methanol production by coal gasification, followed by catalytic conversion of the gas to methane and then methanol. Here the product heating value can be about equal to or somewhat higher than the heating value of the coal fed to the gasification reactor. This depends on how heat is supplied to the carbon-steam reaction, since both fermentation and gasification require heat. In the case of fermentation, the heat requirement is primarily for distillation of the 10 to 14 percent alcohol-water solution produced and for drying the unconverted solids. The heat can be provided by a solid fuel, such as agricultural waste or coal, and is approximately equal to the heating value of the ethanol produced. For methanol production via gasification of coal or biomass, 75 to 100 percent of the heat of combustion of the products must be supplied to the process, generally by burning some of the feed.

Methane can be produced by anaerobic fermentation of organic matter, a process widely used for sewage treatment. Sugars and carbohydrates are easily converted, producing a 50/50 mixture of methane and carbon dioxide. Proteins and lipids produce less carbon dioxide. Since lignin is not effectively converted, wood and other whole plants do well to yield 50 percent of their heating value as methane. The remaining solids can be dried and burned. Drying the sludge, if necessary, is a major consumer of energy.

Gasification of dried biomass is similar to coal gasification, although development of special gasification equipment will be needed. In methane production by gasification, there are no solid fuel byproducts, and essentially all of the feed heat of combustion can be converted to gaseous fuel. Energy amounting to approximately 50 percent of the product heat of combustion must be supplied, however.

Wood responds very well to pyrolysis, with the potential of 70 to 80 percent of its heating value going to liquid and gas. The gas formed can be of good quality, but the liquids are corrosive low-sulfur tars best adapted to direct use as fuel oil. The char remaining can be used to supply the energy for the process. Coal does not respond as well to pyrolysis, yielding only 30 to 40 percent of its heating value in tar and gas. The char remaining is in excess of process heat requirements.

As shown in Table 9, fermentation is slow compared with the alternative high-temperature processes. However, simple tanks and ponds are adequate, in contrast to the high-temperature, high-pressure vessels necessary for competing processes. For large-scale centralized production, the high-temperature processes will probably be more economical and efficient. In the smaller scale operations appropriate to biomass, fermentation becomes competitive and is probably superior for very small scale operations where simplicity is a major requirement. These fermentation processes can be considerably improved through continued R&D; they are expected to find their place in conjunction with energy farming and waste disposal.

## CONVERSION TECHNOLOGY SUMMARY

The foregoing discussion has pointed out the availability of technology for the first-generation conversion processes of major interest. In the case of coal gasification, commercially demonstrated processes are available for converting noncaking coals to low-Btu gas, and both caking and noncaking coals to intermediate-Btu gas. Intermediate-Btu gas is used commercially, here and abroad, as a hydrogen, methane, ammonia, methanol, and gasoline feedstock. The Mobil process for converting methanol to gasoline is approaching the point of commercialization, and production of gasoline from mixtures of carbon monoxide and hydrogen is practiced commercially in South Africa. Production of intermediate-Btu gas from coal or lignite offers strong enough incentives that commercialization requiring no subsidy or loan guarantees is being considered for the Gulf coast area.

Oil shale mining and retorting, while not commercially demonstrated with modern equipment, can be considered to have been carried through the large pilot plant stage and to be ready for pioneer commercialization. Here again, current economics indicate commercial viability, and plans to proceed with commercialization have been announced.

The commercialization status of direct coal liquefaction is less advanced. The cost of producing refined liquid products from coal is believed to be about 10 to 40 percent higher than the cost of producing them from oil shale. There are no current plans for commercial direct liquefaction without substantial government financing. Large pilot plant programs are under way, and demonstration plants with significant government support are planned.

Commercialization of direct coal liquefaction will require establishment of a new industrial activity; the financial risk is high. Technological risks involve the possibilities of encountering lower than expected yields and service factors or higher than planned maintenance; the likelihood of complete inoperability is remote. Pioneer plants designed to minimize these risks will generally be overdesigned, with costs correspondingly higher than those of subsequent plants. Costs can be expected to fall fairly rapidly, even for essentially replicated plants.

Significant improvements can be expected from second- and third-generation technology. For gasification to produce carbon monoxide-hydrogen mixtures, second-generation systems at the point of commercial demonstration, such as the slagging Lurgi and high-pressure entrained-flow systems, can reduce costs by 10 to 15 percent from those of current systems. With further research, development, and optimization, additional improvements of comparable or greater magnitude can be anticipated through techniques designed to meet specific process objectives, such as methods for producing methane during gasification.

In the production of methanol or liquid hydrocarbons from carbon monoxide-hydrogen mixtures, reducing methane yields (typically 40 percent of the feed's heating value) can further lower the cost and improve the efficiency of liquid fuel production. Here, improved catalysts appear to offer major opportunities. Similar improvements (of perhaps 30 percent) can be expected from second- and third-generation systems for shale oil production and direct coal liquefaction.

In a major synthetic fuels industry with very large investment and operating costs, even small efficiency and cost improvements would be important; an overall 30 percent saving would justify a large and sustained R&D program, including parallel development of competing technologies.

Developing a new solids conversion process is very expensive. Typical costs (in 1980 dollars) for developing a new process, beginning with a pilot plant and progressing through a demonstration plant to the point of designing a true commercial operation, are roughly $2 billion. If it is possible to omit the demonstration plant and use just a large pilot plant, the cost may be $500 million. Without the large pilot plant, a process might be developed for $50 to $100 million. The demonstration plant can be considered to be a small commercial plant, differing from a true commercial plant mainly in that its financial support does not require as high a probability of meeting cost and production goals. This very expensive route appears justified only where the ability to scale up from large pilot plant operations is in serious question. Both approaches are being taken in coal liquefaction: a pilot plant and a demonstration plant for solvent-refined coals, and the large pilot plant approach for Exxon Donor Solvent and H-Coal processes.

Major process modifications can be studied in existing pilot plants at reduced cost. However, since the cost of developing new processes or of making major process modifications is very high, a reasonable probability of major industrial use must exist. The high costs and high risks of coal liquefaction development have resulted in cost-sharing for pioneer projects, with the government as a major partner.

As experience is gained from commercial operation, the need for demonstration and large pilot plants will diminish, and smaller pilot plants and lower cost programs will allow individual organizations to develop advances in conversion technology without government funding.

## ENVIRONMENTAL, RESOURCE, AND INDUSTRIAL QUESTIONS

Under the circumstances of declining domestic reserves, establishing a synthetic fuels industry appears obviously desirable. The first steps are being taken; commercial production of alcohol for gasohol and synthetic gas from coal are already under way, and plans for producing liquids from coal and shale are well advanced. The questions of how fast and to what size such an industry should grow will be determined by a set of environmental, resource, and social constraints that are not well understood or predictable at this time, although there are many strongly held opinions. Science and technology can help alleviate or solve many of these major problems.

## Environmental Constraints

***Air Pollution*** The large-scale solids handling and processing activities involved in a synthetic fuels industry have the potential for generating all the currently regulated air pollutants. Emission control will play a major part in the choices of technologies and sites.

The fundamental controls are the National Ambient Air Quality Standards (NAAQS) and the requirement for "prevention of significant deterioration" of air quality in areas that have lower pollutant concentrations than those allowed by the NAAQS. In Class I areas, which include national parks and wilderness areas, air quality must remain virtually unchanged. In Class II areas, some additional air pollution corresponding to moderate industrial growth is allowed. For areas where significant industrial growth is foreseen, states or Indian governing bodies can redesignate some Class II areas to the more liberal Class III. In already industrialized areas, significant deterioration can be avoided by reducing emissions from existing sources to offset new emissions. In most western areas containing coal and shale deposits, there is no established industry to use for offsets. The Colorado shale area, for example, is in Class II; however, the nearby Flat Tops Wilderness Area is in Class I, which under current law must be protected from emissions originating in adjacent Class II areas. Occasionally, natural sources raise hydrocarbon and particulate levels above those set by the National Ambient Air Quality Standards.[5]

Regulations relevant to the production of shale oil have not been fully defined. However, it appears that with the projects, technologies, and regulations tentatively planned for shale oil production, total production might be restricted by air quality regulations to 400,000 barrels per day. If the limitation is not applied, the possible limitation of water supply would need to be addressed. (This is discussed later.) Production at this level would supply regional liquid fuel needs but would make only limited contributions to national energy needs.

Closed systems, such as coal liquefaction or gasification reactors or catalytic conversion units, limit air pollutant emissions to very low levels, since the gases produced can be economically and efficiently cleansed of particles and compounds of sulfur and nitrogen. The combustion equipment used to supply heat, electricity, and steam will produce substantial quantities of these pollutants if coal is burned, and less if clean gas is burned. Gas scrubbing technologies have been developed to meet the requirements of industrial areas, but they cannot meet the requirements for establishing a major industry in a Class II area. Further R&D will probably improve emission control technology.

Another option would be to minimize combustion in critical areas by importing electricity, holding the processing of mined coal or oil shale to a minimum, and choosing technologies that produce relatively small amounts of contaminated gas. A danger is that the first few plants built might preempt allowable emissions, making expansion even with improved technology difficult. Relaxing air quality requirements could, of course, greatly alleviate this problem and will probably be necessary if the western states are to make a major contribution to synthetic fuels production.

Unlike high-quality oil shale, coal and lignite are found in many areas of the country; this obviously alleviates the siting problems imposed by air quality regulations in many parts of the country. In addition, coal can be efficiently transported to industrial locations for further processing to liquid or gas. However, the sheer mass of stone associated with the oil requires that oil shale be retorted on the site of extraction. Unfortunately, production of liquids from coal is estimated to be less efficient and more costly than production of liquids from shale.

***The Carbon Dioxide Problem*** The observed steady increase in the carbon dioxide content of air is attributed to increased use of fossil fuels and to deforestation. Some models of the earth's energy balance predict a rise in temperature of several degrees when the carbon dioxide concentration doubles the present level of 0.03 percent. Models of the world's fossil fuel use and of the natural removal of carbon dioxide from the atmosphere project that this may take place within roughly a century.

Such a large change in the earth's heat balance would result in large climatic alterations, though we cannot predict the exact nature of those alterations. On the conservative assumption that the changes will be undesirable, there is concern about increased use of fossil fuels.

Growing use of coal is of special concern, since coal's low hydrogen to carbon ratio (typically 0.75) implies the release of relatively large amounts of carbon dioxide for a given amount of heat. The best synthetic fossil-derived fuels from this viewpoint are methane and methanol, both of which have hydrogen to carbon ratios of 4. Use of biomass, which merely recycles carbon dioxide drawn from the atmosphere by plant growth, might be especially favored when control of atmospheric carbon dioxide is required. Hydrogen generation from nuclear, solar, or oceanic thermal energy sources is also indicated for the future. This hydrogen could be used as ammonia or added to biomass or fossil fuel carbon to produce liquid fuels.

Minimized use of liquid fuels and maximum use of hydrogen generated from nonfossil sources would significantly reduce carbon dioxide emissions to the atmosphere. The synthetic fuel industry would undergo significant changes under these circumstances. Dwindling petroleum supplies and a continuing need for liquid and gaseous fuels, however, will maintain an important role for this new industry far into the future.

## Constraints of Resources and Financial and Industrial Capacities

***Financial and Industrial Constraints*** Establishing a large synthetic fuel industry will make great demands on the supplies of capital, of construction and manufacturing capacities, and of trained personnel. New capacity of 1 million barrels per day is estimated to require an investment of $55 billion (1980 dollars). This large investment can be put in some perspective by noting that importation of 1 million barrels per day of petroleum at $35 per barrel costs $13 billion per year. With simplification, this investment can be said to be somewhat more than four years' importation. Since imported petroleum prices are expected to continue rising at rates higher than those of general inflation, this ratio can be expected to decrease with time.

One estimate for an 11-million-barrel-per-day synthetic fuel industry attained in 2010[6] indicates that the investment in synthetics would be approximately 30 percent of the total energy investment, and that the total energy investment rate would remain at about 2 percent of gross national product (GNP), the same as for 1976 (assuming 2 percent annual real GNP growth). It is generally believed that these capital requirements, although large, are not beyond the capacities of the private sector.

A large number of new workers will be required for such a large industry. One estimate for a 15-million-barrel-per-day synthetics industry developed over a 30-year period indicates a major direct increase in employment, as shown in Table 10. Employment in manufacturing, transportation, housing, and service industries would also increase. Direct employment would be approximately one million people. Since a population growth of 25 to 50 million is projected over the next 30 years, the number of people entering the labor market should be more than adequate to meet this need. The problem will be one of skills and training. A chronic shortage of experienced and skilled personnel is probable, and special efforts of both private and public sectors would be required to deal with this problem.

**Table 10** Workers needed in a 15-million-barrel-per-day synthetic fuel industry

| Industrial component | Increase over current employment: Number of workers | Percentage |
|---|---|---|
| Mining | 480,000 | 60 |
| Process industries | 390,000 | 55 |
| Construction | 250,000 | 15 |
| Design engineering | 8,400 | 35 |

SOURCE: *The Role of Synthetic Fuels in the United States Energy Future.* Houston, Tex.: Exxon Company, U.S.A., 1980, p. 10.

Municipal growth in lightly populated areas is a more difficult problem, since in many cases there are not adequate bases for rapid growth. In the shale area, for example, with its high resource concentration, the community could grow from the present 100,000 to around one million people at a production rate of 5 million barrels per day. Over 30 years, the growth rate, if constant, would be 8 percent per year. Similar, or worse, problems of high local population growth rates will be found in most western coal-producing areas.

Population growth resulting from large industrial and mining activities in sparsely populated parts of the West will result in major changes in local social and political structures and will inevitably have major and lasting impact on the environment. A significant fraction of the current population of such areas views these changes as undesirable. It can be expected that these problems will have important impacts on the development of the western energy resources.

The conversion of coal to synthetic fuels can be carried out at sites remote from the mines, since the concentrated nature of coal allows it to be transported economically by train, water slurry pipeline, or barge. This can relieve many local problems. It is believed that the growth of transportation facilities can, with adequate planning, keep pace with the growth in coal production, including coal needed for synthetic fuels.

***Water Supply*** On a nationwide basis, and especially in some western states, the ready availability of sufficiently pure and cheap water is becoming limited. Ironically, a large part of the nation's solid fuel resources are in semiarid western regions where agriculture, a major economic activity, is already limited by the supply of low-cost water. A large synthetic fuel and mining industry could become a major user of this scarce resource, and the water supply problem is frequently mentioned as a major growth-limiting factor.

**Table 11** Proposed water use in shale oil production

| | Barrels of water per barrel of oil produced | | |
|---|---|---|---|
| Retorting technology | Paraho direct | Colony (TOSCO) | Modified *in-situ* (Occidental) |
| Mining and handling | 0.35 | 0.44 | 0.14 |
| Power generation | 0.28 | 0.52 | *a* |
| Retorting and upgrading | 1.11 | 2.14 | 1.29 |
| Shale disposal and revegetation | 0.70 | 1.65 | 1.62 |
| Municipal use | 0.28 | 0.25 | 0.35 |
| TOTAL | 2.72 | 5.00 | 2.40 |

[a] Not available.

SOURCE: *An Assessment of Oil Shale Technologies.* Washington, D.C.: Office of Technology Assessment. June, 1980.

Examples of estimated water consumption in the production of shale oil are given in Table 11. In these projects, no water is returned to streams or rivers. About 40 percent is evaporated in retorting, power generation, and refining, and the rest is used for dust control, revegetation, and so on. For mining and handling, there is, of course, less water use (for dust control) in the modified *in-situ* case. In power generation, water is evaporated in a cooling tower to reject heat.

Options for reducing local water consumption include dry heat exchangers for heat rejection, gas turbine power generation, or use of purchased power generated offsite. Heat rejection in retorting and upgrading is large, but retorting actually produces some water from shale. Local consumption for upgrading could be eliminated by transporting the crude shale oil to another area for refining. A large variation is seen in the use of water for shale disposal, depending on whether all the spent shale is wetted (Colony) or whether a large part of it is compacted in dry form (Paraho). Municipal consumption is relatively small and can be held to a low level by recycling municipal water for process use. By carrying out power generation and refining outside the local water supply area, and conserving water in the remaining operations, water consumption can probably be held to around 1 barrel per barrel of oil produced.

At an average water consumption of 3.5 barrels per barrel of oil, water rights held by potential developers could support an industry of 8 million barrels per day, and it is estimated that surplus water adequate for another 2 million barrels per day may also be available.[7] Also, the large quantity of brackish ground water in mine areas could be used, at increased costs for purification. The difficulty is not so much the amounts of water physically present as the allocation of water to the various users.

Since wastewater need not be discharged to surface streams and rivers, this potential source of water pollution need not be a problem. Runoff and drainage water from mines or disposal areas would, in general, be highly saline and might contain boron or other toxic elements. It is possible to contain this runoff by carefully controlled operations. Monitoring and control of the problem on a total watershed basis will probably be necessary.

Other geographical areas offer different sets of problems and other possibilities for dealing with the problems. An example is the choice of water slurry pipelining of coal versus rail transportation or pipelining of gas or liquid products. Because the cost of water is a modest fraction of the cost of synthetic fuel, water conservation technologies like desalination and dry cooling, while too expensive where water is cheap and plentiful, can be used in synthetic fuel production without significantly increasing product costs.

## RESEARCH AND DEVELOPMENT NEEDS

Past support of research and development on synthetic fuels production has had two major components. One is the relatively steady activity in basic chemical and physical research and applied R&D in support of the chemical and petroleum industries. The other is activity specifically aimed at coal, shale, and biomass conversion. Historically, the growth of the latter fluctuating component has occurred when future petroleum supplies appeared in doubt. With major petroleum discoveries—like the East Texas oil field and the very large Middle Eastern oil fields—government and industry support of work specifically oriented to synthetic fuels diminished to low levels. A few special situations—such as the need for coal gasification and coal-to-gasoline conversion in South Africa, and the near commercial viability of oil shale production—maintained moderate but sustained efforts in these fields.

Evidence is strong that petroleum shortage is really here this time, and government, industry, and university research establishments are becoming increasingly involved in work specific to synthetic fuels.

In the past, technology goals were largely determined by economic considerations of conversion efficiency and cost. While these remain the primary criteria, greatly increased requirements—for resource conservation, future land and water use considerations, greatly minimized environmental change, and consistency with local social values—place frequently conflicting demands on synthetic fuels production. New technologies to deal with these problems are needed. Also needed is better basic and quantitative understanding of the chemical and physical processes involved, so that optimal responses to these conflicting requirements are possible.

Technology and research needs cover activities ranging from basic research on the compositions and chemistry of solid fuels to development and demonstration of new mining and handling technologies.

## Mining and Handling Technologies

Mining technology will be challenged by demands for higher levels of resource recovery; for more controlled interaction with aquifers; for predictable and controlled subsidence; for improved waste disposal; and for increased safety, health, and productivity of mine workers. These needs, coupled with a severe shortage of trained personnel, indicate a need for stronger educational and research programs. Basic research on the mechanics of solids and other relevant engineering sciences in the universities is important.

Much that is new in mining technology requires large-scale testing and demonstration. Advances continue to be made, but accelerating the advance may require more sharing of costs and risks by industry and government.

Especially in the mining of shale, the cost of processing the solids involved is large. Open pit and underground mining with high resource recovery rates will call for improved beneficiation techniques whereby the desired organic matter can be separated in higher concentrations. Such a technology, developed enough to make it economically feasible to transport enriched solids to distant refining centers, could relieve much of the environmental and social stress that will otherwise tend to limit shale production.

Solids handling on an unprecedented scale calls for improvements in explosive fracturing, grinding, dust-free transportation, and injection into and removal from high-pressure process equipment. Most of the cost, downtime, and leakage of pollutants in synthetic fuel plants result from solids handling problems. Here, the engineering sciences of solid and fluid mechanics can make important contributions.

## Conversion and Purification

The heart of any synthetic fuel process is the accomplishment of the chemical changes required and the purification of products and effluents. Energy consumption and thermal efficiency are far from their theoretical limits, and advances in the chemistry and engineering of these processes will bring major savings.

There are three general ways to improve a chemical conversion process: improve the selectivity of conversion to the desired product; combine reactions or separations to reduce complexity and energy degradation; and increase reaction rates to reduce equipment size. An example involving all of these approaches is the production of high-Btu gas (methane) from steam and coal. As the reaction temperature is lowered, the conversion selectivity to methane increases. This reduces the cost of the subsequent separation and conversion of the hydrogen and carbon monoxide also produced. Of equal or greater importance is the balancing of heat consumption in the steam-carbon reaction by the heat evolved in the formation of methane. If these processes are carried out separately, heat must be added in one reactor and removed from another, with inevitable loss of useful energy. Unfortunately, as the temperature is lowered, the rate of reaction diminishes to the point of impracticability. The reaction rate can be increased by catalysts, and, while some catalysts have shown considerable promise, there is much room for improvement.

Many other examples could be cited; all of them depend on improved catalysts for major advances. A partial list follows:

- Low-temperature gasification to increase methane production.
- Catalysis of coal liquefaction.
- Reduced methane production in methanol manufacture and in liquefaction.
- Better control of molecular weight in indirect (Fischer-Tropsch) liquefaction.
- Reduced hydrogen consumption in nitrogen and sulfur removal.
- Reduced formation of methane and coke in boiling-range reduction processes.
- Increased and more rapid conversion of biomass to useful products in fermentation processes.

New catalysts for such applications have traditionally come from industrial research laboratories. However, advances in techniques for the study of the physics and chemistry of surface reactions are leading to greater involvement by university and government laboratories all over the world. The sum of these separate efforts may accelerate progress.

Study of the compositions of solid fuels and their synthetic fuel products, and of the chemistry and biological activities of their constituents, is an essential component of research for improving conversion processes. Even the general chemical structure of coal is still a matter of debate, and much remains to be learned of the natures of other solid synthetic fuel raw materials. In addition, liquefaction products from coal and oil shale contain very large numbers of structures and compounds that are not generally known. Full descriptions of their compositions, by groupings of compounds with similar properties, would be useful to chemists and chemical engineers working on synthetic fuels. Modern analytical tools and computer capabilities offer promise for fruitful collaboration in the study of the reactions of these complex mixtures (see Chapter 11, "Chemical Synthesis of New Materials").

It is known that the process of thermal or hydrogenative conversion of solid fuels to liquids can produce polynuclear aromatics capable of producing skin cancer. These compounds are also found in the higher boiling products from petroleum cracking. Such oils have been restricted to use as industrial fuel or in carbon black manufacture, where they can be handled under carefully controlled conditions. Products such as domestic heating oil and diesel fuel have been protected, to a large extent, from these materials by their boiling-point specifications, which tend to exclude these higher boiling materials. Knowledge of these compounds and their biological activities is important, since synthetic fuels can contain a wider variety of active compounds than petroleum, and control by hydrogenation or separation is expensive and energy intensive. Also of great importance is the study of other possible health effects, such as the toxicity of methanol and gasoline components.

Large amounts of the investment and energy consumption in conversion and refining are due to the characteristics of the equipment used to separate such impurities as hydrogen sulfide, ammonia, and carbon dioxide from gases. Distillation and extraction of liquid products to meet consumer requirements, desalination and purification of water, flue gas desulfurization, and removal of nitrogen oxides and particulates are other examples. New demands for completeness of separation and for reduced cost and complexity require renewed chemical engineering research activity in this field.

Synthetic liquid fuels will have compositions different from the fuels produced in the era of plentiful petroleum. End-use equipment must also change to meet higher standards for efficiency, safety, and emissions. It does not seem reasonable to expect that the optimal combination of end-use equipment and fuels will result in exact imitation of the fuels of the past. Arriving at a new optimum will require strong interaction between fuels research and the development of engines and other combustion devices. One important example is in determining the optimal hydrogen contents of jet fuel, diesel fuel, and domestic heating oil. Current combustion devices produce soot, which increases in quantity as the hydrogen content of fuel is lowered. Coal liquids are deficient in hydrogen, and the lowest cost processing of shale and petroleum liquids will also produce products lower in hydrogen than those currently used. Soot control must be emphasized in combustion design so that hydrogen consumption in synthetic fuel production can be lowered. Facilities for the production of experimental fuels will be needed as a part of the effort to optimize the system of fuels and combustion equipment.

Government, industry, and universities are all needed in this effort. University and government laboratories can probably make their best contribution in basic sciences and engineering. Industry is best suited to competitive applied science, engineering, and development. The cost and risk of developing new mining, solid fuel conversion, and emission control processes are high, and joint government-industry participation in funding some major developments, as well as strong support for university research and training in the applicable disciplines, may well be needed.

# Summary and Outlook

Production of synthetic fuels from coal, oil shale, peat, and biomass can draw on a number of commercial technologies. Modern first-generation technologies for coal gasification and conversion of coal gas to methanol and gasoline, and for fermentation of grain to ethanol, are all in commercial use in the United States and abroad. Oil shale mining and retorting have been practiced on large enough scales that pioneer commercial plants could be constructed and operated with acceptable technological risk. Large pilot plant studies of direct coal liquefaction processes are under way, and demonstration plants are planned for other such processes. In addition, much technology from the petroleum refining and chemical processing industries is applicable to synthetic fuels development.

Plans for several commercial synthetic fuel projects have been announced, and the Synthetic Fuels Corporation established by the Energy Security Act of 1980 may stimulate additional commercial application of a variety of synthetic fuel processes. The major first-generation systems will probably have been demonstrated by 1990. Substantial improvements in cost and efficiency will follow as second-generation processes continue to be developed.

## THE CHOICE OF FUELS

The choice of synthetic fuel products will be made on the basis of such considerations as cost and efficiency of conversion; pollution potential; and ease of storage, handling, and combustion. Synthetic liquids and gases will both be demanded. At first, synthetic fuel products will mirror as closely as possible the natural gas and petroleum products they will replace. Over the years, however, the characteristics of fuels and combustion equipment will undergo mutual accommodations. Hydrogenation, for example, is the largest cost item in producing high-quality liquid fuels from coal, and as coal-derived synthetic fuels become more important there will be increasing incentives to adapt engines and other combustion equipment to burn lower hydrogen fuels.

Intermediate-Btu synthetic gas (synthesis gas), with a heat content about one third that of natural gas, will be important because of its relatively low cost, its usefulness as a fuel for local consumption, and its versatility in the production of other fuels and chemicals. Methane (equivalent to natural gas) will be manufactured from intermediate-Btu gas to maintain gas supplies for residential and commercial heating.

Methanol can be efficiently produced from synthesis gas and is potentially a very fine fuel, with the promise of low pollution and very efficient combustion with equipment specially designed for its use. Synthetic gasoline, diesel and jet fuel, and heating oil present greater pollution control problems than do alcohols and are somewhat more costly, but they have the advantage of being adapted to existing commercial combustion technology and will be the major liquid synthetic fuel products for the foreseeable future.

## THE AVAILABLE AND PROSPECTIVE TECHNOLOGY

### Gasification

The dominant commercial coal gasifiers are fixed-bed systems in which the coal moves downward through the reactor in a solid plug, is gasified, and exits as ash, with the gas collected at the top. The Lurgi reactor is such a gasifier. The current restrictions to the use of noncaking coal and low-temperature operation will be removed in second-generation gasification systems during the next decade. These systems include the high-temperature British Gas/Lurgi, the Shell-Koppers system, and the Texaco system, which all involve high-temperature combustion and molten ash removal.

A third generation of gasifiers—using technologies such as fluidized-bed reactors and gasification catalysts—should be ready for commercialization toward the end of this decade.

### Liquefaction

The problem of obtaining high-quality liquid fuels from solid fuels is primarily one of raising the hydrogen to carbon ratio. Carbon can be rejected as char (as in pyrolysis), or hydrogen can be added (as in the dominant direct coal liquefaction systems). In the first generation of synthetic liquid fuel processes, pyrolysis will probably be restricted to use with oil shale and perhaps biomass. Coal's hydrogen to carbon ratio is too low for pyrolysis to produce a high rate of conversion to liquids, and high-conversion processes in which hydrogen, produced separately, is "donated" to the coal receive the most attention in today's development plans. Catalysts can be used to promote hydrogenation. Research on catalysts offers great opportunities for process improvements, as does work on methods for more efficient hydrogen production.

### Shale Oil Extraction

Green River oil shale, found in one small area of Colorado, Utah, and Wyoming, yields liquids relatively high in hydrogen and therefore less needful of costly hydrogenation than coal liquids. This oil shale yields more than 80 percent of its organic matter (as liquids and gases) to simple pyrolysis. Aboveground retorting technology is regarded as ready for commercial use without the intervening steps of demonstration and pilot plants. *In-situ* (underground) retorting is possibly an important alternative in favorable geological conditions.

## ESTABLISHING A SYNTHETIC FUEL INDUSTRY

With the huge resources of solid fuels in the United States and the available and emerging technology, this nation could in principle eliminate petroleum imports and even develop a synthetic fuel export trade. However, an array of constraints, ranging from potential shortages of skilled personnel to the requirements of environmental regulations and state development plans, will limit the industry's rate of growth, and the ultimate size of the industry is therefore difficult to predict. Science and technology will play vital roles in reducing many of these difficulties. Others will require attention on a political and institutional level. Industrywide planning to minimize the local environmental and social impacts of development will be necessary to the establishment of a viable industry.

## REFERENCES

1. *U.S. Energy Supply Prospects to 2010.* Washington, D.C.: National Academy of Sciences, 1979.
2. Ibid.
3. World Coal Study. *Coal—Bridge to the Future.* Cambridge, Mass.: Ballinger Publishing Co., 1980.
4. *Assessment of Low- and Intermediate-Btu Gasification of Coal.* Washington, D.C.: National Academy of Sciences, 1977.
5. U.S. Congress, Office of Technology Assessment. *An Assessment of Oil Shale Technology.* Washington, D.C.: U.S. Government Printing Office, 1980.
6. *U.S. Energy Supply Prospects to 2010.* Washington, D.C.: National Research Council, 1979.
7. U.S. Congress, Office of Technology Assessment. Op. cit.

*"This may well be the last car of its kind produced in America."*

# 16

# *Issues in Transportation*

## INTRODUCTION

To observe that transportation is a critical element in the economic fabric of this nation would be simply to note the obvious. To suggest that there is a need for a long-term strategy for establishing and maintaining the health of the transportation system would elicit no significant debate. To conclude that agreement on the specifics of such a strategy has eluded legislators, government administrators, regulators, suppliers of the services, and users of the system would be painfully correct.

The difficulty in arriving at a national strategy for transportation results, in part, from the difficulty of addressing the overall system. The system is most often viewed as an assemblage of modes (that is, the individual types of transportation), with policies, regulations, and incentives constructed specifically for each mode. While recognizing that the impact of modal policies upon the total system is not easily quantified, the fact remains that the total system is much more than the sum of the individual parts. A national transportation strategy requires dealing with modal issues as well as with the complexity of the many intermodal issues in a way that integrates the total system.

While intermodal issues must be treated appropriately, it should be remembered that the value of each of the various modes depends upon its unique capabilities. Each major type of transportation provides a service that is responsive to a particular demand. Airplanes and trucks transport high-unit-value freight when short delivery time is critical; railroads, barges, and ships mainly carry bulk commodities, for which low cost is the dominant consideration; automobiles, buses, and railroads are used for urban passenger travel where flexibility and convenience are the prime objectives of the traveler; and airplanes are used for long distance passenger service where speed is the principal objective. Because of this segmentation of service, the issues that each mode faces are different and the response of each of the modes will depend on the pressures to which it is subjected.

In the decades before the 1970's, many of the key transportation issues arose from the national desire to improve the mobility of people and goods. In the 1970's, issues shifted to the social and environmental impacts of the transportation system, the availability and cost of fuel, and the safety of the system. In the 1980's, additional issues are already emerging: the growing international competition in vehicle manufacturing and freight transport, the need to reexamine the role of transportation in determining the quality of American life, and the consequent need to address the broadened range of economic factors that affect the viability of the U.S. transportation system.

This chapter primarily considers technological changes in transportation and the possible impact of those changes. While it cannot address the many planning, regulatory, and economic development questions that bear on the future of transportation, it is recognized that transportation services and the technology that supports them are affected strongly by internal social and institutional conditions. Settlement patterns and transportation facilities influence each other, as when a circumferential highway built to serve a suburban population acts as a magnet for additional commercial and residential development. Although the regulatory environment has a strong influence upon what is possible, economics is probably the dominant factor. Once installed, the system is not easily, or quickly, or cheaply altered when new imperatives such as fuel scarcity or pollution problems arise. This is partly linked to the costs and partly to the long development times involved in bringing forth new products. The normal life span of the facilities and many of the vehicles used by carriers can range into the decades. Development of new vehicles also can occupy a decade. Radical departures in vehicle design could require the change or abandonment of supporting investments such as highways, fuel transport and storage systems, ports, and rail lines. Employing new facilities that duplicate or replace still usable national networks could be beyond the capacity of the economy to manage. Thus, much of the new technology will be introduced only gradually, as elements of the existing system live out their useful life.

Just as substantial changes in fuel price or supply can cause a shift in demand that strains the transportation system, so can changes in the needs of those using transport facilities. For example, U.S. coal exports are now limited by the dock facilities available to handle the number of coal ships calling in port. In this case, corrective action hinges not only on the development times but also on the course of U.S. strategy for energy development.

Within the broader social context, uncertainties about such fundamental questions as regulatory policy, pollution control, and national economic policy tend to depress private-sector spending on research and development. While competitive pressures can promote innovation, its realization is likely to be minimized or delayed in an environment in which the basic conditions of operation cannot be predicted.

In spite of the many constraints and uncertainties, new technologies have been developed and introduced. Higher fuel economy, better safety, and lower pollution levels have been achieved in the operation of cars, trucks, trains, airplanes, and ships. Improved containerization technology has helped to integrate freight transportation; improved freight traffic management and materials handling promise further gains. A continuing commitment to transportation-oriented research can be expected to enable still further improvements to be achieved.

Transportation technology will continue to benefit greatly from research in many other areas. Computer applications, composite material developments, and combustion and heat transfer technology are a few examples of technological understanding developed in other contexts having found important uses in transportation. Satellites provide an important supplement to the usual communications links and are having a profound impact on port facility management and on the safe operation of vessels at sea. The optical scanning techniques developed for many areas, including rail, are being used now by the marine sector. This exchange of technology will undoubtedly expand as more sophisticated transportation systems emerge.

## DEMAND PROJECTIONS

In considering the issues that will confront the system in the years ahead, it is essential to take into account the future demands on its capacity. Changing demographic patterns, slowing population growth rates, possible transfers of services among the various modes, and fuel cost and availability all present major uncertainties in projecting demand.

Argonne National Laboratory has developed projections for passenger miles and freight ton-miles to the year 2000. They are shown in Tables 1 and 2. While these projections are in general agreement with most other current projections, it must be recognized that specific numbers can vary substantially among different projections.[1]

## CAPACITY ISSUES

Whether the capacity of the overall system will be able to accommodate the projected demands depends on several issues:

- The extent to which the existing facilities are saturated.

**Table 1** Passenger miles traveled (in billions)

| Mode | 1975 | 1985 | 1990 | 2000 |
|---|---|---|---|---|
| Highway[a] | 2,369 | 3,247 | 3,637 | 4,400 |
| Air | 178 | 356 | 425 | 585 |
| Rail | 21 | 22 | 22 | 23 |
| TOTAL | 2,568 | 3,625 | 4,084 | 5,008 |

[a] Highway mode combines automobiles, personal light trucks, two-wheel vehicles, recreational vehicles, and buses.

SOURCE: R. E. Knorr and Marianne Miller. *Projections of Direct Energy Consumption by Mode: 1975-2000 Baseline.* ANL/CNSV-4. Argonne, Ill.: Argonne National Laboratory, 1979.

**Table 2** Freight haulage (in billion-ton miles)

| Mode | 1975 | 1985 | 1990 | 2000 |
|---|---|---|---|---|
| Highway[a] | 534 | 820 | 946 | 1,210 |
| Air | 7 | 12 | 15 | 20 |
| Rail | 732 | 1,092 | 1,237 | 1,550 |
| Water | 1,047 | 1,481 | 1,597 | 1,730 |
| Pipeline | 905 | 1,037 | 1,131 | 1,040 |
| TOTAL | 3,225 | 4,442 | 4,926 | 5,550 |

[a] Assumes commercial and government vehicles; excludes personal trucks.

SOURCE: R. E. Knorr and Marianne Miller. *Projections of Direct Energy Consumption by Mode: 1975-2000 Baseline.* ANL/CNSV-4. Argonne, Ill.: Argonne National Laboratory, 1979.

- The extent to which technology can extend the capacity of the existing facilities.
- The extent to which competing demands on the system can be accommodated.
- The ability of the economy to support expansion and proper maintenance of the facilities.

While the response to each of these issues is different for the various transport modes, a serious conflict can arise within a mode when the same facilities are used for both passengers and freight. One need only to attempt to travel the downtown streets of most large urban centers while freight is being delivered to recognize this limitation.

## Highway System

There are 3.9 million miles of highway in the United States, with nearly 3.2 million of them classified as rural. The interstate highway system is almost complete, with just over 42,000 miles in operation. The ability of all jurisdictions to maintain this network is of great concern. Repair costs have escalated but revenues for repairs, generally derived from state fuel taxes, have remained constant or even declined. As the efficiency of travel increases, revenues will decrease further unless tax rates are raised or other sources of revenue are found.

Much of the present urban highway system is near saturation. One possible means of increasing vehicle flow, that is, the usable capacity of the highway network, is to improve traffic control. While present control technology can improve traffic flow at low vehicle concentrations, it has been largely unsuccessful at high concentrations. However, as microprocessors are introduced onboard the individual vehicle for engine control, it should be possible to use these computers to couple the vehicle into a local traffic control system to provide an improved demand-actuated system. In the longer term, a more central control, as opposed to local intersection control, would enable the system to adapt to changing traffic conditions and optimize flow. In this circumstance, the driver must be informed continuously of optimum routes. The development of better traffic control systems depends on improving vehicle sensors and communication with the individual vehicle. Research on the development of local controls, on optimal, centralized demand-actuated controls, and on information systems that smooth traffic flow and reduce congestion also should continue.

In the still longer term, there is the possibility of the automated highway with special dedicated vehicles operating totally under computer control. A variation of this, the dual mode, is foreseen as allowing vehicles to be transported on an automated highway and also driven on conventional roads. Neither of these long-term alternatives has been shown to be feasible and thus neither is likely to appear during this century.

## Air System

The air system includes some 15,000 airports that serve civilian aviation. Perhaps the most serious potential problem facing air transportation is congestion; airports overcrowded with flights and passengers can be expected to create costly delays within the next decade. The congestion threat, which began to materialize during the 1960's, abated when the combined effects of a recession and the introduction of wide-bodied aircraft created excess capacity. It now appears that traffic growth will overtake airport capacity again in the near future. The Federal Aviation Administration predicts that 60 airports—including 18 of the nation's busiest—will become saturated during the 1980's. Relief in the form of new airport construction is not anticipated. Further increases in aircraft size, which might alleviate the problem to some degree, will be limited by development costs, the capacity of existing gates and passenger facilities, and the difficulty of providing frequent service with very large aircraft.

The commercial and private use of small aircraft, called general aviation (G.A.), presents both a challenge and an opportunity for air transportation. G.A. comprises over 90 percent of all civil aircraft, ranging from small home-built airplanes to multi-engine corporate jet aircraft. G.A. aircraft are used in a wide variety of transportation operations and carry one third of U.S. intercity passengers. The G.A. fleet is expected to increase by 50 percent in the next decade. Therein lies an opportunity for helping to solve the airport congestion problem. Such aircraft, serving small airports, could siphon off much of the passenger demand otherwise expected at larger terminals. Insofar as G.A. operates on airways and at terminals used by conventional aircraft, its aircraft should possess the same full range of instrumentation as the well-equipped commercial aircraft, including communication, navigation, and control equipment. This is frequently not the case and represents a serious public concern if G.A. is to fulfill an expanded role in passenger service.

Helicopters are being used increasingly in a variety of new transportation tasks in which their unique flight characteristics and cost-effectiveness are important. Business transportation, support of remote or offshore oil operations, forest management, ambulance and other public-service missions, and air-taxi service are among these new applications. The increasing use of helicopters has presented the need, and the opportunity, to improve their performance, reliability, quietness, and smoothness of ride. Rotorcraft technology still lags behind fixed-wing technology, but recent research suggests that higher speed helicopters could be developed to provide competitive passenger and cargo transportation for distances of up to several hundred miles, thereby also helping to counter the anticipated airport congestion.

Two emerging aircraft technologies also could have a significant future impact. The first is supersonic transport. On the basis of research now in progress, it appears that a second generation of supersonic transport could be economically successful and meet the environmental requirements of noise and atmospheric emissions. However, the current level of effort would need to be accelerated in order to support such a development. The second technology is vertical/short take-off and landing (V/STOL). Although V/STOL aircraft are not judged to be cost effective for commercial applications at present, it may be found that they offer a means of reducing the airport congestion problem. The future availability to the civilian sector of supersonic transports and V/STOL aircraft undoubtedly will depend on the level of support provided by defense needs.

Airport capacity could be increased if the minimum safe distance between aircraft, even under adverse weather conditions, could be reduced.

"Although V/STOL aircraft are not judged to be cost effective for commercial applications at present, they may offer a means of reducing the airport congestion problem." [National Aeronautics and Space Administration.]

This would require reducing wake vortex restrictions, minimizing landing time errors, and reducing runway occupancy time (for example, by improving ground maneuverability for earlier turnoffs).

New technology for the future air traffic control system could make possible increased automation of enroute and terminal services, improve systems capacity, reduce delays and consequent fuel waste, and provide a very high level of protection against midair collisions. The changes will entail significant shifts in responsibilities between pilots and controllers and will include automatic decisionmaking and automatic clearance generation. Eventually, advances in electronics will lead to cockpit devices being automatically coordinated with ground-based equipment, permitting pilots to accept more responsibility for aircraft separation. A general upgrading of the airborne and ground electronic controls, surveillance, display, and communications systems can be expected. As in the case of airport capacity improvement, the research leading to these technological advances is in progress, but the pace may not be adequate, particularly in view of the long time typically required to implement new facilities and systems.

Ground traffic flow at airports (and at other intercity passenger terminals) is aided by means such as moving walkways, automated people movers, and the use of television and other devices to inform travelers of schedules. Since these technologies are already in use, research is focusing on increased mobility, improved reliability, and better communications.

## Marine System

The marine system includes the vessels that carry the nation's freight and the dock facilities that support the transfer of goods. It is in speeding the transfer of goods and enhancing marine safety that technological advances can make the greatest contributions to marine transport.

***Merchant Marine*** While the United States is the leading nation in international trade and maritime transportation is the principal mode of this trade, the number of privately owned U.S. vessels and their share of oceanborne commerce have been declining. There are three principal segments to the merchant marine service: liner service (ships operating on a regular schedule over a fixed itinerary), dry-bulk cargo ships, and tankers. In 1970, there were 825 U.S.-flag, oceangoing ships; by the end of 1978, this number had decreased to less than 600. In 1955, the proportion of the total tonnage in U.S. foreign trade carried by U.S.-flag ships was more than 20 percent; by 1978, it was less than 5 percent. Since 1969, the rate of decline in U.S.-flag ship carriage has leveled off. Nevertheless, the long-term trend since World War II has been for a steady increase in U.S. foreign trade but a reduction in the proportion of that trade carried in U.S.-flag ships.

This trend is expected to continue through the 1980's. The reasons for this are many. There is more scope for competition by foreign shippers in the U.S. cargo trade than there is for competition by U.S. shippers in handling non-U.S. cargo; the U.S. tanker and bulk cargo fleet has not been modernized; and the cost of ownership of U.S.-flag vessels is very high. The problems are not likely to lessen as more nations, particularly third world countries, build up their own merchant marines and place additional tonnage in the U.S. trades.

Improvements in the way cargo is transferred between ship and land have made the liner service the one bright spot in the U.S. marine service. The advent of specialized ships using the improved cargo transfer methods—container ships, barge carriers (LASH [Lighter Aboard Ship] and SEABEE [Construction Battalion] vessels), and roll-on/roll-off vessels—has allowed the share of U.S. liner cargo carried in U.S.-flag liners to rise from a low of 22 percent in 1969 to 28 percent in 1977.

Although some increase in tanker tonnage in the U.S.-flag fleet has resulted from the 1970 Amendments to the 1936 Merchant Marine Act, the U.S.-flag tanker fleet generally has lacked the large tankers needed to compete effectively in U.S. foreign trade. United States tanker ship owners rely heavily on foreign flags, especially convenience registries, largely because of the relatively high U.S. costs of ship construction, repair, maintenance, and manning. Construction costs, for example, are up to 250 percent higher in U.S. shipyards than they are in Japan. In the past, higher wage rates accounted for the difference; more recently, as U.S. and foreign shipyards have approached parity in wages, the productivity advantages of the highly automated yards in Japan, Europe, and elsewhere have maintained the gap in construction costs.

Dry-bulk (primarily ore and grain) carriers are active on the U.S. Great Lakes and inland waterways but are almost nonexistent in the U.S.-flag ocean trade. In many ways, this part of the U.S. merchant marine fleet is an anachronism. Compared to foreign-flag ships, these carriers are small, slow, and costly to operate. Operating subsidies for U.S. ships have not offset the advantages of foreign flags, many of which have less costly safety requirements, labor costs, insurance premiums, and overhead charges. As a result, U.S. bulk carriers are used almost exclusively on U.S. government cargoes and on grain cargoes carried under the U.S.-U.S.S.R. Maritime Agreement of 1972.

***Great Lakes Service*** The five Great Lakes and their connecting waterways constitute the world's largest inland fresh water transportation complex. Access to the Atlantic Ocean through the St. Lawrence Seaway enables oceangoing ships to sail around the world from mid-continent U.S. ports. Nonetheless, transportation on the Great Lakes is limited by several major factors and, since 1965, there has been a decline in cargo movements. Key channels and locks freeze during the winter, restricting the navigation season to less than nine months. Efforts to expand the season have proved expensive and only partially effective. The distances between many Great Lakes ports are considerably greater by water than by land, thus limiting the competitive advantage of waterborne trade. Finally, canals and locks limit the ships that can be used in foreign trade to the smaller, less competitive sizes.

Almost 90 percent of U.S. domestic Great Lakes trade is made up of iron ore, coal, and limestone shipments moving in carriers specially designed for this trade, although tanker shipments have been increasing. The average age of U.S.-flag ships in the Great Lakes fleet is nearly 40 years; a program to build vessels of advanced design and large size has begun, but few have been constructed to date.

***Inland Waterways Service*** The weight of cargo carried by the 25,000-mile inland waterways system exceeds the combined shipments of the domestic oceanborne and Great Lakes services. More than 96 percent of the 590 million short tons of dry and liquid cargo that moved on the waterways in 1976 were carried in towed barges. More than half of the tonnage consisted of farm products, coal, nonmetallic minerals, marine products, and miscellaneous goods.

Barge transportation is relatively cheap and efficient. Advances in tow boat and barge design, added to economies of scale, have done much to

"Advances in towboat and barge design, added to economies of scale, have done much to lower and stabilize costs." [EPA Documerica—David Falconer.]

lower and stabilize costs in this largest segment of the domestic trades. The efficiency and low costs of barge transportation are helping inland waterways traffic to expand faster than any other mode. The principal obstacle to the continued expansion of commerce on the upper Mississippi and Ohio Rivers exists in the form of 13 of the existing locks. If these small, aging locks are not expanded or replaced, the waterways' potential for cargo movements will be reduced.

## Rail System

There are about 191,000 miles of main railroad lines and a total track of about 310,000 miles in the United States. Some 28,000 locomotives and 1,700,000 freight cars carry nearly 914 billion ton-miles of revenue traffic annually. The demands on the system, however, are not uniform; while some segments of the track system have experienced substantial new traffic, the demands on other segments have fallen so low that their abandonment is likely.

Investment in track construction, maintenance, and rolling stock has been high. Prior to the 1980 recession, the railroad industry was investing over $1 billion annually in the upgrading of track and about $2.3 billion annually in new cars and locomotives. New yards are being constructed to improve car switching and train distribution. On the average, rail facilities on many railroads are being gradually upgraded.

In spite of the large investments in freight cars and locomotives, car shortages develop periodically. These shortages, and increases in the cost of freight cars, require that cars be handled more efficiently as a part of improving the management of all capital assets. Many studies are in progress on freight car management and several railroads have adopted complex information systems that permit closer control over car and train operations.

Railroads and suppliers also are introducing new equipment to respond to growing intermodal traffic demands. Recently completed demonstration programs have emphasized the importance of rail service frequency and reliability to increased intermodal services.

Other principal technical emphases are upon the reduction of maintenance costs. One approach has been to lessen the stress that trains impose on tracks. In 1972, the railroads, equipment manufacturers, and the Federal Railroad Administration began a ten-year program entitled "Track-Train Dynamics" to develop train-handling techniques and designs for cars that would transmit a minimum of stress to the track. Improper train handling can spread rails, break couplings between cars, or lift cars completely off the track; the occurrence of any of these can cause a derailment. With the help of sensors placed on test tracks and cars and mathematical modeling, this program has supplied the manufacturers with improved specifications for cars and the railroads with new guidelines for train handling under a variety of conditions, including the nature of the terrain, the curvature of the track, and the makeup of the train.

The adoption of radically new train technologies does not appear to be on the horizon. For instance, very high-speed, steel-on-steel (steel wheels on steel rails) operations requiring entirely new rights-of-way appear to involve investments significantly beyond those which can be justified economically. The most revolutionary system under study is the high-speed, magnetically levitated train. The intent is to support the vehicle by magnetic forces a few centimeters above a metallic track, thus making it technically possible to travel at speeds approaching 350 miles per hour. As a result of its own evaluation, the Department of Transportation has discontinued its support of this technology. For reasons that appear unique to their rail transportation needs, Germany and Japan continue to support the development of this technology.

## Intermodal Capacity Improvements

In a general sense, intermodal transportation involves the entire passage of goods or passengers, including passage within individual modes as well as the facilities at which transfer between modes occurs.

Intermodal freight transportation issues are dominated by regulatory policies established by federal legislation and by overlapping jurisdictions of the regulatory agencies. The prohibition by statute or regulation of multimodal ownership is regarded by the rail industry as one of the most significant constraints imposed on the U.S. freight transportation system. Legislation prohibits the ownership of water carriers by railroads and, until recently, the Interstate Commerce Commission's policy discouraged railroad ownership of trucking companies. Most railroads that now own trucking companies acquired them prior to passage of the Motor Carrier Act of 1935.

Competitive multimodal transportation companies providing door-to-door, shipper-to-receiver service would have incentives to use technologies in the most efficient way. The recent deregulation of the air cargo industry provides an excellent example of this increased efficiency. The carrier can now coordinate truck and aircraft schedules to provide a truly premium service for package express service. Deregulation also allowed the package express companies to contract with other carriers to complete trips in areas not covered by their own vehicles.

Overlapping regulatory jurisdictions are another barrier to the increased use of intermodal service. In the 1960's, largely as a result of U.S. innovation, the so-called container revolution began in the international maritime trades. The economic advantages of containers were so great that billions of dollars were invested worldwide to keep national fleets competitive.

The most important technical elements of the emerging intermodal freight system are containers, improved material-handling technology, and overall system management techniques. The widespread adoption and standardization of containers have been directly related to the improved cargo handling capabilities they offer. Containerization also has brought the liner companies directly into cooperation with the railroads,

A roll-on/roll-off carrier of the nineteenth century. Flatcars bearing stagecoaches leave the factory in New Hampshire bound for Omaha and Salt Lake City. [American Antiquarian Society.]

creating a series of shipping linkages involving both ocean and land carriage. These linkages are evident in "land bridge," "mini-bridge," and similar types of intermodal arrangements. They all involve the substitution of joint intermodal land-water service for all-water service over some portion of the total trip. In these cases, the movement of cargo is no longer a disconnected series of events involving the various carriers and the loading, unloading, and storage activities linking each of the carriers; instead, it embodies the integration of all of the elements of ocean and land carriage in order to achieve a maximum economic return.

A good example of this integration is the Port of Seattle. In the 1960's, the port authority began constructing advanced container facilities. The port installed a computerized system that includes location control and warehouse planning; improved cargo transfer and transport equipment; and better ship-handling, mooring, and docking facilities. By 1970, the port was able to capture a large and increasing share of the high-value containerized shipments in the United States-Far East trade. The total weight of cargo handled by the port rose more than 50 percent in ten years, and its value increased by more than 70 percent during the same period. Today, Seattle is the leading container port in the Pacific Northwest, the second leading container port on the West Coast, and the third leading container port in the United States.

Further research can be expected to improve system management techniques. In particular, control procedures that increase the use rate of containers would increase the return on container investment. The technology needed to reduce freight transport documentation exists already, although it cannot yet reduce the procedure to a single waybill. The complete use of this technology must await standardized domestic and international data formats.

## An Emerging Capacity Issue—Coal Transport

Large coal shipments to supply the emerging synthetic fuel industry and export market could generate severe pressure on the U.S. freight system.

***Domestic Shipments*** New facilities and carriers will be needed by the rail, inland waterway, and pipeline systems if they are to meet the projected annual demand of between 600 and 1,000 billion-ton miles of coal movement by the year 2000.[2] Most studies conclude that the railroads can expand their capacity to handle the projected demand; however, the waterways' ability to meet the growth is problematical. It is estimated that coal movements on the waterway system will increase from 156 million tons in 1977 to at least 220 million tons by the year 2000, including coal transferred from rail and movements within ports and harbors. As already noted, expansion of the waterborne movement of coal on the Mississippi-Ohio River system is limited primarily by lock size. It will be necessary to enlarge the lock system if expanded transport on the inland waterways is to occur.

Of the three systems, only pipelines appear to require new technology. Finely divided coal is suspended in a medium, generally water, and pumped through the pipeline. Although the process of dewatering has been developed, some concern still exists about the nature of the resulting effluent and ways of disposing of it. Solvent slurries using fluids other than water are also possibilities and appear to have the potential for reducing the requirements for water and for increasing overall pumping efficiencies.

***Coal Exports*** New port facilities are necessary if the United States is to participate in supplying the rising world demand for coal, which could be enormous. Historically, coke producers have provided the major demand for U.S. coal exports. Each shipload of such coal had to be blended from a variety of coals to yield a mixture with the desired coking characteristics. The railroads serving the East Coast have been installing increasingly complex yards near the docks so that cars from perhaps a score of mines can be put in the sequence required for filling a ship's particular order.

Recently, foreign utilities have sought new sources of steam coal to power their plants. In 1979, the United States exported about 14 million tons of steam coal; by the year 2000, such exports could range between 70 and 200 million tons. At the higher rate, the United States would be one of the world leaders in this market.

Currently, there are no long-term contracts for the export of steam coal, a condition which inhibits investment in the transportation and port facilities required. The low rate of return on dry-bulk cargo operations also has discouraged investment in such facilities by the major ports, which have chosen to improve their financial condition by investing in container facilities that handle high-value cargo. As a result, those ports that have the capability to handle coal exports are now severely congested by container shipments. This leads to increased delays and costs for coal shipment. For example, in Hampton Roads, the largest of the coal ports in the United States, ships must now wait weeks before they secure a berth for loading; demurrage costs are high, with charges for the large vessels exceeding $15,000 a day. Moreover, there is no space at the port to store coal awaiting loading; it is kept in the rail cars which, at times, causes a shortage of cars at the mines in West Virginia and Pennsylvania.

A modest expansion in port facilities is now occurring. Major U.S. coal producers have agreed to build a new coal export terminal at Hampton Roads capable of handling 20 million tons per year; one of those companies also is planning to expand its facilities in Baltimore. U.S. Steel is considering developing some of its reserves in Illinois and shipping the coal by barge to the port at Baton Rouge. This will require an evaluation of the limitations imposed by the small size of the locks on the Mississippi River. If the rise in foreign demand is sustained, arrangements to link further port improvements to long-term contracts probably will become attractive, although the dredging and silt disposal necessary to acccommodate the large coal carriers pose environmental problems that have not been resolved.

## ENERGY

### Historical Situation

Since 1950, transportation has accounted for about one fourth of total national energy consumption, and highway vehicles have demanded about 80 percent of transportation energy.[3] Petroleum's share of transportation energy has grown to about 97 percent today. While the national objective is to reduce our dependence on imported petroleum through the development of synthetic fuels, the synthetic fuels are not likely to supplant significant quantities of traditional fuels until late in the next decade.

Because of the long time required to develop fuels from sources other than petroleum, and despite the fact that the choice of the passenger transportation mode depends on factors other than energy, the near-term emphasis of government and industry has been on conservation. Much attention has been directed toward improving the energy efficiency of new highway vehicles and airplanes and increasing the occupancy of vehicles.

In evaluating the energy efficiency of the various modes that are used for transporting passengers, the key factors are the number of passengers who are carried on each trip and the energy efficiency of the powerplant that propels the vehicle. The vehicle occupancy is conveniently represented as a ratio of the average number of passenger miles traveled to the vehicle miles traveled, PMT/VMT. The average vehicle efficiency is obtained by determining the fuel energy, in British thermal units (Btu), consumed per vehicle mile traveled, and expressing it as Btu/VMT. The transport efficiency is then the ratio of the vehicle efficiency and the occupancy factor, or Btu/PMT. The Btu/gallon content used in these calculations is 124,000 for gasoline, 138,700 for #2 diesel fuel, and 135,000 for jet fuel. The year 1975 was chosen in order that future trends can be compared to a base that existed before major conservation efforts were implemented. The relevant occupancy factors and the passenger transport efficiencies calculated for 1975 are given in Table 3.

For intercity passenger service, the bus was clearly most efficient in 1975. For urban travel, light rail was somewhat better than bus and subway. For both intercity and urban service, automobile transport efficiency ranked well below that of other modes.

While an analysis of the transport efficiency for freight service could be made in terms of Btu's consumed per ton-mile transported, it would tend to obscure the fact that the choice of mode for freight transport depends upon many factors other than energy; thus, low-value bulk freight often moves by water, and high-value freight by air.

Strictly from an energy standpoint, rail and water have the highest efficiency for general freight transport. In 1975, intercity truck transport was significantly more efficient than urban truck hauling because it was usually diesel powered and suffered less from stop-and-go driving conditions.

## Modal Transport Efficiencies of Vehicles Using 1985 Technology

New 1985 vehicles will have substantially improved efficiency over the average vehicle in use in 1975. The full benefit of these improvements will be realized only when the older, less efficient vehicles are replaced by vehicles having the new technology. As noted earlier, this may require

**Table 3** Average modal efficiencies for passenger transport (1975)

| | Intercity | | | | Urban | | | |
|---|---|---|---|---|---|---|---|---|
| | Car | Bus | Heavy rail | Air | Car | Bus | Light rail | Subway (Chicago) |
| Average occupancy (PMT/VMT) | 2.3 | 22.6 | 24.1 | 69.3 | 1.4 | 12.2 | 21.4 | |
| Transport efficiency (Btu/PMT) | 3,990 | 990 | 2,630 | 6,500 | 7,700 | 2,770 | 2,380 | 2,700[a] |

[a] D. L. Green et al. *Regional Transportation Energy Conservation Data Book*. Edition One. ORNL-5435 Special. Oak Ridge, Tenn.: Oak Ridge National Laboratories Energy Division, 1978.

SOURCE: R. E. Knorr and Marianne Miller. *Projection of Direct Energy Consumption by Mode: 1975-2000 Baseline*. ANL/CNSV-4. Argonne, Ill.: Argonne National Laboratory, 1979.

many years. Actions which encourage more efficient vehicle utilization, such as carpooling and reduced movement of nearly empty cars, are capable of making a more immediate impact upon energy efficiency.

***Rail and Marine*** Rail and marine propulsion is predominantly provided by high efficiency diesel engines operated for long periods at the speeds at which they have been designed to have the greatest fuel economy. It appears unlikely, therefore, that substantial improvements in efficiency can be achieved in either of these modes. But some improvement in rail passenger efficiency can be achieved by reducing vehicle weight through the use of new lightweight materials. Rail freight efficiency can be expected to improve through the application of management methods that will reduce the movement of empty cars.

***Air*** Within the past decade, U.S. airlines have achieved an estimated average 50 percent increase in revenue-passenger-miles per gallon of fuel. The improvements have included more conservative scheduling, higher occupancy factors, and improved air traffic control procedures. The introduction of larger, more efficient aircraft in the early 1980's, such as the Boeing 757, will produce further fuel efficiency gains, on the order of 30 percent. Unfortunately, existing airport facilities will, in some cases, limit the use of the new, large-wing aircraft.

For the longer term, significantly greater efficiencies can be achieved. Research being conducted by industry and the National Aeronautics and Space Administration (NASA) in the Aircraft Energy Efficiency (ACEE) programs is directed at a further 40 to 50 percent improvement in the fuel efficiency of airplanes that will be designed in the late 1980's and early 1990's.

The technology objectives of the ACEE program include:

- Improved engine components applicable to existing engines.
- Increased thermodynamic and propulsive efficiencies, and improved durability for new engine designs, with the goal of a 15 to 20 percent reduction in the fuel consumption of current turbofan engines.
- Advanced turboprops providing still further economy improvements, resulting in fuel consumption reductions of 15 to 20 percent beyond the new improved turbofans at equivalent speeds and even greater reductions at somewhat lower speeds.
- Composite small or medium-sized primary structural components to achieve weight savings of about 25 percent compared with metal structures.
- Advances in aerodynamics, active controls, and integrated aircraft design concepts to accomplish fuel savings on the order of 10 to 20 percent.
- Practical technology for laminar flow control—for example, the use of suction to reduce skin friction drag by minimizing turbulence in the airflow adjacent to the surface of the aircraft.

The timely use of some of these technologies may require new certification procedures.

***Car and Light Truck*** The Energy Policy and Conservation Act of 1975 (EPCA) required improved fuel efficiencies for cars and light trucks. This act required that the Secretary of Transportation establish the fuel economy that each manufacturer must achieve for all new vehicles sold during each model year through 1985. These corporate average fuel economies (CAFE) are listed in Table 4.

As indicated, the mandated 1985 CAFE will be nearly double the 1975 level. Further, all of the major U.S. automobile manufacturers have declared their intent to continue to improve fuel economies beyond those required by the EPCA regulations. The technologies that have contributed to these improvements in fuel efficiency are varied. Not all of them will appear on every vehicle. But, as old vehicles are replaced by new ones, the fuel efficiency of the average vehicle will increase.

1. Gasoline engine efficiencies will increase through combustion system changes that allow operation at lower fuel-to-air ratios, the so-called lean burn technology. This will be accompanied by the expanded use of electronic controls. Also, stratified charge

**Table 4** Mandated corporate average fuel economies for new vehicles, by year[a] (in miles per gallon)

| Year | Cars | Light trucks |
|---|---|---|
| 1978 | 18 | |
| 1979 | 19 | |
| 1980 | 20 | 16.0 |
| 1981 | 22 | 16.7 |
| 1982 | 24 | 18.0 |
| 1983 | 26 | |
| 1984 | 27 | |
| 1985 | 27.5 | |

*Note*: The 1975 corporate average fuel economy was 14 miles per gallon.

[a] Corporate average fuel economy is a sales-weighted average of the fuel consumption of all vehicles sold by each manufacturer. A measurement is made of the fuel consumed on two driving schedules, as specified by federal regulation. One driving sequence is representative of urban driving and the other is representative of rural driving. The average fuel consumption is calculated by combining 55 percent of the fuel consumption measured on the urban mode with 45 percent of the fuel consumption measured on the rural mode.

systems are being explored which, if successful, could reduce fuel consumption further. In the future, the fuel may be cut off when the engine is idling and deceleration may be used.

2. Diesel engines have an inherently greater efficiency than do current low-compression-ratio gasoline engines. If the mandated emission levels for passenger cars can be achieved, these engines will find increased use. This topic is addressed below.
3. Transmission improvements will occur through the expanded use of overdrive, mechanical lock-up of automatics in top gear (which reduces fluid coupling losses), and manual transmissions with more gear steps. Work is proceeding on the continuously variable transmission, a means of further increasing overall efficiency.
4. Friction will be reduced in the engine and between the tires and the road; aerodynamic losses between the vehicle and the air also will be lowered. While the improvement of each of these probably will amount to only a few percent, the combination will be important.

Average vehicle weight is being reduced by replacing steel and cast iron with thinner gauge high-strength steel, aluminum, and plastics, by decreasing the size of the vehicle to reduce the passenger capacity, and by redesigning to maximize the effective use of the resulting passenger space.

A new vehicle with a composite fuel economy of 27.5 miles per gallon (mpg), which is typical of the average required by the regulation in 1985, likely will have an approximate city (urban) fuel economy of 23 mpg and a highway (intercity) fuel economy of 34 mph. Assuming no change in the occupancy level between 1975 and 1985 (2.3 passengers per intercity vehicle trip and 1.4 passengers per urban vehicle trip), the average *new* 1985 gasoline-powered car will achieve efficiencies of 1,590 Btu/PMT in intercity trips and 3,850 Btu/PMT in urban trips.

The full benefit of these gains must await the replacement of the older vehicles by the new designs. The other approach to higher transport efficiency—greater occupancy per vehicle mile—has proved more difficult. Incentives have not been found yet that will increase the occupancy of highway passenger vehicles, although the rising cost of fuel certainly has encouraged some movement toward increased car pooling and increased use of buses, trains, and subways.

***Bus*** Reduced vehicle weight and improved efficiency of the engine, transmission and axle (that is, the total powertrain) also have improved the fuel efficiency. Since buses generally have used direct injection diesel engines, the possible improvement in powertrain efficiency would be less dramatic than in cars, but nonetheless important. Improvements in the diesel engine and in the transmission, combined with weight reductions, led to an overall efficiency improvement of between 10 and 15 percent between 1975 and 1980. Further improvement, however, is uncertain because of the constraints imposed by proposed emissions controls. If the proposed 1985 emission standards for heavy duty diesel engines are enacted, the efficiency of these engines probably will be worse in 1985 than in 1980—perhaps by as much as 10 percent.

***Projected Efficiencies*** Table 5 summarizes the effect that the above-described advances will have on the *average new vehicle* being produced in 1985. The 1985 occupancy levels for all modes except air are assumed to be the same as in 1975. The 1985 bus efficiency assumes no deterioration in diesel fuel efficiency because of the more stringent emission standards.

While the transport efficiency of subways in 1985 has not been estimated, there seems to be little reason for it to fall significantly below the Table 3 value of 2,700 Btu/PMT. The efficiency of the BART (Bay Area Rapid Transit) system in San Francisco has been improving but has not yet reached 2,700 Btu/PMT, the level achieved in Chicago.[4]

Commuter car efficiency can be expected to continue to improve after 1985. For example, a vehicle designed for only two passengers could achieve between 2,000 and 2,200 Btu/PMT in urban driving. Efficiencies for intercity trips should improve another 10 to 15 percent over those shown in Table 5, to about 1,400 Btu/PMT.

**Table 5** Energy efficiencies for passenger transport (using 1985 technology)

| | Intercity | | | | Urban | | |
|---|---|---|---|---|---|---|---|
| | Average new 1985 car | Bus | Heavy rail | Air | Average new 1985 car | Bus | Light rail |
| Average occupancy (PMT/VMT) | 2.3 | 22.6 | 24.1 | 89.4 | 1.4 | 12.2 | 21.4 |
| Transport efficiency (Btu/PMT) | 1,590 | 875[a] | 2,500 | 1,800 to 3,000 | 3,850 | 2,400[a] | 2,300 |

[a] Values could be increased by as much as 10 percent if proposed 1985 heavy duty diesel engine emission standards are implemented.

An interesting way to compare relative vehicle efficiencies is to compare the average vehicle weight per vehicle seat. For vehicles with comparable powertrain efficiencies, equal weight per seat should yield roughly equal transport efficiencies. For the passenger car, average weight per vehicle seat decreased from 782 pounds in 1975 to 711 pounds in 1980. It is expected to reach 636 pounds in 1985. For the average transit bus, this value ranges from a low of 579 pounds for the lowest powered 40-foot bus to 700 pounds for the higher powered 35-foot bus. The Chicago and BART subway vehicles weigh 860 pounds and 790 pounds per passenger seat, respectively.[5-6] For comparison, the 747 airplane weighs 850 pounds per passenger seat when not loaded with fuel. This analysis of weight is consistent with the transport efficiencies discussed above.

Three important observations can be drawn from these considerations:

1. Assuming no significant change in passenger occupancy levels, the bus will be the most energy-efficient intercity vehicle, with the automobile a close second.
2. Again assuming no significant change in occupancy levels, bus and light rail can be expected to have the best efficiencies for urban passenger transport in 1985, with the car a close second.
3. Urban transport efficiencies after 1985 for two-passenger commuter cars, intracity buses, and light rail will be equal.

Since it appears possible to achieve generally comparable transport efficiencies with the four principal urban modes—car, bus, light rail, and subway—planning the movement of people should emphasize issues other than energy efficiency. All modes must seek to develop incentives that will increase the average occupancy level of the vehicle, both in urban and intercity trips.

***Freight Efficiencies*** The outlook for improved air, rail, and marine freight fuel efficiencies is the same as that noted above for the passenger version of these modes. Air appears to have the greatest potential. Rail and marine, which already employ highly efficient diesel engine technology, have less potential although, as mentioned previously, improved management systems can reduce empty car movements on the railroads.

The efficiency of urban trucks will improve significantly in the coming years. In 1975, a sizable portion of urban freight was carried in gasoline-powered trucks. There is a growing tendency to replace these with more fuel-efficient, diesel-powered trucks. Diesel engines have predominated in intercity trucks for the past decade but, even in this case, efficiency improvements will occur. The prospects for intercity and urban trucks can be summarized as follows (again, assuming no loss of efficiency through tightened emission standards).

For intercity truck freight, there can be:

1. An increase in diesel engine efficiency of approximately 10 to 15 percent.
2. An increase of up to 15 percent through the combined use of demand-only cooling fans, radial tires, aerodynamic improvement devices, and smooth-sided trailers.

For urban truck freight, there can be:

1. An increase in diesel engine efficiency of approximately 10 to 15 percent.
2. A substitution of diesel engines for gasoline engines. The combined effect should improve urban truck efficiency by about 50 percent.
3. An increase of up to 10 percent should result from the combined use of demand-only cooling fans and radial tires. Aerodynamic efficiency is not as important for urban delivery vehicles as it is for long-haul vehicles.

In addition to the engine and vehicle effects noted, there will be some improvement in efficiency resulting from lighter weight trailers and tractors. Reducing vehicle weight yields two benefits— first, less energy is required to move the vehicle; and second, the freight weight can be increased while the vehicle remains within restrictions imposed by state and federal regulations. While the second benefit is potentially large for intercity vehicles, studies suggest that, generally, most carriers are not able to take advantage of it because only a fraction of their vehicles operate at the legal weight limit.

Overall, in spite of the projected energy efficiency improvements for trucks and airplanes, the most energy-efficient general freight haulers will continue to be trains, barges, and ships.

## Modal Transport Efficiencies—Post 1985

While the spectacular efficiency improvements that are expected between 1975 and 1985 are not likely to be duplicated in the subsequent decade, a further gain of 10 to 15 percent should be well within reach. Much of the potential weight reduction and the direct improvements possible in aerodynamic design of ground-based vehicles will have been achieved. But some improvement in powertrain efficiency will continue to be made. A better understanding of the detailed physical processes that determine engine efficiency will be achieved through expanded research. Through this improved understanding, it will be possible to develop control, combustion, and mechanical systems that minimize friction, reduce heat loss, and optimize combustion, so that the efficiency of the overall system increases while maintaining a low level of emission.

Two new engines—the high-temperature gas turbine and the Stirling engine—are being developed that have the potential of providing good fuel efficiencies. Projections suggest that their fuel economy may surpass that of the best current direct-injection diesel engine by 10 to 20 percent. However, many issues remain to be resolved before it can be concluded that the turbine and Stirling engines will be successful. The fuel economy must be demonstrated, the engines must be shown to be reliable and feasible of large-scale manufacture, and the cost of manufacture must be established. In the case of the turbine, its ultimate success depends upon the development of ceramic components that will allow high-temperature operation. And even if these new developments are technically successful, future diesel engine efficiency—through direct-injection, turbo-compounding, or development of the adiabatic-operating diesel—and future gasoline engine efficiency may equal or surpass the potential of the turbine and Stirling engines. Many other issues must be considered in determining whether a new engine will find widespread use. But one thing is clear. The new alternative must offer significant improvements over the mature technology, a step-function increase, so to speak, or it will not be introduced.

## Alternative Fuels

***Liquids and Gases*** The internal combustion engines, gasoline and diesel, and the external combustion engines, turbine and Stirling, can be modified to operate on a variety of fuels. But only the liquid hydrocarbon fuels—alcohol, gasoline, and diesel—have the high energy content that allows long distance travel. Thus, the water, air, rail, and heavy truck transport systems will continue to rely on liquid fuels. While passenger cars can be made to operate for moderate distances on gaseous fuels, the infrastructure for the delivery of these fuels to vehicles is presently limited.

A national effort is under way to develop sources of alternative liquid and gaseous fuels from coal, shale, tar sands, and biomass. Most observers believe that commercial conversion of oil shale and coal to synthetic liquids will commence in the mid-1980's; by 1990, this source could contribute a few percent of the total U.S. need for liquid fuels and 20 to 30 percent by the year 2000.

The major research efforts center on the effects that alternative fuels will have upon vehicle durability, performance, emissions, and lifetime costs. The National Aeronautics and Space Administration, the Department of Defense, and the aircraft industry are looking for ways to ensure the compatibility of aircraft engines and fuel systems with synthetic fuels, and to determine the system changes that might permit relaxation of fuel specifications. The adaptability of the diesel engine is the primary concern of the rail industry, while the automobile industry is concentrating on the effects on the internal combustion engine of the alcohols and gasoline derived from oil shale and coal and the effects of diesel fuel derived from tar sands. Major concerns are the corrosive nature of some of the fuels, the higher impurity levels which could result from limited refining, and the possible degradation of properties such as octane rating, cetane number, lubricity, and volatility.

Similar issues may arise in the distribution of the fuel. It may be necessary to provide alternative distribution and storage facilities if the synthetic fuel is not compatible with a major part of the existing fuel storage and distribution systems because of its instability, corrosive properties, or energy density. Since the various vehicles used by the major transportation modes have lifetimes between 10 and 40 years, these costly duplicate facilities would be needed for many years. It is therefore essential that the national strategy for providing alternative fuels address carefully and fully all of the implications of the distribution, storage, and utilization of these fuels as well as their production.

***Electricity*** Electricity represents a special form of alternative energy. It can be generated from a variety of basic energy sources such as coal, nuclear fission, and, ultimately, fusion. Although an electrical distribution system is in place and is amenable to expansion, its use in transportation presents some serious and difficult issues.

In the case of rail, it is technically feasible to convert long-haul rail transport from diesel to electric power. But present railroad signaling systems must be changed to accommodate the electromagnetic effects arising from electrical transmission lines over the rails, and the cost of installing the electric power and of modernizing the signaling systems would be large.

Interest in electric cars and vans is increasing. At present, these vehicles are limited to ranges between 50 and 100 miles under typical driving conditions. Recharge times of six to eight hours are expected with existing technology. While some improvements in these characteristics will result from research, it appears unlikely that electric vehicles will be able to achieve the operating characteristics of the internal combustion vehicle. If petroleum is the basic starting fuel, the energy efficiency of an electric vehicle, measured in Btu per passenger mile traveled (Btu/PMT), will be substantially poorer than those shown in Table 5. This is because the vehicle weight per passenger seat of an electric vehicle, with its heavy batteries, will be about 50 percent greater than that of the average new 1985 car. If coal is the starting fuel for generating electricity for the electric car and for the liquid hydrocarbon fuel for the average new 1985 car, the transport efficiency of each, measured in Btu of coal per PMT, will be essentially the same. The low energy density of the storage batteries and the resulting weight penalty reduce the overall efficiency of the electric vehicle by an amount equivalent to the losses that occur in the conversion of the coal to the synthetic liquid fuel that powers the internal combustion car. Thus, factors other than the efficiency of final use in propulsion—considerations such as the adaptability of existing storage and distribution systems—are going to be key elements in ultimately determining the final form of the synthetic energy that is generated.

## SAFETY

### Passenger Safety

Safe operation of the transportation system is of increasing concern to travelers and to citizens who are exposed to the geographic area contiguous to the transportation facilities.

Transportation-related accidents comprise a wide range of events and stem from a great variety of causes. Government regulatory agencies have been active in searching for means to upgrade carrier safety, but a serious stumbling block has been the lack of a generally accepted means of determining the cost effectiveness of proposed regulations. In addition, public opposition defeats important safety proposals that would restrict individual action (such as the mandatory use of seat-belts and the effective control of driving by those under the influence of alcohol). As a result of this opposition, much of the task of improving highway safety has been placed upon the manufacturer of the vehicle whereas, in the case of air and rail, both the manufacturer and the operator of the vehicle share this task.

***Highway Vehicles*** In 1978, motor vehicle accidents caused more than 50,000 deaths and 2 million injuries. About 18 percent of the deaths involved pedestrians. Automobile safety is the subject of vigorous research and development. The safety of passengers involved in accidents is being improved by better passenger restraint systems; by increasing the amount of impact energy that can be absorbed by the vehicle structure; by designing the car to reduce the severity of injury to pedestrians in collisions; and by improving the performance of components such as brakes, steering system, and lights. There is a particular concern regarding the operation of small lightweight passenger vehicles, as they have a higher passenger fatality rate when involved in collisions with larger vehicles. While measures can be taken to improve the safety of the small vehicle, it must be recognized that this generally requires the addition of weight, with a resultant decrease in energy efficiency. Considerable research is under way in the development of new designs that may prove to be more effective in reducing the danger to the passenger; in new materials that may absorb energy more effectively; and in new structures that may reduce damage to the passenger compartment. As noted earlier, an agreed-upon means of determining the cost effectiveness of vehicle safety measures is greatly needed.

***Rail*** In the last five years, the accident rate for railroads increased from 4.15 to 4.50 accidents per billion-gross-ton miles. Although the total cost of these accidents is large, only about 5 percent of the incidents result in injury and about one percent result in death. Almost 75 percent of rail accidents are mainline and yard derailments, usually involving only slight damage to track or equipment; 53 percent of these derailments are attributed to the condition of the track, 21 percent to equipment, and 18 percent to human causes. There are scores of assignable causes within each category, no one of which accounts for more than about 2 percent of all train accidents. Each of the causes is the subject of research supported separately or jointly by government and industry but, given the great diversity of problems being addressed, progress likely will be slow.

***Aircraft*** From 1969 to 1979, the accident rate for U.S. air carriers declined from 1.0 to 0.5 accidents per 100,000 flight hours. The general aviation rate has improved similarly, but from the much higher level of 19 accidents per 100,000 hours to 11. The commuter airline rate of accidents per flight has remained essentially constant at approximately six times that of the trunk carriers.

The differences in safety records reflect significant differences in pilot training and experience, aircraft design and maintenance standards, and operating conditions. The majority of accidents occur in approach and landing; in recent fatal accidents, pilot error was estimated to be the cause in 66 percent of the cases for air carriers, 79 percent for commuter lines, and 88 percent for general aviation.

Current government and industry research should produce better pilot and crew performance, flight-deck management, air traffic control systems, and pilot/traffic-controller interaction. Advanced simulators and flight research vehicles are being used to aid in the study of behavioral factors, crew work load, fatigue, cockpit design, and information systems. Research also is directed at improving the safety of occupants during crashes.

***Marine*** In 1978, marine accidents were reported to have caused 179 deaths and 110 injuries. With personal injury in marine accidents being relatively rare, the principal concern, particularly in the case of large tankers, is the fate of the cargo that is discharged into the sea. This topic is discussed further under the heading of environmental issues.

## Hazardous Materials Transportation

The transportation of hazardous materials poses substantial environmental and safety concerns. These arise from such experiences as chlorine gas released as a result of a derailed tank car or an explosion caused by a highway accident involving a truck carrying propane.

There are some 1,600 materials now classified as hazardous—that is, flammable, explosive, radioactive, or toxic. Such materials are carried about 200 billion-ton miles every year by approximately 600,000 vehicles or vessels. Although the number of fatalities is not very high in this specialized segment of the industry—accidents involving hazardous materials transport caused 31 deaths in 1977 and 46 in 1978—the potential for disaster is large. The response to this has been a substantial amount of legislation that regulates the flow of such traffic.

In 1977, after seven years' study of the behavior of tank cars during derailment, the federal government ordered that three protective devices be installed on 17,600 tank cars: shelf couplers, to resist vertical disengagement and help keep adjoining cars attached in a derailment; head shields, to guard against spearing of tank ends; and insulation, to increase the resistance of tank cars to flames from other cars. All but 338 cars were adapted as of October 1, 1980, and the changes have demonstrated their usefulness already.

Emergency response is also an issue of great importance. Emergency communications are an important aspect of hazardous materials control in the event of an accident. The main problems are developing information systems that can identify the particular material involved in an accident and informing emergency crews quickly on how to handle it. The chemical industry, the railroads, and the federal government are exploring ways in which existing communication systems can be improved.

## ENVIRONMENTAL EFFECTS

Environmental concerns with the transportation system are substantial. They involve the control of undesirable chemical constituents that arise from the combustion of fuel, reduction of the noise associated with the movement of the vehicle along the terrain or through the air, disposal of obsolete vehicles, proper allocation of land to highways, railroads, and airfields, effects of transportation systems upon land use, and the influence of transportation on life style. Some of these, such as emission and noise controls, have been the subject of intense legislative and regulatory action.

### Highway Vehicle Emissions

Highway vehicle emission standards were established by the Clean Air Act of 1970 and the EPCA legislation of 1975. These federally mandated standards for hydrocarbons (HC), carbon monoxide (CO), and oxides of nitrogen ($NO_x$) are given in Table 6. Additional mandated standards for particulate emissions from cars and light trucks of 0.6 grams/mile in 1983 and 0.2 grams/mile in 1985 have been established.

A key to the control of emissions from gasoline engines is the treatment of the exhaust gases with the catalytic converter. The catalyst depends upon a combination of noble metals suspended upon a ceramic substrate. As much as 0.1 troy-ounce of platinum, palladium, and rhodium are used in each converter. In some systems, accurate control of the air-fuel ratio is critical and feedback control is used. Cars and light trucks employ essentially the same control techniques to meet emission standards.

A major issue in controlling the atmospheric emissions of vehicles relates to the emissions from diesel-powered cars and light trucks. While a diesel-powered vehicle must achieve the same emission standards as the gasoline engine, the proposed particulate standard creates a uniquely serious problem for the diesel engine. The much higher concentration of particulates in the diesel's exhaust requires special control techniques. At sufficiently low inertial vehicle weights—below 2,250 pounds—it is expected that the presently stated emission objectives of 1.0 grams/mile $NO_x$ and the proposed 0.2 grams/mile particulate standards can be achieved without exhaust treatment. At higher inertial weights, some form of trapping and disposal of the particulates will be required. This

**Table 6** Federal vehicle emission standards (grams/mile)

| | Passenger car | | | Light truck | | |
|---|---|---|---|---|---|---|
| | HC | CO | $NO_x$ | HC | CO | $NO_x$ |
| 1980 | 0.41 | 7.0 | 2.0 | 1.7 | 18.0 | 2.3 |
| 1981 | 0.41 | 3.4 | 1.0 | 1.7 | 18.0 | 2.3 |
| 1983 | 0.41 | 3.4 | 1.0 | 0.8 | 10.0 | 2.3 |

*Note*: Uncontrolled passenger car emissions (in grams/mile) are taken to be HC = 10.6, CO = 84, and $NO_x$ = 4.1.

has not yet been successfully demonstrated with a production-feasible trap system.

There is also a major question about the carcinogenic risks posed by diesel emissions. If they are found to be large, this highly fuel-efficient engine could be precluded from general use. Further research is required to quantify the cancer risk to the public and to determine the costs and benefits of controls that might be needed to reduce any such risk.

Stringent regulation of emissions from heavy duty diesel engines has been proposed by the State of California for 1983 and by the federal government for 1985. The proposed regulations would require a major reduction in the combined levels of hydrocarbons and oxides of nitrogen, and would change the test procedure from a steady state test to a new transient test. The heavy duty diesel engine manufacturers have indicated that they believe it is presently infeasible to achieve these emission levels.[7] If control procedures are developed, it is expected that a deterioration in fuel efficiency of about 10 percent could result.

## Aircraft Emissions

Research results of the past two decades have been used to reduce aircraft engine emissions considerably. Virtually all transport aircraft meet EPA smoke standards, and other aircraft will be modified to meet the standards in 1981. Present engines do not meet the standards for gaseous emission levels. The technology needed to bring unburned hydrocarbons and CO emissions within the guidelines, using advanced diffusers and mixing geometries to improve combustion efficiency, is being incorporated in combustor designs for advanced versions of the CF6 and JT9D engines. The really difficult area seems to be the achievement of the $NO_x$ standards. NASA and industry are investigating means of providing the two-stage combustion required for major reductions in $NO_x$ emissions, in configurations which will be economically acceptable and reliable in service.

## Rail Emissions

The visible emissions from diesel-electric locomotives have been regulated in a rigorous fashion compatible with available technology and good operating practices. Measures that would require further reductions in $NO^x$ emissions or impose further controls on particulate emissions could threaten the capability of the industry to use diesel-electric locomotives.

## Marine Emissions

Introduction of the large marine tanker has raised serious concern about the effects of a spill on marine life and coastal environments. As yet, there is no effective way to remove or neutralize the large volume of petroleum products that could be released in an accident involving a supertanker. Not only does the design of the tanker influence the probability of avoiding an accident but also the location, type, and management of ports have an influence upon the probability of an accident and the amount and effect of pollution resulting from such accidents. Port facilities, traffic density and patterns, types of vessels, and proximity to environmentally sensitive shore areas are all factors affecting accidental pollution.

Tanker accidents account for an estimated 200,000 tons, or only 3.3 percent, of the estimated 6 million tons of oil that find their way into the oceans each year from land and ships. Most tanker discharges—some 1.3 million tons a year—come not from accidents but from routine operations such as tank washing and deballasting.

Measures to reduce pollution may be characterized as preventive or curative. Considerable research and prototype development have gone into curative procedures aimed at containing the area affected by pollution following a spill and breaking up large spills into smaller ones that can be managed more effectively; development work also has been done on equipment to remove polluted sea water, oil-water separation processes, and demulsifiers that can be applied directly to the oil spills.

In spite of considerable progress on curative measures, most of them cannot be used effectively in the high seas in which accidents frequently occur. Existing procedures are limited to 5-foot seas. Although further work may permit their safe use in 7-foot seas, little can be done safely and effectively in greater than 10-foot seas. For this reason, as well as the long time it usually takes to get equipment and workers to the affected area, there is a growing emphasis on preventive measures. These involve reducing human error through better training, tighter licensing, and new relicensing requirements; improving navigational equipment on both ship and shore; and refining ship routing and ship-to-ship communications.

Consideration is being given to requiring proved equipment, such as gyro-compass, radar, Loran, and Decca, to be installed on all medium and large tankers. More modern systems such as Omega, which tracks shore-transmitted signals in all weather, and computer-assisted collision-avoidance equipment, are available in prototype form and appear to offer potential for reducing accidents.

Preliminary experience with traffic separation and routing schemes in coastal areas with high traffic density, such as the Dover Straits, has demonstrated their value in preventing accidents. The final element in safe traffic routing and control is reliable communication between ships in close proximity. The U.S. government is supporting mandatory bridge-to-bridge communication systems.

## Noise Emissions

Audible noise also is considered a potentially serious emission problem.

***Aircraft*** Although many older, noisier aircraft are still in service, great progress has been made in noise reduction technology over the past 20 years. The modern high-bypass turbofan transport produces at least a 25 decibel lower noise level than did the much smaller 707/DC-8 turbojet. Much of this reduction has taken place in the engine itself, largely resulting from an increase in the bypass ratio adopted initially to improve cruise fuel economy. Nacelle acoustic treatment and noise abatement procedures involving variations in flight path, power, flap setting, and climb or approach angle have brought further gains. As the older aircraft are retired during the early 1980's, it is expected that the transport fleet will meet the existing Federal Aviation Regulation 36 Stage 2 noise rules, and that most of the fleet will comply with the more stringent 1978 Stage 3 levels for certification of new aircraft.

Nevertheless, with increased operations, "stretched" models, and population growth near airports, aircraft noise will remain an issue in the years ahead. Further noise reduction appears possible, but it will be increasingly difficult and may impose considerable economic penalties. One of the difficulties is that during take-off or landing large aircraft may generate airframe noise which, apart from propulsion noise, can be expected to approach Stage 3 levels. Consideration of increasingly more stringent noise regulations must include careful assessment of technical and economic feasibility, safety, and alternative options such as land-use planning to minimize the population near airports.

***Cars and Trucks*** While noise levels for cars are not currently regulated in the United States, the proposed standards will be achieved only through careful design. The situation for trucks is much more critical. Federal regulations preempt state and local noise standards for trucks of over 10,000 pounds gross vehicle weight; however, for trucks 10,000 pounds and under, state and local municipality standards apply. In 1979, these latter standards varied from 83 to 88 decibels, as measured by the procedure developed by the Society of Automotive Engineers; some municipalities specify levels as low as 74 and 75 decibels between 1980 and 1983. These noise levels are unusually stringent, in that a major portion of the noise arises from tire-highway interaction and from the cooling fan, areas which offer very limited opportunities for noise reduction. Major research efforts are under way to determine the prospects for meeting these reduced levels.

***Railroad*** Regulation of railroad noise is evolving slowly. The issues concern the amount of noise permissible from the train itself, at property lines bordering the track, and at other property lines.

***Electrical*** Radio frequency noise is becoming more serious with the increased saturation of the radio frequency spectrum. While the U.S. government does not regulate radio frequency emissions, all manufacturers have agreed that U.S. vehicles will meet the standard established by the Canadian government. The principal source of radio frequency noise from the gasoline engine is the ignition system, with the interruption of high voltage at the distributor, the discharge in the spark plug, and the radiation of energy by the wires. The diesel engine, by the nature of its combustion process, has much less radio frequency emission.

## INTERNATIONAL COMPETITIVE POSITION

The international competitive position of the transportation sector is of tremendous importance to the economic health of this nation. In 1979, U.S. airframe manufacturers exported $7 billion in products (including military exports). The United States imported $17 billion worth of cars, trucks, and buses in the same year. Several key factors are involved in the foreign export-import issue.

The petroleum shortages of 1974 and 1978 created a strong demand for fuel-efficient cars. Since Japanese and European automobile manufacturers have historically served a small-car market, they were able to meet the new U.S. demand more quickly than the U.S. manufacturers. The popular perception then arose that foreign manufacturers were achieving high fuel economy by using advanced technology that was not available to U.S. manufacturers. However, the fuel economy data published annually by the Environmental Protection Agency,[8] indicate that on the average there is no consistent superiority of foreign manufacturers as long as the comparison between foreign- and domestic-manufactured vehicles is made at the same vehicle weight. To be fully competitive with overseas manufacturers, it will be necessary for U.S. manufacturers to improve the reliability and quality of their product.

While the United States still dominates the world commercial airframe market, foreign competition is developing the capability to produce transport, commuter, and helicopter designs as technically advanced as those of our own industry. England, France, West Germany, and the Netherlands, as well as other countries, support substantial aeronautical research and development in government-operated and -funded laboratories in a manner patterned after the NASA approach to aeronautical research. It is significant that, unlike the United States, companies in these countries tend to join forces in civil aircraft development partnerships.

The European A300 large transport aircraft is successfully competing against U.S. models in the world market. The European share of the world helicopter market has grown from 32 percent in the 1970–74 period to 47 percent in the 1975–79 period.

The market for transport aircraft is likely to exceed $10 billion per year over the next decade. In order to compete successfully for this business, industry and government advanced technology programs must be carried on for five to seven years at a cost of hundreds of millions of dollars. These programs will continue to be a prerequisite to private investment of the billions of dollars needed to develop the new aircraft.

## RESEARCH AND DEVELOPMENT SUPPORT

The level of research effort, both in the government and in private industry, varies greatly by sector.

There is little research and development in the private marine sector. Those few companies which had research and development divisions have generally disbanded such operations because of the depressed shipping market. Most research and development are now done by the Maritime Administration, recently transferred to the Department of Transportation.

The highway vehicle industry has established major research and development facilities. Reports to the Securities and Exchange Commission indicate that U.S. industry as a whole spends more than $4 billion annually on research, development, and testing. It is estimated that about one fourth of this relates to advanced product development and research. Support for basic research could be as much as $40 million annually.[9]

The fiscal year 1979 federal research and development budget for highway vehicles was about $250 million, with less than $40 million devoted to basic research on the long-term needs of the automotive industry. This modest federal spending for long-term research represents a serious deficiency in the support of the relevant technology that industry requires. In the rail industry, research facilities and resources are generally very limited. The annual expenditure is about $75 million, almost all of it provided by the federal government. Much needs to be done to increase the level of research and to expand the capability to apply the results of it.

The aircraft industry and the government have a long history of supporting jointly the long-term needs of that industry. Starting with the National Advisory Committee for Aeronautics and continuing under NASA, the federal government has regularly carried on the long-term research that has been necessary to maintain the commercial aircraft industry at the forefront of technology. Major research and development facilities are provided by NASA. While the industry conducts limited research and technology programs, primarily with respect to specific problems encountered or anticipated in current developments, it is principally involved in the effective application of existing technology in the design, development, and production of new products rather than in the generation of new technology itself. The industry also devotes a significant effort to ensuring that it is technologically alert.

# Summary and Outlook

The challenge of sustaining and expanding the transportation system over the coming years exists at two levels: to improve the operating characteristics of each transportation mode—its efficiency, environmental impact, and safety; and to improve the functioning of the system as a whole—including the linking together of each mode to provide the maximum system efficiency. While it has a large role to play at both levels, technology is mandated at each level by social, economic, and political forces. The ability of technology itself to effect radical changes in the national transportation system is limited by the system's immense size, complexity, and cost, and by the time that is required to develop and introduce new vehicles and facilities. No miracle vehicles or fuels promise to revolutionize the way we travel or ship goods in the next five years. However, incremental improvements will continue to be made in vehicles, in the flow capacity of existing facilities, and in the integration of the overall system. Each of these is necessary to meet the rising demand for transportation while minimizing investment in new facilities and lowering the energy required for each passenger mile and ton mile.

## VEHICLE CHARACTERISTICS

### Energy Efficiency

The greatest improvements in energy efficiency by 1985 will occur in the automobile and the commercial airplane; it is expected that the energy efficiency of new cars and airplanes will almost double between 1975 and 1985. Diesel- and electric-powered modes—bus, train, ship, and truck—will realize smaller gains. As a result, whereas the bus was about three times as efficient as the automobile in 1975, it will be only 60 percent more efficient in 1985 (at 1975 levels of occupancy). In the years following 1985, two-passenger cars may be available that will match or exceed the efficiencies of bus and rail in urban transportation. These comparisons are valid internationally; when vehicles of equal weight are compared, the energy efficiency of domestic automobiles matches those of imports. This is expected to continue. Energy efficiency comparisons among the modes will become less of a factor in policy decisions. A key challenge in improving the energy efficiency of passenger transport will be to increase the average number of passengers per trip.

### Alternative Fuels

Various fuels derived from nonpetroleum sources are candidates for use in the transportation system. While it is expected that engines and fuel distribution systems can be developed that will operate on most synthetic liquid fuels, existing systems and engines are not compatible with some of the synthetic fuels, and gaseous fuels offer only a relatively short vehicle range. Electricity is an alternative. The electrical technology exists for trains, but the costs are high. Battery-powered vehicles using electricity generated by coal would be about as energy efficient as 1985 cars powered by coal-derived synthetic liquids. Thus, an overall system analysis is necessary if alternative fuels are to be introduced in the most efficient manner.

## SYSTEM MANAGEMENT AND INTEGRATION

Important opportunities exist to improve the overall performance of the system by smoothing the flow of people and goods at the point where modes intersect, and to increase the capacity of each mode. Better air traffic control technology can help to increase airport capacity by reducing delays in take-off and landing. The extension of commercial aircraft technology to smaller (general aviation) aircraft and better helicopter designs could expand their use on short-haul trips from major airports to smaller ones.

Computerized freight control systems are increasing the fraction of time that railcars carry goods and are helping to expand the integrated use of trucks, railroads, and ships to haul containerized cargo. Greater use of integrated transportation will be aided by the elimination of remaining regulatory prohibitions.

Improved ground and satellite communications are also vital for safer ship navigation, monitoring the transportation of hazardous materials, and improved highway traffic flow.

## RESEARCH NEEDS

The long-term challenges to the transportation sector demand new answers for long-standing problems. Research is a key to this. An improved understanding of the fundamental limitations upon energy efficiency demands that research continue on basic combustion phenomena, on finding new materials and better ways of using materials that will reduce vehicle weight, and on understanding more completely the basic limitations imposed by friction and aerodynamic losses. Exploration of new propulsion systems—that is, engines and their accessories—must continue. Passenger safety will improve through a better understanding of energy dissipation in a collision, through improved management of vehicles in circumstances where collisions are possible, and through improved communication systems. Reducing the environmental impact of the transportation sector in a cost-effective manner will require an improved understanding of the processes that generate pollutants and the relationship of these processes to the operating modes of the transportation system. Reducing the dependency of the transportation sector upon imported petroleum demands careful evaluation of the alternative energy sources and the means by which they can be introduced into commerce. While the transportation sector will continue to benefit from advances originally developed in other contexts, many of the fundamental problems confronting transportation are unique to it and will require the continued support of research by industry and government.

## REFERENCES

1. Joseph R. Wagner. *Brookhaven Energy Transportation Submodel (BETS) Documentation and Results, BNL Report No. 50902, October 20, 1977.* Upton, N.Y.: Brookhaven National Laboratory, 1979. *The National Transportation Policy Study Commission in 1979: National Transportation Policies Through the Year 2000: Final Report, June 1979.* Pub. No. G.P.O. 052-003-00669-3. Washington, D.C.: U.S. Government Printing Office, 1979.

2. Ibid.

3. *The U.S. Department of Transportation, Research and Special Program Administration, Transportation Systems Center in 1979: National Transportation Statistics.* Pub. No. DOT-TSC-RSPA-79-19; G.P.O. No. 050-000-00154. Washington, D.C.: U.S. Government Printing Office, 1979.

4. D. L. Green et al. *Regional Transportation Energy Conservation Data Book.* Edition One. ORNL-5435 Special. Oak Ridge, Tenn.: Oak Ridge National Laboratories Energy Division, 1978.

5. Ibid.

6. *Energy Use Management: Proceedings of the International Conference.* Volumes Three and Four. Edited by Rocco Fazzolure and C. B. Smith. New York: Pergamon Press, Inc., 1978.

7. T. L. Ullman et al. *Effects of Six Variables on Diesel Exhaust Particulate.* American Society of Mechanical Engineers Paper No. 80-DGP-42, presented before the Energy Technology Conference and Exhibition, New Orleans, Louisiana, February 3-7, 1980. New York: American Society of Mechanical Engineers, 1980. E. R. Danielson. "Memorandum to Mobil Source Docket No. OMSAPC78-4." Washington, D.C., 1980. Cummins Engine Company, Inc. *Status of Programs to Meet 1979, 1980, 1983 California Emissions Standards.* Submitted to the California Air Resources Board, December 1978.

8. U.S. Environmental Protection Agency and the U.S. Department of Energy. *1979 Fuel Economy Program, 49-State Test Car List.* February 2, 1979. *1980 Fuel Economy Program, 49-State Test Car List.* October 12, 1979. Washington, D.C.: U.S. Department of Energy, 1979.

9. Based on information contained in the 1979 Form 10-K, Annual Report, required by sections 13 (Periodical and Other Reports) and 15(d) (Broker Registration and Regulation) of the Securities Exchange Act of 1934, and amended (Public Law 95-213), submitted by American Motors Corporation, Chrysler Corporation, Ford Motor Company, and General Motors Corporation.

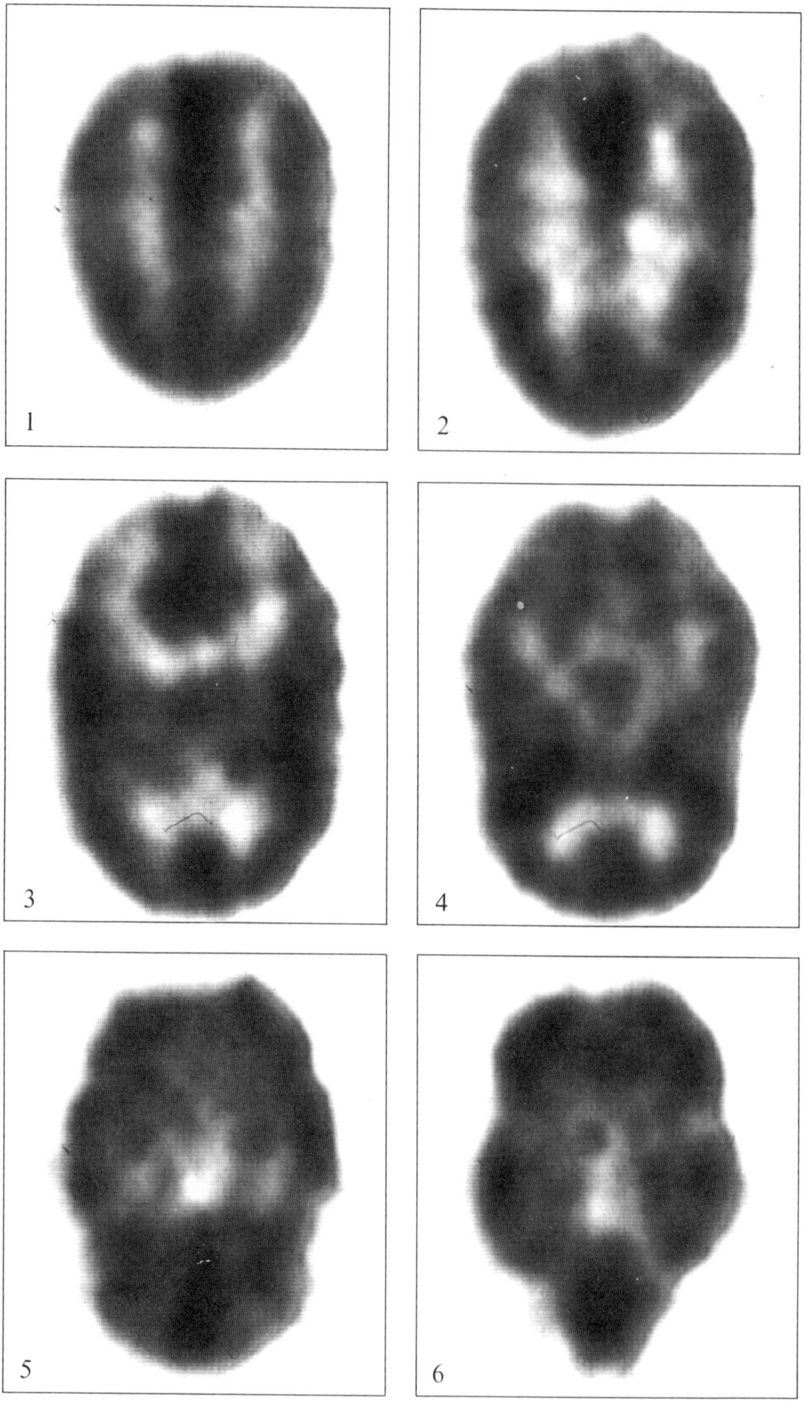
1
2
3
4
5
6

# 17

# *Prospects for New Technologies*

## INTRODUCTION

Many attempts to look far into the future of technology fail in part because events often move faster than all but the rarest of minds can predict. Moreover, unforeseen discoveries may accelerate, inhibit, or divert the course of what seemed to be obvious developments. The invention of the transistor, for example, sidetracked the gradual evolution of vacuum tube technology. The discovery of the relatively simple structure of the molecules that contain the genetic code stored in the cell has accelerated the development of genetic engineering to a point far beyond what one would have guessed as recently as 15 years ago.

But, even though many past predictions have failed, new technology can change our lives so much that we are obliged to speculate as well as we can both on emerging technologies and on emerging situations that have large technological components. This chapter focuses on seven such areas. The topics were chosen without concern for merging them into a seamless whole, but each involves technological issues that could exert major influences on the course of industry and hence on our lives and the course of society.

The first two areas are innovations in medical technology and the uses

---

◀ Positron emission tomography of a normal brain. The images, taken from one scan, record activity at different brain segments, from the bottom up. With this technique, "one can determine patterns of blood flow, blood volume, oxygen perfusion, and various other physiologic, metabolic, and immunologic parameters." [Clinical Center, National Institutes of Health—Rick McCleary.]

of recombinant DNA (deoxyribonucleic acid) methodology. Both reflect the strength of the biological sciences.

Third and fourth are superconductivity and energy storage, intensely active areas in physical science and technology.

Fifth is the prospects for finding deposits of minerals other than fossil hydrocarbons, a subject that is apt to be slighted in this time of great concern with energy resources.

Sixth is the import of the Space Shuttle and particularly of the space telescope.

The seventh area is more long range and speculative than the others. It concerns the extent to which information-processing devices—currently undergoing rapid evolution as a result of dramatic gains in the capacity, speed, and sophistication of microprocessors—can be expected to reach a point where they can be said to think in a manner approaching our perception of thinking in living systems.

## MEDICAL TECHNOLOGY*

Over the years, medical science has benefited increasingly from work in other fields, as is evident in the new technologies that continue to emerge from the laboratory. Those outlined here reflect contributions from electronics, materials science and engineering, nuclear physics, biology, and several branches of chemistry, among other disciplines. The ubiquitous computer also has found a solid foothold in medicine, although its effectiveness suffers somewhat from a lack of standardization both in programs and in protocols. These new technologies should find many uses in diagnosis and treatment during the next five years; in the longer range, they should contribute substantially to our basic understanding of the body and its interactions with its environment. In addition, recombinant DNA methodology (see next section) will be exploited not only in scientific research but also in medical and commercial technologies.

### Microcomputers

The new microelectronics is influencing medical technology profoundly; the trend is evident in developments cited below. Application of the com-

*The following individuals helped to prepare this section: N. G. Anderson, Argonne National Laboratory; F. Dunn, University of Illinois at Urbana-Champaign; M. Goitein, Massachusetts General Hospital, Harvard Medical School; E. C. Gregg, Case Western Reserve University; C. Heidelberger, University of Southern California at Los Angeles; R. M. Krause, National Institute of Allergy and Infectious Diseases, the National Institutes of Health; H. R. Lehneis, Institute of Rehabilitative Medicine, New York University Medical Center; A. R. Margulis, University of California at San Francisco; A. W. Pratt, National Institutes of Health, Division of Computer Research and Technology; G. Rosenthal, National Center for Health Services Research; and H. R. Rusk, Institute of Rehabilitative Medicine, New York University Medical Center.

puter to medicine has spawned a host of new technologies whose development has been accelerated by advances in the manufacture of integrated circuits. Today's microprocessors, which consist of tens of thousands of transistors and related components on silicon chips a quarter of an inch square, have markedly shrunk both the size and cost of computers. The resulting microcomputers are finding an expanding range of uses. One recent survey has projected a market of more than $1.3 billion for microcomputer-based medical devices during the next decade.

The newly developing medical applications of microcomputer technology fall into three broad categories: incorporation into laboratory instruments for control, sequencing, and computation; use in clinical devices to aid in physiological monitoring, diagnostic or therapeutic procedures, and rehabilitation; and use in interfacing older equipment to automated data collection and analysis systems.

In clinical and research laboratories, microcomputers are widely used in analytical apparatus for expediting the acquisition, processing, and storage of data; computations; and printing or plotting results. Clinical applications include monitoring systems for intensive care and cardiac catheterization units, infusion pumps for constant-rate administration of intravenous fluid, and various devices for microscopic and radiographic imaging.

Future applications include systems for improving access to medical information in computerized data bases, assisting in history-taking and diagnosis, monitoring patient care, auditing routine decisionmaking, and expediting communication. Other applications will include increasing use in prostheses (replacement of body parts) of various kinds, including devices for replacing lost physiological or sensory functions.

## Imaging Systems

Rapid progress in means of obtaining images of the body's interior is leading to steadily improving ability to diagnose ailments and study physiological functioning. Promising new imaging systems include X-ray systems, positron emission tomography, ultrasound systems, nuclear magnetic resonance, and the germanium camera.

***X-Ray Systems*** In the short time since its introduction, computerized axial tomographic (CAT) scanning has come to play a key role in the diagnosis and management of many diseases. The technique involves computer reconstruction of X-ray data to give images of plane sections through the body. In the foreseeable future, advances in CAT technology should bring improved resolution and scanners in which electron scanning of crescent-shaped sources produces focused X-rays. The latter development would eliminate moving parts and permit extremely rapid scanning—in the millisecond range—for capturing transient events in the body. Such scanners should facilitate dynamic studies of various kinds, notably in patients with circulatory disorders. Now being designed for portable use in the field—by the military, for example—are CAT scanners that use isotopes such as gadolinium-153 as sources and so do not require a conventional X-ray generator.

Further application of the computer to X-ray imaging systems will result in a gradual shift toward all-electronic computerized detection, storage, and replay of radiographic information. Such a shift ultimately may permit the physician's judgment to be strengthened by computer diagnoses of certain diseases, given improvements in resolution, contrast, computer storage and analysis of information, and the capability for random computer access and retrieval. Such developments also will permit more rapid and accurate communication of information in the course of making diagnoses and caring for patients.

Another important foreseeable advance is the development of new contrast materials with increased X-ray absorption, decreased toxicity, and enhanced specificity for particular organs and disease states.

***Positron Emission Tomography*** The use of positron cameras employing computed tomographic reconstruction techniques makes possible the quantitative measurement of the activity of positron-emitting isotopes throughout scanned sections of the body. (The positron is a positively charged particle with the mass of an electron.) By labeling appropriate chemicals with cyclotron-produced, short-lived, positron-emitting isotopes such as carbon-11, one can determine patterns of blood flow, blood volume, oxygen perfusion, and various other physiologic, metabolic, and immunologic parameters. This technology, still in its infancy, should enjoy increasing use in medical diagnosis and research within the near future, especially in studies of brain and heart function.

***Ultrasound Imaging*** Ultrasound imaging, already an established tool, should be used increasingly in medical diagnosis in the years ahead. In contrast to X-radiation, ultrasound radiation is not known to be mutagenic, carcinogenic, or teratogenic. Thus, it offers the advantages of lower potential risks in mass screening of humans. The further employment of computers in ultrasound imaging, along with the development of techniques for producing focused ultrasonic fields, should lead to major improvements in resolution, bringing the capability for dynamic studies of cardiac function and other physiologic parameters.

***Nuclear Magnetic Resonance*** Another promising imaging technique involves the application of nuclear magnetic resonance (NMR) zeugmatography. Protons in hydrogen atoms, when placed in an external magnetic field, respond to an applied radio frequency by reemitting radio waves of the same frequency. Thus, images can be obtained which are comparable in resolution and contrast with those obtainable with CAT. In principle, many different atomic environments can be imaged, making NMR a particularly fruitful field for investigation.

The method has been used with animals and to obtain images of the human brain. There is some evidence that cancer cells can be distinguished from their normal counterparts by NMR signals. Also, macroscopic movement affects the NMR signal, and an instrument based on this fact has been developed to measure blood flow. The ability to image atoms of fluorine-19, phosphorus-31, and other elements in addition to hydrogen also raises the possibility of using labeled compounds for various tracer studies.

***Germanium Camera*** Among noteworthy advances in imaging techniques is the development of germanium cameras for use in nuclear medicine. The development is promising because of the high efficiency of germanium detectors for photons in useful ranges of energy. Such cameras should yield films of significantly improved resolution and image quality for a given dose, thus reducing the doses involved in diagnosis. The technique is applicable to organs such as the thyroid, liver, and spleen and recently has been used to obtain an image of a wrist that shows every tendon.

## Radiation Therapy of Cancer

Imaging techniques mentioned above, by better localizing and delineating the extent of cancer in a patient, will aid greatly in the choice of appropriate therapy and in monitoring the patient's response to treatment. In addition, although treatment of cancer with radiation is a long-established practice, several newly developing methods of radiation therapy will find increasing application in years to come.

New treatment methods include the use of ionizing particles—such as neutrons, protons, mesons, alpha particles, and heavy ions—of high-linear energy transfer (LET). Experience to date with high-LET radiation therapy has been limited to small clinical trials, primarily with accelerators designed for physics research. Although the results are promising, full evaluation of the potential value of high-LET radiation in cancer treatment will depend on expanded studies now being mounted with new particle accelerators designed expressly for radiotherapeutic use. In principle, high-LET radiation offers the advantages over conventional X- or gamma radiation of more selective localization of the dose to the tumor and greater tumoricidal action for a given dose (although neutrons, the most widely used high-LET radiation, while offering greater tumoricidal action, provide less localization of dose).

Another noteworthy approach is the development of drugs that sensitize oxygen-deficient cells. Such drugs increase the radiosensitivity of poorly vascularized parts of tumors, which otherwise might escape destruction during radiotherapy with X-rays or gamma rays. To date, the most serious limitation of such drugs has been their high toxicity, but research to develop less toxic analogs appears promising.

Also noteworthy is growing evidence that hyperthermia, or heating, can increase the sensitivity of cancer cells to radiation and anticancer drugs. The combined use of hyperthermia and radiation promises to improve the cure rates for certain malignancies.

## Chromosome Analysis

New techniques for chromosome analysis have increased more than 50-fold the number of bandlike markings by which different chromosomal regions are distinguishable from one another. The technique involves microscopic examination of human cells in late prophase (the initial step in cell division). Further refinement and application of this approach should advance greatly the mapping of the human genome—the full complement of genetic information—and our understanding of the role of genetic determinants in health and disease.

## Protein Analysis

High-resolution, two-dimensional electrophoresis using acrylamide gel provides in theory a resolving power capable of separating virtually all of the protein products of the structural genes of man. These genes are estimated to number between 30,000 and 50,000. To attain the resolution needed for such an exacting analytical separation, two-dimensional mapping systems are being developed which can resolve both highly acid and highly alkaline peptides (fragments of proteins). Such systems would reveal, for example, what proteins are present in a tissue sample under differing stresses and, ultimately, even what genes are turned on. The refinement and application of this technology to the analysis of biochemical correlates of health and disease should prove useful in research on normal growth and development, detection of genetic variants, and early diagnosis.

## Prosthetics and Orthotics

Within the near future, advances in biological materials should improve the quality and durability of the two to three million prostheses and orthoses (reinforcements and supports for body parts, such as metal pins and rods) that are implanted yearly in this country. We can now create biological materials with biological and physical properties heretofore unattainable. But, until we know more about the substances that normally control the bonding of cells to each other, it will be difficult to achieve optimal adherence or lack of adherence between implant devices and tissues.

Other advances are expected to come from the expanding use of microprocessors in prostheses. Microprocessor-based devices offer several advantages: the ability to process signals and thus partially compensate for impaired neurological function; adaptability to the special needs of individual patients through modifications in software; and the capacity to be mobile and portable. For example, microcomputer-based prostheses can control motorized artificial hands by responding to myoelectric signals imparted from forearm stumps, allowing the prostheses to be manipulated naturally like normal hands. Among applications under development are aids to hearing and sound perception, synthetic voices, voice-actuated control systems for wheelchairs, gaze-controlled devices, and other aids.

## Synthesis of New Biochemicals

In addition to the use of recombinant DNA technology to expedite the production of selected compounds (discussed in the next section), new chemicals with desired molecular structures are being synthesized using other innovative approaches.

One example concerns the development of new anticancer drugs. Recent advances in enzymology have led to two concepts, among others, that promise to permit the synthesis of increasingly powerful and irreversible inhibitors of enzymes that play key roles in the metabolism of cells. The idea is to use such inhibitors to interfere selectively with the metabolism and growth of tumor cells. One concept involves the design of analogs that mimic the target enzyme, leading to powerful inhibitions. (An example of this type of inhibitor currently under clinical study is PALA, or phosphonoacetyl-L-aspartase.)

The second concept involves what is called a suicide inhibitor, which is chemically inert but resembles the substrate of the enzyme and thus binds to it. Once bound, the inhibitor is promptly converted by the enzyme into a highly reactive compound that irreversibly inactivates the enzyme. Both of these approaches hold promise for the development of new anticancer drugs.

A related approach is the development of chemicals that inhibit the multiplication of viruses by exploiting their ability to code for the biosynthesis of enzymes not present in uninfected cells. The discovery of interferon, which inhibits the multiplication of viruses while leaving host-cell functions relatively undisturbed, indicates the feasibility of this approach. Recent progress in virology is revealing a growing number of virus-coded proteins that should be candidates for specific inhibition by chemical reagents. Application of the same approach to the development of agents for treating parasitic and other infections should be similarly productive.

### Monoclonal Antibodies

Another powerful new technique is a method for producing large quantities of monoclonal antibodies that can be used like well-defined chemical reagents. The method makes use of hybridomas, hybrid cells formed by fusing antibody-forming spleen cells with mouse tumor (plasmocytoma) cells. The hybrid cells so formed produce a single type of antibody molecule—namely, the one elaborated by the parental spleen cells. The cells can be grown as a continuous line from which large quantities of the antibody may be harvested. By immunizing the spleen-cell donors against specific antigens, antibodies of corresponding specificities may be obtained. This technique should have wide application in the development of reagents for diagnosing and treating disease and for biomedical research. (Chapter 11 offers further examples of such chemical tactics.)

## RECOMBINANT DNA

Only a few years ago, recombinant DNA was arousing concern if not fear among the public and some specialists; it seemed to some to be another scientific bogeyman. Today, although the expression recombinant DNA retains a slightly pejorative flavor, it is used commonly to describe a technology for constructing and isolating beneficial molecules using techniques developed by molecular geneticists. This technology evidently is spurring the formation of a new industry. At the same time, intensive research with recombinant DNA is steadily expanding our basic understanding of molecular genetics. The rapid movement of knowledge in this field—only recently considered esoteric and impractical—from the laboratory directly into industry illustrates strikingly how unpredictable are the uses of new knowledge.

### Recombinant DNA Defined

Recombinant DNA is defined in guidelines issued in the early 1970's by the National Institutes of Health to ensure safety in research in the field. The guidelines define recombinant DNA molecules "as either (i) molecules which are constructed outside living cells by joining natural or synthetic DNA segments to DNA molecules that can replicate in a living cell or (ii) DNA molecules that result from the replication of those described in (i) above."

A recombinant DNA experiment has three components: the DNA to be cloned or replicated (the foreign DNA); a segment of DNA (the vector) that can replicate in a host cell; and the host cell itself. The foreign DNA is spliced into the vector DNA, and the combined segments are inserted into the host cell for propagation. The host cell and vector usually are closely related; the DNA to be cloned can come from any organism (hence, foreign DNA).

This technique makes available for study specific segments of the chainlike DNA—the substance of genes—from any organism. The DNA can be analyzed down to the level of its individual subunits, called bases. There are four such distinct bases. The sequence in which they occur along the DNA chains of genes constitutes the coded information that makes up the genetic heritage of organisms. When a cell propagates, most of the genetic information coded into its DNA directs the synthesis of proteins. These long-chain molecules comprise most of the working machinery of the cell. They are formed in several steps, including transcription of the genetic information into the long-chain molecule RNA (ribonucleic acid) and subsequent translation into protein.

In some cases, the function of a specific segment of DNA can be ascertained and its gene product isolated. Recombinant DNA experiments, therefore, provide information on the organization, structure, and function of genes; they also can lead to the making of useful gene products.

## Processing DNA

Molecular geneticists have performed thousands of experiments using recombinant DNA, and the basic technique is well established. Specific fragments of the foreign DNA are produced by enzymes, called restriction enzymes, that recognize certain short sequences of bases in the DNA chain and cleave it where those sequences occur. The desired fragments of foreign DNA are isolated by standard methods for insertion into the vector. The two major classes of vectors are the DNA viruses and the plasmids, naked rings of DNA that occur independently in cells. The vector selected is cut with a restriction enzyme, and the foreign DNA is spliced into the gap using another enzyme, DNA ligase. The host may be any cell in which the recombined vector DNA can replicate; the major limiting factor is the existence of means of inserting the vector into the cell. The most widely used host to date has been a strain of the bacterium *Escherichia coli*. With the vector inserted, host cells are propagated in a suitable medium to produce clone cells containing the identical segments of DNA. At each step in the preparation of recombinant DNA, it is possible to use experimental variations to suit special needs.

## Research Using Recombinant DNA

The recombinant DNA procedure is a powerful experimental tool. At the current stage of development, at least, it is more important to basic science than as the basis of a commercial technology. A few findings obtained with the procedure will illustrate its reach.

***Gene Structure*** In bacterial genes, a continuous sequence of bases codes for a given protein. Research using recombinant DNA has shown, however, that in many genes of higher organisms the sequence is in segments separated by long, noncoding sequences. The entire DNA sequence is transcribed into RNA, and the noncoding sequences are then excised by a mechanism still incompletely understood. Enzymes join the coding sequences into a continuous chain for translation into protein.

One author has described as amazing this discontinuous structure of many eukaryotic genes—that is, genes of organisms whose genetic DNA is bound in the nuclei of cells. The structure is important in two respects. First, since a primary transcript must be processed to form a message, the mechanisms that regulate the expression of genes might work at this level. Second, the segments of genes could be moved around in the genome, thereby providing the raw material for the evolution of new gene functions.

***Antibody Diversity*** Studies with recombinant DNA also have shown that the many theories proposed to explain the enormous diversity of antibodies, important elements of the body's immune system, have been premature. There are several classes of antibodies. An antibody molecule is composed of two light and two heavy chains of protein. Each chain is composed of one region of variable amino acid sequence and one in which the sequence is the same in all antibodies of a particular class. An antibody interacts specifically with one of thousands of antigens, or foreign substances, and the problem has been to explain this diversity in terms of the basic structure of the molecules.

Work with genes for the light chains of antibodies has revealed that the constant (C) and variable (V) regions are on different DNA fragments. A new DNA sequence (J) was found to be associated with the constant region and to code for about a dozen amino acids at the junction of the variable and constant regions. There are about 5 different J regions and 20 different V regions. Sometime during the development of an organism, light chains are made when a V gene is moved near enough to C-J genes to be part of the same transcription unit. Recent findings show that a heavy chain is generated in a similar if somewhat more complicated fashion. These unexpected arrangements and rearrangements of antibody genes provide a strong clue to the mechanisms of generation of antibody diversity. The numbers of combinations made possible by several V and J genes in each antibody chain and in each class of antibody could be sufficient to account for this diversity.

***Functionless DNA*** A third finding of research with recombinant DNA concerns a class of DNA known as intermediate repeat DNA. The fruit fly *Drosophila*, among other organisms, has this DNA dispersed throughout its genome. It has been assumed to play an important regulatory role, providing a simple network by which the functioning of disparate parts of the genome might be coordinated. However, research using the giant salivary gland chromosomes of *Drosophila* has raised considerable doubt that intermediate repeat DNA has a regulatory or any other function, since the same elements are found in different places in different strains. These and similar results have led some to speculate that a considerable fraction of the eukaryotic genome is composed of DNA that is functionless and exists only because it can be replicated and does not harm the organism.

***Fundamental Questions*** The deeper the eukaryotic genome is probed, the more surprises occur. More questions have been raised than have been answered. Still, only with detailed and clever exploitation of recombinant DNA techniques is there any hope of answering the fundamental questions relating to the development of an organism from a fertilized egg.

## Industrial Application

Recombinant DN technology in theory should permit any protein or small set of proteins to be made in convenient hosts. Organisms perform an almost infinite variety of chemical conversions by means of enzymes (protein catalysts); the genes that specify the synthesis of the desired enzymes could be recombined and used to produce the enzymes. The limiting factors are the ability to isolate the particular gene and to arrange matters so that it functions efficiently in a host that replicates well in an inexpensive medium.

The isolation of the desired genes will require great ingenuity. Proper functioning of a gene in its host demands accurate transcription and translation. However, at least for modest rates of synthesis, knowledge of certain base sequences involved in directing the major steps of transcription and translation may suffice. These sequences can be synthesized or isolated from known host genes and added to the foreign DNA inserted into the host cell. For *E. coli*, base sequences that regulate transcription and translation are already known. The amounts of the desired substance produced can be increased by increasing the rates of transcription and translation, increasing the number of copies of the recombinant DNA, or both. The number of different host-vector systems needed for the various syntheses envisioned probably will be small. On the whole, therefore, the special knowledge needed to exploit recombinant DNA technology commercially should be finite and readily transferable to industrial operations.

## Manufacturing Prospects

Products will be made either directly or indirectly. In some instances, the gene product itself, such as insulin, will be the compound sought. In others, the desired product will be an enzyme or enzyme system to be used to perform some subsequent chemical conversion. Production generally will involve fermentation, one of the oldest of man's technologies. It is now highly developed and readily adaptable to the new needs. Sugar, the basic feedstock for many types of fermentation, might become limiting if the industry flourishes. However, recombinant DNA technology itself might help to produce modified organisms that can produce sugar efficiently by fermenting farm, forest, and municipal wastes that contain cellulose, hemicellulose, and lignin. The ability to use such surplus biomass would prevent the new large-scale fermentation processes from competing with the food supply—for example, grain.

What products will be made by recombinant DNA technology? At this early stage in the evolution of the industry, it is difficult to say. It must be recognized that, in general, the ability to make a particular product will be less important than the ability to do so at a profit or at a greater profit than by competing methods. Still, long lists of chemicals to be made have been drawn up; they include products generally associated with the pharmaceutical, chemical, energy, agricultural, and food industries.

The first commercial products of recombinant DNA technology undoubtedly will be pharmaceuticals. A number of them, such as insulin, growth hormone, and interferon, are closest to large-scale production. Vaccines for different viruses should not be far behind. The chemical industry will be considering, if not soon initiating, the production of compounds such as acetone, butanol (butyl alcohol), and ethanol (ethyl alcohol) by this technology. The energy industry is interested in ethanol and methane.

Considerable innovation would probaby be stimulated if biomass could replace the ever more expensive petrochemical feedstocks as a source of chemicals. Moreover, biochemical reactions are highly specific and efficient and have no real byproducts. In a pollution-conscious society, this fact in itself should lend impetus to conversion from chemistry to biochemistry in the chemical process industries. Organisms can perform many of the oxidations, reductions, hydrolyses, group transfers, and other transformations that are important to these industries. The difficulty will be engineering the proper strains of organisms for efficient production.

Food processing and agriculture also are prospects for recombinant DNA technology. At present, a few amino acids are made by fermentation. Others, such as lysine, present in small amounts in food protein, can be made by constructing the right organisms. The ability to grow many plants from single cells has stimulated much thought about genetically modifying plants by means beyond the usual species-limited method of breeding in new germ plasm. Coadaptation of free-living, nitrogen-fixing bacteria and nonleguminous plants would be one way to provide a ready source of nitrogen fertilizer, now largely a product of the chemical industry. Application of recombinant DNA technologies to food processing and agriculture probably will take time because the basic biology of plants is not well known. However, the evident prospects already have stimulated a considerable research effort.

### Unpredictable Future

The future of the nascent industry based on recombinant DNA technology cannot be predicted now. But the industry's size and domain are potentially very large. While the enterprise grows, basic research on recombinant DNA will continue. As new ideas and techniques emerge and develop, application of the technology will grow more and more feasible.

## SUPERCONDUCTOR TECHNOLOGY

Superconductivity was discovered by Kamerlingh Onnes in 1911, but not until the early 1960's did the phenomenon begin to emerge from the laboratory. It has done so along two paths, one leading generally to microdevices, the other to macrodevices. In both areas, substantial applications of new technologies lie not far ahead.

A superconducting material does not resist the flow of electricity and so conducts current without the loss of energy caused by the resistance of conventional conductors. Such a material loses its electrical resistance at a characteristic temperature called the critical temperature, and the critical temperatures of all known superconductors are not far above absolute zero (0 K [Kelvin] or $-273.2°C$ [Celsius]). To reach these temperatures, the materials must be refrigerated with liquefied gases; the higher the critical temperature, the lower the cost of refrigeration.

Microdevices based on superconductivity include microdetectors of electromagnetic energy having unprecedented sensitivity. They also include digital switching elements that permit the design of computers potentially more powerful than can be anticipated by extension of present semiconductor technology. Such thin-film superconducting devices are fabricated by processes already developed for semiconductor integrated circuits.

Superconducting macrodevices potentially can revolutionize the technology of generating, transmitting, and using electrical energy. Ingenious methods have been developed for producing long lengths of composite wire that incorporate superconducting filaments, and commercial quantities have recently become available. As a result, intermediate prototypes of some devices have been built, and the first full-size units are being planned.

## The Final Steps

The most important technical challenges posed by these emerging technologies involve the final steps of the journey from the laboratory to practical use. The superconductors required for switching elements have been found, their intrinsic properties are understood, and they can be fashioned into devices that work as designed, but questions remain: can these devices be assembled into systems having the reliability and manufacturing yield that will make them economically attractive? Similarly, can 5-megawatt macrodevices be scaled up 100-fold? The next five years should provide the answers to these questions.

## Microdevices

Bardeen, Cooper, and Schrieffer developed the basic microscopic theory of superconductivity in 1956. In this theory, conduction electrons of equal and opposite momenta form correlated pairs, all having the same quantum mechanical phase. Superconductors in which the correlation extends over large distances on an atomic scale are called type I; those in which it is relatively short are called type II. The type II materials hold much practical interest because they can be penetrated by lines of magnetic flux.

Brian Josephson noted in 1961 that if two superconductors were weakly coupled—separated by a thin oxide layer, for example—one could have current through such a junction without voltage. The size of the current depends on the phase difference between the two superconductors. The phenomenon is known as the direct-current (DC) Josephson effect, and the devices based on it are called Josephson junctions. The phase difference can be made to vary by placing the junction in a magnetic field and, in this way, the device can be switched from zero to finite voltage and back (it also can be switched by exceeding a critical current). Such devices are called SQUID's (superconducting quantum interference devices) and are basic to superconducting electronics.

An external voltage applied to a Josephson junction leads to oscillations in the supercurrent. Conversely, if the junction is exposed to an alternating-current (AC) signal, the current-voltage characteristic develops a step structure. These effects are called AC Josephson effects and are the basis for oscillators and detectors.

***Digital Electronics*** A Josephson switch is attractive for digital electronics for two reasons: it can switch its state in as little as six picoseconds (a picosecond is $10^{-12}$ second), and its power consumption is exceedingly small. In these respects, Josephson devices are far superior to semiconductor devices. The relevant figure of merit is the power-delay product. The smallest power-delay product for a Josephson device is $10^{-18}$ joule, compared to $3 \times 10^{-14}$ joule for the smallest bipolar semiconductor device.

Within the past few years, all of the functions needed for a computer have been implemented in Josephson technology. These functions include the logic and memory devices and even the power supplies. These various components are being assembled into large systems to study their architecture and design. It seems feasible to construct a computer with a cycle time of four nanoseconds (a nanosecond is $10^{-9}$ second). Since information travels along a superconducting transmission line at about 11 centimeters per nanosecond, such a computer would have to be compressed into a volume of the order of $20 \times 20 \times 20$ centimeters to take advantage of its speed.

***Analog Devices*** The most mature Josephson devices are measurement instruments that employ SQUID's. They are several orders of magnitude more sensitive than competing instruments for detecting magnetic fields and field gradients. Commercial systems have found uses in medical research (for example, monitoring magnetic fields associated with organs like the heart and brain), geophysics (for example, gravimetry and rock magnetism), and prospecting. Since the mid-1970's, the U.S. legal standard for the volt has been maintained using the Josephson voltage-frequency relation.

By combining Josephson sensors, AC/DC converters, and computer circuits, it should be possible to construct very powerful measurement systems. New regimes of real-time analyzers and digital filters appear possible. A fast multichannel spectrum analyzer coupled to a processor to examine data in real time could be used to reduce substantially the amounts of data that need to be stored or transmitted in certain situations.

***Materials and Fabrication*** Circuits of the type described above actually have been realized at the level regarded as characterizing large-scale integrated (LSI) circuits. Indeed, with the important exception of the formation of the tunnel junction, Josephson LSI technology employs processing tools and procedures very similar to those of semiconductor technologies.

The primary materials used at present in Josephson junctions are various lead (Pb) alloys. The underlying ground plane (both electrical and magnetic) is of elemental niobium (Nb). Silicon wafers are used as the substrate. Lead was chosen as the primary superconducting material be-

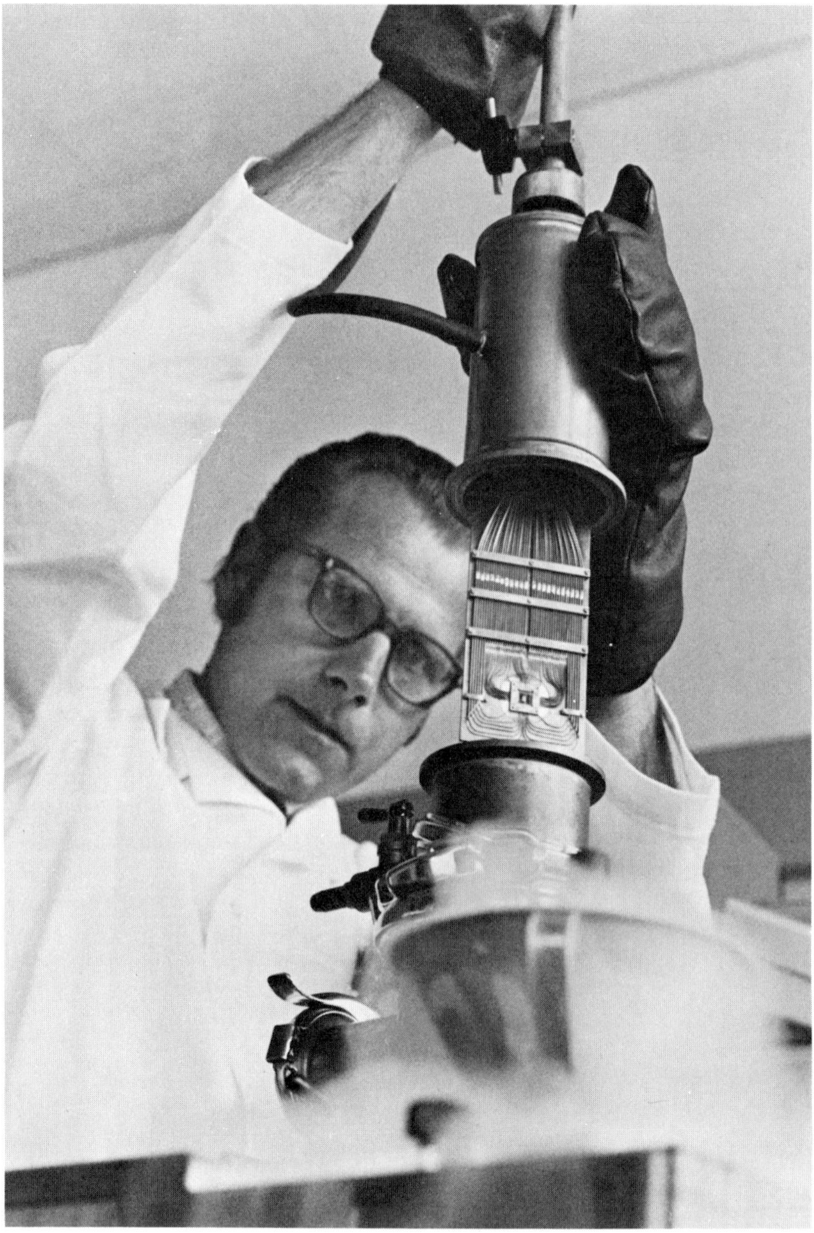

An experimental Josephson memory chip is lowered into a liquid helium bath for testing. "Within the past few years, all of the functions needed for a computer have been implemented in Josephson technology." [International Business Machines Corporation.]

cause of its relatively high critical temperature (approximately 7 K), the insensitivity of its properties to impurities, and the good dielectric properties of its oxides in tunnel junctions.

Tunnel junctions having electrodes of materials with a high transition temperature would be very advantageous. If they could be operated at, say, 15 K, the refrigeration problem would be eased considerably for small-scale applications, since small, portable, Stirling-cycle refrigerators are available in this range. Another advantage is that transition-metal alloys of niobium, tantalum, titanium, and zirconium and intermetallic compounds such as the niobium-tin compound $Nb_3Sn$, whose critical temperature is about 18 K, are very hard and resistant to chemical change. Recently, $Nb_3Sn$ and related superconductors of high critical temperature have been fabricated into Josephson junctions. Undoubtedly, there will be further progress in the synthesis of transition-metal superconducting films, which possibly could lead to a second generation of junctions with the advantages mentioned above.

## Macrodevices

The transformation that superconductor technology could be expected to make in the electric power industry would focus on the replacement of copper-wound, iron-based electromagnets by coreless superconducting coils housed in vacuum-insulated vessels and cooled by liquid helium. There would be two major advantages: less energy would be dissipated as heat, and the available magnetic field would be increased 5-fold, or even more than 20-fold in second-generation devices.

The superconductors that will be used in macrodevices in the next five years probably will be the brittle intermetallic compound $Nb_3Sn$ mentioned earlier and the relatively ductile alloy of niobium and titanium (Nb-Ti). $Nb_3Sn$ has superior superconducting properties, but Nb-Ti is easier to fabricate and is adequate for many purposes.

The diameter of the superconducting strands in a superconducting magnet must be quite small to avoid certain effects of the strong electromagnetic and mechanical forces that develop during operation. With Nb-Ti alloys, strand diameter is reduced by extruding a copper billet containing several rods of alloy. The resulting composite rods are drawn to wire sizes, rebundled, and redrawn until the diameter of the superconducting strands approaches one micrometer. As an extra refinement, the wire is twisted about its central axis to reduce the area of the eddy-current loops between neighboring superconducting filaments. An additional advantage of this type of conductor is the great mechanical strength arising from the reinforcement of the copper matrix by the alloy filaments.

It is more difficult to produce filamentary composites of inherently brittle intermetallic compounds such as $Nb_3Sn$. An ingenious solution for

**Table 1** Some recently built superconducting systems

| | Applications | | | |
|---|---|---|---|---|
| | Particle physics | Magnetohydro-dynamic power generation | Electric power alternator | Materials research |
| Device | Bubble chamber magnet | Duct magnet | Rotor magnet | High field research magnet |
| Location | Geneva, Switzerland | Argonne National Laboratory | Westinghouse | Intermagnetics General |
| Superconductor | Nb-Ti | Nb-Ti | Nb-Ti | $Nb_3Sn$-$V_3Ga$ |
| Maximum central field[a] (tesla) | 3.5 | 5.0 | 5.5 | 17.5 |
| Field volume (cubic meters) | 80 | 1 | $40 \times 10^{-2}$ | $7 \times 10^{-5}$ |
| Maximum stored energy (megajoules) | 800 | 20 | 1 | 1.7 |
| Maximum system power (megawatts) | — | 25 | 5 | — |

[a] Conventional iron-based electromagnets are limited to a maximum field on the order of two tesla.

SOURCE: Adapted from T. H. Geballe and J. K. Hulm. "Superconductors in Electric-Power Technology," *Scientific American,* Vol. 243(November 1980), p. 160, 

the latter, and for a related compound of vanadium and gallium ($V_3Ga$) that is used occasionally, involves a postdrawing diffusion reaction. Both of these systems are in limited commercial production.

***Existing Devices*** Table 1 gives a representative but far from complete list of the many successful superconducting devices developed recently. Two characteristic numbers for every magnet are the maximum field strength and the effective working volume. Table 1 gives these values as well as the maximum stored energy and the power level of the entire system where it is relevant. The CERN (European Organization for Nuclear Research) bubble chamber magnet is distinguished by its enormous stored energy, the maximum so far achieved with superconductors. The magnet for the magnetohydrodynamic (MHD) power generator was built in the United States and shipped to Moscow in a C5A cargo plane for use in a Soviet 25-megawatt MHD system. The rotor magnet for the electric power alternator represents a significant step forward in applying superconductors to rotating machines. Finally, the high-field research magnet, built in this country for use in Japan, holds the record for maximum central field strength using superconductors alone.

## Future Devices

***Fusion Magnets*** Superconducting magnets are essential for containing a thermonuclear plasma by magnetic means. The fusion power research community recently has embarked on a major project, the large coil program, to build very large superconducting magnets for use with tokamak systems. Six contractors, including teams from the United States, Europe, and Japan, will construct six prototype magnets. The project is managed by Oak Ridge National Laboratory, where the coils ultimately will be tested.

The six coils, each D-shaped with a 2.5- $\times$ 3.5-meter bore, will be mounted in a toroidal arrangement at 60° intervals. They will be housed in a vacuum-insulated vessel common to the entire system and refrigerated to 4.2 K by liquid helium. The superconductors in the coils will be multifilamentary composites. These giant experimental coils are merely precursors of the even larger magnets that will be required for practical fusion power reactors.

An alternative magnetic containment for thermonuclear plasma is the mirror machine being developed at Lawrence Livermore National Laboratory. Currently, the laboratory is building a set of superconducting coils for the Mirror Fusion Test Facility. The total weight of the conductor will be about 50 tons, so the system will be comparable in size to the tokamak coils.

***MHD Power Generation*** In magnetohydrodynamic power generators, hot ionized gas is passed through a channel in a magnetic field and the electric current thus generated in the gas is extracted by an array of electrodes along the edge of the channel. There are no moving parts. The efficiency of an MHD generator increases with the temperature of the gas and with the strength of the magnetic field. The maximum gas temperature is limited by the durability of the construction materials in the channel, so a practical generator requires a field strength that can be achieved economically only with a superconducting magnet. Design studies for a full-scale, 600-megawatt MHD electric power plant call for a superconducting magnet weighing about 2,000 tons. A 400-ton magnet is being designed for the MHD Component Development and Integration Facility in Butte, Montana.

***Particle Accelerators*** Particle accelerators have been important research tools for physicists since the 1930's. The largest accelerator in this country, at the Fermi National Laboratory, uses normal electromagnets to accelerate the proton beam. By switching to superconducting magnets, designers at Fermilab hope to reduce power consumption dramatically while doubling the beam energy to $10^{12}$ electron-volts per proton. This new accelerator, the Tevatron, will require 5,000 liters of liquid helium per hour for refrigeration; the helium will circulate in a closed system. An even larger colliding-beam accelerator, Isabelle, is under construction at Brookhaven National Laboratory.

***Generators and Motors*** The simplest use of superconducting magnets in rotating machinery is as replacements for DC magnetic field coils in electric generators and motors. It also may be possible to replace AC coils with superconductors, but technical problems remain to be solved.

To use superconductors instead of copper in the rotor magnet of an alternator, it is necessary to rotate the magnet plus its insulating vessel and to feed the liquid-helium refrigerant into the device through a rotating coupling. These tasks have been accomplished in small machines (see Table 1). A 270-megawatt alternator is being designed and will be constructed by 1984 by Westinghouse in a joint program with the Electric Power Research Institute; plans are being made to test it in an electric utility power plant.

Several advantages are expected to accrue from using superconductors in large alternators. Energy losses in the machines should be reduced by 50 to 65 percent, an attractive saving over the life of the alternators. A superconducting alternator would be more compact than a conventional machine with the same electric output; this should reduce installation costs and, in the long run, first cost. Finally, it is hoped that the use of superconductors will contribute to the overall stability of the power network in which the alternator is situated.

***Power Transmission*** Superconductors may be used in long-distance transmission lines for electric power, but the technology is not far advanced. Such systems will be attractive in locating central power stations away from densely populated areas. Normal underground lines, which in some cases may be the only practical way to bring power into large urban areas, are subject to upper limits on their power ratings, whereas superconducting lines have much less restrictive limits. Studies at Brookhaven National Laboratory, with utility company participation, indicate that typical AC (three-phase, 60-cycle) superconducting power lines would become economically attractive at power levels above 1,000 megawatts.

For both AC and DC lines, the conductor must be fabricated in a form that can cope with the large thermal contractions that occur in any cryogenic power line. The total electric losses must be very low. For short periods—the few cycles necessary to throw control switches—the lines must be able to carry fault currents arising, for example, from lightning. Fault currents can be several times the rated current. The superconductor itself conceivably could be fabricated to handle them, but present designs provide for handling them with parallel copper paths. A 100-meter AC superconducting line is being built at Brookhaven to serve as a flexible coaxial system for demonstration purposes.

### Problems and Opportunities

In addition to the major engineering problems of scaling superconductor devices to commercial size, there is general concern about the efficiency and reliability of refrigerating systems. Most of our experience with helium liquefiers has involved small-scale production in the laboratory, where continuous operation is not a major concern. Currently available refrigerators of about one-watt capacity also are not really appropriate for microscale technology; they are too large for cooling the few chips required for analog devices and too small for large computers. The very large-scale refrigeration required for macrodevices poses an even greater challenge.

It is possible that more efficient superconducting systems using liquid hydrogen at its boiling point (20.4 K) eventually will become a practical option. Reliable operation requires that the superconductor operate well below its critical temperature, so superconductors with critical temperatures above 25 K would be required. None is known at present, although an intensive search is proceeding in some laboratories. It would be an unwarranted gamble at this time to base any technological planning on devices operating above 15 K, where gaseous helium or possibly supercritical pressurized hydrogen can be used as a refrigerant. Nevertheless, many superconductors with critical temperatures above 20 K and with other interesting properties are known, and new ones should still be forthcoming. While Nb-Ti and $Nb_3Sn$ undoubtedly will remain the major materials of superconductor technology for the next five years, research under way to find better superconductors may produce even faster progress toward superconducting technology.

## ENERGY STORAGE

Interest is justifiably high in using energy more efficiently and energy resources more flexibly. One way to improve efficiency is to avoid the need to use energy where and when it is produced. To improve flexibility, one must be able to produce energy from the resource of one's choice and use it in the form of one's choice. The necessary manipulations often require that energy be stored between production and use—for periods ranging from minutes to days, weeks, or longer—and much work is under way on better means of doing so. Major targets are conventional electric power generation and electrically powered vehicles, but there are other applications as well. For example, energy storage mechanisms are essential to any major solar technology.

***Power Systems*** The demand on electric power systems varies greatly with time during the day and week, as well as with the time of year. The base load—the minimum continuous demand plus a suitable margin—typically is met by the system's most efficient generating units. As demand rises to the peak load, it is met by bringing progressively less efficient generating units on line. If the load can be made more uniform, therefore, more of the total energy can be provided by the most efficient equipment. This goal can be achieved, in effect, by storing power from the least-cost equipment when demand is low and using it when demand exceeds the capacity of that equipment.

***Vehicles*** Improved onboard storage of energy is essential to the development of practical electrically powered vehicles. About half of the petroleum we consume is used by vehicles; driving them with centrally generated electric power would permit the mix of fuels used for transportation to be shifted away from petroleum toward coal, uranium, and other resources. It is also possible that electric cars would use energy more effectively (in kilowatt-hours per passenger-kilometer) than would cars powered by gasoline, but the point remains to be resolved. There is no question, however, that vehicles powered by centrally generated electricity would be more efficient than current liquid-fueled cars if the electricity and the liquid fuel were both derived from coal. Electric vehicles also should have an environmental edge: it should be easier to control gaseous emissions from central generating plants than emissions from many individual vehicles using liquid fuels, and this problem does not exist with nuclear or solar sources.

***Other Effects*** Further shifting of the nation's energy economy toward electrification, whether fossil fueled or nuclear, bolstered by better ways to store energy, would provide means of using new energy resources as they become more practical. These include nondepletable resources such as solar energy, wind energy, and perhaps nuclear fusion.

Improvements in energy storage technology can affect many products and technologies. The recent substantial improvement in the storage capacity of small primary (nonrechargeable) batteries, for example, is important to a number of new devices. Among the most visible are the cardiac pacemaker, electric watches, calculators, and myriad small portable electric products.

Energy can be stored in many different ways. Some are well developed, and their future depends primarily on cost and the availability of materials. Others are not so far along; there is good reason to be optimistic about technical improvements, although the degree to which they will materialize is not easy to estimate. The potential market for energy storage devices is immense. One result is growing industrial interest in technological improvements that might lead to commercial products.

## Electrochemical Storage Systems

Of the ways to store energy now being pursued, the electrochemical approach is the most likely to have commercial impact in the next five years. Many electrochemical systems might be used to store energy or convert it from one form to another, but all employ electrodes separated by a solid or liquid electrolyte. The electrodes are of different materials, and a chemical potential exists between them. This potential drives ions through the electrolyte from one electrode to the other and electrons (electricity) through an external circuit. When the external circuit is open, the chemical energy in the cell remains in storage; when the circuit is closed, the chemical energy is converted to electrical energy.

***Batteries*** Secondary, or storage, batteries are those in which the overall chemical reaction can be reversed by applying an external voltage. In this recharging process, electrical energy is converted to chemical energy for storage. Primary batteries are those in which the chemical reaction cannot be reversed simply. On the other hand, the chemical reaction products from some primary cells can be reprocessed externally, so that the cell can be recharged either chemically or mechanically. Fuel cells—which are not storage devices, strictly speaking—produce electricity by the reaction of a gaseous or liquid fuel at one electrode and oxygen at the other. The fuel is stored outside the cell, and the preferred source of oxygen is air. The cell's electrodes and electrolyte are not changed by the reaction.

The largest-volume product of the battery manufacturing industry is rechargeable batteries for lighting, starting, and ignition in automobiles. Essentially all of them are now based on the conventional lead-acid cell. However, production of new types of nonrechargeable primary cells of modest size has been growing rapidly in recent years. These cells reflect considerably more technological progress than does the typical vehicle battery.

Most of the present interest in the large-scale use of electrochemical systems involves electrically rechargeable secondary cells. The commercial technology and market have changed little as yet, but new winds are blowing through this relatively conventional area. A major reason is the recognition that the theoretical limits of performance of a number of new systems are much higher than are those of traditional systems such as the lead-acid cell.

***Materials Dependence*** A hallmark of much of the recent work on electrically rechargeable secondary cells is its direct dependence on new materials for alternative electrochemical systems. New electrolytes include organic liquids, crystalline ceramics, thin films and very fine filaments of ionically conducting polymers and glasses, various molten salts, and alkaline aqueous electrolytes. These electrolytes are being used with a variety of different electrode materials, including alkali metals and alloys that previously could not be considered because they react vigorously with traditional aqueous electrolytes. In addition, new concepts employing liquid electrodes with solid electrolytes, new solid electrodes with novel types of solid-state reactions, and all-solid-state cells have been introduced in laboratory-scale configurations.

Some of this work has reached the point of engineering development; in other cases, the work remains primarily exploratory. Supporting the overall effort is a growing amount of scientific research aimed at elucidating the relevant phenomena and principles.

## Mechanical Storage Systems

Energy can be stored mechanically in three major ways. One is the flywheel, in which kinetic or inertial energy is stored in a rotating mass and (to a much smaller extent) elastic or potential energy is stored in stressed structural members. A second involves storage of potential energy in water pumped from a lower to a higher level. The third is storage of elastic energy in a working fluid, such as water, air, or steam, and (to a lesser extent) in the container that holds the fluid.

***Flywheels*** Flywheels have long been used for purposes such as smoothing pulsed mechanical power, but the energy storage capacity of conventional flywheels is quite limited and the cost is high. Because of considerable progress during the past decade, however, flywheels of modest size with attractive characteristics now seem technically feasible, if still quite expensive. These flywheels employ new design concepts and materials optimized for the purpose. An example of the latter is composite materials in which the reinforcing fibers are oriented preferentially in the direction of maximum stress. Further progress with flywheels depends directly on advances in such materials.

***Pumped Storage*** A number of electric utilities have stored energy in pumped-hydroelectric systems for some years, and the technology is well established. Water is pumped from a lower to a higher reservoir, using relatively cheap, off-peak electric power. When demand on the utility system rises, the water is fed by gravity through turbogenerators that reconvert its potential energy to electrical energy. The major constraints on greater use of such methods are topographical and environmental. The development of underground facilities, involving natural or man-made caverns, is being considered. Such schemes could use greater vertical distances and thus higher pressure differences, giving more power per unit volume of water. This high-head approach is feasible now, but its attractiveness depends partly on the development of improved long-life, pump-turbine technology.

***Compressed Gas*** Storage of energy in compressed gas, usually air, is inherently a more compact approach than pumped-hydroelectric storage facilities. It can be economic in smaller sizes and can use a broader range of geologic formations. However, because of the temperature changes accompanying the compression or expansion of any gas, provision must be made for cooling during the compression part of the cycle and for heating, typically by burning a fuel, as the gas is expanded through a turbine to generate electricity. The first major facility of this type started operating recently near Bremen, in West Germany, so the advantages and disadvantages of this method for utility-related energy storage should become evident soon. Compressed air at the German facility is stored in caverns in a salt dome.

## Chemical Storage Systems

There are a number of schemes for storing energy in the form of readily reconvertible chemicals. The potential advantages include the possibility of coupling the initial conversion to various primary sources of energy, such as solar energy, the relatively low cost of storing and transporting energy in the form of bulk chemicals, and the flexibility of reconverting energy in at least some chemical storage media to other forms, such as heat or electricity.

***Hydrogen*** Some of the most interesting implications of this concept may be illustrated by a scheme in which hydrogen is generated from water, stored or transported (or both) as desired, and subsequently converted to thermal, mechanical, or electrical energy. Each step could be handled in several ways, and the overall approach has attracted sufficient attention to warrant an informal name of its own—the hydrogen economy. Nevertheless, it appears that storing energy in hydrogen would be inherently less efficient and would involve higher capital costs than some other methods of energy storage under conditions in which they could compete directly.

A major challenge of the hydrogen system is to improve the efficiency of generating hydrogen from water by electrolysis. The problem is one of electrocatalysis. We may see substantial progress as we learn more about physical processes that occur on the surfaces of solid electrodes and develop new catalytic materials.

A second aspect of the scheme is the storage of hydrogen in the form of solid metal-hydrogen compounds (hydrides). Such storage is needed to optimize the match between generation and demand. A number of metals and metal alloys react with hydrogen to form hydrides. At the moment, these systems are either expensive, require operation at inconvenient pressures or temperatures, or store too little hydrogen per unit of weight or volume. However, scientific investigations are clarifying gradually the processes involved in forming hydrides and regenerating the hydrogen, and it seems reasonable to expect a good deal of technological progress in this area.

## Storage in Electric and Magnetic Fields

***Electric Fields*** Storage of energy in electric fields is illustrated by the capacitor (or condenser), a common component of automotive ignition systems and many other devices. Energy is stored in capacitors in the form of separated positive and negative charges, often at an insulator-metal interface. As capacitors discharge stored energy, the voltage drops, which is a disadvantage in many applications. Furthermore, the cost per unit of energy stored is quite high, so the devices are used for large-scale storage only where considerable electrical energy is required in a short time. One example is the use of large capacitor storage banks to provide energy for high-powered lasers.

It was demonstrated a few years ago that capacitors using a solid electrolyte and an ionically blocking material can store up to a million times as much energy as those using the typical insulator-metal interface. To date, however, this approach has not been developed commercially.

***Magnetic Fields*** Energy also can be stored in magnetic fields. The use of fields produced by superconducting magnets is potentially attractive because it would permit direct storage of electrical energy at high efficiency; also, high power levels could be handled. But predictions based on current technology indicate that the capital costs of such a system would be very high; to be economically competitive, such installations would have to be very large, with storage capacities of the order of 10 million kilowatt-hours. This requirement is a serious challenge to the designer, since he must avoid the possibility of temperature excursions that could lead to the quick release of immense amounts of energy. The picture could change markedly with the discovery of materials that remain superconducting at temperatures significantly higher than the present maximum.

## Thermal Energy Storage

Various studies have indicated that storing thermal energy in hot water or pressurized steam on a scale suitable for electric utilities is likely to be less efficient and more costly than mechanical methods such as pumped-hydroelectric or compressed-air storage. On the other hand, thermal storage at the point of consumption instead of at the utility can be quite beneficial. For example, electric water heaters and refrigeration units, controlled by timers or signals from the power plant, can be used for space heating or cooling or for refrigeration. Such dispersed storage of energy, which is becoming quite common in Europe, can be used by the utility to reduce peak demand and shift some of the load to the time of day when it can be served by more energy-efficient and lower cost base-load equipment.

Thermal energy generated by solar converters at the home or other local site can be stored in hot water which can be used directly as such or indirectly for space heating and cooling. This approach already is marginally economic in some areas, and its use is growing. Continued growth depends primarily on the provision of economic and institutional incentives, as well as on the development of other alternatives.

Thermochemical storage systems are being explored for several uses, such as the stabilization of local temperatures. These systems, which can be quite large, rely on the absorption and release of large amounts of heat associated with the decomposition of water or phase changes or reactions of certain chemicals. The primary question is cost. There are also problems with lifetime, however, because of corrosion or other kinds of degradation. Furthermore, in schemes where the working material moves, the prevention of clogging is a challenge to the designer.

## Hybrid Systems

Energy storage and conversion methods can be combined in several ways to optimize performance in specific applications. Indeed, such hybrid systems are a central issue in the integration of storage technologies with electric utilities.

Hybrid systems also are especially important in vehicular propulsion. The most prominent possibilities are engine-battery and battery-flywheel combinations. The general idea is to couple a subsystem of relatively high efficiency and specific energy (energy stored per pound) but of low power capability with one that can handle higher power demand for short periods.

No storage system now in development approaches the energy-storage capacity of liquid petroleum fuels. Thus, small engine-generator systems can provide uniquely long range without refueling, and they can be supplemented by flywheels or battery systems to handle transient power loads. However, this approach does not fully avoid consumption of petroleum. One of the chemically rechargeable primary batteries, because of high specific energy, might offer a way to extend the range, but the ultimate practicality of these systems is not known yet. Liquid- or gas-consuming fuel cells also might be used to extend the range. Some of these cells could use fuels, such as ammonia, that can be made from coal.

The battery-flywheel combination would use an electrically rechargeable battery as the primary energy source. The flywheel could meet transient power demands and also could be used with regenerative braking to recover power that would be wasted otherwise.

The future of the many possible hybrid systems depends on improvements in individual subsystems and on matching their characteristics to particular applications. In general, such improvements can be anticipated. Although one may thereby achieve substantial improvement in system efficiency, the inherent complexity of hybrid systems suggests that many of them inevitably will have higher capital costs than nonhybrid energy storage systems.

## Ancillary Problems and Issues

The commercial advent of new energy storage systems would introduce correspondingly new questions, problems, and opportunities. Among them would be issues of resource availability, materials recycling and disposal, environmental impact, and safety.

***Materials*** Some materials in energy storage systems now in development differ from those in commerce today; this raises questions of availability and price. Electrochemical systems based on sodium and sulfur, for example, would pose little difficulty, since both elements are cheap and readily available. For systems involving lithium, on the other hand, the picture is not so clear. This element is plentiful in the earth's crust, but the market has never been large enough to determine the price of the metal if used in quantity. If fusion were to become practical, it would probably create another large demand for lithium, which could make lithium-based batteries more attractive. Extensive new uses of candidate metals such as titanium, vanadium, nickel, or zinc, moreover, would require significant changes in the infrastructure now involved in their supply, processing, and distribution.

***Recycling*** The large-scale use of materials required by some proposed energy storage schemes would raise various questions of disposal and recycling. Any approach involving substantial amounts of scarce or expensive materials surely would require a closed system. In some cases, such as recycling aluminum hydroxide to make aluminum, today's industry would need no appreciable changes in practice but would have to be expanded substantially. In others, requiring the recycling of large amounts of sodium, lithium, nickel, zinc, and organic solvents, current industrial practice would have to change markedly.

***Safety*** Several prospective energy storage systems are inherently hazardous. Any device containing a large amount of energy that might escape rapidly is potentially dangerous. Flywheels pose the hazard of disintegration. Thermal storage devices can leak, chemical storage systems can react with their environment in a harmful manner, and all of the possible electrochemical systems pose characteristic hazards.

In short, the new energy storage systems contemplated would represent major deviations from current technology, and attendant problems can be expected. Such problems are not reason for pessimism, but they must be faced and solved before the new systems can enter the marketplace.

## Near-Term Opportunities

It is evident that improving our methods of storing energy poses substantial challenges to both science and technology. Several major applications not only are important to the national welfare but also will provide extremely large markets if suitable systems succeed technologically. In some cases, there is a good deal of room for improvement, and the incentives are clear.

In recent years, some innovations in both independent storage systems and combinations of energy conversion and storage have opened avenues that have not been fully explored yet; this arena of old technology is thus infused with greatly renewed vigor. A pervasive characteristic is that major progress clearly depends on progress in materials. It is important to learn more about the fundamental phenomena in several areas in order to optimize the use of high-performance active materials. Prominent examples are electrochemical systems that involve complex mass-, heat-, and electron-transport processes. Such systems surely will undergo significant change in the next five years; during that time, as noted earlier, they promise the greatest advances in energy storage.

Questions about new materials and about approaches in areas such as solid electrolytes, organic electrolytes, rapid ionic conductivity, reactions in electrode materials, and catalysts and related matters represent a broad list of new scientific and technological opportunities. Especially important will be the building of closer cooperation in this area between the nation's fundamental research community and industrial developers, for this is an outstanding example of an area of great economic importance in which further understanding and new approaches can have a very large impact.

## RESOURCE EXPLORATION

During the 1930's and early 1940's, the United States produced and consumed about half of the world's tonnage of nonfuel minerals. By the late 1970's, U.S. consumption had doubled, but production and consumption in the rest of the world had risen more than tenfold. Thus, this nation is no longer the dominant producer and, as a consumer, now must compete with a much larger foreign demand.

In dollar value of refined metals, we import about half of our needs (see Table 2). Canada remains our main supplier, but a disquietingly large part of our critical materials—such as manganese, cobalt, chromium, tin, and platinum—comes from remote foreign sources of uncertain dependability. Moreover, environmental policies have diminished the economic viability of some domestic resources, and withdrawal of large tracts of public lands from mineral entry has reduced significantly the accessibility of others.

### A Wealth of Resources

The world still has a wealth of mineral resources, although the supply of readily found, high-grade ores is shrinking inexorably. Certainly, there are more deposits to be found, and some will be of high grade and very large. The question is how to find them. In this context, there are two broad types of deposits: conventional deposits—those generally recognized by the mining industry as possible economic sources of minerals; and unconventional deposits—those that are not so recognized and hence may have been overlooked. It also should be noted that "ore" is generally defined as material that can be mined legally at a profit.

Discoveries of exposed conventional deposits in developed lands have been comparatively uncommon in recent years; indeed, most of Europe's big deposits have been known for centuries or even millennia. Frontiers like Alaska, northern Canada, northern Australia, Siberia, the Antarctic, and the jungle regions of Africa, Asia, and the Americas are the last refuges of exposed but undiscovered conventional deposits. As these frontiers are explored and their deposits developed, the emphasis must shift to unconventional deposits, lower grade conventional deposits, and conventional deposits concealed at varying depths by unmineralized material.

Unconventional deposits may provide a few pleasant surprises but, generally, will be unlikely to meet long-range demands. Lower grade conventional deposits will become ever more important through advances in mining and metallurgy. However, concealed conventional deposits hold the greatest promise for the next few decades and offer the greatest challenge and opportunity to the geologic sciences.

**Table 2** New materials added to the U.S. economy

| Material | 1978 value ($ million) | Percentage of U.S. need produced in United States | Percentage of new imports by nation of source (1977) | | | | | | | | | | | | | |
|---|---|---|---|---|---|---|---|---|---|---|---|---|---|---|---|---|
| | | | Canada | Jamaica | Mexico | Bolivia | Brazil | Chile | Peru | Gabon | Rep. S. Africa | Zaire | Australia | Malaysia | New Caledonia | U.S.S.R. |
| Chromium | 600 | 0 | — | — | — | — | — | — | — | — | 15 | — | — | — | — | 41 |
| Manganese | 200 | 0 | — | — | 7 | — | 18 | — | — | — | 15 | 2 | 2 | — | — | — |
| Cobalt | 150 | 0 | 8 | — | — | — | — | — | — | — | — | 62 | — | — | — | — |
| Tin | 700 | 0.2 | — | — | — | 18 | 4 | — | — | — | — | — | — | 50 | — | — |
| Platinum | 500 | 0.2 | 11 | — | — | — | — | — | — | — | 74 | — | — | — | — | 8 |
| Nickel | 750 | 6 | 56 | — | — | — | — | — | — | — | 3 | — | 3 | — | 9 | 1 |
| Aluminum | 6,400 | 8 | — | 28 | — | — | — | — | — | 28 | — | — | 23 | — | — | — |
| Asbestos | 170 | 16 | 94 | — | — | — | — | — | — | — | 4 | — | — | — | — | — |
| Fluorine | 110 | 18 | 4 | — | 59 | — | — | — | — | — | 17 | — | — | — | — | — |
| Zinc | 680 | 31 | 42 | — | 6 | — | — | — | 3 | — | — | 5 | 4 | — | — | — |
| Silver | 620 | 33 | 46 | — | 24 | — | — | — | 19 | — | — | — | — | — | — | — |
| Tungsten | 160 | 38 | 25 | — | 8 | 19 | 2 | — | 7 | — | — | — | 2 | — | — | — |
| Potash | 450 | 39 | 95 | — | — | — | — | — | — | — | — | — | — | — | — | — |
| Titanium | 230 | 48 | 22 | — | — | — | — | — | — | — | — | — | 48 | — | — | — |
| Iron | 19,000 | 58 | 39 | — | — | — | 4 | 1 | 2 | — | — | — | — | — | — | — |
| Lead | 530 | 76 | 27 | — | 26 | — | — | — | 14 | — | — | — | 3 | — | — | — |
| Copper | 2,300 | 77 | 24 | — | 2 | — | — | 24 | 13 | — | 2 | — | — | — | — | — |
| Sulfur | 550 | 90 | 59 | — | 39 | — | — | — | — | — | — | — | — | — | — | — |
| Phosphate | 590 | 180[a] | — | — | — | — | — | — | — | — | — | — | — | — | — | — |
| Molybdenum | 330 | 200[a] | — | — | — | — | — | — | — | — | — | — | — | — | — | — |

[a] Net export.

SOURCE: Based on information contained in *Minerals in the U.S. Economy: Ten-Year Supply-Demand Profiles in the Non-Fuel Minerals Commodities 1968–1977.* Prepared by the staff of the Bureau of Mines Metals and Minerals Supply/Demand Analysis, published May 1979.

## Understanding of Ore Deposits

Our understanding of ore deposits appears to be developing parallel to, but two or three decades behind, our understanding of global geology. Extensive studies since 1945 of the morphology of ocean basins, global seismicity, and magnetic patterns in the ocean crust, combined with the development and application of radiometric dating methods, culminated in the 1960's in the theory of plate tectonics, which dramatically systematized geologic thought. The groundwork for an analogous culmination in the understanding of nonfuel mineral resources has been laid in the past few years by accomplishments that include:

- The firm establishment of the study of fluid inclusions, which not only reveal the temperatures of mineralization but also produce the best evidence on the nature of ore-forming fluids. Inclusions also demonstrate that many ore-producing hydrothermal systems were boiling, thereby indicating that the depth of formation was relatively shallow.
- An understanding of stable isotopes and other geochemical indicators, which have provided criteria for distinguishing and selecting among alternative sources of metals, sulfur, and ore-forming fluids.
- Improved isotopic geochronology, which has perfected the ability to relate ore-forming processes to regional geologic history.
- The combination of theoretical studies of the stabilities of minerals with elaborate mass- and energy-transfer computer programs. These tools can be used to model ore-forming processes and predict flow patterns and sites of mineral deposition in ground waters set in motion, for example, by intrusions of hot igneous rocks.
- A growing ability to predict the properties of hydrous silicate melts and thus the behavior of magma as it moves through and interacts with the earth's crust.
- The ability to relate the occurrence of many types of mineral deposits to the tectonic-paleogeographic setting.

## New Types of Deposits

A recent example of the prediction of new types of mineral deposits began in the 1960's with the conclusion by the German geologist Adolph Maucher that areas of a certain geologic character in the Alpine chain would be suitable for tungsten and antimony deposits of a submarine hot-spring origin. Exploration based on Maucher's predictions uncovered dozens of tungsten occurrences, including Felbertal, Europe's largest tungsten deposit.

Other novel types of deposits have come into production in recent years. One example is the Carlin type of disseminated, very fine-grained gold deposits now in production in Nevada; a second is the unconformity-vein type of uranium deposits in Saskatchewan and Australia that have proved to be fabulously rich. Both types of deposits are essentially unreported anywhere else. Surely elsewhere on earth there are similar types of deposits that have not been recognized yet.

## Seabed Resources

Only at a few near-shore, shallow-water operations are nonfuel minerals taken commercially from the three quarters of the earth's surface covered by the oceans. The seabed has had much less opportunity to accumulate minerals than the continental masses have had, but still there is a potentially large resource at or near the ocean bottom.

By far the most promising seabed deposits are the sedimentary nodules of manganese oxides on and just below the surface of the deep ocean floor, especially in the near-equatorial regions of the Pacific. These nodules contain up to a percent or so of copper and nickel and a few tenths of a percent of cobalt. The resource is immense, and its use is limited less by technological problems than by the inability of nations to agree on mining rights.

The continental slopes, rises, and shelves contain nodules and pavements of phosphate in certain areas, but these deposits cannot compete economically with the older, higher grade deposits on land. The continental shelves at some points may contain enough placer material to be of interest for tin, platinum-group metals, gold, rutile (titanium dioxide), and a few other inert heavy minerals. But again, except for tin, these deposits generally cannot compete with those on land.

## Submarine Hydrothermal Deposits

One of the greatest successes of science in the past two decades has been the concept of plate tectonics—the demonstration that the earth's crust and mantle interact dynamically. The huge crustal plates atop the mantle separate and converge at net rates of the order of a few centimeters per year. They separate at linear oceanic spreading zones marked by material flowing upward from the mantle. At convergence zones, one crustal plate (usually oceanic) slides under another (usually continental), dragging downward the seafloor plus sediment shed from the continent. Over millions of years, the process renews the floor of the oceans and adds new material to the continents. In some tectonic events, such as occurred a few million years ago in Cyprus, immense slabs of the seafloor are shoved onto the continental mass, and mineral deposits formed deep in the oceans—in this case, copper ore—are exposed for study and use.

***Spreading Centers*** Two types of deposits are recognized to form near oceanic spreading centers. The first is represented in an embryonic ocean, the Red Sea, at the spreading axis between Africa and the Arabian Peninsula. Heat from igneous activity along the spreading axis and water percolating through salt deposited along the margin of the basin unite to create warm pools of brine in low spots on the seabed, two kilometers below the surface. Metal oxides and sulfides precipitate from this brine, and the value of the metals in the sediments below the pools is in the multibillion-dollar range. For a combination of economic, political, and technological reasons, there are no plans at present for immediate development of this resource. Examples of this type of deposit that have been preserved on land have not been clearly identified; nevertheless, some known deposits possibly originated in this way.

***Hot Springs*** A second type of deposit is exemplified by the submarine hot springs discovered in 1978 along the spreading axis of the East Pacific Rise just off Baja California. The springs are fed by seawater that penetrates deep within the underlying basalt, extracts heat and metals from the hot rock, and eventually circulates back to the seabed. As the water reenters the sea and cools, it precipitates chimneylike deposits of minerals around its points of entry. These deposits, if abundant on land, would be very rich zinc and copper ore. Similar but cooler springs with much less mineralization occur along the same spreading axis near the Galapagos Islands. Many more such sites are probable along the 50,000 kilometers of oceanic spreading ridges not yet explored by manned submersible vehicles.

None of the known active deposits of this second type is large enough to excite economic interest. It is encouraging, however, that the famous copper ores of Cyprus apparently represent the fossil result of such an environment. The next few years should see dramatic growth in knowledge of this type of deposit. Ore bodies large enough to exploit may be found, and the knowledge will lead also to more effective exploration for similar deposits on land.

***Convergence Zones*** Seafloor material dragged downward, or subducted, in zones of convergence when one crustal plate slides at an angle under another may be partly remelted at depth and return as igneous intrusions that penetrate the upper plate and may erupt at the surface. Many conventional ore deposits are associated with such volcanic activity. Improving our ability to predict ties between subduction and ore is a limitless opportunity for the future.

An additional type of mineralization occurs underwater where deep marine basins overlie the intrusive-eruptive setting above convergence zones. Seawater is heated by magmatic intrusions below the seafloor, extracts metals and hydrogen sulfide, and finally is discharged to the seafloor to form sulfide deposits containing iron, copper, and zinc with variable amounts of lead, silver, and gold. The deep-sea setting favors the burial of such deposits by sediments and subsequent metamorphism to incorporate them into the continental mass. Such volcanogenic massive sulfide deposits are important suppliers of metals and sulfur. Much has been learned of their origin, especially from the young (12 million years), unmetamorphosed kuroko (black ore) deposits of Japan. We have not observed the formative stage of such deposits yet because preoccupation with oceanic spreading centers has precluded deep exploration of the basins that most favor their development. Ancient volcanogenic massive sulfide deposits on land range in size up to hundreds of millions of tons. If found forming on the seabed, they might well prove to be multibillion-dollar targets.

## Lower Grade Deposits

Over the years, the minerals industry has developed mass mining and concentrating methods for treating relatively low-grade deposits (0.3 percent copper, for example). Such deposits have produced much more total metal than the high-grade deposits mined almost exclusively a century ago. The recovery of byproduct metals—a form of tapping low-grade deposits—has a long and profitable history as well.

Sometimes, the use of low-grade ores has proved advantageous in itself. With the exhaustion of high-grade hematitic iron ore in this country, companies turned to the immense domestic resources of taconite, a magnetite-quartz rock that contains less iron than does hematite. Magnetite is extracted from taconite as a fine dust that must be pelletized before it is fed to the blast furnace. The pellets turn out to be better blast furnace feed than the hematitic ore preferred previously and, despite the extra grinding required to prepare them, reduce the net energy consumed in making iron.

A current issue is the availability of chromium. The United States has little high-quality chromite and imports most of its needs. However, the U.S. Bureau of Mines has shown how to use the abundant low-quality chromite from the Stillwater deposit in Montana. The choice of whether to use the more extensive domestic ore or to rely on imports from the U.S.S.R. and southern Africa will be made on political grounds. Still, the technology available for treating the domestic ore gives political policymakers a strategic option, if not an economic one.

***Solution Mining*** The use of low-grade ores can entail the handling and disposal of vast amounts of rock, but the problem may be addressed by solution mining. Nature emplaces many ores through the agency of solutions, and man might reverse the process by using the proper solvents. Uranium already is being extracted economically from low-grade disseminated ores by solution mining. Attempts to apply this process to copper ores have not yet succeeded. But, as we consider means of mining at depths of perhaps two kilometers or more, where roof support would be a crushing problem, solution mining would appear to be the only feasible approach.

***Byproducts Recovery*** In the recovery of byproducts, new options are appearing continually. One is the possible recovery of rutile from porphyry copper mines in the West. Copper porphyries already rank highly as producers of byproduct molybdenum and gold, and the rutile now going to tailing piles could meet much of our national need for titanium. A second example is the sphalerite (zinc sulfide) present at up to half a percent in some Illinois coals. So much coal is processed that recovery of the sphalerite might meet a substantial part of our need for zinc, of which 69 percent is now imported. Removal of the sphalerite is desirable, moreover, in order to reduce sulfur dioxide emissions when the coal is burned.

## Long-Range Considerations: The Ultralow-Grade Materials

A fundamental question for the long range is the availability of materials from ultralow-grade deposits. This question is much more complex than is recognized by some economic experts.

A critical point is whether more and more of a commodity can be obtained simply by using progressively lower grades of ore (see Figure 1). In other words, if we are willing to raise the price a little, can we have all we want? The alternative, unfortunately a distinct possibility, is that deposits of marginal grade are not extensive and that mining progressively lower grades will thus yield a disproportionately small increase in supply (see Figure 2).

The data needed to determine which pattern is correct are difficult to obtain. In the main, such data are produced by mining companies in a nonsystematic fashion, because companies seek ore, not high-grade waste. The only bright spot in this dimly lit corner is data from the Department of Energy's recent National Uranium Resource Evaluation Program, which suggest (but unfortunately do not prove) that uranium may follow the first pattern—that supply may be limited essentially by the price we are willing to pay. Still, one would expect neither that all commodities follow the same pattern nor that today's cutoff grade is necessarily on or even close to the low-grade side of the small maximum in Figure 2.

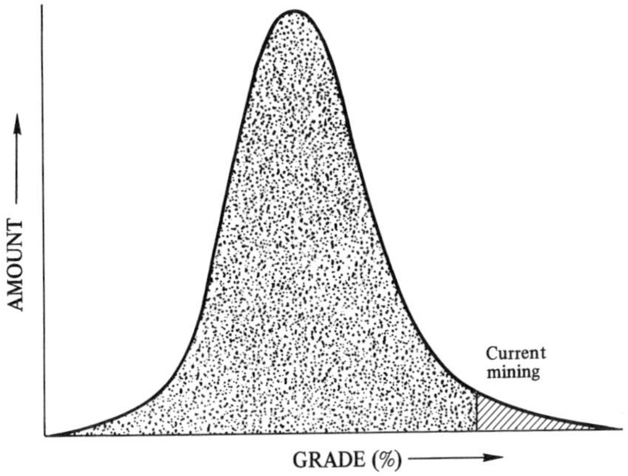


**Figure 1 Typical distribution and mining of selected abundant metals.** The curve represents the typical distribution of geochemically abundant metals such as iron, titanium, and aluminum in the earth's crust. The proportion of the total amount that has been mined up to the present is indicated at the right. The highest grade ores have been mined first but, because the same kinds of minerals occur regardless of grade, the concentrating techniques used in current mining processes to extract metal from high-grade ores also can be used in the future with less rich materials and even common rock. As the grade, or percentage of metal in an ore, declines arithmetically, the curve indicates that the amount of metal available will increase geometrically to a grade corresponding to the peak of the curve. [SOURCE: Brian J. Skinner. "A Second Iron Age Ahead?" *American Scientist,* Vol. 64, No. 3 (May–June 1976), p. 263.]

A second serious aspect of the use of ultralow-grade ores is that the valuable material almost always occurs as higher grade grains in a waste matrix. It also occurs, however, within the crystal structures of the minerals of the waste matrix and, today, this usually minute fraction of valuable material is knowingly discarded when the high-grade grains are separated for refining. If we are forced to ever lower grades, eventually we will reach the mineralogical limit (Figure 2), where all of the metal values lie in the waste matrix. Then, it will no longer be possible to prepare a concentrate without refining the bulk rock; this would increase cost and environmental degradation very considerably.

## Concealed Conventional Deposits

In the next few decades, the greatest hope for discovering new mineral deposits lies in improving our ability to find conventional deposits concealed by unmineralized materials such as lava or sediments. In seeking such deposits, geophysical methods based on density or magnetic or electrical properties are very effective; so is the detection of halos of anomalous concentrations of trace elements that resemble the rings around a

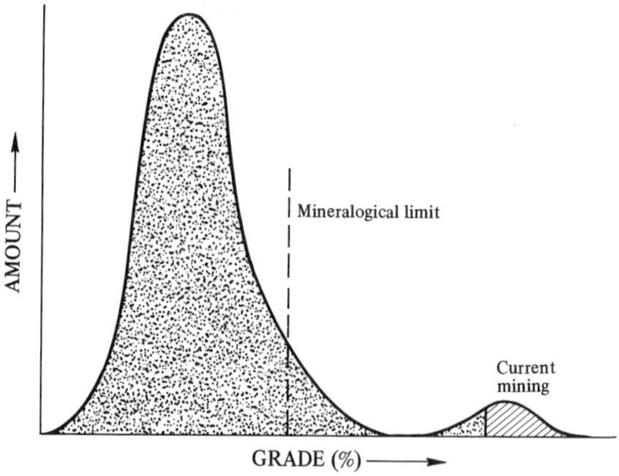


**Figure 2 Probable distribution and mining of scarce metals.** The bimodal curve represents the probable distribution of geochemically scarce metals in the earth's crust. The large peak is the distribution in common rocks, where scarce metals occur not as separate minerals but as atomic substitutes for abundant metals; the small peak represents deposits produced by ore-forming processes such as those resulting from the circulation of brines in the crust. Current mining has reached the point where the distribution curve for easily accessible scarce minerals turns downward. Further reductions in the grade of ore mined will produce declining tonnages of ore. [SOURCE: Adapted from Brian J. Skinner. "A Second Iron Age Ahead?" *American Scientist,* Vol. 64, No. 3 (May–June 1976), p. 263.]

bull's-eye. Nevertheless, there must be many deposits undetectable by current methods, and work is under way on new ones.

Recent studies under COCORP (Consortium for Continental Reflection Profiling), funded by the National Science Foundation, have revealed deep seismic reflectors that may lead us to broad target areas for very deep exploration. Deep discoveries almost surely would have to be mined by solution techniques, however, and both exploration and development would appear to be far in the future.

Mineral exploration and development require extensive analysis of the rocks penetrated by the drill; furthermore, the cost of recovering and assaying samples is considerable. A promising advance in analyzing the walls of drill holes *in situ* is down-hole neutron activation and counting. This technique activates elements with appreciable neutron-capture cross sections and detects radiation emitted by the unstable nuclides as they decay to their stable state. (The two most frequent uses of neutrons in bore holes—in petroleum and uranium exploration—rely not on neutron activation but on the time-dependence of return of scattered or secondary neutrons.)

Studies of the distribution of metal values in the veins of the Julcani district in Peru have provided valuable guides to the directions of flow of the mineralizing solutions. They have led to successful exploration for extensions of mineralization. The method has great promise for many of the base and precious metal deposits associated with the relatively shallow volcanic activity typical of much of the cordilleran regions of North and South America.

***Uses of Models*** One of the most significant trends of recent years is the characterization of mineralization in terms of genetic models and empirical patterns of occurrence. A model of porphyry copper deposits, for example, suggested the location of several hundred million tons of ore at Kalamazoo, Arizona. A molybdenum model developed by industry has led to many molybdenum deposits, including the huge Henderson deposit in Colorado. Models in general, and the more flexible, theoretical genetic models in particular, offer great potential for appraising and exploring land for mineral deposits.

One of the most exciting models being developed deals with the systematization of the occurrence of metal concentrations in mafic and ultramafic rocks. These rocks form immense bodies whose metal content is truly gigantic. The metals of economic interest are nickel, copper, the platinum-group elements, gold, iron, chromium, vanadium, and titanium. These elements are widely disseminated in the parent magmas, but economic concentrations in the rocks occur only when some geologic process concentrates the metals. The model should make it possible, by analyzing trace metals in the rock, to distinguish potentially productive bodies from probably barren ones. The impact on exploration is expected to be significant and should be even more so when the dynamics of magma circulation during crystallization are understood more fully.

Such studies are aimed ultimately at models of entire formative systems. The scale will vary with the type of deposit, from individual magmatic bodies, as in the preceding paragraph, to the relation between the movement of crustal plates and the development of massive metal sulfide deposits on the seabed.

## A Timely Confluence

This nation today is experiencing the disappearance of the sorts of deposits that prospectors could find with a pick and shovel and a lot of hard work; in a few years, the experience will be worldwide. New deposits will be ever more difficult to find. At the same time, demands for materials are rising steadily.

The 1980's will see the timely confluence of the need to find, identify, and produce resources with the scientific and technological opportunities for real progress. Each new exploration technique, each new predictive model, each new type of deposit, and each new commodity found to be useful offers new targets for exploration and the strong chance of providing materials that otherwise would have remained undiscovered and unused.

## THE SPACE SHUTTLE AND TELESCOPE

The Space Shuttle developed by the National Aeronautics and Space Administration (and first flown successfully in April 1981) introduces a new era in space technology. The shuttle, or Space Transportation System, is designed to be a potentially lower cost means of placing payloads in low earth orbit. Being partially reusable, it will avoid the one-mission use of expensive boosters.

The most striking features of the shuttle are that people will be able to do experiments on board and that it will be able to retrieve, maintain, and service free-flying and automated satellites. These capabilities will make it possible to take advantage of opportunities in space that are either too costly or too risky to be pursued using existing expendable unmanned systems. They will permit an unusual variety of payloads to be launched for government agencies, private industry, and national and international organizations. Of the scientific payloads being prepared for launch, the space telescope most aptly matches instrument requirements with the shuttle's capabilities (the telescope's potential is outlined below).

***The Vehicle*** The shuttle itself is a manned orbital vehicle launched vertically by three identical main rocket engines built into the vehicle and two strap-on boosters. The main engines use liquid propellant (hydrogen), and the boosters use solid propellant. The boosters are jettisoned about two minutes after launch, some 25 miles up, but are to be recovered and reused. The liquid fuel for launch is carried in a very large disposable tank that is jettisoned when the vehicle reaches orbit. The shuttle can stay in orbit for about one week, and each operational vehicle will be designed to make at least 100 trips into space. On reentry, the vehicle maneuvers like a glider to land horizontally on a runway.

### Shuttle Technologies

The development of the Space Shuttle has required many advances across a wide spectrum of technology. The most important areas of progress are probably hypersonic aerodynamics, thermal protection during reentry, complex flight-control systems, and very high-pressure, liquid-fueled engines.

***Hypersonic Aerodynamics*** The speed of the shuttle from launch to recovery ranges from very low through subsonic, supersonic, and hypersonic speeds. The maximum Mach number reached is about 25. The shuttle has enough aerodynamic efficiency at reentry speeds to maneuver gently, producing some 1,000 miles of cross range. On reentry, the vehicle is a very large, high-drag glider with difficult aerodynamic problems right down to approach and landing. The aerodynamic characteristics of the shuttle are reasonably well known from analysis and wind-tunnel testing, but, until the first flight, actual control of such a reentry to approach and landing had never been attempted.

***Thermal Protection*** During reentry, the shuttle's skin reaches close to 3,000°F (Fahrenheit) at the leading edge of the wing and on the nose and 500° to 1,500°F at other points. The conventional aluminum structure of the shuttle is protected against such temperatures by means of some 30,000 insulating, silica-foam tiles developed specifically for the purpose and applied with a special new adhesive. This approach is a wholly new technology. Installation of the tiles and the testing of each comprised an enormous program, which was one of the pacing elements of the shuttle's development.

Many other material technologies have been developed for the shuttle. For example, graphite-reinforced epoxy is used for the cargo-bay doors, and boron-reinforced epoxy is bonded to diffusion-bonded titanium members in the thrust structure. While they are not yet in use, low-weight, high-temperature-resistant polyimides for control surfaces are being developed in the Space Shuttle research and development program. Much of this technology will have uses in the future in civilian and military aircraft.

***Flight Control and Avionics*** The shuttle has a digital, fly-by-wire flight-control system. This relatively new system has entered operational service recently on military aircraft. In essence, the pilot controls aerodynamic surfaces by means of electronic signals from the cockpit to remote actuators. Incorporating a digital computer between the pilot and the actuators allows the fly-by-wire system to be operated either manually or automatically.

The electronics system on the Space Shuttle employs five identical computers for sensor computational redundance; it employs computer hardware in addition to software to manage redundancy for controlling thrust vectors and aerodynamic surfaces during reentry. In so doing, the system pushes the state of the art in operational reliability. Moreover, it offers improved and advanced control technology for earth-based applications that require very high reliability.

***Main Engines*** The main engines of the Space Shuttle use liquid hydrogen for fuel and liquid oxygen to burn it. To meet performance requirements, the three engines must operate at pressures much higher than in previous operational hydrogen-oxygen engines. The very high power required for pumps operating at very high revolutions per minute and very high pressure, and associated with valves and intercoolers, entailed a technical development program of great complexity. At the present writing, it appears that the work is approaching successful completion.

## The Shuttle in the 1980's

Perfection of the shuttle system and its evolution to operational status will take place in the early 1980's. Once operational, the shuttle will open the space environment to many activities including:

- Insertion of satellites from the shuttle into earth orbit. This approach will lessen the need for expendable launch vehicles and ease limitations on the weight and size of payloads.
- Retrieving malfunctioning satellites from low earth orbit and repairing or refurbishing them in space or returning them to earth.
- Employing the shuttle as a test bed for instruments and related equipment for use in meteorology, oceanography, and assessment of earth resources.
- Development of major manned and unmanned laboratories to support a range of scientific disciplines. Examples include the Spacelab, the space telescope (discussed below), and the Long Duration Exposure Facility. The possibilities include a life sciences laboratory, a materials research laboratory, an infrared astronomy facility, and a solar optical telescope.
- Carrying deployable satellites on a space-available basis to exploit resources available on the shuttle, such as power, data collection equipment, and thermal conditioning.
- Deployment, assembly, and fabrication of new kinds of large structures in space to serve as economical, multifunction, unmanned, shuttle-serviced platforms and large-aperture communications antennas.
- Launch and in-orbit servicing of upper stages for the interorbital transfer of cargoes and planetary probes. One new-technology upper stage may be the solar electric propulsion system, whose propulsion will be derived from low-thrust, high-efficiency ion engines.

## The Space Telescope

The space telescope being prepared for shuttle launch is the next major step in the revolution in astronomy that has occurred in the twentieth century. The telescope is only 2.4 meters in diameter, compared to the 5-meter telescope at Hale Observatory on Mount Palomar in California, which is second in size only to a 6-meter telescope in the Soviet Union. However, the advantages of making observations from beyond the earth's atmosphere far outweigh the size differences. With the Hale telescope, astronomers can see clusters of galaxies about 2 billion light-years ($12 \times 10^{21}$ miles) away. The space telescope will extend that range sevenfold, bringing into view a volume of space 350 times that accessible to the Hale telescope. Quasars and galaxies 50 times fainter than those which can be seen now will become detectable.

Ground-based astronomers must struggle with the distortion of images produced by fluctuations in the density of the earth's atmosphere. Long-time exposure with the space telescope will produce images ten times sharper than those obtained from the ground. Another great advantage is the broad spectrum of ultraviolet light observable without atmospheric absorption.

The sharper images will appear against a background sky that is free of scattered light and airglow. Concentrating the starlight into a smaller area will enhance proportionately the contrast with the background. In place of conventional photographic film, a televised recording system of much higher sensitivity will be used. Whereas film responds linearly over only a small range of exposure, the photon-counting mode of the television camera will permit very long periods of integration and correspondingly fainter limits of detection.

Finally, the weather in space is always perfect for observation. Exposures of 10 hours may not be uncommon, and all of the sky is accessible almost 24 hours a day. The space telescope should achieve 5,000 hours of operation per year, compared with 2,000 hours, at best, for ground-based equipment.

***Basic Features*** The optical assembly of the space telescope will consist of a 2.4-meter (8-foot) mirror, a meteoroid shield, and a sun shade. The mirror will be made of material with a coefficient of thermal expansion 100 times smaller than that of the Pyrex of the Palomar mirror. Still, it will be necessary to heat the mirror in space to the temperature at which it was polished to preserve the original precision of the surface. The telescope and its focal-plane instruments and guidance module will weigh about 9,000 kilograms (20,000 pounds); the package will be 13.1 meters (43 feet) long and 4.26 meters (14 feet) in diameter. It will be a snug fit in the shuttle's cargo bay.

Because of the great cost of the space telescope, it is essential that it operate for at least ten years. Such long life can be achieved only if the shuttle visits the telescope periodically so that astronaut-scientists can make repairs and replace instrument modules for new experiments. If extensive overhaul is needed, the shuttle can retrieve the telescope and return it to earth.

***Instruments*** The focal-plane structure of the telescope will carry five principal instruments: a wide-field and planetary camera, a faint-object camera, a faint-object spectrograph, a high-resolution spectrograph, and a high-speed photometer. Each instrument represents the most advanced state of the art in optical components and photon detection sensitivity. In addition to these instruments, fine-guidance sensors will be used to make astrometric observations of very high precision. The resulting data will permit the accuracy of the fundamental astronomical coordinate system to be improved by an order of magnitude.

Scientists will be able to communicate with the space telescope at high data rates for about 20 minutes per orbit by means of the tracking and data relay satellite. Low data-rate transmission will be available through the satellite for the remaining 80 minutes per orbit. Most observations will be prescheduled, with commands loaded in the onboard computers well in advance. But, depending on the availability of the tracking satellite, real-time interaction between the telescope and astronomers on the ground will be possible.

***Uses of the Telescope*** Astronomers will use the space telescope to scan every realm of the mysterious universe. They will search for planets orbiting nearby stars and probe the large-scale structure of the universe to the edge of visibility. The key to cosmic evolution is to be found in objects so distant that the time it takes for their light to travel to earth is comparable to the age of the universe.

Ground-based observations reveal an expanding universe, born of a cosmic explosion, the big bang, between 10 and 20 billion years ago. If the rate of expansion is slowing sufficiently so that eventually it will halt and reverse, creating an eventual collapse, that means the universe is closed. It may then oscillate between explosion and collapse on a time scale of a hundred billion years. But, if the expansion continues forever, the universe is open. Creation never repeats; all the stars burn out and their dead remains spread to infinity.

The observed velocities of galaxies to distances of about a billion light-years give no clear indication of the ultimate course of the expansion. The deeper search with the space telescope may finally permit a choice to be made between an open and a closed universe.

***Nearby Galaxies*** Cosmological interpretations of astronomical data can be only as good as our knowledge of nearby galaxies. With the space telescope, every step in the derivation of distance scales will be greatly improved. It is unlikely that individual planets outside the solar system will be distinguished, even when focusing closely on nearby stars. But a star-planet system is on a gravitational seesaw. As the planet orbits the center of mass of the system, the star does the same at a much shorter distance. The apparent oscillatory motion of the star tells us that a planet is present as well as its mass. Barnard's wiggling star is now the only persuasive—if not fully accepted—case for a star-bound planet beyond our solar system. The space telescope, however, may detect planets around as many as 80 stars within 30 light-years of the sun. If such evidence is obtained, the case for abundant cosmic life will be greatly strengthened.

### The Shuttle's Promise

This brief examination of the Space Shuttle and but one of its associated missions can hardly convey a sense of the enormous effort that is going into the program. The shuttle is a great technical and operational departure from the present way of operating in space. Its development is being pursued at the frontier of aerospace technology. Success promises to bring by the mid-1980's a broad new set of opportunities with potentially large scientific and technological payoffs.

## ARTIFICIAL INTELLIGENCE

### Origins

Scientists doing research on artificial intelligence are still far from their ultimate goal—a computer-based analog of the human brain. Nevertheless, their efforts have produced regularly results useful in applied computer science. For example, the design of programming languages in general use has benefited from concepts developed in work on artificial intelligence.

Many branches of science have profited from the reductionist approach—the dissection of complex phenomena into elements whose simple interactions account for these phenomena. This approach is particularly characteristic of computer science, which uses dissection as a universal technique for system design. For example, to produce a full-blown graphic display system, a software designer will first analyze the geometric functions to be provided into vector and matrix operations which can represent these functions. Then he will decompose these vector and matrix operations into standard arithmetical, data-movement, and condition-testing operations and, if necessary, decompose these simple standard operations into absolutely elementary patterned openings and closings of transistorized switches. This method of systematic reduction can be applied now with considerable sophistication; sometimes as many as a dozen intermediate conceptual layers appear between the elaborate functions that certain computer systems provide and the billions of elementary switching steps on which these functions ultimately rest.

From the start, this habit of dissection, coupled with the realization that the computer can manipulate perfectly arbitrary information patterns, has suggested that the broad complex of abilities that constitutes human intelligence might itself be dissected and then reconstructed artificially. To do so would be to build an artificial intelligence, a system of programs that, when run on a sufficiently powerful computer, could imitate all of the intellectual capacities of the human brain: appropriate responses to spoken language, visual perception and manipulation of objects in three dimensions, and even the ability to plan successfully in an environment of varied and complex contingencies, to invent new mathematics and new science generally, to play all games brilliantly, and to converse with wit and verve in many languages.

***Potential Impacts*** The construction of artificial intelligences would affect the circumstances of human life profoundly. The appearance of intelligent beings other than man would surely create a new economics, a new sociology, and a new history. Moreover, if artificial intelligences can be created at all, there is little reason to believe that they could not lead swiftly to the construction of artificial superintelligences able to explore significant mathematical, scientific, or engineering alternatives at a rate far exceeding individual human ability. Such expectations motivate many computer scientists, whether they concern themselves directly with computer realization of some intelligent function or with efforts such as computer systems design, development of programming techniques, or invention of specialized algorithms. The possibility has been mentioned often in wider circles, but public discussion has not begun to reflect this perspective in any adequate way.

The efforts of a growing body of determined researchers in artificial intelligence have produced real successes, and the general scientific and industrial influences of their work are steadily increasing. Nevertheless, the basic techniques available to the worker in artificial intelligence are still modest.

## Current Status

To assess the current situation, we can contrast the techniques of programming with the more ambitious goal of learning. A computer is programmed by supplying it with a carefully composed sequence of instructions which guides its actions in all necessary detail. But to learn, a computer would have to ingest relatively unstructured masses of information, much closer to the highly fragmented information that people deal with routinely, and itself supply the elusive steps of error correction and integration needed to turn this broken material into polished instructional sequences which can be followed literally. Only to the extent that a computer can absorb fragmented material and organize it into useful patterns can we properly speak of artificial intelligence.

Researchers in artificial intelligence have therefore sought general principles which, supplied as part of a computer's initial endowment, would permit a substantial degree of self-organization thereafter. Various principles that allow useful structures to be distilled from masses of disjointed information have been considered candidates in this regard. These include graph search and deduction from axioms.

***Graph Search*** Many problems can be reformulated as one of finding a path between two known points within a graph. Planning and manipulation problems, both physical and symbolic, illustrate the point. Such problems are described by defining (1) an initial condition with which manipulation is to begin, (2) some target state or states that one aims to reach, and (3) a family of transformations that determines how one can step from state to state.

The problem of chemical synthesis is an example: the target is a compound to be synthesized, the initial state is that in which easily available starting substances are at hand, and the allowed manipulations are the elementary reactions known to the chemist. The problem of symbolic integration is a second example: a given formula F containing an integral sign defines our starting state, any formula of the class mathematically equivalent to F but not containing an integral sign is an acceptable target, and the transformations are those that mathematics allows.

In all such problems, the collection of available transformations is a heap of relatively independent items, since transformations can be listed in any order, any arbitrary collection of transformations defines a graph, and a collection of transformations can be freely expanded or contracted without the appearance of any formal obstacle. Thus, the construction of a path through the graph defined by a collection of transformations does represent a situation in which a structured entity, a path, arises via a general principle from something unstructured and indefinitely extensible, a collection of transformations. One can try, therefore, to use this construction as a universal principle for the automatic derivation of structure.

***Deduction from Axioms*** The work of numerous mathematicians in the nineteenth and twentieth centuries has shown that the classical corpus of mathematics can be based on a handful of astonishingly simple axioms and inference rules. The resulting formalism, the logician's "predicate calculus," is a straightforward language of formulas, in which any assertion of classical mathematics can be stated readily. In this system, a simple and easily programmed set of rules suffices to define all possible, mathematically justifiable lines of reasoning. Thus, the predicate calculus gives us a simple but universal formal framework encompassing all logical reasoning.

Arbitrary collections of axioms that express the laws of particular mathematical and nonmathematical domains can be written readily in this calculus; then, if we are willing to search a sufficiently large space of proofs, the consequences of these axioms can be extracted automatically. Since axioms can be listed in relatively unstructured fashion, we have here another situation in which highly structured entities—mathematical proofs—arise via a general principle from something unstructured and indefinitely extensible—a collection of axioms. This construction clearly can be used as a universal principle; a considerable amount of work has been done toward adapting this fundamental logical mechanism to other uses, such as the automatic generation of plans and programs. Generally speaking, such adaptation proceeds without difficulty.

## The Efficiency Problem

Nevertheless, neither the use of this very general logical machinery nor of the graph-searching technique can be regarded as more than a fragmentary key to the problem of automatically developing significant structures out of fragmented masses of information. The reason is the overwhelming inefficiency of these methods. Unless care is taken, any attempt to search the space of all proofs possible in the predicate calculus will founder almost immediately because of the immensity of this space. Similarly, representation of significant problems such as graph searches generally requires consideration of a graph far too large for the largest imaginable computer.

This realization has inspired two decades of effort to prune the searches needed to find either significant proofs in logic or paths or other interesting combinational substructures in graphs—that is, to find ways of distinguishing profitable from unprofitable directions of exploration in order to increase the efficiency of these searches by the very large factors necessary. Several general and quite useful pruning principles have been found, but even the very best general proof-finding and graph-searching techniques known currently are incapable of handling more than small examples.

From this disappointing fact, many researchers have drawn the conclusion that significant progress will be possible only if the sought-after, structure-generating mechanisms somehow use not only very general techniques, such as those reviewed above, but also use larger amounts of information specific to particular domains. However, it is not at all clear how such information can be represented or organized best. Moreover, unless one is careful, the use of such information can be considered a covert retreat from the basic goal: general principles allowing self-organization. Indeed, the presence of enough manually supplied, preorganized information can lead us to classify a system simply as a clever program rather than as a recognizable step toward constructing artificial intelligences.

***Rule Systems*** This objection need not apply to systems to which information can be supplied in the form of separate rules of the kind that an expert in some area might use to transmit his expertise to a beginner. For this reason, such rule systems have begun to attract attention recently. Attempts have been made to use large collections of rules to build artificial expert systems, such as programs able to diagnose human or plant diseases or to recognize clues to the presence of minerals.

Once rules have been supplied to such systems, the systems ingest lists of specific clues or symptoms visible in the situation to be assessed. To each clue or symptom, a numerical indication of the degree of reliability may be attached. The rules specify how these clues are to be combined to yield estimates of the likelihood that particular expert judgments—diagnosis of a specific disease or detection of a mineral—are correct. If the estimated likelihood is high enough, the corresponding judgment is pronounced.

In simple rule systems of this kind, computation of the relevant internal estimate may involve little more than a polynomial combination of the presence/absence/reliability indicators for the clues. When this is so, the rules merely define the coefficients of the polynomials or other functions that will be used. Even in such simple cases, the system may provide a specialized language for defining rules that allows the experts—for example, physicians or mineralogists—from whom the rules are to be acquired to express themselves in a language more familiar to them than the mathematical language of polynomials and functions.

The estimation procedures used in the simpler expert systems of this kind are not very different from some of the mathematical grouping techniques used with limited success for many years to catalog and look up journal articles and other scientific citations automatically. These procedures are admittedly crude, but evidence is beginning to appear that they may not be too unrefined for practical application. (Conceivably, their crudity may reflect that of some of the internal processes that a human expert uses to form a snap judgment of the evidential value of combinations of familiar clues.)

## Links with Applied Computer Science

In one sense, the rule systems, the graph and transformation systems, and even the logical formalism that workers in artificial intelligence have tried to use are simply specialized programming languages. The particularly interesting point about these languages is that they aim, in a more single-minded way than ordinary programming languages, to decompose the programs they express into small, independent fragments. As emphasized above, the ability to deal with fragmented material and integrate it automatically is a fundamental goal of research in artificial intelligence. Languages that make it possible to use small independent fragments to define complex processes are of great interest to programmers even outside the field of artificial intelligence. Such languages eliminate a most troublesome source of programming error and can increase programming speed very considerably. For this reason, research in artificial intelligence has been a fertile source of concepts that have passed into the design of generally used programming languages.

Another connection between work in artificial intelligence and the more technical part of computer science can be noted. To describe and analyze the syntactic structure of programming languages, computer scientists customarily use a kind of formal mechanism called a context-free grammar. Grammars of this kind specify how statements in a programming language are to be parsed (a preparatory operation very close to the sentence diagraming taught in high school English classes a generation ago).

These formal grammars were invented more or less simultaneously by workers in computer science and students of natural language. Although simple, they are remarkably effective in capturing, at least roughly, the general grammatical constructs that people seem to comprehend easily and naturally. Variants of this same formal mechanism have been used in systems for analysis (and, to a limited extent, understanding) of natural language and in systems for decoding spoken utterances. Although context-free grammars are used in many areas and are not always thought of as belonging specifically to artificial intelligence, they do represent a genuine case in which significant aspects of an important human function can be captured by a simple computer model.

## Expectations

Research in artificial intelligence is still immature. As yet, it can boast of few significant general principles and of little theory capable of directing the work of its practitioners. Nevertheless, the field is not without its successes. In particular, the expectation that artificial synthesis of intelligence will be possible has inspired attempts to computerize functions, such as the decoding of spoken utterances and analysis of visual scenes, of a subtlety that might have seemed unreachable otherwise. Full duplication of human sensory and intellectual capabilities, which are rich and subtle indeed, still lies far beyond our reach. But attempts to match them are beginning to be successful enough to furnish items valuable as components of practical systems.

It should be noted that the efforts at artificial vision, while still far short of the full capabilities of human vision, can have a major impact when joined with particular developing industries. An example is the work on robotlike automatic devices—now generally being called industrial robotics—which has much to contribute to industrial productivity. The artificial intelligence work in vision is beginning to add substantially to the capabilities of these devices and will play an important role in the development of the field during the coming decade.

Seen from a more fundamental point of view, the most successful applied programs that have grown out of the work on artificial intelligence may be judged to exhibit only limited degrees of self-organization but nevertheless to represent the application of ingenious programming to unusual areas. Such fragmentary successes continue to accumulate. We may expect that they will cohere gradually, suggesting eventually more adequate general principles of self-organization than are now available.

## Summary and Outlook

The seven emerging technologies described briefly above, although generally disparate, have common threads. An important one is that they arise from a wedding of science and good technology. While much of the basic knowledge that supports most of these technologies was in hand as long as 20 years ago, the rate at which they would emerge was difficult to assess because of the time needed for development. Forecasting remains risky, but the technologies now available have reached a stage at which short-term trends, at least, can be defined with reasonable confidence.

There are more palpable common threads. The importance of the rapidly evolving computer as either a tool or a beneficiary of progress is noted explicitly, if not comprehensively, for each technology except recombinant DNA and energy storage. These two benefit, for example, from computer-based, analytical instrumentation of steadily increasing power.

A third common thread is dependence on the development and refinement of materials. Many emerging technologies rely strongly on our ability to develop and manipulate materials at levels ranging from the bulk or macromolecular to the subatomic. Recombinant DNA has a materials aspect in the sense that it involves the need to develop organisms and associated systems that can efficiently convert biomass to chemicals that could serve as starting points for useful substances.

Energy and the related issue of feedstock enter into the consideration of several technologies. More efficient use of energy resources is the goal of energy storage and, in part, of the practical application of superconductivity. The conversion of biomass to chemicals with the use of the methods of recombinant DNA could ease the pressure on petroleum and other fossil resources now used as fuels and feedstock. The resource assessment functions of the Space Shuttle could contribute to the discovery of both new energy resources and materials.

There is always the question of arousing public uneasiness when new technologies appear. The furor over research on recombinant DNA has faded, for good reason, but it is conceivable that the large-scale use of the technology could rekindle the controversy. Work on artificial intelligence has not stirred dissent yet, but the potential does exist.

Of the technologies considered here, the most uncertain future lies in the ultimate development of artificial intelligence. Although such research is far from its real goals, it can be expected to continue as a central issue for the most innovative minds and will lead in one way or another to results of practical value in computer science and technology.

Advances in medical technology based in part on techniques borrowed from the physical sciences are generating commercial apparatus and techniques useful in diagnosis and treatment. Progress in the field will come steadily and will continue to expand our ability to provide diagnosis and treatment and thereby upgrade the quality of health care.

The use of recombinant DNA technology to make pharmaceuticals is probably close to commercialization, and its use in making other chemicals is not far behind. An important question will center about the economic competitiveness of such processes relative to other means of producing the same products. In a sense, the techniques of recombinant DNA remain to date more important as a source of basic knowledge than as the basis of commercial products.

## BIBLIOGRAPHY

Howard Allaway. *The Space Shuttle at Work*. National Aeronautics and Space Administration. SP-432. Washington, D.C.: U.S. Government Printing Office, 1979.

M. R. Beasley. "Advanced Superconducting Materials for Electronic Applications," *Institute of Electrical and Electronics Engineers, Transactions on Electron Devices*. ED-27. Vol. 2009(October 1980).

T. H. Geballe and J. K. Hulm. "Superconductors in Electric-Power Technology," *Scientific American*. Vol. 243(November 1980).

Fritz R. Kalhammer. "Energy Storage," *Annual Review of Energy*. Volume One. Palo Alto, Ca.: Annual Reviews, Inc., 1976.

Nils J. Nilsson. *Principles of Artificial Intelligence*. Palo Alto, Ca.: Tioga Publishing Co., 1980.

Office of Technology. *The Impacts of Applied Genetics*. Washington, D.C.: U.S. Government Printing Office, 1981.

Brian J. Skinner. "Earth Resources," *Proceedings of the National Academy of Sciences*. Vol. 76(September 1979), pp. 4212-4217.

U.S. Congress. Office of Technology Assessment. "Research and Development of CT and Other Diagnostic Imaging Techniques," *Policy Implications of the Computed Tomography (CT) Scanner: An Update,* Pub. No. OTA-BP-H-8. Washington, D.C.: U.S. Government Printing Office, January 1981, pp. 59-71.

# *Reviewers*

The assistance of the following persons is gratefully acknowledged. The contents of the report are the responsibility of the Steering Committee.

## ENTIRE REPORT

### Governing Board Members and Assembly/Commission Chairmen

*Elkan R. Blout,* Harvard School of Public Health
*Bryce Crawford, Jr.,* University of Minnesota
*W. Kenneth Davis,* Bechtel Power Corporation
*James C. Fletcher,* University of Pittsburgh
*Herbert Friedman,* National Research Council
*Ralph E. Gomory,* IBM Corporation
*Philip Handler,* National Academy of Sciences
*Frederic A. L. Holloway,* Baton Rouge, Louisiana
*Saunders Mac Lane,* University of Chicago
*Thomas F. Malone,* Butler University
*Elliott W. Montroll,* University of Maryland
*Courtland D. Perkins,* National Academy of Engineering
*Frank W. Putnam,* Indiana University
*Frederick C. Robbins,* Institute of Medicine
*Harrison Shull,* Rensselaer Polytechnic Institute
*Herbert A. Simon,* Carnegie-Mellon University
*Mitchell W. Spellman,* Harvard Medical School
*Lewis Thomas,* Sloan-Kettering Cancer Center
*H. Guyford Stever,* National Research Council

*Robert M. White,* University Corporation for Atmospheric Research
*T. Franklin Williams,* Monroe Community Hospital
*Julian Wolpert,* Princeton University

## 1 The Demographic State of the World

*Kenneth E. Boulding,* University of Colorado
*Harrison Brown,* East-West Center
*Leslie Kish,* University of Michigan
*Walsh McDermott,* Robert Wood Johnson Foundation
*William D. McElroy,* University of California, San Diego
*Roger Revelle,* University of California, San Diego
*Theodore W. Schultz,* University of Chicago
*Conrad Taeuber,* Georgetown University

## 2 On Some Major Human Diseases

*Edward H. Ahrens, Jr.,* The Rockefeller University
*Robert W. Berliner,* Yale University School of Medicine
*Bernard D. Davis,* Harvard Medical School
*Robert S. Gordon, Jr.,* National Institutes of Health
*Alfred Knudson, Jr.,* Institute for Cancer Research
*Richard M. Krause,* National Institutes of Health
*Helen Ranney,* University of California, San Diego
*James A. Shannon,* Chevy Chase, Maryland
*James B. Wyngaarden,* Duke University Medical Center

## 3 Directions in Nutrition Research

*Edward H. Ahrens, Jr.,* The Rockefeller University
*Robert W. Berliner,* Yale University School of Medicine
*William J. Darby,* Nutrition Foundation, Inc.
*George K. Davis,* University of Florida
*Gilbert Leveille,* General Foods Corporation
*Nevin S. Scrimshaw,* Massachusetts Institute of Technology

## 4 The Science of Cognition

*Floyd Bloom,* Salk Institute for Biological Studies
*Lee J. Cronbach,* Stanford University
*R. Duncan Luce,* Harvard University
*George A. Miller,* Princeton University

## 5 Ecology and Systematics

*Arthur D. Hasler,* University of Wisconsin
*G. Evelyn Hutchinson,* Yale University

*Paul J. Kramer,* Duke University
*A. Starker Leopold,* University of California, Berkeley
*Ruth Patrick,* Academy of Natural Sciences, Philadelphia

## 6 Plant Disease

*Robert H. Burris,* University of Wisconsin
*Glenn W. Burton,* Georgia Coastal Plain Experiment Station
*Ellis B. Cowling,* North Carolina State University
*James G. Horsfall,* Connecticut Agricultural Experiment Station
*Arthur Kelman,* University of Wisconsin
*Paul J. Kramer,* Duke University
*Anton Lang,* Michigan State University
*Paul Waggoner,* Connecticut Agricultural Experiment Station

## 7 Water Resources

*Myron B. Fiering,* Harvard University
*Luna B. Leopold,* University of California, Berkeley
*Hugh Odishaw,* University of Arizona
*M. Gordon Wolman,* The Johns Hopkins University

## 8 Radioactive Waste Management

*Philip H. Abelson,* Science Magazine
*H. H. Barschall,* University of Wisconsin
*W. Kenneth Davis,* Bechtel Power Corporation
*Earnest F. Gloyna,* University of Texas, Austin
*Charles Hebel,* Xerox Corporation
*Conyers Herring,* Stanford University
*Arthur C. Upton,* New York University Medical Center
*Alvin M. Weinberg,* Institute for Energy Analysis
*Hatten S. Yoder, Jr.,* Carnegie Institution of Washington

## 9 Sun and Earth

*John N. Bahcall,* Institute for Advanced Study, Princeton
*Alan M. Lovelace,* National Aeronautics and Space Administration
*Andrew F. Nagy,* University of Michigan
*Hugh Odishaw,* University of Arizona
*Eberhardt Rechtin,* The Aerospace Corporation

## 10 The Science of Macromolecules

*Robert L. Baldwin,* Stanford University
*Marshall Fixman,* Colorado State University

*Leonard Peller,* University of California Medical School, San Francisco
*John D. Roberts,* California Institute of Technology
*Walter H. Stockmayer,* Dartmouth College

**11 Chemical Synthesis of New Materials**

*Harry B. Gray,* California Institute of Technology
*William S. Johnson,* Stanford University
*Kenneth S. Pitzer,* University of California, Berkeley
*John D. Roberts,* California Institute of Technology
*Frank H. Westheimer,* Harvard University

**12 On Some Recent Developments in Mathematics**

*Hyman Bass,* Columbia University
*Felix E. Browder,* University of Chicago
*Walter Feit,* Yale University
*James Krumhansl,* Cornell University
*Steven Weinberg,* Harvard University

**13 Research in Europe and the United States**

*Pierre R. Aigrain,* Paris
*David Beckler,* Organization for Economic Cooperation and Development
*Robert W. Berliner,* Yale University School of Medicine
*Jacques Bodelle,* Embassy of France, Washington, D.C.
*Sir Allen Cottrell,* Jesus College, Cambridge University
*Heinz Maier-Leibnitz,* Deutsche Forschungsgemeinschaft
*Thomas F. Malone,* Butler University
*Werner Menden,* Embassy of Federal Republic of Germany, Washington, D.C.
*Helen Ranney,* University of California, San Diego
*Charles H. Townes,* University of California, Berkeley
*Mark Wilkins,* Embassy of Great Britain, Washington, D.C.

**14 Research in Industry**

*Michael Boretsky,* U.S. Department of Commerce
*Fred Bucy,* Texas Instruments, Inc.
*John J. Burns,* Hoffmann-La Roche, Inc.
*Albert M. Clogston,* Bell Laboratories
*Edward E. David,* Exxon Research and Engineering Company
*W. Kenneth Davis,* Bechtel Power Corporation
*Carl A. Gerstacker,* Dow Chemical Company
*Ralph E. Gomory,* IBM Corporation

*Julius J. Harwood,* Ford Motor Company
*Wesley Kuhrt,* United Technologies
*Kenneth G. McKay,* New York, New York
*E. R. Piore,* New York, New York

## 15 Fuel Science and Technology

*James Boyd,* Carmel, California
*Robert A. Charpie,* Cabot Corporation
*W. Kenneth Davis,* Bechtel Power Corporation
*Wolf Häfele,* International Institute for Applied Systems Analysis
*Bernard Lee,* The Institute of Gas Technology
*Charles Mankin,* Oklahoma Geological Survey
*Hugh Odishaw,* University of Arizona
*Stanford S. Penner,* University of California, San Diego
*C. Dwight Prater,* Mobil Research Corporation
*Dwain Spencer,* Electric Power Research Institute
*H. Guyford Stever,* National Research Council
*Lawrence E. Swabb, Jr.,* Exxon Research and Engineering Company
*Alvin M. Weinberg,* Institute for Energy Analysis

## 16 Issues in Transportation

*Alan Boyd,* Amtrak
*Paul Chenea,* General Motors Corporation
*William L. Garrison,* University of California, Berkeley
*Aaron Gellman,* Gellman Research Associates
*A. Scheffer Lang,* Gellman Research Associates
*Wilfred Owen,* The Brookings Institution
*Jerrold R. Voss,* U.N. Centre for Human Settlements

## 17 Prospects for New Technologies

*H. H. Barschall,* University of Wisconsin
*Harvey Brooks,* Harvard University
*Albert M. Clogston,* Bell Laboratories
*Edward E. David, Jr.,* Exxon Research and Engineering Company
*James Fletcher,* University of Pittsburgh
*Richard L. Garwin,* IBM Corporation
*Frederick Jelinek,* IBM Corporation
*Arthur Kantrowitz,* Dartmouth College
*Alan M. Lovelace,* National Aeronautics and Space Administration
*John McCarthy,* Stanford University
*Kenneth G. McKay,* New York, New York
*Allen Newell,* Carnegie-Mellon University
*E. R. Piore,* New York, New York

# *Index*